Sustainable Life Insurance

Sustainable Life Insurance: Managing Risk Appetite for Insurance Savings and Retirement Products gives an overview of all relevant aspects of traditional and non-traditional savings and retirement products from both insurers' and policyholders' respective risk appetites. Examples of such products include general accounts, whole life, annuities (variable, fixed and fixed indexed, structured), index-linked products, CPPI-based products, etc.

The book contains technical details associated with both practice and theory, specifically related to modelling, product design, investments and risk management challenges and solutions, tailored to both insurers' and policyholders' perspectives.

Features

- The book offers not only theoretical background but also concrete, cutting-edge "quick wins" across strategic and operational business axes.
- It will be an asset for professionals in the insurance industry, and a great teaching/learning resource for courses in risk management, insurance modelling, and more.
- The book highlights the operational challenges encountered across modelling, product design and hedging.

Chapman & Hall/CRC Financial Mathematics Series

Series Editors

Recently Published Titles

Commodities: Fundamental Theory of Futures, Forwards, and Derivatives Pricing, Second Edition
Edited by M.A.H. Dempster, Ke Tang

Introducing Financial Mathematics: Theory, Binomial Models, and Applications
Mladen Victor Wickerhauser

Financial Mathematics: From Discrete to Continuous Time
Kevin J. Hastings

Financial Mathematics: A Comprehensive Treatment in Discrete Time
Giuseppe Campolieti and Roman N. Makarov

Introduction to Financial Derivatives with Python
Elisa Alòs, Raúl Merino

The Handbook of Price Impact Modeling
Dr. Kevin Thomas Webster

Sustainable Life Insurance: Managing Risk Appetite for Insurance Savings & Retirement Products
Aymeric Kalife with Saad Mouti, Ludovic Goudenège, Xiaolu Tan, and Mounir Bellmane

Geometry of Derivation with Applications
Norman L. Johnson

For more information about this series please visit: https://www.crcpress.com/Chapman-and-HallCRC-Financial-Mathematics-Series/book series/CHFINANCMTH

Sustainable Life Insurance

Managing Risk Appetite for Insurance Savings and Retirement Products

Aymeric Kalife
with Ludovic Goudenège,
Xiaolu Tan, Saad Mouti and Mounir Bellmane

CRC Press
Taylor & Francis Group
Boca Raton London New York

CRC Press is an imprint of the
Taylor & Francis Group, an **informa** business
A CHAPMAN & HALL BOOK

Cover design: © Shutterstock

First edition published 2024
by CRC Press
6000 Broken Sound Parkway NW, Suite 300, Boca Raton, FL 33487-2742

and by CRC Press
4 Park Square, Milton Park, Abingdon, Oxon, OX14 4RN

CRC Press is an imprint of Taylor & Francis Group, LLC

Library of Congress Cataloging-in-Publication Data
Names: Kalife, Aymeric, author. | Goudenège, Ludovic, author. | Tan, Xiaolu, author.
Title: Sustainable life insurance : managing risk appetite for insurance savings and retirement products / Aymeric Kalife ;
with Ludovic Goudenège, Xiaolu Tan, Saad Mouti, Mounir Bellmane.
Description: 1 Edition. | Boca Raton : C&H/CRC Press, 2023. | Series: Chapman & Hall/CRC financial mathematics series |
Includes bibliographical references and index.
Identifiers: LCCN 2022055098 (print) | LCCN 2022055099 (ebook) | ISBN 9781032081557 (hardback) |
ISBN 9781032110929 (paperback) | ISBN 9781003218371 (ebook)
Subjects: LCSH: Life insurance. | Risk management. | Life insurance–Mathematical models.
Classification: LCC HG8771 .K165 2023 (print) | LCC HG8771 (ebook) | DDC 368.32–dc23/eng/20230118
LC record available at https://lccn.loc.gov/2022055098
LC ebook record available at https://lccn.loc.gov/2022055099

ISBN: 9781032081557 (hbk)
ISBN: 9781032110929 (pbk)
ISBN: 9781003218371 (ebk)

DOI: 10.1201/9781003218371

Typeset in Minion
by Newgen Publishing UK

Author Bios

Aymeric Kalife has been the CEO of iDigital Partners since 2019, specializing in insurance and investment advisory while leveraging on digital technologies, and Adjunct Professor in Finance at Paris Dauphine–PSL University. He was Global Head of Savings (Risk Management, Savings Strategy) and Deputy Group Life Chief Actuary at AXA for 11 years, notably supervising the hedging platform of variable annuities ($130 billion assets under management: AuM) and contributing to reshaping the savings business across 20 entities through capital-light structuring innovations. Prior to AXA he worked for eight years in investment banking, as VP volatility strategist at Deutsche Bank, hybrids derivatives structurer at Merrill Lynch, and quant in commodities at EDF and in interest rates at ABN AMRO. His research interests are in market impact and liquidity risk for flow/structured products, fixed and variable annuities, insurance policyholders' behaviour and associated optimal product design, and environmental, social and governance (ESG) asset selection and optimal allocation, developed in a dozen academic publications and awards from the US Society of Actuaries. He holds science (statistics, probability), finance, economics and law master's degrees from Pierre and Marie Curie University/École Polytechnique, ENSAE, Sorbonne and Science Po, an MBA at HEC Paris, and a PhD at Paris Dauphine University and ESSEC Business School.

Ludovic Goudenège has been a researcher in the French national research institute, CNRS (first class), since 2010 and a professor at École CentraleSupélec, and was a consultant in insurance at AXA Group Risk Management. He is a specialist in stochastic partial differential equations, particularly in the presence of strong or singular nonlinearities, for which he develops theoretical results on the convergence and numerical analysis of numerical schemes. He also works with the Inria MathRisk team for the development of numerical methods for finance and insurance. It develops statistical learning methods (machine learning) for solving partial differential equations (PDEs) in large dimensions, nonlinear, dynamic programming, optimal downtime, etc. He holds a PhD in applied mathematics at ENS Cachan (stochastic partial differential equations, numerical simulations, finite elements methods), a master's in probability at University of Pierre and Marie Curie, a master's in mathematical analysis at Université de Rennes and a fifth national rank at the Aggregation of Mathematics. He is now a member of the mathematical federation of the École CentraleSupélec. His research interests are in the theory of stochastic and deterministic partial differential equations, the associated numerical methods and their application in finance and insurance.

Xiaolu Tan has been Associate Professor at the Chinese University of Hong Kong since 2019. He is a member of the editorial board of the journals *Numerical Algebra, Control and Optimization*, *Stochastic Processes and Their Applications* and *Journal of Optimization Theory and Applications*. Previously he was Assistant Professor at Paris-Dauphine University from 2013 to 2019. He was a consultant in insurance at AXA Group Risk Management between 2011 and 2017. He holds a master's and a PhD in applied probability at École Polytechnique. His research interests, translated into more than 30 academic publications, are in stochastic analysis, stochastic optimal control, the numerical simulation of stochastic differential equations (SDEs), the numerical method for nonlinear PDEs, the numerical method for optimal control problems and their application in finance and insurance.

Saad Mouti is an Assistant Professor at UCSB PSTAT Department, after a postdoctoral fellowship at the Consortium of Data Analytics in Risk, part of the Economics and Statistics Department at the University of California – Berkeley. He completed his PhD in financial and actuarial mathematics at Pierre and Marie Curie University (Paris 6, France) in 2017. Prior to that he obtained a financial engineering diploma from the National School of Applied Mathematics and Computer Science of Grenoble in 2010 and an MSc in the statistical processing of information from Paris–Dauphine University in 2011. Saad's thesis was financed by AXA to address different financial risks for the life insurance industry and englobed the pricing and hedging of life insurance products and policyholders' rational behaviour, rough volatility modelling and optimal strategies for buying a book of options under market impact constraints. His research interests are theoretical, applied and engineering approaches to applying data science to risk measurement and management problems. His current subjects are environmental, social and governance investing, asset-pricing theory, volatility modelling and causal inference applications for health science.

Mounir Bellmane has been a Consultant Manager at Towers Watson since 2020, providing advice on asset and liability modelling (ALM), merger and acquisition (M&A) activity, the optimization of ALM models and model point reduction techniques for stochastic models. Previously he was a life and savings specialist at AXA between 2014 and 2020, during which time he managed projects in the following areas: the implementation and monitoring of profitability models, the optimization of run-time calculation, the implementation of Solvency II ALM metrics, capital risk management, product design, the monitoring and calibration of life risks and the design and implementation of dynamic customer behaviour functions. He holds a master's in actuarial sciences from ISFA, a master's in applied mathematics from Paris–Dauphine University and an MBA from ESCP Business School.

Contents

Introduction

After the dramatic developments of the last few years in terms of daunting macroeconomic, demographic and customer challenges, fierce competition and ongoing tech-driven disruption, insurers have had to undertake large-scale change, by investing more than ever to transform their business models and seize subtly different sustainable growth opportunities around the world.

Overall, despite recent slow growth in revenues and intense profit pressures, rising customer expectations, interest rates and ageing populations are expected to drive demand for life insurance products, while ongoing investments in innovations and risk management are the driving factors to meet their appetites in a sustainable way.

Among the biggest challenges is meeting the rapidly changing needs and expectations of consumers within ageing societies; Covid-19 demonstrated why the insurance industry is essential, notably in terms of sustainable financial wellness within the context of savings and retirement. There is an increased demand for more affordable, transparent and customized insurance that better suits evolving needs, and risk appetites that translate into increasingly efficient behaviour.

In this respect, increasingly selective customer risk appetites have been observed since the 1980s, across four risk appetite axes (yields/flexibility/financial guarantees/cost), which can be illustrated through the development of flexible and cheaper savings and retirement policies offered to customers with lower risk tolerance. Life insurers must further expand their value propositions and product sets to address all aspects of financial wellness, which requires, on the one hand, enhancing customers' experience through better advice and, on the other hand, cheap, flexible, performant and capital-light savings and retirement instruments (as illustrated by the rising use of constant proportion portfolio insurance [CPPI] strategies, structured annuities and hybrid general accounts).

Meeting customers' risk appetites has also been developed by centring customer experiences and streamlining their journey as a priority, around flexibility, personalization and relevance.

In this respect, innovations including digital technologies are a key factor in terms of optimizing costs, boosting innovation and driving agility in a sustainable way, shifting the focus from the products to the customers, which reversed the bargaining power between the customer and the asset manager in favour of the former. Such a new paradigm requires switching from a top-down to a bottom-up approach. Being customer-centric is no longer an alternative or a "plus"; it is a requirement for survival. Reputational risk has become the key, with success in the past no guarantee for the future.

The current landscape is also notable for its fragmentation, convergence and intense competition, including from a mix of non-traditional players, which thus requires strong partnerships with asset management, investment banking (for the financial guarantees) and digital tech companies, opening the way to new life insurance archetypes.

These significant changes have strengthened within the context of declining and persistently low interest rates over the past decade, while new regulatory prudential frameworks have increased long-term safety but

DOI: 10.1201/9781003218371-1

also short-term costs. In this respect, life insurers have started to improve the sustainability of both their risk management practices and their business.

As a result, insurers have to be there for their customers and undertake large-scale change quickly so as to make sure that they can serve people in need in a sustainable way, in line with both customers' and insurers' risk appetites, in terms of liabilities and assets alike.

As a result, in order to build sustainable savings and retirement instruments that meet customers' and insurers' risk appetites, Chapter 1 assesses the transformation of the insurance business within the context of increasingly efficient customer behaviours and recurrent equity market crashes, within which specific innovations offer a historic opportunity to grow profitability and resilience, through both new business and in-force management. It then focuses on securing long-term sustainable business growth, from governance to operational risk management, consistent with rising capital prudential regulation, by means of sustainable product approval processes and risk appetite frameworks.

Chapter 2 assesses the drastic changes inherent in the increasingly selective customer risk appetite, in favour of flexible, lower-risk and cheaper guarantees. It then focuses on aligning customers' and insurers' risk appetites through enhanced advisory offerings and product ranges, leveraging digital technologies and banking/asset management skills through long-term partnerships.

Chapter 3 focuses on the measurement of risk appetite, as a sustainable savings and retirement stance requires the evaluation of liabilities to policyholders at their current execution value, consistent with market observations, taking into account every contractual option and financial guarantee included in the contract, notably through the stochastic modelling of both customers behaviour and market volatility, with material impact across the four dimensions of the risk appetite framework; value-wise, it allows the calculation of a best estimate valuation consistent with the market, while, capital-wise, it enables the computation of the solvency capital requirement (SCR) and earnings potential, allowing the insurer to better direct his asset and liability management and be able to set up hedging strategies against these rate variations. Chapter 3 also deals with the computational challenges of the Monte Carlo (MC) simulations involved in such an economic scenario generator (ESG), and proposes some efficient solutions through the use of randomized quasi-Monte Carlo and graphics processing unit (GPU) technology.

Chapter 4 explores ways to screen sustainable savings and retirement liabilities balancing shareholders' against customers' risk appetites, through capital intensity and customer-centric metrics, while innovative liabilities designs can improve the solvency internal rate of return (IRR), mitigate customer behaviour risk and enhance the customer experience. Regarding the asset side, key sustainable innovations reconciling insurer and customer risk appetites lie in the improved risk management of fund management, outperformance in asset selection and allocation while meeting growing demand for ESG assets, and an optimized portfolio rebalancing allowing for a lowering of both transaction costs and financial risks.

Finally, Chapter 5 focuses on the sustainability of hedging, which is apprehended through the decomposition of the hedging profit and loss (P&L) into the sensitivities to the market risk drivers ("Greeks"), the potential control of which is illustrated through sustainable mitigating techniques that capitalize on both structuring expertise and hedging experience: risk measurements (duration/convexity, break-even and par volatilities, hedge effectiveness and cost of carry, performance drag versus tail risk protection measures), mathematical modelling (e.g. optimal control theory, Lévy processes, nonlinear partial differential equations [PDEs] and associated backward stochastic differential equations [BSDEs]), product design innovations and the use of specific hedge instruments, mostly aligned with the customer's risk appetite.

CHAPTER 1

The Transformation of the Risk Appetite for Sustainable Savings and Retirement Policies

After the dramatic developments of the last few years in terms of significant and daunting macroeconomic, demographic and customers challenges, fierce competition and ongoing tech-driven disruption, insurers have to undertake large-scale change at an ever faster pace, by investing more than ever to transform their business models and seize subtly different sustainable growth opportunities around the world.

Among the biggest challenges has been meeting the rapidly changing needs and expectations of consumers within ageing societies; Covid-19 demonstrated why the insurance industry is essential, notably in terms of sustainable financial wellness within the context of savings and retirement. There is increased demand for more affordable, transparent and customized insurance that better suits evolving conditions and can be easily adjusted as the needs change. Insurers have to be there for customers and undertake large-scale change quickly so as to make sure they can serve people in need in a sustainable way.

The new needs and expectations regarding customers' financial wellness have induced drastic and rapid shifts in risk appetite, translating into (i) corresponding suitable and sustainable behaviour in the use of their savings and retirement products (surrender, withdrawals or annuitization), becoming more "efficient" throughout a "multi-stage" life; and into (ii) a demand for more sustainable investments in assets, as illustrated by the significant growth in their ESG (environmental, social and governance) scores.

As a result, life insurers must further expand their value propositions and product sets to address all aspects of financial wellness, centring customer experiences and streamlining their journey as a priority. In this respect, innovations including digital technologies are a key lever to optimize costs, boost innovation and drive agility in a sustainable way.

The current landscape is also notable for its fragmentation; convergence and intense competition, including from a mix of non-traditional players; and widespread collaboration. Carriers will look to partner with or acquire the most promising insurtechs, and banks and asset managers will offer more protection products and seek to differentiate on holistic financial wellness value propositions, forcing insurers to choose between collaboration and competition.

These drastic changes have further strengthened within the context of (i) declining and persistently low interest rates over the past decade, driven by the jump in central banks' balance sheets; and (ii) new regulatory prudential required capital frameworks, which tended to increase long-term safety but also short-term costs. In this respect, life insurers have started to improve the sustainability of both their risk management practices and their business (through lower richness and cost of capital, and in-force initiatives: buyouts, reinsurance, operational cost savings).

DOI: 10.1201/9781003218371-2

Overall, despite recent slow growth in revenues and intense profit pressures, rising customer expectations, interest rates and ageing populations are expected to drive demand for life insurance products, while ongoing investments in innovations and risk management are the driving factors to meet customer appetites in a sustainable way.

As a result, in order to build sustainable savings and retirement policies that meet customers' and insurers' risk appetites, section 1 assesses the transformation of the insurance business within the context of increasingly efficient customer behaviours and recurrent equity market crashes, combined with a persistent and drastic decline in interest rates, within which specific innovations offer a historic opportunity to grow profitability and resilience, through both new business and in-force management. Section 2 then focuses on securing long-term sustainable business growth, from governance to operational risk management, consistent with rising capital prudential regulation (e.g. the European Union's 2009 Solvency II Directive, which codifies and harmonizes EU insurance regulation), by capitalizing on sustainable product approval processes and risk appetite frameworks.

1.1 THE TRANSFORMATION OF THE INSURANCE BUSINESS OVER THE PAST DECADE

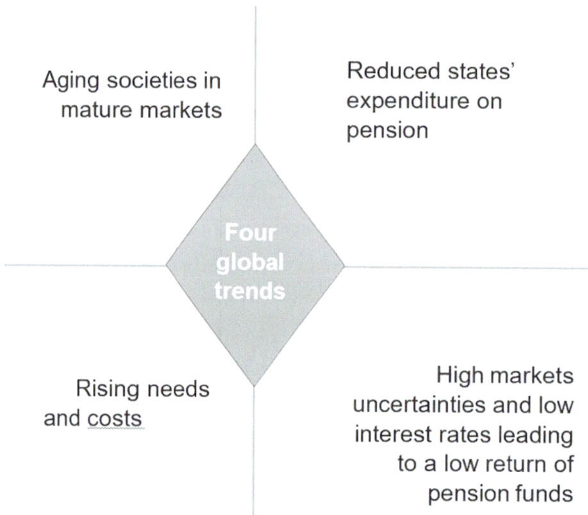

FIGURE 1.1 Four major trends will impact the retirement landscape

1.1.1 Medium- to Long-Term Shifts in Customers' Longevity, Risk Appetites and Behaviours Create Opportunities and Challenges for Insurers, while Decreasing Tax Benefit Calls for Innovations

1.1.1.1 The Demographic Dividend within Ageing Societies

Driven by falling lower birth rates, shrinking workforces and longer life expectancy, the world's older population is growing rapidly, with already more older people than children, and more people of extreme old age than ever before.

Today, 91 out of 195 nations have fertility rates below the replacement level of 2.11. At the same time, all age cohorts in developed countries, from "baby boomers" to "generation Z", can expect, on average, to live to 90 or beyond.

As a consequence, there was a substantial increase of 48% (from 607 to 901 million) of people aged 60 or over between 2000 and 2015, which might reach nearly 2.1 billion by 2050. There are now more people aged over 65 than under 52, accounting for more than 20% of the population within 15 "super aged" countries (including Germany, Italy, France and Japan), and w ithin 44 nations by 2030 and 61 by 2050. Moreover, the "oldest old" (aged 80 or over) population accounted for 14% of old population (aged 60 or over) in 2015, and is expected to triple 2015's value by 2050.

These demographic changes are making today's retirees face an entirely new set of risks and experiences, resulting in new needs and expectations on the part of clients for their pension plans.

For example, in 1939, before America entered World War II, life expectancy was just over 59 for men and slightly above 63 for women. This meant that, when social security was implemented with an eligibility age of 65, the average American would not live long enough to access the benefit. Today, for a married couple aged 65, a 50% chance exists that at least one of the two will reach age 89; a 25% chance exists that one will survive to age 94. In other words, that couple faces a one-in-four chance that one of them will be in retirement for three decades or longer.

And this in a context where costs continue to rise far more rapidly (~ four times) than inflation, putting significant pressure on public social security systems. As noted previously, when social security was created in the 1930s, the average person was not even expected to reach the full retirement age of 65. Today public expenditure on long-term care by 2050, considering only a pure demographic effect, is expected to grow by 1.2 percentage points in relation to gross domestic product (GDP), from the equivalent of 1.1% of GDP now to 2.3% (according to the Organisation for Economic Co-operation and Development). States are increasingly underfunded, leading governments to freeze public pensions despite the rising longevity, with a common trend towards the development of voluntary schemes, offered by private players, in order to ensure clients have a similar standard of living after retiring.

The issue of ageing used to be largely viewed in terms of its economic burden to society, as it tends to focus on the final stages of life. Even so, the benefits of this demographic shift could still outweigh the costs, provided the needs of an ageing population are properly managed.

- Genuinely longer working lives: globally, labour force participation growth is expected to slow from 1.8% per annum (p.a.) over the past 50 years to 0.3% p.a. in the next 50. Improving healthy life expectancy could justify delaying the retirement age across many economies. Employees are working more years of their lives than ever before. Employee preference was the most commonly cited reason that employees continue working while they are eligible to retire. This represents a shift in the mindset of individuals, but also the changing nature of work, which is less physically demanding, the improved health of those age 65 or older and the rise of alternative workforce opportunities.

- Moreover, this demographic shift may go hand in hand with increased consumption: increasing preventive health spending by 0.1 percentage points has been associated with a 9% increase in annual consumption by people aged 60 and over. Given this growing economic power of the elderly, the in-retirement market will be the focus in coming years, with the in-retirement population accounting for nearly a half of total income. Decumulation and income solutions will be the mainstay of retirement business.

- Finally, productivity gains are also expected: rising life expectancy correlates to increased output, thanks to the skills that older workers have accumulated.

Such opportunities open the door to moving towards a multi-stage cycle, with a variety of careers with breaks and transitions. This has the potential to provide societies with more choice and flexibility, and a better balance between work, leisure, family, finances and health.

1.1.1.2 Customers' Risk Appetite and Behaviours Have Become More "Efficient" with respect to Increasingly Defined Needs and Rising Costs versus Reduced State Expenditures on Pensions

As a result of this longer life expectancy and multi-stage cycle, implying increasing lifestyle and health care costs, the idea that individuals and households need to plan for their own retirement is gaining a great deal of attention.

For example, in the United States the sense of insecurity also comes from the worrisome state of the government retirement system, which reached an unsettling milestone in 2010. For the first time since 1983, social security payouts exceeded the amount paid in by workers. Individuals are also coming to terms with the decline of the defined benefit (DB) plan, a once common source of guaranteed lifetime income (with 28% coverage rate across employers in 1980). During the last 30 years it appears that most employers have determined that a defined benefit plan is too costly and carried too much long-term financial risk for the company (translating into a 3.5% coverage rate in 2006). They moved instead to tax-advantaged defined contribution (DC) savings plans (e.g. 401(k) plans in the United States) for their workers (31% coverage rate in 2006 versus 8% in 1980), thereby shifting the responsibility for retirement security and longevity risk from corporations to their employees. Unfortunately, most Americans have not yet truly comprehended this new risk that they now own. Perhaps even more worrisome is the fact that the percentage of Americans who are not covered by either a DB or a DC plan has remained constant at around 50%.

Those in the Americas are saving the least (14.5%), followed by Europeans (14.9%). People in Asia are saving the most, with an average of 15.9%. At a regional level, Russians are saving the least (11.1%), followed closely by Spaniards (11.2%). People in Austria and Switzerland are saving the highest proportion of their income (21.6% and 21.3% respectively).

In fact, although people are overly optimistic about how long their retirement savings will last (considering that they will be able to take out each year 10% without running out of money), they are still worried about outliving their money at retirement (90% in Spain and Japan, 80% in France, 70% in the United States, Germany and the United Kingdom), and their concerns about retirement risks have significantly increased since the 2008 financial crisis (for reasons such as market fluctuations, low interest rates, lack of guaranteed income, lack of social security benefits, paying down existing debt, lack of emergency savings, inability to meet monthly payments).

In this new context individuals are wishing to save more for their pension (75% to 90% across France, Germany, Italy and the United States), with millennials (18- to 37-year-olds) leading the way (16% of their current income, versus 13% for the "baby boomers"). Yet savings amounts are still low (only 50% to 66% actually save for retirement), while few among those who do save believe in the sufficiency of their savings (less than 15 to 30% of them in the United States, Europe and Japan).

As a result, private pension plans penetration is still too limited, mainly due to these factors.

- Current products and services are not adapted to clients' needs, notably in terms of simplicity, security and flexibility. A third (34%) of people take less risk with their retirement savings than with their personal savings. Insurers and pensions do not provide enough advice for the clients to invest efficiently for their pensions. The most important criterion by which clients choose retirement solution providers is by their quality of advice, followed by expertise and reputation.
    - For example, despite this being a top concern of employees entering the workforce, and total student debt outstanding eclipsing $1.5 trillion in the United States, only 1% of plan sponsors offer student debt repayment and refinance programmes integrated with their defined contribution plans. However, 38% of plan sponsors report considering this integration, and, even when it is offered, use of these programmes is low (6.3% and 2.0% respectively).
    - Even so, surveys show that significant efforts in terms of caring have been made by defined contribution plan sponsors (with respect to electing a deferral percentage that works best for them, choosing their investments and, ultimately, the portability of their retirement savings).
- Clients lack awareness: the majority of individuals over 45 do not feel financially well prepared for retirement, and do not know the extent to which they need to save for retirement, while most individuals have other savings priorities than retirement (which is ranked only fifth in the United States), such as day-to-day living expenses, their childrens' education, health insurance (30 to 40% in France, Germany and Switzerland), protecting the family in the event of death or paying their own debt. And many cannot afford to invest for their pension (e.g. 50 to 60% in France, Germany and Switzerland). Almost all

non-retired people think that access to information about how much money they need for their desired lifestyle and their likely living costs in retirement, or visualizing how life might be once they have retired, could persuade them to save more for retirement.

- The complexity of products: 80% of US consumers, for example, feel under-informed, and nearly a half do not understand the different pension options.

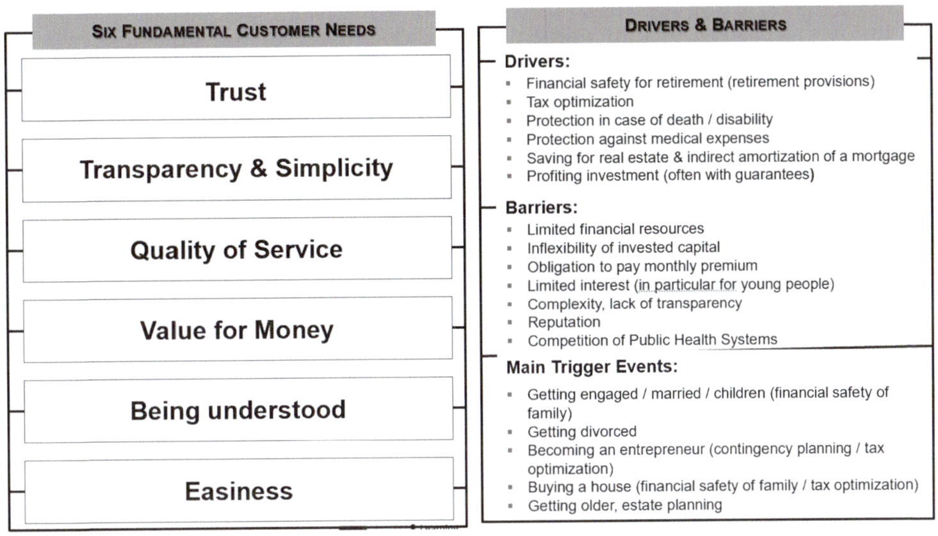

FIGURE 1.2 Customers' needs, drivers and barriers

As a result, insurers and pension providers need to help customers convert their financial worries into actions. In order to better apprehend the changes in customers' risk appetite, and thus their behaviour, let us focus on the individual life insurance products in the United States, with two major kinds of products: so-called "life products" (term life/whole life/universal life); and then "annuity products":

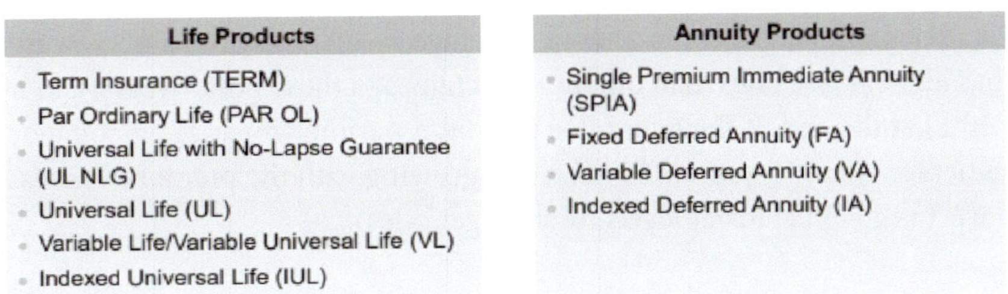

FIGURE 1.3 Major US savings and retirement products

Let us first review the generic structure of term life/whole life/universal life insurance policies.

- Term policies (massively developed in the context of high interest rates during the 1970s and 1980s) are issued for a given period, such as ten, 20 or 30 years. Their sole purpose is to provide financial protection (a death benefit) for your dependants in the event of your death. But, if you do not die during the term, the policy will expire when the term ends, and no benefits will be paid. One of the biggest benefits of term life insurance is that premiums remain the same throughout the term ("level premium period") of the policy (although they are reset at each renewal), which provides cost certainty, and it is usually much less expensive than whole/universal life if it is bought at a young age. The biggest drawback is that, if you do not die during the term of the policy, neither you nor your beneficiaries will receive any benefit. As a result, it is usually smart to choose a term that will cover the years when you are young and working, and your beneficiaries will need income if you pass away.

- Unlike a term insurance policy, a whole life/universal life insurance policy has value other than the death benefit, and lasts for your whole life. With this extended period, premiums are considerably more expensive, but with the major benefit that the premiums can grow (tax-deferred) as a guaranteed cash value (expressed as a percentage of the policy's face value) and be accessed over the life of the policy. And, if you do not touch this cash prior to passing away, it is disbursed to your beneficiaries in addition to the face value of the policy. These excess whole life premiums are invested at the insurance company's discretion. Most people will be better served by purchasing this product at a younger age (~ 20) to secure lower premiums.
- Universal life insurance (launched in the early 1980s) adds flexibility to whole life: you can make premium payments at any time you wish and in any amount you wish; and excess premium payments can be anything from stocks to bonds to mutual funds/annuities (albeit potentially restricted by the insurance company), thus with higher market risk exposure.

The following points can be made with regard to lapse behaviour.

- After the 2008 financial crisis the overall annual policy lapse rate (policy cancellation rate) decreased 4.0% annually (6.2% for term life, 4.3% for universal life, 2.9% for whole life), down from 4.5%, 4.2% and 4.9% in the 2007–2009, 2005–2007 and 2001–2002 studies. This applied across all three product lines and all policy years, although decreases occurred most significantly in the first three policy years, stemming from the economic downturn.
- Whole life lapse rates were noticeably higher in the early policy years relative to the other study years, caused by the recessionary economic conditions. It also appears that younger issue ages were more likely to react to the economic conditions than older issue ages.
- However, the aggregated average lapse rates across product lines increased in absolute levels in policy years 11 to 20 after the financial crisis, due to a change in the mix of business, with more exposure to term, which has higher lapse rates than other product lines in those policy years, and in particular a lapse shock (30 to 60%) at the end of the guarantee premium period, which often coincides with a premium increase (whether ten-year, 15-year or 20-year, and growing with the premium increase), and this across issue age cohorts (although at lower levels for younger ones).

Second, let us review the generic structure of annuities with living benefits.

- The policyholder receives at maturity a lump sum or an income (lifetime or fixed term) stream, after a "waiting"/"deferral" period during which savings are invested in financial assets (70 to 80% equities for variable annuities [VAs], mostly fixed income for fixed annuities [FAs] and fixed indexed annuities [FIAs], though also some equity-indexed annuities [EIAs]), and benefits from a minimum "guaranteed" rate (a 1 to 2% so-called "fixed credited rate" in FAs, a "credited rate linked to an index, up to a monthly cap that is reset on a yearly basis" in FIAs, a 4 to 6% "roll-up" rate in VAs).
- During the waiting/deferral period the insured is still allowed to withdraw up to a certain percentage of the guarantee on a yearly basis (with a penalty if larger than the guarantee times the roll-up rate), although she is usually incentivized with a "deferral" bonus to delay such withdrawals (usually by ten years), while potential surrender is penalized during a "surrender charge" period (usually seven years).

- In addition, if the contract contains a death benefit, then a certain amount is paid to the beneficiaries in case the policyholder dies during the term of the contract. Naturally, this death benefit is offered only if the income stream has *not* been activated.

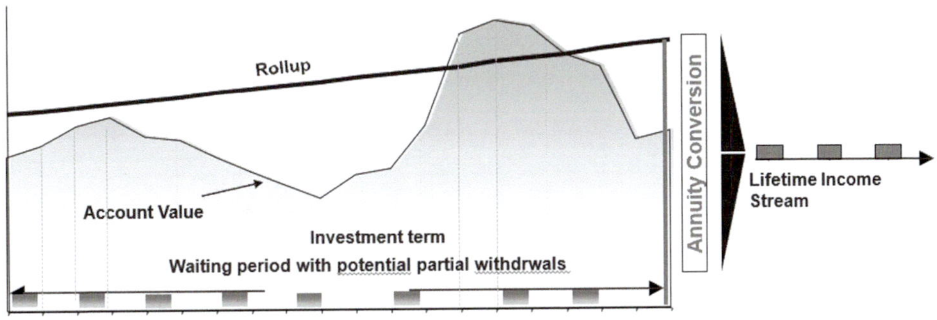

FIGURE 1.4 Retirement savings annuities mechanisms

As a result, behaviour risks mainly lie in three areas – surrender (or lapse) risk, withdrawal risk and annuitization risk (annuity election) – each of which is concentrated within specific phases of contract life: the withdrawal risk timing mostly occurs in the early stages, during the "ten-year withdrawal deferral bonus period"; the lapse risk timing is mostly concentrated after the "seven-year surrender charge period"; and the annuitization risk usually occurs far later.

Regarding policyholders' behaviours, there are still some key structural patterns, for instance regarding surrender behaviour.

- The surrender experience involves three regimes.
  - Maximum levels occur at the shock duration – the year following the end of the surrender charge period (mostly seven years: the "shock lapse"); although this stood at 30% at the onset of the 2008 economic crisis, shock rates below 10% were observed during 2016, and otherwise a post-crisis regime has prevailed, with shock rates in the range of 12 to 16% from 2009 through mid-2015 and in 2017/18. The 2016 dip is believed to be the result of uncertainty surrounding the US Department of Labor's proposed fiduciary rule and other political factors.
  - Lapse rates fall significantly in the quarter just after the shock lapse, and then decline gradually thereafter. In the quarter immediately after the shock lapse, lapse rates decrease by over 50%. In the five years following the shock lapse, rates fall by approximately 30% from the highs experienced in the first quarter after the shock lapse.
  - Surrender behaviour decreases with living benefits, and falls even lower with income utilization, and when guarantees are more valuable (e.g. with higher credited rates for FIAs, or with higher "in-the-moneyness" [ITM] for living benefits.

ITM is defined as follows.

- Nominal "in-the-moneyness" = current guarantee level/account value – 1.
- Actuarial "in-the-moneyness" = discounted future guaranteed income over the life of the product/account value – 1. Such an actuarial measure incorporates mortality and discounting effects.

- Although "out-of-the-money" (OTM) contracts experience surrender rates three times higher than ITM contracts under both measures, the "in-the-moneyness" is very much wider for the nominal measure than for the actuarial measure (80% versus 12% for the guaranteed lifetime withdrawal benefit [GLWB]).

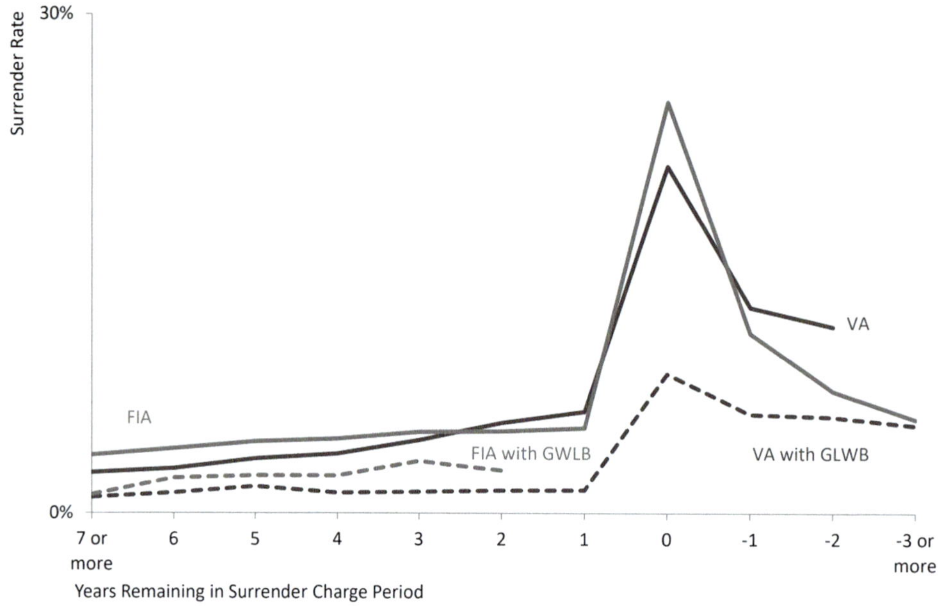

FIGURE 1.5 US pattern of customers' surrender rates

The following points can be made regarding customers' withdrawal trends.

- Withdrawals vary in terms of tax qualification status (a notable portion of nonqualified policies will not withdraw, whereas all qualified policies will withdraw by some ultimate duration) and by tax regime, with large withdrawals occurring just after some tax penalties disappear (59.5 years of age in the United States) or appear (71.5).
- Withdrawals increase with age, and this in an accelerated way.

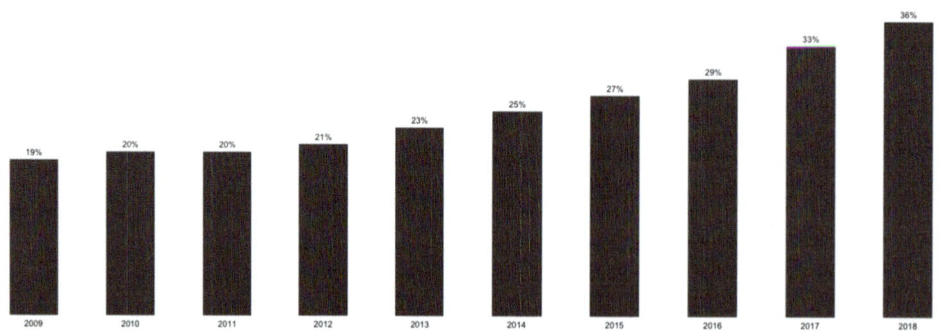

FIGURE 1.6 US pattern of withdrawals behaviour

Regarding annuitization rates (within the guaranteed minimum income benefit [GMIB]), they are de-incentivized by tax penalties before the policyholder is 59.5 years old, then jump before the 71-year-old required minimum distribution (RMD) tax penalty, but still stay low. The annuitization rate then increases until the last eligible year to annuitize. Part of the persistently low albeit rising annuitization rate lies in the usually associated guaranteed minimum death benefit (GMDB) to the GMIB, which is activated at death only if the annuity has *not* been activated. Many GMIB holders indeed consider this GMDB as sufficiently lucrative to justify buying the combo GMIB/GMDB.

Yet the dynamic sensitivity of lapsing, withdrawals and annuitization has grown over the past decade, becoming increasingly "efficient" in terms of the income utilization of the economic value (induced by the "moneyness", the ratio of income present value [PV] to annuitization PV or the roll-up product feature) for the policyholder, as illustrated by the following.

- Lower lapses, decreasing with the actuarial "in-the-moneyness" of policies after the surrender charge period.
- Income utilization has increased and become more efficient (with 65% at the maximum allowed for the GLWB), with withdrawals highest at issue and after bonuses expire after the guaranteed base roll-ups end (usually year 10 for both VAs and FIAs, specifically since 2008 with the so-called 200% ten-year deferral bonus, with a five-point uptick in the 11th contract year, lower before the 11th year [6 to 7%] and afterwards [9%]). The withdrawals frequency is then dynamic, increasing with age and with the "in-the-moneyness" of the policy for both FIAs and VAs (more than tripling, up to a 30% rate, for the GLWB).
- The annuitization rate also depends on the economic value of other benefits, such as continued income utilization, and so decreases as the ratio of income PV to annuitization PV increases.

In addition, there has been the emergence of some significant increasing connections between lapse and past withdrawal behaviours, as a contract's prior partial withdrawal history influences its persistence.

First, policyholders who withdraw more or less than they are contractually allowed show a higher tendency to lapse.

- Those who consistently withdraw moderately *more* than the maximum allowed withdrawal amount are about three times as likely to lapse in the subsequent year as policyholders withdrawing amounts equal to the maximum (i.e. efficiently).
- In contrast, those taking moderately *less* than the maximum allowed are about 1.25 times as likely to lapse as those withdrawing efficiently.
- Furthermore, if that same policyholder were to take extreme excess withdrawals in the year before the end of the surrender charge period, reducing the benefit base by at least 70%, then she tends to lapse at a rate more than six times higher than those withdrawing efficiently. A policyholder who continues large excess withdrawals during the last surrender charge ultimate year is more than ten times as likely to lapse.

Second, in contrast, policyholders with longer withdrawals are associated with a lower lapse rate.

- Policyholders who take systematic withdrawals on lifetime benefit riders experience very low surrenders in the first systematic withdrawal year, increasing thereafter.
- A policyholder who has been withdrawing efficiently for five years (ten years) tends to lapse at a rate equal to about 80% (70%) of that of an otherwise identical policyholder who has been withdrawing for just one year.
- These effects are diminished if the policyholder has a recent history of inefficient utilization – i.e. either under- or over-withdrawing the maximum amount. At the shortest end of the withdrawal history spectrum, a policyholder who is still deferring (by definition, withdrawing efficiently for zero years) is more likely to lapse than a policyholder who has been withdrawing efficiently for any positive length of time.

- The effects of inefficient prior utilization on lapse behaviour dissipate once a policyholder begins to take efficient withdrawals: policyholders who moderately over- or underutilized in prior years, and then subsequently began to take efficient withdrawals (withdrawals equal to the maximum allowable amount), exhibit very similar lapse behaviour as exclusively efficient policyholders

Overall, there is evidence of a rising anti-selection feature within customers' behaviour, with the emergence of three types of savings use behaviour patterns, characterized by the frequency and amounts of withdrawals, and strongly influenced by age.

- Before the age of 60: one-off withdrawals of heterogeneous sizes, very often massive (greater than 200% of the annual guaranteed withdrawal); illustrative of the consumption of expensive durable goods, real estate purchases, financing of education and vocational training, children's studies, the management of periods of unemployment, etc.
- From 60 to 70 years of age: regular withdrawals, of moderate and homogeneous sizes, during retirement; illustrative of the entry into annuity.
- After 70 years of age: mostly the entry into annuity, with regular, moderate and homogeneous withdrawals; but also some excess withdrawals (for tax reasons or donation).

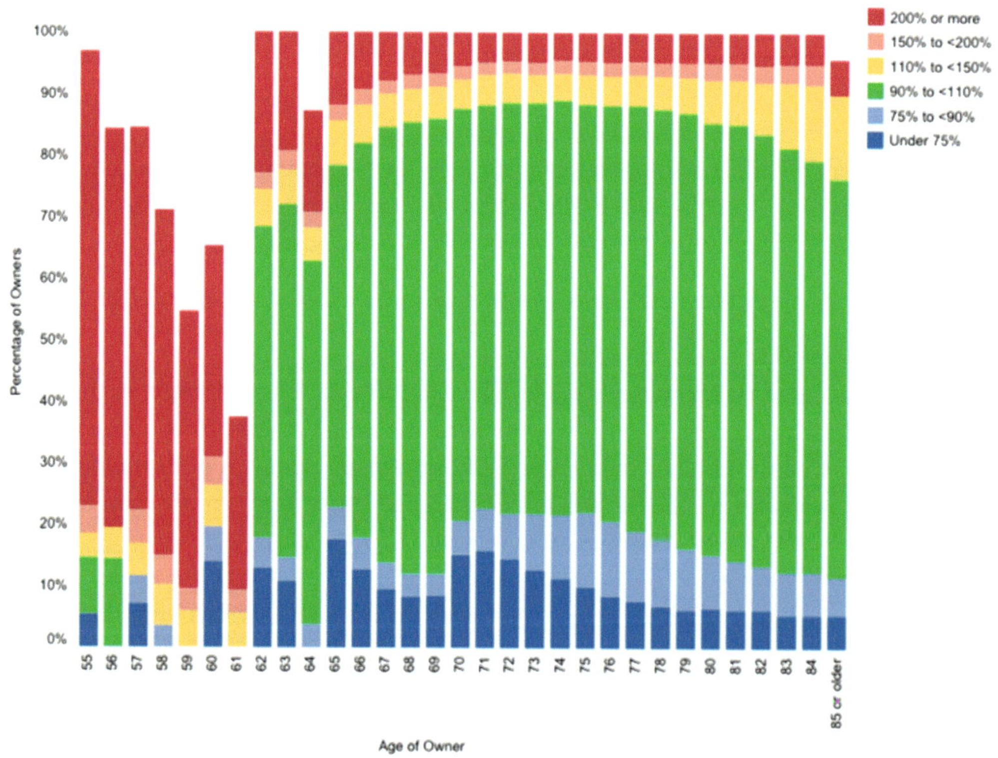

FIGURE 1.7 US withdrawals pattern

This segmentation of customer profiles reflects strong differentiation in terms of saving behaviours, which calls for an optimization of the product design tailored to its uses for an enriched customer experience.

For this multi-stage life cycle to be realized, insurance will need to design innovative products and services so as to address older-age appetites, at key events throughout their lives (e.g. marriage, divorce, new house, setting up of new business, retirement). And there will also be increased demand for more affordable, flexible, transparent and customized insurance that better suits such evolving conditions and can be easily adjusted as the needs change.

The customer experience has thus become companies' prime focus, with consumers now aware of their demands and rights more than ever. Consequently, insurers have had to raise the level of their offerings in order to engage customers.

	Upcoming clients needs…	…Require a specific answer from providers	Customers' appetite: Low income indiv.	Customers' appetite: Middle class indiv.	Customers' appetite: HNWI / Large corp.
Technical needs	■ Longevity risk coverage	Providing financial solutions for retirees living longer than expected / their savings product			
	■ Interest rates coverage	Guaranteeing principal with an above average interest rate and a fair return on investment			
	■ Inflation coverage	Covering against inflation peaks with annuities indexed on inflation	●	●	●
	■ Inheritability	Providing retirement products whose funds can easily be inherited by designated beneficiaries			
New set of expectations	■ Holistic products including health & LTC	Offering retirement products with additional health and/or protection coverage	◑	●	●
	■ Simplicity / Transparency	Offering simple, accessible and easily understandable products	●	●	◕
	■ Flexibility and funds availability	Tailoring flexibility solutions to adapt to the clients risk profile, and withdrawal needs etc.	●	◕	◑
	■ Accumulation/ Decumulation flexibility	Tailoring accumulation and decumulation flexibility solutions to adapt to the clients wishes to invest or withdraw money at all time, pre and post retirement	●	◕	◕
Additional services required	■ Advice	Advising customers on products and amounts to save	◑	●	●
	■ Online self servicing	Offering self planning and monitoring of retirement accumulation and decumulation online	◑	●	●
	■ Education and awareness	Educating customers in existing savings strategies through digital tools for instance	●	◕	◑
	■ Assistance services	Offering services combinations: daily assistance, transportation	◑	◕	●

FIGURE 1.8 New clients' needs and expectations for their pension plans

Notes: HNWI = high-net-worth individual; LTC = long-term care.

Within such a context, there is a lot of room for improvement in the use of data and digital technologies by insurance enterprises. Insurers that commit to gathering and analysing data for effective customer profiling and targeting will find themselves better equipped across the full value chain: differentiating their offerings (e.g. underwriting), strengthening their customer relationship management (e.g. claim management) and improving risk management and fraud prevention (through pattern identification). Moreover, the access to data has been drastically improved: whereas customer behaviour had formerly been centred on a reluctance to share data, and heightened concerns about the safety of such data, freely shared data from consumers has become widespread with data capture cookies being phased out on digital platforms.

1.1.1.3 Sustainability Offers a Historic Opportunity to Lead, Innovate and Grow Purposefully

Climate change and sustainability have re-emerged atop board and C-suite agendas as the direct impacts of the Covid-19 pandemic have receded. Previous discussions about sustainability were largely theoretical and centred on making public pledges of support. However, today leading insurers are taking tangible steps and adopting hard metrics to address the full range of environmental, social and governance issues and opportunities.

In fact, ESG ratings aim to measure a company's resilience to long-term relevant risks, through good sustainability practices within the company. Looking at the ESG factors one by one, the following points emerge.

- The "E" factors refer to the company's behaviour on environmental issues such as resource depletion, climate change, pollution.
- The "S" factors relate to the company's treatment of people, workers and local communities, including health and safety aspects.
- The "G" factors refer to corporate policies and governance, including tax strategies, corruption, the company structure and remuneration.

As a result, companies that score well should benefit from improved terms across lines of insurance – the liability of employers, directors and officers, product liability, public liability – given their focus on mitigating risk factors that cause reputational harm.

A high ESG score is thus likely to help a company to (i) attract more investors, (ii) lower the cost of capital and (iii) result in better operational performance, thus stock performance.

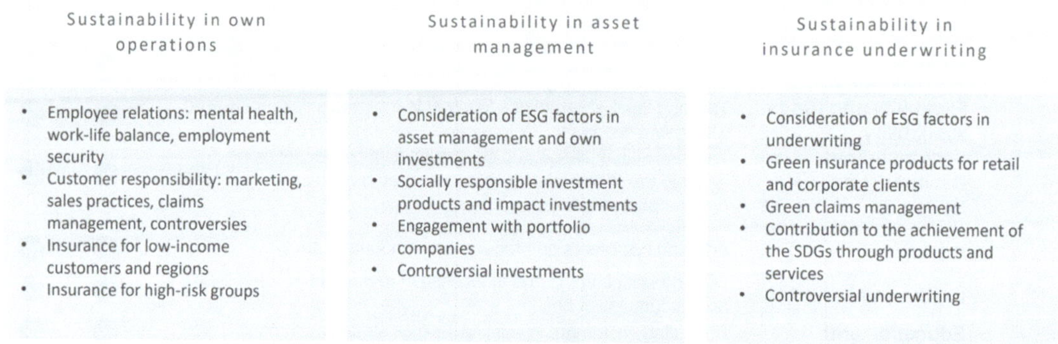

FIGURE 1.9 ESG issues in insurance

Note: SDGs = the United Nations' Sustainable Development Goals.

For most insurers, the focus is, first, on the "E" in ESG, as climate change will have the biggest and most immediate impact on the industry's financial performance. But, over the long term, this will also affect the health of the population, and therefore have an impact on the cost of health and sickness benefits and, ultimately, the value and cost of annuity policies, translating into a growing focus on the "S" in ESG.

The "S" is key, as insurers offer intangible (as opposed to physical) products that address the important social functions of a safety net: retirement savings and protection from financial losses. Properly designed and marketed insurance products are meant to incentivize socially desirable behaviours such as "financial wellness" or "retirement preparedness" for life insurance customers. Besides, negative publicity around "S" considerations may restrict a firm's ability to attract and retain talent if there is a "social stigma" associated with the firm, while compliance may boost employee motivation and attract talent through social credibility.

Another driver of the "S" screening is that insurance companies and pensions have been committed to responsible investing for years, dedicated to a long-term view. Investment strategies range from disciplined asset liability management to impact investing to diversity inclusion and investing to mitigate climate change. Based on polls, ESG is also a significant consideration for 80% of the companies in risk transfer, reinsurance and insurance-linked securities (ILS), though where they factor in their underwriting deliberations is less clear.

Therefore, the ESG score is becoming an increasingly key factor in companies' portfolio asset allocation, within the context of not only growing customer demand but also regulatory requirements. As a matter of fact, the value of ESG assets under management (AuM) is expected to reach some $50 trillion by 2025 – a third of global AuM.

Finally, the focus on the "G" factor translates into different governance and policies across the various departments (risk and cybersecurity, notably).

Many insurers produce regular ESG/sustainability reports and/or corporate responsibility filings, which typically discuss their approach to ESG, their accomplishments and their overall strategy in addressing the issue. Conversely, other firms seem to produce far less in the way of materials on the matter, and do not discuss the topic in calls and presentations with investors.

A number of companies produce in-depth materials or have dedicated websites on their ESG efforts, which highlight targets for emissions, their overall environmental impact and diversity and mobility opportunities within their workforce, and have committees and standards for their corporate governance structure. Conversely, some firms simply produce a standard code of conduct, with very limited mention of ESG in public forums or comprehensive ESG-related reports from Brighthouse Financial or Equitable Holdings.

However, it is difficult to determine the extent to which ESG is incorporated into most insurers' business practices. To be able to accurately benchmark, track and measure results of ESG business transformation efforts, insurers must identify the right key performance indicators (KPIs). Among the key axes are these:

- define clear and short-term objectives;
- clearly articulate what you want to achieve;
- work out how to measure progress, so establish a baseline to start from;
- identify KPIs to act on to improve performance;
- report on results at regular intervals; and
- be discerning and select solutions that are scalable, adaptable and capable of helping you achieve your outline goals.

Insurers can take many meaningful steps in the near term to help advance the transition to a greener economy. Mapping action plans to specific targets and establishing quantifiable performance metrics relative to sustainability are two ways that insurers can live their purpose. Within a broader ESG strategy, insurers must identify priority focus areas, clarify why they are allocating resources to them and determine the benefits they expect to achieve.

A clear road map must also reflect the impacts on different parts of the business and how ESG strategies will be executed. In tracking performance against sustainability targets, insurers should monitor risk exposures, value creation and progress towards specific goals. As reporting and disclosures become standardized, the most transparent companies will benefit from easier access to capital, increased customer loyalty and better share price performance.

1.1.2 Key Lessons from the 2008 Financial Crisis and the Persistently Low Interest Rates Environment regarding Sustainability

1.1.2.1 The 2008 Financial Crisis Was a Stress Test

We lived through a stress test in the financial crisis (–51% over 17 months), above the 99.5% percentile – a crash in equity markets, an increase in volatility, a drop in interest rates, a perfect storm in financial markets – beyond previous market crashes (–43% in 1973/74, –41% in 2000–03).

With the onset of the financial crisis some of the most successful players skidded into trouble. For example, between October 2007 and March 2009 six of the ten largest publicly listed VA issuers in the United States lost about 90% of their market capitalization, 50% of their sales and 25% in their AuM. Most products became unprofitable in economic terms, while the hedging strategies proved insufficient, leading to write-offs and capital injections. The industry's profitability sagged under rising guarantee values, the collapse of earnings from mutual fund fees, negative hedging results and exploding hedging costs. As a result, many companies' credit ratings were downgraded by the rating agencies. VA sales, having reached a pre-crisis high of $184 billion, have decreased by a half since then.

> AXA Group announced a US$3bn capital injection into its US subsidiary AXA Financial to bolster its solvency. One factor in this decision was variable annuity hedging losses projected at between €350 and €450 million for the fourth quarter of 2008 …
>
> In its report for the quarter ending September 30 2008, Dutch insurer AEGON announced a €48 million loss on US VA earnings.
>
> OLD MUTUAL
> *August 2008, investor Q&A*

Hartford first suspended sales in the United Kingdom, exited the VA business in the United States and Japan, then sold its new business capabilities to a strategic buyer. Mitsui halted sales of VAs in Japan. ING stopped selling VAs and variable whole life policies in Japan. MetLife divested Brighthouse in 2016. ING, which had $12.3 billion in sales in 2008, ranking fourth among sellers, exited the VA business and separated its legacy policies in fixed and variable annuities from its ongoing business – now VOYA – which it sold to Apollo Capital in 2016. Ameriprise discontinued wholesale sales at the end of 2010.

The best-in-class players managed to navigate relatively well through the crisis thanks to higher diversification, the focus on their core businesses, disciplined and timely monitoring of all asset class exposure (full reporting on "toxic" exposures, temporary measures on investment, stringent derivatives policies with daily or weekly margin calls) and active risk management (a dynamic hedging programme aligned with risk appetite, and a liquidity emergency plan).

The market turmoil of 2008 and early 2009, followed by the persistent decline in interest rates over the past decade, cemented a deep-seated crisis of confidence about how to truly create retirement security. What has occurred is nothing short of a decisive shift in the financial mindset for all Americans, but especially for those in or near retirement.

1.1.2.2 High Market Uncertainties and Persistently Low Interest Rates Have Had a Drastically Negative Impact on the Sustainability of Savings and Retirement Products

Since the 1990s recurrent unexpected severe volatility spikes (as illustrated by recurrent crashes: 1994 bond crash; 1997 Asian crisis; 1998 Russian crisis; 2001/02 Internet bubble crash; 2008 crash; 2010, 2012, 2014 European sovereign debt mini-crashes; August 2011 mini-crash; 2016 Brexit mini-crash; February and October–December 1998 mini-crashes; 2020 Covid-19 crash), combined with persistently low interest rates, have made most savings products less attractive and sustainable, translating into earnings losses for the insurance companies and lower yields for policyholders.

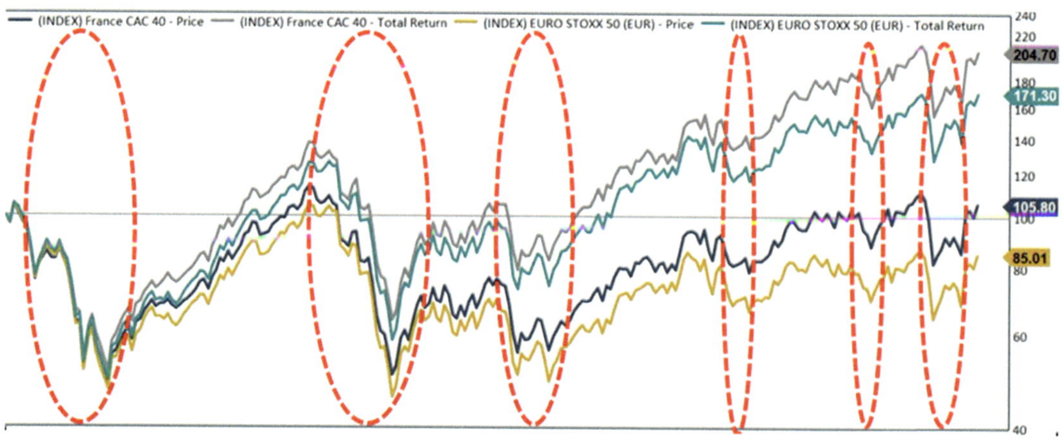

FIGURE 1.10 Recurrent equity market crashes and volatility spikes over the past 20 years

Market volatility certainly had a strong impact on returns sustainability, as illustrated by the extent to which the sequence of returns affects a portfolio: despite an average 8% return of the S&P 500 over the past 30 years, the range has varied dramatically from 37% in 1995 to –37% in 2008. Of course, this is not an issue for consumers focused on accumulation who do not draw down on their investments, and so weather the volatile times and ultimately achieve the average return for any given holding period, assuming the dollars stay invested. But, for those drawing down assets or taking income from their holdings, the variability of performance can have a huge impact, as illustrated below.

Let us consider the case of a retired couple with a $1 million nest egg and plans to withdraw 5% per year ($50,000), adjusted for inflation (3.5%) each year.

- If the portfolio generates a steady annualized return of 6% per year, the couple could count on their distributions lasting for 30 years.
- What happens if the portfolio moves up and down every year, while keeping the same 6% average return over 30 years and withdrawing $50,000 per year until the value hits zero? It then depends on the sequence of returns.
    - Four distinct rate-of-return values cycling every four years would translate the portfolio earnings into 27% in year 1, 9% in year 2, 7% in year 3 and –15% in year 4.
    - Consider the reverse-sequence rotation of the above, starting with –15%, then 7%, 9% and 27%: the different sequencing of returns represents a 13-year difference in how long the money will last.
    - In fact, the money could run out much more quickly than anticipated, or could be extended if investments perform better in the early years. As a result, avoiding early losses is key for performance; as any potential future gains would now accrue off a smaller base, you may not have ample time to benefit from a market recovery, particularly if you need to make additional withdrawals from your retirement account.

The problem is the unpredictability of variable investments in the short run, when it is already hard enough to accurately project an average annual return over an extended period. No matter how accurate you may be in projecting an "average" return, it provides little assurance that retirement income goals will be met if the sequence of returns happens to work in an unfavourable way.

It may seem that a simple solution for reducing a retiree's sequence of returns risk would be for you to reduce your equity holdings in your portfolio in favour of fixed-income investments. However, this approach compromises the portfolio's upside potential and may lead to a quicker – and premature – reduction of your long-term savings.

The severity of the financial crisis and the longer-than-usual time needed to benefit from a market recovery (2009–13) have undermined people's once steady faith in the equity markets as a solution to their retirement problems, forcing the next generation of retirees to look for solutions different from those that worked for their parents: among the entire baby boomer group, a desire to lock in guarantees, in lieu of higher returns, has emerged, with 80% preferring an investment with a 4% return and a guarantee against losses compared to one that pays 8% interest but is subject to market downturns. In short, the new buzzwords for ageing baby boomers are "lifetime income" and "guarantees".

Actually, whether you're in your 40s or 50s, and looking for upside investment potential or some guaranteed returns, annuity solutions can address a range of long-term retirement planning needs. Income annuities are a useful hedge against sequence-of-returns risk for two main reasons: (i) they provide a guaranteed source of lifetime income that is not correlated to market ups and downs or interest rate fluctuations; and (ii) the annuity regular income lowers the withdrawals that you might need to take to cover expenses. This is particularly good news should the market perform poorly in the early years of one's retirement, because these solutions can help retirees avoid selling at the bottom.

However, the cost of those annuities guarantees has significantly jumped, not only with the rising equity volatility but, even more, with the declining interest rate environment throughout the last decade, triggered and driven by the rise in central banks' balance sheets since the 2008 financial credit crisis (in the United States

$4.6 trillion in 2019, then $9 trillion after the Covid-19 crash). Persistently low interest rates have indeed been the major driving factor behind the lower sustainability of savings and retirement products.

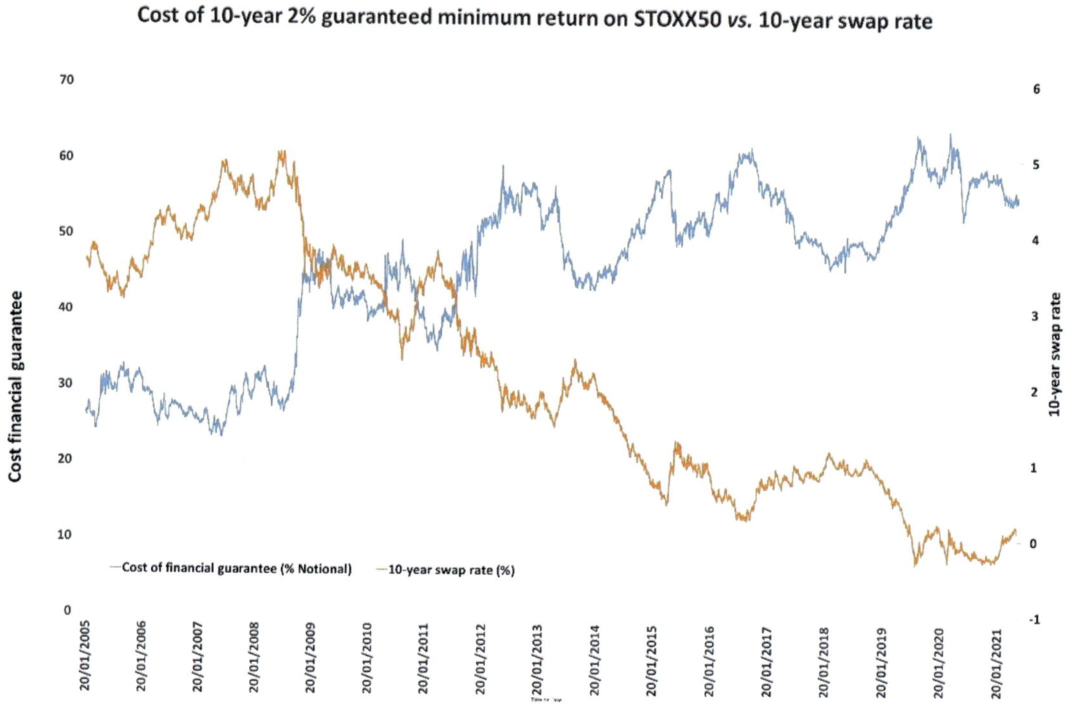

FIGURE 1.11 The cost of guarantees has jumped in line with declining interest rates

Despite some attempts by the Fed to reduce temporarily the balance sheet by $600 billion and a few interest rates hikes from 2017 to 2019 and the recent jump in consumer inflation (up to 7% annually in 2021/22), interest rates have not recovered to pre-financial crisis levels. The normalization from artificially low levels is unlikely before a few years given the huge remaining balance sheets of central banks.

Such a historically persistent low interest rate environment has contributed to a drastic increase in both premium levels and the risks inherent in major savings and retirement products, translating into a significant sales drop. Although large fixed-income allocations and greater amounts of hedging have helped reduce volatility, it has become extremely challenging for assets to grow at guaranteed rates in a prolonged low-rate environment.

- On the one hand, lower interest rates increase reserves (using lower discount rates, resulting in a higher present value of expected convex claims – in contrast with usual linear fees, which are hardly impacted), capital requirements (which are related to regulatory shocks as a percentage of those higher reserves) and the cost of hedging (through higher options prices induced by lower discount rates).
- On the other hand, lower interest rates, and thus yields, increase difficulties in supporting minimum guaranteed rates, both in equities for VAs (due to higher costs of guarantees) and in fixed income for FA/FIA/term and whole life and universal life and general accounts (lower credited rates due to credit spread compression), and decrease return on equity (ROE) due to lower net income from higher reserves levels.

Unfortunately, excessively early expectations after 2009 that interest rates would rise again have pushed most writers to consider growing further and developing products that are still generous, and thus risky, at the same price, as illustrated by high minimum guaranteed rates and withdrawals (e.g. 5 to 7% roll-up and guaranteed withdrawal rates, annual/quarterly/daily ratchets within US variable annuities by Jackson, Metlife and Prudential).

Even so, premiums were increased (e.g. a rise in hedge fees from 0.80% to 1.15 to 1.35% p.a.), while the implicit technical rates embedded within the guaranteed income were drastically reduced, indirectly, through much less generous mortality tables (i.e. by discounting the official mortality tables by a significant factor, which implied significantly higher premiums).

Such combined de-risking, through offering lower product richness while raising fees, has also impacted policyholders' behaviour into lower lapse levels, as new savings and retirement products became less and less attractive, with surrenders at the shock duration going down from 30% in 2008 to 10% in 2016 and 12 to 16% in 2018.

In contrast with the situation prior to the 2008 financial crisis, when the vast majority of annuity guarantees were in-the-money, as interest rates have remained persistently low and volatile while policies have aged within a slow market recovery (from 2009 to 2013), a greater percentage of the book went ITM with rising cumulated fees, which implied lower lapses that accounted for the major part of the earnings loss arising from customers' behaviour, while partial withdrawals went down in line with rising past durations.

Estimates of the lapse response to the "moneyness" of guarantees vary widely, as insurers are making very different assumptions with regard to behaviours in their annuity books. For business that is 30% in the money – as with much of the business in force today – estimates of the reduction in lapse propensity due to the "moneyness" vary hugely, from 20 to 75%, thus dropping down to less than 1%. This led to the so-called "double trouble": the most expensive policyholders (those with "in-the-money" contracts in which the benefit base is larger than the assets) are lapsing at a lower-than-expected rate while more profitable policyholders (those with "out-of-the-money" contracts) are lapsing at a higher-than-expected rate. The impact can be especially costly to insurers, as more benefits will be paid out than priced for, while margins are compressed – at a time when hedging costs are increasingly expensive because of low/volatile equity market returns and low interest rates.

Behavioural risk, and in particular lapse risk, has one similarity to market risk: the magnitude of its potential impact on balance sheets, which translates into significant rising reserves, capital increase and earnings losses. But, in contrast to market risks, policyholder behaviour risk is virtually impossible to hedge for, and can be transferred only through reinsurance; this is why a company that has misestimated policyholder behaviour may have little choice other than to recognize the adverse experience in its financials.

As a result, additional fundamental de-risking measures were implemented by most life insurers from 2013 to 2016.

- Reserves were strengthened to reflect updated policyholder behaviour assumptions, to (i) lower lapse (level and floors) for ITM policies, while (ii) higher for OTM policies; withdrawals assumptions were reduced beyond a certain policy past duration or whenever the long-term interest rates go below a certain interest rate level; lower lapse levels were assumed for remaining policies after a buyout, with a lengthened time for full lapse recovery. Reserve charges of several billion dollars have been incurred, notably by AXA Equitable Life Insurance, Metlife and ING US's life insurance companies, while John Hancock, Sun Life Assurance Company of Canada (US) and Prudential Insurance Company of America took lesser charges.
- Buyout initiatives were launched targeting the richest policies to lapse or convert into alternative insurance policies (e.g. offering between 35 and 70% of International Financial Reporting Standards [IFRS] value in cash, floored at two years of premiums; in the form of an increase in annuity account value in the base contract in exchange for terminating the riders, clients then being able to withdraw the cash or remain in the contract with no benefits). From customers' perspective, they could be interested in monetizing part of their account value/benefit base in the form of an accelerated guaranteed surrender value to, potentially, subsequently reinvest their proceeds in higher-return products (the Buoni del Tesoro Poliennali five-year yield stood at 2.9% in Italy in 2013). From the insurer's perspective, the buyout offers

the opportunity to redeploy capital to less capital-intensive products or increase dividend, achieving a positive P&L by not fully paying out the IFRS guarantee reserve.

- Product designs were amended to incorporate de-risking and lower richness:
    - higher fees (monitoring and evaluation [M&E] and riders); possibility for the insurer to increase fees at his discretion;
    - age restrictions (higher withdrawals at higher attained ages);
    - lower benefits/guarantees (lower credited rates in FAs and FIAs, lower with-profit crediting and bonus in general accounts, lower guaranteed rates for term/whole life and universal life policies;
    - lower roll-up rates (cut from 7 to 4%) or floating roll-up rates in line with long-term Treasury rates; lower guaranteed withdrawal rates; lower frequency of ratchets; limiting maximum annual growth on resets/bonus features;
    - stricter asset allocation limits (higher mandatory minimum allocations to fixed/balanced accounts); restrictions on the number of equity funds (focusing on large indices such as S&P 500 with lower volatility); increased use of "tracker" funds; stricter governance in product approval process;
    - restrictions in the ability of policyholders to "time the market";
    - premium suspension (no additional contributions into legacy); and
    - increased hedge coverage.
- Most life insurers have also improved their risk management practices by increased use of volatility hedging, adding macro hedges to protect against steep interest rate drops.

Even so, despite the generous buyouts to lapse rich old policies and significant incentives to convert into new ones (equipped with the 200% ten-year deferral bonus), and the improved risk management techniques and culture, all these initiatives still could not fully mitigate the impact of the persistently low interest rate environment with lower lapses, the higher cost of hedging and lower credited rates within FAs, FIAs and general accounts (due to credit spread compression).

With the introduction in the European Union of quantitative easing in 2015, followed by the Brexit referendum in 2016, interest rates have persistently declined. Despite reducing guarantees on new business and taking action over asset liability management (ALM), low yields have created a hostile environment for the traditional life insurance industry.

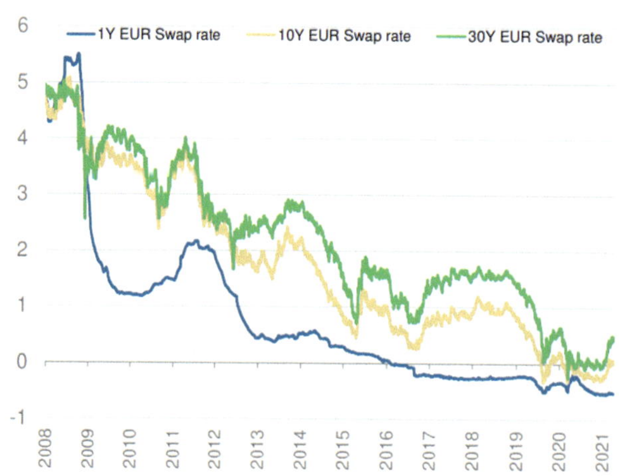

FIGURE 1.12 Interest rates declined over the 2010s

In Europe, German life is the most exposed, with yields falling more sharply than new business guarantee rates, translating into duration. In this way, a disconnect between the proportion of capital allocated and the earnings – and, crucially, cash – was generated.

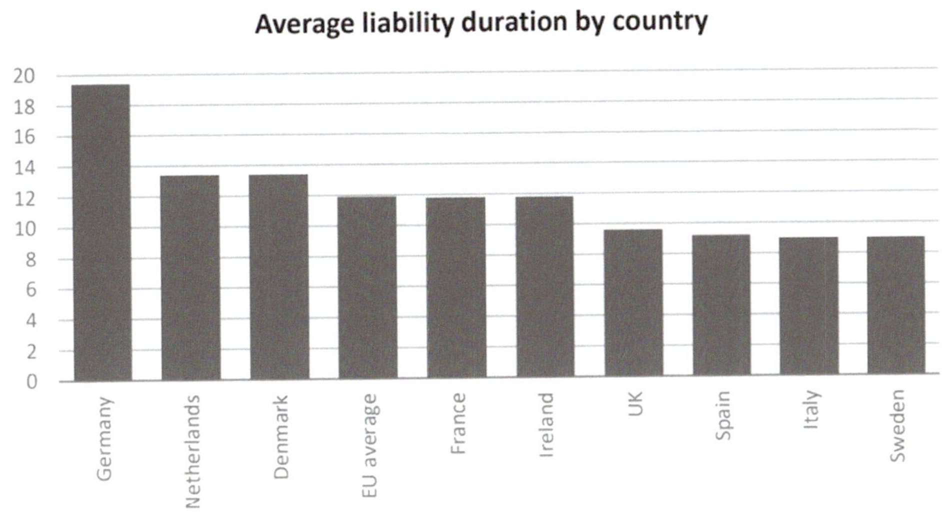

FIGURE 1.13 Germany has the longest life liability duration in Europe

As bond yields have continued to trend lower, even the lower guarantee rates on the new generation of products have become increasingly unaffordable. Companies are pushing further to reduce the cost of guarantee through various ways, such as offering a guarantee for only a proportion (typically 60 to 90%) of the premium instead of 100%, or charging fees higher than the "nominal" guarantee rates.

Further solutions have been developed since 2016 across the world.

- Decelerating the distribution of variable annuity and general account sales until interest rates mean-revert to sustainable long-term levels.
- Reducing the capital guarantees of the general account, "net of fees", and so with a level of 80 to 92% of the initial amount.
- The establishment of a hybrid euro fund that incorporates a portion of general account (e.g. 30 to 50%) while the residual is invested in non-guaranteed unit-linked (UL) assets.
- As persistent low interest rates have increased customers' concerns about fees and triggered drastic changes in their behaviour (lower lapse and partial withdrawals), retrieving resilience in profitability requires the development of customer-centric designs within a "rational" framework that balances benefits against fees.
- The reinsurance of businesses: fixed and variable annuity businesses embedding rich guarantees, such as Jackson in 2019, Equitable Financial (2006–2008 legacy block, accounting for 33% of the fixed-rate GMIB and 13% of total variable annuity in force) in 2020 and Prudential (block before 2010) in 2021 to Apollo. Such reinsurance initiatives enabled the generation of approximate value and capital release while reducing risks (respectively $1.2 billion and $800 million for Equitable Holdings on a statutory basis, so –64% statutory capital requirements and about –50% of net amount at risk).
- The exit and disposal of businesses. MetLife divested Brighthouse in 2016. ING, which had $12.3 billion in sales in 2008, ranking fourth among sellers, exited the VA business and separated its legacy policies in fixed and variable annuities from its ongoing business (now VOYA), sold to Apollo Capital in 2016. Ameriprise discontinued wholesale sales at the end of 2010.

- Further buyouts, notably "lump sum payment options" that provide certain contract holders with a percentage of the PV of the lifetime payments in a one-time lump sum payment at the time their annuity account value falls to zero (either due to a withdrawal or surrender).

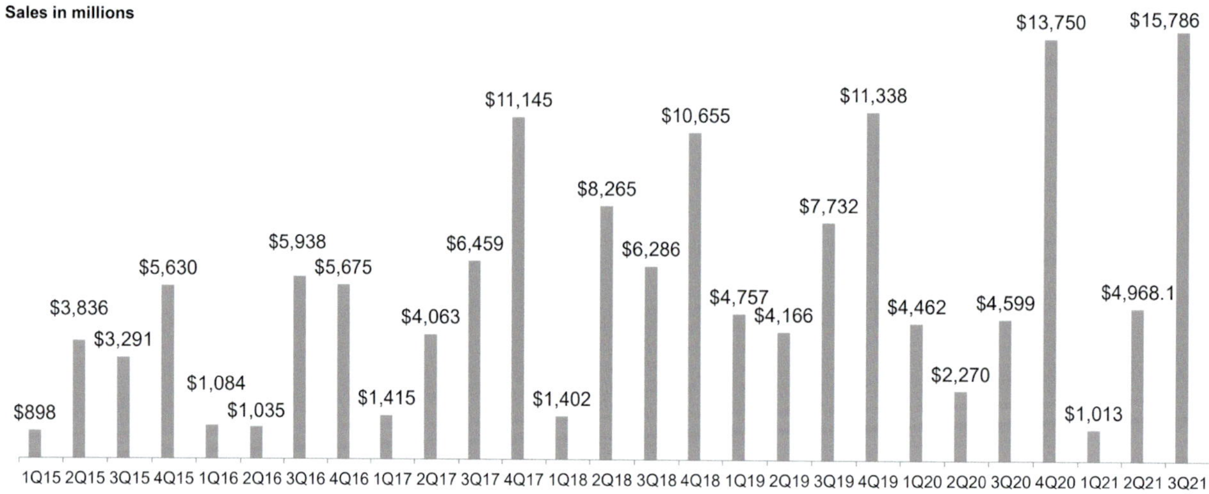

FIGURE 1.14 Buyout sales in the United States

- Develop alternative products more resilient to persistently low interest rates, such as:
  - fixed index annuities, which not only suffer less from guaranteed rates, thanks to market value adjusters and resets of guaranteed rates after short guarantee periods, but also benefit from the equity upside potential, as illustrated by the 75% S&P 500 growth since 2013;
  - asset management retail products (individual constant proportion portfolio insurance [iCPPI] policies), offering best-effort-only capital protection (versus capital guarantee within fixed or variable annuities);
  - structured index annuities (so-called registered index-linked annuities: RILAs) with third-party guarantees, leveraging on partnerships with investment banks to hedge the market risks;
  - "eurogrowth" funds, which hold a partial capital guarantee (e.g. 80 to 85%) only at maturity (and no more at all times); and
  - German, Swiss and Italian life insurers launched "hybrid" contracts, which combine unit-linked and traditional savings features; these hybrid products typically offer lower guarantees and have much lower interest rate risk, and therefore are more capital-efficient.

Even so, the higher market volatility and further lowering of interest rates triggered by the Covid-19 market crash has strengthened the increasingly "efficient" behaviour of policyholders, illustrated by lower lapse and withdrawal rates, among contracts with the highest propensity to exercise the benefit (e.g. in-the-money contracts following the end of the deferral bonus period or the surrender charge period).

1.1.3 In-Force Management Opportunities and Challenges: a Strategic Asset for Sustainable Savings and Retirement

Until recently insurers have focused on acquisitions, and resources and technology were deployed to generate growth, and sometimes returns, from new customers. In particular in emerging Asia, with its strong economic growth, increasing urbanization has increased incomes, contributing to an increase in life insurance penetration rates. Even so, in recent years the ratio of the new business premium to the in-force premium has decreased, with worldwide premiums +2% over the 2008–18 compound annual growth rate (CAGR) p.a., vs. +7% CAGR p.a. for the 1998–2008 period.

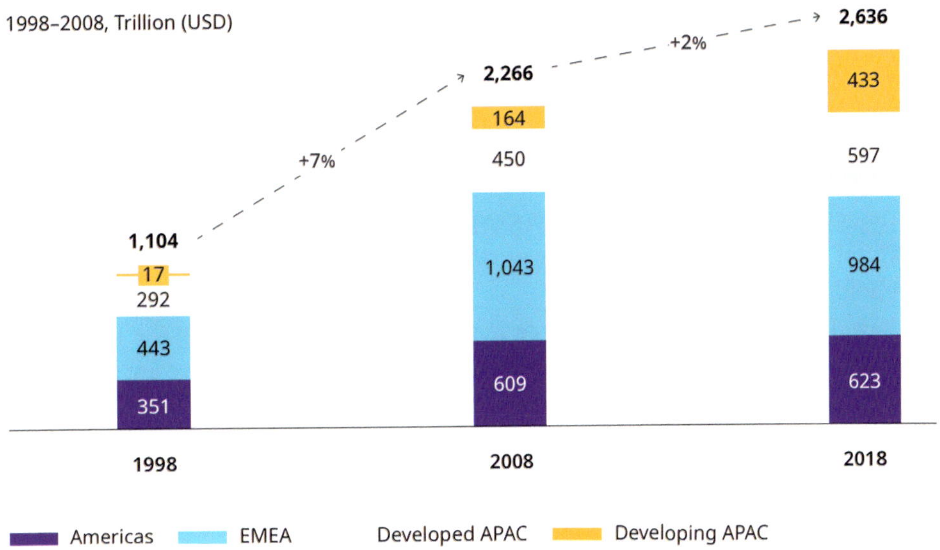

FIGURE 1.15 Premium growth of life insurers

Notes: EMEA = Europe, Middle East and Africa; APAC = Asia–Pacific.

In-force portfolios have traditionally provided life insurers with a steady and reliable stream of earnings that has driven stable industry margins and contributes a relatively large share of the profit and revenue. However, with low interest rates here to stay, inducing both lower yields and higher reserves of increasingly costly guarantees, and increasing volatility in investment markets and risk-based capital regimes becoming prominent in most markets, margins have been under pressure while customers' risk appetite and the needs of both financial protection and market upside are less satisfied. Thus insurers need to re-examine how they can create value for their customers while delivering attractive returns for shareholders, translating into KPIs such as stronger profitability and improved capital efficiency.

Furthermore, operational costs have continued to increase, hiding within them a worrying trend around the slow pace of automation and digitization of back-end processes in the sector. The in-force book is hardly homogeneous, comprising many product lines, vintages and customer sub-segments with varying degree of profitability. Therefore, an opportunity exists to focus on one of their most valuable and underexploited assets – their in-force portfolio – while unlocking significant basis points (bps) of reserves in additional operating profits.

Because of the scale of insurers' in-force portfolios, a relatively small incremental improvement in in-force performance can significantly impact bottom-line earnings and the value of the business as a whole.

These performance gains can be significantly improved if insurers use a targeted approach to in-force portfolio management, given the wide variety of customer and policy characteristics that exist within life insurance portfolios. There are many ways insurers can improve the performance of their in-force portfolios:

- improved customer experience, notably through customers' retention management combined with cross-/upselling and conversions of legacy products to alternative offerings;
- capital optimization;

- earnings generation and increased resilience;
- expense control;
- strategic rationalization of the asset allocation without compromising value for customers;
- liability management through more sustainable and flexible new business; and
- risk transfer.

Any project to improve performance must also be grounded in granular and robust analytics. High-quality informative analytics, combined with deep business experience and local market knowledge, tell us where to find and how to extract the value from an in-force portfolio that can give a much-needed boost to sustainable earnings and shareholder value.

Some insurers have implemented major in-force management projects during the last decade to enhance value using the activities outlined above, while more insurers will follow suit. Here are some examples.

- AXA Group's in-force optimization programme, which was composed of over 300 projects between 2010 and 2015, has managed to release €2.6 billion of capital for life and savings business, delivered €500 million in additional underlying earnings, with a 12% ROE, and delivered significant cost savings (lower administrative costs and acquisition costs through higher distribution efficiency and commercial productivity). Among the key initiatives are:
  - reshaping the capital-intensive general account savings offer (14% of the annual premium equivalent [APE] during the in-force programme) through reduced guarantees on new business, or combined with UL/protection riders (using customer incentives such as "Bonus Euro+" and distribution commission schemes), into less capital-intensive hybrid savings that are more resilient within the persistent low interest rate environment, and a selective growth focus;
  - closing (if unprofitable), repricing, redesigning (e.g. passive assets or floating-rate liability guarantee) and de-risking new business in variable annuities, and externalizing guarantees on market risks to third parties (asset managers, investment banks such as Structured Capital Strategies in the United States, Valore Performance in Italy, Sérénité in France or in Switzerland);
  - actively upselling protection and health within savings and retirement products (17% of APE during the in-force programme) – benefiting both from new clients' needs and lower exposure to interest rates risks;
  - diversifying the unit-linked product offer (34% of APE during the in-force programme), based on customers' profiles, with alternative asset classes;
  - expanding the ALM duration gap strategy to lower the sensitivity to interest rates risks;
  - introducing buyout initiatives (Belgium in 2016);
  - identifying growth opportunities in specific countries:
    - group pensions in France (defined benefits schemes for high-level executives and defined contribution schemes to all employees until annuity exit, with a high small and medium-sized enterprise [SME] market penetration rate) and in the United States (municipal workers and SMEs, through leveraging a leading position and expertise from the educational profession business, with direct-to-consumer distribution capabilities);
    - individual retirement income solutions in the United States (leveraging on a dedicated proprietary tied agent network, the strength of the existing wholesale model, cross-selling opportunities with

US property and casualty [P&C] carriers, expanding the 403 (b) K-12 education market, while decreasing market risks exposure through floating-rate guarantees/passive asset management and selective buyout offers), in Japan (partnerships with a national tax advisory company to educate customers), Italy (online tools to educate retirees and generate leads) and France (a "PERP" savings plan, characterized by tax-deductible lump sums blocked until annuity exit, through the development of the eurogrowth product, a higher share of unit-linked, modernization of IT systems and reinsurance);

    - high growth savings markets in Thailand (bundling with health and protection riders, a successful proprietary network and a bancassurance channel);

  - enhancing the distribution (41k tied agents specialized in savings) through new digital entry points, raising adviser expertise (sales tools and processes, regular training and news flow, technical support to sales force), partnerships with major banks, and social-media-linked customer relationship management; and
  - the disposal of pension businesses.

- In 2014 Zurich embarked on phase 1 of its in-force management programme and aimed to deliver $80 to 100 million in additional operating profits in two years. By the end of 2016 the programme had managed to deliver more than $150 million. Phase 2 of the programme, from 2017 to 2019, also exceeded the $100 million target, delivering $172 million in additional operating profits p.a. by 2019. Initiatives to simplify operations also have managed to reduce operating expenses, by $600 million.

1.1.3.1 Assess and Prioritize Opportunities for Earnings, Value and Capital Enhancements

The in-force book is hardly homogeneous, consisting of many product lines, vintages and customer sub-segments, all with varying degrees of profitability. Given the range of performance improvement opportunities by type and by portfolio, insurers will need to prioritize their actions based on the financial impact, the risks involved and the implementation timescales.

Insurers must first understand the profitability of their own in force by conducting a detailed review of what is in force and determining its profitability by sub-segments, thus providing the appropriate prioritization and allocation of resources to the most impactful actions. Additional investigation will probably be required to ensure the appropriate allocation of revenue, expenses and capital items at a more detailed level, including consideration of the impact of marginal changes in the portfolio, including any potential diversification effects.

In this respect, leveraging the efforts of a dedicated team to institutionalize a focus on the in-force book across the enterprise makes sense. Such a team must streamline processes for the capture of financial data for analysis, establish a knowledge store (i.e. a comprehensive inventory of all data, metrics, analysis and reporting that currently exist for the life in force, to be able to pinpoint the source of variances and perform statistical analysis to identify the impact if current trends continue), partner with service centres and customer analytics to identify key operational and customer metrics and then integrate the financial and operational data into a holistic view for the management board.

The financial impact of each opportunity will typically examine a range of accounting bases (local regulatory, Solvency II, local generally accepted accounting practice [GAAP], IFRS and embedded value [EV]); in fact, for some opportunities such as retention management, different accounting metrics can produce very different results and, potentially, drive different management actions. Depending on the specific business, the focus may be different across the four axes, earnings/value/capital/liquidity: on earnings volatility and capital burden, to support cash generation and new business for capital-intensive business (general account, variable annuities), but also on efficiency of capital usage and value creation, including customer behaviour (single-premium immediate annuity [SPIA] and deferred annuity).

Even so, the selection of most economic metrics, such as embedded value, combined with a metrics illustrative of earnings cash distributions to shareholders, will be the usual key drivers of decision-making to align actions more closely with both customer and shareholder values.

Once the most adequate metrics has been selected, both the short- and long-term implications of decisions will need to be taken into account (e.g. not just the current value of customers' in-force policies but their customer lifetime value, including allowance for any market cyclicality), as well as other material business issues that affect daily operations (e.g. how the cash flow position and the profit of the business will be affected), and balance sheet and business strategy impacts.

Within such a process, insurers also need to look again at their existing end-to-end processes around in-force servicing, such as claim processing and payments, which are often plagued with inefficiencies and low-value manual tasks, which could be expedited and automated through process re-engineering and automation, so as to reduce not only cost but also claim leakage, and improve the customer experience (e.g. improved response time due to straight-through processing, faster turnaround times for customer requests, more predictable outcomes and experiences, and an increased level of personalization).

In this respect, insurers will need to look at their existing IT infrastructure and assess whether there are opportunities for cost optimization through outsourcing, such as to potential software-as-a-service (SaaS) providers and cloud-based policy administration systems that not only reduce complexities associated with in-house customizations but also have many downstream benefits, such as faster and more seamless customer response and more autonomy to agents through self-service portals.

With a suite of approved business cases, opportunities can be prioritized and an overall programme can then be established, developed and implemented to improve in-force portfolio performance.

The in-force management initiatives can take advantage of the levers depicted in Figures 1.16 and 1.17 to make an impact.

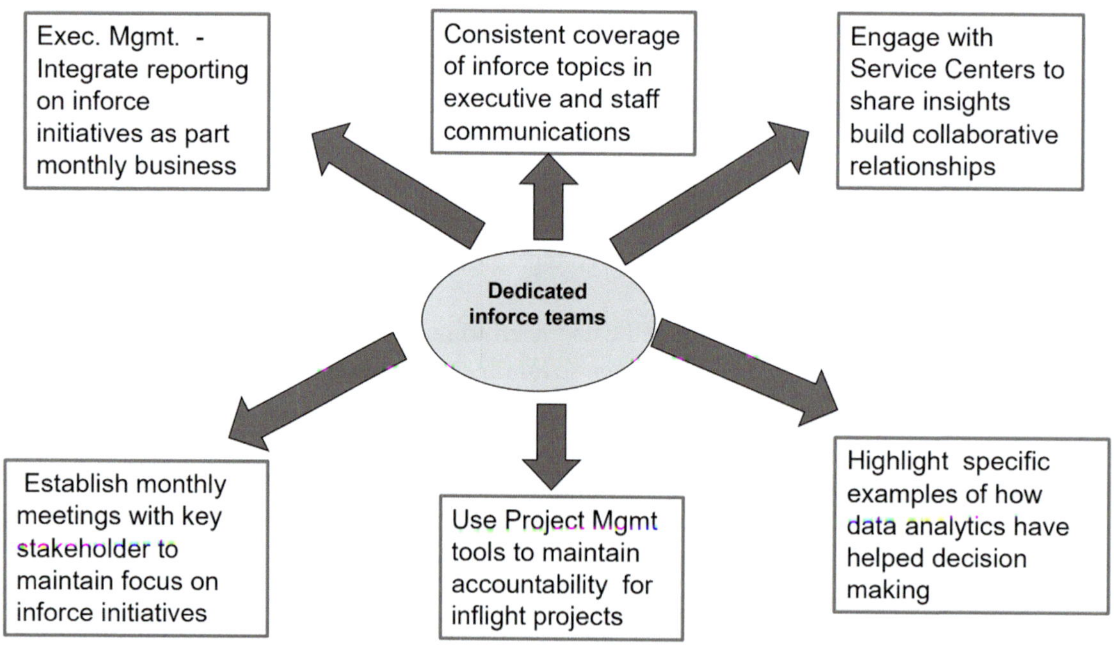

FIGURE 1.16 Dedicated in-force management team

Lever	How can it impact profitability?	Potential issues
Cross-Sell and Up-Sell	• Additional product sales • Improve inforce economics	• Impact on distribution • Cost to execute
Customer / Advisor Behavior	• Lapse • Utilization of benefits • Additional premium contributions • Asset allocation • Execute options against the company (e.g. utilization) • Trigger lapse/rollover	• Very difficult to influence • Waking of 'sleeping dogs' with any proactive outreach • Ensuring a fair deal for consumers • Willingness and ability to pay compensation • Ensuring tri-party alignment of interests • Ability to execute any programs at sufficient scale
Financial management	• Direct costs (e.g. hedging) • Overall risk profile and financial flexibility (both hedging and reinsurance)	• Limited supply of capital currently • Many options may be cost prohibitive or contain limited benefits
Funds	• Manage separate account economics and risk profile	• Legal and compliance
Pricing and Offers	• Manage our fee levels (product, rider, convenience) as appropriate	• Many products are limited • Legal and compliance • Cost of execution
Rates	• Managing crediting rates to improve spreads	• A large portion of the business is at minimum • Crediting strategy as environment changes

FIGURE 1.17 Levers for in-force management impact

Retention management and asset and liability management are two of the key areas for in-force portfolio performance improvement, which we develop and illustrate further below.

1.1.3.2 Better Understand Customers' Retention Drivers and Tailor Interactions to Improve Portfolio Profitability

For life insurers with a cross-regional footprint, lapse rates vary radically across markets, for market-specific and product structure differences. Given the long-term nature of most life insurance contracts and the high acquisition costs associated with getting new customers, even basis point improvements in lapse rates can unlock millions in profit and embedded value.

Retention management is the process of measuring the value of customers at a granular, segmented level, of identifying policyholders' behavioural characteristics that drive lapse rates and subsequently implementing measures aimed at retaining positive or high-value customers. Among the key stages of such a process are the following.

- Developing a rigorous understanding of individual customers' value across their life cycle, which reflects the value of future earnings of existing policies and the potential value of additional policy sales (a customer lifetime value assessment). As their life circumstances change, very few customers actively seek out and reassess if their insurance needs are adequate.
- Analyses of the drivers of policyholder behaviour and their quantitative impact on withdrawal rates, allowing for the interactions and correlations between factors: product class, distribution channel, socio-economic characteristics, policy riders, policy duration (current and remaining), the "moneyness" of guarantees, benefit amount, policyholder age and gender, number of recent contacts with the insurer.

For instance, what makes a customer use systematic small and regular withdrawals as opposed to using irregular but very significant ones or lapsing?

- Developing more effective customer lifetime value models requires bringing together multiple internal data sources, supplemented with relevant external data sets benchmarked with competitors, and deploying more sophisticated analytical and optimization predictive techniques, including propensity-to-lapse models, to estimate the impact of changes in features. Deploying these models effectively requires not only embedding this information within the operations and servicing teams as well as the distribution teams but also effectively training them on how to use this information.
- This approach also allows insurers to segment customers into homogeneous groups that respond similarly to changes in behavioural drivers. Insurers can then consider how to influence policyholder behaviour in each segment so as to improve overall portfolio profitability and to estimate the degree to which each segment will react to changes in each behavioural driver.
- Insurance/pensions can then focus their retention management activities on higher-value customers, by, for instance, tailoring written communication to the high-value policyholders to emphasize the merits of their policy, particularly over time; scripting specific responses to inbound telephone surrender enquiries from high-value policyholders (with messages to persuade them to maintain their policies) yet delivering a different message to less valued customers (which may direct them to an alternative product that would benefit both parties, subject to regulatory requirements); offering loyalty programmes (e.g. non-insurance-related products or voucher bundles) that are proactively offered to high-value customers and timed to coincide with high-withdrawal external-factor changes (e.g. significant changes in investment market returns).

Models are an excellent starting point, but human behaviour is unpredictable – so it is best to learn from an initial pilot. These activities can then be incorporated into policyholder data sets, allowing subsequent analyses to analyse the effect on retention rates and assess their effectiveness and/or cost efficiency.

1.1.3.3 Manage In-Force Assets to Enhance their Resilience to Lower Interest Rates and Yields

One of the ALM challenges facing insurers is the lack of assets of sufficiently long duration. Insurers can leverage derivative instruments to extend the duration of the asset portfolio to enhance the ALM position.

Furthermore, insurers may not always seek to precisely match asset and liability positions, because such a strategy could be inflexible and expensive. Insurers could deliberately choose to create a temporary mismatch if such a strategy allows them to generate incremental returns.

Lastly, under the current and persisting low interest rate environment, there has also been increasing interest from insurers in alternative assets, which could provide good risk-adjusted returns and low correlation to traditional assets. Insurers can invest beyond real estate and private equity, to those assets such as infrastructure, natural resources and private debts. In this respect, in-force blocks that have longer terms and low cash value are more suitable to being backed by alternative assets.

1.1.3.4 Manage In-Force Portfolio Liabilities to Enhance the Runoff Value

Liabilities can also be managed after the sale, and insurers have a variety of ways to influence and enhance the runoff of life insurance portfolios. Three levels can be distinguished.

- The first level relates to the ability of insurers to reprice/restructure the liabilities within the current legal and regulatory framework – i.e. when the insurer is voluntarily offering better terms and conditions than are legally required. Typically, contractual requirements are compared with actual practice, invoking potentially unused insurers' rights or providing the right incentives to distribution channels to promote the desired behaviour. Examples would be with-profit policies, using the discretion of the insurer within the profit-sharing rules; using the discretion of the insurer within the asset allocation through fund substitution (e.g. in the United States during October–December 2008 by switching all funds into major large indices; capping the weight in asset allocation to less liquid funds), or within the modelling of liabilities (discount rate reference – swap or govie [government] rates – for the liabilities value); forbidding additional future premiums; and letting fees increase under extreme market conditions. Most often, as in Europe, policyholders individually must agree to policy changes before they can be applied; but sometimes, in the United States, customers just need to be notified by a letter at least 90 days in advance, with changes then implicitly being approved in the event of no explicit refusal. In markets such as Australia, where protection policies are priced annually, significant value can be unlocked by looking more granularly at the emerging experience and better differentiating the pricing of the portfolio.

- The second level relates to treating customers differently depending on the underlying financial attractiveness of their policy to the insurer – such as, for example, launching unit-linked targeted campaigns to encourage additional premium payments and extending the policy term, whereby distribution partners and customer service staff are encouraged and incentivized. In contrast, for policies with high guarantees or any other onerous policy terms with significant capital consumption, insurers consider reviewing policyholder communication and targeting surrender.
- The third level relates to the conversion of an entire portfolio into new policy types or a lump sum (instead of the systematic lifetime payments embedded in the annuity policy), translating into cash or credit the base contract.

This often happened during the last decade as interest rates persistently declined to historic lows, which generated so-called "lapse-supported" in-force-embedding onerous guarantees.

Within these buyout and conversion initiatives, policyholders receive compensation for the loss of these features through higher reserves translated into cash. Individual conversions generally diversify in-force portfolios, as seen in programmes in Germany and Austria, whereby clients' policies were converted into unit-linked CPPI policies with a reduced guarantee in exchange for higher upside potential. Most countries allow conversion exercises only if policyholders individually agree to policy changes.

	AXA EQ	Transamerica	Hartford
Rider	Greater of Roll or Annual Ratchet GMDB (Standalone)	Family Income Protector, Managed Annuity Program, Managed Annuity Program II or Guaranteed Minimum Income Benefit.	Lifetime Income Builder II (GWBL)
Number of contracts	~22000	~25,000	?
Regulatory Filings	• State filing in all jurisdiction where the rider was sold • SEC Filing	• SEC Filing	• State filing in all jurisdiction where the rider was sold • SEC Filing
How offer amount is calculated	~70% of IFRS reserves or 2 years of rider fees whichever is greater.	80% of minimum annuitization value or income base under Eligible rider minus the cash surrender value.	a) Contract value on the full surrender date. Any applicable fees are waived upon surrender or b) Contract value on the Surrender date plus 20% of Payment Base subject to a cap of 90% of Payment Base
Impact to base contract	GMDB rider is terminated and the Offer amount is added to AAV; client have option of maintaining base contract with a Return of AAV as the new death benefit	Once offer is accepted, client must surrender contract and transfer to Transamerica or external carrier's product	Once offer is accepted, client must surrender contract and transfer to external carrier's product
Impact to surrender charge and other fees	Existing surrender charges remain as is and if applicable are not waived if client surrenders contract. Prorated death benefit charges are waived.	Surrender charges are waived however rider charges are prorated	Surrender charges and other fees are waived
Impact to other riders	• If the Earnings Enhancement Benefit (EEB) is elected, it is also terminated. • The standard death benefit (ROP) is also terminated.	Any death benefit is terminated	Any death benefit or other living benefit is terminated
Marketing Support	Client letters, prospectus supplement, Brochure; microsites; dedicated service desk	Client letters, prospectus supplement	

FIGURE 1.18 Buyout examples

Note: ROP = return of premium.

These are the lessons learned.

- A simple valuation method is helpful, as advisers feel they need to understand the calculation to better explain it to their clients.
- Clients and representatives also want to see the "number", through an offer amount in letters as well as via a website; over-communication is a necessity, and reduces the opportunity for misinformation. Clear and transparent client–adviser communication is also a key requirement; finally, engaging the distribution organization is critical to success.
- The success of a buyout depends on its content (whether immediate annuity or lump sum), the offer amount (the "fair" amount being the SPIA premium generating the equivalent future income), the policyholder's age, the interest rate level (lump sum buyouts take up rates rising with interest rates, particularly for account value depletion buyouts).
- Although the first buyout initiatives may be very successful (e.g. with a take-up rate of up to 60%), subsequent campaigns have been be far less successful (often a 3 to 10% take-up rate: 8% in the event of a 50% IFRS reserve offer, 11% in the event of a 70% IFRS reserve offer), as the interested customers have already left.
- Assessing the recoverability of the customer behaviours after the buyout to a long-term lapse rate is extremely hard. In addition, the anti-selection effect may be very significant, with the risk that the targeted policies (the richest ones) may not be the actual ones or may not be efficient (e.g. policyholders who would have left without the buyout incentive).

The different in-force management initiatives launched over the past decade have led to several key lessons.

- **One size does not fit all.** In-force management approaches need to be tailored carefully to each portfolio's specifications, as well as the values and objectives of individual companies, country-specific laws and the need to manage reputational risk.
- **There is no silver bullet.** Generally, in-force management is driven by many small steps, and only rarely do individual actions have a significant impact on overall performance. Often only the combination of the three levels outlined above helps companies improve their situation in a meaningful way.
- **Everyone should benefit.** Conversion offers need to be financially attractive to both the policyholders and the insurer to attract sufficient participation. For instance, low-payment lifetime annuities could be converted to higher-payment annuities with a fixed term, translating into more meaningful annuity payments for customers, while substantially reduced administrative costs, a reduced term and reduced longevity exposure are the benefits for the insurer.

1.2 SECURING LONG-TERM SUSTAINABLE BUSINESS GROWTH: FROM GOVERNANCE TO EFFECTIVE RISK MANAGEMENT

1.2.1 Risk Is Embedded within the Insurer's Business Model

Risk management is not a by-product but, rather, is at the core of the business model of insurance companies. Risk lies at the heart of the insurance business. Risk is the insurer's raw material, as insurance is the business of managing risk:

- taking risks off clients' balance sheet;
- taking these risks on their balance sheet;

- transforming it and/or transferring it; and
- generating profit from this activity.

The risk management objective, therefore, is that risks are taken, in a controlled and disciplined manner:

- risks are taken consciously;
- risks are adequately measured;
- risks are limited and frequently monitored; and
- risks are at the forefront of management actions.

The risks across the life insurance industry are multiple and diverse: financial (equity, interest rates, basis, foreign exchange [FX]); policyholder behaviour and longevity risks; operational, counterparty and liquidity risks.

Specifically, life products are long-term products, and profitability and value creation must be measured over the life cycle of the contracts: contracts quite often generate a negative result in year 1 (acquisition cost), but they usually generate positive margins in the following years.

As mentioned above, life insurance risks come from a number of sources.

- The biometric cover given to policyholders – e.g. the death risk for a term insurance, for which the death frequency can be higher than expected, and value creation can be at risk, though the longevity risk is the risk that mortality is lower than expected, which implies a longer duration (and cost) of the lifetime income period within an annuity.
- Customer behaviours – e.g. lapses for an investment product, when customers may lapse earlier than expected (higher lapse rates) and value creation can be at risk. Lapses include cancellations, total and partial surrenders and changes in paid-up status (regular premiums). The lapse mass is the risk of adverse unanticipated lapses in the coming year as a result of a mass phenomenon due to an additional lapse driver that is observable only in extreme scenarios. Among other customer behaviour risks there is the risk that policyholders are more rational than expected and exercise more embedded options in a way that adversely affects the company (e.g. guaranteed annuity options, as illustrated in the subsection above).
- The company and its ability to manage its expenses correctly, which may be higher than expected, putting value creation at risk

These risks translate into billions of dollars in exposures and associated capital requirements – but also business opportunities, such as growing needs for financial protection and yields, and care for sustainability and responsibility.

1.2.2 From Risk as a Constraint to a Culture: Risk Management as Sustainable Business Partner and Risk Factor Control

> There are known knowns. There are things we know that we know.
> There are known unknowns. That is to say, we know there are some things we do not know.
> But there are also unknown unknowns, the ones we don't know we don't know.
>
> DONALD RUMSFELD
>
> *Former US Defense Secretary, 12 February 2002*

Attitudes towards risk are often ambivalent, from threat perception to opportunity realisation.

Our classic knowledge of risk has its origin in our perceptions, and so it is put at risk today. We fear dangers that we are familiar with or that we can directly perceive with the senses: when confronted with a wide array of hazards, we select and identify them as risk sources. Risk is thus defined by checking whether and how it threatens to jeopardize our way of life. Risk is also a measure of danger:

$$\text{Probability of occurrence} \times \text{Magnitude of harm.}$$

However, as many dangers today can no longer be perceived with the senses, we do not know how high the risk is, and everybody may say something different that we cannot verify. Thus we may fear risks that can cause only moderate harm while we ignore significant risk scenarios.

Moreover, as the world is increasingly closely linked (due to globalization, increasing interconnectedness and the maximization of efficiency of systems), we become more vulnerable to external disturbances, as risk factors influence one another, such that effects can lead to a long chain of mutual dependences, which may not be apparent at first glance and develop stealthily -> unable not only to assess the probability but also to foresee the possible nature of harm.

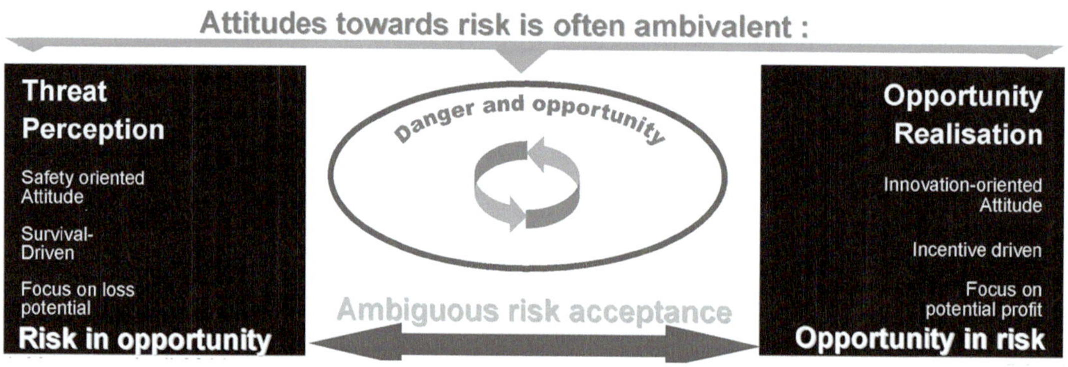

FIGURE 1.19 Attitudes towards risk

Without taking risk there would be insufficient appetite for innovation and limited opportunities for growth. Risk cannot be eliminated without forgoing profitability -> assessing how we can reduce risks and when it is worth becoming more committed.

Making decisions on risks is about striking a balance between potential gains (rewards) and potential losses (risks) -> maximizing potential gains while being constrained by maximum tolerable losses.

Risk management is about decreasing uncertainty, getting as many events as possible in the "known" category and limiting unexpected volatility. The key is in being able to move swiftly and quickly in doing so, and that starts with identifying opportunities and threats and effectively monitoring them in a consistent way that people can adapt and get used to.

The objective of risk management is not to prevent people from taking risks but to allow people to understand and manage in the optimal way the risks that they take. The objective of risk management is not to avoid taking risk but to promote a culture of selective risk taking by providing advice in risk taking.

Risk management should not focus simply on analysing "cold" figures; for example, value-at-risk – i.e. the amount lost on positions if market values move adversely by a certain amount – assumes continuous liquidity for sufficient amounts to close all positions, which may not be possible if all market participants hold similar positions and try to execute closing trades at the same time. The same limitations apply regarding stress tests, when the maximum relevant historic experience might not be sufficient.

Risk management should understand in depth the market, the underlying exposure, underwriting and claims processes to be able to "read" the figures, assess the level of capital required through realistic scenarios and help in building an appropriate action plan. The risks analysed will be specific to the product types and to their respective market. In particular, the size and distribution of market participants need to be understood, and the size and concentration of positions held by these participants and the way they interact with each other and with exogenous events must be anticipated and evaluated.

Regarding the content of risk management advice, it should not be based only on figures and models: sound business judgment should also be used. Assessments, models, etc. are always wrong to some degree … A risk management framework may not be sufficient to solve all real-time risks (e.g. systemic risks with feedback loops). There might be multiple varied interpretations of novel and unfamiliar risks that materialize in unexpected ways. The usual normalized prospective approach based on prediction might not be possible. Mathematical analysis may offer little guidance in the short term when action is required. Controlling practices such as top-down commanding may not respond well "online".

Rules of thumb erring on the side of caution make sense in the event of a lack of reliable data. Common sense, sound governance, expert judgment, intuition and out-of-the-box thinking ("Think the unthinkable!") are essential. Additionally, a pragmatic approach is often better than an in-depth study that comes too late.

> Any observed statistical regularity will tend to collapse once pressure is placed upon it for control purposes. And remember that, when a measure becomes a target, it ceases to be a good measure.
>
> PROFESSOR CHARLES GOODHART FBA
>
> *Chief adviser to the Bank of England*

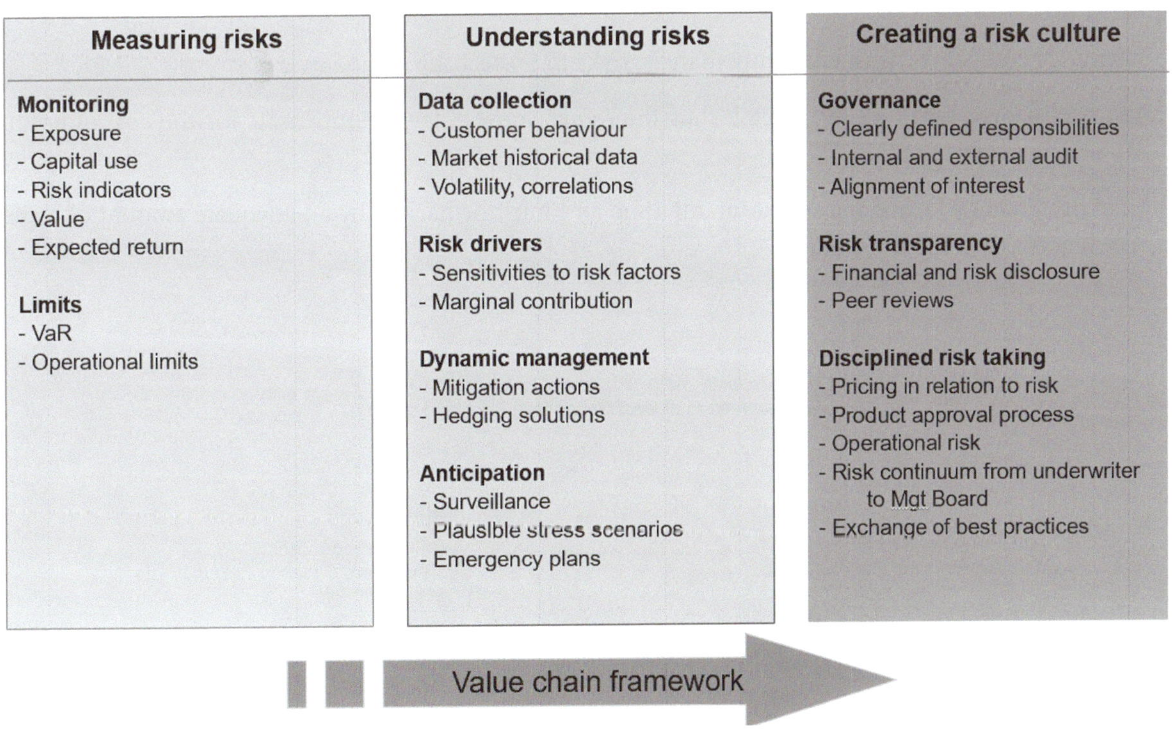

FIGURE 1.20 From risk measure to risk culture across the value chain of risk management

Note: VaR = value at risk.

What Risk Management promotes

✓ **Independence**
Organization that ensures independence from business and adequate governance to express Risk Management opinion

✓ **Courage**
Ability to ring the bell when necessary, based on mutual trust and acknowledgment of second opinion value

✓ **Innovation**
Use cutting-edge technics to ensure risks are properly modeled with most advanced findings

✓ **Role model in the insurance sector**
Always strive to align with best practices and try to be instrumental on most decisive debates (e.g. Solvency II or Systemic risk)

✓ **Business partner**
Remain close to business, and try and find the right balance between customer centricity and long term protection of the company

What RM does NOT promote

✗ **Risk reporting**
Added value comes from analysis, alerts and action plans, not only from risk dashboards and reportings

✗ **Experts in an ivory tower**
We seek dialogue and constant exchanges with business, in order to stick to our strategy and our products

✗ **"Moscow's eye"**
Trust does not exclude control. Risk Management studies intend to challenge, not criticize, and to bring a different angle from the business

✗ **Stuck to models**
All models are wrong, some are useful. We try to build on our business expertise to take distance from the models and bring a value-adding analysis

✗ **Arrogant**
We try to be open to dialogue and bring a constructive challenge

FIGURE 1.21 Risk culture

To play its role efficiently, risk management must be closely embedded into business processes and decision-making. Taking selective risk is part of a triptych: risk, return, capital.

- Insurance companies must efficiently manage risks to protect the company.
- Taking risks requires holding capital; thus the return must at least compensate for the cost of holding capital ("cost of capital").
- Unexpected large claims may occur at any time and must be backed by an adequate amount of capital, which acts as a company's buffer.

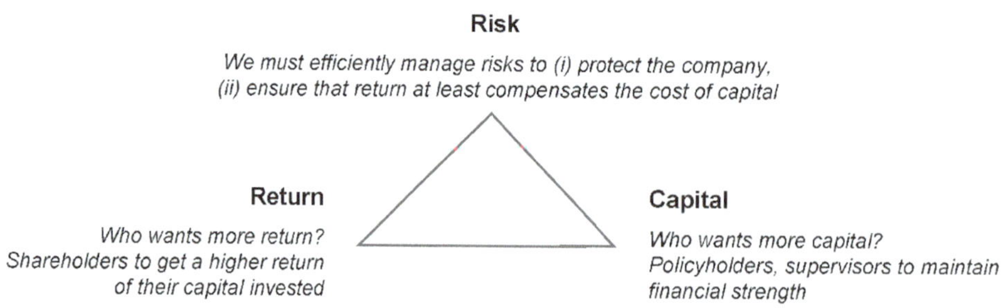

FIGURE 1.22 Triptych risk/return/capital

In this respect, an attractive risk provides "good" return on capital (short-term and long-term), has a positive contribution to the solvency ratio and is manageable over time (in terms of guarantees with the ability to reprice or mutualize losses; in terms of expertise in underwriting and medical selection; in claims management and general knowledge of the risks).

High capital requirements do not necessarily, therefore, mean "unattractive risk" if they are sufficiently compensated for by high profitability (e.g. lapse up in unit-linked or health and protection).

Capital scarcity adds a new level of complexity in the running of the business. Working together on these subjects could give the insurance company a competitive advantage. As a result, risk management becomes a sustainable business "partner" and not just a risk factor control function, with three main missions.

- To define and formalize a framework to help in choosing consciously the risks we want to retain.
  - Governance and processes: *not* to annoy people and be bureaucratic *but* to ensure that proper information is considered and challenged to base decisions on. On level 1 or 2 (see Figure 1.23), operational decisions are mainly taken at entity level and controlled through group processes/ initiatives. When level 3 is reached (i.e. impacts and damages are increasing), the group moves temporally towards more centralization of decision-making, with local or central execution. In the event of a crisis scenario, decisions to mitigate the impacts are recentralized (e.g. equity-hedging programmes).

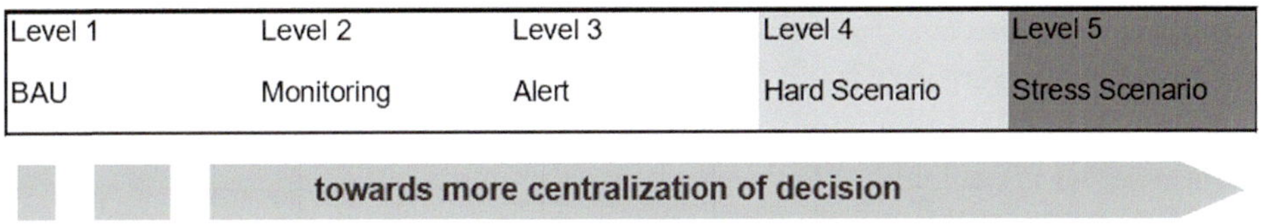

FIGURE 1.23 Governance of risk escalation

Note: BAU = business as usual.

  - Defining metrics: *not* to create new complicated indicators *but* to provide a common measurement basis across all risks and consistent across the group.
  - Setting limits: *not*, sadly, to decrease the bottom line *but* because the company's capacity to take risks is limited and should be clearly allocated -> define alert levels and "What if?" action plans at group and operating unit levels to detect a crisis in its early phases and mitigate the impact (hedging, BCP, communication plans …).
  - Control and challenge the level of risk assumed in the different departments.
  - Report to senior management on the level of risks.
- To aim for the long-term survival of the company ("Still be here in 50 years").
  - Why purchase an insurance product if you are uncertain about the insurer's ability to pay claims when it occurs in the future?
  - Insurers have a key social responsibility to play in providing long-term solutions (pensions, long-term care).
  - Insurers need to be prepared against adverse events: *not* because they are so pessimistic *but* to quicken their reaction and make a difference when it matters most -> analyse the consequences of extreme events through description of the scenarios and quantification of the impacts.
- To promote a culture of selective risk taking, based on the identification, evaluation and management of risks.
  - People in the chief investment officer (CIO) function, underwriting, etc., are all risk takers.

- The professionals in the risk management community cannot manage the risks of a large company purely by themselves.
- Everybody should therefore feel as though they are a risk manager, beyond the risk teams themselves.
- Risk awareness will then rise, as they feel more "in charge" of delivery.

In conclusion, the risk management function is there to ensure that insurance companies do only the business they want to do.

1.2.3 From Prudential Regulation Frameworks to Effective Sustainable Insurance Risk Management

1.2.3.1 Why Do We Need Regulation?

The aim of prudential regulation in insurance is to protect policyholders from a company's insolvency (long-term pensions, savings), something that has been reinforced in the wake of the crisis: financial stability is considered the number 1 priority.

Since insurance obligations are very far in the future, companies need to start saving early to pay out future claims, and so are required by the regulator to hold reserves.

The purpose of reserves is twofold:

- to help insurance companies honour commitments to policyholders and promote solvency ("prudential reserves"); and
- to help determine company earnings for stockholder markets ("earnings reserves").

In addition to statutory reserves, regulators require insurance companies to hold an additional layer of capital, the solvency margin (capital requirements), against unexpected events and hence provide policyholders with further protection.

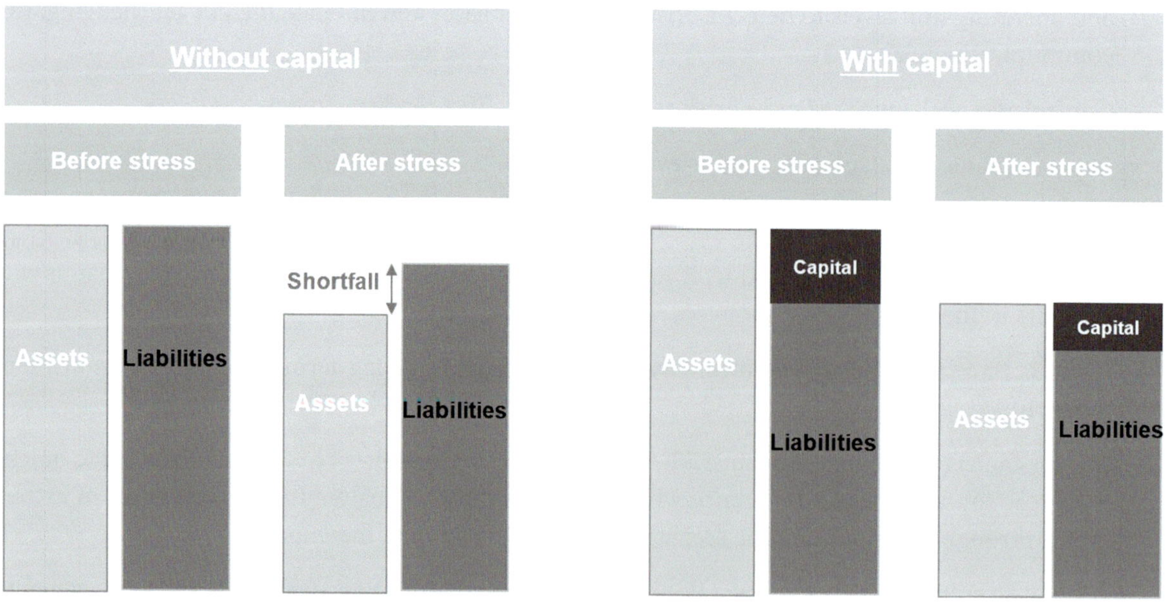

FIGURE 1.24 Asset versus liability management with and without capital

Players from various horizons usually contribute to the setting of such prudential regulation (see Figure 1.25).

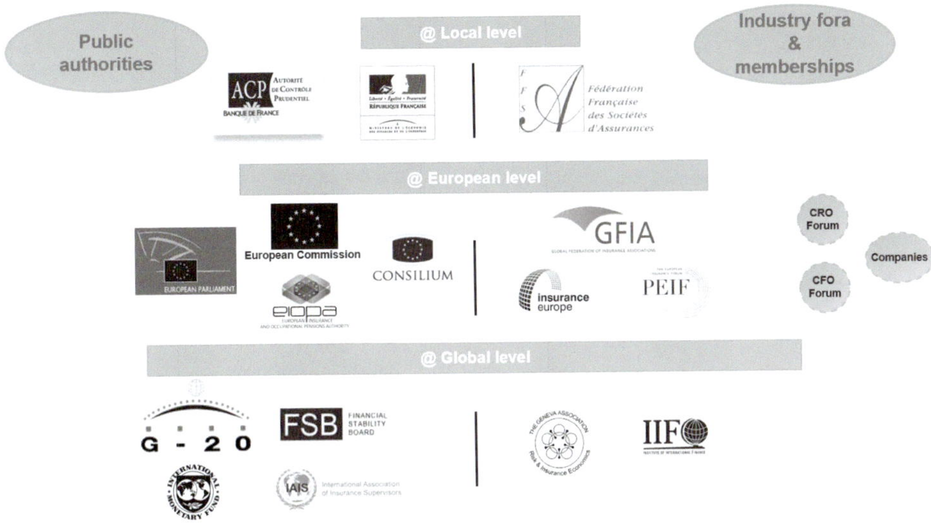

FIGURE 1.25 Major players around prudential regulation

1.2.3.2 *The Structure of Capital Requirement Prudential Regulation: Focus on the European Solvency II Framework*

There is a trend of modernizing the required capital frameworks across the globe, with significant changes towards market-consistent, principle-based and risk-based frameworks: the United Kingdom's ICA/ICG, Swiss Solvency Test, Solvency II in the European Union since 2016 and the United States' National Association of Insurance Commissioners (NAIC) statutory framework from 2020 to 2023.

Overall, these new required capital frameworks are designed to:

- provide greater protection to policyholders through an enhanced prudential regime;
- better align capital requirements to the firms' asset and liability profiles and enhance the quality of capital;
- improve firms' own risk management with the aim of reducing the probability of firm failure; and
- reform the public disclosure of firms' financial condition, to encourage the effective exercise of market discipline.

These prudential regulations usually rely on three pillars, defined as follows.

- Pillar 1 covers the capability of an insurer to demonstrate that it has adequate financial resources in place to meet all his liabilities and that they consist of the quantitative requirements, such as the amount of capital an insurer should hold. It is all about the calculations, models and capital requirements.
- Pillar 2 sets out requirements for the governance and risk management framework that identify and measure the risk against which capital must be held, as well as for the effective supervision of insurers.
- Pillar 3 focuses on the disclosure, reporting and transparency requirements around these risks and capital requirements.

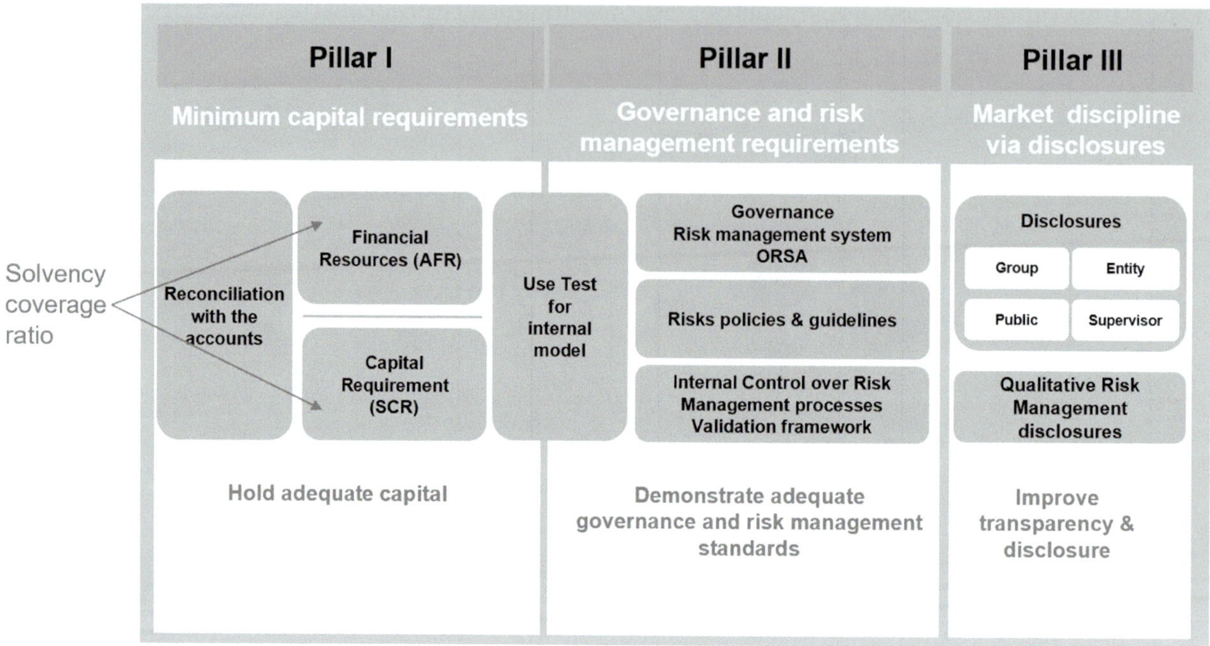

FIGURE 1.26 Solvency II pillars

Notes: AFR = available financial resources; ORSA = own risk and solvency assessment.

Given that the NAIC framework reform is under way, in the following (Figure 1.27) we illustrate with respect to the European Solvency II prudential regulatory framework, characterized by the following changes from the former Solvency I: under Solvency I, there is no difference between risks and ALM is not considered, nor the diversification of risks nor advanced risk transfer techniques. In addition, there is a conservative approach to hybrid debt capital, putting insurers at a disadvantage with regard to banks.

FIGURE 1.27 Key milestones in Solvency II implementation

Note: QIS = quantitative impact study.

- In terms of the balance sheet, there was a switch under Solvency II from a mostly accounting view to an economic view (see Figure 1.28).

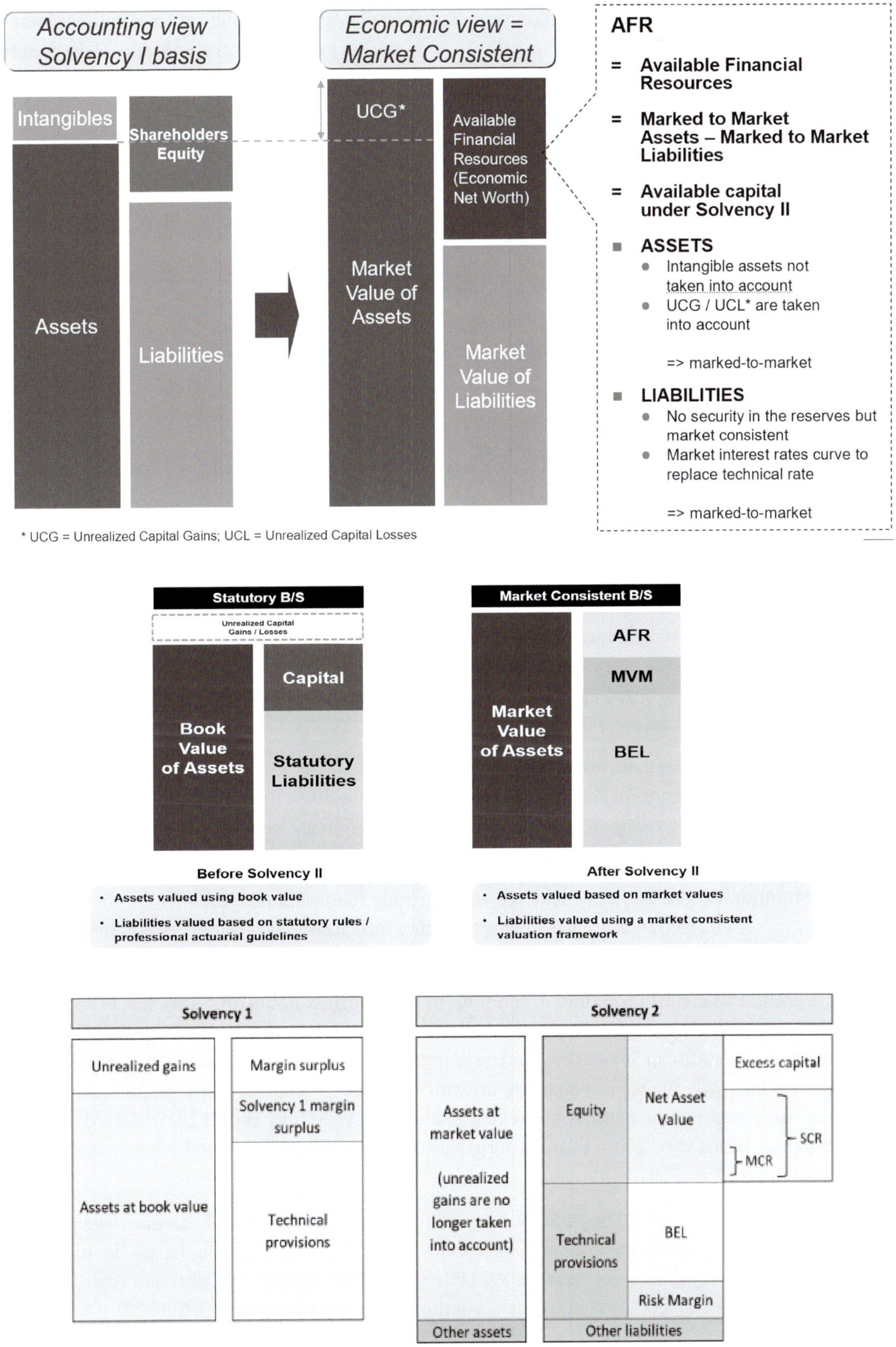

FIGURE 1.28 Solvency II versus Solvency I balance sheet

Notes: UCG = unrealized capital gain; UCL = unrealized capital loss.

AFR (available financial resources) is the amount of capital available to cover the capital requirements – i.e. to absorb losses in stress events. AFR is the surplus from a market value balance sheet, and it changes from Solvency I into Solvency II, as shown in Figure 1.29.

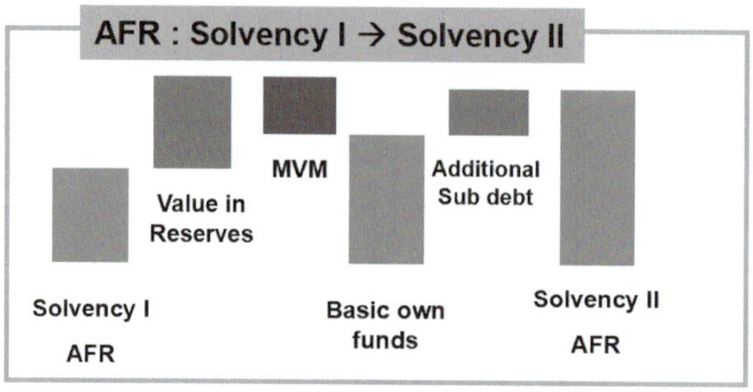

FIGURE 1.29 Solvency I to Solvency II AFR

Notes: MVM = market value margin; BEL = best-estimate liability; MCR = minimum capital requirement.

- In terms of the value of liability, there has been a switch from a statutory accounting value to a "market value", using two specific components relative to the market (thus hedgeable) and technical (life, thus non-hedgeable) risks under Solvency II: the best-estimate liabilities, defined in article 77 of the EU directive, correspond to the expected current value of future cash flows required to settle the insurer commitments towards policyholders, using the relevant risk-free interest rate term structure (potentially adjusted for counter-cyclical measures). The cash flow projections should, ideally, be performed on a policy-by-policy basis. However, approximations are permitted, and grouped model points can be used provided certain conditions are met, including validation of accuracy.

  For with-profits businesses, discretionary benefits need to be allowed for. When calculating the best-estimate liability figure for businesses that incorporate a discretionary element, the BEL must be calculated separately for the guaranteed and future discretionary benefits. The future discretionary benefits should relate to normal expected bonus distributions only.

  The cash inflows and outflows considered must include the premiums, the insurer's benefits and all types of expenses due to the insurance contracts, taking into account every option inherent to the contract, which are financial guarantees or rights granted to the policyholder.

  Future premiums can be taken into account up to the "contract boundary", which is broadly defined as the point at which a company can unilaterally terminate the contract, refuse to accept a premium or change the premiums or benefits in such a way that they fully reflect the risks. For life insurance business, this normally means the maturity or expiry date of the contract, or in some cases an earlier date at which premiums or benefits are reviewable so that they fully reflect the risks. Allowance for future expenses needs to take into account both overheads and directly attributable expenses, and future expense inflation.

  All BEL assumptions should be genuine best estimates, with no prudential margins. The projections should allow for all expected decrements and policyholder actions, including lapses. Insurance companies must take into account all relevant available data, both internal and external, when arriving at assumptions that best reflect the characteristics of the underlying insurance portfolio.

  The risk-free discount rates are published by the European Insurance and Occupational Pensions Authority (EIOPA) on a monthly basis. The rates are based on swap rates when there is a sufficiently deep and liquid swap market, or government bond rates otherwise. These rates are then adjusted (by EIOPA) to reflect the risk of default of the counterparty (i.e. credit risk adjustment). For longer maturities for which data are not available, the yield curve is extrapolated to a defined long-term equilibrium rate.

When insurers have long-term predictable liabilities, and can hold matching assets to maturity, they are not exposed to the risk of changing spreads on these assets (they are not exposed to the liquidity component of spread risk, although they are still exposed to default risk). In such cases, insurers are allowed to adjust the risk-free discount rate in line with the spread movements of their assets, called "matching adjustment", derived by taking the spread on the portfolio of matching assets and deducting the "fundamental spread", an allowance for the credit risks retained by the insurer. EIOPA publishes the fundamental spreads that insurers must use.

When insurers have liabilities that are not eligible for use of the "matching adjustment", they can alternatively add a "volatility adjustment" to the risk-free discount rate. The purpose of the volatility adjustment is to reduce the risk of forced sales of assets in the event of extreme bond spread movements. The volatility adjustment is based on the spreads on a representative portfolio of assets for each relevant currency (or for each national insurance market when the assets held by insurers in that market are under particular stress). Risk-free discount curves including the addition of a volatility adjustment are also published by EIOPA for firms to use.

Regarding savings and retirement contracts, the best examples of options and financial guarantees inherent in the contract are as follows.

✓ Surrender: during the term of the contract the policyholder has rights on the mathematical provisions of the contract, as she may claim her right on it during the contract and thus terminate it before the initial settlement date. This choice may be motivated by the difference between the rate paid by the insurer and those of competing insurers.

✓ Minimum guaranteed rate, fixed at the subscription that must be respected by the insurer during all the life of the contract, regardless of the evolution of its portfolio.

✓ Profit-sharing mechanism: life insurers have to share with their policyholders a portion of their technical and financial benefits, depending on the evolution of the assets held by the insurer. The profit-sharing mechanism can be contractual or discretionary.

- The risk margin called the market value margin (MVM): such as to ensure that the total value of the technical provisions is equivalent to the amount a potential buyer would be expected to require in order to take over and meet the insurance and reinsurance obligations of the undertaking without any additional financial risks. It is thus intended to increase the technical provisions to the amount that would have to be paid to another insurance company in order for it to take on the best-estimate liability. It therefore represents the theoretical compensation for the risk of future experience being worse than the best-estimate assumptions, and for the cost of holding regulatory capital against this.

  The risk margin is determined using the cost of capital (CoC) method, which takes the perspective that sufficient capital is needed to be able to run off the business – i.e. based on the cost of holding capital to support those risks that cannot be hedged. These include all insurance risk, reinsurance credit risk, operational risk and "residual market risk". For Solvency II, it is a fixed rate of 6% per annum.

  The MVM is estimated as the present value of the expected cost of current and future SCR for non-hedgeable risks to support the complete run-off of all liabilities.

The MVM calculation thus involves first projecting forward the future capital that the company is required to hold at the end of each projection period (e.g. year) during the run-off of the existing business. For Solvency II, the projected capital requirement is a subset of the SCR, consisting of those risks that cannot be hedged in financial markets. These projected capital amounts are then multiplied by the CoC rate.

The product of the cost of capital rate and the capital requirement at each future projection point is then discounted, using risk-free discount rates, to give the overall risk margin.

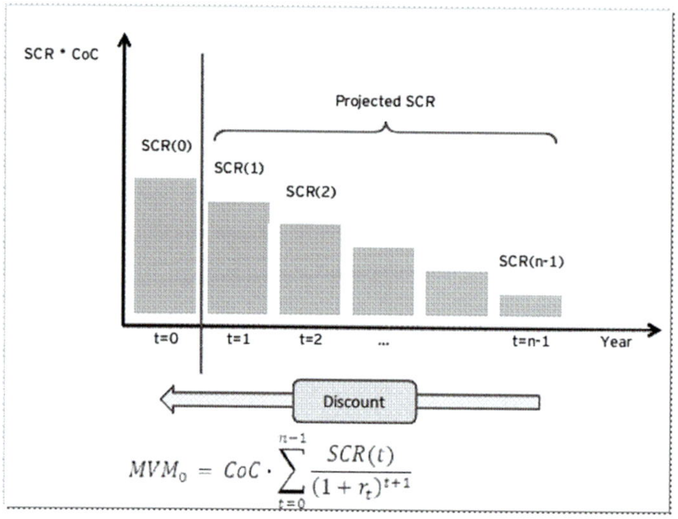

FIGURE 1.30 MVM computation

Since the projection of the SCR is potentially complex, various simplified approaches can be used. For example, this could involve selecting a driver (e.g. reserves or sum at risk) that has an approximately linear relationship with the required capital or its components. The initial capital requirement can be expressed as a percentage of that driver, and the projected capital is then approximated as the same percentage of the projected values of the driver.

Although the risk margin must be disclosed separately for each line of business, it can be reduced to take into account diversification between lines up to legal entity level. The allocation of diversification benefit can be approximated by apportioning the total diversified risk margin across lines of business in proportion to the SCR calculated on a standalone basis for each line, or by other approximate methods if appropriate given the materiality of the results.

- Also based on a market-consistent approach, technical provisions should represent the amount that the insurance company would have to pay in order to transfer its obligations immediately to another insurance company. The technical provisions consist of the BEL and the MVM. The calculation should be segmented by homogeneous product type.

- In terms of the solvency capital requirement, this is translated into a more realistic measure as an economic and risk-based capital requirement: in Europe, under Solvency 1, there is no link between capital requirements and the level of risk

when the "expected result" is based on the "baseline reserving assumptions"

From Solvency I …

The required capital is defined by simple formulas

Life (without death cover):
- 4% of life mathematical reserves
- 1% of unit-linked reserves

Shortcomings:
- Born in the 1970s…
- Capital as % of volumes / accounting results, not risks (no diversification)
- No risk assessment of asset side
- No level playing field on reserves
- No incentives for risk management

… to Solvency II

- **SCR (Solvency Capital Requirement) a "Target level" required to operate with a low probability of failure (1/200 year over 1 year based on VaR risk-measure), calculated with two options**
  - Standard formula (defined by EIOPA* via Quantitative Impact Studies)
  - Internal models (upon validation by supervisors)
- **MCR (Minimum Capital Requirement), an "Absolute Minimum level", i.e. trigger level for ultimate supervisory action:**
  - Simple (factor based or as a % of SCR)
  - Absolute minimum

*EIOPA: European Insurance and Occupational Pensions Authority

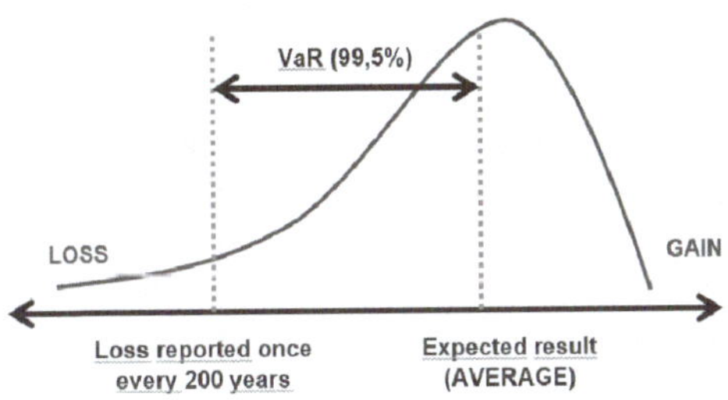

Market	Credit	Life	P&C	Risk op.
Interest Rates	Mortgages	Cat	Reserves	Risk Op.
Spread	Fixed Income	Mortality	Premium	
Equity	Reinsurance	Longevity	Cat	
FX		Disability		
Volatility		Lapses		
Real Estate		OCB		
Hedge Funds		Expense		
Private Equity		Health		

FIGURE 1.31 Solvency I to Solvency II SCR

Note: OCB = other customers' behaviours.

As a value-at-risk metric over a one-year time horizon, the SCR measures the change between the average "value" of a product and a specific percentile of the loss distribution (e.g. 99.5th percentile).

Under Solvency I, the capital requirement is based on year-end accounting results (premiums, reserves) linked to the volume of business (4% of life mathematical reserves, 1% of unit-linked reserves), with no consideration of underwritten risks or risk diversification.

In contrast, under Solvency II the SCR is a function of risks underwritten, e.g. market or life.

- The main drivers of the market SCR are spread, equity and interest rates. The life business (savings and retirement) tend to consume the vast majority of the SCR (~80%). Let us also mention that the counterparty default risk is distinct from the market risks.
- The main drivers of the life SCR are the technical risks, such as customer behaviour and expense risks (~60%) and biometric risks (~40%).
- Among products, savings and retirement are mostly geared towards market risks, while protection and health are more geared towards technical risks.

The Solvency II capital requirement considers diversification of risks through an aggregation risk matrix, including both geographical and business diversifications (Bourgeois 2014).

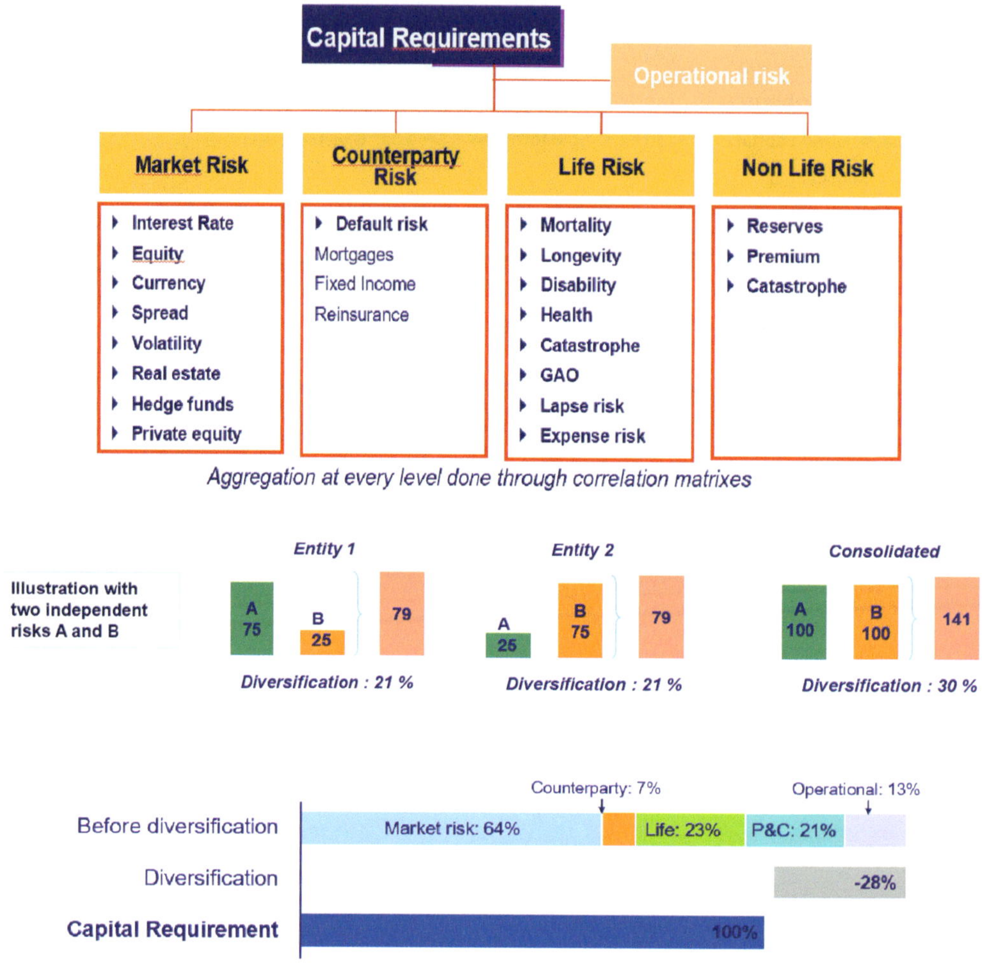

FIGURE 1.32 Risk diversification under Solvency II

Note: GAO = guaranteed annuity option.

Risks that diversify well are usually mortality, disability and medical expense risk, and some customer behaviours (lapse up, mass lapse) that are highly profitable while being mitigated via hedging, reinsurance and retention actions. While risks that do not diversify well (so-called "cumulative") are usually sensitive to the financial market environment and are difficult to hedge (thus requiring costly reserve strengthening or buyouts), such as some customer behaviours (lapse down, annuitization, fund switches) and longevity (risks influenced by the market environment) and market risks within guarantees. Thus, these risk definitely need prudent assumptions.

As a result, protection and health and unit-linked without guarantees are generally well diversified, while savings with rich guarantees (whether UL or general account) or longevity are mostly concentrated in the cumulative risk category, which is usually very dependent on the market environment (and the associated moneyness of embedded options).

It is then possible to draw and monitor the SCR risk signature for each product per type of risk (see Figure 1.33) and illustrate the impact of both market conditions (increase or decrease in interest rates or volatility) and mitigating actions (additional hedge instruments, improvement in product designs, reinsurance …) (Bourgeois 2014).

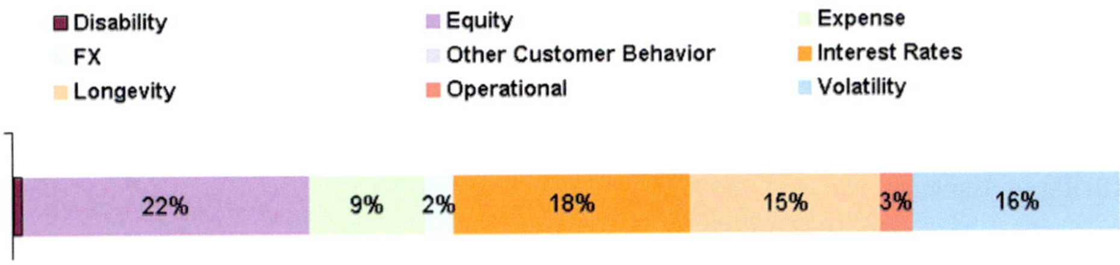

FIGURE 1.33 SCR risk signature

Given that AFR is the amount of capital available to cover the SCR – i.e. to absorb losses in stress events – it makes sense to produce and monitor the solvency coverage ratio, which is AFR/SCR.

The insurer usually targets a specific coverage ratio level, for the following reasons:

- to allow the business to absorb a good degree of volatility;
- to allow early action to stop a deteriorating capital position;
- to maintain solvency after a number of adverse, but plausible, events;
- to make allowance for non-quantifiable risks, not measured by the SCR; and
- to be prudent.

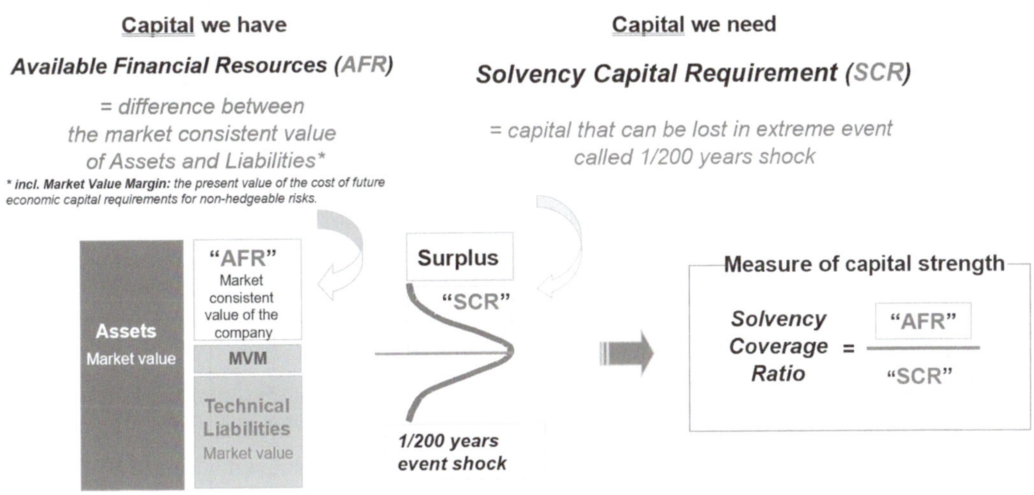

FIGURE 1.34 AFR and SCR

Solvency II prudential regulation is about making sure that the initial available financial resources are high enough to remain positive, after one year, when confronted with a combination of the most adverse conditions that can occur every 200 years.

The supervisor can oblige the undertaking to retain an additional amount of capital (capital add-on) to cover insufficiently controlled risk. If the level of capital SCR is not reached, it will probably result in regulatory intervention and require remedial action. The minimum capital requirement is the minimum amount of capital the insurer needs to cover his risks. If an insurer's risk capital falls below the MCR, he will be prohibited from writing any further business.

In the event of this happening, among the management (corrective) actions considered are:

- a capital injection;
- a restructuring of assets;
- a reduction in expenses;
- a review of reinsurance structure;
- a cut in dividends;
- an increase in fees; or
- closing/reducing new business.

✓ **Quick win**: coverage ratios are more volatile under Solvency II than Solvency I, so a close monitoring of the Solvency II coverage ratio is required, specifically within the context of market volatility and low interest rates, as experienced over the past decade. The impact is shown in Figure 1.35.

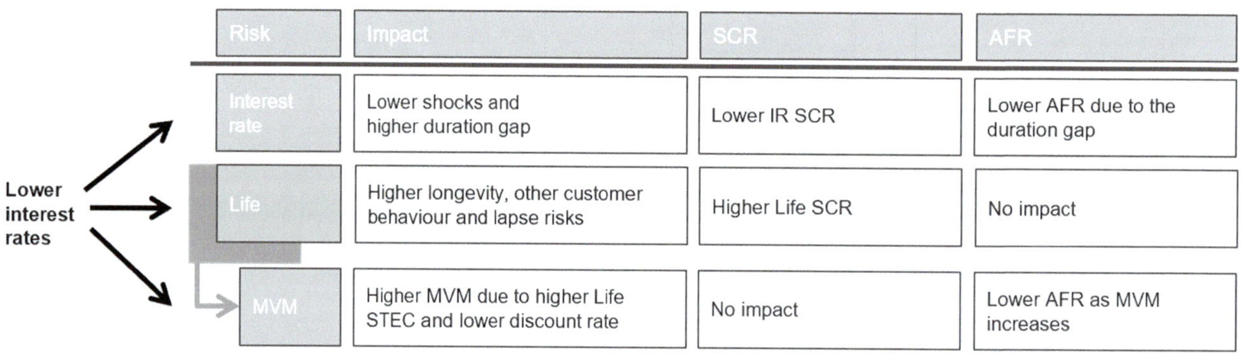

FIGURE 1.35 Impact of risks on SCR and AFR under Solvency II

Linked to the SCR is the notion of "required capital", which can be defined as the capital on top of other resources (notably VIF: value in force – i.e. the value of the business already sold, and still with the company) such that the target Solvency II ratio (e.g. 150% or 200% of the SCR) is achieved (floored by 0):

NB: The value in force is the present value of local regulatory (statutory) profits projected over the entire future duration of existing liabilities, with provision for the cost of financial options and guarantees, and net of the cost of holding additional capital required to support the business.

The VIF is thus composed of three elements: CE PVFP + TVOG + CoC/NFR

✓ The certainty-equivalent present value of future profits (CE PVFP) corresponds to the deterministic value of new business without taking credit for any future investment risk premiums -> discount rate = risk-free rate.

It is before the time value of option and guarantee (TVOG) costs and the cost of capital (CoC): cash flows are projected using a risk-free rate of interest then discounted at the constant risk-free rate. Effectively, this approach assumes that all assets are risk-free and, as such, the insurer has no financial risk. The next two components of value allow for this risk.

✓ The time value of options and guarantees is the risk-neutral value minus the CE PVFP – i.e. the difference between the risk-neutral value calculated with stochastic models and the certainty-equivalent value calculated with a deterministic model (in practice a stochastic model run on a single CE scenario is used) at current risk-free rates -> products with guarantees that are correlated with financial markets will effectively be discounted at a rate higher than the risk-free rate.

This is significant when worthy guarantees/embedded options are offered: interest rate guarantees on traditional products (guaranteed cash values and guaranteed annuity options), profit sharing, such as bonus rates, credited interest rates and policyholder dividends, guaranteed benefits on unit-linked products (GMDB, accumulated benefits, income benefits, withdrawal benefits …), "no lapse" guarantees on life insurance products, dynamic policyholder behaviour, such as full or partial surrender, annuitization and premium discontinuance.

✓ The cost of capital divided by non-financial risks (NFR) is the cost incurred through the payment of investment expenses and taxes on investment income of assets held in excess of the policyholder reserves, for operational and non-financial risks.

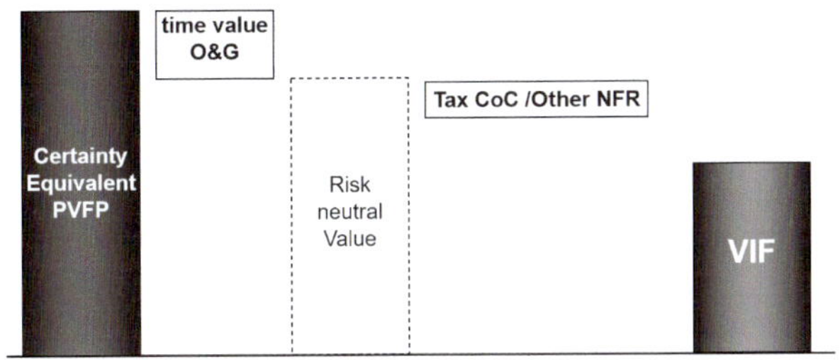

FIGURE 1.36 Decomposition of the VIF

Then the sum of the VIF and the ANAV (adjusted net asset value = free surplus + required capital, which represents the economic net assets, derived by taking reported shareholder net assets, adding unrealized capital gains and then adjusting for intangible assets and other items) gives the so-called "statutory balance sheet" attached to the local statutory accounting framework, which thus defines the profit and loss – i.e. the sum of different margins (financial, technical, expense), or premium – claims + investment – Δreserve – expense.

Ultimately, EEV (European embedded value), AFR and SCR all share the same framework, as shown in Figure 1.37.

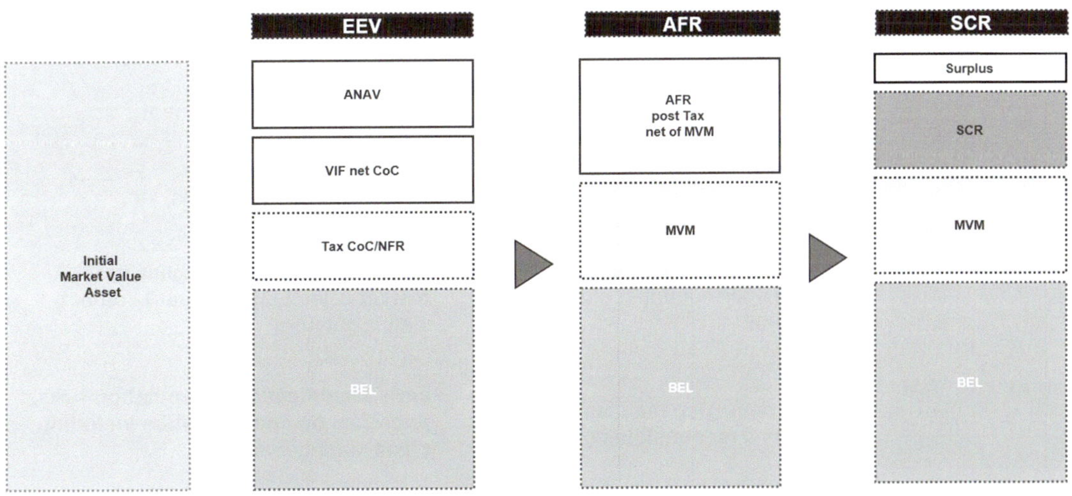

FIGURE 1.37 EEV/AFR/SCR balance sheets

1.2.3.3 From Regulation Concept to Sustainable Corporate Risk Management Practices: Enterprise Risk Management (ERM)

Despite the diversity of risks, there is one integrated ERM (see Figure 1.38).

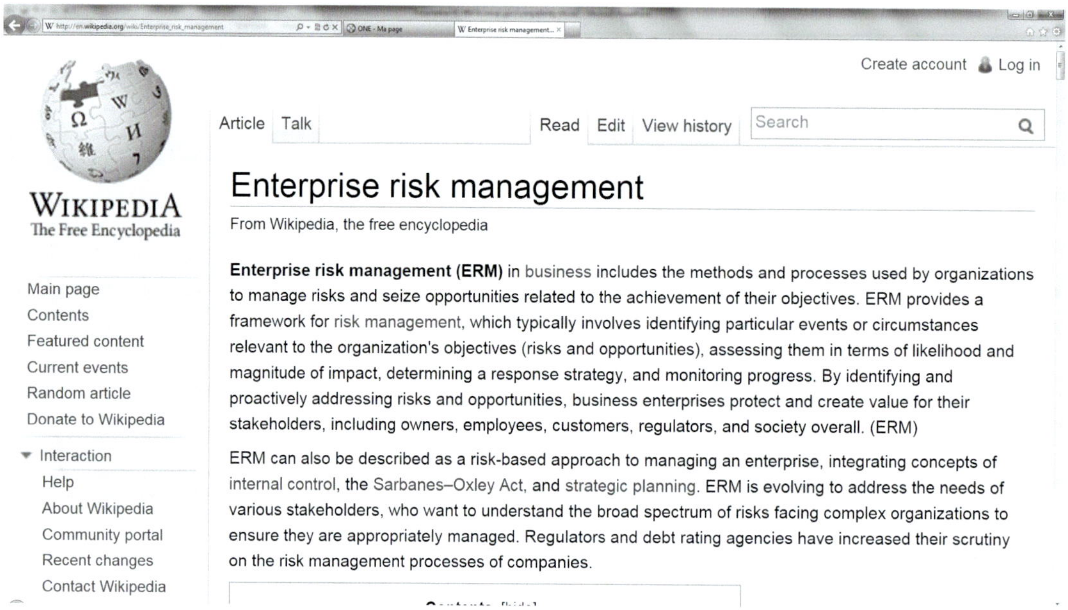

FIGURE 1.38 ERM definition

In terms of governance, the ERM must be compliant with Pillar II, which deals with the system of governance, rules and guidelines, and internal control (Figure 1.39).

FIGURE 1.39 Risks embedded within ERM

In particular, the own risk and solvency assessment (ORSA) is defined by the US National Association of Insurance Commissioners as "an internal assessment conducted by [the] insurer of the material and relevant risk associated with an insurer's current business plan and the sufficiency of capital resources to support those risks". The two primary goals of an ORSA, per NAIC, are as follows.

- To foster an effective level of enterprise risk management at all insurers through which each insurer identifies, assesses, monitors and reports on his material and relevant risks, using techniques that are appropriate to the nature, scale and complexity of the insurer's risks, in a manner that is adequate to support risk and capital decisions.
- To provide a group-level perspective on risk and capital, as a supplement to the existing legal entity view.

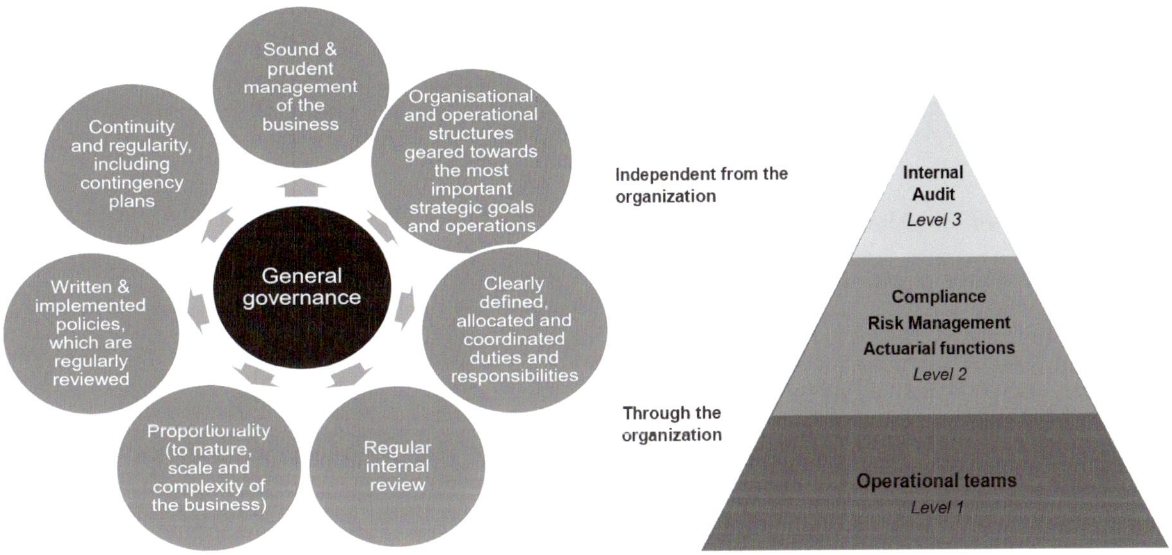

FIGURE 1.40 Governance and internal control of ERM

The ORSA is to ensure that insurers:

- identify and assess all risks they are (or could be) exposed to;
- maintain sufficient capital to face these risks;
- develop and better use risk management techniques; and
- monitor and manage these risks.

It is the key linkage between the company's risk capital modelling capability and the way the business operates and should be an integral part of the business strategy. It provides a holistic view of short- and long-term risks in relation to current and future capital and solvency needs. It is, therefore, the process by which the company demonstrates how the SCR will continue to be met while executing its business plan.

It is a management tool for the senior executives to undertake strategic decision-making within appetite, considering risk and capital profile and sensitivity to future stressed conditions. Thus it helps to strengthen the culture of risk management (see Figure 1.41).

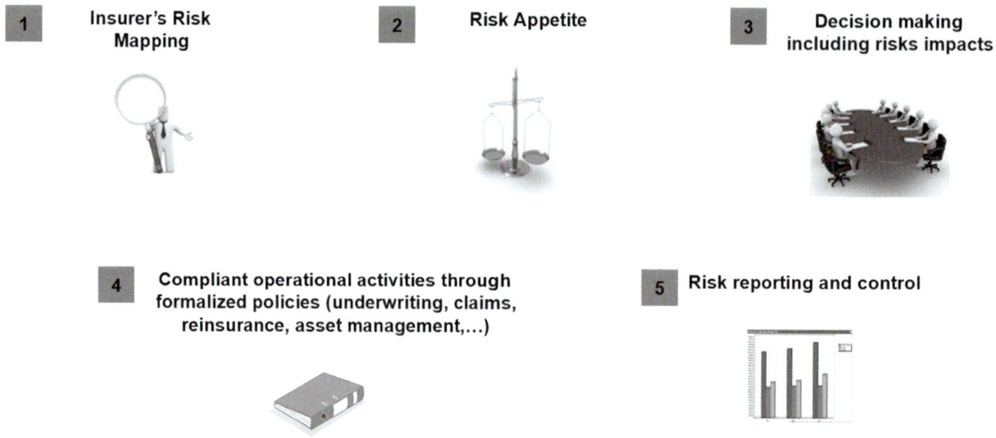

FIGURE 1.41 Risk management strategic axes

In terms of reporting, the ORSA translates into an annual summary report that includes information regarding:

- a description of the risk management framework;
- an assessment of risk exposures;
- capital planning, including balance sheet projections and "reverse stress tests";
- a capital and prospective solvency assessment; and
- evidence of the effectiveness of the ORSA process for the internal and external stakeholder review coverage ratio.

The ORSA report includes three sections that are strongly and directly linked to the ERM framework:

- the description of the insurer's risk management framework;
- the insurer's assessment of the risk exposure; and
- the risk capital and prospective solvency assessment.

In terms of governance, the ERM is based on the "subsidiary principle": subsidiarity is an organizing principle of decentralization, stating that a matter ought to be handled by the smallest, lowest or least centralized authority capable of addressing that matter effectively. The Oxford English Dictionary defines subsidiarity as the idea that a central authority should have a subsidiary function, performing only those tasks that cannot be performed effectively at a more immediate or local level (source: Wikipedia).

There are two axes.

- Dual global and local risk governance: the local chief risk officers (CROs) report to both the group CRO and the local chief executive officer (CEO), while the group CRO reports to the group CEO.
- Organization by risk factors at group level: life versus financial risks.

Decisions should be made by appropriate committees based on thorough studies -> it is a structured way to handle conflicts that naturally arise and reduce individual psychological biases in making decisions in a risk environment.

In particular, a fit and proper policy must ensure that all persons who effectively run the undertaking or are responsible for other key functions are at all times fit and proper; it has to:

- demonstrate the competence and integrity of the decision-makers to the regulator;
- implement a mechanism for preventing conflicts of interest; and
- in compliance with the EU directive, establish separate responsibilities across the key functions within the risk management system.

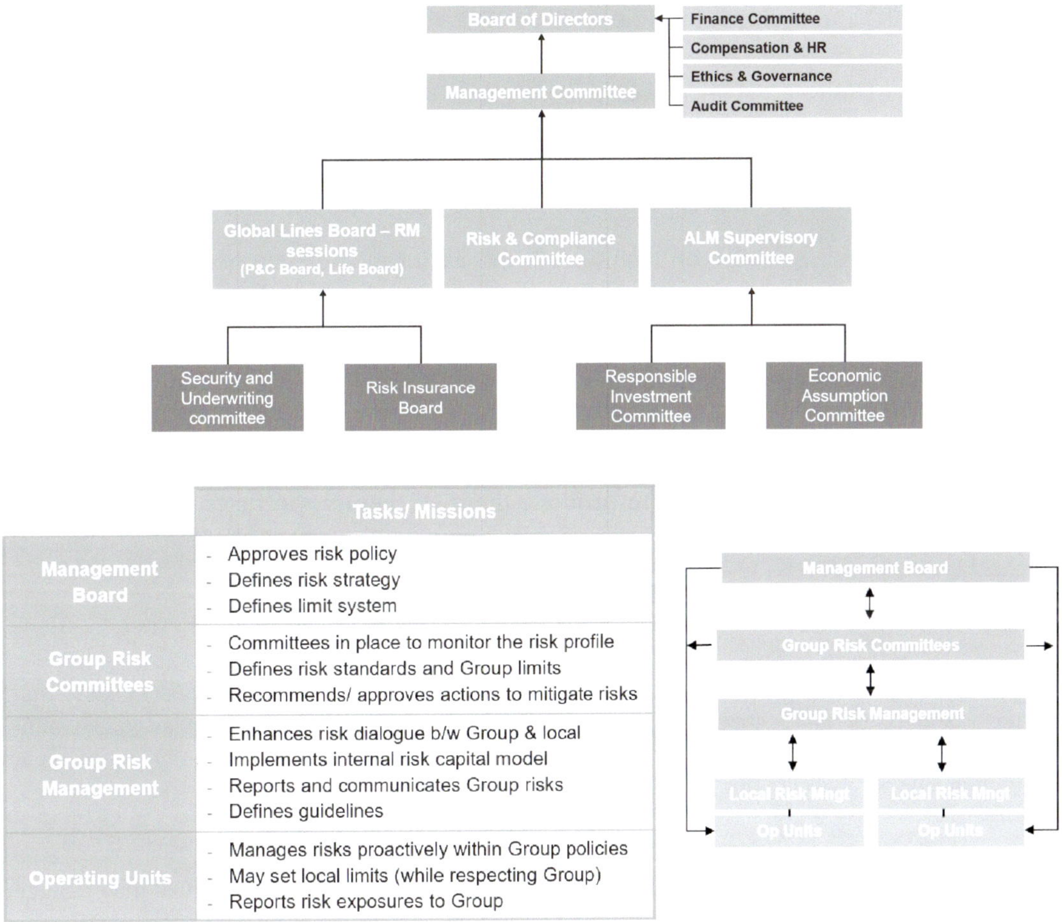

FIGURE 1.42 ERM governance

In addition, appropriate "checks and balances" and second opinions should be in place.

Overall, much of the insurance industry has a well-documented ERM framework, but evidence of use and effectiveness remains inconsistent regarding the following.

- Risk culture documentation and evidence.
- Risk appetite and limits frameworks: insurers have struggled to directly link risk appetite and risk limits across key risk types. Including quantitative measures in the risk appetite requires the adoption of a consistent quantitative measure of risk.
  - Limited focus on "group risks" (e.g. fungibility of capital between legal entities in stressed conditions), when accounting differences across geographies and product lines increase difficulty.

  - Difficulties due to a lack of exposure data; significant approximations are used within the quantitative metrics.
  - Difficulties for the projection of future income statements and balance sheets (including new business) under stressed conditions for key risks.
  - Limited comprehensive views of future balance sheets under various accounting regimes (e.g. statutory, GAAP, economic) and the capital framework.
- Evidence of risk management's role in the business and strategic decision-making process: Consistent implementation of an ERM framework across an insurance group can be a challenge across countries (e.g. multinationals) and/or lines of business.

The specific impacts of Solvency II are as follows.

- Regarding the implementation of Solvency II and ORSA within the insurance industry, the challenges have been diverse, numerous and very significant, among them the following.
  - An increased reporting burden on insurance firms, as Solvency II requires insurance companies to supply numerous reports a year to supervisors, many of which are produced quarterly.
  - There are concerns with the significant ongoing cost
  - Technology is crucial to the implementation.
  - Flexibility and automation are essential, given the volume and frequency of reports, and the speed with which companies have to produce them (it takes up to three weeks just for the SCR number).
  - The use of different data formats makes aggregation a challenge

✓ **Quick win**: one solution is to collect all the data in a single central data warehouse (a "data lake"), which will simplify the process of data governance, meaning companies can apply controls or rules quickly and consistently to all data rather than seeking to replicate rules across separate databases. Virtualization digital technologies may be very useful in this respect.

- Regarding the impact of Solvency II on capital management, using market values may exaggerate the true exposure of the balance sheet to temporary market volatility and therefore force the company to hold significant extra capital buffers to cope with the large volatility.
- Regarding the impacts of Solvency II on the business strategy, they also have been diverse, numerous and very significant.
  - Consistent with the drastic and persistently low interest rates, combined with recurrent volatility spikes, there has been a shift in business, as illustrated by the drastic reduction in general accounts and most unit-linked ones with rich guarantees, together with the emergence of new "capital-light" savings products with lower guarantees to policyholders.

    In this respect, the volatility adjustment (a spread on top of the risk-free rate, specifically introduced to alleviate the burden on long-term guarantees when there are persistently low interest rates) failed to achieve its designed purpose during the recent market turmoil, with limited and artificial effects.

Besides, the risk margin is very sensitive to long-term interest rates, meaning that, in periods of record low interest rates, it can cause insurers to hold a high a level of capital, particularly hitting insurers that are writing longer-term annuities or savings products with any form of guarantee.

There may be a move to increase the last liquid point for the euro swap rate curve from 20 to 30 years, while the ultimate forward rate (UFR) decreased (from 4.2% to 3.6%), which could then significantly increase the value of certain long-term liabilities.

Given the structural long-term nature of the life insurance business, the industry expresses the wish to work with the regulator to ensure that opportunities are taken to develop a more effective long-term approach to the treatment of insurers holding long-term assets, so as to match long-term liabilities.

- There is also the danger that the sheer volume of the rules and guidance acts to limit understanding and effectiveness of Solvency II.
- Consistent with the very high SCR for equities (linked to significant 39/49% downward shocks), there has been a drastic reduction in equities, from as high as 35% in 2000 down to 3 to 5% in major insurers such as Allianz and AXA. Moreover, due to their higher capital requirements, insurers may choose or be forced to sell equities in falling markets, which causes markets to fall further, with the risk of a vicious circle.
- Consistent with the mitigating effect of hedging, there has been an increase in the use of hedging assets (whether FX, fixed income or equities).
- Growing M&A activity: since Solvency II rewards well-diversified insurers with lower capital requirements, larger insurers may look to diversify further, and smaller insurers with specialized businesses (and less capacity to handle all the new analysis and reporting that Solvency II requires) may seek to combine.

1.2.4 Effective Sustainable Insurance Risk Management Implementation

Even though it is not sufficient to solve all real-time risks, an effective risk management framework is definitely necessary: Enterprising businesses need to accept calculated and appropriately mitigated risks in exchange for measurable and commensurate rewards.
Among the key elements are:

- a measurable tolerance and appetite for risk, including the trade-off between risk and reward;
- systematic risk identification as anything that could prevent an organization from achieving its goals and objectives (whether strategic, operational or financial);
- assessing the likelihood, severity and correlation of identified risks;
- developing targeted risk response strategies (internal control, risk mitigation initiatives); and
- creating a system of ongoing review and monitoring that uses key risk indicators (KRIs) and trigger points for the timely identification and escalation of organizational threats.

1.2.4.1 Sustainable Operational Risk Management

The risk management function shall "[p]ut in place strategies, processes and reporting procedures necessary to identify, measure, monitor, manage and report on a continuous basis the risks, at an individual and at an aggregated level, to which the company is or could be exposed, and their interdependencies" (article 44, Solvency II).

The risk management function should:

- assist the board, management and other functions in the effective operation of the risk management system;
- monitor the risk management system;
- maintain an entity-wide view on the risk profile of the undertaking;
- report in detail on risk exposures and advise the board and management on risk management matters;
- identify and assess emerging risks;
- regularly evaluate the design and effectiveness of the risk management system and report findings;
- design, implement, test, validate and document the internal model;
- assess the appropriateness of external credit rating assessments; and
- cooperate closely with the actuarial function.

In terms of processes, see Figure 1.43.

The risk management process involves:[3]

1. **Establishing Context:** This includes an understanding of the current conditions in which the organization operates on an internal, external and risk management context.
2. **Identifying Risks:** This includes the documentation of the material threats to the organization's achievement of its objectives and the representation of areas that the organization may exploit for competitive advantage.
3. **Analyzing/Quantifying Risks:** This includes the calibration and, if possible, creation of probability distributions of outcomes for each material risk.
4. **Integrating Risks:** This includes the aggregation of all risk distributions, reflecting correlations and portfolio effects, and the formulation of the results in terms of impact on the organization's key performance metrics.
5. **Assessing/Prioritizing Risks:** This includes the determination of the contribution of each risk to the aggregate risk profile, and appropriate prioritization.
6. **Treating/Exploiting Risks:** This includes the development of strategies for controlling and exploiting the various risks.
7. **Monitoring and Reviewing:** This includes the continual measurement and monitoring of the risk environment and the performance of the risk management strategies.

FIGURE 1.43 Risk management processes

Operationally, the above cornerstones translate into the following five pillars of the ERM.

- Independent and comprehensive control.
  - Second line of defence: always somebody in charge – i.e. to ensure that monitoring and resolving procedures are in place;
  - ORSA process.
  - Governance and standards (who/how to make decisions/on what information).
  - Within this framework, operational entities are autonomous; outside this framework, prior approval from risk management is required (e.g. large case sizes, such as over €20 million; business not in line with underwriting guidelines; distribution channel compensation system; innovative marketing campaign).
- How much risk to take and which: the risk appetite, with monitoring across the four dimensions of earnings, value, capital and liquidity, across all risks, including stress scenarios and emergency plans.

- A systematic second opinion: challenge to business decisions (on reserves, ALM and asset allocation, reinsurance).
- Rigorous underwriting and pricing:
  - underwriting guidelines; and
  - systematic product approval process (PAP) prior to launch, including proactive updates of product features and pricing in response to market conditions.
- Unified metrics to measure exposure: economic capital (additionally, a measure of profitability), leading the approval process with the regulator, reverse stress scenarios.

FIGURE 1.44 ERM processes

1.2.4.2 *The Product Approval Process for Sustainable Savings and Retirement*

The PAP aims at ensuring that an adequate risk–return trade-off is met before any product launch and formalizes the approval before significant investments are made. It helps the sharing of best practices of product development across the group (design, pricing and risk management). It provides the basis for wider business action on poor profitability and/or highly capital-intensive products.

The PAP makes sure that, for the launch of a new product, there is an appropriate decision-making process in place in local entities.

- It measures value creation: risk analysis based on minimum requirements (product features, pricing, hedging, profitability).
- It ensures consistency throughout the group: pricing report (description, analysis of profitability and sensitivities, capital efficiency, risk mitigation strategies), legal/compliance/regulatory aspects.

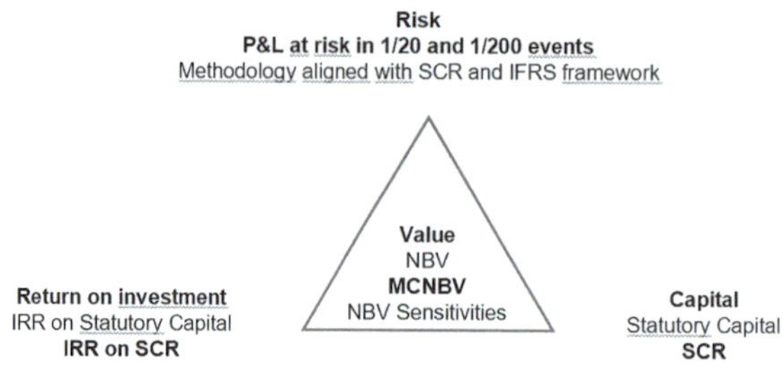

FIGURE 1.45 KPI and KRIs

✓ Among the profitability metrics are the following.

- NBV (new business value) = CE PVFP – strain + TVOG + CoC/NFR

  NBV is the value of new business issued during the current year, after deducting the cost of capital. The corresponding ratio is the NBV margin (denoted NBVM), defined as NBV / PVFP (where PVFP is the present value of future profit).

✓ **Quick win**: the NBV is usually high for protection and health, but low for investment and savings (I&S) (excluding the accumulator) and servicing profitability.

- 1st year AFR = CE PVFP – strain + TVOG + MVM
  where
  - CE PVFP is certainty-equivalent present value of future profits.
  - Strain is the initial cost of selling the business.
  - TVOG (< 0) is the time value of options and guarantees.
  - CoC is the cost of holding capital (< 0) and NFR is non-financial risks.
  - MVM is the market value margin (cost of life insurance, operational risks and reinsurance default risks) < 0
- The sensitivities of these two metrics with respect to equity, interest rates and volatility moves.
- ROI = statutory return on investment (internal rate of return).
- ROE = GAAP return on equity.
- Profit margin = present value of profits as a percentage of present value of premium.
- ROA = present value of profits as a percentage of present value of reserves/fund.
- VNB = present value of one year of new business as a percentage of first-year premium.
- The free cash flow = statutory earnings – Δrequired capital.
- The decomposition of change in value from the previous year, through a "roll-forward" waterfall chart (Bourgeois 2014).

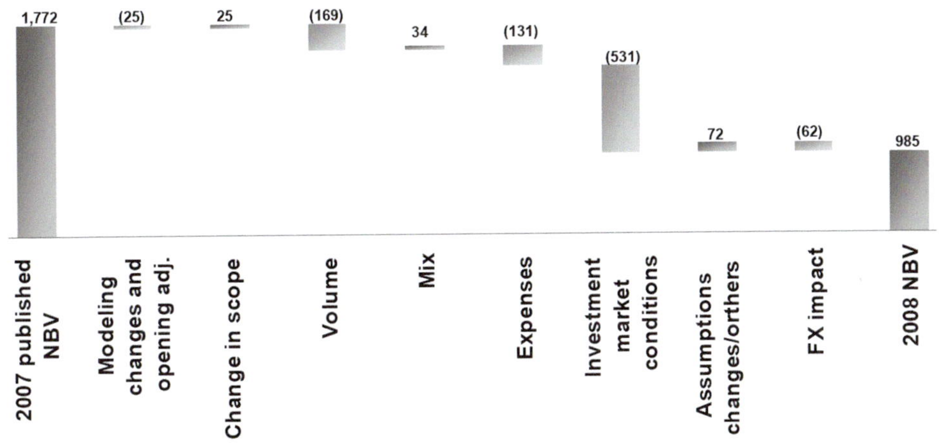

FIGURE 1.46 Roll-forward waterfall of NBV

✓ Among investment efficiency metrics is the investment performance ratio = investment performance / SCR

✓ Among the capital efficiency metrics are:

- Coverage ratio = AFR / SCR (or risk-based capital [RBC] ratio in the United States).
- Solvency II return = ΔAFR / STEC.
- Hard capital = 150% or 200% SCR – VIF (value of in force) + MVM
- $\textbf{Required capital}_t = \textbf{Max}\left(150\%\,\textbf{STEC}_t - \textbf{VIF}_t + \textbf{MVM}_t, 0\right)$

✓ **Quick win**: the MVM is to be compared with VIF (thus, looking at the difference, VIF – MVM, which makes a *significant* portion of the AFR). MVM is significant for general account savings, translating into potential negative AFR. Even so, a significant MVM can be compensated for by high VIF, resulting in a positive AFR, as in protection and health and UL policies.

✓ **Quick win**: in terms of capital leverage, the MVM is to be positioned against the SCR: high MVM relates to high SCR, the maturity of business run-off (over 15 years) and the projection of future premiums (e.g. Switzerland, Hong Kong, Germany, Japan).

- Capital consumption, which measures the total amount of free surplus consumed for new business = 1st year AFR – 150% or 200% SCR.

✓ **Quick win**: the capital consumption metric is usually low for health protection and servicing, while traditional life and savings tend to remain the highest for capital consumption.

- The internal rate of return (IRR), the discount rate at which the present value of distributable earnings is equal to the strain and initial required capital invested to support the business underwritten; defined as the solution to the following equation:

$$\sum_{t=0}^{n} \frac{Distributable\ earnings_t}{(1+IRR)^t} = 0$$

$$\sum_{t=1}^{n} \frac{Distributable\ earnings_t}{(1+IRR)^t} = strain + initial\, required\, capital$$

$$\text{Distributable earning}_0 = \text{Statutory earning}_0 - \text{required capital}_0$$

$$\text{Distributable earning}_t = \text{Statutory earning}_t - \text{required capital}_t + \text{required capital}_{t-1}.$$

The amount and timing of distributable earnings are based on four key components: the first year's investment (i.e. strain to acquire new business, including acquisition and administration costs, plus initial required capital); statutory earnings (SE; including investment return on capital); release of required capital (RC); and financial and non-financial assumptions underlying the projection.

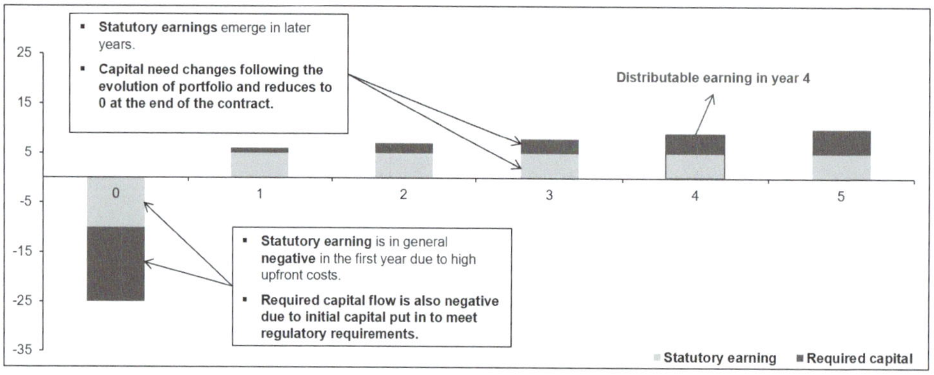

FIGURE 1.47 Timing of distributable earnings

The IRR may be computed on a base case scenario or on a stress scenario (e.g. market crash or economic deflation).

The impact of a change in the regulatory capital requirement framework may be significant, depending on the magnitude of risks.

- When there is no additional capital requirement, SII IRR = SI IRR.
- Due to the low-risk nature of a pure unit-linked product, the AFR is sufficient to cover the majority of the SCR, implying a lower hard capital requirement under Solvency II, while the Solvency II IRR is higher than the Solvency I IRR.

✓ Among profitability/capital efficiency combined metrics are:

- short-term return = earnings / SCR; and
- long-term return = (VIF – CoC × PVSCR) / PVSCR; this reflects the long-term margin adjusted for the cost of capital required to finance the SCR; the present value solvency capital requirement (PVSCR) provides the average view of the capital used over the maturity of the product.

Screening and benchmarking the quality of businesses are then possible through the following charts of solvency IRR versus coverage ratio, or NBV versus SCR in favour of life and savings without guarantees (or with guarantees externalized to a third party) and protection products (see Figure 1.48) or NBV versus capital consumption, also in favour of life and savings without guarantees (or with guarantees externalized to a third party) and protection products.

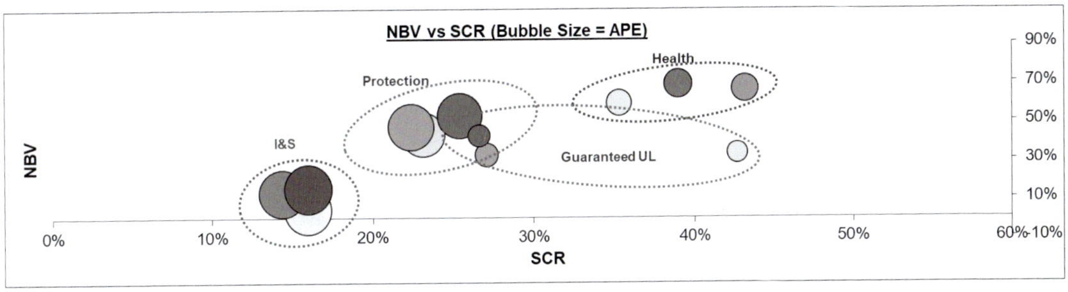

FIGURE 1.48 NBV versus SCR

- Improve formalization: all new products must have a formal sign-off required by the chief financial officer (CFO), senior actuary and head of product management.
- Launch of specific products requires group management board approval.

In terms of governance, the PAP is local responsibility with group oversight (Bourgeois 2014).

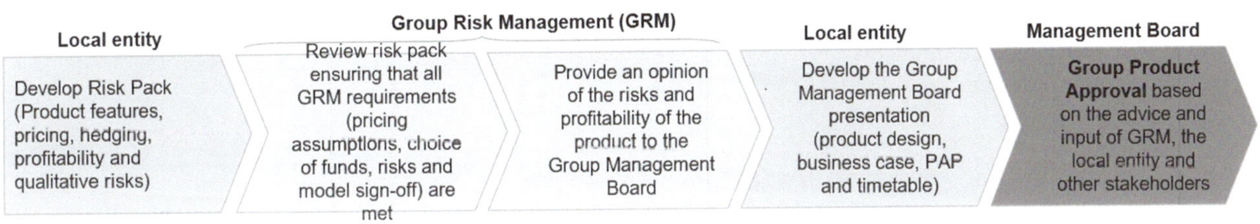

FIGURE 1.49 PAP process

For products that do not meet the targeted PAP threshold for the country, sign-off by the local CEO and group business line is required.

PAP reports must have a formal sign-off at least by the local heads, usually the CFO/CIO/CRO/chief actuary or pricing actuary (if no chief)/chief marketing officer (CMO)/the head of product management. In addition, the compliance must receive a formal sign-off before launch by the head of legal or compliance officer or senior legal officer and the head of product management. In addition, the chief corporate responsibility officer is usually consulted.

In terms of reporting, the PAP report is composed of five elements:

- product design;
- pricing (e.g. fees/loadings, GMxB pricing, protection, …);
- profitability analysis (base case + sensitivities);
- risk analysis; and
- risk pack required in some cases (e.g. GMxBs, long-term care) (guaranteed minimum death and living benefits are collectively known as GMxBs).

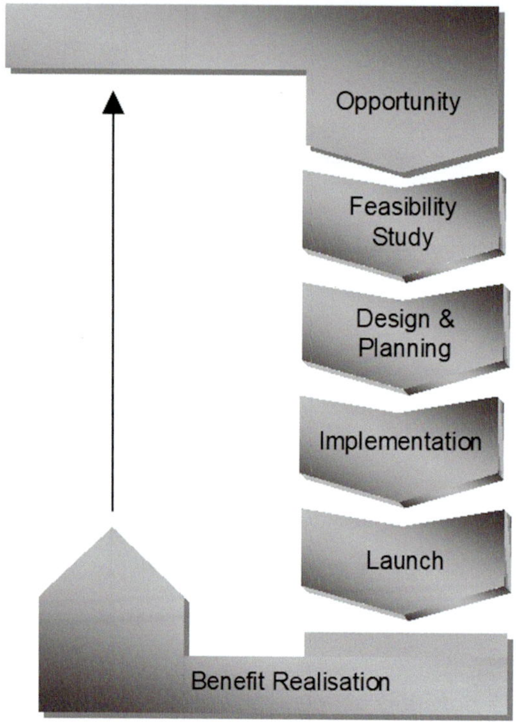

FIGURE 1.50 PAP report components

1.2.4.3 The Risk Appetite Framework for Sustainable Savings and Retirement

The risk appetite is a risk governance tool to ensure that risks are limited, their consequences on all dimensions understood and reviewed on a regular basis. Risks are therefore taken on the basis of management choices. The aim of the risk appetite framework is to ensure that appropriate governance, reporting, limits and decision processes have been set up to drive risk management decisions. It helps the group

- have a clearly stated risk appetite that defines the tolerance of senior management to a negative variation in earnings, value, capital and liquidity;
- monitor the accumulation of risks through the positions versus the risk limits; and
- manage its own exposure through interactions with local operating units.

In terms of governance, the risk appetite is a structured process between the group and local units to ensure consistency of risk tolerance, related governance and decision processes – empowering local units to actively manage their risk exposures through their own choice of local risk limits. In this respect, it both empowers local units and manages the group. At group level, the management committee is responsible for the definition of the group limits, while the different committees manage the group's exposure. Local operating units have defined their governance and set their own choice of local risk limits (consistent with but not equal to the group's) and respective action plans. The product launch requires local and regional risk management review and formal approval.

The coordination is increased in periods of crisis, with a temporary ban applied across all business lines and all countries.

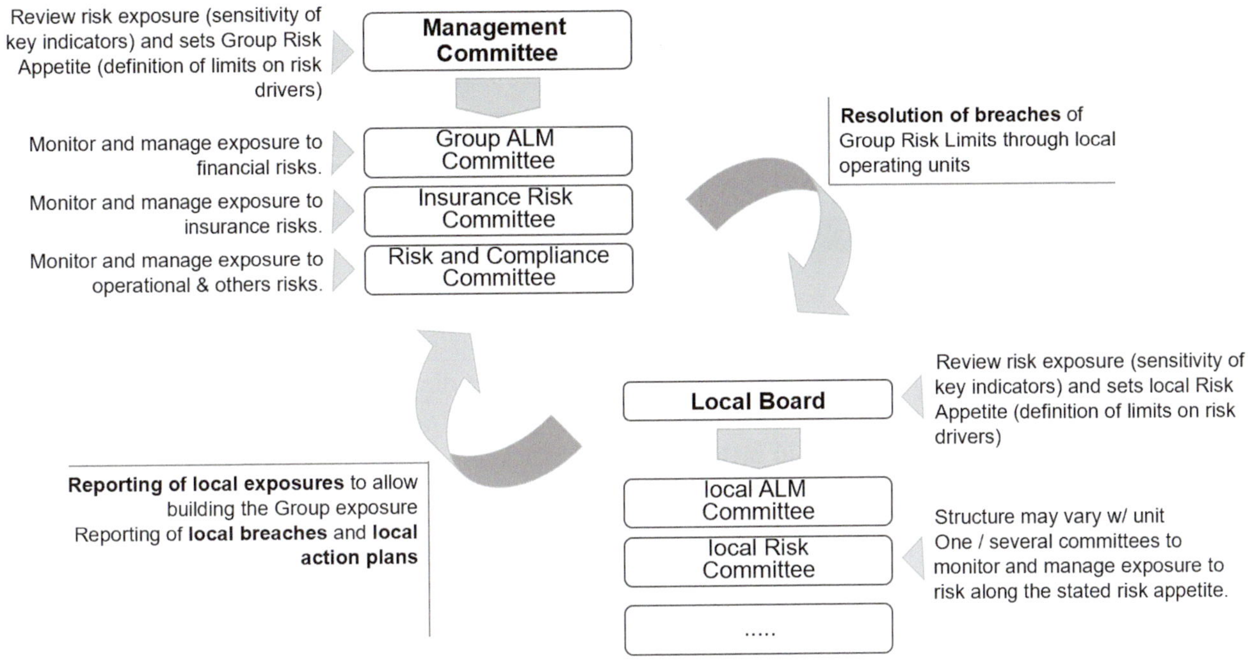

FIGURE 1.51 Risk appetite governance

It is an operational framework with alerts and limits to actively manage risk exposure. In this respect, the risk appetite considers the potential impact of risks on four dimensions across several key performance and risk indicators to set up limits on functional risk indicators (see Figure 1.52).

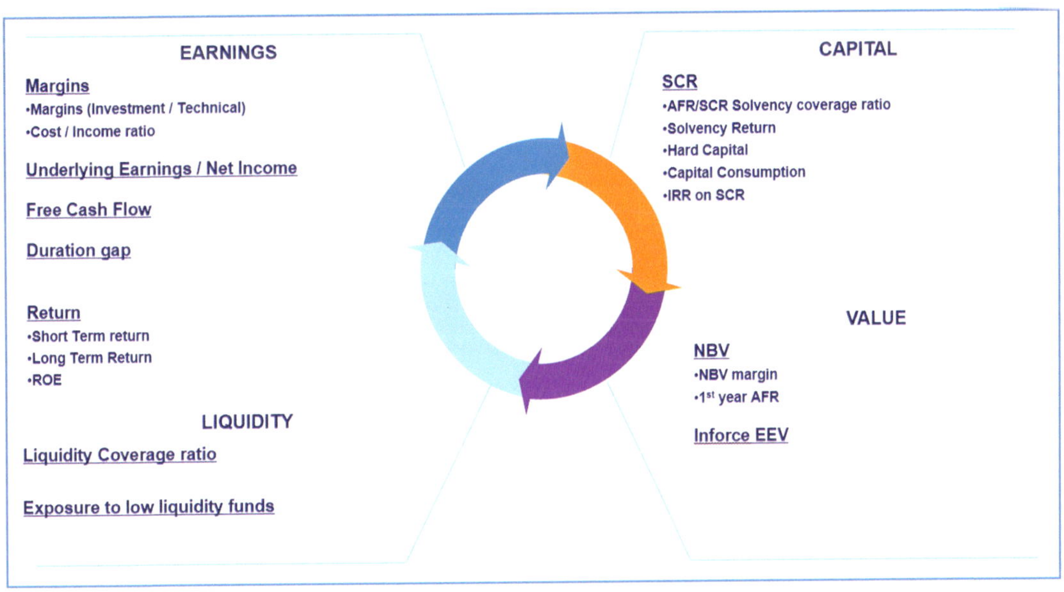

FIGURE 1.52 Risk appetite dimensions and corresponding KPIs/KRIs

The functional risk indicators are simple indicators (no complex calculations), used day to day to manage the risk exposures; they are easily and frequently monitored, reported and understood by people in the operations. They are needed to ensure that risk appetite will translate operationally, as actions are taken to manage risk depending on the area of the reported exposure.

For each functional risk indicator, the three areas of the exposure are defined by two thresholds (see Figure 1.53).

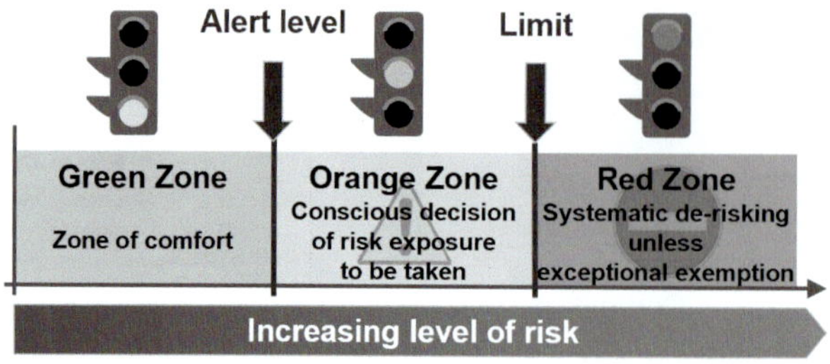

FIGURE 1.53 Traffic light system

These thresholds are called the risk appetite limits across the four dimensions (earnings, value, solvency, liquidity), usually consistent with a 1/200-year (extreme scenario) or a 1/20-year event (adverse scenario).

The risk limits are then specifically set to each category of risk (financial, life, operational) (Bourgeois 2014).

✓ **Quick win**: regarding the market risk appetite, the total exposure (P&L) can be decomposed into a Taylor expansion of so-called "Greeks", with respective risk limits for both adverse and extreme scenarios.

The above computation is made for both liability option value (LOV) and hedge assets. The SCR is based on the net hedging position (hedge assets change minus the LOV change).

Such decomposition is linear, offering a direct comparison between the categories of risks, offering a direct and transparent P&L attribution.

P&L derived for each Risk Factor	Adverse	Extreme
Equity level	-0.9	-1.40
Equity implied vol.	-4.6	-9.25
Equity realized vol.	-12.3	-31.01
Interest rate level	0.3	0.52
Interest rate implied vol.	-0.6	-1.17
Interest rate realized vol.	-1.7	-3.60
T-Error: Index fund	-6.7	-16.80
T-Error: Managed fund	-28.8	-57.60
T-Error: Other	0.0	0.00
Invt Grade Credit risk	-0.1	-0.15
Sub-Invt Grade Credit risk	-0.3	-0.60
P&L derived on a Combined Basis	-5.4	-12.08

vs. Normalized Earnings

Breaches to Risk Limit	Adverse	Extreme
Equity level	-0.14x	-0.20x
Equity implied vol.	-0.65x	-1.32x
Equity realized vol.	-1.76x	-4.43x
Interest rate level	0.05x	0.07x
Interest rate implied vol.	-0.09x	-0.17x
Interest rate realized vol.	-0.24x	-0.51x
T-Error: Index fund	-0.96x	-2.40x
T-Error: Managed fund	-4.11x	-8.23x
T-Error: Other	0.00x	0.00x
Invt Grade Credit risk	-0.01x	-0.02x
Sub-Invt Grade Credit risk	-0.05x	-0.09x
Breaches to Combined Risk Limit	-0.77x	-1.73x

FIGURE 1.54 Risk appetite framework

As the methodology is aligned with the capital requirement methodology, the risk appetite framework directly assesses the impact of management decisions on the solvency ratio.

In terms of processes, the following are entailed.

- Monthly monitoring of the risk appetite metrics across the four dimensions.
- A quarterly release of a group risk report, with the sensitivities of the seven key indicators to various risk factors (financial and insurance) in case of adverse events (1 in 20 years) and extreme events (1 in 200), and the translation into risk limits at group level, to be validated by the management committee.
- Regular reporting and management of the group risk exposure versus limits, specifically monitored by the group ALM supervisory committee for financial risks, the risk insurance supervisory committee for insurance risks and the risk and compliance committee for operational risks.
- This process mirrored in local units: quarterly reporting, risk limits, ongoing management of exposure, with ongoing discussions with the group to ensure alignment.

In the event of these limits being breached, mitigating actions are required, in particular to secure the risk appetite in a sustainable and cost-efficient way – i.e. at a pace function of opportunistic views versus a systematic approach.

- Hedging for market risks, which involves the following:
  - minimizing the interest rate risk by adjusting the duration and convexity using swaps and different options (caps, floors, swaptions);
  - minimizing the equity volatility risk using options, based on opportunistic volatility or interest rates triggers;
  - using total return swaps to reduce basis risk; and
  - "pre-hedging" to secure profitability.
- The credit risk is also actively managed through:
  - having a well-diversified credit portfolio across all sectors and with good ratings;
  - the disciplined use of repos and credit default swaps (CDSs);
  - more stringent derivative exposure management, with collateralization of all trades being implemented; and
  - the reallocation of some funds with corporate exposure to governments.
- Liquidity risk is closely monitored, and the necessary mitigation measures are put in place so that liquidity resources exceed liquidity requirements under stress conditions.
- Enhanced product designs are introduced:
  - restriction of features such as revisions to guarantee features, including flexible roll-ups, the ability to increase rider fees, the waiting period for dollar-for-dollar withdrawals and resets, reduced equity exposure and penalties on deferral period and fund switches;
  - adjusting profit-crediting rules for traditional with-profits liabilities;
  - premium suspension (no additional contributions into legacy);
  - higher fees; and
  - more prudent assumptions (e.g. a lower lapse floor, higher volatility or correlation assumption, more efficient customer behaviour).
- Reinsurance for life risks (longevity, normality, lapses).
- In-force actions are taken:
  - reduction of basis risk through changes in the fund line-up (e.g. from 10/90 to 70/30 passive versus active funds, allocating cap to funds with low liquidity);
  - the roll-out of managed-volatility funds to reduce exposure to realized volatility;
  - selective partnering in third-party activity, incentivizing sales through commissions;
  - buyout initiatives; and
  - the disposal of structurally unsustainable businesses.

- Strengthening reserves through assumption changes (e.g., for non-hedgeable or non-reinsurable risks such as behaviour risks, a two-sided dynamic lapse function and "efficient" withdrawal assumptions).
- Stopping selling new businesses.

These mitigating actions can be categorized across the four dimensions of risk appetite. The associated costs of the measures above need to be balanced with the expected future savings under a 1/20- or 1/200-year event. Given that the risk appetite framework is structured across these four dimensions (earnings, value, solvency, liquidity), the mitigating actions are expected to produce a significant improvement in the alignment between capital and earnings targets (see Figure 1.55).

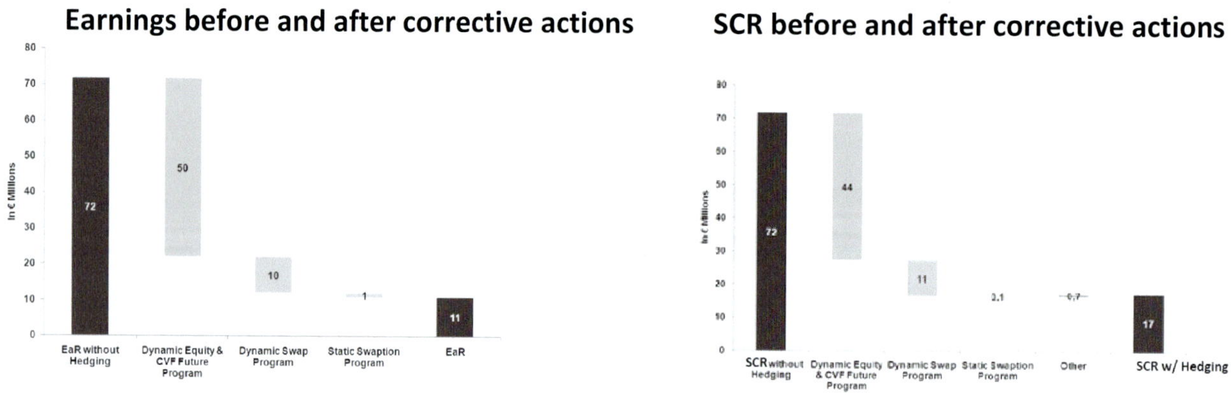

FIGURE 1.55 Impact of corrective risk management actions on SCR and earnings at risk

It is also possible to assess the impact of the mitigating actions through the evolution of the SCR risk signature.

- Swaption hedges reduce the interest rates convexity risk, while equity options hedges reduce the implied volatility (IV) risk.
- A new shock calibration, with a lower dynamic lapse floor, may in contrast increase the lapse risk.

Product Risk Appetite for Sustainable Savings and Retirement

Customers' needs may change often, quickly and drastically, so it requires agility at minimum costs for insurance companies to stay afloat. In this respect, an increasingly selective customer risk appetite has been observed since the 1980s, across four risk appetite axes (yields/flexibility/financial guarantees/cost), which can be illustrated, for example, through the development of most savings and retirement blockbusters, such as universal life and the variable and fixed index annuities, in favour of increasingly flexible and cheaper guarantees to customers with lower risk tolerance.

Moreover, the development of digital technologies has further shifted the focus from the products to the customers, which reversed the bargaining power between the customer versus the asset manager in favour of the former. Such a new paradigm requires switching from a top-down to a bottom-up approach. Being customer-centric is no longer an alternative or a "plus"; it is a requirement for survival. Reputation risk has become the key, with success in the past providing no guarantee for the future.

As a result, sustainable life insurance that aligns insurer and customer risk appetites in an era of such increasingly efficient customer behaviour and an economically adverse environment requires, on the one hand, enhancing customers' experience through better advisory practices (e.g. by pairing the right funding strategy with the right policy structure) and, on the other hand, cheap, flexible, performant and capital-light savings and retirement products, illustrated by the rise in CPPIs, structured annuities (RILAs) and hybrid general accounts.

It also requires strong partnerships with asset management, investment banking (for the financial guarantees) and digital tech companies, opining the way to new life insurer archetypes.

In the quest to build sustainable savings and retirement, section 1 assesses the drastic changes in an increasingly selective customer risk appetite, in favour of flexible, lower-risk and cheaper guarantees. Section 2 then focuses on aligning both customer and insurer risk appetites through an enhanced advisory and product range, leveraging digital technologies and banking/asset management skills through long-term partnerships.

DOI: 10.1201/9781003218371-3

2.1 THE INCREASINGLY SELECTIVE CUSTOMER RISK APPETITE ACROSS THE FOUR AXES: YIELDS/FLEXIBILITY/FINANCIAL GUARANTEES/COST

2.1.1 Case Study: US Life Insurance Market, where Customer Risk Appetite Switched between Variations of Universal Life depending on Relative Trends of Equities versus Interest Rates

2.1.1.1 The Fast-Growing Customer Risk Appetite for Universal Life in the 1980s, in line with the Sustained High Interest Rate Regime

Prior to the introduction of universal life (UL) insurance in the early 1980s, the life insurance available to policyholders consisted of term and whole life insurance.

- Term policies (massively developed in the context of high interest rates during the 1970s and 1980s) are issued for a given period, such as ten, 20 or 30 years. Their sole purpose is to provide financial protection (a death benefit) for your dependants in the event of your death. But, if you do not die during the term, the policy will expire when the term ends, and no benefits will be paid. One of the biggest benefits of term life insurance is that premiums remain the same throughout the term ("level premium period") of the policy (although they are reset at each renewal), which provides cost certainty, while the biggest drawback is that, if you do not die during the term of the policy, neither you nor your beneficiaries will receive any benefit.
- Unlike term insurance policy, a whole life insurance policy has value other than the death benefit, and lasts for your whole life. With this extended period, premiums are considerably more expensive, but with the major benefit that the premiums can grow (tax-deferred) as a guaranteed cash value (expressed as a percentage of the policy face value) and be accessed over the life of the policy. And, if you do not touch this cash prior to passing away, it is disbursed to your beneficiaries in addition to the face value of the policy. These excess whole life premiums are invested at the insurance company's discretion. Most people will be better served by purchasing it at a younger age (20) to secure lower premiums.

  As a result, a whole life insurance requires a policyholder to pay a fixed, given amount of premium each year, in exchange for having a fixed, given death benefit, and a progression of fixed cash surrender values over time (all of which are guaranteed).

  The result is a policy that gives a "reasonable" rate of growth on the cash value, and a "reasonable" internal rate of return for those who held the policy until death (at least, relative to interest rates at the time).

  Unfortunately, though, by trying to give clients a return on both these elements – the cash value and the death benefit – the implied returns are often a lot lower for either than what could be achieved by focusing individually on just one element or the other.

In this context, the significance of the introduction of more "flexible premium" universal life products was that it gave policyholders the flexibility to more fully maximize either the internal rate of return on the death benefit portion of the product or the investment return on the investment account in the product.

Moreover, universal life (launched in the early 1980s) adds not only guarantees but also flexibility to whole life: you can make premium payments at any time you wish and in any amount you wish; and excess premium payments can be anything from stocks to bonds to mutual funds/annuities (albeit potentially restricted by the insurance company), and so with higher market risk exposure.

In addition, although traditional life insurance was something of a black box to the purchaser, universal life took an "unbundled" approach and was transparent to the policyholder with respect to policy expense charges, cost of insurance charges, policy level activity and interest rate credits, to name a few.

In the early 1980s market interest rates were very high. The prime rate hit an all-time high of 21.50% on 19 December 1980, and this provided a solid launching point for universal life sales, as UL credited a market-like interest rate in the double digits. As universal life continued to dominate the market in the 1980s, more companies began to enter this market, until most of them had at least one UL insurance product on the books. The ongoing potential seemed endless, specifically for universal life or variable life (VL) plans including credited market interest rates.

Definitely during most of the 1980s customers' risk appetite for UL was mostly driven by such high single- or double-digit market credit rates, including a minimum 4 to 6% guaranteed yield, with which traditional life plans (both participating – dividend paying – and non-participating) were not able to keep up. Quite a number of the UL sales of that era came from insured individuals exchanging their traditional whole life plans for the newer, interest-sensitive life plans.

2.1.1.2 The Switch in Customer Risk Appetite to Variable Life in the 1990s, in line with Falling Interest Rates and the "Roaring Bull" Equities Market

As a result of this newly focused customer risk appetite, the falling credit rates illustrated on many of the universal life plans led to their downfall in the 1990s as interest rates started their aggregate drift downwards. As a result, there was a spate of class action litigation in the 1990s relating to interest-sensitive life insurance illustration performance. Insurance companies began to rethink their UL product line, especially with respect to interest rates that were projected on illustrations.

As a result, the "roaring bull" market in equities, in addition to the litigation and model regulation, saw numerous sales moving away from universal life and into variable life products (including variable universal life: VUL). Variable universal life did not have the same issues with respect to interest rate crediting that non-variable UL did, because VUL interest rate crediting in the separate account was based on the underlying performance of the subaccounts. This contrasted with non-variable UL, for which the interest rate credit was generally in the control of the insurance company. Furthermore, variable life provides the option to grow cash value, whereas in a universal life insurance policy the value grows at a rate set by the insurer.

Therefore, such a move was driven by a shift in customers' risk appetite from fixed-income yields to more uncertain but higher equities yields, combined with the search of more flexibility in growing the cash value.

2.1.1.3 The Growing Customer Risk Appetite for Fixed Indexed and Secondary Guarantee Universal Life in the 2000s, in line with Higher Risk Aversion versus Downside Risk

Even so, universal life has seen a comeback of sorts in the 2000s, as its market share as a percentage of life insurance premiums has grown from 27% in 2002 to 42% in 2011 and 38% in 2013.

There again, such a major move was driven by a shift in customer risk appetite.

- The significant equity market corrections led to a flight from the perceived riskiness of variable products.
- These were further fuelled by the flight to quality caused by the uncertainty in the post-9/11 financial market.
- The higher risk aversion versus the downside risk often associated with variable life also played a role, met in part by changes in product design, as illustrated by the launch of "fixed indexed universal life" (FIUL) policies.

FIUL products are indeed structured to participate in the upside of the equity markets while avoiding/limiting the downside risk of variable products, through a combination of fixed-income instruments (bonds) to provide the underlying interest guarantee while using call options on stock market indices to provide the upward potential over the interest rate guarantee. Like other UL products, there is a minimum interest rate guarantee (2%, for example). Although FIUL products, like variable products, shift the burden of determining the interest rate credited from the insurance company to the financial markets, there are still significant issues as a result of the current low interest rate environment.

- The lower the interest rate, the higher the price of the bond in general, and the lower the residual available money to invest in call options to take advantage of the upside in the equity market.

- Stock market volatility has been more the norm in the new millennium, which causes the prices of call options to increase.

FIUL products have made a definite splash in the 2000s by offering the potential for upward appreciation while limiting the downside risk often associated with variable products. The FIUL is more in line with mainstream UL products, in that the potential to accumulate significant cash values is the draw.

Another major growth area in the 2000s, perhaps even more so than fixed indexed UL, has been the growth and proliferation of "secondary guarantee UL" products, commonly referred to as "shadow account UL". This is a product line in which the emphasis is on securing insurance protection over a prolonged period of time, rather than trying to accumulate significant cash values.

The insurance protection is guaranteed to stay in force as long as the underlying "shadow" account is positive, even if the cash values of the normal UL accumulation fund turn and stay negative. Premiums are lower than a normal UL product and, at older ages, are attractive enough to lure buyers, given that term products are not available or not feasible at those older ages. Nonetheless, such shadow account products are popular as the result of this solid lifetime guarantee. This product is for those insureds who favour insurance coverage over cash value accumulation.

Although most, if not all, life insurance companies have jumped on the shadow account wagon (much like life insurance companies that turned their focus over to selling UL in the 1980s),

- the amount of capital and reserve needed to support a shadow account UL product is quite large in relation to the reserves held for other non-shadow account UL products, due to the length and nature of the coverage guarantee. So insurers grapple with the capital-intensive nature of these products due to the high amount of statutory reserves that companies selling these products are required to hold.
- Another risk for the insurance company selling shadow account products is lapse risk. Most shadow account UL products are lapse-supported, meaning that more lapses mean more profit for the insurance company. This is in contrast to most life insurance products, with which the opposite situation exists (i.e. fewer lapses lead to more profits). Therefore, it is very important for the insurance company to be very judicious in its estimate of future lapse rates during the pricing process. Most insurers try to be prudent; some even factor a 0% lapse rate in pricing once the shadow account is "in the money" (some consider this to be at the point when the shadow account is positive and the regular fund cash values are zero or negative). Lapsing at that point would provide no benefit to the insured, as there is no cash value to be had. The only lapses would be from those who no longer can afford (or desire to) pay a premium or from those who feel that they no longer need life insurance protection.
- Yet another risk for the insurance company is older age risk, which comes when a disproportionate number of policies issued are to older age individuals. Since term insurance for older age individuals is not offered or is so costly as to not be an option, the only way to affordably maintain life insurance protection for this group is through a shadow account UL product. The average issue age of the shadow account UL product is generally in the 60s. Part of the issue with respect to mortality risk is that there is still limited experience regarding mortality rates and the mortality curve slope at older ages. Given that several shadow account products on the market can last until maturity (age 121, for example), the limited amount of older age mortality experience data that exists at ages 90 and up can be a deterrent to entering the shadow account market for some life insurance companies.

Even so, while there are numerous risks for the insurance company in the shadow account UL market, as well as capital intensiveness and limited profitability, these products have been a hit, with the consumer looking to maintain life insurance protection for a long timeframe at a reasonable price.

2.1.1.4 Strategies for Universal Life Products in the Current Market of Persistently Low Interest Rates

There has been no life insurance company that has been insulated or immune to the low interest rate environment when it comes to the manufacturing and issuing of universal life products. All have had to grapple with the issues of how to develop products that will provide their insureds with a solid value proposition while having to deal with a less than optimal interest rate market.

While UL holds the dominant market share with respect to premiums versus competing life insurance products, the share could erode in a bull market (as variable life sales increase) or in a volatile/apprehensive market (as insureds move to whole life and term products in a flight to quality).

Moreover, there is always the potential for a new product type to enter the market and shift the focus away from UL products, much as UL products shifted focus away from more traditional life products in the 1980s.

For our old friend UL, whose foundation was to provide cash value accumulation, the fixed index UL and even VUL products provide an alternative for that market. Even so, with regard to the type of UL that has been in the market the longest, which is both non-variable and non-indexed in nature, it may be quite some time before interest rates rise to such a level as to make these products competitive in the market again.

Although the interest rate spread earned by insurance companies on UL has been compressed and is no longer quite the profit driver it used to be, insurance companies continue to try and find ways to make cash accumulation UL both appealing to the general life-insurance-buying public and profitable to sell. These include managing the other sources of profit, such as mortality, expense and surrender, even more closely than beforehand.

- For example, as mortality rate studies continue to show a downward trend in mortality rates as more people engage in healthier lifestyles, some insurance companies may factor this directly into their pricing (as well as future mortality improvements) in attempts to earn a larger spread on expected versus actual mortality. With respect to expenses, some insurance companies have lowered agent commissions, which, while not exactly pleasing to the sales force, can help increase the profitability of a product. Becoming more effective at overall insurance expense control has been another way companies have tried to cope.
- Others have used increased policy charges to allow for the crediting of higher rates. Some have increased their earned rates by lowering the quality of the new asset portfolio or going out longer on the yield curve.
- Features such as new riders (LTC, for example) have also been a way to make UL more attractive in light of a low interest rate scenario. Moreover, market-value-adjusted products have been introduced.

Whatever method is in use, insurance company creativity continues to show through as one of the strong points of the industry. However, cash accumulation UL will come back into the forefront one day, when interest rates begin their inevitable climb upwards. Until then companies will continue to exercise their vast creativity to continue selling these products and provide value for their clientele.

2.1.2 The Surge of Traditional Variable Annuities in the 2000s and Their Significant Decline since 2015 with Persistently Low Interest Rates and Recurrent Financial Market Volatility

Life insurers began offering enhanced guarantees for variable annuity contracts in the early 2000s, especially guaranteed minimum living benefits (GMLBs), in addition to the existing death benefit guarantees (GMDBs). Guaranteed minimum death and living benefits are collectively known as GMxBs.

The US Securities and Exchange Commission (SEC) defines variable annuities like this:

> [A] contract between you and an insurance company, under which you make a lump-sum payment or series of payments. In return, the insurer agrees to make periodic payments to you beginning immediately or at some future date. You can choose to invest your purchase payments in a range of investment options, which are typically mutual funds. The value of your account in a variable annuity will vary, depending on the performance of the investment options you have chosen.

Variable annuities are life insurance products with unit-linked or managed fund investment characteristics and optional guaranteed benefits for the policyholder. With a variable annuity, the insurance company agrees to make payments to you in the future. You can purchase a variable annuity contract by making either a single purchase payment or a series of purchase payments.

A variable annuity has two phases: the savings (or "accumulation") phase and the payout (or "annuitization" or "income") phase. During the savings phase you make purchase payments into the contract and the earnings accumulate on a tax-deferred basis. The payout phase starts when you begin receiving regular payments from the insurance company by electing a variable annuity income option. Many contracts include a variable annuity commencement date, generally between ages 85 and 95, at which variable annuity owners are required to select a payout option (also known as "forced annuitization").

The insurer invests those funds in stocks, bonds or money market instruments, directly or via a contracted party (asset manager or bank, for instance). The innovation is that the policyholder is guaranteed a regular payment on a certain predetermined date fixed in the contract and is served it for the rest of her life (living benefit option). Those products are particularly well fitted for retirement purpose, because they couple guaranteed coupons indexed with the participation in an index (typically the S&P 500 in the United States).

Payments above the guaranteed amount depend on the performance of the investment portfolio. The policyholder is therefore protected against downside risk but able to benefit from upside potential. Differently from a traditional annuity product, the policyholder may have more choices of the funds invested in and may withdraw them earlier, although some penalties can apply to any returns that have accumulated.

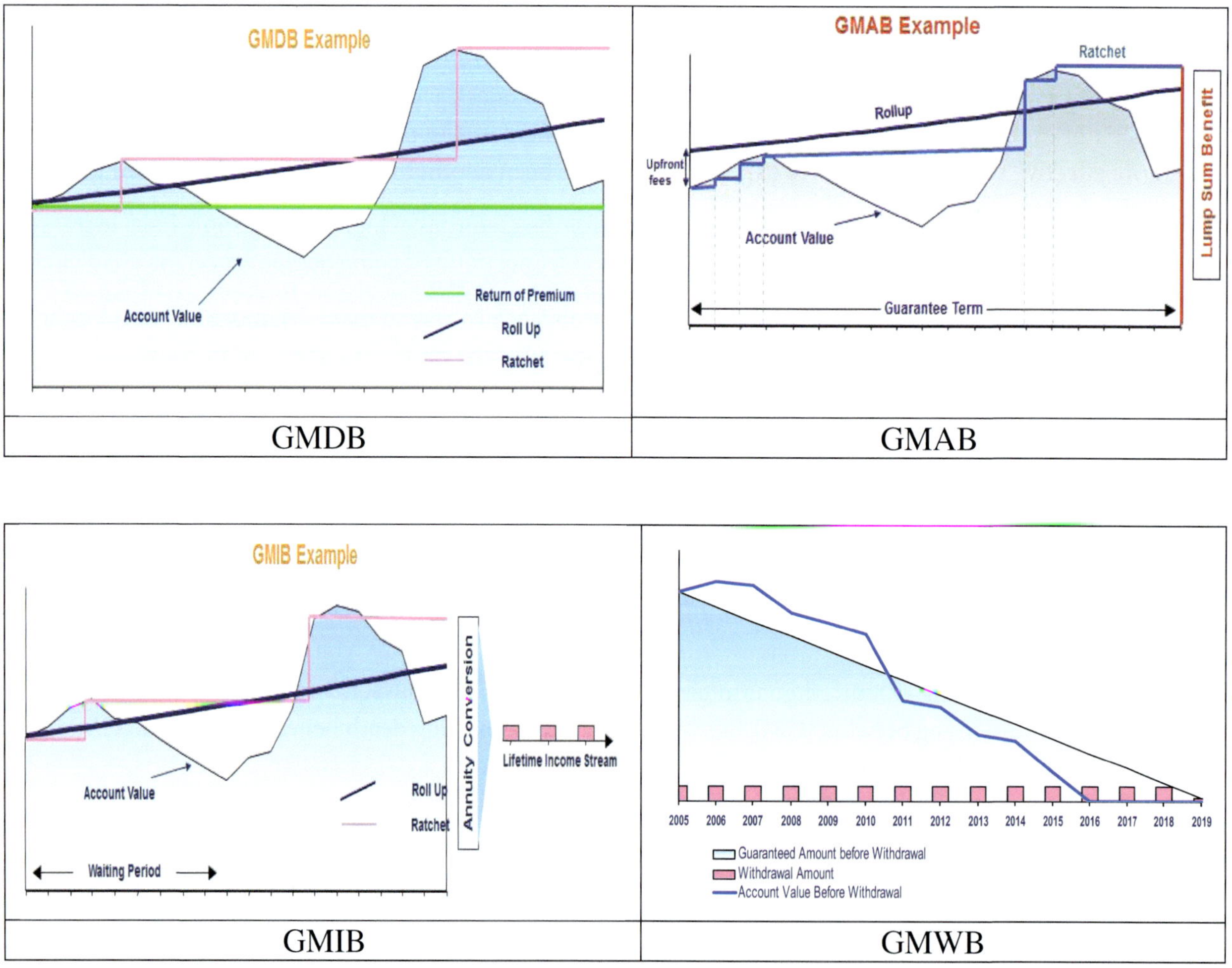

FIGURE 2.1 GMxB cash flow patterns

Note: GMAB = guaranteed minimum accumulation benefit.

In contrast to conventional annuity products, in which the policyholder has limited or no decision power over investments and may not be able to cancel the policy once the policy has been purchased, variable annuities allow the policyholder to make decisions after the purchase and even cancel the policy, often with the full return of the original investment guaranteed.

These products have a long term, so therefore there are risks that need to be monitored and reported continuously over a long period of time. Variable annuities have been causing various and contrasting reactions, mainly because, in the past, design features lacked robustness. Many facts abound in this direction. Variable annuities expose insurers to a number of risks. The main ones are market risk and insurance risk (among those are longevity, mortality and lapse risks). Insurers use a variety of measures to manage these risks, from product design features and fees and charges to risk pooling, reinsurance and hedging strategies (we will see those techniques in details in the second part).

From a client point of view, VA offer investment flexibility and tax-deferred growth, since the policyholder has the freedom to make additional deposits to the annuity and change investment allocations based on a number of investment options that may generate substantial returns. Taxes on the account growth are deferred until the policyholder begins taking money via annuity payouts.

The reasons why the VA succeeded in the United States are the following.

- The US ageing population is growing, and with it there is a growing need for accumulation and retirement products.
- Market volatility increases probability of ruin at the wrong time – generally just before or after retirement.
- There has been a continuing reduction in classical and traditional pension plans.

Market sales stand at ~$100 billion annually. Therefore, variable annuity remains one of the dominant guaranteed income products

Unfortunately, the persistently low interest rates, combined with recurrent volatility in the market environment during the 2010s, has contributed to a drop in sales of 30% (from $150 billion to $100 billion annually in the 2010s), as it becomes more challenging for assets to grow at guaranteed rates in a prolonged low rate situation.

- On the one hand, lower interest rates increase the premium, the reserves and the cost of capital, stemming from the higher present value of expected claims and the cost of hedging (as put options become more expensive).
- On the other hand, lower interest rates increase the difficulty in supporting minimum guaranteed rates, and decrease ROEs due to lower net income from higher reserve levels.

Unfortunately, excessively early expectations after 2009 that interest rates would rise back up pushed most companies (except AXA Equitable, whose sales plummeted due to management efforts to limit volumes before de-risking the VA) to consider growing further and developing more generous – thus riskier – products at the same price, as illustrated by the increased then sustained minimum guaranteed rates and withdrawals (e.g. 5 to 7% roll-up and guaranteed withdrawal rates, quarterly or daily ratchets for some VAs at Metlife or Prudential US).

For example, MET gained a significant share of the market in 2011 as weaker competitors retrenched, but its sales plummeted in 2012 due to management efforts to limit volumes. Jackson National's market share continues to increase (Lifeguard Freedom Flex becoming the best-selling GLWB product in the market), while Lincoln has been more aggressively selling VAs since 2013, driving up its market share.

Nevertheless, as interest rates remained persistently low and volatile through the 2010s, fundamental de-risking measures were implemented by most life insurers.

- Product design de-risking:
  - higher fees (M&E and riders; rider fees linked to the Volatility Index [VIX] of the Chicago Board Options Exchange [CBOE]; withdrawal percentages linked to Treasury rates, as at SunAmerica and Allianz) and lower benefits (lower roll-up rates or linked to the interest rate level: AXA Equitable); lower frequency of ratchets; lower guaranteed withdrawal rates; stricter asset allocation limits (higher mandatory minimum allocation to fixed/balanced accounts); reduced benefits and higher fees to reduce GAAP reserve.
  - restriction in ability of policyholders to "time the market";
  - restriction in number of equity funds and increased use of "tracker" funds; automatic reallocation of funds within volatility managed funds (AXA, Aegon, Metlife) or use of CPPIs (Prudential, Nationwide) to mitigate volatility in the GAAP reserve;
  - stricter governance in product approval processes;
  - possibility for the insurer to increase fees at discretion;
  - the introduction of more prudent "efficient" modelling of customer behaviour risks; and
  - the addition of macro hedges to protect against steep interest rate drops.

  Such product de-risking has increased US GAAP ROEs to the low to mid-teen levels for new business.

- Most life insurers have also improved their risk management practices, such as the increased use of volatility hedging and, as already mentioned, adding macro hedges to protect against steep interest rate drops.
- Many life insurers still curbed the distribution of VA sales until the interest rate mean had reverted to long term levels.
- AUM growth in the VA business tended to be less volatile due to allocations to fixed-income investments (28% in fixed income and cash, while an increasing portion of the residual was allocated to dynamically balanced funds).

Although the de-risking has provided healthier margins for new business, it could not fully mitigate the impact of the persistently low interest rate environment, as illustrated by the higher cost of hedging variable annuities and the multiple increases in reserves (AXA Equitable Holdings, Metlife, Prudential, Voya) due to updates of policyholders' behaviour (e.g. in line with significantly lower surrender and partial withdrawal rates than expected).

Even so, further solutions had to be implemented in order to survive and thrive in the low interest rate environment in place since 2015.

- Reducing premium payment options when possible, and reducing guarantees if policyholders deviate from their scheduled premium pattern.
- Decelerating the distribution of sales until interest rates mean-revert to sustainable long-term levels is another reasonable response.
- As persistently low interest rates have increased customers' concerns about fees and triggered lower lapse and partial withdrawals, insurers have also developed more customer-centric designs within a "rational behaviour" framework that balances benefits against fees.
- Introducing buyout initiatives.

Even so, the situation became more difficult.

- It became more challenging for assets to grow at such high guaranteed rates (~5% roll-up rate; JNL's LifeGuard Freedom Flex is 7%) in a prolonged low rate environment; a 5% lifetime withdrawal rate starting at age 65 is the industry norm.
- Maintaining the same new business volumes became onerous on the balance sheet.
- There was no stable relationship between reserves and capital, according to the NAIC: although statutory reserve (AG43 using a locked-in rate specified by the carrier, constant throughout the projection) levels are relatively interest-rate-insensitive, lower rates increased the level and the volatility of required capital requirement (C-3 Phase II, based on a ten-year US Treasury + 50 bps – constant throughout the projection).
- Lower rates also increase the GAAP reserves (Financial Accounting Standard [FAS] 157) from claims being discounted at lower rates, implying lower a net income.
- The higher cost of hedging as options become more expensive reduces the incentive to fully hedge market risks.
- VA writers are not appropriately risk-adjusting new business ROEs, since the cost of hedging is not incorporated in net income.
- ROEs are lower due to lower net income from higher reserve levels.
- In conjunction with regulatory uncertainty, low interest rates have helped drive the divestiture strategies of VA blocks.

As a result, many VA carriers have decelerated VA sales and pivoted their distribution network to alternative products such as indexed annuities, which not only suffer less from guaranteed rates, thanks to market value adjusters and a reset of guaranteed rates after short guarantee periods, but also benefit from equity upside potential, as illustrated by the massive S&P 500 growth since 2013.

2.1.3 The Growing Customer Risk Appetite for Indexed Annuities (FIAs or EIAs) during the 2010s, Providing Flexible and Cheaper Guarantees to Customers with Lower Risk Tolerance

Traditional life insurance products have lost much of their attractiveness with the persistently low interest rates. First of all, there is the stock of old contracts for which high interest rates are still guaranteed. For example, in the case of financial statements in the years 1994 to 1999, 4% was usually guaranteed in Germany. And, because the contracts usually run for several decades, this return must still be generated today.

The German insurance supervisory authority requires companies to create an additional savings pot in order to continue to serve such guarantees from the past. This is the so-called additional interest reserve. Insurers provide this reserve at the expense of the surpluses with which they have wooed customers in the past.

So the insurance companies themselves are slowly taking the product out of their programme, because interest rates are low and costs are too high, while low interest rates forced advisers to seek out alternative products to meet their clients' capital preservation needs.

This is why many insurance companies now offer a different product: index-linked pension insurance, or "index policy" for short. This promises the preservation of the contributions and, at the same time, a chance of high returns.

With an index-linked pension insurance, the insurance company first invests the customer's contribution in its security assets – in particular safe investments that have virtually no risk of loss. Then the surpluses are determined. These consist of the return on savings within one year and the participation in the cost gains.

However, the provider does not put the surplus aside but puts it into risky investments, which should bring more return in the long term than fixed-income forms of investment. This is possible because the insurance guarantees only the contributions themselves, not a fixed interest rate on the contributions. The insurance company invests the money from the surplus in a stock index. The investment experts of the insurance company do not buy shares in index funds but, instead, acquire so-called options on the index. Basically, the customer indirectly pays the price for the options that the insurance company buys from an investment bank.

The leverage effect of such an embedded option is needed to achieve a high profit with the little capital invested that comes from the surplus. The profit thus generated is credited up to a certain "cap" to the accumulated capital, which has achieved a return in this way.

In the United States traditional VA providers entered or put new focus on this market during the 2010s (Nationwide, PacLife, ING, Lincoln, Genworth, etc.), and industry marketing efforts will continue to drive market growth, while private equity firms have entered into the market by buying carriers.

The first FIA was issued in 1995 in the United States, where sales grew to $20 billion by 2005 with the launch of death and living benefits combo. In that year the vast majority of FIA sales came from foreign subsidiaries such as Allianz Life, ING, Old Mutual and Sun Life. This trend was magnified when Aviva PLC acquired AmerUS in 2006. In 2013 the only significant foreign subsidiaries that remained were Allianz Life and Jackson National. Solvency II requirements and new Canadian regulatory requirements made FIAs less attractive to the parent companies of foreign subsidiaries, contributing to divestitures and strategic realignments.

A number of acquirers then gained a solid market foothold through their acquisitions. Many acquirers view their asset management and structuring capabilities as a way to generate value from existing blocks. In many cases these carriers have continued to issue new business via competitive products, resulting in increased assets under management and market share.

The FIA market was considered by many industry participants as a "niche" market in 2005. Today this perspective is largely reversed, and the universe of participants is much broader. In 2005 sales through independent distribution accounted for approximately 90% of FIA sales. In 2013 ~25% of FIA sales came from outside the independent channel. Most of this change is explained by the growth and success of new entrants with a proven track record in these alternative channels.

Growing sales and a flurry of multi-billion-dollar M&A deals (by Athene, Global Atlantic, Guggenheim and Resolution) have put a brightening spotlight on fixed indexed annuities. From 2007 to 2013, despite near-record low interest rates, FIA annual sales growth averaged nearly 8%, and their share of the overall annuity market grew from 9.7% to 17.1%. Also drawing attention were the numerous M&A deals that have taken place since 2010. 2021 sales stood at $63 billion after a peak at $74 billion in 2019 – a half of all fixed annuity accounts and a half to three-quarters of variable annuity sales since 2016.

The FIA market has been well positioned since 2010 relative to the traditional fixed annuity and variable annuity markets.

- The minimum guarantee is enough to meet the non-forfeiture laws in most US states: the products are sold as fixed annuities and not as variable annuities.
- There is no need to be registered with the SEC or the (National Association of Securities Dealers (NASD) to sell them.
- There is greater account value growth potential compared to traditional fixed-rate annuities, due to the equity-indexed crediting rate.
- Low interest rates make indexed annuities more attractive than regular FAs to conservative investors.
- There are significantly lower fees than with variable annuities (~1% versus ~3%).
- Most FIA carriers' in-force blocks are composed of recent sales, and thus have surrender charge protection, which reduces disintermediation risk.

- Approximately half the in-force FIAs feature a market value adjustment (MVA).
- In contrast to their VA counterparts, FIA GLWB riders benefit from stable statutory and US GAAP accounting.
- There is a more stable risk profile than VAs for insurance carriers, due to the high-quality fixed-income assets, as well as the hedged exposure to equity markets.
- Significant growth in banks and broker dealer distribution will probably continue thanks to the low commission and short surrender charge designs with competitive indexing features.
- Certain FIA carriers compensated for the declining yields in the 2010s by seeking additional liquidity and credit risk premiums.
- Recent rising rates and market volatility since 2021 have had a positive impact on sales and reduced pressure on new business profitability.

Even so, as the FIA market matures, operating costs and service to consumers and distributors will become more important differentiators.

2.1.3.1 Product Basics

An indexed annuity (whether a fixed indexed annuity or an equity-indexed annuity: EIA) is a way to participate in the attractive returns of a market-driven investment with a low downside risk thanks to the guaranteed return. They are called indexed annuities because the performance of such annuities is tied to an index.

The interest earned is based on the performance of the index, subject to a cap (as opposed to being based on rates specified by the insurer for fixed annuities), with a guaranteed minimum rate of return, the only condition being to hold the indexed annuity until the end of the term. An EIA holder receives a percentage of the index's gain, via a participation rate.

Participation rates generally range from 50% to 90%, and do not exceed 100%. For instance, a participation rate of 80% and an 8% return from the index would result in a 6.4% gain by the investor. Some insurance companies tend to set caps on products to avoid upside risks. Some of them have also decided to add contract "riders", as living benefits (in the same way as on variable annuities). For instance, a guaranteed living benefit (GLB) guarantees a regular payment (the frequency ranges from monthly, quarterly to annual) during the client's lifetime. As expected, the older you are at inception, the more you get as a GLB rider.

This is in contrast to fixed annuities, for which some excess asset returns are "credited" to the policyholder (the "fixed amount" specified by the insurer) and the remaining spread is manufacturing revenue. Instead of a fixed amount being paid to the policyholder, the credited rate of the FIA is used to purchase options to create the cap, the floor and index crediting payoffs (the "options budget").

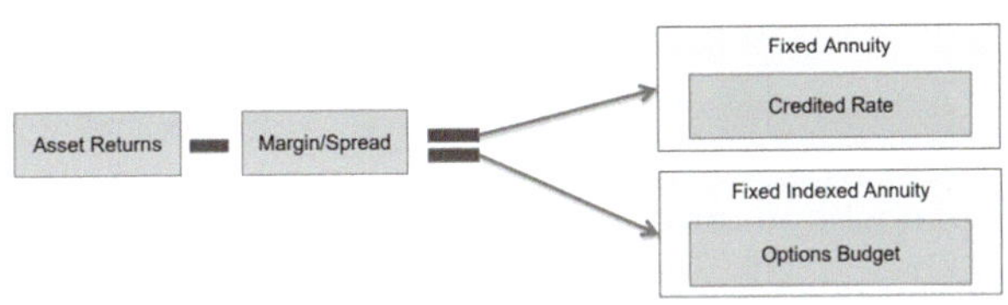

FIGURE 2.2 FIA components

They are short- to medium-term products (five to ten years), with the guarantee in the form of a call option (in contrast to variable annuities' guarantees, which can be seen as a put option) on the stock index, with potential additional guaranteed income as a benefit rider. Moreover, an EIA guarantee is usually in-the-money at the maturity (while a VA guarantee is rarely in-the-money at maturity).

What has to be noted is that this payment occurs even if the account balance does not support the income stream. For example, suppose a client who was 61 years old invested $100,000 in an EIA with a 5% GLB annual rider. In the event of the rider being activated immediately, the client is guaranteed a withdrawal of $5,000 per year for the rest of her life even if the account balance drops and cannot support the income stream!

As a key remark, the equity indices used to link EIA contracts are not total return, which means that dividends are not reinvested, and so they accumulate more slowly than the total return versions of the indices.

Product designs are multiple, and include the following.

- A death benefit rider is often included.
- Several crediting methods are used for indexed annuities: point-to-point (PTP), annual ratchet and high-water mark (i.e. earns the highest return since the inception of the index attained during the whole life of the policy).
- Some insurance companies keep the right to change participation rates, cap rates and other fees, either annually or at the start of each contract term. This could potentially affect the client return.
- In most cases, there are surrender charges during the early years of the contract. Moreover, an investor should be aware that withdrawals prior to age 59½ may be subject to a 10% federal income tax penalty.
- Annual insurer fees are specifically low (~0.5 to 1%, vs. ~3 to 4% for variable annuities), although they are not charged whenever the index return is negative, while commissions range from 5 to 8%.

Among the latest product design innovations of FIAs are:

- customized, rules-based volatility control indices, with a higher potential upside;
- liquidity-focused features (e.g. shorter surrender charge periods);
- index-linked income, with "stacking" roll-ups or growth potential in income, even after the policyholder begins withdrawals; and
- a fee-based version (which still accounts for a marginal portion of sales), offering higher customer value (higher cap, lower surrender charges) with less strain to the insurer (as the company will not pay an up-front commission).

The targeted customers have the following characteristics:

- conservative investment behaviour;
- ages 50 to 70 (available from 45 to 80);
- average age 62 years;
- with investable assets of up to $2 million;
- in need of growth to maintain their lifestyle;
- typically aware of the major risks and challenges along the retirement path; and
- average policy size $75,000 premium (versus around double for VAs).

There again, as with ULs and VAs, the level of risk tolerance in retirement also drives the product selection.

Performance-wise, fixed indexed annuities tend to perform relatively well from a risk appetite perspective, offering an inferior best performance but also a significant reduction in the worst performance compared to EIAs or VAs.

	Best	Average	Worst
S&P 500	17.2%	8.6%	-6.2%
4.5% Annual Cap	4.5%	3.3%	1.7%
3.0% Annual Cap	3.0%	2.3%	1.3%
1.8% Monthly Cap	5.8%	2.8%	0.3%

Annualized client return net of fees
Daily starting points, 10 year hold, using 30 years of S&P data

FIGURE 2.3 Annualized returns net of fees

Overall, this is a summary of the key features and benefits of indexed annuities.

- There is full capital preservation upon death or surrender at the end of the surrender penalty period.
- There is a higher upside potential than many current products using a different design structure.
- They have a simple, balanced crediting strategy without complicated formulas and modifiers.
- There is a lock-in privilege that can be triggered at any time.
- They have a unique free partial withdrawal feature that eliminates any crediting penalty.
- They feature a return of the premium benefit, which provides protection in extended low or negative equity scenarios, when strategy fees reduce accumulation values below the initial premium.
- There is a two-year indexing period.
- They provide alternative indices that give greater diversification and upside potential.
- There is flexibility to choose the fee level and the equity participation level.
- They have low fees.
- Unlike the VA market, ~81% of FIAs are distributed through independent financial planners and banks; the bulk of the distribution is via non-registered channels.

2.1.3.2 Crediting Mechanism

2.1.3.2.1 Examples Let us consider a two-year indexing term, 50% equity participation, 1% declared rate, starting index value = 1,000, $100 starting average, 1% annual strategy fee, no partial withdrawals, no lock-in selected.

- In the event of positive index returns, the index at the end of two years is 1,400. The index return is 40%.
  - Equity component gain/loss (G/L) = 50% × 40% = 20%.
  - Declared rate component G/L = 50% × (1.012 – 1) = 1.005%.
  - Combined rate = 20% + 1.005% = 21.005%.
  - Index crediting rate = 21.005%, above contractual floor of 0%.
  - Ending AV after two years = $100 × .992 × 121% = $118.60.
- In the event of negative Index returns: the index at the end of two years is 950. The index return is –5%.
  - Equity component G/L = 50% × –5% = –2.5%.

- Declared rate component G/L = 50% × (1.012 − 1) = 1.005%.
- Combined rate = −2.5% + 1.005% = −1.495%.
- Index crediting rate = 0%, due to contractual floor of 0%.
- Ending AV after two years = \$100 × 0.992 × 100% = \$98.01.

2.1.3.2.2 Setting the Crediting Strategy Let us consider the following situation.

- Product: strategy A, 2% strategy fee, 0.74% product spread, 1% declared rate.
- Economic conditions: 4.1% corporate bond yield, two-year annuity factor = 1.883, two-year swap rate = 0.52%, two-year at-the-money (ATM) implied volatility = 19.16%, implied volatility skew = −17.86%.

As a result, we can compute the following.

- Revenue = 2% + 0.74% = 2.74%.
- Annual option budget = 4.1% − 0.74% = 3.36%.
- Total option budget = 1.883 × 3.36% = 6.33%.

All coupons collected less the spread constitute the amount spent on buying options.

Setting the crediting strategy requires an iterative process in order to find the equity participation rate (EQ) and the call option strike (K), which generate positive returns while remaining compliant with the call total option budget.

- Condition 1: EQ × (1 − K) + (1 − EQ) × (1.012 − 1) = 0%.
- Condition 2: EQ × call option price (K, 2 years) = total option budget

→ EQ = 65.94%, K = 98.96%.

The EQ participation is guaranteed for only two years. After two years steps 1 to 3 will be repeated based on current portfolio earned rate and option prices. If the portfolio experiences high defaults or low reinvestment rates, new EQ participation will be set so that the target revenue is still made in the next two years.

2.1.3.2.3 Embedded Guarantees and Market Value Adjustment

2.1.3.2.3.1 Embedded Guarantees

- Financial guarantees.
  - 0% minimum return on initial investment – i.e. return of principal. Supported by the insurer purchasing a long bond and using its coupons to purchase equity call options. Strategy fees are applied after 0% minimum.
  - Minimum guaranteed value (MGV): 87.5% of initial premium increasing at 1%. Cost to the insurer upon lapse only.
  - Minimum 5% equity participation guarantee.
- ROP death benefit (DB), reduced by withdrawals. ROP DB cannot be less than the MGV.

2.1.3.2.3.2 *Market Value Adjustment*

- The equity indexing term is only two years, but the average maturity of the bond portfolio at issue is ten to 11 years.
- In the event of lapse while interest rates rise, bonds would need to be sold at a loss. There is a charge discouraging early lapses (0.25% × 7.5 = 1.9%, at issue), while MVA is an additional charge meant to recoup these bond losses.
    - MVA = 75% × remaining surrender charge term × (Treasury at issue – current Treasury – 0.25%), assuming that the interest rate duration of a ten-year corporate bond is 7.5.
- MVA effectiveness is reduced by the MGV = 87.5% of premium -> the most we can collect at issue is 12.5% (included 10% clawback).
- MVA effectiveness is increased by the potential positive equity performance and the account value (AV) growing above the initial contribution.

2.1.3.2.4 Free Partial Withdrawal Treatment The contract pays interest credits on the withdrawal amount, up to a 10% free corridor. Here are two examples.

- If there is a 20% gain in the combined index after one year, AV0 = 100, and free withdrawal 1 = 10, the client collects $10. However, the AV is reduced only by $10 / 1.2 = $8.33. In other words, the full 20% interest is credited on the $8.33 withdrawal.
- If equity decreased and combined index return is 0% at time 2, the ending balance is 100 – 8.33 = 91.67.

2.1.3.3 *Pricing: Focus on Equity-Indexed Annuities*

Three types of design are considered below: point-to-point (the simplest one), ratchet ("cliquet" in French) and lookback (the "no regret" one).

- PTP is a design in which the policy earns the realized return on the index (or some other risky asset) over a certain period of time at a prescribed participation rate, with a minimum guarantee.
- Ratchet is a design in which the interest rates are credited annually on the higher of the index return during the prevailing year and the minimum guaranteed rate. The resulting interest credits are locked in. The values of the policy never decrease.
- The lookback design is a contract that earns the highest return on the index attained during the life of the policy. The interests are credited at the end of the term based on the difference between the highest level reached by the index during the term and the index value at the beginning of the term.

2.1.3.3.1 Point-to-Point Version A point-to-point equity index annuity is the simplest design, providing protection against market falls while the cap rate reduces the cost. The value of the PTP financial guarantee depends on the guaranteed return and the fixed participation rate, which can vary significantly.

The Esscher transform offers closed-form formulas for EIAs. Let $S_1(t)$ and $S_2(t)$ be two asset prices at time t, for $t \geq 0$, and δ_1 and δ_2 denote the respective dividend yield rates for assets S_1 and S_2. Let us define now:

$Y_i(t) = \log\left(\frac{S_1(t)}{s_1(0)}\right)$ for $i = 1,2 \cdot Y_i(t)$ is the rate of return on asset i over $[0,t]$. $Y(t) = \left(Y_1(t), Y_2(t)\right)'$

Define now $M(z,t;h)=\mathbb{E}\left(\exp(z'\times Y(t))\times\frac{\exp(h'\times Y(t))}{\mathbb{E}(\exp(h'\times Y(t)))}\right)$ to be the moment-generating function of Y under the Esscher transform with parameter h. $\frac{\exp(h'\times Y(t))}{\mathbb{E}(\exp(h'\times Y(t)))}$ is the Radon–Nikodym derivative used to define the new probability measure, under which the discounted asset price is a martingale; therefore, it is a risk-neutral probability.

Gerber and Shiu (1994, 1996) have shown that, when the assets follow a jointly multidimensional geometric Brownian motion (GBM) or a compounded Poisson process, there exists a vector h' called the risk-neutral Esscher parameter vector such that $\mathbb{E}_s\left(exp\left(-\left(r-\delta_i\right)\times t\right)\times S_i(t);h'\right)=exp\left(-\left(r-\delta_i\right)\times s\right)\times S_i(s)$ for $\leq t$, with $\mathbb{E}_s$ denoting the conditional expectation up to time s. Assuming that the market is complete, one can show that the risk-neutral measure and h' are unique. Assuming also that $Y(t)$ is normally distributed with mean $\mu t=\left(\mu_1 t,\mu_2 t\right)'$ and covariance matrix $Vt=\left(\sigma_{i,j}t\right)_{i,j=1,2}$ under the real-world measure, then h' can be determined analytically via the following formula: $\mu+Vh'=r*1-\begin{pmatrix}0.5\times\sigma_{11}+\delta_1\\0.5\times\sigma_{22}+\delta_2\end{pmatrix}$ and $1=(1,1)'$.

The payoff of a point-to-point EIA at time t is represented by $\cdot P=max\left[min\left[1+\alpha\cdot R(t),(1+\zeta)^t\right],\beta\cdot(1+g)^t\right]$ with α the participation rate, ζ the cap rate, g the minimum guaranteed rate on β the premium, and $R(t)=\frac{S(t)}{S(0)}-1$ is the return on the index. As we can notice, there is a cap on the contract $(1+\zeta)^t$, which reduces its cost, since it introduces an upper bound. There is also a protection against a downside market, represented by $\delta\times S(t)dt$.

Let $S(t)=S(0)\times exp(Y(t))$ for $t\geq 0$ be the asset price paying out dividends $\delta\times S(t)dt$ between t and $t+dt$. $Y(t)$ is a random variable denoting the compounding rate of return on the asset over time interval $[0,t]$.

Now suppose that we have a policy that, given an initial premium of one, gives at T the maximum of $exp(\alpha\times Y(t))$ for $a>0$ and of a fixed exercise price $K=\exp(lnK),K>0$.

We can then express the policy, via the fact that K is seen as a guaranteed minimum, as

$$e^{-r\times T}\mathbb{E}\left(\max(exp(\alpha\times Y(t)),K);h'\right).$$

If we decompose the maximum via max(exp(α × Y(t)), K) = exp(α × Y (t)) × $\mathbb{I}$(α × Y (t) > K) + K × $\mathbb{I}$ (α × Y (t) < ln K, with $\mathbb{I}$(.) the indicator function.

We then use the linearity of expectation. Remember that we have assumed that Y is normally distributed with means $\mathbb{E}\left(Y(T);h'\right)=\left(r-\delta-0.5\times\sigma^2\right)\times T$ and $\mathbb{E}\left(Y(T);h'+\alpha\right)=\left(r-\delta-0.5\times\sigma^2+\alpha\times\sigma^2\right)\times T$ under Esscher transforms for parameters h' and $h'+\alpha$ respectively, and variance $\sigma^2\times T$ (note that it is unaffected by Esscher transforms).

Then, using Esscher transform on our expectation, we get that the point-to-point EIA policy price is

$$P_{pp}=e^{\left[(\alpha-1)\times r-\alpha\delta+\frac{1}{2}\alpha(\alpha-1)\sigma^2 T\right]}\times\varphi\left(\frac{\left(r-\delta-\frac{1}{2}\times\sigma^2+\alpha\times\sigma^2\right)\times T-\frac{\ln K}{\sigma}}{\sigma\sqrt{T}}\right)+e^{-rT}K\varphi\left(\frac{\frac{\ln K}{\sigma}-\left(r-\delta-\frac{1}{2}\times\sigma^2\right)\times T}{\sigma\sqrt{T}}\right)$$

with φ denoting the cumulative distribution function (CDF) of a standard normal random variable.

We can also change the design of the strike price by assuming that the client gets a guaranteed minimum maturity value that is a percentage β of the original premium compounded at a minimum guaranteed rate of return g for the duration of the policy. Thus, via this hypothesis, we get that $K = \beta\exp(g \times T)$.

In general, the price is an increasing function with respect to variables K, σ and α; as for r and T, the situations are not that clear-cut. The price can be increasing or/and decreasing with respect to r and T, depending on the level of other variables, α in particular, as illustrated below (Pinault 2013).

- When α is small, say less than 0.8, the price is almost a perfectly linearly decreasing function of T.
- If α is larger, say between 0.8 and 1.2, the price increases in T for the first five to eight years, and then decreases linearly thereafter. We can also show that, for T less than 10, the price is increasing linearly in α for α between 0.5 and 1.1

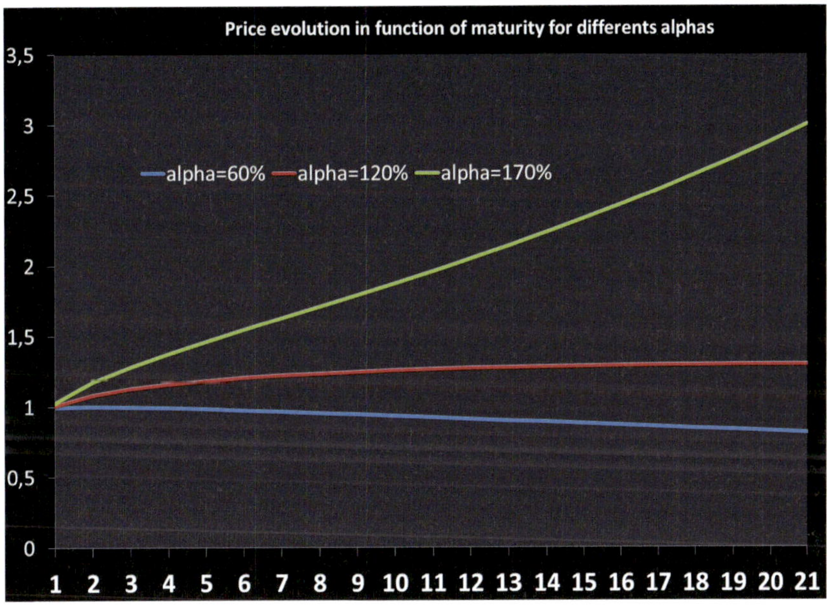

FIGURE 2.4 Sensitivity of FIAs to participation rates

Remark: in case α is smaller than 80%, the price of the EIA point-to-point decreases quasi-linearly in function of T. For α larger than 80% and smaller than 120%, the price increases in T for the five to eight years and then decreases linearly thereafter.

Hypotheses are – 30%, $r = 4\%, g - 3\%, \beta = 90\%$.

Using an α that is less (greater) than the break-even, α is going to produce a price that is less (greater) than one. Note that these participation rates are considerably higher than those being offered in the market under the same economic conditions.

The price is an increasing function of the participation rate alpha (see Figure 2.5).

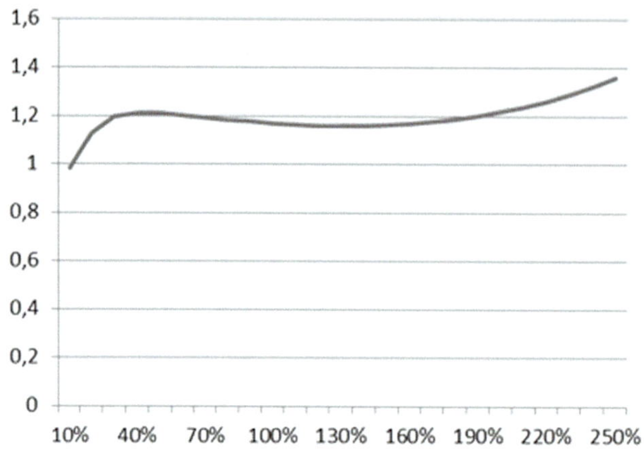

FIGURE 2.5 Price versus alpha

2.1.3.1.2 Ratchet Version With this design, we are assumed that the annuity, at the end of every period, will receive the maximum between the return on an asset with a participation rate of α and a guaranteed minimum rate g. With these product features, performances are locked in and will never decrease. Let us assume that the asset pays out dividends $\delta S(t)\,dt$ between t and $t + dt$ and that the payoff is made at the end of the n periods. The price of this contract at inception is

$$\mathbb{E}\left(e^{-r\times n}\prod_{j=1}^{n}\max\left(\exp(g),\exp\left(\alpha\times Y_j\right)\right);h'\right).$$

where $Y_j = Y(j)-Y(j-1)$ is the rate of return on the asset in period j, $j=1,2,3,\ldots,n$.

If we assume that $\{Y_j\}$ is a sequence of independent and identically distributed (i.i.d.) random variables, then we can simplify the expression by getting out of the expectation the product, since those are independent variables, and, using the identical distribution of those, we can assess that the product of the expectations is in fact the same expectation but at the power n.

Thus we get the following expression for the price:

$$P_R=\left[e^{\left[(\alpha-1)\times r-\alpha\delta+\frac{1}{2}\alpha(\alpha-1)\sigma^2\right]}\times\varphi\left(\frac{-\frac{g}{\alpha}+r-\delta-\frac{\sigma^2}{2}+\alpha\sigma^2}{\sigma}\right)+e^{-(r-g)}\varphi\left(\frac{\frac{g}{\alpha}-r+\delta+\frac{\sigma^2}{2}}{\sigma}\right)\right]^n.$$

Now let us suppose that we impose a cap c on the periodic return, with $c > g$. Thus, it will be the maximum interest rate the policy can earn in each period. If we keep the same assumptions as above, here is the price of the ratchet EIA:

$$\mathbb{E}\left(e^{-r\times n}\prod_{j=1}^{n}\max\left(\exp(g),\min\left(\exp\left(\alpha\times Y_j\right),\exp c\right)\right);h'\right).$$

Using classical arguments of the i.i.d. random variable, as before, since the $Y_j's$ are i.i.d. with variance σ^2, the equation above becomes

$$P_{\text{Cap}} = \left[e^{-(r-g)} \left(\frac{\frac{g}{\alpha} - r + \delta + \frac{\sigma^2}{2}}{\sigma} \right) + e^{(c-r)} \varphi \left(\frac{-\frac{c}{\alpha} + r - \delta - \frac{\sigma^2}{2}}{\sigma} \right) + e^{\left[(\alpha-1)\times r - \alpha\delta + \frac{1}{2}\alpha(\alpha-1)\sigma^2\right]} \right.$$
$$\left. \times \left\{ \varphi \left(\frac{\frac{c}{\alpha} - r + \delta + \frac{\sigma^2}{2} - \alpha\sigma^2}{\sigma} \right) - \varphi \left(\frac{\frac{g}{\alpha} - r + \delta + \frac{\sigma^2}{2} - \alpha\sigma^2}{\sigma} \right) \right\} \right]^n.$$

Let us now study a little more the sensitivity of the capped-ratchet EIA to the participation rate α.

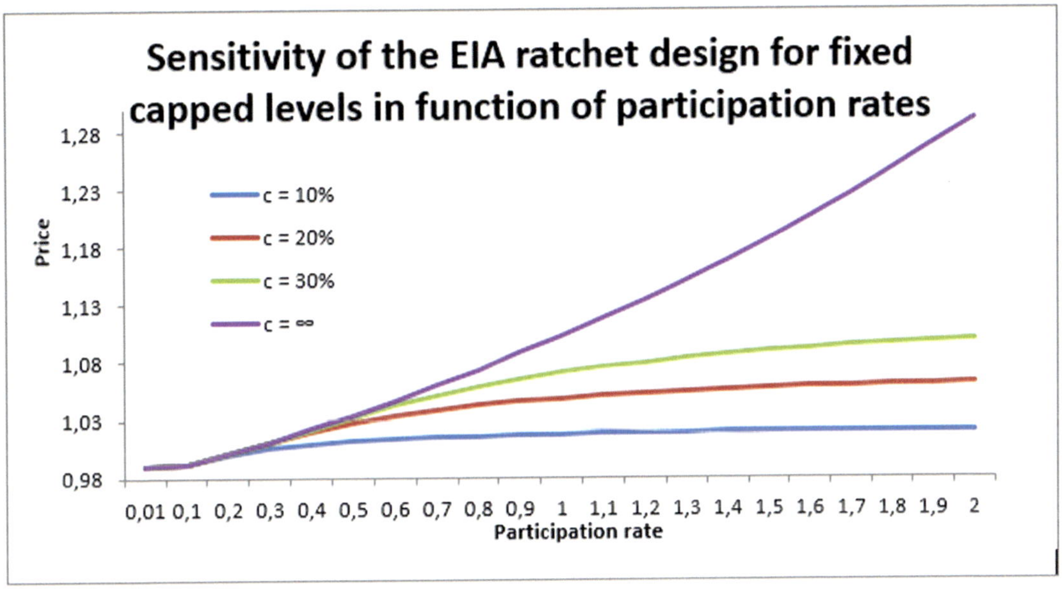

FIGURE 2.6 Price versus participation rate

For finite cap levels, the price of the capped-ratchet EIA is increasing, then it flattens, and eventually it converges when α goes to infinity. The reason for this phenomenon is the presence of caps; therefore, even though the participation rate explodes, with this design the policy cannot earn more than the rate c in each period. The richer the policy in c, the steeper the slope of the price, and therefore the slower the convergence.

The break-even participation rates are considerably lower than those of the point-to-point design; hence, the higher prices for cliquet EIAs. Notice how sensitively these participation rates respond to the risk-free force of interest r when c is low. Moreover, all the break-even participation rates listed below are decreasing in σ.

2.1.3.3.3 High-Water Mark Version The high-water mark product consists in looking at the index level at each policy anniversary, taking the highest of these and considering it as the index level at the maturity date.

For our calculation, we will consider the continuous lookback case, bringing a closed-form formula that is easy to calculate.

Let $M(T) = \max_{0\le t\le T} Y(t)$ be the maximum return rate on the index over the time period $[0, T]$. We will assume that the payoff of the policy at time T is the maximum between $\exp(\alpha M(T))$ and a guaranteed amount K.

Thus, at inception (time 0), the value of the contract is $e^{-r\times T}\mathbb{E}\left(\max(\exp(\alpha\times M(T)), K); h'\right)$.

Using the same formulas as before, and the same properties, we get the following expression of the price of the continuous lookback EIA:

$$
\begin{aligned}
P_{tb} &= e^{-rT+h(\alpha)}\Phi\left(d_1(\alpha)+\alpha\sigma\sqrt{T}\right)\\
&+e^{-rT+m(\alpha)}\Phi\left(d_2(\alpha)+g(\alpha)\sigma\sqrt{T}\right)\\
&-e^{-rT}\frac{2(r-\delta)-\sigma^2}{g(\alpha)\sigma^2}\cdot\left[-e^{g(\alpha)\kappa(\alpha)}\Phi\left(d_2(\alpha)\right)+e^{m(\alpha)}\Phi\left(d_2(\alpha)+g(\alpha)\sigma\sqrt{T}\right)\right]\\
&+e^{-rT}K\left[\Phi\left(-d_1(\alpha)\right)-e^{\frac{\{[2(r-\delta)-\sigma^2]\kappa(\alpha)\}}{\sigma^2}}\Phi\left(d_2(\alpha)\right)\right]
\end{aligned}
$$

where

$$\kappa(\alpha)=\frac{\ln K}{\alpha},$$

$$g(\alpha)=\alpha+\frac{2(r-\delta)}{\sigma^2}-1,$$

$$h(\alpha)=\alpha(r-\delta)T+\frac{1}{2}\alpha(\alpha-1)\sigma^2T,$$

$$m(\alpha)=-g(\alpha)(r-\delta)T+\frac{1}{2}g(\alpha)(g(\alpha)-1)\sigma^2T,$$

$$d_1(\alpha)=\frac{\left(r-\delta-\frac{1}{2}\sigma^2\right)T-\kappa(\alpha)}{\sigma\sqrt{T}},$$

$$d_2(\alpha)=\frac{-\left(r-\delta-\frac{1}{2}\sigma^2\right)T-\kappa(\alpha)}{\sigma\sqrt{T}}.$$

The formula is quite long, but it is nonetheless a closed-form formula, and it is more practical to handle.

The price of the lookback is an increasing function of the participation rates and tends to be linearly decreasing in function of maturity for α lower than 100%. For $\alpha = 120\%$, we can notice that the price increases during the five years, and thereafter is quite converging.

2.1.3.4 Pricing and Profitability Issues

2.1.3.4.1 Market Assumptions Fixed indexed annuity products tend to experience unfavourable profitability under most European NBV frameworks.

- Market-consistent framework: risk-neutral modelling makes fixed products unfavourable, due to the fact that crediting rates could be in excess of risk-free rates and that there is insufficient sharing of the downside credit risk with the client.
- Expenses: commissions and maintenance expenses create additional drag on the realization of future profits.

Even tweaking the product design by employing an MVA or reduced commissions could be either ineffective or infeasible. Most US insurers do not use a market-consistent framework for pricing, and others probably use different liquidity premium assumptions and customer behaviour assumptions.

There are a few actions that could be implemented to improve the profitability.

- Drop crediting rates after the initial guaranteed period; this could lead to undesirable consequences for the company's reputation.
- Assume higher liquidity premiums in the NBV discount rate; this might possibly not be consistent across product lines.

In addition, from a risk management perspective, because of this and the "fixed income/book value lenses" of many FIA carriers, many companies primarily view GLWBs as a source of insurance risk, which is consequently left mostly unhedged. However, GLWB riders impact both the duration and convexity of the insurance liability, and its sensitivity to index returns. Additionally, GLWBs can make FIA statutory reserves insensitive to changes in the index, which in turn can cause important statutory accounting volatility as the hedge-crediting P&L emerges.

2.1.3.4.2 Customer Behaviour Risks Even so, the potential added living benefits compound the problem. Fixed indexed annuities differ from VAs in that they are much less sensitive to equity market fluctuations. Nonetheless, there are important ways in which policyholder behaviour on this product, too, responds to volatility in the market.

Credited interest rates are a key driver of persistency on FIAs. This is a matter not only of the absolute rate – surrenders are especially high when credited rates are below 2% – but also of the availability of attractive market alternatives. When competitive market rates fall, all else equal, surrenders fall too.

Beyond interest rates, we must also consider equity index performance. An FIA is distinct from a traditional fixed deferred annuity in that credited rates reflect participation in index-crediting strategies. Policyholders accept lower fixed returns in exchange for the potential of market upside. But experience suggests that, when that upside does not materialize, policyholders become dissatisfied, and surrender at greater rates. Equity market drops lead to higher surrender rates. An exception, though, is observed on policies with a guaranteed lifetime income benefit (GLIB). When a GLIB is present, presumably the policyholder's main objective is lifetime income, so there is a sensitivity to the actuarial moneyness – the relationship of account value to the present value of guaranteed income: the more valuable the guarantee is in economic terms, the more likely the policyholder is to persist. In contrast to VAs, the FIA actuarial moneyness is not driven by account value fluctuations but by the interest rates, as it goes further in the money as interest rates fall, increasing the present value of future payments.

A secondary way in which financial markets affect surrender behaviour is through a market value adjustment. An MVA is designed to reduce the incentives and costs of disintermediation – that is, when increases in market interest rates make it advantageous for a policyholder to surrender the annuity and invest the proceeds in a higher-yielding instrument. Although a properly designed MVA should neutralize the gains or losses, making the policyholder indifferent to market interest rates, we observe that policyholders are not indifferent. Rather, surrender rates for policies with a positive adjustment exceed those for policies with a negative adjustment or no adjustment. In the aggregate, policyholders act as though a positive MVA is a bonus rather than a mechanism to make both parties whole. Note that a positive MVA is generated when interest rates fall relative to the interest rate at the time of the initial investment.

Overall, we can say the following.

- Lower credited rates lead to higher surrenders, somewhat offset by the fact that lower competitive interest rates reduce surrenders.
- Lower interest rates make living benefits more valuable, leading to greater GLIB persistence.

- Equity market drops lead to higher surrenders on policies without GLIB.
- Equity market drops lead to higher surrenders on policies with a positive market value adjustment. There will be more of these as interest rates fall.
- Lower interest rates make living benefits more valuable, so we should expect to see greater rates of income utilization in response to declines in interest rates.

Fortunately, significant in-force blocks are starting to exit the surrender charge period, which will give FIA carriers a wealth of data on surrender behaviour for application in assumption setting and customer retention. Most FIA designs hold significant surrender charges so as to provide a certain level of protection against early surrenders and to recoup some acquisition expenses.

2.1.3.4.3 Product Transparency, Premium Costs and Performance Net of Fees There is a comparatively high risk of a total loss: if things go badly, the money used to buy the options is also lost. Moreover, the profit is also limited in particularly good months with a fixed cap (called the cap). Thus the return can hardly be estimated in advance, because each insurance company sets the cap anew every year. Many providers have already had to lower the cap in recent years because there were only low surpluses due to the low interest rates.

Index policies also incur high ongoing administrative costs, which greatly reduce returns. In addition, there are the usually high closing costs, which the customer pays off with the contributions in the first few years. Moreover, about 15% of these costs are so-called risk costs. They secure the repayment of all contributions in the event of death during the term.

The amount of the costs ultimately affects the return. The value by which the average annual return is reduced due to costs is called the effective cost ratio. On average, if the company generates a 4% return on the contributions paid, the saver is effectively left with only 2.64%; 1.36% may sound little but it makes a huge difference.

These returns are very dependent on the selected period: for example, in the period from 1994 to 2013, an average German index policy would have generated 2.8%. If this period had been postponed by only two years – i.e. to 1996 to 2015 – it would have been only 1.9%, even though the average price increase of the Euro Stoxx 50 during the two periods was similar. Responsible for this is the strong dependence of the policy on the type of fluctuations during the term. If the index rises sharply at the beginning, the effect on the contract is very low, due to the initially low policy value. However, if the index is still gaining well in the recent years of maturity, this will bring much more. Then the value of the policy is high, and a return on it brings in more. With an end of the term in December 2015, for example, the last two years were far less profitable than with an end in December 2013.

The systematics of calculating returns via cap and quota also makes the relationship between price development and return non-transparent. High fluctuations in the index do not have the same effect on the development of the index policy due to the limitations. Even if you knew that the index would perform well in the long run, vagueness would remain about the index policy return.

Providers often make forecast calculations in their product descriptions. They show the payout that would result at the end of the contract period with an annual return through the index of about 4, 8 or 10%. But are such returns realistically achievable?

In a simplified statistical model, we simulated 10,000 theoretical histories of the index over 20 years (using the volatility of the Euro Stoxx 50 from 1996 to 2015 as a basis) and taking into account the observed average price increase of these 20 years. As a result, the effective returns after deducting the costs (~1.5%) of more than 4% are extremely rare. As a very likely range to be estimated, values of 0.5 to 2.5% occur in almost 60% of scenarios, with an average of 1.54%. This is consistent with the observed interest rate granted purely via the bonus participation over the period.

2.2 MEETING BOTH SHAREHOLDERS' AND POLICYHOLDERS' APPETITES THROUGH SUSTAINABLE CAPITAL-LIGHT PRODUCTS?

Because they provided healthy returns with no risks in the previous era of higher and decreasing rates, traditional with-profit products could be deemed the single and proper answer to a variety of customer needs. However, the situation now requires a much broader range of savings solutions to address specific time horizons, levels of risk appetite, liquidity profiles and investment objectives.

Many life insurers are at a strategic crossroads, facing new options and decisions about where to invest and grow the portfolio, where to harvest and which areas to exit. To successfully execute these choices, they must identify the capabilities that it is critical to develop and master, and those that can be "good enough".

Whatever their choices, insurers also must clearly communicate a compelling story to investors, so that investors understand why and how the business is performing and can gauge the model's sustainability. Choosing the right metrics to manage and communicate will give investors confidence that an insurer has made the right strategic and operational decisions through and through.

As a result, life insurers must take into account sustainable growth and solvency constraints, while meeting both customers' and shareholders' interests, through sustainable capital-light savings and retirement products, across an increasingly adverse economic environment (e.g. persistently low interest rates and economic volatility) and increasingly efficient customers behaviours.

This includes investment products with capital protection solutions, including partial capital guarantees, structured policies that comply with the European Union's Undertakings for the Collective Investment in Transferable Securities (UCITS) framework and commercial package policies that cover short- and long-term needs.

2.2.1 Evolving the Range of Sustainable Savings and Retirement Products Aligning Both Customers' and Insurers' Risk Appetites

2.2.1.1 An Increasingly Adverse Environment

With Solvency II having become effective on 1 January 2016 and the low interest rate environment persisting, European life insurers face tremendous economic and regulatory pressure due to the structure of their books. This is exacerbated in some countries where companies sold products on a large scale with high, but often completely unhedged, guarantees in times of more benign market conditions.

Most traditional savings and retirement products hold guaranteed interest rates that are usually fixed for the whole term of the contract (sometimes this means accumulation plus the annuity phase); although guarantees are often not adequately hedged, the level of the guaranteed rates is high in comparison to market rates, and with the interest rate guarantee an annual bonus is provided.

Indeed, one of the key challenges lies in the contrast between low nominal interest rates and high long-term interest rate guarantees granted to contracts that have been sold when interest rates were higher. The market-based valuation of insurance liabilities therefore leads to high volatilities in terms of the company's P&L, while recent higher capital charges introduced by new regulatory frameworks, such as Solvency II and NAIC, expose some of the inherent risk of traditional products that had not been considered previously.

All the non-traditional business constitutes a big portion of new business in many markets. This holds in particular for recent years, when insurance companies have provided higher interest rates than banks and attracted large volumes of single-premium business.

Moreover, in managing the in-force book, mostly made up of traditional (non-linked) business, insurers need to reinvest their maturing bond assets into lower rates in order to meet their long-term liabilities – a huge challenge for both the company's profitability and its financial stability under Solvency II.

Switching from the shareholders' to the customer's perspective, there are, essentially, two main ways falling interest rates impact European life insurers.

- When interest rates fall, the overall willingness of people to enter into long-term insurance contracts declines, since customers are less willing to potentially lock their money and their future premiums into contracts offering an expected low return.

  Most customers drift to alternative investment approaches, such as short-term savings vehicles (the "wait and see approach"), or directly invest in other asset classes, such as real estate.

  Thus, insurers' new business volumes are progressively coming under pressure, in particular in traditional insurance markets within Europe that have historically favoured bond investments to back policyholder liabilities.

- The second challenge is the value proposition. Many European markets still want insurance guarantees even though returns on traditional products are less than zero after the deduction of underlying costs.

  Hence providing insurance products that include guarantees by simultaneously offering the potential for attractive returns is impossible.

A turnaround in product strategy was indeed noted in 2015, and the offerings of many large life insurance companies have changed. For instance, in Germany, major life insurers have announced plans to exit the traditional life insurance market by no longer selling classical guarantee products and by revamping their product line-ups to reduce exposure to financial guarantees.

2.2.1.2 Increasingly Efficient Customer Behaviour Is Required to Reshape Customer Journeys for an Enhanced Customer Experience

2.2.1.2.1 Increasingly Efficient Customer Behaviour Savings and retirement behaviour has undergone significant change in the direction of increasingly efficient use over the last decade, as illustrated by the emergence of increasingly differentiated retirement savings patterns, in line with new needs and for the subsequent life phases.

To this end, there are three types of savings use behaviour patterns, illustrated by the frequency and amounts of withdrawals, all strongly influenced by age.

- Before the age of 60: one-off withdrawals of heterogeneous sizes, most often massive (over 200% of the annual guaranteed withdrawal), illustrative of the consumption of expensive durable goods, real estate purchases, financing education and vocational trainings, children's studies, management of periods of unemployment.
- From 60 to 70 years of age: regular withdrawals, of moderate and homogeneous sizes during retirement, illustrative of the entry into annuity.
- Over 70 years of age: mostly the entry into annuity, with regular moderate and homogeneous withdrawals, but also excess withdrawals (for tax reasons or donation).

This segmentation of customer profiles reflects the strong differentiation of savings behaviours. Meeting the expectations of the insured then requires an optimization of the product design, through product features tailored to the savings profiles, for an enhanced customer experience.

This underlines the necessity of developing segmented solutions that address different parts of the market. By segmenting customer groups, insurers can increase the number of personalized offerings that might address, and even predict, a specific need. As such, life insurers can better respond to customer needs and expectations around flexibility, personalization and relevance.

The shift towards a more diverse range of value propositions designed to answer the variety of customer needs – and away from the simplicity of guaranteed products – represents a major challenge for distribution.

It requires a careful rethinking of sales and advisory processes, including an adoption of hybrid digital sales formats and greater personalization.

Our research finds that remote and video-based advice platforms can be two and one-half times more productive than traditional face-to-face meetings. In addition, while agent interactions are still at the core of selling complex products, chatbots and other automation tools can add tremendous value to the overall hybrid online and offline customer journeys.

Among agents and sales representatives, specializations – such as in retirement – will allow for further efficiencies. As these agents and representatives develop deeper expertise that is focused on specific customer segments and offers, they can provide more personalized advice.

Technology will probably play a crucial role in driving these changes and fostering efficiency across the advisory process. Moreover, technology could enable growth in the part of the market in which small-ticket sales and low margins make traditional, physical advisory nonviable.

Fundamentally, a digital front end and connected tech capabilities must provide the backbone that upholds various distinctive customer journeys and helps ensure compliance with the consumer protection regulations (e.g. the European Union's 2016 Insurance Distribution Directive: IDD). Providing visual aid and explanations to agents and sales representatives could increase their productivity as well as their advisory effectiveness.

2.2.1.2.2 Reshaping Customer Journeys from Design to Advice, for Enhanced Customer Experience: Pairing the Right Funding Strategy with the Right Policy Structure on Universal Life

2.2.1.2.2.1 Background Before the development of universal life insurance in the early 1980s, the only permanent insurance available to policyholders was whole life insurance, which required a policyowner to pay a fixed, given amount of premium each year, in exchange for having a fixed, given death benefit, and a progression of fixed cash surrender values (CSVs) over time (all of which were guaranteed).

Permanent life insurance was therefore relatively straightforward: fixed premiums, fixed death benefits and fixed interest rates, generating a consistent growth of cash surrender values over time. Thus, determining the internal rate of return on policies was relatively easy, resulting in a "reasonable" rate of growth on the cash value, and a "reasonable" IRR for those who held the policy until death (at least, relative to interest rates at the time).

However, as "flexible premium" universal life products were developed in the early 1980s, providing a return on both elements – the cash value and the death benefit – policyowners gained flexibility to structure premium payments differently. This introduced the possibility of placing greater emphasis on either the IRR of death benefits or the IRR of cash surrender values – resulting in a lot lower IRR for both than could have been achieved by focusing individually on just one element (death benefit IRR or cash surrender value IRR). Financial advisers and policyholders thus started to improve the way to structure flexible universal life premiums in order to fully optimize either one of these objectives.

2.2.1.2.2.2 Mechanics of the Interplay between the Cash Surrender Value, Death Benefit and Cost of Insurance Charges As illustrated below, optimizing requires understanding the dynamics influencing both death benefit and cash surrender value IRRs, and in particular the interrelationships between cash value, death benefits and cost of insurance (COI) charges, and the ways that changes in the net amount at risk (NAR) compound over time.

In this respect, the COI charge is the main expense deducted from a universal life policy to compensate (i.e. pay) for the actual risk of needing to pay a death benefit if the insured dies (the other expenses include premium loads, monthly policy charges and surrender charges).

There are two main components to this COI charge.

- The COI rate accounts for the risk (i.e. probability) that the insured dies, which increases each year given the higher chance of dying with age, in a convex way (see Figure 2.7).

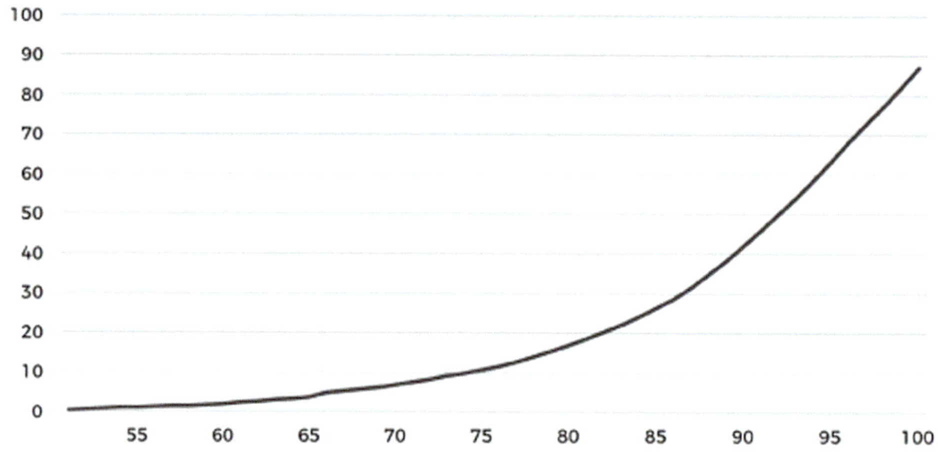

FIGURE 2.7 COI versus age

- The NAR reflects the amount at risk if a death benefit does need to be paid. Since the insurance company already has a certain amount of cash value associated with the policy, it is only the difference between the death benefit and the accumulated amount of cash value.

  As a result, the full charge for the COI for the year = (COI rate for year) × (NAR for year).

Paying a level premium to endow the policy at age 100 (i.e. account value = death benefit at age 100) therefore results in the account value increasing each year, and thus the difference between the death benefit and the account value – the net amount at risk – decreases in turn, which effectively reduce the COI charge in the later years ... even though the actual COI rate is increasing in those later years. Since the COI rate increases as the insured gets older, the only way to keep the COI charges low is through an offsetting decrease in the NAR in the later years.

This means how "funded" (i.e. how much cash value has been accumulated) a universal life policy is can itself have a material impact on the COI charges of the policy.

2.2.1.2.2.3 The Three Flexible Premium Options The flexible premium option with universal life allows the paying of premiums with three alternative "funding options".

- "Pay a level premium" each year to keep the policy in force, which results in paying more than the COI in the early years so that the account value accumulates over time and results in a low NAR in the later years, so that the overall cost is still low. This was traditionally how whole life insurance policies worked.

- "Pay a minimum premium" just to keep the policy in force over time, which results in the premiums being low in the early years, when the COI rate is low, but very expensive in the later years, when the COI rate is higher. While this results in the risk of the premiums in the later years being very high (as COI rates increase and the NAR is still large), the policyowner may be trying to optimize the time value of money, as she has a lower chance of living to the later years, maximizing her death benefit IRR.

- "Pay a hybrid" of the two above, which involves paying a level amount of premium to a point in the future, and then a minimum premium (just to maintain the policy) after that.

The "minimum premium" strategy is significantly cheaper in the early years but escalates dramatically in the later years: the carrier mandates a certain level of premium for the first few years, while, thanks to flexible premium policy, the policyowner can elect to skip premiums over the next few years. Hence the following dilemma: although the minimum-funded policyowner (minimizing the policy premiums in order to maximize the IRR of the death benefit) may have saved in premiums up to that point by minimally funding the policy, going forward she will pay drastically higher COI charges, which will result in significantly higher premiums going forward than their maximum-funded counterpart (maximizing the cash surrender value – i.e. maximizing premiums to minimize the COI charges, which in turn maximizes the tax-deferred growth in the cash value) even though the COI rates are the same.

When is it better for the policyowner to pay lower premiums now, at the risk of paying very high premiums later?

2.2.1.2.2.4 Maximizing the Death Benefit IRR Not surprisingly, the IRR on a life insurance death benefit is very favourable for those who die within 15 to 25 years (i.e. before life expectancy), but tails off in later years.

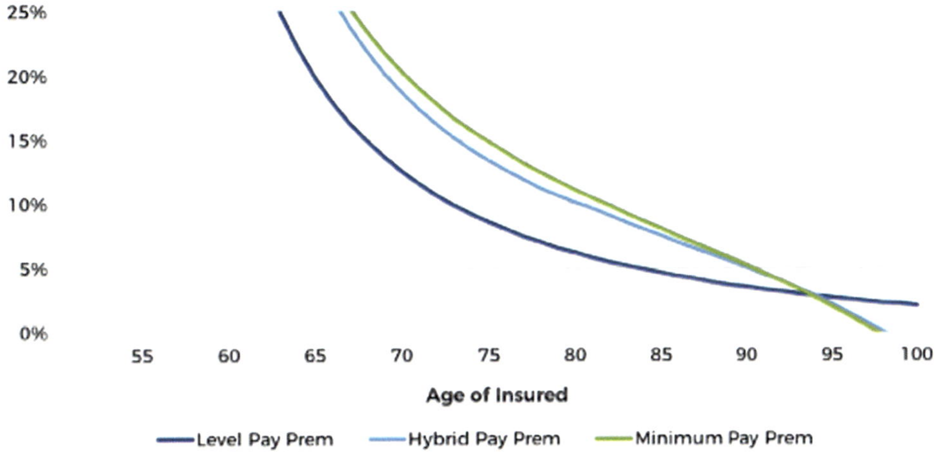

FIGURE 2.8 Death benefit IRR assuming death at end of year

Although the "hybrid pay" and "minimum pay" options essentially overlap, the "level pay" premium notably lags behind at each age in terms of IRR. Indeed, the caveat of the IRR comparison above is that not all the potential IRRs have the same probability of occurring: the very high IRRs are improbable because the odds of death well before life expectancy are not very high. Nor are the IRRs in later years very likely, because at that point the cumulative odds of death would be very high.

As a result, to properly compare strategies, it is not enough just to compare the IRRs; it is also necessary to account for the fact that the insured has to pay the future life insurance premium only if she is still alive. This means there is a need to determine what the chance is of the insured surviving to a given point in time in the future (using actuarial mortality tables), and discount the value of both the potential cash flows (both premiums going out and the death benefit coming in) in a net present value computation for each strategy.

As a result, the actuarial present value of the later premium payments in terms of today's money is actually smaller than the first-year premium payment owed on the policy. If the insured is alive during a policy year, then she will have to pay the scheduled premium (an expense), whereas, if the insured dies in a policy year, the policyowner will receive the death benefit (income). Thus, by using each year's survival and death probabilities, and the corresponding premium and death benefit schedule, the expected cash flows for the year can be determined and an IRR calculated.

As expected, the "pay a level" premium strategy has the highest actuarial present value of premiums (i.e. is the most expensive) and, as a result, has the lowest expected death Benefit IRR. The "minimum" premium strategy is the least expensive and, correspondingly, has the highest death benefit IRR.

2.2.1.2.2.5 Maximizing the Cash Surrender Value IRR Whereas maximizing the IRR of the insurance policy's death benefit requires minimizing the premiums, maximizing the IRR of the cash surrender value requires maximizing the premiums while minimizing the death benefit, so that the policyowner can minimize the total COI charges (which in turn maximizes the tax-deferred growth in the cash value).

In the event of a whole life policy, in which the premiums and associated future cash value growth were all guaranteed, it was not flexible enough to proactively change the strategy. However, the flexible premium structure of universal life insurance can use the tax-deferred growth build-up of large amounts of premium accumulated to accumulate assets while not paying taxes on them.

In order to limit this, Congress in the United States enacted legislation – the Tax Equity and Fiscal Responsibility Act (TEFRA) of 1982 and the Deficit Reduction Act (DEFRA) of 1984 – that capped the amount of premiums (to avoid unduly shrinking the net amount at risk in a manner that would have otherwise virtually eliminated paying any COI charges). In order for the policy to qualify as life insurance – allowing tax-deferred build-up on the cash value – the policy must satisfy one of the following two tests (the choice of the test has to be made at issue).

- The guideline premium test (GPT), which requires that the death benefit at all times be a certain multiple above whatever the then current accumulated cash value may be (known as GPT corridor factors). It ensures that, even if the policyowner pays the maximum premium, the death benefit itself will eventually rise as well, ensuring that the NAR stays at a certain level – and that the insurance policy remains an insurance policy (with some death benefit). In addition, the GPT limits the premiums, based on an actuarial calculation that solves for the maximum premium to fund the death benefit obligation (known as guaranteed charges).
- The cash value accumulation test (CVAT): unlike the GPT, this test does not limit the premium payments but requires only that the death benefit at all times be a certain multiple above the cash value (known as CVAT corridor factors). These CVAT corridor factors are indeed significantly higher than the corresponding GPT factors, in order to make up for the lack of a premium limitation requirement.

At first sight the GPT seems the better option, because the corridor factors are significantly lower than the CVAT corridor factors, making it easier under the GPT to limit the NAR, which in turn limits the amount of COI charges and allows for greater tax-deferred growth. However, since the CVAT does not have a premium limit, it has significantly higher corridor factors for a given age, and so causes the death benefit to increase at a much earlier age.

This of course increases the NAR, which, when combined with the higher COI rates in the later years, results in the CVAT policy paying significantly higher COI charges than the corresponding GPT policy. Accordingly, the cash surrender value of the GPT policy quickly outpaces that of the CVAT policy in the later years: since the CVAT corridor factors increase the death benefit to a larger extent than the GPT corridor factors, the higher resultant NAR increases the COI charges, which drags down the CVAT cash surrender values in the later years when compared to the GPT, which inevitably results in lower CSV IRRs of the CVAT policy.

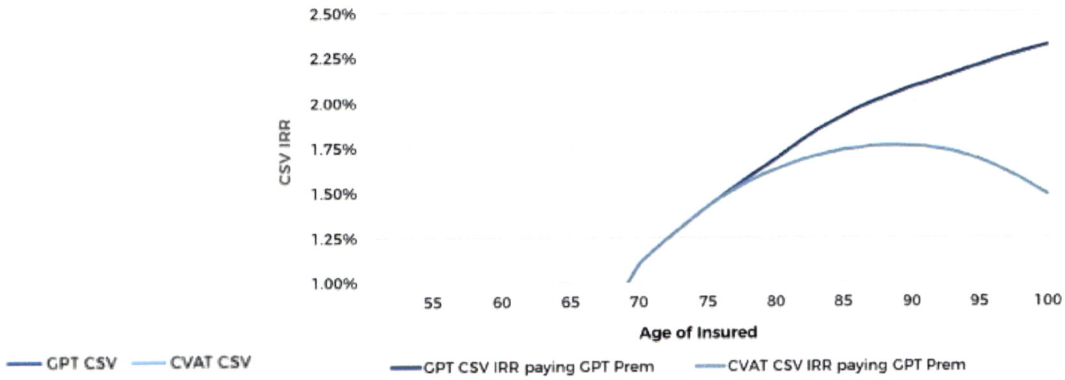

FIGURE 2.9 GPT versus CVAT CSV

A GPT policy therefore results in higher long-term CSV IRRs than a corresponding CVAT policy (the IRRs in the first 25 years are the same, since the corridor factors have not come into play for either the GPT or the CVAT). However, for the first 20 years the return for both the max-funded GPT and the CVAT is extremely low, because of the high initial policy charges in the early years of the policy (which drag down those cash value returns), due to agent commissions (notably 100% of the first-year premium), through a combination of premium load (2 to 15% of each premium dollar), the regular policy fee and the surrender charge (extremely high in the early years, but decreasing over time up to 15 to 20 years). These cumulative costs significantly impact the cash value of the life insurance policy in the early years of a GPT policy – e.g. 21% of the value is eaten up by policy expenses (premium load, policy charges and cost of insurance) and another 68% is eaten up by surrender charges, leaving only 11% of the premiums paid available as the end-of-year cash surrender value. As a result, the universal life insurance policy as an investment does not even break even until year 15.

In this context, CVAT policies can be more appealing, because they do not limit the premiums that are paid into the policy, now or in the future. So the policyowner can make a minimum premium payment when the initial policy expenses are high (i.e. in the early years), and then make a large payment after the high-expense period has ended, which would significantly increase the IRRs on the investment (a strategy that many advisers and especially agents neglect, as lower up-front premium payments with such a strategy could dramatically reduce the commissions paid to the agent on an otherwise equivalent death benefit policy). CVAT simply forces the death benefit to potentially increase more rapidly (by having a larger corridor value).

This strategy to leverage a CVAT policy can bump up the cash surrender value IRR dramatically. The CVAT policy also comes with the added advantage that the policyowner is, essentially, buying a policy when she is young and in good health, so that years later she can pay in very large amounts of "premium contributions" into the policy at low COI rates – even if the health of the insured has declined (which makes it an even better "deal").

Buying a small-face-amount CVAT policy allows a large pay-in to be made in the later years, when policy expenses are lower. This large lump-sum contribution causes the death benefit to increase because of the CVAT corridor factors. This bumps up the NAR initially and increases the COI charges assessed to the policy. However, the large contribution also allows interest to accumulate in the policy. Over time, the interest in the policy accumulates more rapidly than the corridor factors are lifting up the death benefit, which causes the NAR over time to decrease, in turn reducing the COI charges assessed on the policy and maximizing the growth of the cash value (see Figure 2.10).

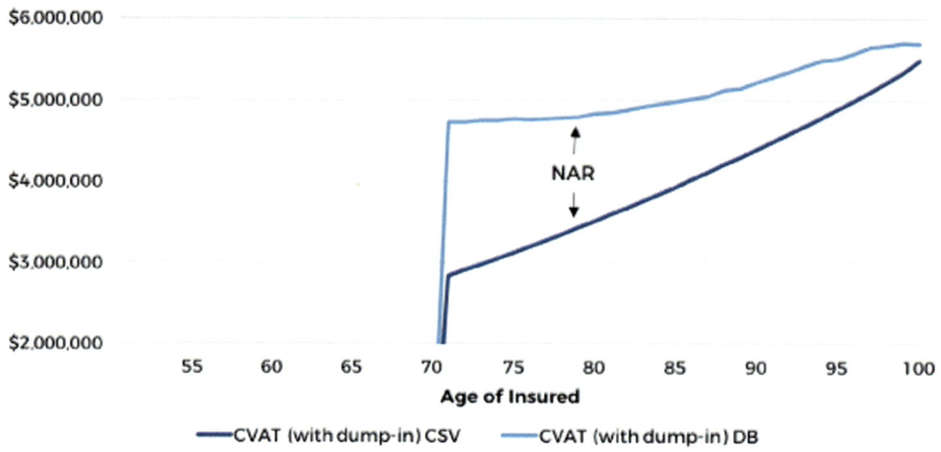

FIGURE 2.10 NAR versus age of insured

Although making a large lump-sum premium contribution in year 21 initially causes both the DB and CSV to increase (and is permissible even without additional underwriting, because the policyowner is not increasing the face amount, per se, but simply allowing the premium payments to cause the death benefit to increase indirectly), as the insured gets older the cash surrender value grows at a larger rate than the DB, because of two reasons.

- The CVAT corridor factors decrease as the insured gets older. The narrowing CVAT corridor permits the death benefit to get closer to the cash value over time, which in turn permits a reduction in the NAR.
- The large dump-in causes interest to accumulate in the policy more rapidly than the COI charges are growing. Even though the COI rate is increasing as the insured gets older, the actual COI charge is not growing that quickly (because of the narrowing NAR). In fact, the interest the policy is accruing each year is significantly larger than the COI charges allowing the CSV to grow over time, even in the later years when the COI rates are otherwise very high.

  By dumping significant premium payment in the later years, the policyowner completely avoids the high policyowner expenses and surrender charges that drag down returns in the early years of a life insurance policy. As a result, at the end of the year the cash surrender value is close to 95 to 100% of the total premium cost paid into the policy. All the policy expenses are now a small fraction of the total premium cost for the policy. As a result, two years after the dump-in, 95 to 100% of the policyowner's total premium cost is available as cash surrender value.

This is a drastic contrast to the max-funded GPT policy, in which after two years some 90% of the total premium cost paid into the policy is eaten up by expenses. Even though the policyowner is 70 years old at the time of the large dump-in, the cost of insurance charges are still only 2% of the total premium cost – which is the same percentage for the 50-year-old with the GPT policy – thanks to the fact that the narrower corridor (due to the higher age) permits a much smaller NAR. With the GPT max-funded strategy, the policyowner pays the maximum annual premium (e.g. $110,000/year on a $5 million policy) starting from age 50. However, with the CVAT dump-in strategy, the policyowner pays the minimum annual premium (e.g. $550/year) into a $100,000 policy, and then does a large dump-in at the start of the 21st year.

2.2.1.2.2.6 How Do We Compare the CSV IRRs from the GPT Max-Funded Strategy to the CVAT Dump-In Strategy? The best way to do this is to treat the first 20 years of payments into the CVAT dump-in strategy as an option price for the right to dump in a large premium later. At the end of the 20th year the accumulated value from paying 20 years' worth of annual premium amounts of $550 amounts to $14,947 (after adjusting for the terminal cash surrender value and the time value of money). We can then treat this accumulated cost, and $3 million large dump-in premium, as a collective cost in the first year of the policy. This way we can compare year by year the CSV IRRs for each strategy to each other. The only difference is that, for the GPT strategy, year 1 is on a 50-year-old, while for the CVAT dump-in strategy year 1 is on a then 70-year-old. Using the CVAT dump-in premium strategy drastically increases the early year CSV returns over the GPT strategy. While the GPT strategy does not break even until year 15, the CVAT dump-in strategy does so in year 3.

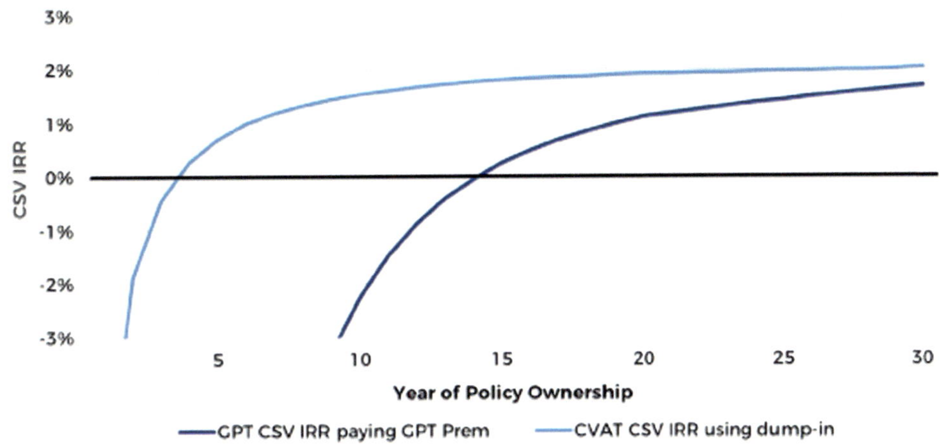

FIGURE 2.11 CSV IRR versus year of policy

By using the CVAT dump-in strategy, the policyowner dramatically increases the early years' CSV IRRs – a lead that holds throughout the subsequent life of the policy as well. The boost in early-year CSV IRRs is of extreme importance, given that 50% of all purchasers of permanent life insurance policies still cancel their policies within the first ten years. As shown earlier, cancelling a life insurance policy early can result in the loss of hundreds of thousands of dollars in high early-year expenses that the policyowner cannot recoup. The embedded option of the CVAT strategy to lock in an underwriting class early, and dump in large premiums later, allows the policyowner more liquidity and better early and long-term CSV returns.

Although the CVAT dump-in strategy allows numerous advantages, it is important to be aware of the three following issues.

- Doing a large dump-in in the later years of the policy is the best way to maximize the CSV IRR. However, if the dump-in is "too large", it runs the risk of making the policy a modified endowment contract (MEC), which causes any loans or partial withdrawals from the policy to be taxable (unlike the standard treatment, whereby withdrawals come from principal first and loans are non-taxable), while also subjecting withdrawals to a 10% early withdrawal penalty prior to age 59½ (similar to annuities and individual retirement accounts).
- The goal here is to show how to exclusively maximize the CSV IRR of a flexible premium universal life insurance product; in other words, it is presumed that the goal is to maximize the cash value, and that the life insurance is not for death benefit purposes (as the optimal funding strategy for maximizing the death benefit is entirely different!).
- Perhaps the most obvious critique is that the best way to increase the CSV IRR is to invest in a permanent life insurance product that is investing in high-yield assets, but these returns are now subject to ordinary income rates, which can be as high as 37% for federal tax. (State income taxes may apply as well, but would typically apply equally to both.)

2.2.1.2.2.7 Conclusions and Takeaways Ultimately, the key point is to understand that strategies that maximize the long-term cash value of universal life insurance – which are all about minimizing the net amount at risk, to reduce COI, while managing the tax code requirements that there must always be at least "some" amount of death – are fundamentally different from the strategies to maximize the IRR of the death benefit of term life insurance (which is driven by minimizing the amount of cash in the policy so as to maximize the net amount at risk, and therefore the amount of money the insurance company must come to the table with in the event of death).

This means it is crucially important to pair the right funding strategy with the right policy structure in order to accomplish the long-term goals of a client with a particular flexible-premium universal life insurance policy.

The bottom line is that, while the traditional life insurance illustration typically shows steady premiums paid to age 100, not only is it not necessary to do so with a flexible-premium universal life policy but also it is often not beneficial to do so. This is both because (i) the IRR of the death benefit can be better maximized with a minimum funding strategy from the start and (ii) even the IRR of the cash value with a "maximum funding" strategy is often better from starting with a minimum funding solution, using a CVAT-style policy and then maximizing the contributions later.

Given the increasingly efficient behaviour of customers, all the more so within an increasingly digital society, it is important to understand the mechanics of how best to maximize a life insurance policy, and why the funding strategies to maximize cash value accumulation are fundamentally different from maximizing the IRR on the death benefit itself. This means that it is especially important to properly pair the right funding strategy and policy structure with the actual long-term goals of the client!

In order to accomplish the long-term goals of a client, in line with her risk appetite, and anticipate her increasingly efficient observed behaviour, an agent-based modelling approach of policyholder behaviour is developed in Chapter 3, through behavioural simulations that combine individual decision-making rules and artificial intelligence (AI). Such agent-based modelling is used for a variety of business applications, such as customer retention, pricing and fee structuring, product design and risk management.

2.2.1.3 Evolving the Product Range of Sustainable Savings and Retirement Instruments, Aligning Both Customers' and Insurers' Risk Appetites, through Cheap, Flexible, Performant and Capital-Light Financial Guarantees

Life insurance companies should review their product strategies accordingly.

Life insurers have been responding to the challenges of persistently low interest rates squeezing already tight margins and making it difficult to cover rising capital requirements from prudential regulation by rethinking their product offerings to less capital-intensive products. New ("capital-light") investment solutions that provide yield in accumulation and income and low volatility in retirement are essential. These require new investment capabilities in multi-asset solutions, real assets and novel solutions that smooth volatility and generate income in retirement.

They are gradually shifting towards "capital-light" products, such as unit-linked products or products with lower or no guarantees, or hybrids as a combination of traditional savings and unit-linked features, thus transferring part of the investment risk to policyholders, in an attempt to consume fewer capital requirements.

As guaranteed products remain in vogue in the European life insurance market, with customers recognizing the value of guarantees within an investment environment characterized by low yields and high volatility, while the costs involved can be prohibitive for the insurer, insurers developed over the 2010s a generation of long-term savings that keep some financial protection while being cheap, flexible and capital-light (i.e. consuming less capital under Solvency II).

There are a range of approaches, and a range of terms are used to describe these new products: index-linked products, unit-linked products with guarantee mechanism, constant proportion portfolio insurance in underlying funds, individual CPPIs, static and dynamic hybrids, and "select" products (an index-linked surplus) are just a few examples, and we expect this trend to expand in the coming years.

These new products tend to align both customer and insurer interests.

- From the customer's perspective, they offer a similar guaranteed maturity benefit to the current traditional product. Only in very bad scenarios are credited rates below the technical rate possible.
- From the insurer's perspective, in most scenarios the minimum annual guaranteed rate of interest is 0%, with a significant reduction of solvency capital requirements.

Remark: as "select" products are index-linked products, see the previous section on indexed annuities. The returns for index select are indeed linked to two components: the stock index, which provides the policyholder with participation in the European equity market, and a guaranteed annual interest rate.

The policyholder can control the composition of the returns annually, as she can choose every year if she wants her accrued account value to participate in the general assets (surplus participation) of the insurer or in some formula-based participation in the equity index. The client's surplus is used to purchase an option on this index participation.

At the beginning of each year a cap will be determined, which specifies the maximum level of monthly returns (e.g. 4%) the policyholder can receive from the development of the equity index. Negative returns will be floored at zero. But, even in a case of one year of total loss on the markets, the policyholder can receive positive returns by choosing to receive her returns not only from the index but also from a guaranteed interest rate.

Initially very successful, it was not understood by many distributors and clients, as it was often marketed like a normal unit-linked product without a guarantee (i.e. that, if you tick a box, you get a guarantee that you will get your premiums back – with no explanation that ticking the box leads to a significant investment in traditional life insurance). In addition, the product's fund exposure has decreased due to low interest rates.

2.2.1.4 Capital-Light CPPI Savings Products

The sustainability of savings and retirement products is also needed to secure both growth and solvency constraints for insurers, notably through reasonable amounts of risk and capital requirements. This change in approach must therefore be meeting both customers' and shareholders' interests.

In Europe (e.g. Germany, Switzerland, France, Italy, Spain) and Asia (Japan, Hong Kong, Thailand) such innovations launched during the 2010s are represented by dynamic hybrid products, which are typically offered as deferred annuity contracts. Their market share has significantly increased since 2016, climbing as high as 30% of new life savings annual sales in Germany in 2021.

Hybrid products are insurance policies designed to combine potentially higher returns of investments on equity funds with a certain degree of security required by policyholders. They combine the merits of traditional participating life insurance and unit-linked products by periodically shifting funds between the policy reserve stock of an insurer (with an interest rate guarantee), a guarantee fund and/or an equity fund.

They are mostly characterized by the use of a dynamic asset management approach to vary the participation in a range of risky assets in response to these assets' estimated future volatility.

- Applied to a traditional equity/bond fund, for instance, the overlay would dial down the allocation to equities into cash or cash-like instruments, if the expected fund volatility exceeds the predefined target.
- If the expected fund volatility falls below the target, an overlay would use leverage to increase the fund's equity holdings and increase its volatility exposure. This leverage is typically capped at 100% of the fund value.

The adoption of a sophisticated, dynamic risk-managed approach would invest heavily in equities, so as to maximize returns, during periods of relatively calm markets, but then limit exposure during periods of volatility.

Such portfolio insurance strategies have long been used by banks and have by now become important for the product design in life insurance. This is done with the aim of combining the stability of traditional life insurance policies with benefiting from positive capital market returns.

2.2.1.4.1 Value Proposition On the one hand, capital should work actively and achieve the highest possible returns. On the other hand, the portfolio should also include components that secure the capital – i.e. maintain its value. In this respect, the CPPI is a risk management trading strategy that determines a portfolio's allocation between two pools of assets:

- risky (or growth) assets that are intended to provide the returns; and
- protection assets (usually fixed income assets; here the insurer's general assets as riskless assets), intended to provide some level of predefined capital protection.

In principle, the CPPI model strives for an optimal distribution between opportunities and risks, as well as security. The allocation between the protection assets and growth assets becomes dynamic, as follows.

- In rising markets, the asset allocation mechanism (algorithmic or deterministic) is designed to either (i) maintain the maximum exposure to the growth assets or (ii) increase the exposure to the growth assets when the maximum permitted exposure has not been reached.
- In falling markets, the asset allocation mechanism will allocate more to the protection assets, in extreme markets allocating 100% of the portfolio to the protection assets.

In managing in this way, the CPPI asset allocation mechanism aims (but does not guarantee) to provide returns via the growth assets subject meeting the predefined capital protection constraints.

A variation on the regular CPPI strategy is time-invariant portfolio protection (TIPP), within which the protection assets are usually money market assets; rather than providing protection at a distant point in the future, TIPP solutions typically provide a certain level of protection the next day (always less than 100% and usually 70 to 90%), allowing solutions to be structured in which the value of the portfolio is protected from falling below, say, 80% of its highest value since inception.

It was designed in 1987 by economists Robert Jones and Fisher Black. It is a financial mathematical model that pursues two objectives. The available capital is to be protected against falling prices on the capital markets in order to never fall below a predetermined minimum value. In addition, the part with higher risk is to be used, so that active capital accumulation is carried out at the same time.

To make this work, the model takes into account some individual parameters that the insurer adjusts depending, in Germany, on the Riester contract. These are, for example,

- the form of the chosen warranty;
- the contract;
- the amount of capital; and
- the investment intervals.

Individualized CPPI (iCPPI) employs the same mechanisms as CPPI but, rather than applying to a pooled set of assets (many policyholders invested in the same CPPI fund), it applies this at the individual policyholder level (hence the term "individualized"), thereby allowing a life insurance company to provide protected solutions fully tailored to the individual policyholder. Furthermore, the market risk associated with providing such guarantees can be precisely and fully hedged out to a third party (e.g. an investment bank), leaving the life company with the residual actuarial risk that it is best positioned to manage.

The iCPPI has been launched across Germany, Switzerland, Italy, France and Thailand. The special feature of iCPPI is shown by how the capital is managed by the provider. On the one hand, this must pursue an increase in capital, but, on the other hand, the floor must not be undercut. For this reason, when share prices rise, more and more is shifted to risky positions in order to increase the multiplier and exposure. When stock prices fall, a so-called bear market, more and more is shifted to safe stocks, such as bonds. Since they are usually less affected by it, or even record gains in a bear market, the previously defined floor is preserved.

Savings products with iCPPI are particularly interesting for young and middle-aged people, as illustrated in the German market with the so-called Riester iCPPI, which usually invests for at least two decades (or much longer), during which time fluctuations due to the long investment period can not only be compensated for but can even be actively exploited. Since the Riester pension is paid into regularly and continuously, the Riester provider will automatically take advantage of periods of weakness on the capital market. Therefore, the Riester insured receives lower entry prices and, thus, equally, later a high return.

2.2.1.4.2 The CPPI Mechanism Given a capital amount to be guaranteed at maturity, a typical floor strategy at time t is a fixed-rate floor, which is given by the discounted value of the guaranteed amount at maturity. The most common floor strategy is to allocate the minimum amount of assets into the risk-free fund.

But the CPPI not only guarantees a fixed payoff at maturity but chases the upside potentials via dynamic trading using leverage as well. In this respect, the cushion – the difference between the portfolio value and the floor – will be invested in the mutual fund. Moreover, the CPPI portfolio usually leverages its cushion C to chase higher return, and its leverage ratio m stays constant as a "constant proportion". The cushion value C fluctuates with the market; once it approaches zero, all the fund will be invested in the risk-free asset till the maturity.

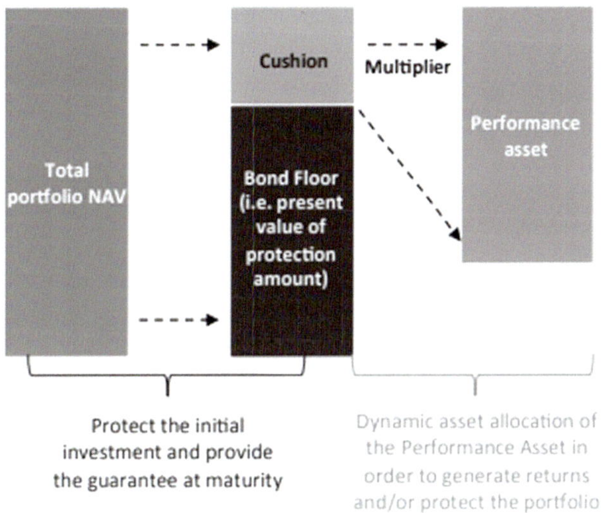

FIGURE 2.12 CPPI components

Note: NAV = net asset value.

The CPPI then controls for each period the percentage of total assets invested risk-free and risky, respectively.

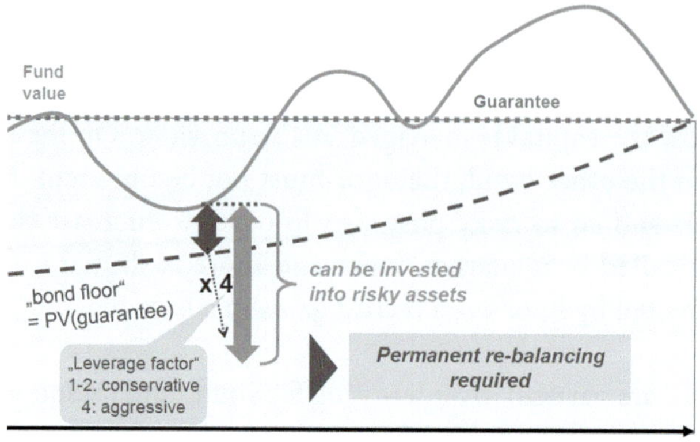

FIGURE 2.13 CPPI dynamic rebalancing mechanism

Let us illustrate with a basic example.

- Assume that, at t = 0, the initial value of the fund is V0 = 100, which is also the capital guaranteed at maturity T = 5yrs. m = 4 and the risk-free rate r is 5%.
  - The initial floor is thus F(0) = 100(1 + 5%) – 5 = 78.35.
  - The initial cushion is C(0) = V(0) – F(0) = 100 – 78.35 = 21.45.
  - With m = 4, the initial risky exposure is E(0) = m*C(0) = 4 × 21.45 = 85.8.
  - The initial non-risky exposure is thus M(0) = V(0) – E(0) = 100 – 85.8 = 14.2.
- After one day (assuming a daily rebalancing frequency), if the market goes 4% up,
  - E(1) = E(0)*(1 + 4%) = 89.23;
  - M(1) = M(0)*(1 + r) = 14.2(1 + 5%/365) ≈ 14.2;
  - V(1) = E(1) + M(1) = 89.23 + 14.20 = 103.43;
  - F(1) = 100*(1 + 5%)5–1/365 = 78.36; and
  - C(1) = V(1) – F(1) = 103.43 – 78.36 = 25.07.
  - The implied multiplier is m(1) = E(1) / C(1) = 3.56, which is different from the target m = 4. The manager will thus need to buy m × C(1) – E(1) risky assets in order to keep the right risky exposure.

The partitioning of an account value AV with a guaranteed value G is given by

$$B_t = (1-\alpha_t)\cdot AV_t,$$
$$S_t = \alpha_t \cdot AV_t = AV_t - B_t,$$

where B denotes the risk-free or low-risk part of bonds and S provides the portion to invest riskily – i.e. in stocks. The percentage α is calculated as follows:

$$\alpha_t = \min\left(\max\left(m\cdot\frac{AV_t - (1+r_f)^{-\Delta t}\cdot G_{t+\Delta t}}{At}, 0\right), \alpha_{\max}\right).$$

with $m \geq 0$ indicating a multiplier to control for risk aversion. The numerator is referred to as a cushion, since it specifies the fraction that is not required to meet the guarantee and that can be invested in the risky asset.

If we are using time-continuous rebalancing (the Black–Scholes model), the CPPI strategy is equivalent to:

- buying a zero coupon with nominal G to guarantee the capital at maturity (super-hedging the floor); and
- investing the remaining sum in a risky asset that has m times the excess return and m times the volatility of S and is perfectly correlated with S.

Within such a continuous-time modelling framework, the CPPI fund value never falls below the guaranteed floor.

- The portfolio value at maturity is explicitly given by

$$V_T = G + \left(V_0 - Ge^{-rT}\right)C_m\left(\frac{S_T}{S_0}\right)^m$$

$$C_m = exp\left((1-m)\left(r + \frac{m\sigma^2}{2}\right)T\right).$$

- At each time t the account value is

$$V_t = F_t + C_0\exp\left(rt + m(\mu - r)t + m\sigma W_t - \frac{m^2\sigma^2 t}{2}\right)$$

$$\mathbb{E}\left(V_t\right) = F_t + C_0 e^{rt + m(\mu - r)t}$$

$$\mathbb{V}\left(V_t\right) = C_0^2 e^{2(rt + m(\mu - r)t))}\left[e^{m^2\sigma^2 t} - 1\right].$$

⇒ The manager has no risk of breaching the floor in this framework.

However, as the underlying may experience non-continuous dynamics (e.g. jumps), this guarantee is then no longer assured whenever the risky asset experiences an extreme variation below the "bond floor" (required to guarantee the capital at maturity), thus experiencing the so-called "gap risk". Technically, the gap risk happens whenever the risky assets falls by more than $1/m$, where m is the multiplier of the CPPI as already defined. Although such gaps are assumed to be negligible in most financial models used in practice, their occurrence and magnitude have drastically increased over the past ten years as interest rates decreased while volatility increased, which means that selling more risky assets therefore pushes the multiplier m (and market performance) down. This significantly decreased the appetite for CPPI types of product during the 2010s, notably in Italy and Belgium.

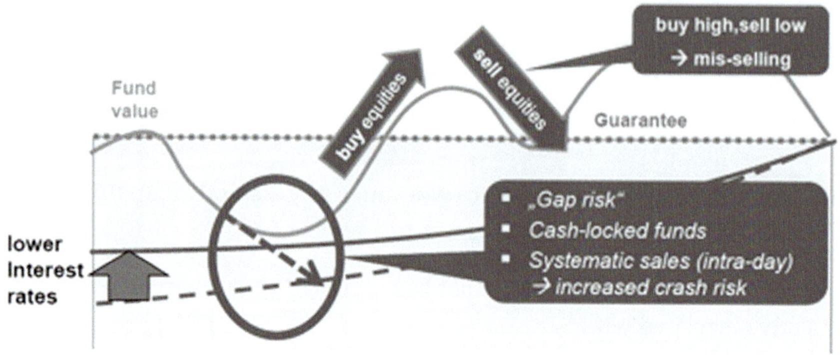

FIGURE 2.14 CPPI gap risk

In order to mitigate such gap risk, improvements in modelling, product design and hedging strategies need to be devised and implemented, notably through the use of risk metrics (e.g. value-at-risk) constraints (see Chapter 5 regarding the risk management of the gap risk).

2.2.1.4.3 The Different Versions of the CPPI

2.2.1.4.3.1 Static Two-Fund Hybrid Product Here the growth assets are funds.

This version is referred to as "static" as there are no reallocations of the contract balance between the protection assets and the growth fund investment during the entire term of the contract. The advantage of static grant validation lies in the clearly defined investment pattern. Regardless of the development of stock market prices, there is no mixing between the guarantee component and the fund component. This guarantee concept is particularly interesting for savers who prefer a balanced investment ratio.

Static hybrid products combine a traditional life insurance policy with a fund investment. In doing so, static hybrid products split the premium into two investment parts. One part is invested in the premium reserve stock (PRS) of a life insurer, which resembles an investment in a traditional policy with a guaranteed rate of return and the right to receive surplus, whereas the second part is invested in an equity fund. The initial partitioning of the premiums is maintained throughout the contract term and is given by (see Bohnert & Gatzert 2104)

$$PRS_t = \frac{G_{t+\Delta t}}{(1+r_G)^{\Delta t}},$$
$$EF_t = AV_t - PRS_t.$$

If we consider a 40-year time horizon, in the first year as much as ~70% of the contributions are invested to comply with the contribution guarantee given the low interest rate environment. This means that 30% flow into freely selected index funds. With each additional year, the contribution part that is invested in the protection assets increases. The contribution portion that is invested in funds, on the other hand, decreases over the course of the contract period. However, the interest surplus is invested in funds over the entire term.

If we consider a 30-year time horizon, ~75% of the contributions are invested to comply with the contribution guarantee. This means that fewer than 25% flow into freely selected index funds. If we consider a 20-year time horizon, close to 85% of the contributions are invested to comply with the contribution guarantee. This means that only ~15% flow into freely selected index funds. If we consider a 15-year time horizon, close to 90% of the contributions are invested to comply with the contribution guarantee. This means that just ~10% flow into freely selected index funds.

As a result, in the case of static two-fund hybrids, a significantly high proportion of the payments flows into protection assets, especially for short to medium contract periods. Investors aiming for higher investment ratios with significantly shorter maturities should prefer a dynamic three-fund hybrid guarantee model, as illustrated below.

Moreover, at a certain time during the contract term the investment in the policy reserves might be higher than required to meet the set guarantee level – i.e. a part of the investment in the policy reserve stock could, theoretically, be transferred to the risky investment with a higher return on average. This leads to the dynamic version of hybrid products.

2.2.1.4.3.2 Dynamic Two-Fund CPPI/TIPP Products (Germany, Switzerland, Italy, France) Dynamic hybrid products merge the concepts of CPPI and the static hybrid product. In contrast to splitting the premiums, in the case of a static hybrid product, a contract's total account value is split and invested in the policy reserve stock of an insurer and an equity fund (and/or guarantee fund), and it is periodically (e.g. monthly or daily) reallocated between these pots.

A two-fund shifting mechanism that invests the maximum proportion of the account value in an equity fund or guarantee fund, along with ensuring the guarantee, is given by

$$PRS_t = \begin{cases} \dfrac{G_{t+\Delta t} - (1-\lambda)\cdot AV_t}{(1+r)^{\Delta t} - 1 + \lambda}, & \text{if } \dfrac{G_{t+\Delta t}}{(1-\lambda)\cdot AV_t} > 1 \\ 0, & \textit{otherwise} \end{cases}$$

$$F_t = \begin{cases} AV_t - PRS_t, & \text{if } \dfrac{G_{t+\Delta t}}{(1-\lambda)\cdot AV_t} > 1, \\ AV_t, & \textit{otherwise} \end{cases}$$

where F cannot lose more that λ% per period and can be provided by a third party or by the insurer (see Kochanski & Karnarski 2011).

2.2.1.4.3.3 The Eurogrowth Fund, where Customers' and Insurers' Interests and Risk Appetites Are Still Mostly Not Aligned In fact, the two-fund CPPI mechanism is also embedded to some extent within the so-called "eurogrowth" fund, which has been developed and distributed by major life insurance companies in France.

Introduced by the Berger/Lefebvre report (see Berger & Lefebvre 2013), the "eurogrowth" product guarantees a part of the prime at maturity only. Even though it was supposed to become the third pillar of life insurance with the euro funds and "units of account", it has not been successful. It distinguishes itself from the euro funds by a higher investment capacity in risky assets due to the suppression of the guarantee at all times.

The persistently low interest rate environment has deprived the diversification provision, which is built beyond the mathematical provision as the growth generator. Indeed, the fall of interest rates mechanically increases the mathematical provision (equal to the guaranteed amount discounted with 90% of the so-called TEC [targeted excess capital] rate, and invested in bonds), at the expense of the diversification provision (individual, equal to the difference between the contract value net of charges and the mathematical provision, invested in risky assets) and, ultimately, the performance. As a matter of fact, four years after its launch only €2.3 billion of the outstanding €4 billion have been collected on this type of product – far from the €1,800 billion of outstanding assets recorded on euro funds in the same period.

Indeed, it appears that the initial distribution between the mathematical provision and the technical diversification provision has considerable importance on the performance of the contract; in fact, on the former eurogrowth policies, a 0% return was observed for periods when the discount rate was negative, as the contract was subscribed. This is explained by the initial provisioning, which is therefore composed 100% of the mathematical provision, because the guaranteed amount is discounted at a rate of 0%. The initial diversification provision is therefore nil, which implies that no share of it will be allocated to this contract. Throughout the life of the latter, even in the event of a rise in rates or sharp rises in the equity market, the diversification provision will therefore be zero, which explains the return of 0% in the end.

As a result, in France the PACTE (Plan d'action pour la croissance et la transformation des entreprises) law of 2019 deeply simplified the product by suppressing the use of TEC discount rates by removing the distinction between the diversification provision and the mathematical provision, in favour of a reinforced diversification provision.

What is the impact of the removal of the mathematical provision on the choice of asset allocation, the capital guarantee and the maturity of the contract for an insurer and insured compromise in terms of profitability and risk taking? Are customers' and insurers' interests and risk appetites aligned within such a design as a result?

The removal of this mathematical provision has multiple very positive consequences for the insurer, but it has some negative consequences for the customer, which makes the alignment between customer and insurer interests and risk appetites less than obvious, and often contradictory or adversarial.

- First, despite its removal, the insurer has to continue calculating a theoretical mathematical provision for provisioning purposes. The advantage, for him, is that this theoretical mathematic provision is no longer attached to the "TEC" discount rate, so he may decide to discount this notional provision at a less conservative rate in order to increase the risky portion, which offers more flexibility in his provisioning. At maturity, if the diversification provision is lower than the guarantee, then the difference will be financed by the insurer; otherwise, the only performance of the asset will honour the guarantee.
- Second, in the event of early redemption, the insurer will no longer be required to reimburse the insured up to his mathematical provision; he will do so according to the valuation of the diversification share at the time of redemption. This device mitigates the insured's guarantee for the benefit of the insurer at first, and to the detriment of the customer.
- Third, there are lower capital requirements (by ~10 to 20% for a 40–60 equity–bond asset allocation), mainly driven by the removal of the mathematical provision, although the impact is marginal in the event of a low (e.g. 10 to 20%) equity allocation, and it also decreases as the capital guarantee decreases and as the maturity increases. Thus the worst here for the insurer occurs for the highest equity allocation, the longest maturity and the highest level of capital guarantee, improving both the insurer's IRR and capital requirements, which is mostly the opposite for the customer to a large extent (i.e. except for a very high equity allocation, whose potential high volatility might be detrimental to the return). This means that customers' and shareholders' interests and risk appetites are, again mostly not aligned here.

  In practice, the maturity tends to be beyond ten years, while the capital guarantee level lies between 80 and 85% (due to the small negative impact on the customer's IRR as the capital guarantee decreases from 90 to 80% – i.e. the initial premium net of 1 to 2% annual fees over ten years).
- Fourth, this also made its mechanism closer to a unit-linked policy, and thus much easier for the insurer; indeed, the removal of the mathematical provision makes the provision and the returns the same for all policyholders, in contrast to individualized ones within the former version of the eurogrowth fund (depending on the date of subscription, the duration of holding, the guaranteed amount at maturity and the customer's risk appetite). This simplifies the operation of these funds and makes them clearer for the insured, and in particular for neophytes: the insurer can display a single return that will be applied to each contract, regardless of its characteristics.
- Fifth, such a removal offers a greater yield generator for the "average" within the low interest rate environment, albeit to the detriment of the policyholders already in the portfolio, who will have to share the

wealth with these new policyholders. This is in contrast to the former version, within which the new contribution would have increased the total return of the fund while nothing would have been redistributed to these new insureds. This would have increased only the performance of contracts already in the portfolio, subscribed at higher rates. Now the value created will therefore be shared among all policyholders, which allows a transfer of wealth from the old policyholders, who subscribed at a higher rate, to the new ones, who subscribed at a lower rate.

Such a common return to all policyholders has a direct impact on policyholders' behaviour: within the old eurogrowth system, with low interest rates, no insured exhibiting rational behaviour invests her money, since the expectation of return is accordingly very low, and will wait for rates to rise or turn to other more attractive products. The removal of the mathematical provision, and the expectation of a yield that it brings, should therefore make it possible to win new customers and thus increase the outstanding amounts of this product. In contrast, if interest rates rise, the new eurogrowth product could prove harmful for policyholders. Indeed, by offering lower returns than those of the old product, if the insured exhibit rational behaviour, there should therefore be fewer subscriptions.

In addition, since parameters such as the percentage of guaranteed capital, or the date of subscription, will no longer affect the individual return, this should lead policyholders to move rationally towards the most secure options with the fewest constraints for them – i.e. the shortest contracts with the best possible capital guarantee. The insurer would then be forced, for reasons of prudence, to change the allocation of its assets to increase the share of bonds compared to that of risky assets. This change in asset management would lead to a decrease in the expectation of returns for the customer while limiting the risks taken by the insurer. But this would also reduce the fund's performance compared to a fund in which savings would on average be held for a longer period of time and there would be lower capital guarantees. This again illustrates that customers' and insurers' interests and risk appetites are not aligned.

- Sixth, when compared to the traditional CPPI, the leverage is much lower, as the multiplier m = 1; this, on the one hand, removes the gap risk for the insurer but, on the other hand, makes the upside potential performance rather low (albeit higher than for the euro fund over the past few years) and mostly volatile for the customer, especially within a persistently low interest rate environment (~1%) with a very small portion of the premium allocated to the risky asset.
- Finally, the value proposition appears complex and unclear not only to customers but also to insurers from an operational management perspective; indeed, since the Pacte law the mechanism also includes the so-called "deferred collective diversification provision", which is a non-individual (in contrast with the diversification provision) and non-compulsory provision for smoothing financial returns in the market, which partially reduces the volatility of the diversification provision by absorbing some of the performance when markets are rising and reinjecting performance when they are low (which are bounded by regulatory constraints: the allocation to the deferred collective diversification provision is possible only if the amount of the deferred collective diversification provision does not exceed, after the allocation, 8% of the maximum between, on the one hand, the amount of the mathematical provisions of the auxiliary allocation accounts that would have to be entered if the discount rate used for their calculation had been zero and, on the other hand, the value of the assets of the auxiliary allocation accounts).

Thus, it gives the insurer the opportunity to keep part of the past performance for redistribution when the performance of the funds deteriorates in the future. It can be compared with the provision for participation in surpluses for the eurofund. It therefore makes it possible to fund the profit-sharing provision account balance and subsequently to defer the profit sharing when the return on the asset is high and distribute it when the return falls.

Like the PPE of the traditional eurofund, this provision has to be redistributed after eight years, since it belongs to the insured. Here again the policyholder who stays longer will benefit all the more from the stock of collective provisions of deferred diversification, which once more means the customers' and insurers' interests are not aligned.

Overall, such a euro growth fund design does not seem to meet the diversity of customer's risk appetites and mostly favour the insurer's interests. It is definitely the case that customer and insurer interests and risk appetites are still mostly not aligned.

2.2.1.4.4.4 Dynamic Three-Fund Adjusted CPPI/TIPP Products (Germany, Switzerland) The majority of German life insurers now rely on dynamic three-fund CPPI rather than two-fund arrangements.

This is a self-financing strategy, which dynamically rebalances the portfolio amount on the risk-free asset and two risky funds: the "value protection" fund, as within a traditional CPPI; and a growth fund without guarantee.

The contract balance is thus distributed across three investment layers based on a dynamic redeployment mechanism, the aim of which is to invest as much as possible in the growth funds while securing the guarantee capital.

The dynamic redeployment mechanism ensures at all times that the total contract balance is at least equal to the present value of the guaranteed capital at the start of the pension. This is done through a dynamic interaction between the value protection fund and the protection assets.

As long as the guarantee of the value protection fund, together, if necessary, with the fixed income assets, reaches the present value of the guarantee capital at the beginning of the pension, it is guaranteed at all times that the necessary guarantee capital can be generated at the beginning of the pension. The additional credit can thus be invested in growth assets without guarantee, with the aim of hedging against specific risks, such as inflation.

- Investment layer 1 is made of the fixed income assets and always offers a fixed money market interest rate according to the actuarial interest rate valid at the time of conclusion of the contract (0.90% guaranteed interest plus surplus participation). Goal: an investment in the protection assets comes into play only if the contract balance is not sufficient to ensure the guaranteed performance solely through the value protection fund (investment layer 2).

- Investment layer 2 is made of the value protection fund (guarantee fund; ideally, an active exchange-traded fund [ETF] portfolio) and guarantees a minimum value of 80% of the share value of the last valuation date of the previous month for one month at a time. Goal: the value protection fund (if necessary, together with the protection assets) is at any time in a position to secure the currently required guarantee capital.

- Investment layer 3 is made of up growth assets (active funds, managing funds and ETFs). Goal: this layer is not needed to secure the guarantee; it is an extra "return engine", in addition to the value protection fund.

A three-fund dynamic hybrid product's mechanism with the aim of investing as much as possible in risky funds results in

$$PRS_t = \begin{cases} \dfrac{G_{t+\Delta t} - (1-\lambda)\cdot AV_t}{(1+r)^{\Delta t} - 1 + \lambda}, & \text{if } \dfrac{G_{t+\Delta t}}{(1-\lambda)\cdot AV_t} > 1 \\ 0, & \textit{otherwise} \end{cases}$$

$$GF_t = \begin{cases} AV_t - PRS_t, & \text{if } \dfrac{G_{t+\Delta t}}{(1-\lambda)\cdot AV_t} > 1 \\ \dfrac{G_{t+\Delta t}}{1-\lambda}, & \textit{otherwise} \end{cases}$$

$$EF_t = AV_t - PRS_t - GF_t$$

(see Kochanski & Karnarski 2011 and Bohnert & Gatzert 2014).

While, in the case of static two-fund hybrids, a comparatively small proportion of the contributions flows into the growth investment, as illustrated above, especially for short to medium contract periods, the dynamic three-fund hybrid guarantee model creates significantly better chances to invest in the growth fund investment. Thus, the guarantee concept is particularly interesting for savers who prefer higher equity quotas.

All dynamic hybrid systems apply a periodical rebalancing process between the used funds. However, the time intervals between these shifts differ for the various concepts.

- On the one hand, short intervals are necessary to be able to immediately react to market movements and to reallocate capital to less risky investments, such as the insurer's policy reserve stock or a guarantee fund, for falling markets, or vice versa in the event of an upward trend for the investments.
- On the other hand, costs are incurred for every reallocation of funds, which thus constitute a trade-off between minimizing transaction costs and maintaining the capacity to act as quickly as possible. Here, we can observe two different implementations – i.e. the considered concepts either conduct shifts (if necessary) every day or once per month. The two-fund concepts usually coincide with the daily shifting, while nearly all three-fund approaches use the monthly shifting, due to the fact that the three-fund products use a guarantee fund in addition to the policy reserve stock to provide the guarantee.

In most cases, the dynamic three-fund CPPI strategy outperforms CPPI strategy in terms of preventing downside risk (in terms of the protection ratio, 1% value at risk, and 1% expected shortfall), offering higher returns with the higher volatility (see Chen et al. 2019), and also in terms of capital requirements (~1% AuM), while their operational management is under control. Moreover, the use of money market funds for investment layer 1 and the 80% maximum value for investment layer 2 very significantly decrease the "gap risk" as experienced by the traditional two-fund CPPI.

Even so, high volatilities and low interest rates lead to low equity allocation, and so low returns for the customer. In addition, the fee, although lower than in traditional two-fund CPPIs, remains significant (~3%).

As a result, such a three-fund CPPI design improves the alignment between customers' and insurers' interests and risk appetites, albeit not completely.

2.2.1.5 *Capital-Light, Flexible and Cheap Structured Variable Annuities (Registered Index-Linked Annuities: RILAs) in the United States*

RILAs are deferred equity-linked annuity products that can be purchased in the United States by individual investors as part of both qualified and non-qualified retirement accounts. RILAs are intended as long-term investments, and, if used as such, they enjoy the same preferential tax treatment as traditional variable annuities

Carriers such as AXA (2010), MetLife (2013), CUNA Mutual (2013) and Allianz Life (2013), followed by most US life insurers, have launched these VA/FIA hybrids as a new and innovative way to attract VA assets without offering rich guaranteed living benefits.

The customer appetite lies in the offer of better accumulation opportunities (relative to indexed annuities and fixed annuities) for pre-retirees and retirees open to modest, tailored amounts of downside risk. They are very similar to certain structured retail products (which have total annual sales north of $45 billion per annum). They are sold primarily through wirehouses (e.g. Morgan Stanley, Goldman Sachs, Merrill Lynch) and financial institutions (e.g. Wells Fargo, Chase).

Rapid sales growth began in 2015, with the volume quadrupling from about $0.5 billion to about $2 billion per quarter, reaching $9 billion in 2017, $24 billion in 2020 and $39 billion in 2021. They thus constituted 40% of the traditional VA (TVA) market, while also vastly outpacing growth in fixed annuities – both fixed-rate and indexed. Annual sales are expected to reach $50 billion by 2026.

Like other deferred annuity products, a RILA has an accumulation phase followed by a payout phase (sometimes called the "distribution phase"). The latter entails various ways to annuitize the accumulated account value at a specified time. Since insurers do not guarantee the interest rate used to determine these annuity payouts, the payout phase boils down to an immediate fixed annuity and can be considered in isolation from the accumulation phase.

Moreover, after an initial three- to ten-year period with surrender charges (during which free withdrawals are typically limited to 10% of the account value), the investor can usually surrender her contract at the end of any guarantee-crediting term and receive the full account value. As a result, many RILA contracts may never reach a meaningful "payout phase", as investors opt for a lump-sum payout.

The accumulation phase is divided into crediting terms of usually one year, two years, three years or six years. During this time the account value is credited a return based on the performance of a common market index over a given term of usually one to six years, subject to minimum return guarantees in the form of a floor or buffer for downside protection in exchange for a cap on potential gains or a limited participation in the underlying index. By retaining some of the downside risk, investors are able to benefit from relatively generous rates on the upside.

At the end of the accumulation phase term, the insurer credits the RILA account with a return that is equal to the index return, subject to restrictions on the upside – through a cap on the annual performance (i.e. a maximum credited return – e.g. 8% or 16%) and/or reduced participation in the gains of the index – and some form of downside protection.

In terms of payoff, they essentially consist of a zero-coupon bond, a European ATM call spread that is bought, and an OTM European put that is sold, with the three main indices crediting (see Figure 2.15).

- SVA 1
  - ✓ Upside gain up to a cap
  - ✓ Downside loss in excess of a "buffer" rate
- SVA 2
  - ✓ Upside gain up to a cap
  - ✓ Downside loss floored at predetermined level
- SVA 3
  - ✓ Fixed upside gain if market return is not negative
  - ✓ Downside loss in excess of a "buffer" rate

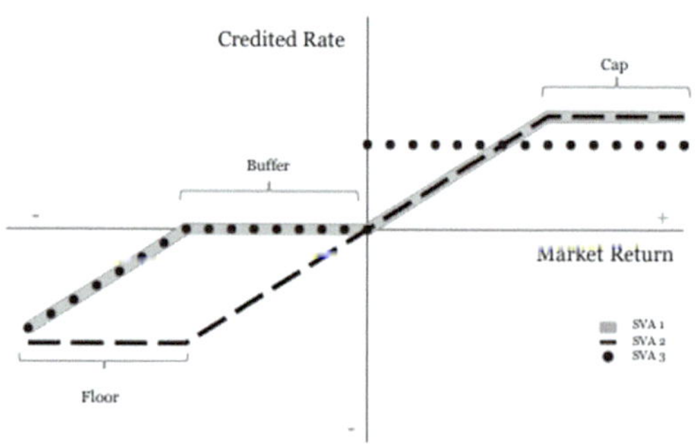

FIGURE 2.15 RILA versions

Note: SVA = structured variable annuity.

As a result, RILAs offer a moderate upside equity potential (achieved through a combination of buying and selling call options) and some downside protection (achieved through a combination of buying and selling put options), which takes three main versions.

- A "buffer" version (i.e. losses to the index are credited only above a certain threshold – e.g. 10 or 30%), in which the insurance company takes the first dollar loss, and the policyholder is fully exposed beyond that limit, so it can be characterized as "full protection in scenarios with small losses" and "moderately reduced downside in scenarios with large losses".
- A "floor" version (i.e. a maximum loss percentage – e.g. 10%) in which the policyholder is exposed to losses up to a maximum, while the insurer becomes fully exposed beyond that limit, so it can be characterized as "no protection in scenarios with small losses" and "substantially reduced downside in scenarios with large losses".
- A downside participation rate – e.g. the loss credited to the account is only 50% of the loss on the part of the index.

The fundamental characteristics of the crediting mechanism are the downside protection structure and rates, which remain fixed, and the upside crediting rates, which the carrier can change over time, for both new policies and renewals at the end of a crediting term. This grants insurers the flexibility to adjust their pricing to changing market conditions – a desirable feature that is not available to them with TVA policies. In fact, during the Covid-19 pandemic market crash in 2020 carriers adjusted the cap rates (larger for buffer versions, lower for floor versions) to reflect the lower interest rates and increased market volatility, which particularly affected the floors of the downside protection: while floor and buffer contracts produced very similar levels of profitability pre-pandemic, the floor contracts yielded over 100 basis points less in profitability in mid-2020.

Even so, many carriers of floor contracts chose to absorb some of the financial consequences of the current conditions (i.e. retained a significant part of the resulting increase in hedging costs), as they did not decrease the rates to the full extent and saw their estimated surplus decline by over a percentage point on average. They were still able to pay their expenses and make a steady profit, and increased their earnings through investing the RILA premiums into long-term and fixed-income assets that earn them a credit risk premium. In practice, RILAs have so far generated a relatively consistent surplus for the insurers, most of which comes almost entirely from the credit spread earned from investing the RILA account balance in fixed-income securities.

Their risk profile suggests that long-run target profitability should lie between FIA profitability and VA profitability, while their financial risks are significantly lower due to the similar profile of guarantees available in the financial market. In terms of asset and liability management, the following can be said.

- The bond component is hedged by investing in fixed-income securities, which generate yield that accrues to the insurer (who may take some credit risk, interest rate risk and liquidity risk).
- Meanwhile, the insurer hedges interest crediting by purchasing a call option (or a call spread option in case of a cap) and selling an OTM put option (for the buffer version) or buying a put spread (for the buffer version) simultaneously, on an exchange-traded or over-the-counter (OTC) basis.
- There is an ongoing need for highly rated, short-duration fixed-income instruments, to be set aside as collateral to back derivatives traded to fund index-linked crediting.

From the insurer's risk appetite perspective, these hybrids can also be designed in such a way as to balance the risk profile of existing VA blocks, while reducing transaction costs.

- RILA asset liability management involves the sale of put options while variable annuity risk management involves the purchase of put options.
- Insurers with traditional VA books could therefore save on derivatives transactions costs via the internal offset of such supply and demand for put options.

RILAs offer providers other potential benefits, which could help explain why many are willing to sell the product with such generous guarantees.

- First, it may not be necessary or optimal to fully hedge the RILA guarantees with financial options, because they can provide a partial natural hedge to TVAs
  - If the financial market performs well during a given period, insurers tend to profit from their TVA offerings – because they collect more fees while the value of the embedded guarantees declines – but they have to credit a return to their RILA accounts.
  - Conversely, if the market generates a negative return, then the RILA accounts (i.e. the insurer's liabilities) decline in value, while the insurer's assets (which are invested in fixed-income securities) probably will not. Under this financial scenario, the profit from the RILA contracts can help offset the increase in the insurer's TVA liabilities. Therefore, the presence of RILA products can reduce the hedging costs and the capital requirements for the insurers' TVA offerings – both of which are often substantial due to the long-term nature of the TVA guarantees. This might also explain why insurers are offering RILAs with longer-term guarantees (three to six years), even though they cannot easily hedge their resulting liabilities, since the financial markets do not trade options with those maturities.
- Second, even if they fully hedge their RILA liabilities, insurers may benefit from pooling them with their in-force TVA policies due to the diversification between the fixed-income risk of the former and the equity risk of the latter. This could potentially reduce the firm's reserve and/or capital requirements.
- Third, and related to this, providers may try to make RILAs particularly attractive in order to entice current TVA policyholders to exchange their contract for a RILA product or to attract potential investors to RILAs instead of TVAs. Since the upcoming changes to GAAP accounting principles are likely to have a major negative impact on insurers who have a lot of long-term risk exposure from TVAs on their books, the short-term nature of RILAs may make them (at least temporarily) preferable for insurers, even if they may be less profitable on their own. In addition, increasing the assets under management from RILA contracts would produce a better balance with their TVA assets, so that they can take greater advantage of the natural hedge (or the diversification benefits) between the two product types, as described above.

More importantly, as long as the embedded guarantees can be approximated by combined financial options payouts (sold and managed by investment banks), the life insurer can near-perfectly hedge the embedded equity risk by trading the corresponding options in the financial markets. This way, the insurer is able to transfer away his entire exposure to equity risk from this guarantee at a relatively low cost, and so does not experience any more financial risks nor capital requirements for financial risks (except for the counterparty risks related to the financial options transactions with the investment bank). Thus RILAs are capital-light savings products.

This is particularly true given that the guarantees are written on only a few popular market indices (most commonly the S&P 500, the Russell 2000 and the MSCI EAFE) that have very liquid option markets. In contrast, the long-term guarantees embedded in TVA policies have to be hedged "manually" using instruments such as futures contracts and exchange-traded funds (since the underlying mutual funds cannot be shorted), which exposes insurers to substantial levels of hedging errors and basis risk, while index annuities often include multi-term guarantees that resemble exotic options and that cannot be hedged with commonly traded derivatives. Therefore, RILAs are perfectly suited for insurers who want to offer their customers equity-linked savings opportunities without themselves being exposed to excessive equity risk.

RILAs offer a very cheap alternative, as most annual fees lie between 0 and 1.25%, with an average annual cost of only around 0.2% of the RILA investment amount, which means that investors can expect to receive 99.8% of their RILA investment back.

As RILA products do not include any fees or charges to the policyholder, providers benefit from a favourable imbalance between the guarantee's downside protection and its upside cap, the ability to invest the RILA funds into, say, corporate bonds and earn a credit risk premium, and in some cases from using a price-based index that does not account for dividend payments.

The missed dividends act effectively as an implicit fee paid by the policyholder. This is comparable in magnitude to the charges of typical TVA products – which account for expenses, administrative costs and the mutual fund fees – except that it is not stated explicitly in the RILA contract. However, RILA contracts are usually priced much more favourably for investors than TVAs, despite this implicit "fee". And, of course, insurers can just as easily offer a total return index instead of a price index and reflect the difference in a less generous cap rate.

Finally, a return-of-premium death benefit guarantee (GMDB) is embedded in most RILA products (without any explicit charge). Some RILA carriers have also begun to offer GLWB riders (for a separate fee) attached to the investor's RILA account value.

Naturally, there remain withdrawals and redemption behaviour risks.

RILAs and TVAs have much in common, including their regulatory framework, their use as equity-based retirement savings vehicles with downside guarantees, and (to a large degree) their target clientele. RILAs offer investors an upside potential that is comparable to TVA policies, while their downside protection limits the likelihood of a financial loss. However, there are crucial differences between the two product types.

- For instance, TVAs entail a direct investment of the premium(s) in one (or several) of many available mutual funds, while for RILAs the investor can choose between only a few mainstream market indices, and the premium is not directly invested in the index.
- Most RILA do not distribute dividends, in contrast to traditional VAs.
- RILA fees (0 to 1.25% p.a.) are far lower than TVAs (~3% p.a.).
- Moreover, the financial guarantees embedded in TVA policies are long term and are applied only when the policy terminates – either at maturity or due to the policyholder's death – or if the account value reaches zero due to guaranteed withdrawals. This makes the TVA guarantees difficult to value and hedge due to their very long-term nature (possibly 25 years and more), strong dependence on policyholder behaviour and exposure to basis risk.

  In contrast, RILA guarantees are short term and are tied to the performance of a major stock index, while the liabilities can be decomposed into short-term European options on popular market indices, so that insurers can hedge the embedded liabilities near-perfectly through the financial markets. Thus RILAs are much easier for insurers to manage.
- As a result, RILAs can provide carriers with a highly predictable and stable profit, permitting the product to be offered at a low cost. The absence of fees in most RILA products – and the overall pricing, which is quite favourable to investors – should make the contracts more marketable compared to TVAs, which many financial advisers consider overpriced.
- Furthermore, RILA guarantees appear to be much less dependent on policyholder behaviour than their TVA counterparts. For instance, TVA policyholders often have strong incentives to surrender or exchange their policy when the embedded guarantee moves "out of the money". RILAs provide investors with the

same flexibility in accessing their money when needed, but since the guarantee is always "at the money" (at least for single-term guarantees) there is little incentive for investors to anti-select against the insurer.

The incentive structure for agents and brokers – who receive a new commission when they can persuade a policyholder to such an exchange – may still lead to significant lapse rates, but probably lower than for TVAs. In addition, RILA carriers can further reduce lapse incentives by offering (slightly) more generous renewal rates compared to the cap rates for new policies. In fact, my review of current market offerings suggests that many carriers are already doing this, even when contracts are still subject to surrender charges.

- A particular motivation for TVA providers to transition to RILAs is the upcoming regulatory change in US GAAP, known as ASU 2018-12 or "targeted improvements to the accounting for long-duration contracts" (LDTI), which will require life insurers to reflect their liabilities from TVA guarantees under fair (i.e. risk-neutral) market values on their income statements. This will generally lead to an upward correction of the TVA liabilities and diminish the profit projections of these products. RILAs, in contrast, will be largely unaffected by this regulatory change, because insurers typically hedge the embedded equity risk.
- Additionally, RILAs are formally categorized and regulated as "variable annuities" and are pooled with TVAs for accounting and regulatory purposes. Both product types can, therefore, potentially reduce the reserves and risk-based capital requirements of their variable annuity portfolios. This is because RILAs can act as a natural hedge for the TVA guarantees and because the premiums from RILA policies are generally held in fixed-income securities, which helps diversify the providers' exposure to equity risk from their TVA offerings. Therefore, in the years to come life insurers may be motivated to develop RILA products and to promote them more aggressively, and possibly in favour of TVAs.
- Finally, RILAs also bear resemblance to FIAs. In particular, one can view an FIA as a special case of a RILA with a floor of 0%. That is, the insurer provides full protection on the index return in exchange for a low cap rate of currently between 2 and 4%. In other words, relative to FIAs, RILAs provide considerably more exposure to the underlying market index with respect to both upside and downside risk.

Regarding simulations of payouts, they reveal that over the long term the payout profiles of RILA contracts are comparable to TVA policies with long-term guarantees. While RILAs allow for temporary losses on the investment, this risk is largely diversified away through time, leaving only a small chance (around 5%) of a long-term financial loss. On the other hand, in terms of the potential to take advantage of strong equity markets, RILAs can easily compete with standard TVA products. Thus, for individuals who have a long-term investment horizon and are looking for equity exposure with some downside protection and preferential tax treatments, RILAs may well be the product of choice.

In addition, some weakness lies in the lack of liquidity for financial options beyond six years, which structurally limits the maturity of the equity upside embedded within the RILA.

Overall, such RILAs further bridge the gap between customers' and insurers' interests and risk appetites, but they still suffer from capped performance and limited downside protection, further penalized by a limited maturity.

2.2.1.6 The Hybrid General Account with Embedded Structured Guarantees

2.2.1.6.1 Trailblazing a Path beyond the traditional General Account Insurance Product, Leveraging on Customers' Experiences and Digital Technologies

Persistently low interest rates have created a dilemma for those planning for retirement.

- The most famous traditional insurance product, called "general account", characterized by full guaranteed capital and 80 to 90% low-risk debt investments (broadly half govie bonds, half corporate bonds), has reached strong limits of sustainability, as illustrated by unattractive yields for the customer and the expensive cost of capital for the insurance company. Several insurance companies CEOs have recently announced the expected death of this major insurance product. Even so, alternative equity unit-linked products remain inherently risky, given the multiple unexpected volatility spikes and drawdowns since 2015.

- General account products are still today based on a top-down approach, not caring enough about customers' risk appetite, product usage and time horizons.

As a result trailblazing a path beyond the general account relies on a multiple-pillar integrated strategy, meeting the risk appetites of shareholders (sustainable growth and higher profitability, lower cost of capital, controlled risks) and customers (higher yields combined with some guarantees, tailored to emerging behaviour changes) alike.

1. Be "customer-centric" in designing the liabilities, from "profitability/risk/cost" profiles tailored to specific customers' segments, while ensuring recurrent revenues to the insurance company for a resilient/sustainable business growth (e.g. mitigating lapse and withdrawal behaviour risks).

   - Such "capital light" hybrid insurance policy designs meet diverse and evolving customer yields appetites and protection needs, translating into some dynamic usage of life insurance product features over their full duration (full or partial guaranteed capital, punctual lump sums to support child education payments, personal training, unemployment periods), at reasonable costs in line with the persistent low interest rate environment (from a 1% to a 2.5% all-inclusive annual fee).
   - This initiative takes advantage of customers' experience through building client profile segmentation (current and expected), extracted from integrated internal and external data (market environment, underwriting, the in-force book, customer support, claims management, social networks) and the use of digital technologies ("data virtualization", chatbots, artificial intelligence).
   - In particular, the AI digital technology provides an "efficient behaviour" approach, which is a prudent forward-looking approach, based on individual policyholder's optimization of net profit over the full life of the contract duration, depending on key drivers such as the moneyness of the guarantee or the level of interest rates. An "efficient" modelling framework is a useful benchmark that can be used in pricing, once key drivers are also considered, to get realistic prices consistent with practice: market conditions, policy durations, past decisions (potential partial withdrawals), the tax regime. Such a modelling framework can also be used in product designs to stir policyholder behaviour while also fitting client appetite. For instance, the design of the fees opens a way to improve both competitiveness and the clients' risk appetite fit, ensuring sustainable growth.

2. Design a cheap asset investment mix, a source of sustainable returns and tailored to specific customer segment risk appetites, while mitigating ALM risks and meeting regulatory constraints for the insurance company.

   - Such a design initiative offers more resilience and generates significantly higher returns leverages on (i) a high fundamental bottom-up selection of quality stocks selection that is more resilient to crashes (healthy and stable profitability, strong free cash flows, low debt and shareholder-friendly practices, above-average dividend payout, low net equity issuance) and technical (with a focus on skewness and kurtosis); (ii) an asset allocation using capital requirements and transaction-cost-optimized portfolio rebalancing, forward-looking volatility mitigating mechanisms, and some derivatives for performance protection to mitigate drawdowns or boost returns within sideways markets and to manage "gap risks"; (iii) long-term partnerships with asset management firms and investment banks.

- Such an initiative can leverage on digital technologies to merge and visualize data (data virtualization and business interruption), forecasting returns/risk and customizing the utility function of customers' (artificial intelligence), transaction cost modelling (AI technologies using "optimal control" and "feedback loops" modelling for devising the rebalancing strategy depending on the risk tolerance relative to a target allocation and the level of transaction costs).
- In particular, "big data" and artificial intelligence provide more resilient managed-volatility funds, which do not overreact to and exacerbate non-fundamental sell-offs but consistently, only those with sound economic concerns (e.g. by introducing macroeconomic and supply and demand indicators into the volatility control mechanism).

3. Reduce the "time-to-market" to much lower costs than today, through the use of the robotic process automation (RPA) digital technology, providing automated processes and reporting.
4. Lower the cost of distribution channels via better customer targeting and product positioning, while offering high standards of customized advisory on product selection and usage, by developing online digital platforms (including educational tools) that dynamically integrate customers' savings behaviours (via "machine learning" technology) and help "cross-selling" (via chatbots).
5. Improve agility and governance by leveraging on internal dashboards with real-time 360° views based on digital technologies (data virtualization and visualization) for the regular monitoring of results internally by the management board to help define long-term goals and manage critical situations (dashboards integrating business, actuarial and risk management views; sales volumes; profitability; solvency ratios; "capital intensity"; "wealth consumption").

2.2.1.6.2 The Hybrid Structured General Account The general account experienced a golden age for both insurers and customers during the 1990s by offering the triptych of "high returns, capital guarantees and free access to money at all times". Unfortunately, the persistently low interest rate environment, combined with more stringent regulatory frameworks, since the 2008 crisis has made this product unsustainable for insurers and customers alike, offering low value for money for the customer, while being capital-intensive for the insurer.

This "hybrid general account" product trailblazes a way to rebuild this golden age triptych in a sustainable way by customizing a combination of the general account (i.e. the availability and security of money at every moment), though only to a minor extent (but consistent with the actually empirically observed use of savings by customers – e.g. 15 to 40% of premiums), with enhanced-yield capital guarantees for the major part (60 to 85% of premiums, and this time only at maturity).

As a result, such a "hybrid general account" retains the best of the alternatives developed previously.

- It offers free access at any time to a significant portion (15 to 40%) of the savings (as was the case in the traditional general account).
- There is unlimited downside protection (in contrast to products structured for yield enhancement), with a capital guarantee of at least 100% (raw of fees) of the premium at maturity, with potentially richer guarantees (minimum 2 to 5% roll-up rates, thus offering a minimum 120 to 150% capital guarantee at maturity), and this despite the persistently low interest rates. Let us highlight that such downside protection cannot be deactivated if some lower market level is breached (as is usually observed in structured products and euro medium-term notes such as autocallable or reverse convertibles).
- There is an unlimited boosted upside potential (in contrast to eurogrowth products, RILAs and EIAs) offered by the enhanced-yield capital guarantees, such as diverse long-dated (ten-year) equity options,

offered by investment banks. These offered a 4 to 9% average annual net return (notably through the 2 to 3% minimum return + dividends) from 2005 to 2021 (and with a rather narrow corridor) to the customer through all market conditions so far experienced (including persistent low interest rates and market crashes). Naturally, the yield can be further boosted by tailoring the equity option design to the customer's risk appetite (e.g. using light exotic options or flooring the downside protection so that it starts at –50%).

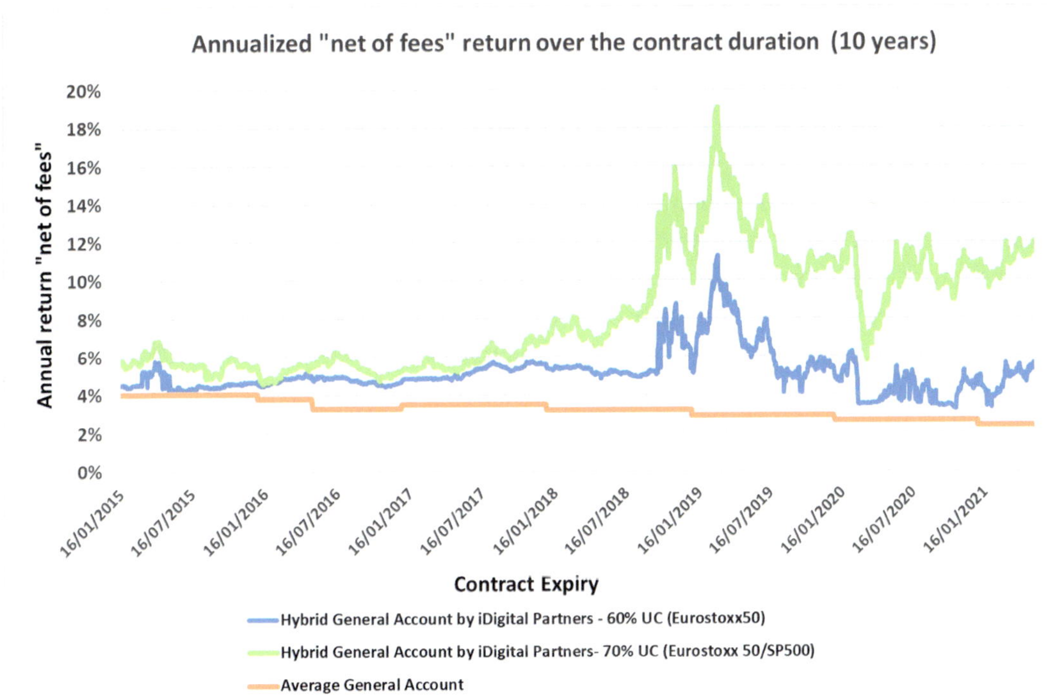

FIGURE 2.16 Annual returns net of fees after a ten-year holding period, 16 January 2015–16 April 2021

- The fees are low (total fees lower than 2%) and, in addition, resilient to market turmoil (as the fees decrease when volatility increases), which are perfectly suited to the uncertain market conditions as experienced today.
- There are very low capital requirements.
  - The capital requirement of the general account has been reduced by 60 to 90%, in line with the decreased investment.
  - There is almost zero capital requirement for the financial risks of the equity part, since the options are managed by the investment bank; only the bank's default risk remains.
  - The customer's behaviour risk is also limited, given the resilient attractive yield of the product (most of the surrender risk is here not linked to financial market conditions, and so rather stable and predictable, which can be confirmed by the "agent-based efficient modelling" framework, as detailed in Chapter 3).

Overall, such a product seems to align both customers' and insurers' interests and risk appetites, as it delivers, on the one hand, resilient high returns guaranteed at low cost to the customer, while offering flexible designs tailored to the customer's risk appetite, coupled with a capital-light and low-risk structure attractive to the insurer.

2.2.2 Reducing Financial Risks and Capital Requirements through Banking, Asset Management and Digital Partnerships

2.2.2.1 Leveraging on Digital Technologies through Partnerships with Assurtechs: Illustration of the Data Governance within IFRS 17

Technology will probably play a crucial role in driving these changes and fostering efficiency across the advisory and reporting process. Thus, technology could enable growth.

More specifically, insurers can leverage on digital technologies to offer an end-to-end solution for enhanced reporting and decision-making within International Financial Reporting Standards 17 (IFRS 17).

Most insurers' IFRS 17 implementation is already in flight and the accounting and reporting concepts are already quite well understood. It is the perfect time to get the data governance stream up and running – all the more so as IFRS 17 involves high requirements in consolidating, analysing and disclosing granular levels of data across a wide scope, while current systems often work in silos with largely file-based exchange.

To design a strategy for data governance, a roadmap is key to ensuring that it moves beyond the attainment of compliance towards more strategic initiatives, such as enhanced risk management, better-informed underwriting, improved customer experience, stronger data security and innovative new product design.

Such a programme relies on a technology platform that must be scalable, agile, customer-centric and cost-efficient and interoperable with this existing estate. Modern data governance approaches are all about collaboration between business users, the business and IT, and consumers and owners of data.

The end-to-end process orchestration is achieved by leveraging on digital innovations such as RPA and data virtualization.

By joining forces, such an ecosystem offers an end-to-end solution to the data governance issue, with a rich blend of domain knowledge, tools and execution capabilities spanning finance, actuarial and operations.

2.2.2.1.1 Why Is Data Governance Needed for IFRS 17? IFRS 17 looks to bring new levels of transparency to insurance contracts, to ensure "high quality, understandable, enforceable and globally accepted financial reporting standards based upon clearly articulated principles" to improve the comparability of insurers' financial statements, regardless of sector, geography or products.

In this respect, IFRS 17 requires:

- current estimates at each reporting date of each contract, then pooled by similar risks profiles, reflecting up-to-date information about cash flows and the timing and risk associated;
- information on sources of profit and losses through underwriting activity and investing premiums from customers, and the extent and nature of risks arising from the contracts;
- the recognition of profit emergence patterns over the duration of the insurance contract, and information about the expected future profits; and
- accounting policies to be standardized and auditable.

Such higher requirements in consolidating, analysing and disclosing granular levels of data, across a large scope, and the need to maintain a full prior history will imply four data challenges:

- data quality improvements involved within the numerous new computations;
- massive data management to reduce both run-time and analytical lags in a cost-effective manner;

- managing the implementation process, including workflow automation, communications and stakeholder management; and
- aligning and automating numerous financial statement reporting systems and documentations across the entire actuarial modelling and financial reporting chain, at both the group and business unit/entity levels.

2.2.2.1.2 How Can Data Governance for IFRS 17 Be Achieved with Data Virtualization?

2.2.2.1.2.1 From Compliance to Strategic Governance A roadmap is key to ensuring that there is movement beyond the attainment of compliance towards more strategic initiatives such as enhanced risk management, better-informed underwriting, an improved customer experience, stronger data security and innovative new product design.

- Solvency II also requires insurers to invest in data quality, control and management; however, there are differences in the detail. A key challenge will be to ensure that these different types of data are available and that systems have the flexibility to accommodate such differences.
- As a result, a specific data management and governance policy becomes a central foundation for successful and consistent IFRS 17 implementation, all the more so given the fragmented nature of existing actuarial systems.

2.2.2.1.2.2 The Data Governance Objectives In order to manage the four above challenges, the proposed data governance scheme is expected to:

- provide consistent uniform guidelines on actuarial, finance and risk data management, leveraging on Denodo's technology;
- develop data quality checks and correction mechanisms, in order to avoid data redundancies, discrepancies and inconsistencies, leveraging on data virtualization's technology; and
- drive and monitor an end-to-end process of orchestration.

2.2.2.1.2.3 An End-to-End Process of Orchestration Leveraging on Digital Technologies
This will consist of the following:

- establishing harmonized data architecture, leveraging on data virtualization's technology;
- defining procedures that align data management across departments (including documentation), leveraging on Denodo's technology;
- identifying potential opportunities for the automation of repetitive/time-consuming tasks, leveraging on RPA technology; and
- standardizing financial/management reporting templates, leveraging on RPA technology.

2.2.2.2 Combining the Skills of Insurance and Banking/Asset Management within the Context of Drastic Changes in Financial Markets, Regulations and Policyholders' Behaviour Patterns

The two industries each come with distinct challenges. Therefore, insurers must invest in specific capabilities and assets aligned with their strategic and operational decisions.

2.2.2.2.1 Strengthening Long-Term Partnerships with Clients to Fit Their Risk Appetites and Constraints Since the 2008 financial crisis institutional market participants have become much more heterogeneous: some have higher risk tolerance and a longer time horizon and are willing to take on an illiquidity premium (e.g. by investing in real estate, infrastructure assets); others are intolerant of any large drawdowns and are looking for some downside protection at a reasonable cost. The rapidly changing regulations also had a drastic impact on the asset selection or ALM strategies (e.g. the variety of treatments across asset classes under Solvency II). Moreover, the drastic and persistent decrease in interest rates significantly increased the differentiation across insurance companies in the management of their liability risk profiles and wealth consumption over time.

As a result, institutional investing needs to know clients' risk appetites and liability structures to customize adequate asset selection, asset allocation and risk mitigations strategies, whether they seek income or yields or some downside protection at a reasonable cost. In this respect, customers' investment decisions are strongly influenced not only by their profitability (earnings and value)/risk/capital/liquidity targets under various accounting frameworks (IFRS, Solvency II) but also by the evolving risk profile of their liability portfolios, which depends on their design combined with policyholders' behaviours and longevity.

In fact, sudden and significant changes in policyholders' lapse and use (partial withdrawals, election rate) of variable annuities and general accounts since the 2008 crisis have had drastic negative impacts on earnings for most insurers, which required proper identification to activate mitigating actions through customized asset selection/allocation and ALM strategies. Knowing the specific risk appetite and time horizon of insurers was also key to properly fine-tune the design of the managed-volatility funds and CPPI funds, and thus to mitigate at least partially the market spikes during February and October–December 2018. Knowing the structure of the liability cash flows and their embedded policyholder optionalities has also been necessary to properly select the hedge assets portfolio and set the portfolio rebalancing triggers based on ALM and market volatility criteria.

Different metrics may be used in this respect: liability capital intensity and capital requirements, liability financial wealth consumption, liquidity stress tests, the duration gap, the degree of rationality in policyholders' behaviours and stress tests for liability assumptions. At the same time, regular teach-ins would raise clients' awareness about new investment/protection assets opportunities.

Ultimately, comprehensive "products with advice and strategies" solutions that fit clients' needs, time horizon and constraints (e.g. capital requirements, liquidity, policyholders' behaviours) are required in order to build long-term partnerships across the rapidly changing regulatory, investment and policyholder behaviour environments.

Such initiatives will leverage on a customer-dedicated digital platform to strengthen long-term partnerships through investment teach-ins while offering asset and liability analytics, which will enable insurers not only to monitor their customers' actual portfolios but also to customize their offers to their clients' actual needs and risk appetites.

2.2.2.2.2 Enhancing Research Monetization through Customer-Centric Solutions Institutional investors refer to a collection of demanding investors with specific portfolio structures and durations or risk types (strongly nonlinear market risks, liquidity, policyholders' behaviours, longevity) who are likely to use "sophisticated" portfolio management metrics/techniques for their asset selection and asset allocation (factor analyses, asset cross-correlation and co-integration).

As the fee for institutional investors is based on the value of the client's account, risk management/ALM techniques and derivatives can be used to protect the generated revenues, all the more so given the rapidly changing gap between assets and liabilities within the persistently low interest rate environment and the frequent market corrections occurring at multiple times throughout the year. There again, the expertise and experience of the liability product designs, including their evolving risk profiles and wealth consumption patterns, combined with the policyholders' changing behaviour, are key to mitigating the risks in a robust way, using smoothed cash flow matching or duration methods.

In addition, tailored nonlinear strategies using combinations of cash and derivatives, leveraging on both fundamental research and technical indicators, enable the extraction of alpha from event-driven opportunities across a range of indices/sectors/corporates: macroeconomic trends or events, relative value trades between asset classes or peers, M&A activity, changes in sector regulations, credit ratings downgrades, sectors or corporates earnings/dividends surprises, share buybacks, changes in corporate management.

Such initiative will leverage on a customer-dedicated digital platform to monetize "nuanced" market views through tailored cash and derivatives portfolio (ALM or investment) strategies, capitalizing on fundamental research.

2.2.2.2.3 Growing Portfolio Managers' Market Timing Capacity and Agility in the Current Uncertain Market Environment while Retaining Their Long-Term Fundamental Investment View

Governments and central banks are running out of fiscal and monetary tools to stimulate growth, which means that business cycles are likely to be shorter and more severe. At the same time, clients have shorter time horizons and increased aversion to risk, and are less likely to accept short-term underperformance in the hope that the manager will perform well in the long run.

In fact, markets are driven by swings in investor sentiment as well as by changes in fundamentals, particularly in the short to medium term. The market volatility pattern has drastically changed over the past ten or so years, influenced by the increasing synchronicity between markets (equity rates, but also with derivatives markets) and between the asset management industry and investment banking trading activity. In addition, the "reflexivity" of the markets – the impact that market moves have on investor portfolio preferences, which in turn impact the markets – will continue to amplify market volatility for the foreseeable future.

As a result, market timing becomes key, without losing sight of long-term investment goals and valuation perspectives. Beyond improving their investment outcomes through active asset allocation and broadening their portfolios with additional asset classes, more sophisticated tactical and adaptive asset allocation (rolling optimization, momentum rotation and target volatility, tail risk analyses) and derivatives momentum/investor sentiment prospective indicators will improve market timing ability and agility. Finally, as institutional investors trade in large blocks, which can significantly affect the price, liquidity management becomes key to avoiding growing overreactions in asset allocation subsequent to non-fundamental sell-offs (e.g. April 2006, August 2011, August 2015, June 2016), as experienced by the poor performance of most wealth management flexible/managed-volatility funds, due to their tendency to lag behind the trend during volatility spikes, thus missing the upside during market recovery periods.

Such initiatives will leverage on a customer-dedicated digital platform to grow the market timing capacity and agility with "stractical" (strategic/tactical) analyses and tools, including some behavioural finance.

2.2.2.3 Towards a New Type of Life Insurer?

A focus on underwriting and manufacturing requires product innovation and actuarial chops. Alternatively, a distribution-intensive business requires knowing how to use technology to enhance advisers' productivity in sales and assets under management. Insurance companies can choose one of the five types that we see standing out in the market, each of which entails clear choices of where to play and how to win.

- Product-focused manufacturers emphasize product design and capital management, such as Jackson National, which focuses on variable annuities in the United States, offering varying investment options, benefit options and fee structures to address different investor needs, such as annuities targeted to ultra-high-net-worth clients. Jackson has a focused product strategy and a clear sense of how to win: adviser-centred distribution with product leadership, backed by the industry's largest and most productive wholesaling force.

- Risk-focused investment managers tend to succeed through M&A activity, joint ventures and other forms of partnership. A unique partnership with Apollo, for example, allows Athene to invest in more specialized, higher-margin securitized assets and, eventually, to move to direct origination.
- Distribution-focused providers' archetypes come in two forms.
  - The first form is a career agency model focused on protection. For instance, MassMutual has customer relationships developed through a captive force of over 8,000 advisers, allowing it to excel in cross-selling, and, with its alternative distribution method through its affiliate Haven Life, it is an online direct-to-consumer provider.
  - The second form focuses on broad financial planning, as exemplified by Ameriprise. The company started as a wealth manager and entered insurance through M&A. It distributes in the United States through a nationwide network of high-productivity captive and franchisee advisers. In recent years Ameriprise has explicitly shifted its focus away from insurance and annuities, growing through a series of well-timed acquisitions in an extended bull market.
- Data and analytics platforms, supplemented by product design, M&A and partnership skills. Discovery, based in South Africa, is perhaps the most successful example, through its digitally enabled platform called Vitality, a wellness programme that rewards policyholders with discounts and perks based on healthy living. Its execution hinges partly on broad cross-selling, whereby healthy behaviours bring rewards across the product categories. Investors have benefited with an average 16% total shareholder return over the past five years.
- Platform retirement players: here a company has to master customer service and relationship management, data and analytics, and sustainable scale. Voya, for instance, divested its annuities business and exited life insurance in order to emphasize retirement products for companies of all sizes. Now the company is using its employee benefits business and focus on financial wellness to expand the platform of services offered in the workplace market. The strategy has reduced Voya's risk profile and increased free cash flow.

Each archetype poses distinct challenges. For example,

- For example, some insurers who rely on agent networks become detached from customers' priorities. Their cultures traditionally emphasized actuarial and distribution capabilities, and the companies invested to keep agents happy. Developing close customer relationships will require new muscles and habits.
- Archetypes that depend on strong product design, by contrast, will have to step up their continual innovation by adopting recent technologies such as the internet of things, artificial intelligence to optimize risk profiling and price selection, and chatbots for client interactions. John Hancock has shifted to interactive policies that track fitness and health data through wearable devices and smartphones, in partnership with Discovery's Vitality platform. By creating a programme based on preventive health and accrued benefits and rewards, Hancock increases engagement and contact with customers.

Regardless of archetype, insurers must contend with the digital disruption sweeping through their industry, as the number of insurtechs in life insurance has grown rapidly in the past five years. Incumbents have woken up to the need to simplify and digitalize their businesses, to benefit both consumers and advisers.

CHAPTER 3

Measuring Risk Appetite for Sustainable Savings and Retirement

As part of the pillar 1 requirements of prudential regulatory frameworks, insurers have to evaluate their liabilities to policyholders at their current execution value, which is the best valuation, consistent with market observations, of the current value of future cash flows required to settle insurer commitments to policyholders, taking into account every contractual option and financial guarantee included in the contract. These include the following.

- Surrender: during the term of the contract the policyholder has rights on the mathematical provisions of the contract, as she may claim her right on it during the contract and thus terminate it before the initial settlement date. This choice may be motivated by the difference between the rate paid by the insurer and that offered by competing insurers.
- The minimum guaranteed rate, fixed at the subscription that must be respected by the insurer during the whole life of the contract, regardless of the evolution of its portfolio.
- A profit-sharing mechanism: life insurers have to share with their policyholders a portion of their technical and financial benefits, depending on the evolution of the assets held by the insurer. The profit-sharing mechanism can be contractual or discretionary.

To estimate the price of the uncertainties inherent in the multiple sources of optionality of a life insurance product, a deterministic model is not adequate, especially in view of the nonlinear structure of the financial guarantees. Therefore, life insurers resort to stochastic models to make a sufficiently robust evaluation that takes into account the peculiar risks implied by the volatility over time of the market and the underlying assets.

The complexity of the estimation of these inherent options, especially for a life insurance contract, requires the use of a simulation approach, based on the Monte Carlo method. The use of such an economic scenario generator is significant, and various across three of the four dimensions of the risk appetite framework.

- In terms of value, it allows the insurer to evaluate the economic value of the future cash flows relative to his obligations, in order to perform a best estimate valuation consistent with the market.

DOI: 10.1201/9781003218371-4

- In terms of capital, it also enables the insurer to compute the SCR and measure the impact of an instantaneous change in the economic and financial environment, as required by European regulation (stress test scenarios used in the SCR evaluation).
- In terms of earnings, the scenarios are also used as inputs for an ALM model that models the interactions between assets and liabilities, asset and liability management strategies and policyholder behaviour, which allows the insurer to better direct his asset and liability management in order to foresee market evolutions and their potential impact on the balance sheet and be able to set up hedging strategies against these rate variations.

As a result, in order to build sustainable savings and retirement, taking into account a few considerations on the risk-neutral versus real-world pricing framework, section 1 focuses on hybrid financial modelling (interest rates, then equity), including enhancing both the precision and computational time of simulations, before dealing with market crashes through comparisons between local, stochastic and jump models. Section 2 then focuses on customers' behaviours (surrender, annuitization, withdrawals) by introducing an "efficient forward-looking" modelling framework, which proves to be both prudent and consistent with empirics. Finally, section 3 illustrates the computational challenges of Monte Carlo simulations, and proposes some efficient solutions through the use of randomized quasi-Monte Carlo and GPU technology.

3.1 CONSIDERATIONS ON RISK-NEUTRAL VERSUS REAL-WORLD PRICING FRAMEWORKS

In the context of the valuation of liabilities in the insurance framework, two approaches are possible with an ESG: evaluations based on an ESG can be conducted either with the risk-neutral approach or the real-world one, provided that a deflator is used to ensure the consistency between the two projection methods. A deflator is a stochastic discounting function that includes a time component and measures the market value of a risk from a projection of the flows under the real-world probability.

3.1.1 The Risk-Neutral Approach

With this approach, we suppose that there is no arbitrage opportunity and that all economic agents are risk-neutral. As a result, all risk premiums are null, which avoids the need of determining their complex form, and valuations can be conducted by discounting future cash flows at the risk-free rate.

The neutrality assumption for economic agents is widely used for the pricing of derivative products in finance and of optionalities embedded within an insurance policy, because it simplifies the valuation computations. Indeed, under the risk-neutral probability, the expected returns of financial assets are equal to the risk-free rate and their discounted prices are martingales. It is important to note that the generated scenarios do not aim to reflect future evolutions expected by the insurer.

The use of the risk-neutral probability is also recommended by the first pillar of Solvency II, since it facilitates market-consistent evaluation of the insurer's liabilities, in particular contracts with financial options and guarantees. Models have to be calibrated to market prices quoted on the evaluation date, so that the generated scenarios respect the property of consistency with the observed market prices.

3.1.2 The Real-World Approach

With this approach, the aim is to replicate the historical behaviour of the market data. The simulations project economic variables under the real-world probability as closely as possible to the observable market environment. Models have to be calibrated to historical data representative of the economy over the projection horizon in order to project the actual behaviour of these variables as observed in the historical data.

Note that models can also be calibrated to anticipations of the future economic environment, so that the projected variables replicate these anticipations. This approach considers the real returns, and therefore involves a risk premium in asset returns – i.e. a return in excess of the risk-free rate – which greatly complicates asset valuation, since risk premiums are in practice very difficult to model.

A real-world ESG allows the generation of a valid real-world distribution for all the risk factors. The generated scenarios are then especially used for optimizing the strategic asset allocation, performing computations in the ORSA framework or computing the SCR.

✓ **Quick win**: let Nt be the numeraire of the real-world measure P. Let r be the risk-free rate and P* be the risk-neutral measure. Then, from the standard change of measure results

$$\frac{dP^v}{dP} = \frac{e^{rt}N_0}{N_t},$$

so that we get the general result for the numéraire of the real-world measure P:

$$\boxed{N_t = N_0 e^{rt} \frac{dP}{dP^*}}.$$

Now, in the particular one-dimensional Black–Scholes (GBM) case, we have for the stock price under P

$$\frac{dS_t}{S_t} = \alpha dt + \sigma dW_t,$$

and, again, standard results tell us that

$$\frac{dP^*}{dP} = e^{-\frac{\lambda^2}{2}t - \lambda dW_t},$$

where $\lambda = \frac{\alpha - r}{\sigma}$ is the Sharpe ratio, and so

$$N_t = N_0 e^{\left(r + \frac{\lambda^2}{2}\right)t + \lambda W_t}.$$

Now, if we differentiate that formula, we obtain for the dynamics of the real-world numéraire N

$$\frac{dN_t}{N_t} = \lambda dW_t + \left(r + \lambda^2\right)dt = \frac{\lambda}{\sigma}\frac{dS_t}{S_t} + r\left(1 - \frac{\lambda}{\sigma}\right)dt.$$

Thus we see that the numéraire N is the value of a portfolio that invests a proportion λ/σ of its value in stock and $1-\lambda/\sigma$ of its value in the risk-free asset.

Remark: if one assumes a capital asset pricing model (CAPM) (or, more generally, an arbitrage pricing model [APT] framework), the risk premium (or premiums, if many factors) should be defined with respect to the undiversifiable risk – i.e. the market portfolio (respectively, APT factors). The rationale here is that the idiosyncratic risk (a.k.a. diversifiable risk) of stocks can be diversified away, so there is no risk premium to remunerate against assuming this risk.

In this case, the real-world numéraire in this market is the so-called market numéraire Mt. It is the process starting at M0 = 1 and, with volatility λt, the risk premium

$$\frac{dM_t}{M_t} = \lambda_t dW_t^{\mathbb{P}}.$$

Under this numéraire, the expected return of a given asset is given by the risk-free rate and instantaneous covariance between the market numéraire and the asset,

$$\mu_t^S = r_t + \left\langle \frac{dM_t}{M_t}, \frac{dS_t}{S_t} \right\rangle_t$$

$$\mu_t^S - r_t = \frac{\sigma_t^{S,M}}{\sigma_t^{M,M}}\left(\mu_t^M - r_t\right),$$

and claims can be priced by taking the expectation under the real-world measure:

$$\text{Price}_t = \mathbb{E}^{\mathbb{P}}\left[\text{Payoff}_T \frac{M_t}{M_T} \mid \mathcal{F}_t\right].$$

✓ **Quick win**: the choice to use a real-world rather than a risk-neutral pricing framework depends on a few rationales.

- In Europe, the risk-neutral framework is the reference across the four dimensions: capital requirements (Solvency II), and so for the cost of capital; value (EEV, market-consistent embedded value [MCEV] ...); and liquidity, which has a significant impact on the sustainability of savings and retirement. However, regarding the earnings dimension, (IFRS) risk premiums are still taken into account, in favour of a real-world pricing framework.
- In the United States and Asia, risk premiums are usually taken into account across the four dimensions, thus favouring a real-world pricing framework. Only the hedging is aligned with a risk-neutral framework, in order to be in line with the way hedge assets are priced and managed. Nonetheless, let us mention that a risk-neutral hedging framework is still prudent as opposed to a real-world one (see Chapter 5).
- Regarding behaviour risks that usually significantly impact many savings and retirement products, such as annuities (e.g. ~35% of capital requirements: see later), the real-world framework is naturally more realistic, albeit less prudent than the risk-neutral framework.

For the sake of presentational clarity, the following developments are made under the risk-neutral measure.

3.1.3 The Different Steps in Constructing an Economic Scenario Generator

When constructing a risk-neutral ESG, regulation gives insurers a certain amount of freedom for developing a model that complies with regulatory requirements.

- First, select the most appropriate models to project the macroeconomic variables of the market. The insurer must question the characteristic and market phenomena (e.g. yield curve shape) that he wants to be reproduced by the models, since they do not have the same theoretical properties.
- The insurer must also put these choices into perspective with the current economic environment. For example, interest rate models that can generate negative rates were until recently criticized for this characteristic. However, in the current context of low interest rates, the use of these models may seem more attractive to insurers, and even be explicitly required by the regulator.
- The calibration of an ESG represents another very delicate step in the development process and will greatly influence the correct market-consistent valuation of the portfolio. The insurer must first select the market data to use to calibrate the ESG models, ensuring that they reflect in a certain manner the market situation at the evaluation date and that they are adapted to the considered models. The imperfection of stochastic models means that the latter cannot simultaneously reproduce all the observed market prices. Hence it is necessary to resort to an optimization problem, which nonetheless depends heavily on operational constraints, especially the available computing power. The results obtained can be very sensitive to the method retained (optimization algorithm) and to the market data used to calibrate the model parameters.
- Finally, it is necessary to check the reliability of the generated scenarios by verifying that the theoretical properties of the underlying models and the regulatory requirements for an ESG are respected. In the event of a risk-neutral ESG, two main theoretical properties in particular can be tested:
    - the market consistency criterion, the aim of market-consistent scenarios being to reproduce market prices and implied volatilities; and
    - the martingale criterion, as, if there is no arbitrage opportunity, the discounted prices of financial assets are martingales under the risk-neutral probability.

3.2 HYBRID FINANCIAL RISK MODELLING FOR SUSTAINABLE SAVINGS AND RETIREMENT

Savings and retirement product prices may be decomposed as the net difference between discounted financial cash flows as follows:

- inflows: the charges (fees paid by the policyholders); and
- outflows: the claims (withdrawals, the annuity, the surrender amounts).

As these savings and retirement products cash flows, spread over a medium- to long-term horizon, are discounted, the price and sensitivities to both interest rates (implying both expected future returns *and* discount rate) and market volatility (both equity and rates) are expected to be very significant.

Indeed, as mentioned in Chapter 1, the risk signature of savings and retirement products is usually dominated by financial risks (at least around 50%), specifically when the savings and retirement contracts include financial guarantees.

✓ **Quick win**: considering devising savings and retirement products without financial risks is just an illusion, given their structural long maturity, which supports a contribution by financial risks to a minimum of 50% in practice. Only the expertise and experience through the triptych "modelling, product design and hedging" is a way to ensure their sustainable development. In this respect, the insurer may naturally decide to share the risks with third parties, such as investment banks and asset management agencies.

Given that the assets of savings and retirement policies are mostly invested in equities and bonds, the corresponding interest rate and equity modelling is detailed below, giving rise to a hybrid modelling framework.

Moreover, financial drivers over the past 30 years have been increasingly characterized by lower interest rates combined with recurrent equity market crashes and volatility spikes (as illustrated by the structural instability of the "option implied volatility" – a proxy for future expected market volatility – across all time horizons). These have made savings and retirement prices both more expensive and volatile.

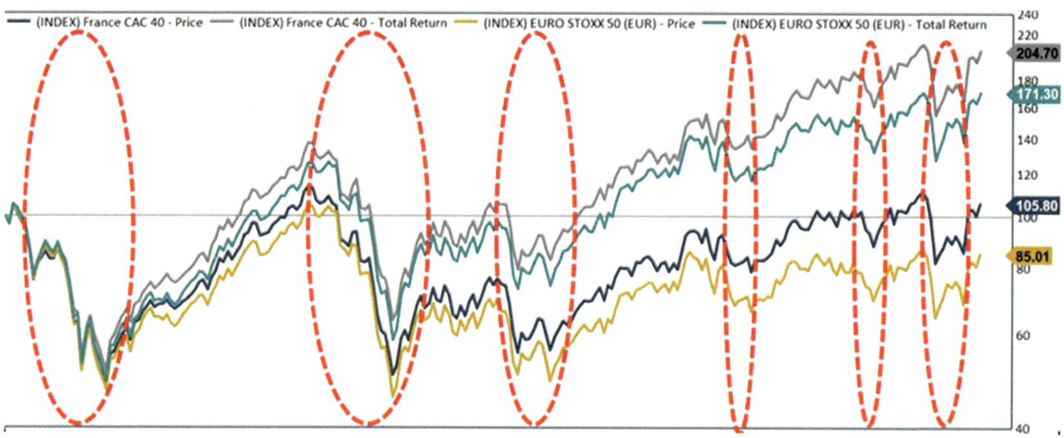

FIGURE 3.1 Recurrent equity market crashes

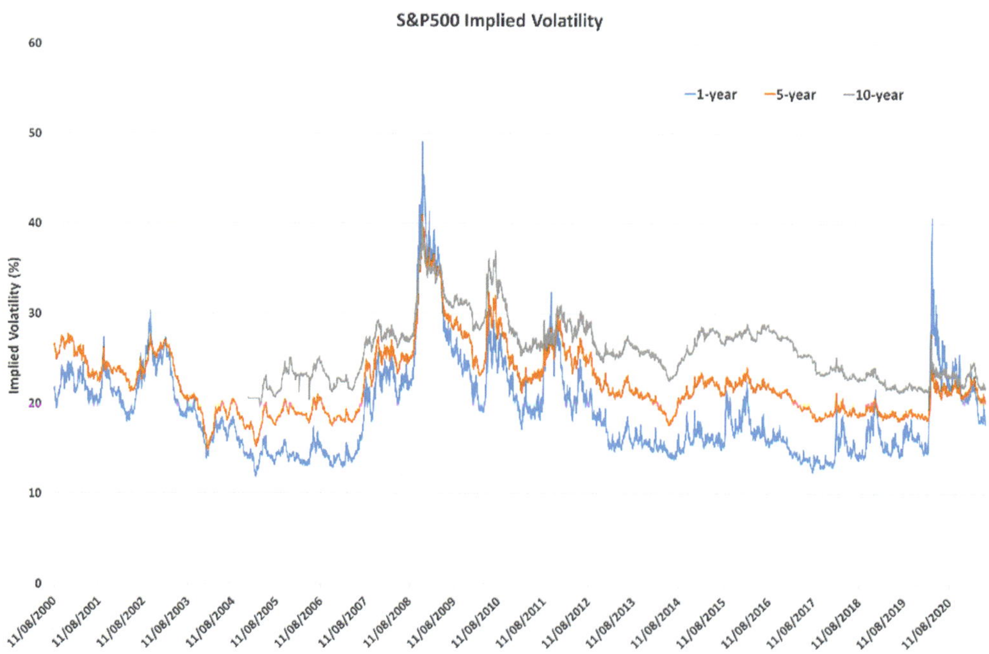

FIGURE 3.2 Recurrent market volatility crashes over the past 20 years

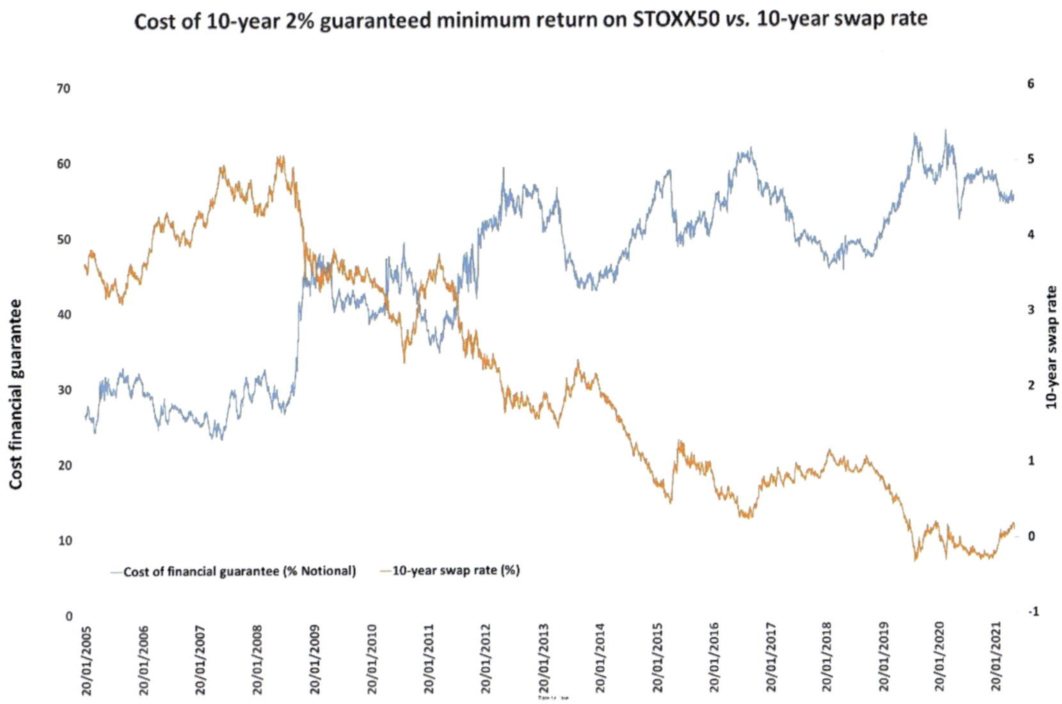

FIGURE 3.3 Recurrent market volatility crashes over the past 20 years: rising and increasingly unstable cost of financial guarantees

As a result, modelling these risks becomes a key challenge regarding their sustainability. Selecting the best potential modelling framework depends on three major criteria:

- the quality of fitting (low pricing errors in terms of calibration to market quotes);
- the robustness of parameters, which are expected to change in magnitude in line with the market environment (i.e. with the potential changes in the yield curve for interest rate models) while staying stable in the event of no change in the market environment; and
- the interpretability of parameters, which need to be intuitively linked to the empirical behaviour: a model must remain intuitive, simple to understand and justified by observations.

It is, naturally, illusory to think that a single model can account for all possible market situations. In addition, having simple templates is essential. Subtlety should be introduced into models only when necessary, because complicating models generally leads to two negative consequences:

- there is a longer computation run-time, due to higher discretization schemes to simulate the asset; and
- the intuition and understanding are usually harder to develop.

3.2.1 Hybrid Financial Modelling for Sustainable Savings and Retirement Liabilities

3.2.1.1 Interest Rate Dynamics Modelling for Sustainable Savings and Retirement

Interest rate variations can impact the financial result of an insurer through the rate guarantees to policyholders.

- A decrease in interest rates generally leads to an increase in the medium- and long-term financial liabilities, even if it may also lead to short-term unrealized gains beneficial to financial communication.

- Conversely, when interest rates increase, the rates paid by the insurer may not be as high as those of the market or competition, and so cause a wave of surrenders.

As a result, a sustainable modelling of interest rate dynamics allows the insurer to better direct his asset and liability management in order to foresee market evolutions and their potential impact on the balance sheet and be able to set up hedging strategies against these rate variations.

The interest rates are usually modelled in the following ways.

- In Heath–Jarrow–Morton (HJM) models, as an integral of the instantaneous short rate r (with its specific dynamics) over the maturity:

$$\int_s^t r_u du.$$

The short rate dynamics is either Gaussian (Hull–White), or log-Gaussian (Black–Karasinski) or square root Gaussian (Cox–Ingersoll–Ross: CIR), all three including mean reversion as empirically observed:

$$dr = \left[\theta_t - a r\right] dt + \sigma dz,$$

$$d\ln r_t = \left[\theta_t - a \ln r_t\right] dt + \sigma dz.$$

$$dr_t = \kappa\left(\theta - r_t\right) dt + \sigma\sqrt{r_t}\, dw_t.$$

Then a deterministic function of time φ is used, allowing the reproduction of the initial curve of the zero coupon and ensuring the required no-arbitrage condition.

- In the LIBOR market model (LMM or Brace–Gatarek–Musiela [BGM] model), where each holds its own dynamics and is then correlated through a correlation matrix.

In both cases, the parameters of the dynamics are then calibrated to interest rate product prices (bonds, options).

Regarding the economic scenario generators for savings and retirement liabilities, despite their shortcomings the HJM models (no-arbitrage short-term models) hold the following significant practical advantages over LIBOR market models.

- They are tractable, in particular for the long-dated and complex features of savings and retirement contracts. They are also easy to understand and implement.
- With their additional forward rate framework, they provide freedom in choosing the volatility term structure consistent with different maturities.
- However, LIBOR market models often imply high instantaneous volatilities (through highly calibrated volatility parameters) for the LIBOR forward rates, which make them unusable in practice with an ALM model. This caveat occurs less in HJM (Hull–White) models.
- The ability of some HJM models (e.g. the one associated with the Hull–White dynamics – so-called G2++) to generate negative rates, which makes them more suitable models in the current economic environment.

- Even so, due to a lack of observations of the underlying (instantaneous forward rate), they prove difficult to calibrate, and so require some specific algorithms and/or modelling add-ons (e.g. volatility term structure; see later).

As a result, HJM models are mostly used by practitioners regarding interest rates modelling embedded within savings and retirement pricing. In contrast, the market models are mostly used by practitioners for interest-rate-hedging assets (e.g. swaptions), as illustrated in Chapter 5.

In the next section, dedicated to liabilities, the HJM models are the focus of the interest rate modelling framework.

3.2.1.1.1 The Impact of Interest Rates Is Significant on Savings and Retirement The basic guarantees embedded within savings and retirement may be proxied with a vanilla put option, which is usually considered as only slightly impacted by interest rates, as illustrated in textbooks by the small negative rho ("short rho") hedging sensitivity. In fact, the low impact observed on the market stems from the fact that most traded vanilla puts are short- to medium-term options (three months to two years). However, as their maturities grow, their negative rho sensitivity increases significantly.

In contrast, the put premiums are positively impacted if the volatility level ("long vega") is positive, but as maturity grows its positive larger contribution is progressively dampened by the negative rho sensitivity. In other words, the longer the maturity, the higher the sensitivity of put premiums to rates compared to their sensitivity to volatility.

As a result of those two opposite effects, put premiums ultimately start to decrease beyond a certain maturity (around seven to ten years); overall, the net marginal impact of a decrease in interest rates level by 100 bps on a 20-year at-the-money put goes up to an additional 40% premium. Long-dated puts therefore present a key distinct feature from short-dated puts.

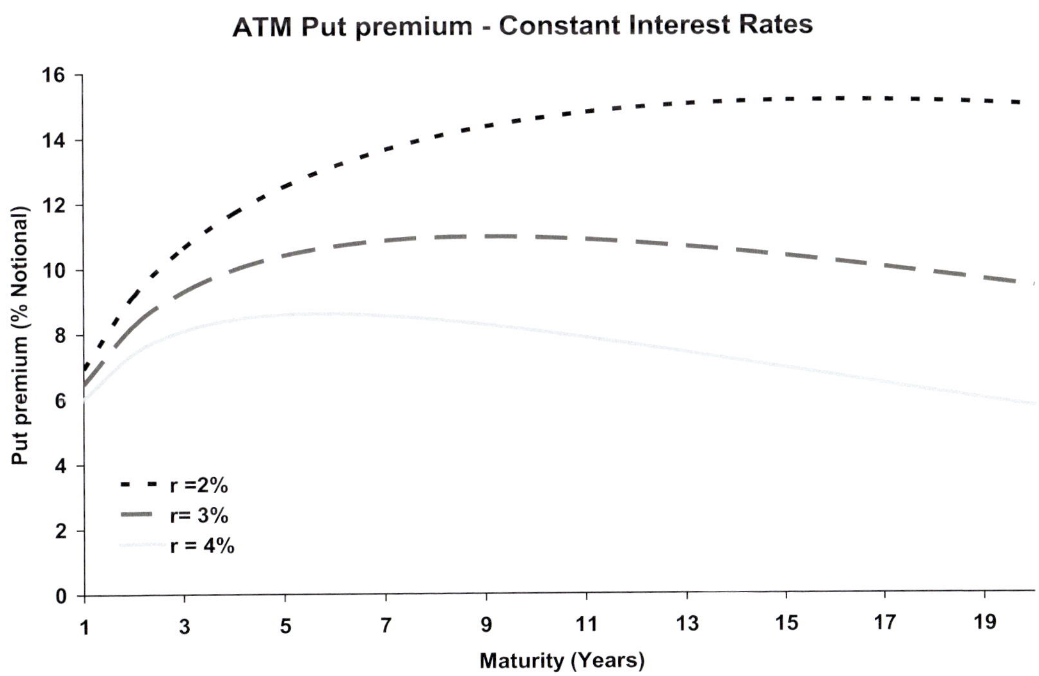

FIGURE 3.4 Recurrent market volatility crashes over the past 20 years: interest rate level impact is significant on long-dated puts

Besides the growing marginal impact of the interest rate level as maturity grows, there are two additional impacts, stemming from (i) the volatility of interest rates and (ii) the correlation between equity and rates, whose contribution increases as the put maturity grows. In fact, stochastic rates and a positive equity–interest rate correlation increase the put volatility, and thus the price.

As a result, the interest rate modelling of savings and retirement requires a specific hybrid modelling, by introducing a non-constant interest rate (below we use the Hull–White one-actor model) with a non-zero correlation between the equity level and the interest rate level. In this respect, the longer the maturity and the higher the interest rate level, the larger the impact of those combined elements; there is an impact of around 30% and upwards for 20-year ATM put prices due to the hybrid modelling.

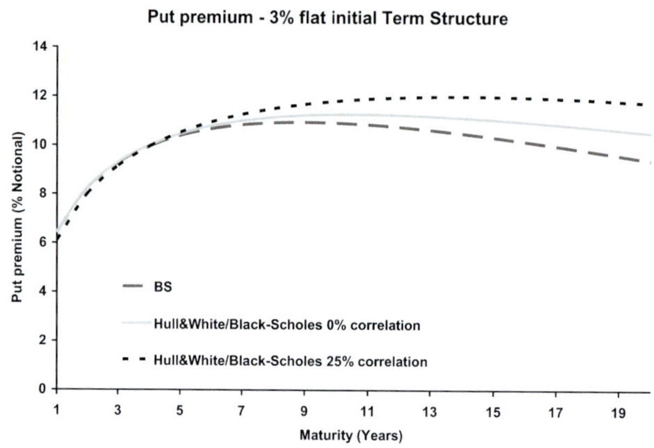

FIGURE 3.5 Recurrent market volatility crashes over the past 20 years: impact of correlation

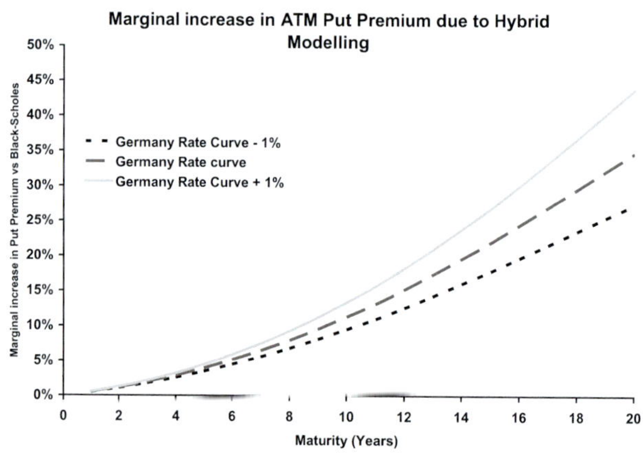

FIGURE 3.6 Hybrid modelling impact on put premium

In particular, the relative additional impact of the equity interest rate correlation on pricing increases with interest rates levels: e.g. a 25% correlation adds as much as switching from constant to stochastic interest rates (using a 0% equity–interest rate correlation).

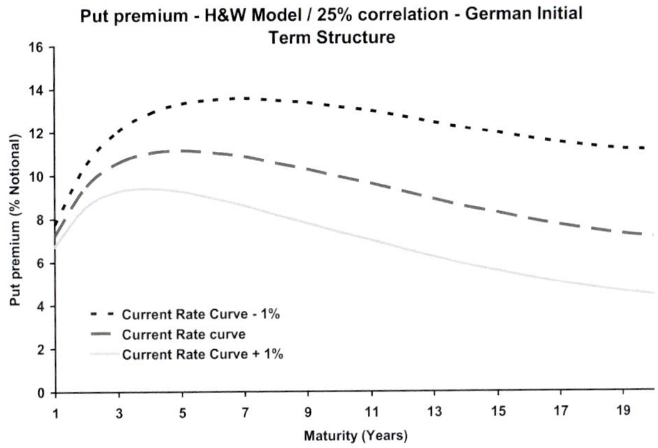

FIGURE 3.7 Equity–rate correlation impact

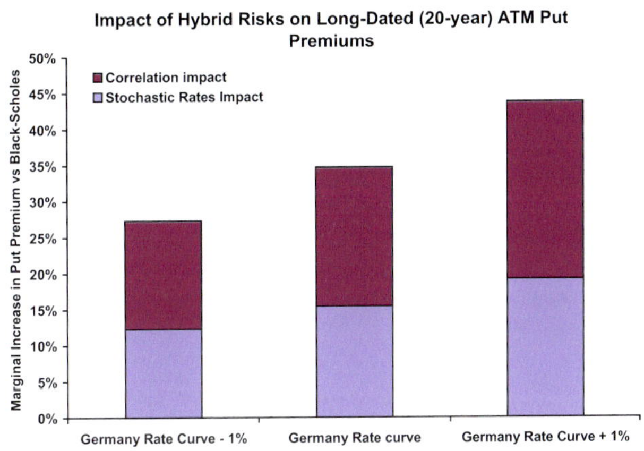

FIGURE 3.8 Correlation contribution grows with rates and maturity

Regarding sensitivities to rates, the interest rate level and the impact of hybrid modelling go the same way.

- The interest rate level impact on the put delta is very significant.
  - There is up to 40% additional delta for a 20-year put for a decrease by 1% in the interest rate level.
  - The longer the maturity, the lower the interest rate level, the larger the impact on the delta.
  - In contrast, the impact on the gamma is not significant, as expected given the very low gamma for long-dated puts.

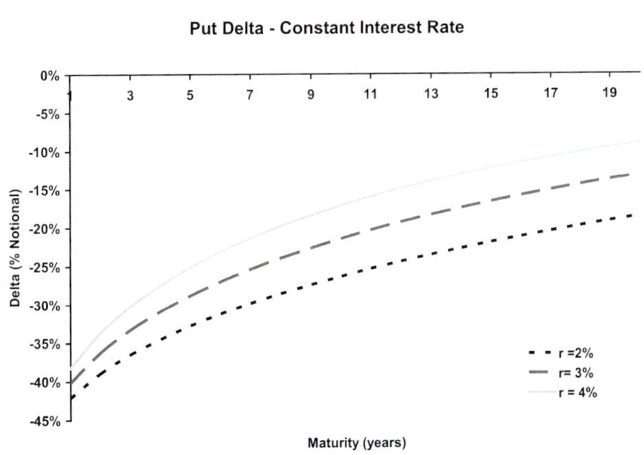

FIGURE 3.9 Interest rate impact on delta is significant

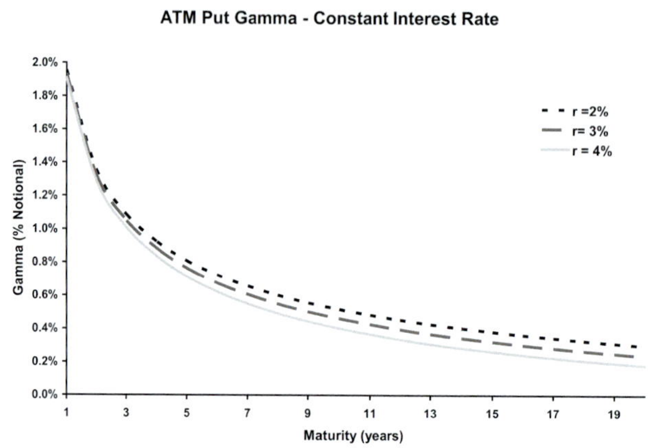

FIGURE 3.10 Interest rate impact on gamma is not significant

- The impact of hybrid modelling goes the same way, to a lesser extent.
  - It grows with both rising maturity *and* the interest rate level.
  - +16% absolute delta for a 20-year ATM put due to hybrid modelling.
  - Correlation contribution grows with interest rate level and maturity.

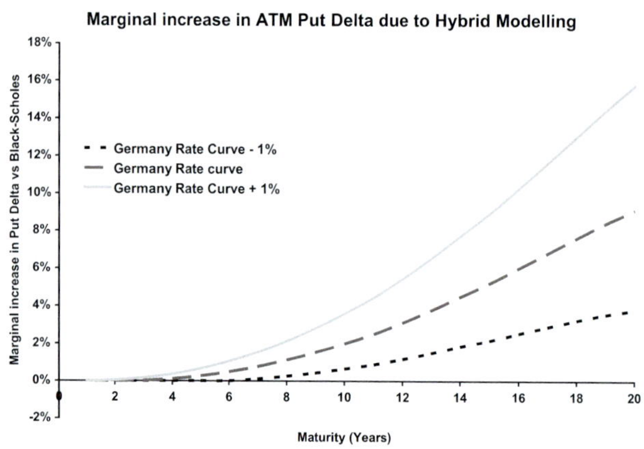

FIGURE 3.11 Hybrid modelling on delta goes the same way

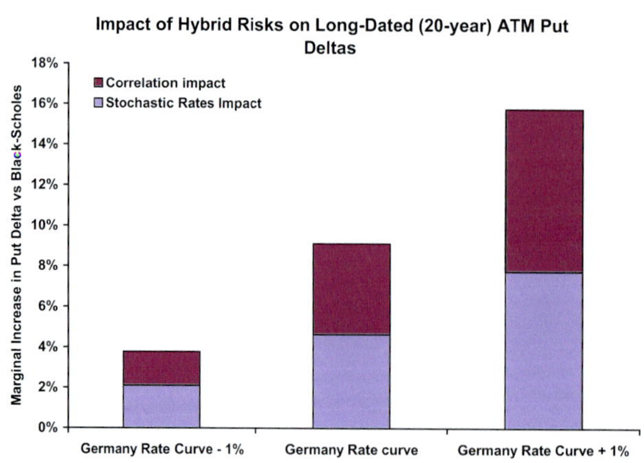

FIGURE 3.12 Correlation contribution

Combining these two influences implies more than 50% additional absolute delta for a 20-year put and a decrease by 1% in the interest rate level.

These impacts of interest rates on pricing and hedging are larger in the context of savings and retirement, given their longer maturities, potential numerous claims and specific payoff exotic features.

> ✓ **Quick win**: it is strongly recommended to take the cross-sensitivity delta–rho Greeks into account, given their significant impact in both pricing and hedging, and to further extend hybrid modelling (interest rates, equity–interest rate correlation), not only for improved pricing and hedging.

3.2.1.1.2 Key Axes for Sustainable Interest Rate Dynamics Modelling

3.2.1.1.2.1 Major Features and Drivers of Interest Rate Modelling Frameworks Interest rate models are characterized by three main features:

- the distribution of interest rates, whether a Gaussian, log-normal or square root process;
- the number of factors within the interest rate level dynamics; and
- the volatility of interest rates.

A specific model may be more suitable depending on four major interest rate drivers:

- the interest rate level: whether it is high or low, a Gaussian linear (e.g. Hull–White), a square root (e.g. CIR) or a log-normal model (e.g. Black–Karasinski) may be more adequate
- the shape of the interest rate curve: its curvature, illustrative of the correlation interest between rates across maturities, whether it is very positive or negative, a one-factor or two-factor model may be more adequate;
- calibration issues: computational run-time, ease of use (tractability), stability; and
- the interest rate volatility level (high or low) and term structure (steep or flat): a constant flat volatility, deterministic volatility term structure or interest rate stochastic volatility may be more adequate.

3.2.1.1.2.2 Interest Rate Levels Interest rate levels have a bearing on how insurers respond.

- High interest rates facilitate being modelled by Gaussian processes while keeping the probability of negative rates negligible -> a Gaussian linear mean-reverting Ornstein–Uhlenbeck (Hull–White) stochastic rate model:

  For a one-factor Hull–White model in which the instantaneous short rate r is written as the sum of an Itô process x and a deterministic function of time φ, allowing reproduction of the initial curve of the zero coupon and ensuring the required no-arbitrage condition,

  $$r_t = x_t + \varphi_t\,,\; dx_t = -ax_t dt + \sigma dW_t\,,\; \varphi_t = f(0,t) + \frac{\sigma^2}{2a^2}\left(1-e^{-at}\right)^2.$$

- Low interest rates: these must not be allowed to become negative with too much probability.

For a one-factor Hull–White model, the probability of rates going negative is

$$\mathbb{P}(r_t \leq 0) = \Phi\left(-\frac{\alpha(t)}{\sigma\sqrt{\frac{1}{2a}\left[1-e^{-2at}\right]}}\right), \quad \forall t \in [0,T],$$

where

- Φ denotes the distribution function of a standard normal distribution, and a and σ are, respectively, the model's mean reversion speed and volatility parameters.
- $$\alpha(t) = f(0,t) + \frac{\sigma^2}{20}\left(1-e^{-at}\right)^2, \quad \forall t \in [0,T],$$

  with $f(0,t)$ the instantaneous forward rate at time 0.

This is indeed very small within high interest rates. However, if the level of rates is low, then $\alpha(t)$ is small, which increases the probability of having negative rates, all other things being equal.

The CIR square root process model used is to be preferred, as it provides strictly positive rates while being more stable than Ornstein–Uhlenbeck processes.

$$dx_t = a(b-x_t)dt + \sigma\sqrt{x_t}\,dW_t, \quad \varphi_t = f(0,t) - \frac{2ab\left(e^{th}-1\right)}{2h+(a+h)\left(e^{th}-1\right)} - x_0\frac{4h^2e^{th}}{\left[2h+(a+h)\left(e^{th}-1\right)\right]^2}$$

$$h = \sqrt{a^2+2\sigma^2}.$$

NB: let us note that this is true only in theory: if we simulate this process using a discretization scheme such as Euler's, this property is no longer true; however, this operational caveat can be mitigated through a more adequate discretization scheme than Euler's, such as the Milstein scheme (see later for details).

- Even so, the growing and persistent occurrence of negative (including long-term) interest rates since 2015 (e.g. across Japan, Switzerland, Germany, France) has reduced this comparative advantage of the CIR model. The ability of the Hull–White model to generate negative rates makes it indeed a more suitable model in the current economic environment. Moreover, so-called "shifted models", such as the "shifted Hull–White model", also provide a way to cap the high probability of generating negative rates.

3.2.1.1.2.3 The Shape of the Interest Rate Curve The curvature of interest rates also plays a major role. Indeed, if it is low, as illustrated by perfectly correlated rates along the term structure, then a single-factor model may be very suitable. Naturally, this means that the one-factor model is suitable only for some constrained interest rate curve shapes. As a result, it is suitable for products that depend on one single rate only.

However, it may generate significant pricing errors whenever interest rates are imperfectly correlated (as already illustrated with swaptions), illustrative of a stronger curvature, which leads to a two-factor models to take into account the imperfect correlation between short- and long-dated rates while allowing for potential twists across the yield curve.

Indeed, correlations between interest rates across the term structure are usually highest for adjacent maturities and decrease as the maturity difference between the rates increases. Correlations also tend to be higher between longer rates than between shorter rates: the one-year rate has the lowest correlation with the rest of the system, because this is a money market rate that is more influenced by government policies than the longer rates.

In fact, principal component analysis (PCA) studies applied to yield curves usually suggest the existence of three factors that, jointly, account for 99% of the changes in the yield curve.

- The first factor, rather "flat" across maturities, captures the common trend component of all rates across the yield curve, illustrative of potential parallel moves in the yield curve; if this first principal component (PC) changes at a time when the other principal components are fixed, then the returns all move by roughly the same amount.
- The second factor, decreasing across maturities, usually captures a change in the slope of the yield curve, illustrative of a steepening or flattening move in the yield curve (a "tilt" component); if this second PC changes at a time when the other components are fixed, then the returns move up at one end of the term structure and down at the other end.
- The third factor, decreasing then increasing across maturities, usually captures a change in the curvature of the yield curve; if the third PC changes when the other components are fixed, then the returns move up (or down) at both ends of the term structure and down (or up) in the middle of the yield curve.

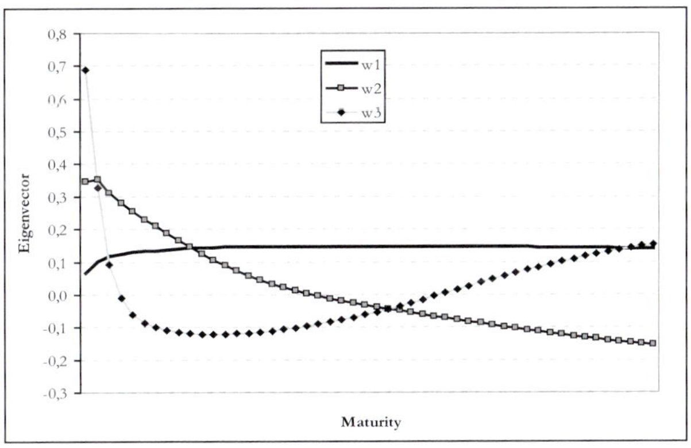

FIGURE 3.13 Behaviour of principal components of a yield curve

The correlation factor does indeed have a significant impact on prices (see Figure 3.15).

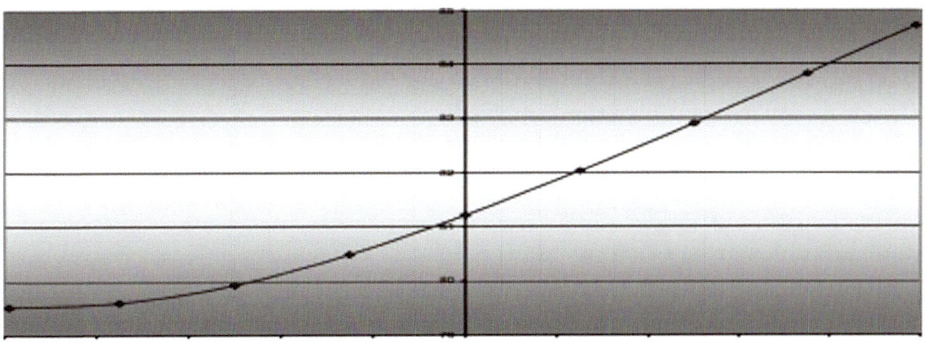

FIGURE 3.14 Impact of the average pairwise correlation between rates on the swaption price

Moreover, as a two-factor model embeds more parameters (seven, as opposed to three in the case of Hull–White models), it usually has greater explanatory power. In a two-factor model, the instantaneous short rate r is written as the sum of two Itô processes x and a deterministic function of time φ:

$$r_t = x_t + y_t + \varphi_t,$$

where both processes x and y follow a Hull–White

$$dx_t = -ax_t dt + \sigma dW_t^x$$
$$dy_t = -by_t dt + \eta dW_t^y$$

or a CIR process

$$dx_t = a(b - x_t)dt + \sigma\sqrt{x_t}\, dW_t^x$$
$$dy_t = c(d - y_t)dt + \eta\sqrt{y_t}\, dW_t^y$$

and are, potentially, non-perfectly correlated through the correlation parameter ρ.

However, the two-factor-CIR model hardly manages hump-shaped rate curves and offers poor tractability in the event of negative correlation between the two factors or if the yield curve gets inverted (as is empirically experienced during market crashes or scares).

3.2.1.1.2.4 Calibration Issues In order to generate meaningful and market-consistent economic scenarios, an interest rate model must be able to replicate market prices. Calibration can be defined as the process of estimating model parameters consistently with market information.

In the context of savings and retirement contracts, the purpose of market-consistent calibration is to reproduce the prices at which liabilities would be traded if they were really traded on the market. Accordingly, models should be calibrated to financial instruments that can reflect the nature and term of these liabilities. In practice, we calibrate (in general with a least squares minimization problem) ESG models to market prices of assets that have similar characteristic to insurers' liabilities, such as put options or swaptions for savings and retirement with embedded optionalities.

This is why calibration to given market data is such a crucial step when it comes to modelling. With regard to historical calibration, the purpose is to estimate parameters such that the model replicates an empirically observed history of market prices. Here we focus only on risk-neutral calibration processes.

Calibration Algorithm Selection

When calibrating a market-consistent model to given market data, we seek for the parameters that minimize the difference (or error) between the prices obtained with our model and market prices. Precisely, this problem can be reduced to a least squares minimization problem:

$$\theta^* = \arg\min_{\theta} \sum_i \left[\omega_i (\text{ Model price } (i,\theta) - \text{ Market price } (i)) \right]^2.$$

Depending on which function we want to minimize, we can choose weights ωs to be

- all equal to 1; we then consider squared absolute errors, which gives more weight to instruments with high prices,

- equal to 1/market price(i); we then consider squared relative errors, which gives the same weight to all instruments,
- equal to specified values chosen by the user in order to give more weight to specific instruments.

We are then left with two problems: choosing (i) the derivatives to calibrate the model (caps, swaptions as very liquid assets and considered vanilla by market participants, their market price can be reasonably supposed to be correct) and (ii) the optimization algorithm for finding the best fit to market data.

In practice, it is neither possible nor meaningful to match exactly the observed prices. Therefore, the problem of calibrating the model is formulated as an optimization problem. The aim is to minimize the pricing error between the model prices and the market prices for a set of traded options. A common approach to measure this error is to use the squared difference between market and model prices; this approach leads to the nonlinear least squares method:

$$\inf_{\Theta} G(\Theta), \quad G(\Theta) = \sum_{i=1}^{N} w_i \left| C_i^{\Theta}\left(t, S_t, T_i, K_i\right) - C_i^{*}\left(T_i, K_i\right) \right|^2,$$

where N denotes the number of options used, wi is a weight and C is the market price of the call option observed at time t. CΘ denotes the model price computed using a vector of model parameters $\Theta = (\nu 0, \kappa, \theta, \sigma, \rho)$.

Among alternative measures used in calibration are the following.

$$\textit{Maximum absolute relative error}: \mathrm{MARE}(\Theta) = \max_i \frac{\left|C_i^{\Theta} - C_i^{*}\right|}{C_i^{*}}, \quad i = 1, \ldots, N,$$

$$\textit{Average absolute relative error}: \mathrm{AARE}(\Theta) = \frac{1}{N} \sum_{i=1}^{N} \frac{\left|C_i^{\theta} - C_i^{*}\right|}{C_i^{*}},$$

$$\textit{Root mean square error}: \mathrm{RMSE}(\Theta) = \sqrt{\frac{1}{N} \sum_{i=1}^{N} \left(C_i^{\Theta} - C_i^{*}\right)^2}.$$

With regard to weights, wi stands for a particular weight. In practice, the weights are put where the most liquid quotes are on the market. This is usually around ATM. We aim to have the model prices as close to the mid-price as possible. Thus we put weight on the small spreads. The bigger the spread, the less weight is put on the particular difference between the model price and the market price during the calibration process.

Thinking of G as an objective function of Θ, the optimization problem is not an easy one. Function G is not convex and is not of any particular structure. Moreover, it may have more than one global minimum, and, even if it did not, it is not clear whether the unique minimum can be reached.

There are two families of optimization methodologies, deterministic (or local) and stochastic algorithms, between which one must find a balance between time and efficiency.

- On the one hand, deterministic or local algorithms do not guarantee a global optimum (only a local one), but they are much faster.

  These algorithms need an initial point to start their routines (unlike most stochastic algorithms). Moreover, starting from a random initial point will most likely converge to a local minimum -> one has to choose an initial guess, which is "good enough", to find starting values (by taking the latest calibrated parameters, for example), or generate a large number of initial points and then run the optimization process from different best guesses, in order to converge to the desired global minimum. Nevertheless, even with this approach, the probability that the deterministic algorithm converges to a local minimum (rather than the global one) remains high.

Among those local algorithms are: the Newton method or quasi-Newton method (which uses a gradient descent to find the local minimum and requires a differentiable objective function); the built-in Excel Solver (which uses the generalized reduced gradient method); and the Nelder–Mead (or downhill simplex) algorithm (which operates in multidimensional space and which does not require the objective function to be differentiable, but does require the feasible region to be small enough).

Local optimization algorithms may be useful when minimizing smooth functions with appealing properties (monotonic, convex). Otherwise, stochastic algorithms should be preferred to perform global optimization, despite their longer computation time.

- On the other hand, stochastic algorithms have a higher probability of convergence to the global optimum, but they are much slower.

  Indeed, contrary to deterministic ones, stochastic algorithms introduce some randomness in the optimum research, which increases the probability of convergence to the global optimum, and can even guarantee it if we suppose that the algorithm runs for an infinite length of time. In particular, they appear especially useful because the corresponding objective functions generally present many strong local minimums (due to their complexity and the high number of parameters) and are highly sensitive to small variations in input data (market prices, the initial zero-coupon curve).

  Simulated annealing and genetic algorithm are two examples of stochastic algorithms. A simulated annealing algorithm works as follows: at each iteration, it generates a random new point whose position from the previous one is determined by a probability distribution (known as the acceptance probability, and given by the Boltzmann distribution). This acceptance probability has to be parametrized such that points that minimize the objective function have high probabilities. It then converges to new points that reduce the value of the objective function but can also move to states that increase the value of the target function with a specific probability (also specified by the user). As a result, the probability of being stuck into a local minimum is reduced, since the algorithm is able to consider more combinations of parameters. As a matter of fact, the convergence to the global minimum is theoretically guaranteed.

  Even so, a simulated annealing algorithm requires an initial point to start. In contrast, a genetic algorithm avoids this need for starting values. However, it is known to be very computationally intensive, and it does not guarantee a convergence to the global minimum – not even to a local one.

✓ **Quick win**: eventually the best practical method for a complex optimization problem would probably be to first call a stochastic algorithm to at least converge to the region of the global minimum, and then use a local deterministic algorithm in order to refine the solution.

Even so, some optimization problems may be too time-consuming to use a stochastic algorithm, in which case running a local algorithm from different initial points provides a reasonable trade-off – i.e. a solution that can reasonably be considered as a global optimum.

Benchmarking the Different Interest Rates Models regarding Calibration

CIR-type interest rate models do not lead to closed formulas – or, at least, semi-closed ones – for the price of swaptions. In addition, it is not possible to introduce a correlation between (Brownian) factors within a two-factor CIR model if one wants to keep the no-arbitrage condition for the φ function, translating into

$$\phi^1(t) = \frac{2ab\left(e^{th_1} - 1\right)}{2h_1 + (a+h_1)\left(e^{th_1} - 1\right)} + x_0 \frac{4h_1^2 e^{th_1}}{\left[2h_1 + (a+h_1)\left(e^{th_1} - 1\right)\right]^2}$$

$$\varphi_t = f(0,t) - \phi^1(t) - \phi^2(t),\ \phi^2(t) = \frac{2cd\left(e^{th_2} - 1\right)}{2h_2 + (c+h_2)\left(e^{th_2} - 1\right)} + y_0 \frac{4h_2^2 e^{th_2}}{\left[2h_2 + (c+h_2)\left(e^{th_2} - 1\right)\right]^2}$$

$$h_1 = \sqrt{a^2 + 2\sigma^2}$$

$$h_2 = \sqrt{c^2 + 2\eta^2}$$

The log-normal Black–Karasinski model suffers from calibration issues in terms of both computational run-time and stability of calibrated parameters.

- It does not yield analytical formula for bonds and interest rate options.
- Mapping short rates into long rate dynamics needs time-consuming numerical replacements.
- Every "up" move in rates proportionally inflates volatility, making the duration artificially longer, which may requantify the rho hedging needs.
- Empirically, long-term rates are not log-normal and have lower volatility than short rates; therefore, calibration to long-dated products can be complicated.
- In the event of a steep yield curve coupled with sharply falling proportional volatility, the volatility modelled as proportional to the rate level cannot be recommended.
- Numerical issues arise when rates are high for Black–Karasinski calibration processes.

Hull–White outperforms globally in terms of calibration.

- Calibration errors are overall lower, in particular on the key rate buckets (ten to 20 years) most important for savings and retirement (see Figures 3.16 and 3.17).

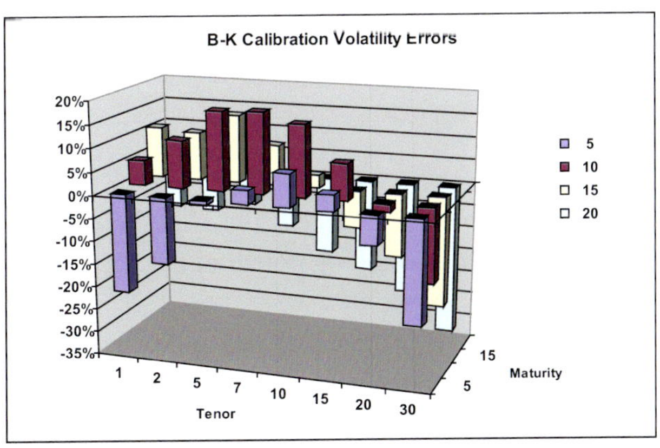

FIGURE 3.15 Black–Scholes calibration volatility errors, as a percentage of the volatility level

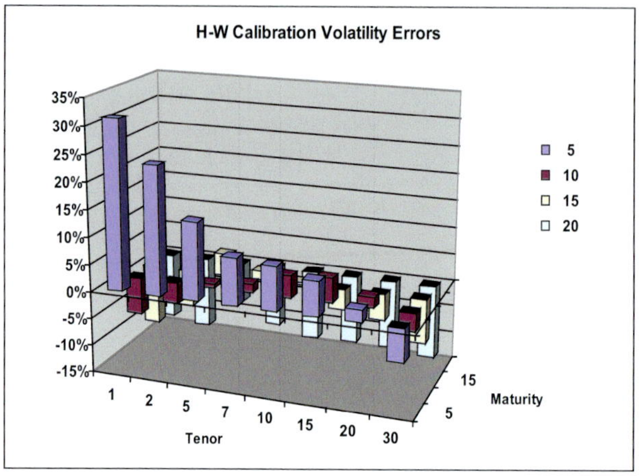

FIGURE 3.16 Hull–White calibration volatility errors, as a percentage of the volatility level

These discrepancies are further confirmed by most empirical analyses.

- Black–Karasinski calibrates only by trees, which requires a very large number of parallel runs.
- Run-time is far lower for Hull–White.
- Hull–White's calibrated "sigma" is rather responsive to the slope and the shape of the volatility surface.

3.2.1.1.2.5 The Volatility of Interest Rates Interest rate volatility may be constant or may depend on the maturity (term structure volatility) – i.e. exhibit a piecewise linear deterministic function (e.g. constant) for the volatility across maturities (see Figure 3.18).

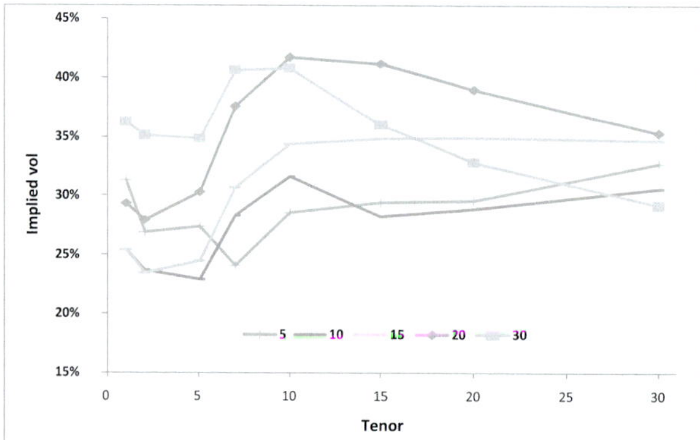

FIGURE 3.17 Implied volatility versus tenor

3.2.1.1.3 The Overall Benchmark Among the HJM models, the Hull–White model holds many advantages, which make up for the small drawback of not precluding potential negative interest rates.

- This is, in fact, rather harmless in terms of (average) pricing, and manageable for the VaR metrics through introducing constraints on the mean reversion parameter a.
- There is a weak if not negligible empirical relation between the rate volatility and the rate level, which is a sign of Gaussian rather than log-Gaussian behaviour. As illustrated in Figure 3.19, interest rate volatility

(the blue line) has remained very much independent of single-digit Treasury rates, in contrast with double-digit rates, where volatility grows with the rate level; overall, it has been remained independent of both Treasury and swap rates since the late 1980s.

This is further confirmed by calibration analyses of a constant elasticity of variance (CEV) model (which encapsulates the three potential alternative models: log-normal if $\gamma = 1$ versus normal if $\gamma = 0$) on swaption prices.

- It provides rather more conservative prices versus Black–Karasinski.
  - Both distribution dispersions grow with maturity with a further shift on the right.
  - The Black–Karasinski one experiences a fatter right tail, coupled with an increasing peak around zero.
  - By contrast, the Hull–White one presents an increasing negative area.

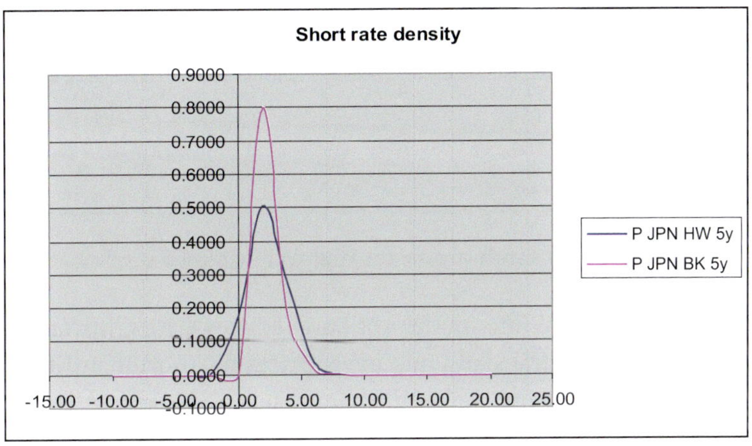

FIGURE 3.18 Short rate density over time

This is further confirmed by empirical analyses (see Figure 3.20).

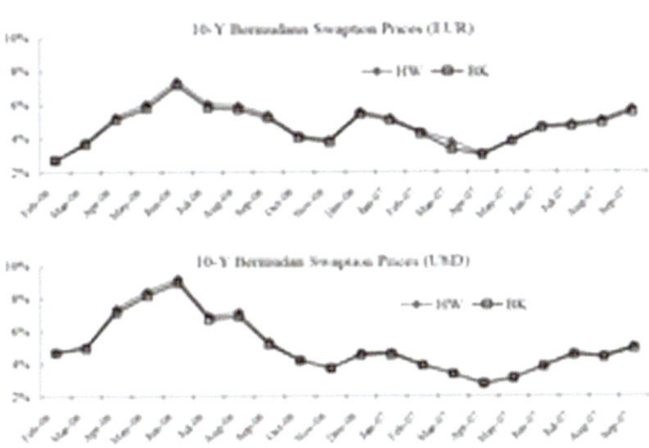

FIGURE 3.19 Hull–White swaption prices are slightly higher

- It holds a great deal of analytical tractability, with a closed-form formula for bonds and options on bonds, cap and floor, and a quasi-closed-form formula for swaptions.
- It allows for "quasi-parallel" shocks with rate deviations gradually depressed along the curve, as consistent with the behaviour of implied volatility (IV) falling with the swap maturity.

- The explicit formula for the standard deviation (stdev) of the zero-coupon rate can be directly compared with the Black quoted implied volatility.

However, given the low interest rate environment (potentially combined with high volatility, or with an inverted rate term structure), the additional constraint on the mean reversion a parameter not only implies significant instability in the a parameter (despite the use of interpolation/extrapolation smoothing techniques) but also makes the parameter somewhat insignificant (i.e. close to zero). This makes interpreting it very odd, specifically for a so-called "mean-reverting" model.

ATM IV error absolute increase/decrease in errors by using 2F model	1	5	7	10	15	20	30
5	-2.6%	-0.7%	-0.4%	-0.1%	0.1%	0.3%	0.5%
10	1.6%	-0.2%	0.1%	0.3%	0.6%	0.8%	0.9%
15	1.0%	-0.2%	-0.5%	-0.7%	-1.0%	-1.1%	1.2%
20	0.7%	-0.5%	-0.8%	-1.0%	-1.2%	-1.4%	1.4%
30	0.2%	-0.9%	-1.1%	-1.3%	1.3%	1.5%	1.5%

FIGURE 3.20 Model error on short maturities and tenors

2FHW vs 1FHW prices	1	5	7	10	15	20	30
5	-3%	-1%	0%	0%	0%	0%	1%
10	-1%	0%	0%	0%	1%	1%	1%
15	-1%	0%	0%	1%	1%	1%	1%
20	-1%	0%	1%	1%	1%	1%	2%
30	0%	1%	1%	1%	1%	2%	2%

FIGURE 3.21 Model errors on US swaptions with two-factor Hull–White model

In this respect, the two-factor Hull–White model not only reduces errors, further leveraging on the simplex algorithm (see Figure 3.21), which translates into lower prices for short maturities yet is more expensive for long maturities (see Figure 3.22), but also makes the parameters no longer insignificant (see Figure 3.23).

2-Factor HW:

Parameters	
a	9,73%
b	0,14%
sigma	0,83%
eta	1,05%
rho	-81,06%

FIGURE 3.22 Calibrated parameters

However, this is to the detriment of the much greater instability of the calibrated parameters (see Figure 3.24).

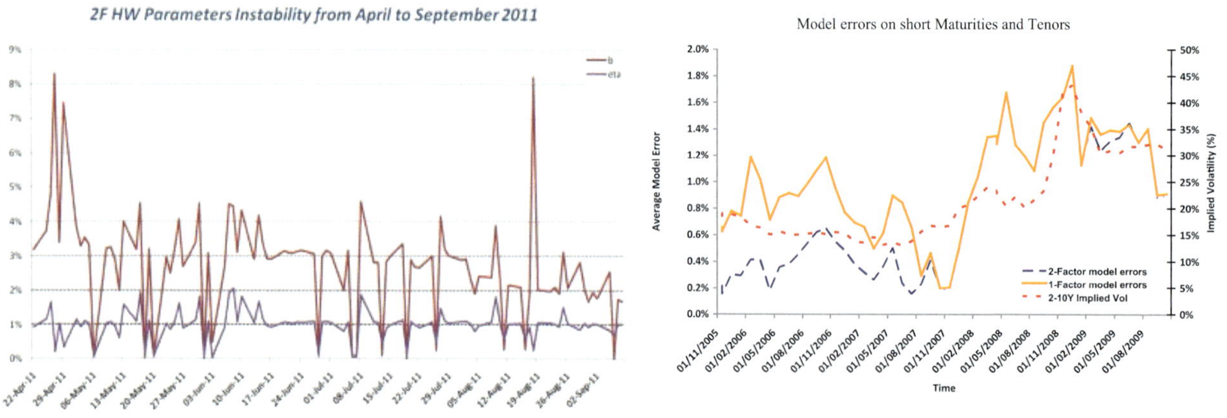

FIGURE 3.23 Two-factor Hull–White calibrated parameters

Such instabilities in calibrated parameters lead to the introduction of the volatility term structure model (as illustrated empirically in Figures 3.25 and 3.26).

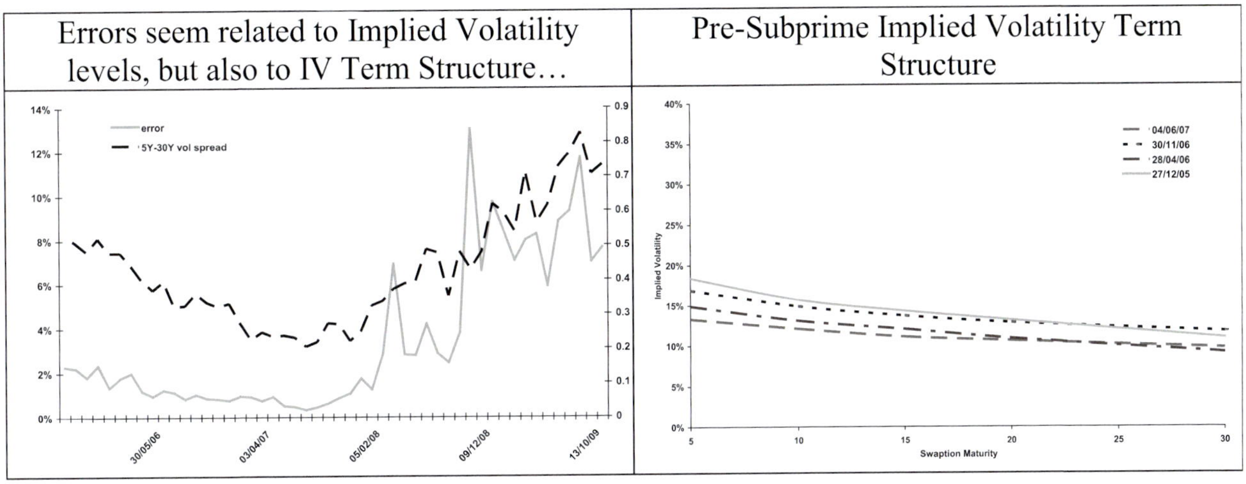

FIGURE 3.24 Historical analyses illustrate the impacts of volatility levels

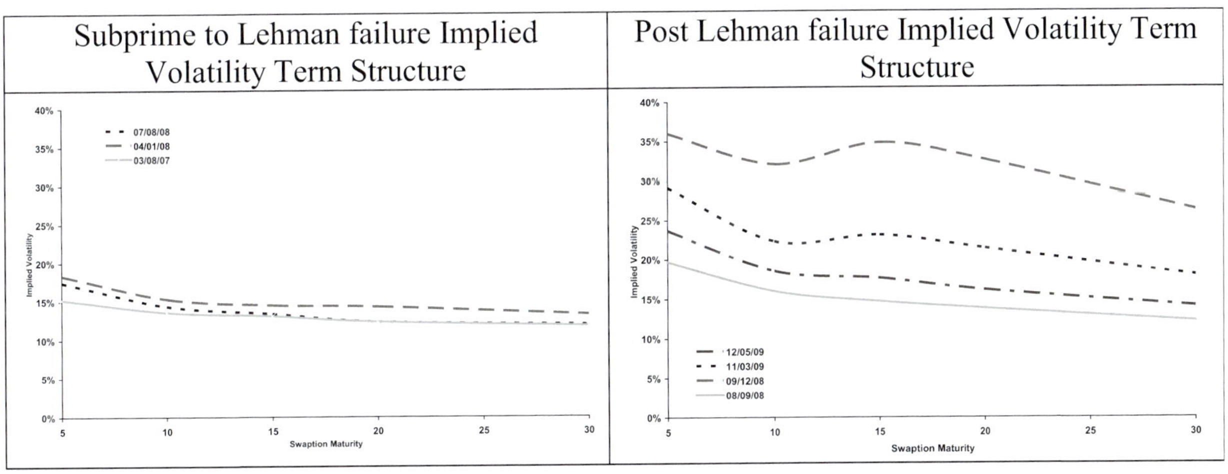

FIGURE 3.25A Volatility term structure steepened during the crisis

The Hull–White volatility term structure model provides numerous and significant advantages.

- Error rates are significantly reduced.
- The average relative error rate and maximum error rate are both significantly reduced.
- The maximum improvement is for long-term maturities, in which the highest risks are concentrated.
- Flexibility is significantly improved.
- The model is consistent with the distortion of the interest rate curve shape, offering realistic interest rate behaviour, illustrated by more hectic short-dated volatilities but more stable long-dated volatilities.
- Last but not least, only Hull–White with the volatility term structure provides interpretable results: the calibrated parameters are not just more stable (illustrated in Figure 3.26 for both European and US markets) but they also move along consistently with the market conditions, with mean reversion a and correlation both evolving within a much tighter and sensible range.

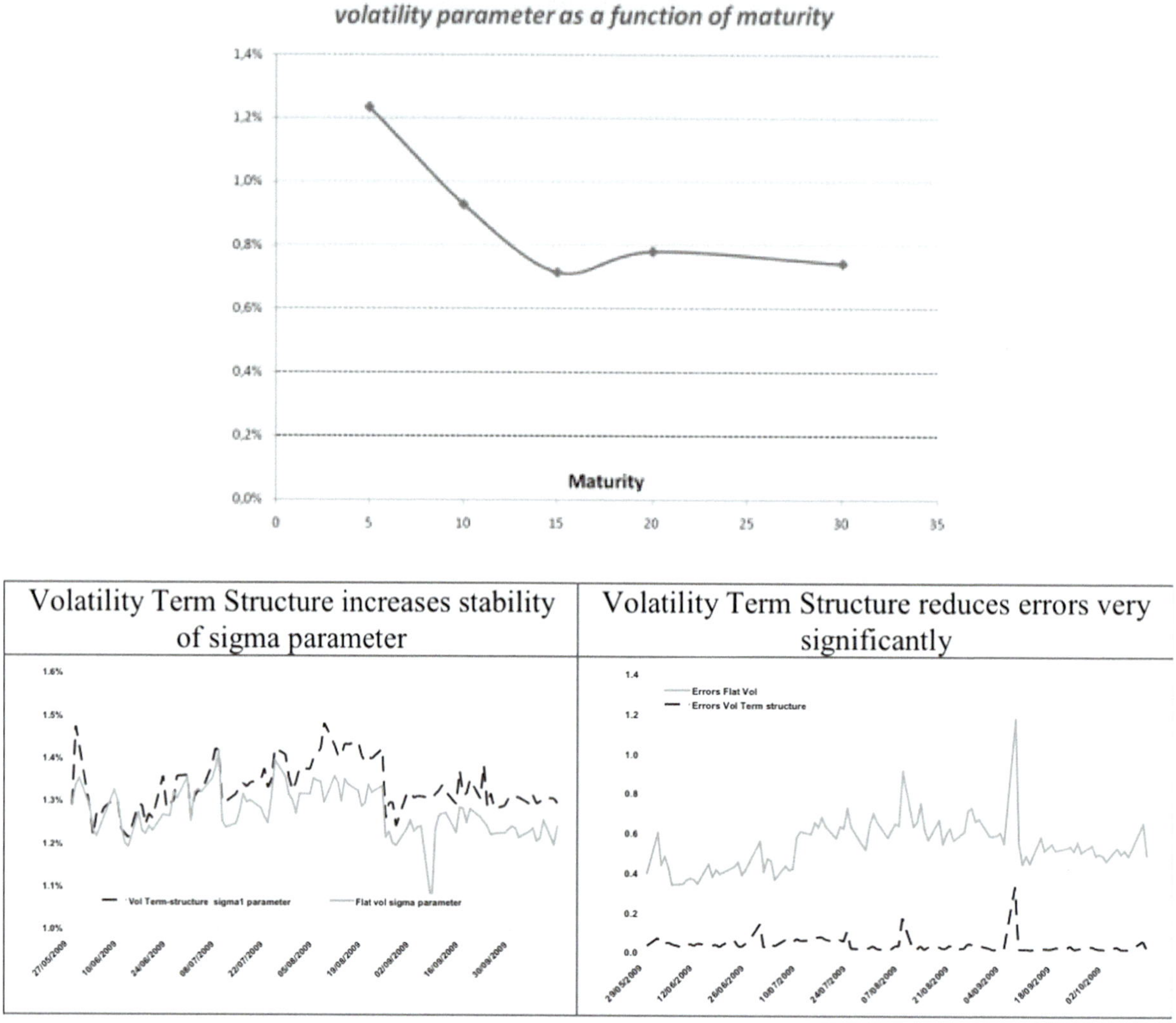

FIGURE 3.25B Historical calibration of a and volatility parameters

When two factors and the volatility term structure are both involved, then the instability introduced by the classical two-factor Hull–White model disappears, while the error is further improved:

3.2.1.2 Equity Hybrid Dynamics Modelling for Sustainable Savings and Retirement

As discussed regarding the potential best interest rate model, there is no "one size fits all" solution, as it depends on the market environment: whether equities and/or interest rates are low or high.

For instance, in the event of high equities, a high level of equities generally goes with low equity volatility and low equity volatility of volatility, and so in favour of a constant equity volatility. As a result, a Black–Scholes model could be considered as sufficiently realistic.

In contrast, in the event of low equities, an expected higher equity volatility and volatility of volatility may require an equity stochastic volatility. As a result, a stochastic (Heston) equity volatility model may be more adequate.

It is interesting to see that, at any given moment, there are four possible situations:

- volatility and the instant short rate are above those of Black–Scholes;
- volatility and the instantaneous short rate are below those of Black–Scholes;

- volatility is above that of Black–Scholes and the short rate is below; and
- volatility is below that of Black–Scholes and the short rate is above.

How do these two quantities work?

- The interest rate impacts the drift of the asset. Thus, if rates are above the level of the Black–Scholes model, growth will be faster.
- Volatility affects the range of possible movements for the source of risk associated with the asset. If volatility increases, asset movements will be greater, whether in the event of an increase or a decrease.

In the following, two-factor hybrid modelling is developed:

- a two-factor Hull–White and Black–Scholes model; and
- a generic Itô process rate and equity stochastic volatility (Heston) model.

3.2.1.2.1 The Black–Scholes and Two-Factor Hull–White Hybrid Model We consider the equity Black–Scholes constant volatility model:

$$\frac{dS_t}{S_t} = r_t dt + v dW_t^S$$

We also consider a two-factor Hull–White model (with potential constant or term structure volatility),

$$r_t = x_t + y_t + \varphi_t,$$

where the deterministic function of time φ allows us to reproduce the initial zero-coupon curve (and the instantaneous forward rates curve, since $f(t,T) = -\frac{\partial \ln P(t,T)}{\partial T}$) and ensure the required HJM no-arbitrage condition.

We consider the two-factor Hull–White model with the correlation ρ between the two factors (Brownian motions):

$$\begin{cases} dx_t = -ax_t + \sigma dW_t^x \\ dy_t = -by_t + \eta dW_t^y, \quad dW_t^x dW_t^y = \rho_{xy} dt \end{cases}$$

$$\begin{cases} x_t = x_s e^{-a(t-s)} + \sigma \int_5^t e^{-a(t-u)} dW_u^x \\ y_t = y_s e^{-b(t-s)} + \eta \int_s^t e^{-b(t-u)} dW_u^y \end{cases}$$

$$r_t = x_s e^{-a(t-s)} + \sigma \int_s^t e^{-a(t-u)} dW_u^x + y_s e^{-b(t-s)} + \eta \int_s^t e^{-b(t-u)} dW_u^y + \varphi_t.$$

This model allows us to obtain an *exact simulation for the dynamics of the asset, and thus not to introduce discretization bias in our simulations.* Its only drawback is the constant volatility of rates and equity.

$$\ln\left(\frac{S_{t_{i+1}}}{S_{t_i}}\right)=\int_{t_i}^{t_{i+1}} r_s ds-\frac{v^2}{2}+v\left(W_{t_{i+1}}^S-W_{t_i}^S\right),\ dW_t^S dW_t^y=\rho_{Sy}dt,.dW_t^S dW_t^x=\rho_{Sx}dt.$$

$$\int_{t_i}^{t_{i+1}} r_s ds=\int_{t_i}^{t_{i+1}} x_s ds+\int_{t_i}^{t_{i+1}} y_s ds+\int_{t_i}^{t_{i+1}} \varphi_s ds \text{ with}$$

$$\int_{t_i}^{t_{i+1}} x_s ds=x_{t_i}\int_{t_i}^{t_{i+1}} e^{a(st_i)}ds+\sigma\int_{t_i}^{t_{i+1}}\int_{t_i}^{u} e^{a(uv)}dW_v^x du,$$

where the stochastic version of Fubini's theorem allows us to write

$$Z_x=\int_{t_i}^{t_{i+1}}\int_{t_i}^{u} e^{-a(u-v)}dW_v^x du=\int_{t_i}^{t_{i+1}}\int_{v}^{t_{i+1}} e^{-a(u-v)}du dW_n^x.$$

This follows a Gaussian distribution with variance

$$V_{Z_x}=\int_{t_i}^{t_{i+1}}\left(\int_{v}^{t_{i+1}} e^{-u(u-v)}du\right)^2 dv=\frac{1}{a^2}\left[t_{i+1}-t_i-\frac{2}{a}\left(1-e^{-u(t_{i+1}-t_i)}\right)+\frac{1}{2a}\left(1-e^{-2u(L_{i+1}-t_i)}\right)\right].$$

As a result,

$$\int_{t_i}^{t_{i+1}} x_s ds=x_{t_i}\frac{1-e^{-a(t_{i+1}-t_i)}}{a}+\sigma Z_x,\ \int_{t_i}^{t_{i+1}} y_s ds=y_{t_i}\frac{1-e^{-b(t_{i+1}-t_i)}}{a}+\eta Z_y$$
$$Z_y\sim\mathcal{N}\left(0,V_{Z_y}\right)\ et\ V_{Z_y}=\frac{1}{b^2}\left[t_{i+1}-t_i-\frac{2}{b}\left(1-e^{-b(t_{i+1}-t_i)}\right)+\frac{1}{2b}\left(1-e^{-2b(t_{i+1}-t_i)}\right)\right].$$

How to simulate our quantities step by step in this model? First of all, let us note that, in addition to the annual return of the stock, we need the factors for the rates, as well as the entire short rate used for the discount factor. We combine the different equations

$$\begin{cases} x_{t_{i+1}}=x_{t_i}e^{-a(t_{i+1}-t_i)}+\sigma X \\ yt_{i+1}=y_{t_i}e^{-b(t_{i+1}-t_i)}+\eta Y \\ \ln\left(\frac{S_{t_{i+1}}}{S_{t_i}}\right)=A\left(x_{t_i},y_{t_i}\right)+\sigma Z_x+\eta Z_y+vZ_s \end{cases}$$

with

$$A\left(x_{t_i},y_{t_i}\right)=\int_{t_i}^{t_{i+1}}\varphi_s ds+x_{t_i}\int_{t_i}^{t_{i+1}} e^{-a(s-t_i)}ds+y_{t_i}\int_{t_i}^{t_{i+1}} e^{-b(s-t_i)}ds-\frac{v^2}{2},$$

$$X=\int_{i_4}^{t_{4+1}} e^{-\theta(t_{+1}-s)}dW_w^\pi,.Y=\int_{l_4}^{t_{4+1}} e^{-b(t_{i+1}-w)}dW_w^y,\ Z_s=\left(W_{t_{i+1}}^S-W_{t_4}^S\right).$$

This system is therefore composed of three equations with five correlated Gaussian random variables.

Let U be the Gaussian vector composed of these five random variables: $U = (X, Y, Z_x, Z_y, Z_s)$.

If we pose Σ_U the variance-covariance matrix of this vector, then there is a single lower triangular matrix L_U verifying the following relationship (Cholesky decomposition of the variance-covariance matrix Σ_U):

$$L_U L_U^{'} = \Sigma_U.$$

Overall, to simulate our assets (i.e. the matrix U), five independent reduced centred normal laws N_i need to be simulated. Then we apply the formula

$$U = L_U \begin{pmatrix} N_1 \\ N_2 \\ N_3 \\ N_4 \\ N_5 \end{pmatrix}.$$

All that remains is to calculate the matrix Σ_U (the Cholesky decomposition L_U being computed by a classical algorithm). To that extent, let us calculate the covariance between Z_y and Z_s:

$$\mathrm{Cov}\left(Z_y, Z_s\right) = < Z_y, Z_s > = < \int_{t_i}^{t_{i+1}} \int_{v}^{t_{i+1}} e^{-b(u-v)} du dW_v^y, \int_{t_i}^{t_{i+1}} dW_v^S >$$

$$\mathrm{Cov}\left(Z_y, Z_s\right) = \rho_{S_y} \int_{t_i}^{t_{i+1}} \int_{v}^{t_{i+1}} e^{-b(u-v)} du dv,$$

$$\mathrm{Cov}\left(Z_y, Z_s\right) = \frac{\rho_{S_y}}{a} \left(1 - \frac{1 - e^{-a(t_{i+1} - t_i)}}{a} \right).$$

As long as the equity volatility is deterministic, there is here an exact simulation for the dynamics of the asset, and thus no introduction of discretization bias in our simulations, as illustrated by the identical value of the expected stock value E[S], whether based on the Monte Carlo simulations or on the closed-form formula (see Figure 3.27).

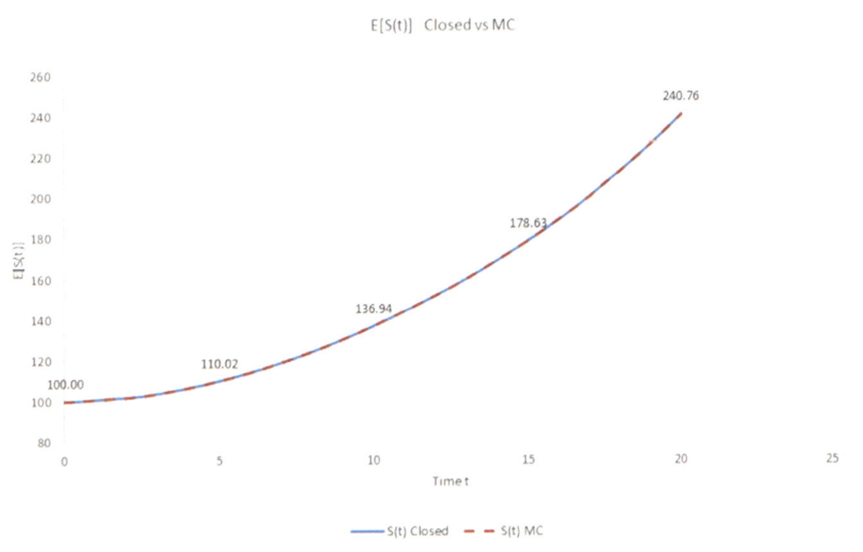

FIGURE 3.26 Monte Carlo versus exact stock level

Under the Black–Scholes one-factor Hull–White model, both European and American option prices are more expensive.

3.2.1.2.2 The Heston Hybrid Model

3.2.1.2.2.1 The Multidimensional Dynamics

$$\begin{cases} dS_t = r_t S_t dt + \sqrt{V_t} S_t dW_t^S \\ dV_t = a(b - V_t)dt + \eta\sqrt{V_t} dW_t^V \\ dr_t = \mu(t, r_t)dt + v(r, r_t)dW_t^r, \end{cases}$$

$$\begin{cases} dW^S dW^V = \rho_{sv} dt \\ dW^S dW^r = \rho_{sr} dt \\ dW^r dW^V = \rho_{rv} dt, \end{cases}$$

$$d\begin{pmatrix} S_t \\ V_t \\ r_t \end{pmatrix} = \begin{pmatrix} r_t S_t \\ a(b - V_t) \\ \mu(t, r_t) \end{pmatrix} dt + \begin{pmatrix} \sqrt{V_t} S_t & 0 & 0 \\ 0 & \eta\sqrt{V_t} & 0 \\ 0 & 0 & v(t, r_t) \end{pmatrix} d\begin{pmatrix} W_t^S \\ W_t^V \\ W_t^r \end{pmatrix}.$$

Since Brownian motions are correlated, it is possible to return to a model with three independent Brownian motions using the Cholesky decomposition of the instantaneous correlation matrix.

$$\begin{pmatrix} 1 & 0 & 0 \\ \rho_{sv} & \sqrt{1-\rho_{sv}^2} & 0 \\ \pi_1 & \pi_2 & \pi_3 \end{pmatrix}, \begin{cases} \pi_1 = \rho_{sr} \\ \pi_2 = \dfrac{\rho_{rv} - \rho_{xw}\rho_{sr}}{\sqrt{1-\rho_{sv}^2}} \\ \pi_3 = \sqrt{1 - \rho_{sr}^2 - \left(\dfrac{\rho_{rv} - \rho_{sw}\rho_{sr}}{\sqrt{1-\rho_{sv}^2}}\right)^2}, \end{cases}$$

$$d\begin{pmatrix} S_t \\ V_t \\ r_t \end{pmatrix} = \begin{pmatrix} r_t S_t \\ a(b - V_t) \\ \mu(t, r_t) \end{pmatrix} dt + \begin{pmatrix} \sqrt{V_t S_t} & 0 & 0 \\ \rho_{sv}\eta\sqrt{V_t} & \sqrt{1-\rho_{sv}^2}\,\eta\sqrt{V_t} & 0 \\ \pi_1 v(t, r_t) & \pi_2 v(t, r_t) & \pi_3 v(t, r_t) \end{pmatrix} d\begin{pmatrix} W_t^1 \\ W_t^2 \\ W_t^3 \end{pmatrix}.$$

The model is presented here in its version in which interest rates have only one factor, in order not to overload already heavy ratings (a stochastic two-factor rate just adds a degree of uncertainty – i.e. a dimension – in the model; therefore, instead of dealing with 3 × 3 matrices, we would be dealing with 4 × 4 matrices for variances, but the approach remains the same).

Now, how to simulate in this model the evolution of the asset? *Unlike the previous model, there is no exact simulation method, so we will have to implement a discretization scheme.* Three schemas are available to us: the Euler schema, the 1D Milstein schema and the 2D Milstein schema. We focus on these three schemes to the extent that discretizing a stochastic differential equation introduces a bias into the simulations.

3.2.1.2.2.2 Dealing with Discretization Bias within Multidimensional Simulations In terms of simulations, although the Heston model was introduced back in 1993, the research into discretization methods for its dynamics has appeared only recently.

An exact scheme was developed by Broadie and Kaya (2006), which simulates the Heston processes from its exact distributions, but suffers from extremely slow computational run-time.

Different stochastic Taylor expansions have been devised, such as Euler schemes and higher-order Milstein schemes, which provide increasing accuracy, but unfortunately they have potential negative values in the variance process.

In order to deal with such an issue, combined with further improved run-time, other schemes such as quadratic exponential (QE) schemes by Andersen have been introduced. These can be further accelerated by the use of multi-level Monte Carlo (MLMC) simulation schemes.

3.2.1.2.2.2.1 The Exact but Significantly Slow Scheme (Broadie and Kaya)

$$dX(t) = \left(r - \frac{1}{2}V(t)\right)dt + \sqrt{V(t)}dW_X(t),$$
$$dV(t) = \kappa(\theta - V(t))dt + \sigma\sqrt{V(t)}dW_V(t).$$

Broadie and Kaya composed a bias-free (exact) method to simulate the Heston model as follows.

The volatility at time t and stock price, given the values of v_u and for u < t, can be written as

$$v_t = v_u + \kappa\theta(t-u) - \kappa\int_u^t v_s ds + \sigma\int_u^t \sqrt{v_s}\,dW_s^{(1)},$$

$$\ln(S_t) = \ln(S_u) + r(t-u) - \frac{1}{2}\int_u^t v_s ds + \rho\int_u^t \sqrt{v_s}\,dW_s^{(1)} + \sqrt{1-\rho^2}\int_u^t \sqrt{v_s}\,dW_s^{(2)},$$

where $W^{(1)}$ and $W^{(2)}$ are uncorrelated Brownian motions.

Thus

$$\ln(S_t) = \ln(S_u) + r(t-u) + \left(\frac{\kappa\rho}{\sigma} - \frac{1}{2}\right)\int_u^t v_s ds + \frac{\rho}{\sigma}\left(v_t - v_u - \kappa\theta(t-u)\right) + \sqrt{1-\rho^2}\int_u^t \sqrt{v_s}\,dW_s^{(2)}.$$

The exact simulation method uses sampling *exactly* from the distribution of v_t.

Conditional on v(t), the cumulative distribution function of v(T) is a non-central chi-square distribution:

$$F_{x^2}(z;v,\lambda) = e^{-\frac{1}{2}}\sum_{j=0}^{\infty}\frac{\left(\frac{1}{2}\right)^j}{j!2^i + j\Gamma\left(\frac{y}{2}+j\right)}\int_0^z x^{\frac{y}{2}+j-1}e^{-\frac{1}{2}}dx.$$

$$d = \frac{4\kappa\theta}{\sigma^2}$$

$$F_{V(T)}^{Q}(z \mid V(t)) = F_{\chi^2}\left(\frac{zn(t,T)}{e^{-\kappa(T-t)}};d,V(t)n(t,T)\right), \qquad n(t,T) = \frac{4\kappa e^{-\kappa(T-n)}}{\sigma^2\left(1-e^{-\kappa(T-t)}\right)},T>t.$$

This can be done given v_u for some u < t by using sampling from the noncentral chi-squared distribution (which is a special case of the gamma distribution with scale parameter equal to two).

The next step having v_t and sample from v_u is generating a sample from the distribution of the integrated variance over time – i.e. the integral

$$\int_u^t v_s ds,$$

whose cumulative distribution function is

$$F^Q_{IV(t,T)}(z \mid V(t), V(T)) = \frac{2}{\pi}\int_0^\infty \frac{sin(\phi z)}{\phi} Re(\Xi(\phi))\, d\phi,$$

where

$$\Xi(\phi) = \frac{\gamma(\phi) e^{-\frac{1}{2}(\gamma(\phi)-\kappa)(T-t)}\left[1-e^{-\kappa(T-t)}\right]}{\kappa\left[1-e^{-\gamma(\phi)(T-t)}\right]}$$
$$\times \exp\left\{\frac{V(t)+V(T)}{\sigma^2}\left(\frac{\kappa\left[1+e^{-\kappa(T-t)}\right]}{1-e^{-\kappa(T-t)}} - \frac{\gamma(\phi)\left[1+e^{-\gamma(\phi)(T-t)}\right]}{1-e^{-\gamma(\phi)(T-t)}}\right)\right\}$$
$$\times I_{\frac{d}{2}-1}\left(\sqrt{V(t)V(T)}\frac{4\gamma(\phi)e^{-\frac{1}{2}\gamma(\phi)(T-t)}}{\sigma^2\left[1-e^{-\gamma(\phi)(T-t)}\right]}\right) /$$
$$I_{\frac{d}{2}-1}\left(\sqrt{V(t)V(T)}\frac{4\kappa e^{-\frac{1}{2}\kappa(T-t)}}{\sigma^2\left[1-e^{-\kappa(T-t)}\right]}\right)$$

with

$$\gamma(\phi) = \sqrt{\kappa^2 - 2\sigma^2 i\phi},$$

where $I_v(\cdot)$ is the modified Bessel function of the first kind with v degrees of freedom and the corresponding probability distribution function (PDF) is

$$f^Q_{IV(t,T)}(z \mid V(t), V(T)) = \frac{2}{\pi}\int_0^\infty \cos(\phi z)\, \mathrm{Re}(\Xi(\phi)) d\phi.$$

This is *not* computationally friendly. The trapezoidal rule is used to compute the probability distribution. The value of the integral is then simulated by the inverse transform method, by first generating a uniform standard normal variable U and then looking for the value of x for which the $P(V(u, t) \leq x)$ is equal to U.

The only unknown term, $\int_u^t \sqrt{v_s} dW_s^{(1)}$, can be easily obtained from the dynamics of v_t. Furthermore, the fact that v_t is independent of the second Brownian motion is used, and therefore the distribution of $\int_{f'}^t \sqrt{v_s} dW_s^{(2)}$ given the path generated by v_t is normal with mean zero and variance $\int_u^t v_s ds$. This scheme gives an unbiased estimator of the derivative price:

$$C_{heston} = \mathbb{E}\left[BS\left(S_0\xi, \sqrt{1-\rho^2}\sqrt{\frac{1}{T}\int_0^T V_s ds} \mid \int_0^t V_s ds, \int_0^t \sqrt{V_s} dW_s\right)\right].$$

However, the algorithm is very time-consuming (around 100 times the run-time of traditional discretization schemes), because of the application of the numerical inversion of a cumulative distribution using the characteristic function involving a modified Bessel function of the first kind, which makes it almost impossible to apply in real-world situations.

3.2.1.2.2.2.2 Accelerating the Broadie and Kaya Scheme: the QE Scheme by Andersen Andersen's goal was to speed up the first step of Broadie and Kaya: the QE scheme uses two different approximations to the non-central chi-square distribution, depending on the values of the variance. A non-central chi-squared random variable with a high non-centrality parameter can be represented by a power function applied to a Gaussian variable; then Andersen proposes an approximation based on moment-matching techniques.

Instead of sampling the integral of the variance he uses an approximation:

$$\int_u^t \sqrt{v_s} dW_s^{(2)} \approx \sqrt{\gamma_1 v_n + \gamma_2 v_{n+1}} Z_S \sqrt{\Delta_n},$$

which yields the following discretization:

$$\ln(S_{n+1}) = \ln(S_n) + K_0 + K_1 v_n + K_2 v_{n+1} + \sqrt{K_3 v_n + K_4 v_{n+1}} \cdot Z_S,$$

$$K_0 = -\frac{\rho\kappa\theta}{\sigma}\Delta_n,\ K_1 = \gamma_1\Delta_n\left(\frac{\kappa\rho}{\sigma} - \frac{1}{2}\right) - \frac{\rho}{\sigma},\ K_2 = \gamma_2\Delta_n\left(\frac{\kappa\rho}{\sigma} - \frac{1}{2}\right) + \frac{\rho}{\sigma},$$
$$K_3 = \gamma_1\Delta_n(1-\rho^2),\ K_4 = \gamma_2\Delta_n(1-\rho^2)$$

$$\text{and } \gamma_1 = \gamma_2 = 0.5.$$

However, we also include the drift, and so the resulting scheme that we use is

$$\ln(S_{n+1}) = \ln(S_n) + r\Delta_n + K_0 + K_1 v_n + K_2 v_{n+1} + \sqrt{K_3 v_n + K_4 v_{n+1}} \cdot Z_S.$$

This scheme generally yields good results, even when the Feller condition is violated.

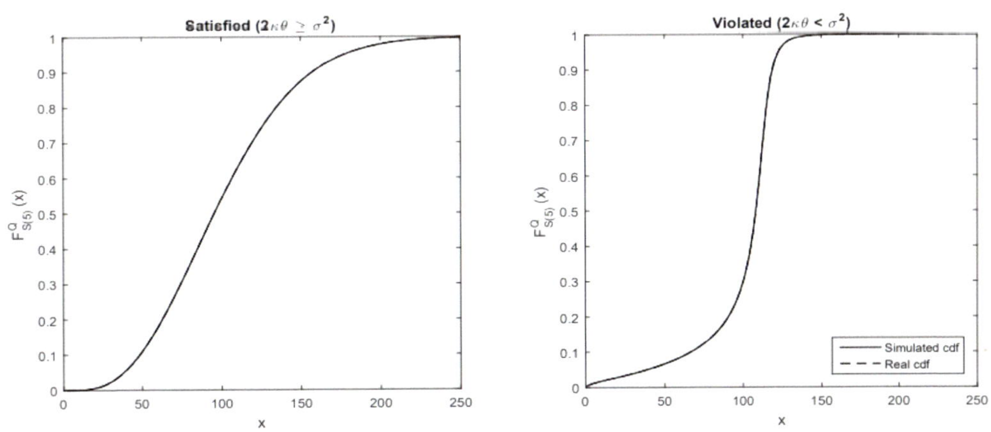

FIGURE 3.27 Feller condition test

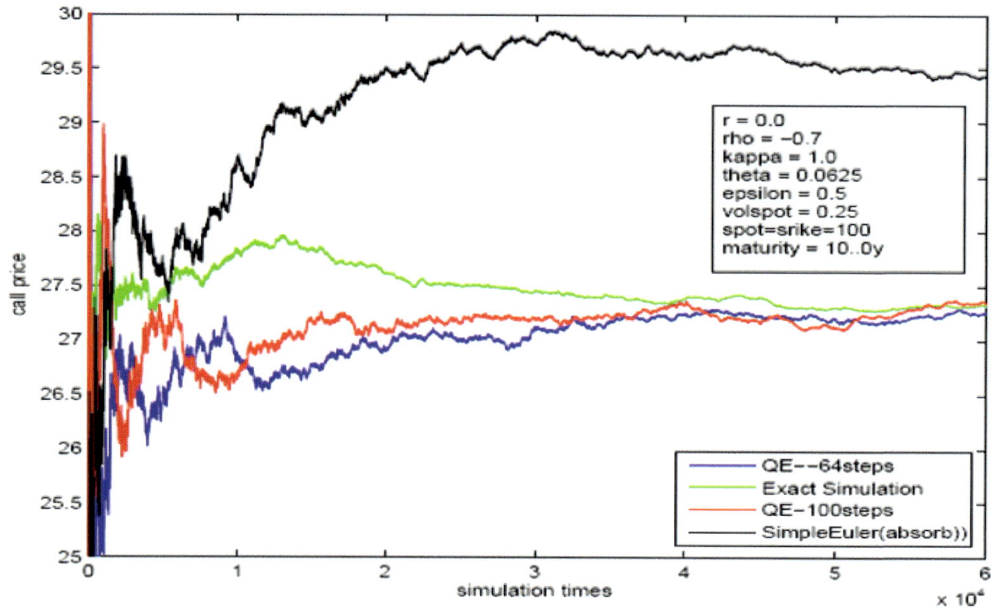

FIGURE 3.28 Convergence versus simulation times

3.2.1.2.2.2.3 Increasing the Order of Convergence of the Discretization Schemes The performance of simulation schemes sometimes depends on the satisfaction of a specific condition, known as the Feller condition, for the square root process.

If $2\kappa\theta \geq \sigma^2$, then the variance process variance cannot reach zero. When the Feller condition is violated (i.e. $2\kappa\theta < \sigma^2$), then the origin is accessible and strongly reflecting; in such cases, the likelihood of hitting zero is often quite significant. In Figure 3.30, the CDFs of V(T) given V(t) for two different sets of parameters illustrate the Feller condition. It can easily be seen that the cumulative distribution function is non-zero near the origin if the Feller condition is violated (dashed line); however, this is not the case when the condition is satisfied (solid line).

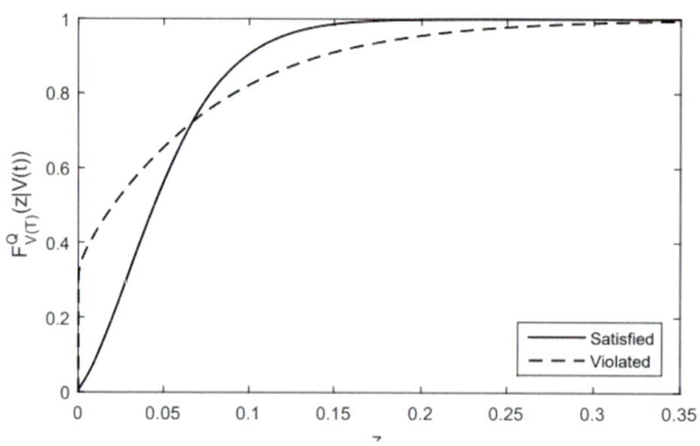

FIGURE 3.29 Feller condition test

Appropriate stochastic Taylor expansions can give consistent numerical schemes of an arbitrarily desired higher order. These expansions are derived through an iterated application of the stochastic chain rule (the Itô formula).

Stochastic Taylor Series (Schmitz-Abe 2007; Schmitz-Abe & Shaw 2005; Schmitz-Abe & Giles 2008)

$$\begin{aligned} dX_t &= A(Z_t)dt + B(Z_t)dW_{1,t} \\ dY_t &= C(Z_t)dt + \rho D(Z_t)dW_{1,t} + \hat{\rho} D(Z_t)dW_{2,t} \\ Z_t &= [X_t, Y_t, t], \quad \langle dW_{1,t}, dW_{2,t} \rangle = 0 \end{aligned}$$

$$X_t = X_{t_0} + \int_{t_0} A(Z_s)ds + \int_{t_0} B(Z_s)dW_{1,s},$$
$$Y_t = Y_{t_0} + \int_{t_0}^{t} C(Z_s)ds + \rho\int_{t_0}^{t} D(Z_s)dW_{1,s} + \hat{\rho}\int_{t_0}^{t} D(Z_s)dW_{2,s}.$$

The first integrals are deterministic Riemann integrals and the rest are Itô stochastic integrals.

As a reminder, if f is a differentiable function of Z, one obtains Itô's lemma:

$$df(Z_t) = \left(\frac{\partial f(Z_t)}{\partial t} + A(Z_t)\frac{\partial f(Z_t)}{\partial X} + C(Z_t)\frac{\partial f(Z_t)}{\partial Y}\right)dt$$
$$+\left(\frac{B^2(Z_t)}{2}\frac{\partial f^2(Z_t)}{\partial X^2} + \rho B(Z_t)D(Z_t)\frac{\partial f^2(Z_t)}{\partial X \partial Y} + \frac{D^2(Z_t)}{2}\frac{\partial f^2(Z_t)}{\partial Y^2}\right)dt,$$
$$+\left(B(Z_t)\frac{\partial f(Z_t)}{\partial X} + \rho D(Z_t)\frac{\partial f(Z_t)}{\partial Y}\right)dW_{1,t} + \hat{\rho}D(Z_t)\frac{\partial f(Z_t)}{\partial Y}dW_{2,t}.$$

$$f(Z_t) = f(Z_{t_0}) + \int_{t_0}^{t}(L_{de}f)(Z_s)ds$$
$$+\int_{t_0}^{t}(L_{W_1}f)(Z_s)dW_{1,s} + \int_{t_0}^{t}(L_{W_2}f)(Z_s)dW_{2,s},$$

where L_{de}, L_{W1} and L_{W2} are deterministic operators.

Applying the Itô lemma to $f(Z_t) = A(Z_t)$ and $f(Z_t) = B(Z_t)$ we get

$$A(Z_t) = A(Z_{t_0}) + \int_{t_0}^{t}(L_{de}A)(Z_u)du + \sum_{i=1}^{2}\left(\int_{t_0}^{t}(L_{W_i}A)(Z_u)dW_{i,u}\right).$$

Thus

$$X_t = X_{t_0} + A(Z_{t_0})\int_{t_0}^{t} ds + B(Z_{t_0})\int_{t_0}^{t} dW_{1,s}$$
$$+\left[\tilde{I}_{(W,W)} + \tilde{I}_{(dt,W)} + \tilde{I}_{(W,dt)} + \tilde{I}_{(dt,dt)}\right]_{t_0}^{t},$$

where the four double Itô integrals are defined by

$$\left[\tilde{I}_{(W,W)}\right]_{t_0}^{t} = \int_{t_0}^{t}\sum_{i=1}^{2}\left(\int_{t_0}^{s}(L_{Wi}B)(Z_u)dW_{i,u}\right)dW_{1,s} = O(\Delta t),$$
$$\left[\tilde{I}_{(dt,W)}\right]_{t_0}^{t} = \int_{t_0}^{t}\int_{t_0}^{s}(L_{de}B)(Z_u)du\,dW_{1,s} = O\left(\Delta t^{\frac{3}{2}}\right),$$
$$\left[\tilde{I}_{(W,dt)}\right]_{t_0}^{t} = \int_{t_0}^{t}\sum_{i=1}^{2}\left(\int_{t_0}^{s}(L_{W_i}A)(Z_u)dW_{i,u}\right)ds = O\left(\Delta t^{\frac{3}{2}}\right),$$
$$\left[\tilde{I}_{(dt,dt)}\right]_{t_0}^{t} = \int_{t_0}^{t}\int_{t_0}^{s}(L_{de}A)(Z_u)du\,ds = O(\Delta t^2).$$

Here

$$\left(L_{W_1}\cdot\right)(Z_t)=B\frac{\partial\cdot}{\partial X}+\rho D\frac{\partial\cdot}{\partial Y},$$
$$\left(L_{W_2}\cdot\right)(Z_t)=\hat{\rho}D\frac{\partial\cdot}{\partial Y}.$$

The Euler Scheme

The traditional Euler approximation is essentially equivalent to ignoring all the last double integrals, leading to

$$\widehat{X}_{t+\Delta t}=\widehat{X}_t+A\left(\widehat{Z}_t\right)\Delta t+B\left(\widehat{Z}_t\right)\Delta W_{1,t}.$$

The Euler scheme is the simplest schema to implement. It is very effective for products with short maturities, insofar as the time step dt can converge to zero without the computational run-time exploding.

Application to the Heston Equity Hybrid for Savings and Retirement

$$\begin{cases} S_{t+\Delta t}=S_t+r_tS_t\Delta t+\sqrt{V_t}S_t\Delta W_t^S \\ V_{t+\Delta t}=V_t+a(b-V_t)\Delta t+\eta\sqrt{V_t}\Delta W_t^V \\ r_{t+\Delta t}=r_t+\mu(t,r_t)\Delta t+v(t,r_t)\Delta W_t^r. \end{cases}$$

Generally speaking, practitioners cannot rely on the Euler algorithm to obtain paths of the variance process. When the Feller condition proposition is violated, the variance may take negative values, since the origin is then accessible and strongly reflecting (see Figure 3.31). If a better approximation is required, one needs to approximate one or more of the four double integrals.

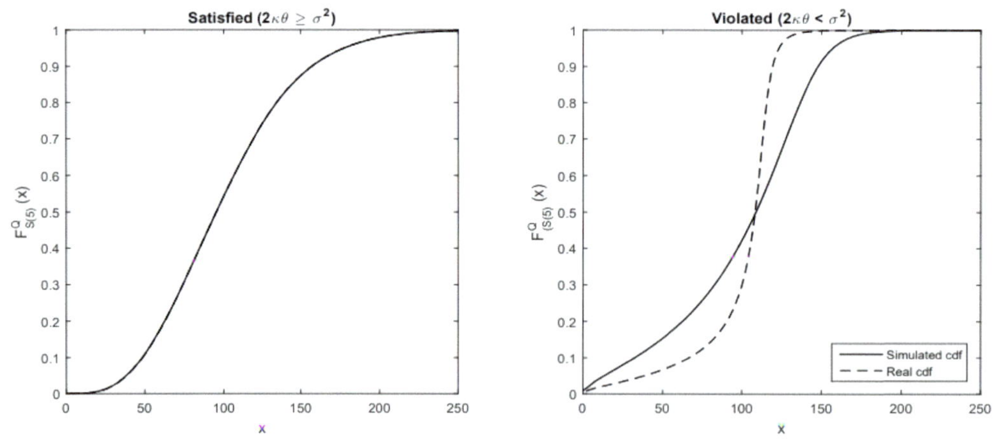

FIGURE 3.30 Feller condition test

Note that the essence of the method is to use the substitution repeatedly to obtain constant integrands in higher-order terms. For example, if one repeats this argument to

$$f(Z_t)=\left(L_{W_1}B\right)(Z_t)$$

$$f(Z_t)=\left(L_{W_2}B\right)(Z_t),$$

and we use the following approximation,

$$\left[I_{(i,i)}\right]_t^{t+\Delta t} = \int_t^{t+\Delta t}\int_t^s dW_{i,u}dW_{i,s} = \int_t^{t+\Delta t} W_{i,s}dW_{i,s} - W_{i,t}\int_t^{t+\Delta t} dW_{i,s}.$$

Using the Itô lemma,

$$d(W)^2 = 2WdW + dt, \ the \ WdW = ^{\frac{1}{2}\left(dW^2 - dt\right)}.$$

Thus

$$\begin{aligned}\left[I_{(i,i)}\right]_t^{t+\Delta t} &= \frac{1}{2}\int_t^{t+\Delta t}\left(dW_{i,s}^2 - ds\right) - W_{i,t}\left(W_{i,t+\Delta t} - W_{i,t}\right) \\ &= \frac{1}{2}\left(W_{i,t+\Delta t}^2 - t - \Delta t\right) - \frac{1}{2}\left(W_{i,t}^2 - t\right) - W_{i,t+\Delta t}W_{i,t} + W_{i,t}^2. \\ &= \frac{1}{2}\left(\left(W_{i,t+\Delta t} - W_{i,t}\right)^2 - \Delta t\right) = \frac{1}{2}\left(\Delta W_{i,t}^2 - \Delta t\right)\end{aligned}$$

Then the first double Itô integral above can be approximated as a result of the Itô lemma:

$$\left[\tilde{I}_{(W,W)}\right]_{t_0}^t = \int_{t_0}^t \sum_{i=1}^2 \left(\int_{t_0}^s \left(L_{W_i}B\right)\left(Z_u\right)dW_{i,u}\right)dW_{1,s} = O(\Delta t)$$

$$\left[\tilde{I}_{(W,W)}\right]_t^{t+\Delta t} = \frac{1}{2}\left(L_{W_1}B\right)\left(Z_t\right)\left(\Delta W_{1,t}^2 - \Delta t\right) + \left(L_{W_2}B\right)\left(Z_t\right)\left[I_{(2,1)}\right]_t^{t+\Delta t} + O\left(\Delta t^{\frac{3}{2}}\right).$$

This is called the Milstein correction; it is a stochastic effect.

The double Itô integral has the following mean and variance:

$$\begin{aligned}E\left[I_{(1,2)}\right] = E\left[I_{(2,1)}\right] &= \frac{\Delta W_{1,t}\Delta W_{2,t}}{2}, \\ \mathrm{Var}\left[I_{(1,2)}\right] = \mathrm{Var}\left[I_{(2,1)}\right] &= \frac{\Delta t}{12}\left(\Delta t + R^2\right) \\ R^2 &= \left(\Delta W_1\right)^2 + \left(\Delta W_2\right)^2,\end{aligned}$$

So, by truncating X_t at $O(_t)$, the 1.0 strong order of convergence, usually called the Milstein scheme, is

$$\widehat{X}_{t+\Delta t} = \widehat{X}_t + A\left(\widehat{Z}_t\right)\Delta t + B\left(\widehat{Z}_t\right)\Delta W_{1,t} + \frac{1}{2}\left(L_{W_1}B\right)\left(Z_t\right)\left(\Delta W_{1,t}^2 - \Delta t\right) + \left(L_{W_2}B\right)\left(Z_t\right)\left[I_{(2,1)}\right]_t^{t+\Delta t}$$

$$\left[I_{(2,1)}\right]_t^{t+\Delta t} = \int_t^{t+\Delta t}\int_t^s dW_{2,u}dW_{1,s}.$$

Now, using $d\left(W_1W_2\right) = W_2dW_1 + W_1dW_2$,

$$\left[I_{(2,1)}\right]_t^{t+\Delta t} = \int_t^{t+\Delta t}\int_t^s dW_{2,u}dW_{1,s}, \quad \left[I_{(2,1)}\right]_t^{t+\Delta t} = \int_t^{t+\Delta t} W_{2,s}dW_{1,s} - W_{1,t+\Delta t}W_{2,t} + W_{1,t}W_{2,t},$$

while

$$\begin{aligned}\left[I_{(1,2)}\right]_t^{t+\Delta t} &= \int_t^{t+\Delta t}\left(W_{1,s}-W_{1,t}\right)dW_{2,s}\\ &= \int_t^{t+\Delta t} W_{1,s}dW_{2,s} - W_{1,t}W_{2,t+\Delta t} + W_{1,t}W_{2,t}\\ &= -\int_t^{t+\Delta t} W_{2,s}dW_{1,s} + W_{1,t+\Delta t}W_{2,t+\Delta t} - W_{1,t}W_{2,t+\Delta t},\end{aligned}$$

$$\left[I_{(1,2)}\right]_t^{t+\Delta t} + \left[I_{(2,1)}\right]_t^{t+\Delta t} = \left(W_{1,t+\Delta t}-W_{1,t}\right)\left(W_{2,t+\Delta t}-W_{2,t}\right) = \Delta W_{1,t}\Delta W_{2,t}.$$

Using the definition of the Lévy area,

$$\left[\underline{L}_{(1,2)}\right]_t^{t+\Delta t} := \int_t^{t+\Delta t}\int_t^s dW_{1,u}dW_{2,s} - \int_t^{t+\Delta t}\int_t^s dW_{2,u}dW_{1,s},$$

$$\left[\underline{L}_{(1,2)}\right]_t^{t+\Delta t} = \left[I_{(1,2)} - I_{(2,1)}\right]_t^{t+\Delta t}, \qquad \begin{aligned} E\left[\underline{L}_{(1,2)}\right] &= E\left[I_{(1,2)} - I_{(2,1)}\right] = 0\\ \operatorname{Var}\left[\underline{L}_{(1,2)}\right] &= \frac{\Delta t}{3}\left(\Delta t + R^2\right).\end{aligned}$$

The first double Itô integral can therefore be written as

$$\begin{aligned}\left[I_{(2,1)}\right]_t^{t+\Delta t} &= \frac{1}{2}\left[\left(\left[I_{(1,2)}\right]+\left[I_{(2,1)}\right]\right)-\left(\left[I_{(1,2)}\right]-\left[I_{(2,1)}\right]\right)\right]_t^{t+\Delta t}\\ &= \frac{1}{2}\left(\Delta W_{1,t}\Delta W_{2,t} - \left[\underline{L}_{(1,2)}\right]_t^{t+\Delta t}\right).\end{aligned}$$

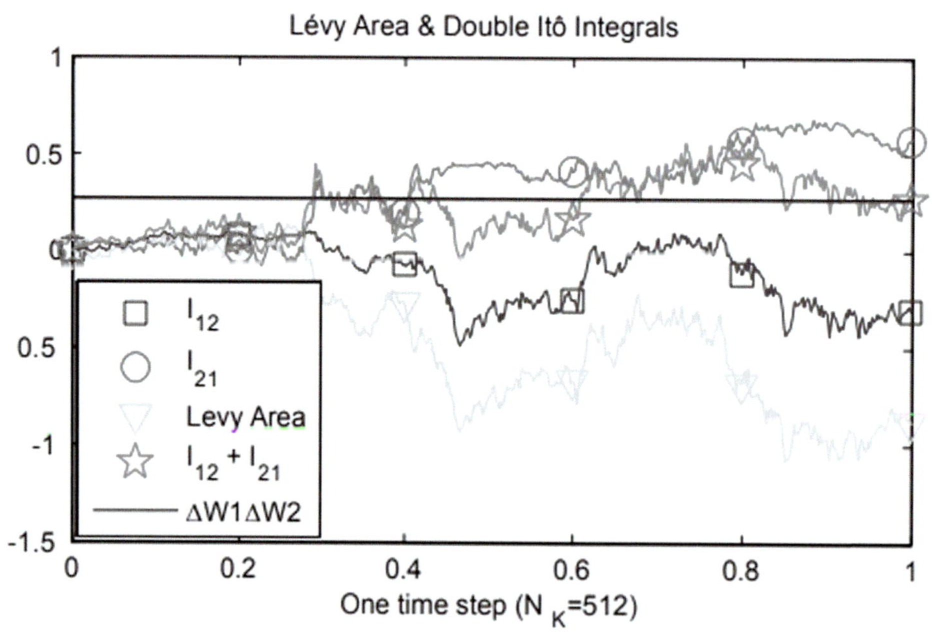

FIGURE 3.31 Lévy area and double Itô integrals

Then the Milstein scheme leads to

$$\widehat{X}_{t+\Delta t} = \widehat{X}_t + A\left(\widehat{Z}_t\right)\Delta t + B\left(\widehat{Z}_t\right)\Delta W_{1,t} + \frac{1}{2}\left(L_{W_1}B\right)\left(\widehat{Z}_t\right)\left(\Delta W_{1,t}^2 - \Delta t\right)$$
$$+\frac{1}{2}\left(L_{W_2}B\right)\left(\widehat{Z}_t\right)\left(\Delta W_{1,t}\Delta W_{2,t} - \left[\underline{\bar{L}}_{(1,2)}\right]_t^{t+\Delta t}\right).$$

Doing the same argument above for Y (t), the discrete time strong approximation of order 1.0 (Milstein scheme) for Y (t) is

$$\widehat{Y}_{t+\Delta t} = \widehat{Y}_t + C\left(\widehat{Z}_t\right)\Delta t + D\left(\widehat{Z}_t\right)\Delta\widehat{W}_{2,t} + \left[\rho\,\tilde{I}_{Y,1} + \hat{\rho}\,\tilde{I}_{Y,2}\right]_t^{t+\Delta t}$$

$$\left[\tilde{I}_{Y,1}\right]_t^{t+\Delta t} = \frac{1}{2}\left(L_{W_1}D\right)\left(\widehat{Z}_t\right)\left(\Delta W_{1,t}^2 - \Delta t\right) + \left(L_{W_2}D\right)\left(\widehat{Z}_t\right)\left[I_{(2,1)}\right]_t^{t+\Delta t}$$
$$\left[\tilde{I}_{Y,2}\right]_t^{t+\Delta t} = \frac{1}{2}\left(L_{W_2}D\right)\left(\widehat{Z}_t\right)\left(\Delta W_{2,t}^2 - \Delta t\right) + \left(L_{W_1}D\right)\left(\widehat{Z}_t\right)\left[I_{(1,2)}\right]_t^{t+\Delta t}.$$

Overall,

$$\widehat{X}_{t+\Delta t} = \widehat{X}_t + A\Delta t + B\Delta W_{1,t} + \frac{1}{2}B\frac{\partial B}{\partial X}\left(\Delta W_{1,t}^2 - \Delta t\right) + \frac{1}{2}D\frac{\partial B}{\partial Y}\left[\Delta W_L^-\right]_t^{t+\Delta t},$$
$$\widehat{Y}_{t+\Delta t} = \widehat{Y}_t + C\Delta t + D\Delta\widehat{W}_{2,t} + \frac{1}{2}D\frac{\partial D}{\partial Y}\left(\Delta\widehat{W}_{2,t}^2 - \Delta t\right) + \frac{1}{2}B\frac{\partial D}{\partial X}\left[\Delta W_L^+\right]_t^{t+\Delta t},$$

where

$$\left[\Delta W_L^{\pm}\right]_t^{t+\Delta t} = \rho\left(\Delta W_{1,t}^2 - \Delta t\right) + \breve{\rho}\left(\Delta W_{1,t}\Delta W_{2,t} \pm \left[\underline{\bar{L}}_{(1,2)}\right]_t^{t+\Delta t}\right).$$

Milstein integration includes all the O(Δt) terms, neglecting higher-order terms.

Milstein's scheme is a more refined method of discretization than Euler's: one adds here a term of zero average (the Lévy area is supposed to be zero) derived from a finer approximation of the stochastic integral of the asset, which allows us to approach more faithfully the real simulation of the asset under consideration (see Schmitz-Abe 2007). Here we can write this diagram in the following form.

Application to the Heston Equity Hybrid for Savings and Retirement

The 1-D Milstein Scheme

The 1-D Milstein scheme ignores the stochastic variation v_t in the S_t equation, and thus the first double Itô integral:

$$\begin{cases} S_{t+\Delta t} = S_t + r_t S_t \Delta t + \sqrt{V_t}\,S_t \Delta W_t^S + \frac{1}{2}V_t\left[\left(\Delta W_t^S\right)^2 - \Delta t\right] \\ V_{t+\Delta t} = V_t + a\left(b - V_t\right)\Delta t + \eta\sqrt{V_t}\Delta W_t + \frac{1}{4}\eta\left[\left(\Delta W_t^V\right)^2 - \Delta t\right] \\ r_{t+\Delta t} = r_t + \mu\left(t, r_t\right)\Delta t + v\left(t, r_t\right)\Delta W_t^r + \frac{1}{2}v'\left(t, r_t\right)v\left(t, r_t\right)\left[\left(\Delta W_t^r\right)^2 - \Delta t\right] \end{cases}.$$

When the Feller condition is not satisfied, it still is possible to show that the occurrence of negative variances is greatly reduced as compared to the Euler–Maruyama scheme. However, this algorithm still yields poor results when $4\kappa\theta < \sigma^2$, as Figure 3.33 shows.

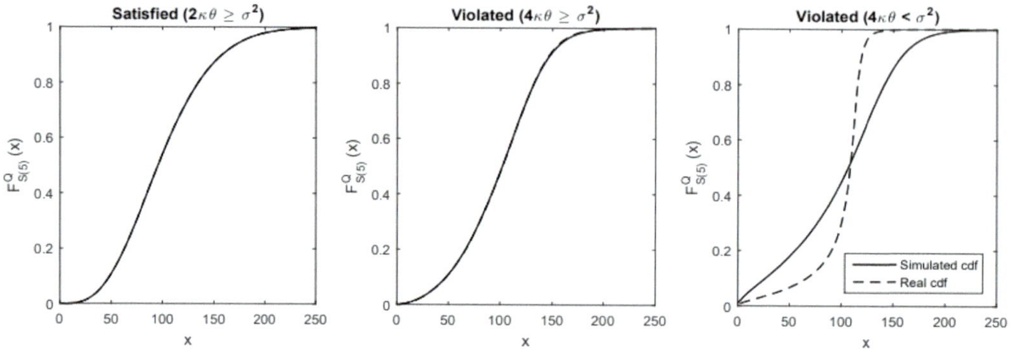

FIGURE 3.32 Feller condition test

The 1-D Milstein scheme has the same order of "weak" convergence 0.5 (i.e. for expected values – e.g. option prices) as the Euler scheme without having to calculate the first double Itô integral. The reason for this is that the scheme was applied without taking into account the correlation between the two systems.

<u>The 2-D Milstein Scheme</u>

To obtain a 1.0 order of convergence with the Milstein scheme one needs to apply the scheme to the vector form of, use independent Wiener processes and compute correctly the double integral or Lévy area, as follows.

Considering this term leads to the 2-D Milstein scheme:

$$\begin{cases} S_{t+\Delta t} = S_t + r_t S_t \Delta t + \sqrt{V_t} S_t \Delta W_t^S + \left(\frac{1}{2} V_t + \frac{1}{4}\rho_{sv}\eta\right)\left[\left(\Delta W_t^S\right)^2 - \Delta t\right] + \frac{1}{2}\sqrt{1-\rho_{sv}^2}\,\eta \mathbf{I}_{(2,1);t} \\ V_{t+\Delta t} = V_t + a\left(b - V_t\right)\Delta t + \eta\sqrt{V_t}\Delta W_t^V + \frac{1}{4}\eta\left[\left(\Delta W_t^V\right)^2 - \Delta t\right] \\ r_{t+\Delta t} = r_t + \mu\left(t, r_t\right)\Delta t + v\left(t, r_t\right)\Delta W_t^r + \frac{1}{2} v'\left(t, r_t\right) v\left(t, r_t\right)\left[\left(\Delta W_t^r\right)^2 - \Delta t\right] \end{cases},$$

$$\mathbf{I}_{(2,1);t} = \int_t^{t+\Delta t} \int_t^u dW_v^2 dW_u^1.$$

Figure 3.34 shows the price of puts of different maturities for Black–Scholes and hybrid Heston models with different discretization schemes, in which the price differences may be significant:

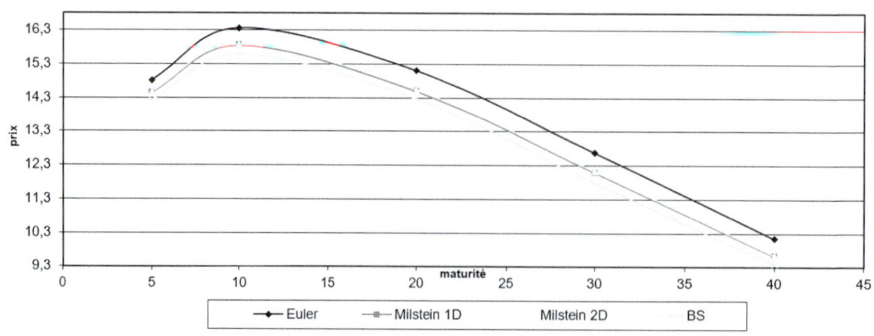

FIGURE 3.33 Put prices depending on the discretization scheme: equity–rate correlation 25%; equity volatility–level correlation –70%; rate–equity volume correlation 0%

Note: Number of simulations: 50,000.

This confirms the results on Heston call and put option prices (see Figure 3.35).

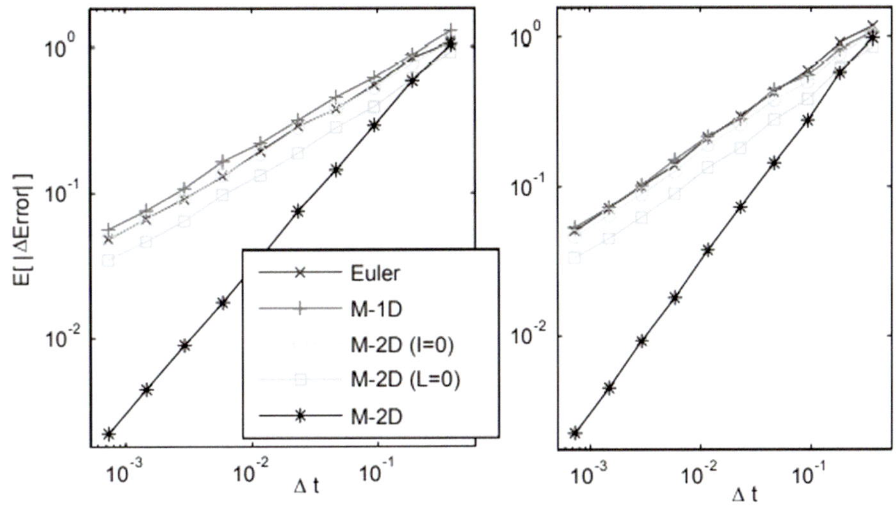

FIGURE 3.34 ΔError versus Δt

We also consider a combination of the exact scheme from Broadie and Kaya for the underlying asset process with the Milstein scheme for the volatility process, denoted "E+M".

An integration scheme that is more accurate than the accuracy of the Milstein scheme requires the evaluation of the three residual double Itô integrals. Unfortunately, these integrals and the one from the Milstein scheme cannot be accurately expressed in terms of the under laying random process ΔWt alone. Rather, accurate evaluation of these integrals requires the generation of additional random numbers.

In Figure 3.36 we illustrate the mean absolute relative errors for the different schemes, for 100,000 simulated paths.

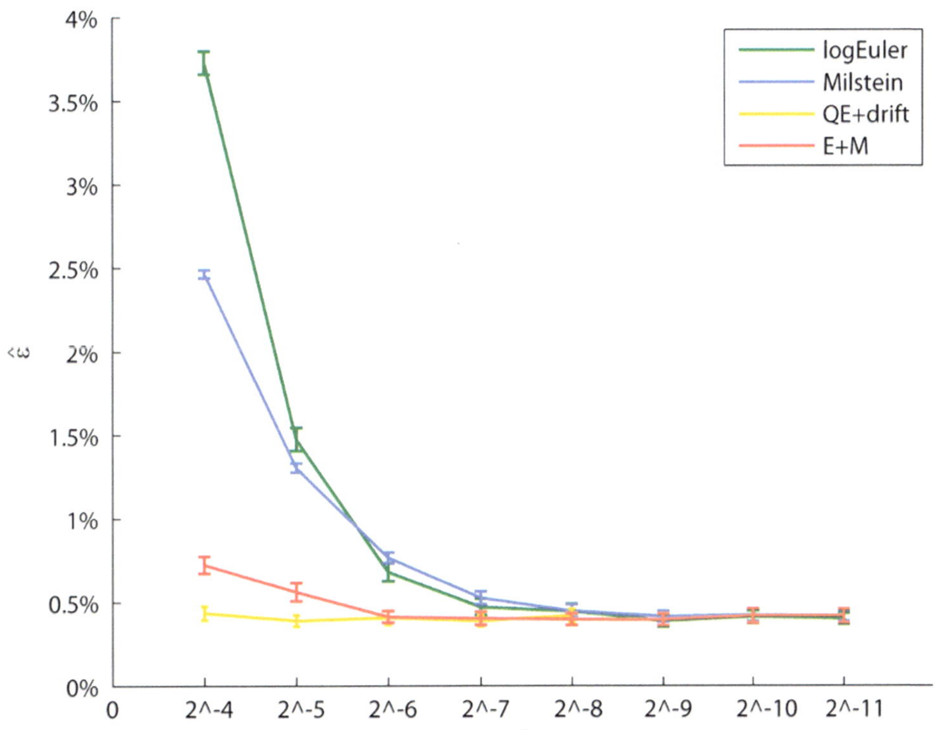

FIGURE 3.35 Mean absolute relative errors

3.2.1.2.2.2.4 *Reducing the Computational Cost of Simulations: the Multi-Level Monte Carlo*

$$dS(t) = a(S,t)dt + b(S,t)dW(t), \quad 0 < t < T,$$

to compute the expected value of f(S(T)), where f(S) is a scalar function with a uniform Lipschitz bound – i.e.

$$| f(U) - f(V) | \leq c \| U - V \|, \quad \forall U, V.$$

The simplest estimate for $E[f(S_T)]$ is the mean of the payoff values with N independent path simulations:

$$\hat{Y} = N^{-1} \sum_{i=1}^{N} f\left(\hat{S}_{T/h}^{(i)}\right).$$

It is well established that, provided a(S,t) and b(S,t) satisfy certain conditions, the expected mean-square error (MSE) in the estimate can be split into two elements,

$$\begin{aligned} \text{MSE} &= \text{E}\left[(\hat{Y} - \text{E}[P])^2\right] \\ &= \text{E}\left[(\hat{Y} - \text{E}[\hat{Y}])^2\right] + (\text{E}[\hat{Y}] - \text{E}[P])^2, \end{aligned}$$

where the first term is the variance of the estimator and the second term, $(E[Yb] - E[P])^2$, is the square of its bias due to discrete approximation.

It is therefore asymptotically of the form

$$\text{MSE} \approx c_1 N^{-1} + c_2 h^2,$$

where c_1, c_2, are positive constants. The first term comes from the variance due to the Monte Carlo sampling, and the second term comes from the O(h) bias introduced by the discretization scheme.

To compute Y we need to discretize {St}. If we divide a time interval [0, T] into D ∈ N subintervals [0 = t0 < t1 < t2 < · · · < tn < · · · < tD = T] by setting a time step interval Δt = T/D for n = 1, 2, . . ., D, we get the standard Monte Carlo (SMC) method.

To make the MSE $O(\varepsilon^2)$, so that the root-mean-square error (RMSE) is O(ε), requires that $N = O(\varepsilon^{-2})$ and h = O(ε), and hence the computational complexity (cost) is $O(\varepsilon^{-3})$:

$$\begin{aligned} MSE &\approx c_1 N^{-1} + c_2 dt^2 \\ &= O\left(\varepsilon^2\right) + O(\varepsilon). \\ &= O\left(\varepsilon^3\right) \end{aligned}$$

And this decides whether the Euler or Milstein discretization scheme is used.

The MLMC method reduces the first term, and the variance reduction effect leads to reduction in computational complexity.

Using MLMC, the computational complexity can be reduced to $O(\varepsilon^{-2} (\log \varepsilon)^2)$ for the Euler discretization and $O(\varepsilon^{-2})$ for the Milstein discretization, which reduces the variance, leaving unchanged the bias due to the discretisation scheme.

Computation cost	Standard Method		ML-MC
$MSE \approx c_1 N^{-1} + c_2 dt^2$			
$= O(\varepsilon^2) + O(\varepsilon)$	*Euler*: $O(\varepsilon^3)$	$\rightarrow$	$O\left(\varepsilon^{-2}(\log\varepsilon)^2\right)$
$= O(\varepsilon^3)$	*Milstein*: $O(\varepsilon^3)$	$\rightarrow$	$O(\varepsilon^{-2})$

FIGURE 3.36 Computation cost

The MLMC is a multi-grid method that uses a geometric sequence of grids, each typically twice as fine in each direction as its predecessor. If one were to use only the finest grid, the discretization error would be very small, but the computational cost of iteration would be very large. On a much coarser grid, the accuracy is much less, but the cost is also much less. A multi-grid approach solves the equations on the finest grid, by computing corrections using all the grids, thereby achieving the fine grid accuracy at a much lower cost.

Set D = M^L. Let Pbℓ, ℓ = 0, 1, . . . , L denote the approximation of $P(S_T)$ on level ℓ; Pbℓ is the discretized version of $P(S_T)$ with a time step interval hℓ = T/ M^l. Under the MLMC framework, the expected value on the finest level L can be uniquely constructed by

$$\mathrm{E}\left[\hat{P}_L\right] = \mathrm{E}\left[\hat{P}_0\right] + \sum_{\ell=1}^{L} \mathrm{E}\left[\hat{P}_\ell - \hat{P}_{\ell-1}\right].$$

Define

$$\hat{Y}_\ell = \begin{cases} N_0^{-1} \sum_{i=1}^{N_0} \hat{P}_0^i, & (\ell = 0) \\ N_\ell^{-1} \sum_{i=1}^{N_\ell} \left(\hat{P}_\ell^i - \hat{P}_{\ell-1}^i\right) & (0 < \ell \le L). \end{cases}$$

Then

$$\hat{Y} = \sum_{\ell=0}^{L} \hat{Y}_\ell,$$

and the MLMC implementation is as follows (see Giles 2008).

The expectation of a payoff *P* with maturity T is calculated by

$$E\left[P_L\right] = E\left[P_0\right] + \sum_{l=1}^{L} E\left[P_l - P_{l-1}\right]$$

where *L* is the level in the algorithm that simulates the scheme or time approximation with different time steps dt:

$$dt_L = \frac{T - T_0}{M^L} \quad \text{for}: \quad 2 \le M \in \mathbf{Z}.$$

For an error = ε, the algorithm will stop when it reaches convergence:

$$\max\left\{M^{-1}\left|P_{L-1}\right|, \left|P_L\right|\right\} < \frac{1}{\sqrt{2}}(M-1)\varepsilon.$$

For each level L, we need only a certain number of Monte Carlo paths:

$$N_L = \left[2\varepsilon^{-2} \sqrt{V_L dt_L} \sum_{l=1}^{L} \sqrt{\frac{V_l}{dt_l}} \right].$$

So, for a given ε =| exact solution – approximation |, these are the steps.

1. Start with L = 0.
2. Estimate the variance V_L using an initial N_L = 10^4 samples:

$$V_L = \sum_{l=1}^{L} N_l^{-1} \sum_{i=1}^{N_l} \left(P_l^{(i)} - P_l^{(i)} \right).$$

3. Define the optimal N_L:

$$N_L = \left[2\varepsilon^{-2} \sqrt{V_L dt_L} \Sigma_{l=1}^{L} \sqrt{\frac{V_l}{dt_l}} \right].$$

4. Evaluate extra samples at each level as needed for the new NL.
5. If L > 1, test for convergence:

$$\max\left\{ M^{-1} \left|P_{L-1}\right|, \left|P_L\right| \right\} < \frac{1}{\sqrt{2}}(M-1)\varepsilon.$$

6. If L = 1 or not converged, set L: = L + 1 and go to step 2.

Here we apply it to various options, while using the basic Euler and 1-D Milstein (with Lévy area = 0) discretization schemes, and we confirm the results by Giles (2008) for both European options (embedded within endowment contracts and term and whole life) and lookback options (which are often embedded within EIAs, variable annuities and structured annuities).

For ε = 1e-4 (1 bps), the MLMC method is more efficient than the standard method for

- European options: 21.68 times for the Euler scheme and 56.57 times for the Milstein scheme; and
- lookback options: 520.58 times for the Euler scheme and 537.72 times for the Milstein scheme.

3.2.1.3 Equity Market Crash Modelling for Sustainable Savings and Retirement

3.2.1.3.1 Recurrent Equity Market Crashes Since the 1990s recurrent unexpected severe volatility spikes (as illustrated by recurrent crashes: 1994 bond crash, 1997 Asian crisis, 1998 Russian crisis, 2001–2 Internet bubble crash, 2008 crash, 2010/2012/2014 European sovereign debt mini-crashes, August 2011 mini-crash, 2016 Brexit mini-crash, February and October–December 2018 mini-crashes, 2020 Covid-19 crash), combined with persistently low interest rates, have made most savings products less attractive and sustainable, translating into earnings losses for insurance companies and lower yields for policyholders.

Market volatility had a strong impact on the return's sustainability, as illustrated by the extent to which the sequence of returns affects a portfolio: despite an average 8% return of the S&P 500 over the past 30 years, the range has varied dramatically from 37% in 1995 to –37% in 2008.

Comparing the actual market crash empirical distribution with the theoretical probability induced by the Gaussian distribution that drives the equity Black–Scholes model clearly illustrates that the likelihood of extreme events is underestimated in terms of frequency.

Date	Closing	Log Return	Average frequency under Normal law
19-Oct-87	1738.74	-0.2563	once in 10^{53} years
26-Oct-87	1793.93	-0.0838	once in 72503 years
15-Oct-08	8577.91	-0.0820	once in 41318 years
01-Dec-08	8149.09	-0.0801	once in 21725 years
09-Oct-08	8579.19	-0.0762	once in 6068 years
27-Oct-97	7161.15	-0.0745	once in 3402 years
17-Sep-01	8920.70	-0.0740	once in 2914 years
29-Sep-08	10365.45	-0.0723	once in 1798 years
13-Oct-89	2569.26	-0.0716	once in 1405 years
08-Jan-88	1911.31	-0.0710	once in 1173 years

Ten largest down moves of the Dow since 1954.

FIGURE 3.37 Ten largest down moves of the Dow Jones since 1984

Index	Mean	Std Deviation	Skewness	Kurtosis
SP500 (1970-2001)	0.0003	0.0099	-1.6663	43.36
SP500 (1977-1999)	0.0009	0.0119	-0.4409	6.94
Nasdaq-Composite	0.0012	0.0154	-0.5439	5.78
DAX	0.0015	0.0157	-0.4314	4.65
SMI	0.0009	0.0141	-0.3584	5.35
CAC-40	0.0013	0.0143	-0.2116	4.63

FIGURE 3.38 Statistical moments

Statistically, the empirical distribution of asset returns is leptokurtic: the empirical fourth moment (so-called kurtosis) around the mean is greater than for a normal distribution with the same variance, which implies a large likelihood of extreme returns and fewer midrange returns than would be expected under a Gaussian distribution. This translates chart-wise into a higher peak and fatter tails. These fat tails and associated kurtosis are further emphasized by the rising volatility of volatility during market crashes, as Figure 3.40 shows).

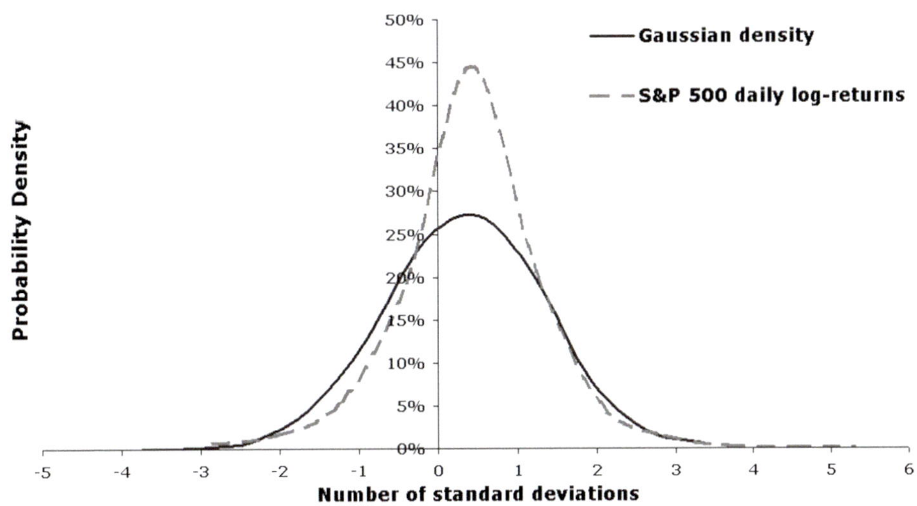

FIGURE 3.39 S&P 500 versus Gaussian distribution

Additional key empirical features of the return's volatility are as follows.

- Volatility clustering and mean reversion: periods of high and low volatility often appear in "blocks", such that large moves follow large moves and small moves follow small moves, implying that volatility (or variance) is auto-correlated. One clearly sees that there are periods with high absolute log-returns and periods with lower absolute log-returns. This is in contrast with the absolute value of simulated normal random variables (with the empirical standard deviation of the Standard & Poor's 500 Index [SP 500 or SPX]), where we can see a more homogeneous picture, often referred to as white noise.

  These observations motivate the introduction of models for asset price processes in which volatility is itself stochastic and mean-reverting.

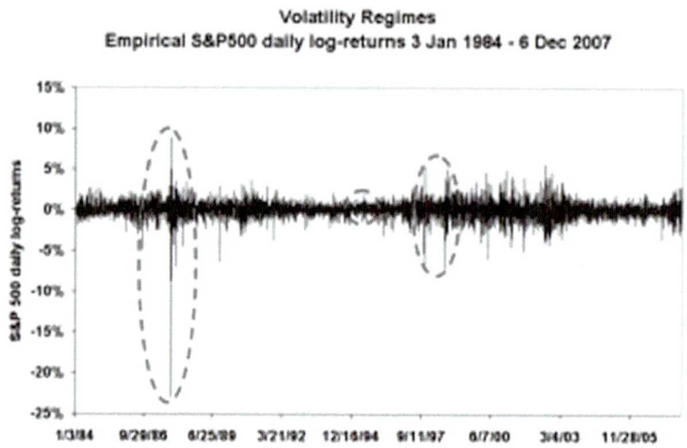

FIGURE 3.40 Volatility regimes, 3 January 1984–6 December 2007

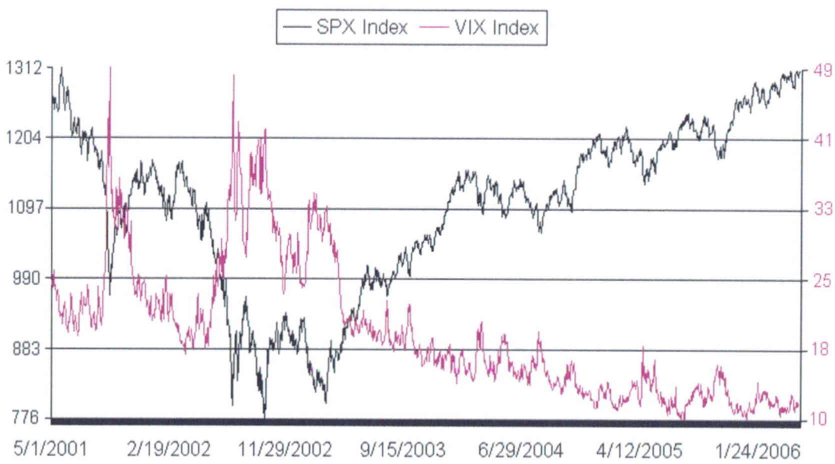

FIGURE 3.41 Leverage effect, 1 May 2001–31 March 2006

- Leverage effects: falling share prices increase the debt-to-equity ratio of firms, which leads to higher uncertainty, and thus volatility, in terms of the share price; hence price movements and volatility are negatively correlated, as empirically observed below

These two additional behaviour features of return volatility are also provided by the use of a random volatility within the return's modelling.

Another feature lies in the expected shortfall risk measure behaviour, defined as the expected excess over a given (high) level, conditionally on this level being exceeded. The sample version of the expected shortfall over a certain level is simply the average of the excesses over that level.

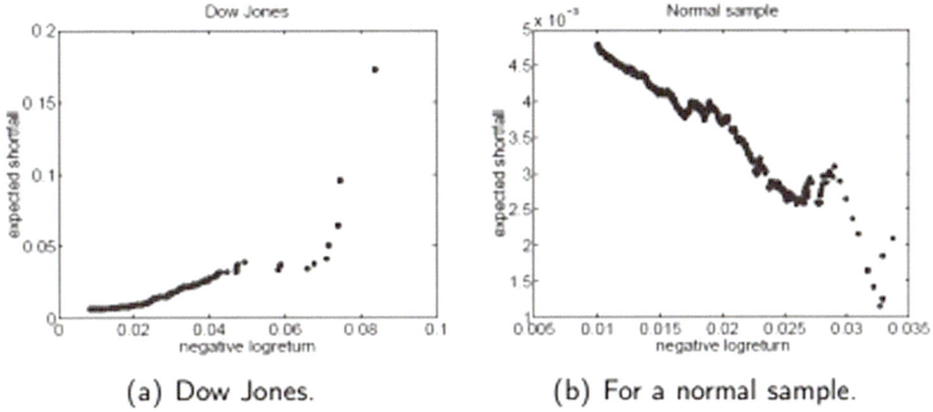

FIGURE 3.42 Expected shortfall over 1,000 largest negative daily log-returns of the Dow Jones and of a normal random sample with the same mean and variance

Note that the expected shortfall increases with the level: the higher the level being exceeded, the higher the excess by which it will be exceeded! Once more, this is in sharp contrast with the normal sample panel of the figure; for a normal (Gaussian) distribution sample with the same mean and variance, the expected shortfall decreases rapidly (note the different axes). In a light-tailed world, given that you exceed a high level, you hardly exceed it at all. But in a heavy-tailed world, once you know you will get hit, you may get hit much harder than expected!

Because of the thicker tail on the left, expressing the fear of a market crash, illustrated by the negative relationship between the volatility and the market – and thus between the volatility and the strike of the option – we expect the probability that an out-of-the-money put option finishes in the money to be higher than that for a normal distribution. This results in volatility skew, whereby implied volatility is much higher at very low strikes.

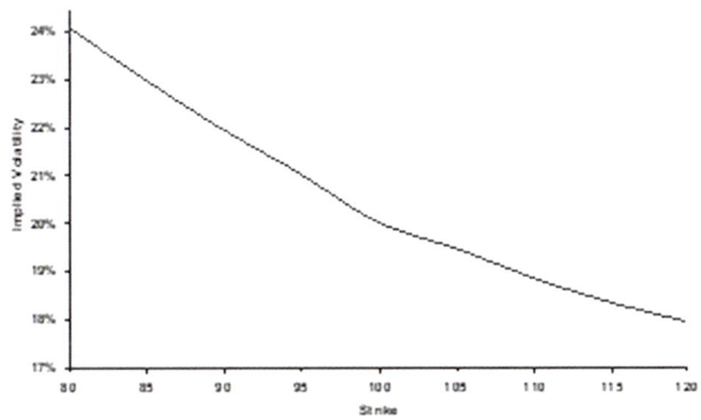

FIGURE 3.43 Implied volatility skew

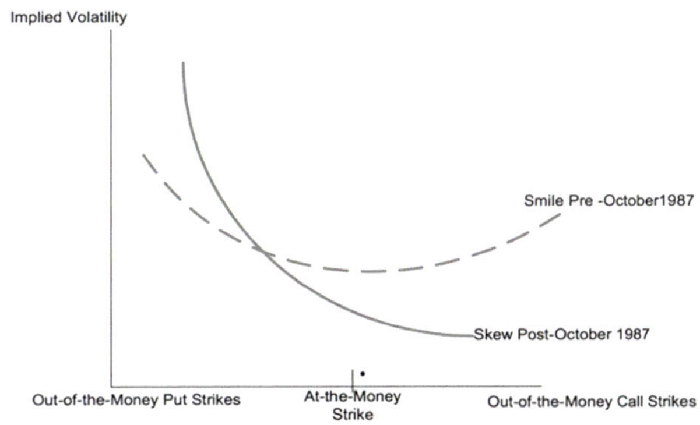

FIGURE 3.44 Movement of the volatility skew

The volatility skew appeared after the 1987 crash and was connected in some way with the visceral shock of discovering, for the first time since 1929, that a giant market could drop by 20% or more in a day or two.

Indeed, the markets began to observe a skew structure after the stock market crash of 1987; operators realized that, in the event of a sharp decline in the market, volatility increases and liquidity decreases, and, conversely, volatility decreases with a rising trend. Thus, they gave a higher probability to bearish events. To account for this phenomenon in the price of the options, they introduced a skew, which can make options more expensive for a lower strike and cheaper for a larger strike, as we can see in Figure 3.46.

This can be proxied with the following decomposition into a slope factor (so-called "skew" and a convexity factor):

$$V(K,T) = V_{ATMf} - \frac{skew}{\sqrt{T}} \ln\left(\frac{K}{F}\right) + convexity * \ln\left(\frac{K}{F}\right)^2 .$$

All these empirical features are part of the volatility skew behaviour across maturities.

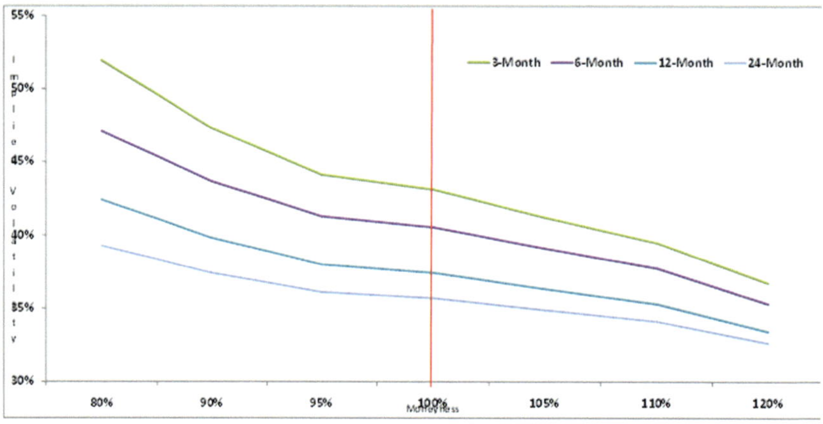

FIGURE 3.45 Skews across maturities

3.2.1.3.2 From Random Volatility to Pure Jump Process Modelling As demonstrated by Renault and Touzi (1996), any random volatility model is able to give rise to a "volatility skew".

Indeed, thicker tails and higher kurtosis can be generated by random volatility, as illustrated below, where we consider the most basic random volatility model using a binomial distribution (50% likelihood to be 10%, 50% likelihood to be 30%, versus a constant 20% volatility), providing a higher peak and larger kurtosis to the returns distribution (see Figures 3.47 and 3.48).

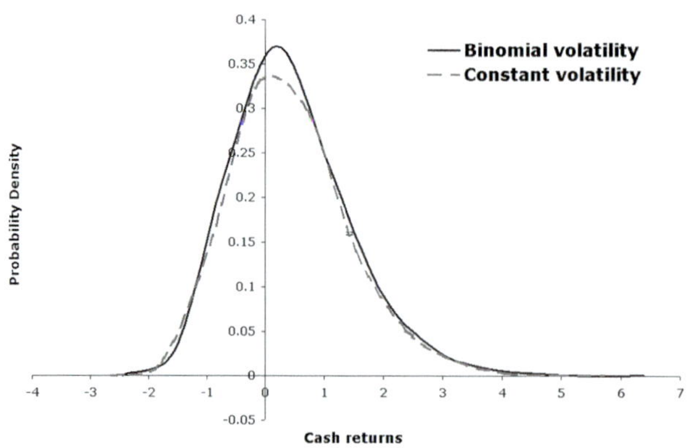

FIGURE 3.46 Returns distribution

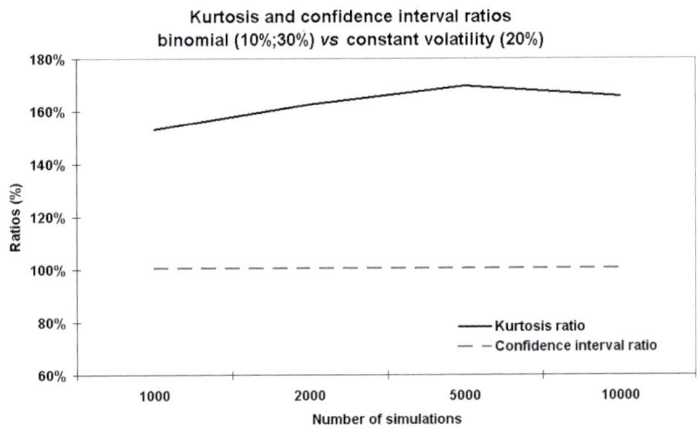

FIGURE 3.47 Kurtosis and confidence interval ratios

The asymmetry of the volatility skew can be derived from a negative correlation between the stock price and its volatility (it is usually around –70%).

Those two features can be modelled, by any of these methods.

- Implicitly, through a "local" volatility, defined as the expected instantaneous volatility at particular exercise price St = K and maturity t = T (this associates a local deterministic function σ to each pair of strike and maturity of the option (K,T), so with no additional random risk of its own; the corresponding asset dynamics is:

$$\frac{dS_t}{S_t} = r(t)dt + \sigma(S,t)dW_t^Q).$$

- Explicitly, through a so-called "stochastic volatility model" (e.g. Heston, stochastic alpha, beta, rho [SABR]), in which we assume that the volatility itself follows a stochastic process, with its specific source of risk, and so the Brownian motion driving the volatility process is different from the asset's Wiener process (remark: any stochastic volatility model has to be consistent with the local volatilities in expectation):

$$E\left[\sigma_T^2 \mid S_T = K\right] = \sigma^2(K,T).$$

Let us mention the two most well-known stochastic volatility models.

- Heston with mean reversion and a CIR process:

$$\begin{cases} \dfrac{dS_t}{S_t} = rdt + \sigma_t dW_t \\ d\sigma_t^2 = b\left(\sigma_\infty^2 - \sigma_t^2\right)dt + \beta\sigma_t dZ_t. \end{cases}$$

- SABR:

$$dF_t = \alpha_t F_t^\beta dW_t$$
$$d\alpha_t = v\alpha_t dZ_t.$$

Both have correlation ρ between volatility and the forward rate level.

In the event ρ= 1, it gives way to normal and log-normal models (depending on the value of β).

- Explicit pure jump dynamics (Merton, variance gamma, Carr–Geman–Madan–Yor [CGMY] models).

3.2.1.3.3 Local Volatility Models Are Fine for Life Products (Endowments, Term, Whole Life, Universal) but Face Limitations for Complex Savings and Retirement The local volatility can be proxied as

$$\sigma(T,K)=\sqrt{2*\frac{\frac{\partial C}{\partial T}(K,T)+rK\frac{\partial C}{\partial K}(K,T)}{K^2\frac{\partial^2 C}{\partial K^2}(K,T)}}.$$

Expressed in terms of implied volatility:

$$\sigma_{loc}^2(K,T)=\frac{\frac{\sigma_{imp}}{T-t}+2\frac{\partial\sigma_{imp}}{\partial T}+2r(T)K\frac{\partial\sigma_{\text{imp}}}{\partial K}}{K^2\left(\frac{\partial^2\sigma_{imp}}{\partial K^2}-d_1\sqrt{T-t}\left(\frac{\partial\sigma_{imp}}{\partial K}\right)^2+\frac{1}{\sigma_{imp}}\left(\frac{1}{K\sqrt{T-t}}+d_1\frac{\partial\sigma_{imp}}{\partial K}\right)^2\right)}.$$

Remark: the local volatility skew is about twice as big as the implied volatility skew.

As an example of a local volatility model, we select below the CEV process, which is the solution to the following SDE for $t \geq 0$: $dXt = \mu Xtdt + \sigma X^{\alpha/2}\, t\, dWt$, $X0 = x$, with instantaneous variance of dXt/Xt being $\sigma^2 X^{\alpha-2}{}_t$. Note that, when $\alpha = 2$, this is just the Black–Scholes model.

As the local random volatility model (CEV here) is more realistic than the constant volatility (Black–Scholes) model, the gap between their respective sensitivities (Greeks) tells us about the risks taken by assuming a constant volatility, as illustrated on a put (see Figure 3.49) across moneyness and maturities and depending on the volatility regime: the constant volatility (Black–Scholes) model tends to underestimate the risks for long-term deep in-the-money-guarantees (with more negative delta), in low volatility regimes (2005) and high volatility regimes (2009).

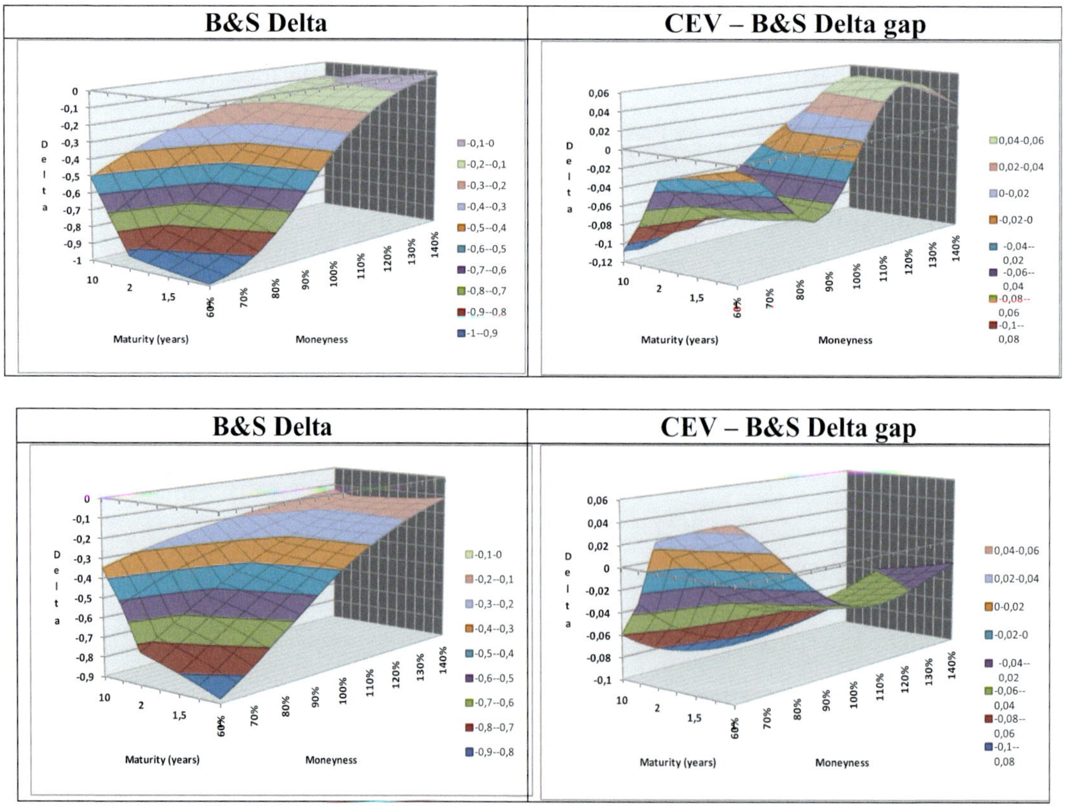

FIGURE 3.48 B–S delta versus gap between CEV and B–S deltas, 2005 & 2009

A local volatility model can fit the initial skew in normal market conditions, but faces limits in hectic markets, either through poor fitting (as illustrated in Figure 3.50 for 3 June 2009).

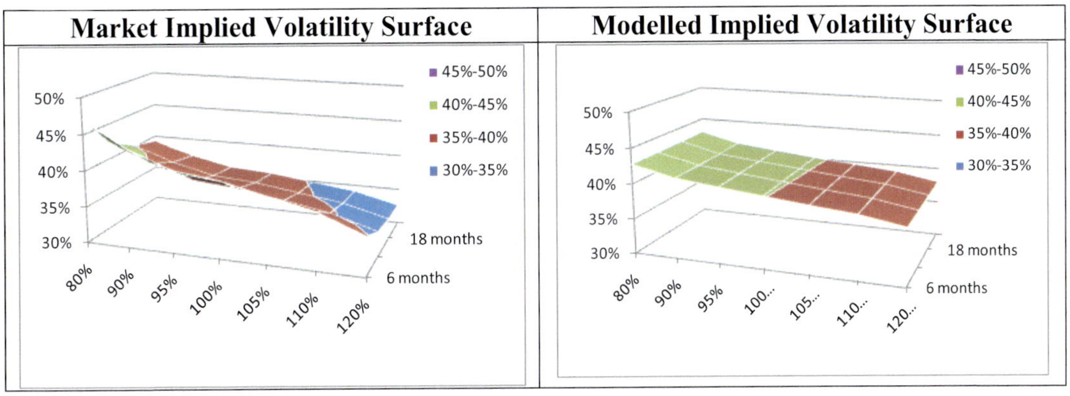

FIGURE 3.49 Fitting implied volatility surface, 3 June 2009

Local volatility models are self-consistent and arbitrage-free and can be calibrated to precisely match observed spot skews: the model correctly simulates the market smile at any strike and for any maturity T starting from the present date.

Local volatility models are indeed easy to calibrate statistically:

$$V(K,T) = V_{ATMf} - \frac{skew}{\sqrt{T}} \ln\left(\frac{K}{F}\right) + convexity * \ln\left(\frac{K}{F}\right)^2.$$

Thus, using the Dupire formula for local volatility,

$$V_{\text{ATMf}} - \frac{skew}{\sqrt{T}} \ln\left(\frac{K}{F}\right) + convexity * \ln\left(\frac{K}{F}\right)^2 = \frac{1}{K}\sqrt{2\frac{\frac{\partial C}{\partial T}(K,T)}{\frac{\partial^2 C}{\partial K^2}(K,T)}}.$$

We can then calibrate the three different factors (ATM, skew slope and skew convexity).

- Volatility at the money: using straddle market quotes using the following first-order approximation:

$$\frac{Vol_{impl}\left(Call_{ATMF}\right) + Vol_{impl}\left(Put_{ATMF}\right)}{2} = V_{ATMF}.$$

- Skew: using call spreads and put spread market quotes:

$$\frac{Vol(110\%) - Vol(90\%)}{2} \sim skew.$$

- Convexity: using butterfly quotes:

$$\frac{Vol(90\%) - 2 * Vol(100\%) + Vol110\%)}{2} \sim convexity.$$

Then interpolation methods (e.g. cubic splines) are used to ensure the continuity of the volatility surface …

$$\sigma_{imp}(K,T) = a(K) + b(K)T + c(K)T^2 + d(K)T^3$$

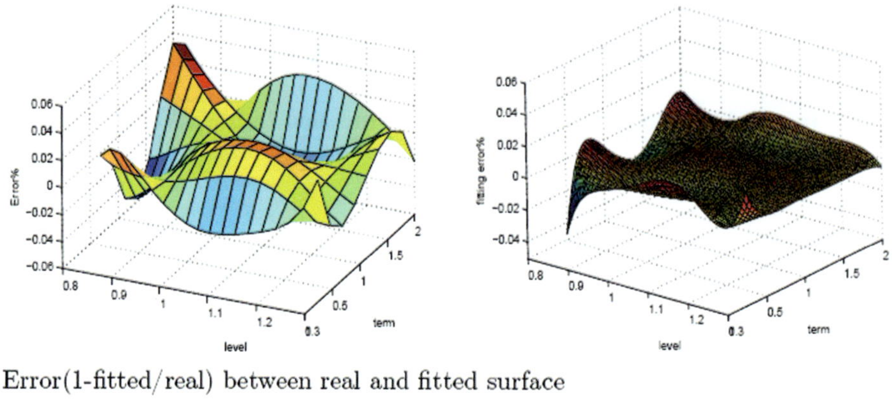

FIGURE 3.50 Error between real and fitted surfaces

… combined with no-arbitrage inequality constraints, illustrative of positive spreads and butterfly option premiums:

In terms of sensitivities, we have the following relationship:

$$\Delta = \Delta_{BS} + Vega_{BS} \frac{\partial \sigma_{imp}}{\partial K}.$$

As long as the savings and retirement contract payoff depends on the maturity date only (European payoffs), such as endowment, term, whole life products, only the spot skew is at risk, which implies that local volatility is fine.

However, as illustrated below, whenever the savings and retirement contract payoff depends on intermediary dates, such as annuities (equity-enhanced annuities, fixed or variable annuities), which very often involve ratchets and flexible withdrawal optionalities, then the forward skew is at stake, and the dynamic behaviour of skews need to be modelled. And there those skews dynamics predicted by local volatility models are exactly the opposite of the behaviour observed in the marketplace.

- When the price of the underlying asset decreases local volatility, models predict that the smile shifts to higher prices; when the price increases, these models predict that the smile shifts to lower prices. In reality, asset prices and market smiles move in the same direction.

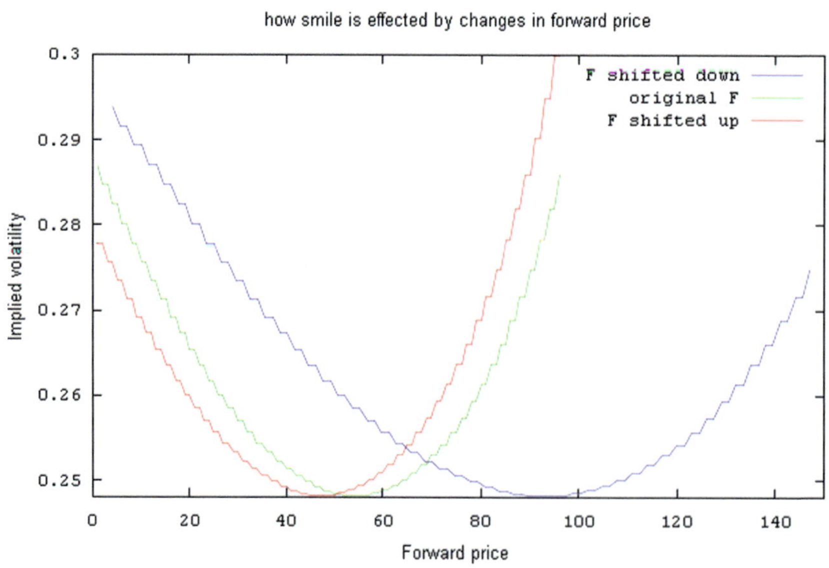

FIGURE 3.51 How the smile is affected by changes in the forward price

> This contradiction between the model and the marketplace tends to destabilize the delta and vega hedges derived from local volatility models, and often these hedges perform worse than the naive Black–Scholes hedges.

Moreover, other behaviours of the skew modelled by a local volatility model are not realistic.

- As the price of the underlying asset increases, the slope of the curve decreases. The changes in the slope of the smile are way too high for the local volatility model.
- The curvature of the smile stays roughly as it is. Note that this is due to the fact that the curvature of the smile is roughly equal for all points on the curve. From market data we might see a small decrease in the curvature for an increase in the forward swap rate.
- Local volatility may give rise to a skew with quite a steep slope, but with a small curvature, in contrast with the observed short-dated implied volatility skews of swaptions.

In addition, experience has demonstrated that, although the model is extremely accurate for evaluating products that start today and end at time T, it is not accurate for forward products that start at a future time U and end at T. This is very unfortunate, as most savings and retirement contracts are forwards products.

Finally, the local volatility model implies that the volatility at a given time depends only on the underlying S(T), whereas experience has demonstrated that this dependence is not really constant.

The failure of local volatility means that we cannot use a Markovian model based on a single Brownian motion to manage our skew risk. This implies the need for a further parameter to introduce another stochastic source, translating into a two-factor model (forward rate level and volatility), leading to the popularity of stochastic volatility models.

3.2.1.3.4 Stochastic Volatility Models Are Suitable for Annuities (Enhanced, Fixed, Variable)

3.2.1.3.4.1 Stochastic Volatility Models Improve the Calibration Error The generic formulation is as follows:

$$dS = S(\mu - D)dt + S\sqrt{v}dW_1,$$
$$v = \sigma^2, \quad E[dW_1 dW_2] = \rho dt.$$

These are the most usual specific cases.

Hull and White (1987):

$$d\sigma = \sigma\left[\bar{w}dt + \bar{\xi}dW_2\right] \quad or^3 \quad dv = v\left[wdt + \xi dW_2\right].$$

Stein and Stein (1991):

$$d\sigma = (\bar{w} - \bar{\zeta}\sigma)dt + \bar{\xi}dW_2 \quad or^5 \quad dv = (\sqrt{v}w - \zeta v)dt + \xi\sqrt{v}dW_2.$$

Heston (1993):

$$dv = (\bar{w} - \bar{\zeta}v)dt + \bar{\xi}\sqrt{v}dW_2 \quad or^6 \quad d\sigma = \left(\frac{\bar{w}}{\sigma} - \bar{\zeta}\sigma\right)dt + \bar{\xi}dW_2.$$

Generalized autoregressive conditional heteroscedasticity (GARCH):

$$dv=(w-\zeta v)dt+\xi v dW_2 \quad \text{or}^{7} \quad d\sigma=\left(\frac{\bar{w}}{\sigma}-\bar{\zeta}\sigma\right)dt+\bar{\xi}\sigma dW_2.$$

3/2 model:

$$dv= v(w-\zeta v)dt+\xi v^{3/2}dW_2 \quad \text{or}^{8} \quad d\sigma=\sigma^2\left[\left(\frac{\bar{w}}{\sigma}-\bar{\zeta}\sigma\right)dt+\bar{\xi}dW_2\right].$$

Their respective steady-state distributions (based on Monte Carlo simulations) are shown in Figure 3.53.

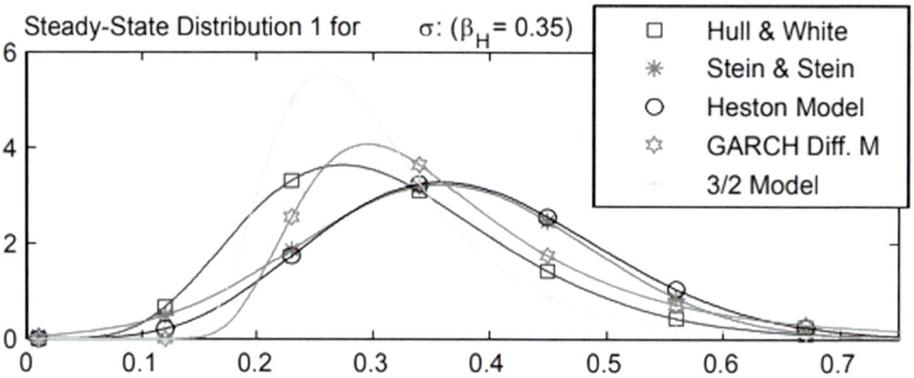

FIGURE 3.52 Steady-state distribution of volatility

In the following we focus on the Heston model. For small d_t, this model keeps the volatility positive, and is the most popular from among them because of its two main features: it has a semi-analytical pricing formula, which is easy to implement; and the solution is typical (it displays the same qualitative properties that one expects in general time-homogeneous cases):

$$\frac{dS_t}{S_t}=\mu_t dt+\sqrt{V_t}\,dW_t^1$$
$$dV_t=k\left(\theta-V_t\right)dt+\sigma V_t^{\alpha}dW_t^2.$$

In this case, the parameter σ is interpreted as the volatility of the variance, or better known as "vol of vol" k is the rate of mean reversion and θ is the long-term average of the variance. The Heston model, which provides a quasi-analytical formula similar to Black–Scholes, is a special case of the above model with α = 0.5.

Compared to local volatility models, the vol of vol parameter (epsilon) has a non-negligible impact specifically on long-term options: increasing the volatility of the flight has the effect of increasing the convexity of the skew, as illustrated. This effect is mitigated as maturity increases, the latter having the opposite effect on convexity.

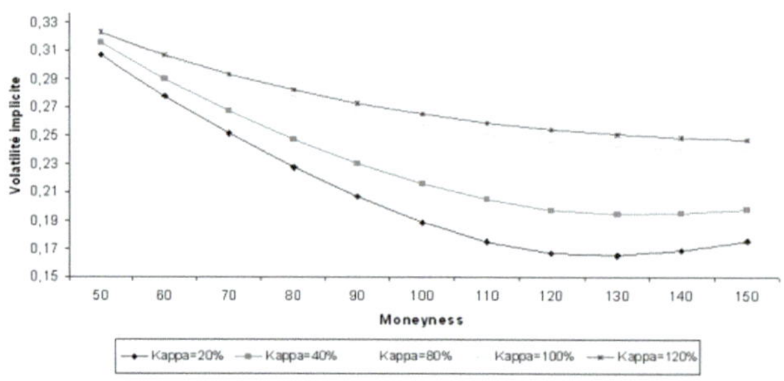

FIGURE 3.53 Impact of kappa on implied volatility across moneyness

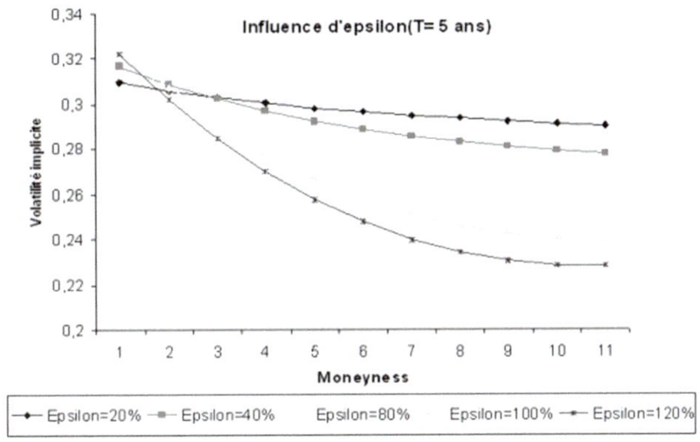

FIGURE 3.54 Impact of epsilon on implied volatility across moneyness

In contrast, increasing the rate of the mean reversion parameter is equivalent to concentrating the variance more strongly around its long-term average, and so reduces the variance of the variance process and the skew accordingly. Intuitively, decreasing κ must therefore increase the convexity of the smile.

Regarding the correlation parameter, a more negative correlation of the model between the asset and its variance process tends to give more weight to bearish events: when the value of the asset decreases, its variance will be more likely to increase.

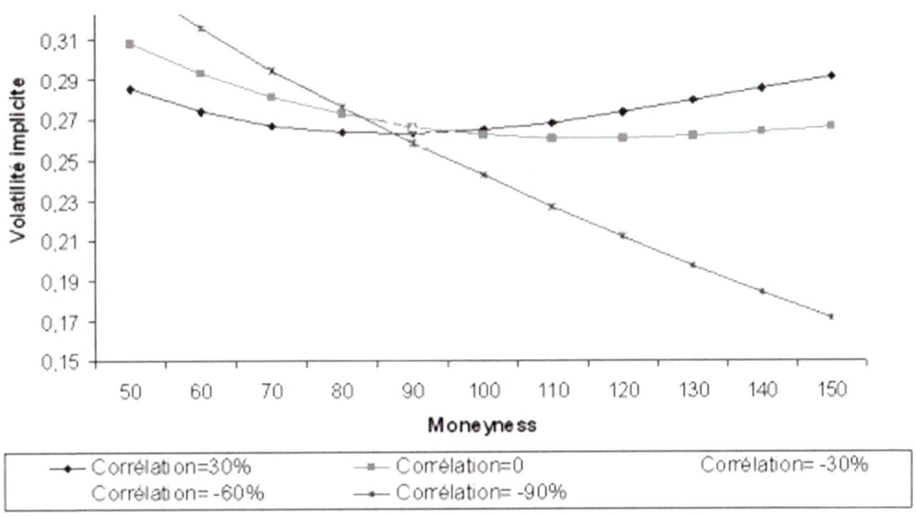

FIGURE 3.55 Impact of correlation on implied volatility across moneyness

Overall, in terms of options prices versus Black–Scholes, short-term ATM calls are undervalued compared to the BS model, and this undervaluation is an increasing function of the volatility of the stochastic variance; in contrast, short-term OTM Calls are overvalued compared to the Black–Scholes, and this overvaluation is an increasing function of variance volatility stochastic; both models converge in premium when variance volatility stochastic tends towards zero.

If Heston overvalues the price compared to Black–Scholes in the short term, in the long term the differences are greatly reduced, linked to a flattening of the skew, albeit at a lower pace than local volatility models.

Regarding the calibration, stochastic volatility models improve the calibration error, as illustrated in the Heston model during the financial crisis that started in 2008, on 3 June 2009 (see Figure 3.57).

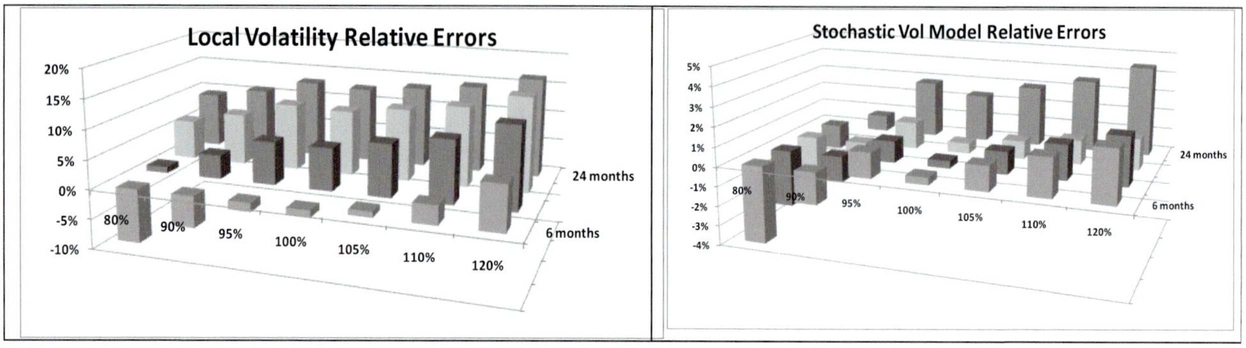

FIGURE 3.56 Local volatility and stochastic volatility models' relative errors

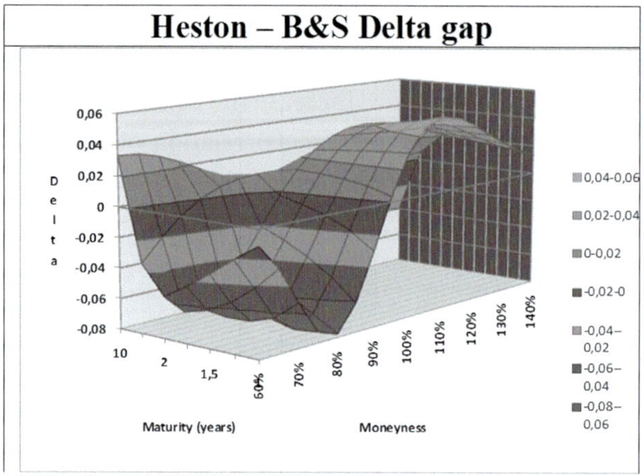

FIGURE 3.57 Gap between Heston and B–S deltas

Such a model enables not only a better fit to the market volatility level but, more importantly, adaptation to a broad range of market volatility regime dynamics. More specifically,

- an increase in the forward swap rate is a decrease in the ATM level, but there is also an increase in the slope as well as curvature; and
- an increase in the instantaneous volatility directly affects the ATM level of the skew, with a secondary effect of a slight steepening.

Therefore, stochastic volatility models model have the capability of capturing the volatility skew in various market conditions and predict the correct skew dynamics, which leads to stable hedging of our skew exposures. We then plug in the above volatility to compute European options prices.

3.2.1.3.4.2 The Heston Model Faces Multiple Calibration Issues In terms of calibration, one of the difficulties in solving the calibration problem is that the information observed from market data is often insufficient to exactly identify the parameters. Several sets of parameters may perform well and produce model prices that are consistent with the prices observed on the market, causing the awkwardness of the problem.

Specifically to the Heston model, the Feller condition, which ensures that the variance process is positive ($2\kappa\theta > \sigma^2$), is hard to satisfy in practice when calibrating to market data.

Regarding the Heston model, GA and ASA are usually used as global optimizers and MS Excel's Solver and MATLAB's lsqnonlin as local optimizers. Global optimizers can find a solution on their own even without any initial guess, but local optimizers, on the other hand, need a suitable starting point. Hence the combinations of global and local optimizers also follow, with the function tolerance stopping criterion to be TolFun = 1e−9.

3.2.1.3.4.3 The SABR Model Provides Significant Tractability with respect to Skew Modelling This is the SABR model:

$$dF_t = \alpha_t F_t^{\beta} dW_t$$
$$d\alpha_t = \nu \alpha_t dZ_t.$$

α is the stochastic volatility whereas β specifies the relationship between the future spot price and the ATM volatility and is dependent on time to maturity.

From liquid European options prices, the equivalent implied log-normal volatility for Black's model is then obtained through an analytical, closed-form approximate solution (see Hagan et al. 2002):

$$\hat{\sigma}(K, f, T) = A \cdot \left(\frac{z}{\chi(z)}\right) \cdot B$$

with

$$A = \frac{\sigma_0^T}{(fK)^{\frac{1-\beta}{2}} \left[1 + \frac{(1-\beta)^2}{24} \ln^2 \frac{f}{K} + \frac{(1-\beta)^4}{1920} \ln^4 \frac{f}{K} + \ldots\right]},$$

$$B = \left[1 + \left(\frac{(1-\beta)^2}{24} \frac{\left(\sigma_0^T\right)^2}{(fK)^{1-\beta}} + \frac{\rho\beta\nu\sigma_0^T}{4(fK)^{\frac{1-\beta}{2}}} + \frac{2-3\rho^2}{24}\nu^2\right) \cdot T + \ldots\right],$$

$$z = \frac{\nu}{\sigma_0^T} (fK)^{\frac{1-\beta}{2}} \ln \frac{f}{K},$$

$$\chi(z) = \ln\left(\frac{\sqrt{1 - 2\rho z + z^2} + z - \rho}{1-\rho}\right).$$

Thus the SABR model predicts the correct smile dynamics, which leads to stable hedging of our skew/smile exposures; it predicts that, whenever the forward price changes, the implied volatility curve shifts in the same direction and by the same amount as the forward price.

As a result, it leads to stable hedging of our skew/smile exposures.

Remark: the formula simplifies quite considerably for ATM (K = f) options, in line with the middle factor:

$$\frac{z}{\aleph}(z)_{=1.} \sigma_{ATM}(\tau) = \sigma_B(f, f, \tau) = \frac{\sigma_0^T}{F^{1-\beta}} \left(1 + \left(\frac{(1-\beta)^2 \sigma_0^{T^2}}{24F^{2(1-\beta)}} + \frac{\rho\beta\sigma_0^T}{4F^{1-\beta}}\nu + \frac{2-3\rho^2}{24}\nu^2\right)\right)_{\tau}$$

✓ **Quick win**: SABR volatility can be used as a way to smooth the implied volatility surface instead of using one of the usual non-parame tric techniques, such as cubic splines, thus making calibration easier with other models such as Hull–White on the liability side.

The relative impacts of the four parameters on fat tails on the skew, level, slope and curvature of the skew are assessed in Figure 3.59.

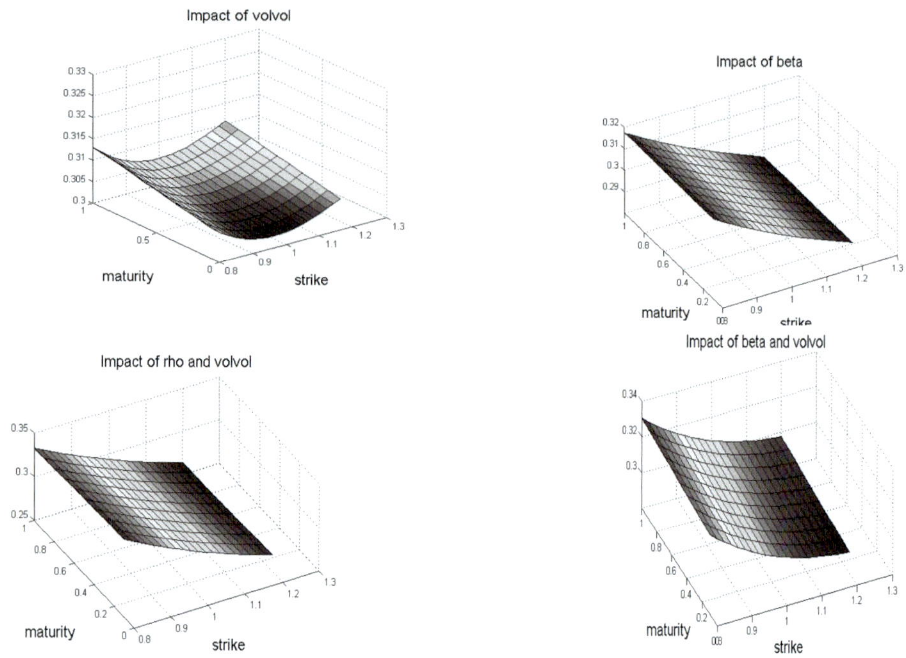

FIGURE 3.58 Impact of the different parameters of the SABR model

The β Parameter

The β parameter controls how pronounced the volatility skew is. It has three main effects on the skew, of decreasing importance.

- There is a progressive steepening as it goes from one to zero, which is more pronounced for OTM options.
- There is a lowering of the level of the smile as it goes from zero to one, which is not surprising, because the "total" volatility of the change in forward rate is given by $f\beta\sigma(t)$.
  - Everything else remaining the same, as β increases from zero towards one the term $f\beta$ decreases in magnitude (it is equal to one for $\beta = 0$ and equal to the forward rate for $\beta = 1$), thereby reducing the total volatility seen by the change in the forward rate.
  - However, ceteris are never paribus with fitting. For a given market smile, changing β changes both the slope and the level (and, to a lesser extent, the curvature) of the smile. Therefore, we adjust the σ_0 parameter to maintain the level of total volatility: $f^{\beta}\sigma_0 = f^{\beta'}\sigma'_0$.
- There is the introduction of a curvature to the skew as it goes from one to zero.

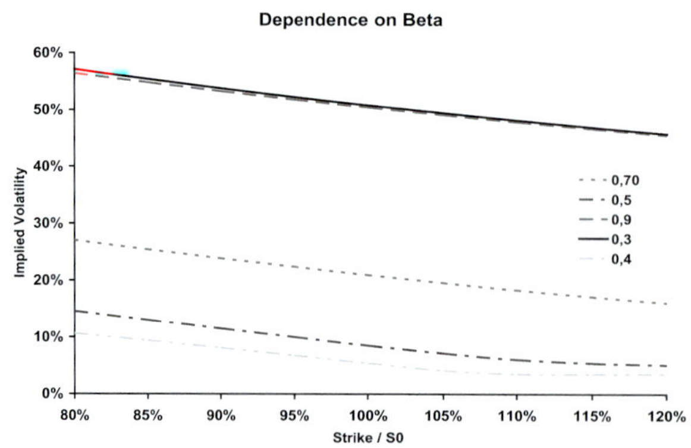

FIGURE 3.59 Implied volatility depending on the beta level

An increase of β therefore not only decreases the steepness but also decreases the curvature; this is obvious by noting that the implied volatility skew at low strikes falls much more rapidly than the high strike increase.

These mixed effects stem from the fact that the β parameter controls the distribution of the forward rate.

- When β = 1, the volatility distribution is log-normal. Thus we get a stochastic log-normal model:

$$\sigma_B(K) = \sigma_0^T\left(\frac{z}{\aleph(z)}\right)\cdot\left\{1+\left[\frac{\rho v\sigma_0^T}{4}v+\frac{\left(2-3\rho^2\right)}{24}v^2\right]T\right\}$$

$$z=\frac{v}{\sigma_0^T}\ln\frac{f}{K}.$$

- When β = 0, the volatility price distribution is Gaussian, producing a steeper skew:

$$\sigma_B(K) = \sigma_0^T\frac{\ln\frac{f}{K}}{f-K}\left(\frac{z}{\aleph(z)}\right)\cdot\left\{1+\left[\frac{\left(\sigma_0^T\right)^2}{24fK}+\frac{\left(2-3\rho^2\right)}{24}v^2\right]T\right\}$$

$$z=\frac{v}{\sigma_0^T}\sqrt{fK}\ln\frac{f}{K}.$$

- When β = ½, the volatility distribution is a CIR process (as in Heston).

To use a value of one implies that the at the money volatility moves horizontally as the market increases or decreases. A value of less than one would indicate that the volatility decreases/increases as the market increases/decreases.

With the value of β quite stable over time, empirically speaking, then it can be selected in two ways: selecting from "aesthetic considerations" based on market experience, since market smiles can be fit equally well with any specific value of β; or estimating from the market data, since the ATM volatility proxy is dominated by the first term, thus $\ln\sigma(F,F)=\ln\alpha-(1-\beta)\ln F+\ldots$.

This naturally suggests a time-weighted regression, very much as for exponentially weighted moving average (EWMA) volatility calculations.

<u>The Three Other Par ameters</u>

The α parameter controls the level of the skew. It is a "volatility-like" parameter: it is not equal to the volatility, but there will be a functional relationship between this parameter and the at-the-money volatility. It can be calibrated to the market smile by inverting the ATM volatility proxy:

$$\sigma_{ATM}=\frac{\alpha}{F^{1-\beta}}\left(1+\left(\frac{(1-\beta)^2\alpha^2}{24F^{2(1-\beta)}}+\frac{\rho\beta v\alpha}{4F^{1-\beta}}+\frac{2-3\rho^2v^2}{24}\right)\tau\right).$$

And, to find the "smallest positive root" of the following third-order equation,

$$\frac{(1-\beta)^2\tau\alpha^3}{24F^{2(1-\beta)}}+\frac{\rho\beta\tau v}{4F^{1-\beta}}\alpha^2+\left(1+\frac{2-3\rho^2v^2}{24}\tau\right)\alpha-\sigma_{ATM}F^{1-\beta}=0.$$

The correlation factor ρ between the underlying and the volatility controls the slope of the skew, thus producing a similar effect to β. It has a secondary effect of a small decrease in curvature.

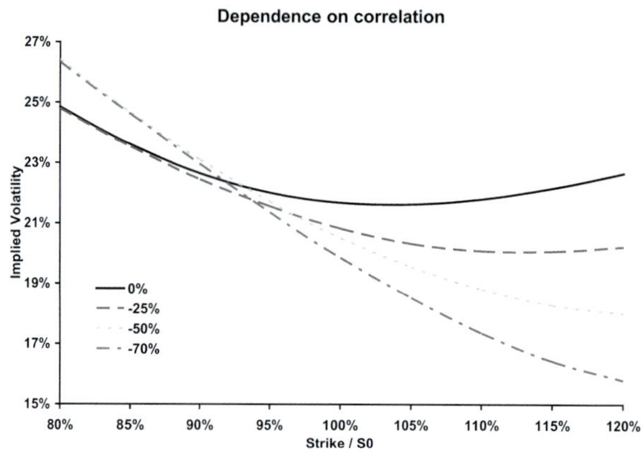

FIGURE 3.60 Dependence on correlation

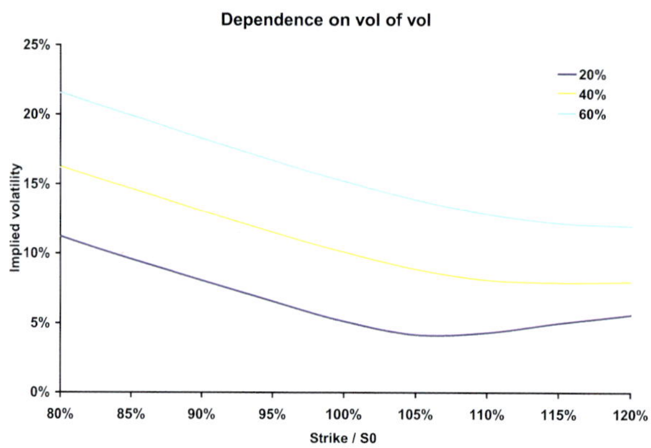

FIGURE 3.61 Dependence on vol of vol

The vol of vol ν parameter controls how much the smile displays in the skew. Therefore, when carrying out a fit to market implied volatilities, β – together with ρ – mainly picks up the burden of recovering the slope, while ν accounts for most of the curvature.

Calibration

- Under usual market conditions, the Levenberg–Marquardt algorithm can be used in order to calibrate the three parameters – a useful nonlinear least square fitting method, more robust than the Gauss–Newton (embedded in Excel), because in many cases it finds a solution even if it starts far off the final minimum.

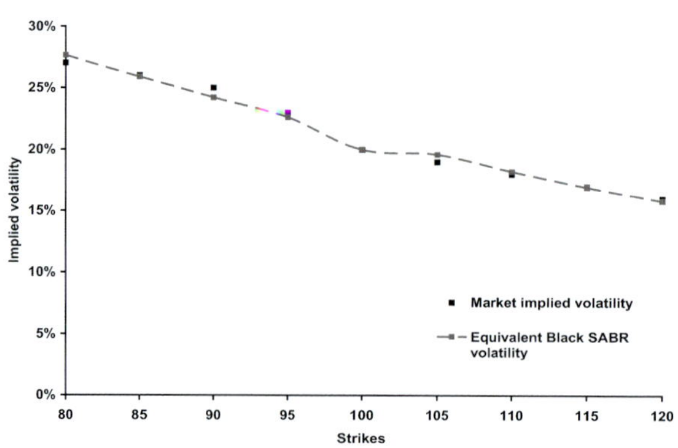

FIGURE 3.62 Implied volatility versus strikes

We may give greater weights to the values that are closest to the at-the-money strike, whereas more distant values do not have such a great impact in determining the form of the SABR.

- For flat short maturities, the caplet curve has a pronounced skew, meaning that the normal skew model β = 0 adapts better – with less undershoot. As for the weighting, it seems as if the weighted parameters act better for short maturities.
- For flat long maturities, both the β curves are capable of adapting relatively well, although the β=1 flat log-normal tends to be slightly better.

- Within illiquid market conditions, there are steps to calibrate, where β can be fixed in advance.

- The α parameter can be calibrated to the level of the ATM volatility, where the above formula simplifies to

$$\sigma_{ATM}(\tau)=\frac{\alpha}{F^{1-\beta}}\left(1+\left(\frac{(1-\beta)^2\alpha^2}{24F^{2(1-\beta)}}+\frac{\rho\beta\alpha}{4F^{1-\beta}}v+\frac{2-3\rho^2}{24}v^2\right)\tau\right).$$

The inversion is numerically easy. Thus, as the at-the-money volatility has a term structure and changes frequently, so too does the value of α.

This leads to the following third-order equation,

$$\frac{(1-\beta)^2\tau\alpha^3}{24F^{2(1-\beta)}}+\frac{\rho\beta\tau v}{4F^{1-\beta}}\alpha^2+\left(1+\frac{2-3\rho^2v^2}{24}\tau\right)\alpha-\sigma_{ATM}F^{1-\beta}=0,$$

in which we are assuming that we have already solved for r and v. For typical parameter inputs, this cubic has only one real root, but it is perfectly possible for it to have three real roots, in which case we seek the smallest positive root.

- To calibrate ρ and ν to the skew, the two-dimensional Nelder–Mead algorithm, a no-analytical-search method, is used, in order to minimize the following:

$$\underset{\rho,v}{Min}\sum_{K,T}\left(\sigma_{SABR}(K,T)-\sigma^*(K,T)\right)^2.$$

Calibration results show an interesting term structure of the model parameters, as a function of the maturity and underlying.

- β – together with ρ – mainly picks up the burden of recovering the slope, whereas ν accounts for most of the curvature.
- Typical is the shape of the parameter α, which starts out high for short-dated options and then declines monotonically as the option maturity increases.
- The vol of vol parameter ν, which controls curvature, is also more marked for short expiries, which is consistent with a strong convexity for short-dated options and smaller for long-dated options: when markets become more excited, the left and right tails of the distribution become fatter, and the skew correspondingly more curved.

- In contrast, our experience with correlations is less clear. In some markets a nearly flat skew for short maturity options develops into a strongly downward-sloping skew for longer maturities. In other markets there is a strong downward skew for all option maturities, and in still other markets the skew is close to zero for all maturities.

A good criterion for reverse-engineering the parameter β is to see which choice of the exponent makes parameters such as the volatility or the correlation most stable on most days.

✓ **Quick win**: The ATM volatilities, or, equivalently, α, need to be updated frequently, say once or twice a day, while the smiles and skews, or, equivalently, ρ and ν, need to be updated infrequently, say once or twice a month.

3.2.1.3.4.4 The SABR Model Provides a Tractable Decomposition of Tail Risks This takes place through skew components. Up to the fourth moment risk can be analysed analytically within such a modelling framework. The SABR model allows the market risks to be obtained immediately from Black's formula

(i) Skew Risks
To make the model's qualitative behaviour apparent, the implied volatility is further approximated with

$$\sigma_B(f,K)=\frac{\alpha}{f^{1-\beta}}\left\{1-\frac{1}{2}(1-\beta-\rho\lambda)\log\frac{K}{f}+\frac{1}{12}\left[(1-\beta)^2+\left(2-3\rho^2\right)\lambda^2\right]\log^2\frac{K}{f}+\ldots\right\},$$

where $\lambda=\frac{\nu}{\alpha}f^{1-\beta}$.

This approximate expression is very accurate as long as one does not look at strikes that are too out of the money or expiries that are too long.

- The main term in the approximate formula for the volatility is $\sigma_B(f,f)=\frac{\alpha}{f^{1-\beta}}$, which represents the value ATM, almost entirely determined by the exponent β, with the exponent β = 0 (a stochastic Gaussian model) giving a steeply downward-sloping skew, and the exponent β = 1 giving a nearly flat skew.

- The second term, $-\frac{1}{2}(1-\beta-\rho\lambda)\log\left(\frac{K}{f}\right)_1$, represents the overall skew – the slope of the implied volatility with respect to the strike K. We can decompose it into two principal components:

  - the beta skew, $-\frac{1}{2}(1-\beta)\log\left(\frac{K}{f}\right)$, which is downward-sloping, since $0 \le \beta \le 1$; it arises because the local volatility, $\frac{\alpha}{f^{1-\beta}}=\frac{\alpha f^{\beta}}{f^{1-\beta}}$, is a decreasing function of the forward price.

  - the vanna skew, $-\frac{1}{2}\rho\lambda\log\left(\frac{K}{f}\right)$, is the skew caused by the correlation between the volatility and the asset price; typically the volatility and asset price are negatively correlated, so, on average, the volatility α would decrease (increase) when the forward f increases (decreases); it thus seems unsurprising that a negative correlation ρ causes a downward-sloping vanna skew.

- The last term, $\left\{\frac{1}{12}\left[(1-\beta)^2+\left(2-3\rho^2\right)\lambda^2\right]\log^2\left(\frac{K}{f}\right)+\ldots\right\}$, also contains two parts.
   - $\frac{1}{12}(1-\beta)^2\log^2\left(\frac{K}{f}\right)$ appears to be a smile (quadratic) term, but it is dominated by the downward-sloping beta skew, and, at reasonable strikes K, it just modifies this skew somewhat.
   - $\frac{1}{12}\left(2-3\rho^2\right)\lambda^2\log^2\left(\frac{K}{f}\right)$ is the smile induced by the volga (vol-gamma) effect. Physically, this smile arises because of "adverse selection": unusually large movements of the forward F happen more often when the volatility α increases, and less often when α decreases, so strikes K far from the money represent, on average, high volatility environments.

(ii) Delta and Vega Risks

The SABR model offers more accurate and interpretable hedging sensitivities in terms of convexity through refined computations of gamma and vega, which take into account not only the vol of vol impact but also relationships between volatility and forward rate dynamics.

- The delta hedge is computed by shifting the current value of the underlying while keeping the current value of α fixed, leading to the following change in the option value:

$$\Delta V=\left\{\frac{\partial V}{\partial f}+\frac{\partial V}{\partial \sigma}\frac{\partial \sigma}{\partial f}\right\}\Delta f.$$

So the delta is computed as $\Delta=\frac{\partial V}{\partial f}+\frac{\partial V}{\partial \sigma}\frac{\partial \sigma}{\partial f}$,

where the first term is the original Black delta, and the second accounts for the systematic change in the implied volatility as the underlying changes.

As discussed above, the predicted change consists of a sideways movement of the volatility curve in the same direction and by the same amount as the change in the forward price. So there may also be minor changes to the shape of the skew due to changes in the forward price f.

✓ **Quick win**: theoretically, one should use the above delta to risk manage option books. However, in many markets, it may take several days for volatilities to change following significant changes in the forward price f. In these illiquid markets, as the volatility σ_{ATM} does not immediately change following a change in the forward price f, one would be forced to re-mark α to compensate, using α as defined implicitly in the calibration process above; expressed in terms of σ_{ATM} and f, the delta risk is given by

$$\Delta=\frac{\partial V}{\partial f}=\frac{\partial B}{\partial f}+\frac{\partial B}{\partial \sigma}\left(\frac{\partial \sigma}{\partial f}+\frac{\partial \sigma}{\partial \alpha}\frac{\partial \alpha}{\partial f}\right).$$

The delta risk is now the risk to changes in f with σ_{ATM} held fixed. The last term is just the change in α needed to keep σ_{ATM} while f changes. Note that this term is zero when β = 1 (normal model). Thus re-marking caused by market delays would change the hedges and can prove to be costly.

- Similarly, the vega risk is computed as $\Delta V = \left\{ \frac{\partial V}{\partial \sigma} + \frac{\partial V}{\partial \sigma} \frac{\partial \sigma}{\partial \alpha} \right\} \Delta f$.

So the vega is computed as $\Lambda = \frac{\partial V}{\partial \sigma} \frac{\partial \sigma}{\partial \alpha}$.

It is traditional to scale vega so that it represents the change in value when the ATM volatility changes by a unit amount.

- Vanna $= \frac{\partial B}{\partial \rho} = \frac{\partial B}{\partial \sigma} \frac{\partial \sigma}{\partial \rho}$ expresses the risk to the skew increasing.
- Volga $= \frac{\partial B}{\partial v} = \frac{\partial B}{\partial \sigma} \frac{\partial \sigma}{\partial v}$ expresses the risk to the skew becoming more convex.

(iii) Modified Delta and Vega Risk Formulas

Here we take the implicit interaction between market level and volatility into account.

- In the SABR model, the underlying forward f rate is correlated with the SABR volatility α. Thus, whenever the forward rate f changes, the volatility α also changes – at least, on average. In fact, the time evolution of α can be decomposed into two independent components:
    - one due to the change of the forward rate f; and
    - one due to the idiosyncratic change in α: $\frac{\partial \alpha}{\partial f} \Delta f = \frac{\rho v}{f^{\beta}} \Delta f$.

Thus, accounting for this average change in α caused by movements in the forward f leads to a new term in the formula for the delta risk, a term not contained in the original SABR paper. One can hedge delta more precisely by adding this term into the delta risk:

$$\Delta V = \left\{ \frac{\partial V}{\partial f} + \frac{\partial V}{\partial \sigma} \left(\frac{\partial \sigma}{\partial f} + \frac{\partial \sigma}{\partial \alpha} \frac{\rho v}{f^{\beta}} \right) \right\} \Delta f.$$

So the new delta is computed as $\frac{\partial V}{\partial f} + \frac{\partial V}{\partial \sigma} \left(\frac{\partial \sigma}{\partial f} + \frac{\partial \sigma}{\partial \alpha} \frac{\rho v}{f^{\beta}} \right)$.

Similarly, for the vega risk, where we take into account the average change in the forward rate f caused by the change in SABR volatility α: $\frac{\rho f^{\beta}}{v} \Delta \alpha$. Thus

$$\Delta V = \left\{ \frac{\partial V}{\partial \sigma} \left(\frac{\partial \sigma}{\partial \alpha} + \frac{\partial \sigma}{\partial f} \frac{\rho f^{\beta}}{v} \right) \right\} \Delta \alpha,$$

$$\Lambda = \frac{\partial V}{\partial \sigma} \left(\frac{\partial \sigma}{\partial \alpha} + \frac{\partial \sigma}{\partial f} \frac{\rho f^{\beta}}{v} \right).$$

- The new term in the delta risk is minimal when one hedges both vega risks and delta risks but can be substantial when only delta is hedged.

In fact, the new term in the delta risk $\frac{\partial\sigma}{\partial\alpha}\frac{\rho v}{f^{\beta}}$ is just $\frac{\rho v}{f^{\beta}}$ times the classic SABR vega risk. In a vega weighted portfolio this term is zero, so, if the classic vega risk and delta risk are both hedged, then the new delta risk is also hedged.

- Suppose one specifies β and then selects ρ to match the observed volatility skew. The delta risk then depends on the β chosen. With the new term introduced above, the delta risk is less sensitive to the β chosen, depending mainly on the slope of the implied volatility skew.

3.2.1.3.4.5 Even So, the SABR Model Suffers from Specific Caveats However, this more sophisticated and flexible modelling framework might be burdened in some market conditions, with a series of limitations and pitfalls regarding calibration and hedging practices.

- There is no mean reversion in the process for the volatility. Since correlation gives the short-term skew, and mean reversion level determines the long term value of volatility, we might lose some flexibility in fitting the skew.
- As the SABR model is a stochastic-volatility enrichment of the local volatility model called "CEV", it has a unique solution only if β > ½.
- The implied volatility smile must flatten at large maturities; however, this is violated by the Hagan et al. proxy above. The result is that butterflies with low strikes can have negative value when priced in this approximation. The origin of the problem is that the Hagan approximation is derived using a short-time asymptotic approximation, whereas interest rates traders need to calculate prices and risks for options with maturities as high as 30 years.
- The proxy implied volatility expression above, obtained using perturbation theory, has an internal flaw: it is inconsistent, as β→1 – i.e. does not converge to the known value as β→1. More recently the subject has been treated in a rigorous analytical manner, whereby somewhat tedious computations led to a new, corrected version of the original formula. The new term provides convergence as β→1.

Specifically in the case of equivalent "normal" implied volatility:

$$\frac{z}{\aleph(z)}=\frac{vx\frac{\zeta}{z}}{\ln\left(\frac{\sqrt{1-2\rho\zeta+\zeta^{2}}+\zeta-\rho}{1-\rho}\right)}$$

instead of

$$\frac{z}{\aleph(z)}=\frac{vx}{\ln\left(\frac{\sqrt{1-2\rho z+z^{2}}+z-\rho}{1-\rho}\right)},$$

where

$$\zeta=\frac{v}{\sigma_0^T}\frac{f-K}{(fK)^{\frac{\beta}{2}}},$$

$$z=\frac{v}{\sigma_0^T}\frac{f^{1-\beta}-K^{1-\beta}}{1-\beta}$$

- Furthermore, the above formula is known to produce wrong prices in the region of small strikes for large maturities: it assigns a negative price to a structure with a positive payoff, or, equivalently, it implies a negative probability density for the stock price process in some region. However, if one uses the formula recently derived recently, this problem either appears in yet lower strikes or disappears completely.
- Regarding the calibration issues.
  - The "overloading" of the skew shape on the exponent, β, is chosen based on practice or estimated statistically, before applying a calibration algorithm on the other three parameters. In contrast to the "CEV" local volatility model, it is not easy to choose the β, due to the additional stochastic volatility: here, as the forward swap rate changes, so will the instantaneous volatility. Since the two driving stochastic processes (forward rate and volatility) are correlated, a similar behaviour will occur, no matter which β one chooses. Thus, the SABR model is somewhat over-parameterized, which in part explains his success in fitting the skew.
  - Very similar quality skew fits can be obtained with different combinations of β and ρ. Increases in β and ρ induce (primarily) a decrease in out-of-the-money Black-implied volatility and an increase in in-the-money Black-implied volatility – i.e. a change in the slope of the smile. However, the changes are not entirely symmetric across the smile (changes at in- and out-of-the-money strikes are not symmetric in their magnitude): parameters β and ρ have a secondary effect on curvature, just as ν has a secondary effect on steepness. This is undesirable. To "fix" this problem, it is market practice to fix either of the two parameters, and optimize over the other.
  - The skew curvature can be obtained with different combinations of β and ν. In fact, β is responsible not only for the steepness of the skew but also for the fatness of the left tail: for instance, a β = 0 will give more probability to very low values of the forward rate than a β = ½ or 1, and so induce a steeper skew through a wrong correlation ρ, but, most importantly, it also "invents" a considerable amount of volatility of volatility that is absent in the true model.

Warning: thus a fit with the wrong β is often as good as the fit obtained using the true process, but it will induce a fat left tail through a stochastic volatility. This is a worrying feature, because such an important feature as the presence or absence of stochasticity in the volatility – a feature that, in our example, can be conjured up or made to disappear by a suitable choice of β – will considerably change the hedging. So, is stochastic volatility a reality or an artefact of fitting a wrong model? This is a topic we tackle in another note.

  - The fitted implied volatility surface is nearly flat, floating above the actual implied volatilities at long maturity. The ATM volatilities at the shortest maturity are much higher than those at longer maturity. Two types of unpleasant shape may occur depending on parameters:
    - when the shortest maturity contracts are close to expiry the volatility may experience an extremely large ATM volatility, and
    - when the time to maturity is large enough the ATM volatility may decline below zero.
  - There are also calibration inconsistencies. The asymptotic behaviour of implied volatilities for very large strikes turns out to flatten out to a constant, as opposed to diverging rapidly, as SABR would predict. Finally, some pricing inconsistencies may emerge with SABR, since the model is solved by means of asymptotic expansions with a finite, irreducible error.
  - SABR claims to dissociate volatility dynamics from skew fitting. However, if two different SABR models fit the same skew, they also fit the same volatility dynamics (as illustrated), taking the two extreme cases as a log-normal process and a normal volatility process.

Thus we get the results shown in Figures 3.64 to 3.66.

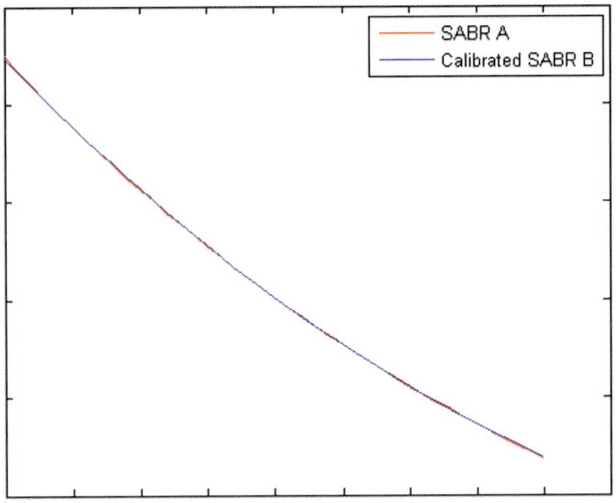

FIGURE 3.63 Same skew

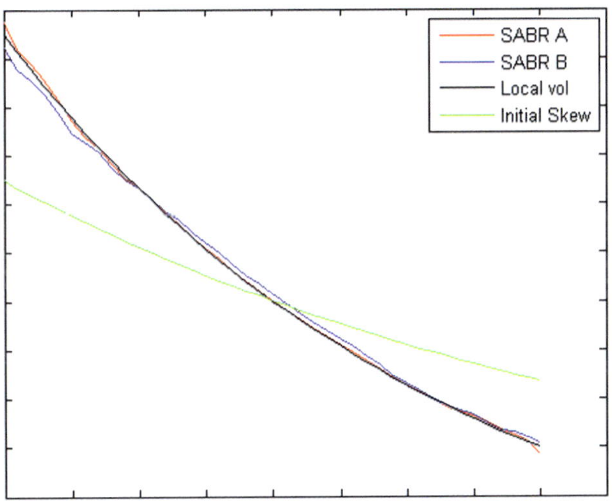

FIGURE 3.64 Similar volatility dynamics

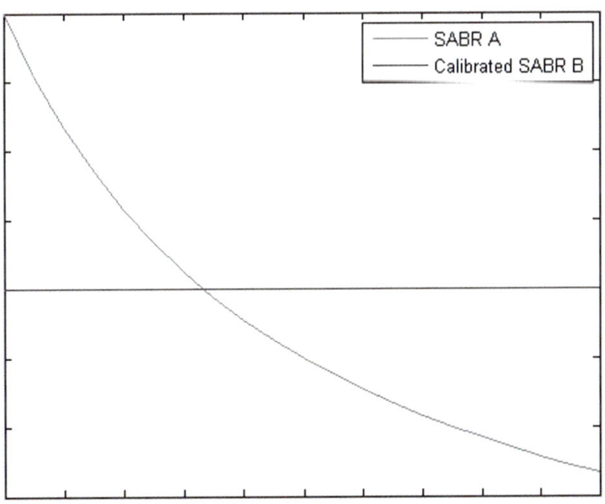

FIGURE 3.65 Different ATM volatility

Overall, SABR is therefore unlikely to be the definitive solution, and is perhaps rather just yet another stepping stone in a long chain.

3.2.1.3.5 Introduction to (Lévy) Pure Jump Modelling for Recurrent Equity Market Crashes

3.2.1.3.5.1 Why Pure Jump Processes? Given the recurrent (and at an increasing pace) equity market crashes, it makes sense to have a look at the introduction of infinite jump processes.

In fact, Brownian motion has continuous sample paths, whereas in reality prices are driven by jumps. The Brownian motion needs a substantial amount of time to reach a low level, whereas jumps can cause an almost immediate move to that level.

Short options tend to be underpriced by the Black–Scholes formula. As the time to maturity T approaches zero the probability of an out-of-the-money option moving into the money (or vice versa) is very small with a diffusion. In fact, it tends to zero as T does. But, with jumps, this probability does not drop to zero, meaning that the option values will be higher, as observed in markets. This feature is present in all models with jumps.

In terms of tail modelling, Lévy models can provide very flexible density structures, and accommodate a wide range of skewness and kurtosis.

In Figure 3.67 we show the comparison on both fat tail modelling and hedging between the Black–Scholes constant volatility models, stochastic volatility models (so-called "diffusion models" in the figure) and jump models.

Empirical facts	Black-Scholes	Diffusion models	Jump models
Large sudden price movements	Once in 10^{53} years	Difficult	Generic and controlable
Heavy tails	No	Possible	Generic and controlable
Risks with options	Can be handled without risk	Can be handled without risk	Cannot be handled: risky
Market incomplete: some risk can not be hedged	Market complete	Market complete or can be completed	Incomplete
Losses concentrated to few large downwards jumps	Losses concentrated to long period of trend movements	Large sudden jumps don't occur	Jumps can give large losses
Some hedge strategies better than others	All hedge strategies are perfect	All hedge strategies are perfect	Optimization problem
Exotic options hedged by standard options	Options redundant: each options can be hedged	Options redundant: each options can be hedged	Options not redundant: standard options can reduce hedge risks

FIGURE 3.66 Models' behaviours

3.2.1.3.5.2 What Are the Main Lévy Pure Jump Processes? In terms of modelling, jumps processes are modelled by Lévy processes (see Figure 3.68).

Lévy processes recover a large class of processes, characterized by the so-called Lévy measure ν, which is defined in Figure 3.69.

Definition

A process $(X_t, t \geq 0)$ in $\mathbb{R}^d$ is a Lévy process on a filtered space $(\Omega, \mathcal{F}, \mathbb{P}, (\mathcal{F}_t, t \geq 0))$ iff :

- $X_0 = 0$, $\mathbb{P}$-as,
- for all $n \in \mathbb{N}^*$ and for all $0 \leq t_0 < t_1 < \cdots < t_n$, the random variables $X_{t_0}, X_{t_1} - X_{t_0}, \ldots, X_{t_n} - X_{t_{n-1}}$ are independent,
- the law of $X_{s+t} - X_s$ does not depend on s,
- the process $(X_t, t \geq 0)$ is continuous in probability,
- it exists $\Omega_0 \in \mathcal{F}$ such that $\mathbb{P}(\Omega_0) = 1$ and such that for all $\omega \in \Omega_0$, $X_t(\omega)$ is right continuous for $t \geq 0$ and left limited for $t > 0$.

FIGURE 3.67 Lévy process definition

Let $(X_t, t \geq 0)$ be a Lévy process in $\mathbb{R}$, then it exists :

- $b \in \mathbb{R}$,
- $\sigma^2 \in \mathbb{R}^+$,
- a Lévy measure ν on $\mathbb{R}$ such that

$$\nu(\{0\}) = 0 \text{ et } \int_{\mathbb{R}} (x^2 \wedge 1)\, \nu(dx) < +\infty,$$

and such that for all $u \in \mathbb{R}$:

$$\mathbb{E}\left[\exp\left(iuX_t\right)\right] = \exp\left\{iubt - u^2\frac{\sigma^2}{2}t + t\int_{\mathbb{R}}\left(e^{iux} - 1 - iux\mathbb{1}_{|x|\leq 1}\right)\nu(dx)\right\}.$$

Moreover this characterization is unique and the family (b, σ, ν) completely characterizes the process.

FIGURE 3.68 The Lévy measure

The Lévy measure ν and the volatility σ carry useful information about the structure of the jump process, notably whether it embeds a finite or infinite number of jumps – so-called finite and infinite activity (see Figure 3.70).

	Finite activity	Finite variation
$\nu(\mathbb{R}) < \infty$	YES	YES
$\nu(\mathbb{R}) = \infty$	NO	?
$\sigma = 0$ and $\int_{\|x\|\leq 1} \|x\|\nu(dx) < \infty$	?	YES
$\sigma \neq 0$ or $\int_{\|x\|\leq 1} \|x\|\nu(dx) = \infty$	NO	NO

FIGURE 3.69 The Lévy measure and the structure of the jumps

Let us mention the most usual Lévy jump processes.

- Merton (1976): it has a diffusion component, and so is a jump diffusion process. The measure ν is the measure of a compound Poisson process with intensity λ, while the law of the jumps is a Gaussian normal law with mean μJ and variance σJ. Its characteristic function is

$$\exp\left\{iubt - \frac{\sigma t u^2}{2} + t\lambda\left(e^{i\mu_J u - \sigma_J^2 u^2/2} - 1\right)\right\},$$

$$\mathbb{E}[X_1] = b + \lambda\mu_J \ and \ \mathrm{Var}[X_1] = \sigma + \lambda\mu_J + \lambda\sigma_J^2.$$

- Variance gamma (1990): this Lévy process has no diffusion component and has infinite activity.

  $\log(R_t) = \log(S_{t+1}/S_t)$ is distributed with mean and random variance $\sigma^2 V$, where μ and σ^2 are constants and V is a gamma random variable with parameters c, γ. The characteristic function of Xt = log(Rt) − μ is

$$\phi_X(u) = \left(1 + \frac{\sigma^2 u^2}{2c}\right)^{-\gamma t} = \left(1 + \left(\frac{\sigma^2 v}{m}\right)\left(\frac{u^2}{2}\right)\right)^{-\frac{m^2 t}{v}},$$

  where m = γ/c and v = γ/c² are the mean and variance.

  The variance of this Xt process is σ² and the fourth central moment is 3 σ² (1 + v), resulting in a kurtosis of 3 (1 + v).

- The CGMY model as a generalized version of the variance gamma model.
  Let X ~ gamma law (C, M) and Y ~ gamma law(C, G); X − Y then follows a "variance gamma law" VG(C, G, M), with three dedicated parameters, C, G and M, with the following characteristic and density functions, and statistical moments:

$$\phi_{X-Y}(u) = (1 - iu/M)^{-C}(1 + iu/G)^{-C} = \left(\frac{GM}{GM + (M-G)iu + u^2}\right)^C,$$

$$f_{VG}(x; C, G, M) = \frac{(GM)^C}{\sqrt{\pi}\Gamma(C')}\exp\left(\frac{(G-M)x}{2}\right)$$
$$\left(\frac{|x|}{G+M}\right)^{C-1/2} K_{C-1/2}((G+M)|x|/2),$$

$$\begin{array}{ll} & VG(C,G,M), \\ \textit{mogenne} & C(G-M)/(GM), \\ \textit{variance} & C\left(G^2+M^2\right)/(GM)^2, \\ \textit{skewness} & 2C^{-1/2}\left(G^3-M^3\right)/\left(G^2+M^2\right)^{3/2}, \\ \textit{kurtosis} & 3\left(1+2C^{-1}\left(G^4+M^4\right)/\left(M^2+G^2\right)^2\right) \end{array}$$

where Kλ is the Bessel function of the third kind of index λ and Γ(·) is the gamma function.

The law is extremely flexible, and its structure can easily be understood with the help of parameters.

✓ If G = M, the distribution is symmetric. If G < M, then skewness is negative, and vice versa.

✓ The C parameter mainly controls the kurtosis.

3.2.1.3.5.3. How to Price Contracts Using Lévy Pure Jump Processes

(i) Pricing by Simulations

Using pure jump processes (such as VG or CGMY), there is no single risk-neutral measure, but an infinity. We therefore choose a simple correction transformation of the mean, whereby the process in VG is translated to get a martingale.

Assume

$$\log\left(S_t / S_0\right) = \mu t + Y(t),$$

where Y(t) is a two-parameter variance gamma process.

Then the price of a product under the risk-neutral measure is, as usual, given by the expected discounted payoff, where the expectation is taken under a risk-neutral measure. Under this risk-neutral measure, the drift in the log price must be the risk-free rate r equal to $\frac{1}{t}\left(\log\left(\mathbb{E}\left[S_t / S_0\right]\right)\right) = \mu + \frac{1}{t}\left(\log\left(\mathbb{E}\left[Y_t\right]\right)\right),\quad r = \mu - \frac{1}{v}\log\left(1 - \frac{v\sigma^2}{2}\right).$

Then a European call option can be priced as follows:

$$C_{VG} = \int C_{BS}\left(s, K, \sqrt{\sigma^2 v}, r\right) g(v) dv,$$

where g(v) is the density of the gamma distribution.

(ii) The FFT Pricing Method for European Optionalities

As long as savings and retirement products embed European options, an efficient pricing alternative goes through fast Fourier transform (FFT) methods (Carr & Madan 1999), which combine both run-time efficiency and high accuracy. However, they are not efficient for path-dependent optionalities.

The only assumption is that the characteristic function of the underlying is given analytically. The basic idea of the method is to develop an analytic expression for the Fourier transform of the option price and to get the price by Fourier inversion.

Let C(T,K) denote the price of a European call with maturity T and strike K. If we denote K = exp(k), then

$$C(T,K) = \int_k^{+\infty} e^{-rT}\left(e^s - e^k\right) q_T(s) ds,$$

where qT(s) is the risk-neutral density of s(T) = log[S(T)].

As the Fourier transform and its inversion work only for square-integrable functions, we do not consider directly the option price but a modification of it,

$$c(T,k) = \exp(\alpha k) C(T, \exp(k)),$$

which is square-integrable for a suitable α > 0. The choice of α may depend on the model for St. The Fourier transform of c is then defined by

$$\psi(T,v) = \int_{-\infty}^{+\infty} e^{ivk} c(T,k) \mathrm{d}k.$$

Under the classical Fourier transform computational approach,

$$\begin{aligned}\psi(T,v) &= \int_{-\infty}^{+\infty} e^{ivk} \int_{k}^{+\infty} e^{\alpha k} e^{-rT} \left(e^{s} - e^{k}\right) q_T(s) \mathrm{d}s \; \mathrm{d}k \\ &= \int_{-\infty}^{+\infty} e^{-rT} q_T(s) \int_{-\infty}^{s} \left(e^{\alpha k + s} - e^{(\alpha+1)k} e^{ivk}\right) \mathrm{d}s \; \mathrm{d}k \\ &= \int_{-\infty}^{+\infty} e^{-rT} q_T(s) \left(\frac{e^{(\alpha+1+iv)s}}{\alpha + iv} - \frac{e^{(\alpha+1+iv)s}}{\alpha+1+iv} \right) \mathrm{d}s \\ &= e^{-rT} \frac{\phi(T, v - (\alpha+1)i)}{\alpha^2 + \alpha - v^2 + i(2\alpha+1)v},\end{aligned}$$

where φ(T, ·) is the Fourier transform of q(T, ·).

The European call price therefore becomes

$$C(T,K) = \frac{\exp(-\alpha k)}{\pi} \int_{0}^{+\infty} e^{-ivk} \psi(T,v) \mathrm{d}v. \approx \frac{\exp(-\alpha k)}{\pi} \sum_{j=0}^{N-1} e^{ivjk} \psi\left(T, v_j\right) \eta.$$

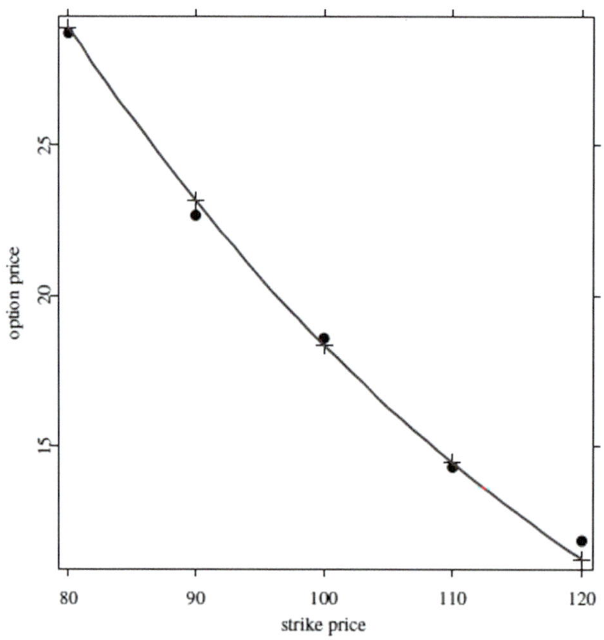

FIGURE 3.70 Analytic of Monte Carlo versus FFT prices

Different alternative methods are then available to compute such a Fourier transform, which we benchmark in Figure 3.72 regarding key criteria: model flexibility, easiness of implementation, computational run-time and accuracy.

Method	Payoff and Model flexibility	Implementation	Runtime	Accuracy
Classical Fourier transform	+++	+++	- - -	+++
Carr-Madan FFT	++	+++	+	+
Carr-Madan alternative	+	+++	+	++
Matsuda alternative	-	+++	++	++
Optimal Carr-Madan FFT	++	-	-	+++
Voltchkova-Tankov alternative	++	++	++	+++
Goudenège alternative	+++	++	++	+++

FIGURE 3.71 Comparison of different pricing methodologies

(iii) Pricing by PIDE for Exotic Optionalities

An alternative to transform methods for pricing options is to derive and then solve numerically the partial integro-differential equation (PIDE) that the option price satisfies.

The advantage of PIDE methods is that complex and exotic payoffs can be treated easily; the limitations are the slower speed in comparison to transform methods and the computational complexity when handling options on several assets.

Here, we derive the PIDE corresponding to the price of a European option in a Lévy-driven asset, using martingale techniques.

Let us denote by f(Lt,t) the time t price of a European option with payoff function g on the asset S; the price is given by

$$f(L_t,t) = e^{-r(T-t)}\mathbb{E}\left[g\left(S_0 e^{L_T}\right)\right] = V_t, \quad 0 \le t \le T.$$

By arbitrage theory, we know that the discounted option price process must be a martingale under a martingale measure. Therefore, any decomposition of the price process as

$$e^{-rt}V_t = V_0 + M_t + A_t,$$

where M ∈ Mloc and A ∈ Aloc, must satisfy At = 0 for all t ∈ [0,T].

This condition yields the desired PIDE using the Itô lemma:

$$\begin{aligned} d\left(e^{-rt}V_t\right) &= d\left(e^{-rt} f\left(I_{t^-},t\right)\right) \\ &= -re^{-rt} f\left(L_{t^-},t\right)\mathrm{d}t + e^{-rt}\partial_2 f\left(L_{t^-},t\right)\mathrm{d}t \\ &\quad + e^{-rt}\,\partial_1 f\left(L_{t^-},t\right)\mathrm{d}L_t + \frac{1}{2}e^{-rt}\partial_1^2 f\left(L_{t^-},t\right)\mathrm{d}\left\langle L_t^c\right\rangle \\ &\quad + e^{-rt}\int_{\mathbb{R}}\left(f\left(L_{t^-}+z,t\right) - f\left(L_{t^-},t\right) - z\partial_1 f\left(L_{t^-},t\right)\right)\nu_L(\mathrm{d}z,\,\mathrm{d}t), \end{aligned}$$

$$
\begin{aligned}
d\left(e^{-rt}V_t\right) &= d\left(e^{-rt}f\left(L_{t^-},t\right)\right) \\
&= e^{-rt}\left\{-rf\left(L_{t^-},t\right)dt + \partial_2 f\left(L_{t^-},t\right)dt + \partial_1 f\left(L_{t^-},t\right)b\, dt\right. \\
&+ \partial_1 f\left(L_{t^-},t\right)\sigma dW_t + \frac{1}{2}\partial_1^2 f\left(L_{t^-},t\right)\sigma^2 dt \\
&+ \int_{\mathbb{R}} z\partial_1 f\left(L_{t^-},t\right)(J_L - \nu_L)(dz,\, dt) \\
&+ \int_{\mathbb{R}} \left(f\left(L_{t^-}+z,t\right) - f\left(L_{t^-},t\right) - z\partial_1 f\left(L_{t^-},t\right)\right)(J_L - \nu_L)(dz,\, dt \\
&+ \left.\int_{\mathbb{R}} \left(f\left(L_{t^-}+z,t\right) - f\left(L_{t^-},t\right) - z\partial_1 f\left(L_{t^-},t\right)\right)\nu_L(dz)dt\right\}
\end{aligned}
$$

$$
\begin{aligned}
d\left(e^{-rt}V_t\right) &= d\left(e^{-rt}f\left(L_{t^-},t\right)\right) \\
&= e^{-rt}\left\{-rf\left(L_{t^-},t\right)dt + \partial_2 f\left(L_{t^-},t\right)dt + \partial_1 f\left(L_{t^-},t\right)b\, dt\right. \\
&+ \partial_1 f\left(L_{t^-},t\right)\sigma dW_t + \frac{1}{2}\partial_1^2 f\left(L_{t^-},t\right)\sigma^2\, dt \\
&+ \int_{\mathbb{R}} \left(f\left(L_{t^-}+z,t\right) - f\left(L_{t^-},t\right)\right) \qquad (J_L - \nu_L)(dz,\, dt) \\
&+ \left.\int_{\mathbb{R}} \left(f\left(L_{t^-}+z,t\right) - f\left(L_{t^-},t\right) - z\partial_1 f\left(L_{t^-},t\right)\right)\nu_L(dz)dt\right\},
\end{aligned}
$$

$$
\begin{aligned}
d\left(e^{-rt}V_t\right) &= d\left(e^{-rt}f\left(L_{t^-},t\right)\right) \\
&= e^{-rt}\left\{-rf\left(L_{t^-},t\right)dt + \partial_2 f\left(L_{t^-},t\right)dt + \partial_1 f\left(L_{t^-},t\right)b\, dt\right. \\
&+ \partial_1 f\left(L_{t^-},t\right)\sigma dW_t + \frac{1}{2}\partial_1^2 f\left(L_{t^-},t\right)\sigma^2\, dt \\
&+ \int_{\mathbb{R}} \left(f\left(L_{t^-}+z,t\right) - f\left(L_{t^-},t\right)\right) \qquad (J_L - \nu_L)(dz,\, dt) \\
&+ \left.\int_{\mathbb{R}} \left(f\left(L_{t^-}+z,t\right) - f\left(L_{t^-},t\right) - z\partial_1 f\left(L_{t^-},t\right)\right)\nu_L(dz)dt\right\}.
\end{aligned}
$$

Hence, the price of the option satisfies the partial integro-differential equation,

$$
\begin{aligned}
0 &= -rf(x,t) + \partial_2 f(x,t) + \partial_1 f(x,t)b + \frac{\sigma^2}{2}\partial_1^2 f(x,t) \\
&+ \int_{\mathbb{R}} \left(f(x+z,t) - f(x,t) - z\partial_1 f(x,t)\right)\nu_L(dz),
\end{aligned}
$$

for all $(x,t) \in \mathbb{R} \times (0,T)$, subject to the terminal condition

$$
f(x,T) = g\left(e^x\right).
$$

(iv) Calibration Issues and Mitigation

Under the "risk-neutral" measure, the flexibility of the generating distributions allows the implied volatility smiles produced by a Lévy model to accurately capture the shape of the implied volatility smiles observed in the market

However, a major drawback of Lévy-based models is that they are i.i.d., and therefore cannot capture the term structure of implicit volatilities. We can have a perfect fit for each maturity, but all maturities cannot be calibrated simultaneously; every maturity implies a different set of model parameters.

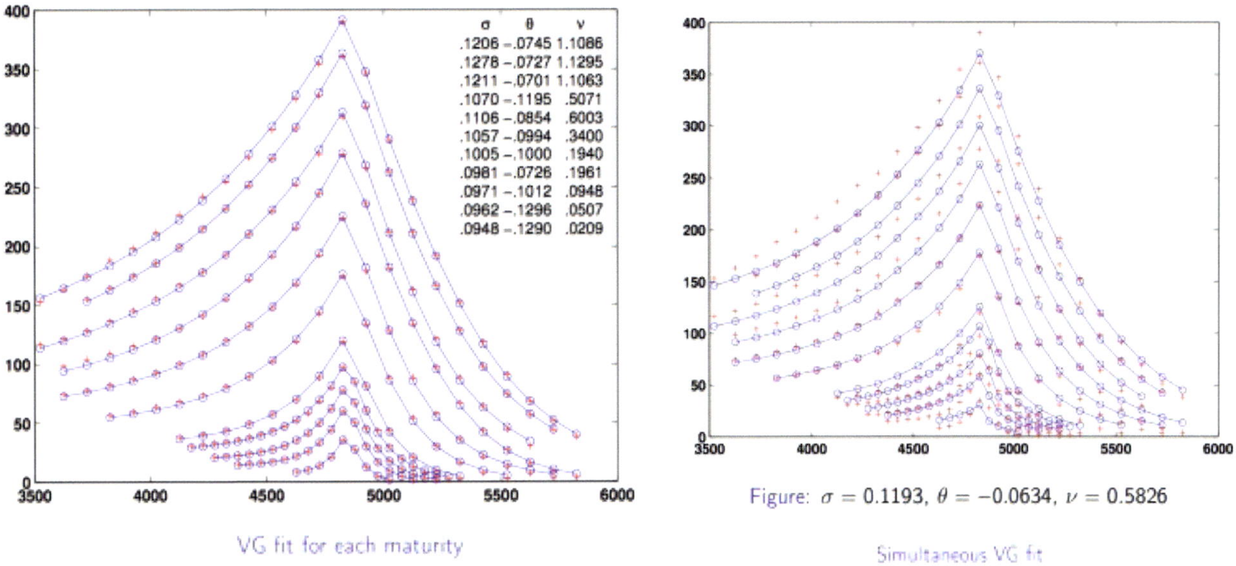

FIGURE 3.72 Quality of fitting implied volatility distribution across maturities by variance gamma model

The improvements in fitting are shown in Figure 3.75.

The empirical success of pure jump Lévy processes is not maintained when one considers the variation of option prices across maturity.

In fact, these homogeneous Lévy processes impose too strict conditions on the term structure of the risk-neutral variance, skewness and kurtosis.

- Specifically, the variance rate is constant over the term, skewness is inversely proportional to the square root of the term and kurtosis is inversely proportional to the term. In contrast, the data suggest that these risk-neutral moments are often rising with the term.
- Economically, the usual supposition that investor uncertainty is increasing with the term suggests that return distributions should spread out as the holding period is increased. On the other hand, risk premiums in markets are likely to display mean reversion, which would impact the term structure of skewness and kurtosis implicit in the risk-neutral distribution.
- In addition, volatilities estimated from the time series are usually clustered, which is commonly referred to as volatility persistence across maturities, which is inconsistent with homogeneous Lévy processes, and possibly explains the failure of such processes to explain option prices across the maturity dimension.

Collectively, these considerations suggest that it may be desirable to incorporate a richer behaviour across maturity than is implied by homogeneous Lévy processes.

In this respect, Lévy models can be extended by incorporating stochastic and mean-reverting volatilities. Below we consider the VG Lévy process and generate the desired volatility properties by subordinating it to the time integral of a Cox–Ingersoll –Ross process. The randomness of the CIR process induces stochastic volatility, while the mean reversion in this process induces volatility clustering. We term the resulting processes VGSV in recognition of their synthesis with stochastic volatility:

$$\begin{aligned} d\log S_t &= f(v_t)dt + \sqrt{v_t}\left(\beta dW_t^S + \sqrt{1-\beta^2}\; dL_t\right), \\ dv_t &= \theta_v(\bar{v} - v_t)dt + \phi_v\sqrt{v_t}\; dW_t^v, \\ dW^S\, dW^v &= \rho dt, \quad \text{var}(L) = 1. \end{aligned}$$

- This model allows correlation for $\rho \neq 0$.
- Heston for $\beta = 1$.
- Pure jump process for $\beta = 0$.

These processes are tractable, in that analytical expressions can be derived for their characteristic functions. On employing their exponentials to describe stock prices, European options can be priced via Fourier methods.

	Heston	*Jump*	*Full*
$\sqrt{v_0}$	0.0912	0.0992	0.0908
θ_v	0.9363	1.3509	1.3270
$\bar{\sigma}$	0.1949	0.1732	0.1781
ϕ_v	0.3380	0.3282	0.3551
ρ	−0.5570		−0.9870
σ		0.8604	0.3474
θ		−1.2153	6.4040
β			0.6708
rmse	4.806	11.095	4.242

FIGURE 3.73 Calibrated parameters

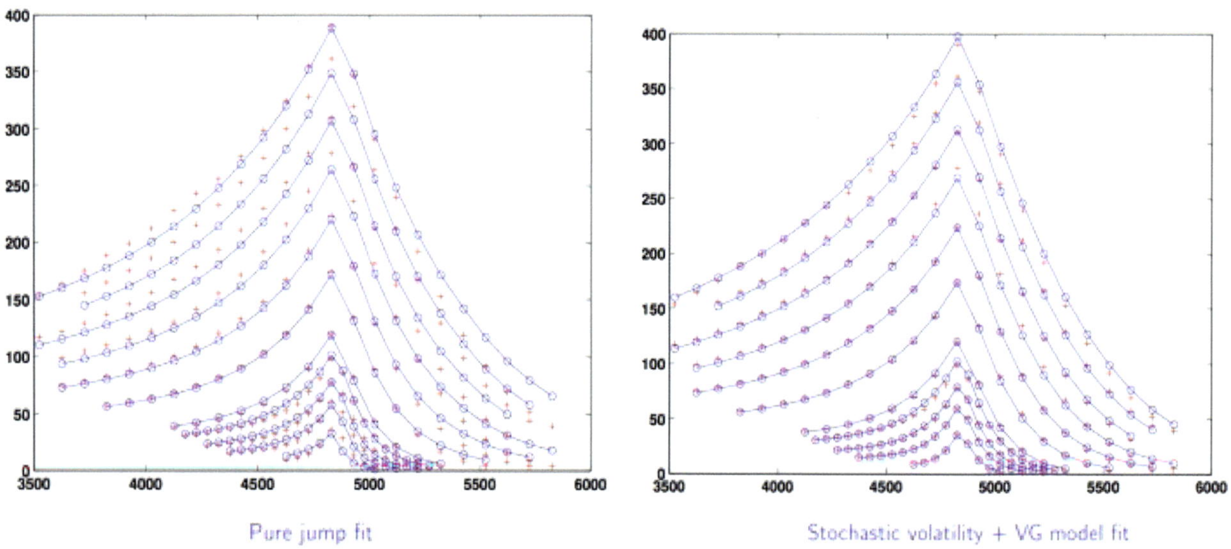

FIGURE 3.74 Quality of fitting implied volatility distribution across maturities by variance gamma model with a CIR volatility process

Such a new process provides us with a flexible family of option-pricing models, capable of being calibrated to market option prices varying across both the strike and maturity dimensions.

The calibrated process may then be used for pricing complex savings and retirement contracts. These processes can also be used in a simulation to evaluate the quality of hedging strategies, and to enhance our structural understanding of risk management issues. With the resulting models, one may assess the impact on market values of changes in reasonably intuitive parameters, such as the speed of adjustment, the level of long-run volatility and the variance of volatility. Thus, the parameters of the models synthesize the information content of option prices in a concise manner and open the door to interesting investigations into the economic determinants of asset pricing as inferred from an analysis of derivative markets.

(v) Summary

In Figure 3.76 we benchmark the different modelling frameworks across key criteria: flexibility, computational run-time, (average) pricing accuracy, density fitting.

Method	**Payoff flexibility** Implementation	**Runtime**	**Pricing**	**Calibration** Good Density	Global Summary
Black-Scholes / Carr-Madan FFT	+++	+++	-	- -	0
VG / Carr-Madan FFT	++	++	+	+	4+
Variance Gamma / Voltchkova-Tankov-Goudenège FFT alternative	++	++	++	+	5+
Black-Scholes / Monte-Carlo	+++	+	-	- -	- -
Black-Scholes + Sto Vol / Monte-Carlo	++	- -	+	++	+
VG / Monte-Carlo	++	0	+	+	2+
VG + Sto Vol / Monte-Carlo	++	- -	+	++	+
CGMY / Monte-Carlo	++	0	++	+	3+
CGMY + Sto Vol / Monte-Carlo	++	- -	+	++	+
Black-Scholes / PIDE	+++	++	+	- -	+
VG / PIDE	+++	++	+	++	5+
VG + Sto Vol / PIDE	++	+	++	+++	6+

FIGURE 3.75 Comparison of pricing methodologies

3.3 CUSTOMER BEHAVIOUR RISK MODELLING FOR SUSTAINABLE SAVINGS AND RETIREMENT

As illustrated in Chapter 1, behavioural risks have been one of the major sources of losses over the past decade, in particular for savings and retirement financial guarantees, leading to multiple earnings losses, reserve strengthening and reinsurance deals. The impact is in the billions of dollars for the companies with the largest exposure (as already experienced for the variable annuity business in the 2010s, notably at Brighthouse and AXA Equitable regarding surrender and partial withdrawals), and potentially long-term solvency.

> Though equity-market declines are generally seen as the biggest risk in VA contracts, most insurers effectively hedge that risk via derivatives. That leaves the less-easily hedged and more unpredictable policyholder behaviour, and particularly lapses, as a key driver of the profitability of these popular products.
>
> Retention by policyholders was much greater than expected, causing insurers to take significant, unexpected earnings charges and write-downs over the past year and a half.
>
> Recent experience for these guarantees provides [the takeaway that] … companies tend to retain customers that cost them the most and lose those that cost them the least.
>
> Moody's Investor Service (2013)

Thus incremental improvements to our understanding of customer behaviour can have enormous dollar impacts.

Policyholder behaviour is the result of a complex process that depends on the overall economic environment (equity and interest rates levels in particular), the financial and personal situation (attained age, tax rate), the contract characteristics (policy past and residual duration, policy year, guarantee options, policy size), the insurer's product compared to the competition and the effectiveness of its financial communication. There is enormous uncertainty around how policyholder behaviour will emerge over the lifespan of business currently on the books of companies.

Fortunately, as time passes the available data grow: for companies that have been consistently present in the variable annuity marketplace, there is now over a decade of experience. In addition, there are valuable data on customers available from third-party vendors. While the experience data have some meaningful limitations in forecasting future experience, the industry could probably gain significant value by using the available data to develop better forecasting tools.

In order to better apprehend the different key behavioural risks, let us remind ourselves of the generic savings and retirement product mechanics (see Figure 3.77).

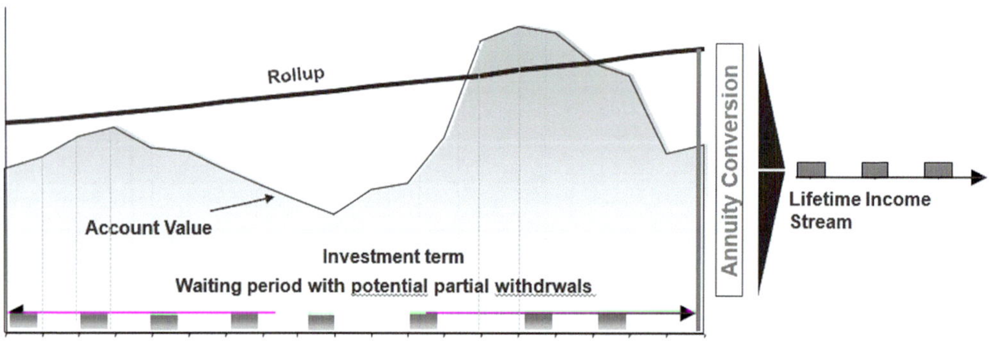

FIGURE 3.76 Retirement savings annuity mechanics

As a result, the behaviour risks mainly lie in four risks: withdrawal deferral period risk, surrender risk, withdrawal risk, and annuitization risk (annuity election), which are each concentrated within specific phases of contract life: the withdrawal risk timing mostly occurs in the early stages during the "ten-year withdrawal deferral bonus period", the lapse risk timing is mostly concentrated after the "seven-year surrender charge period" and the annuitization risk usually occurs far later.

As illustrated in Chapter 1, the most essential change in behaviour over the past decade has been the drastically rising efficiency of customers' behaviour (whether surrender or withdrawals in particular), translating into several anti-selection effects notably experienced during buyout initiatives.

Below we present the different approaches and then illustrate them through the four main customer behaviour risks: withdrawal deferral period risk, surrender risk, withdrawal risk and annuitization risk.

3.3.1 Traditional Experience Study versus Predictive Modelling Approaches

Traditional tabular descriptive analysis uses one-way or two-way splits of the data to analyse the impact due to the limited number of variables, and suffers from several limitations.

- Aggregating data fails to control for confounding effects, which may result in spurious correlation.
- Assumptions are typically formulated to best fit the most recent experience.
- Validation is typically performed on the entire dataset rather than a holdout set.
- The credibility measure is based on exposure rather than a probabilistic measure of the parameters.
- It is easy to use and implement but lacks statistical rigour.

In contrast, the predictive model approach has these advantages.

- It captures a greater number of drivers without sacrificing credibility.
- It uses all available data by effectively accounting for correlations in the model.
- Interactions between variables can be fully explored without splitting the data.
- There are safeguards against overfitting by training models on a subset of data and validating the model on a holdout set.
- Credibility can be estimated based on the statistical significance of parameter estimates, model performance on a holdout data set or the consistency of either of these measures over time.

3.3.2 From Backward-Looking to "Reasonable" Forward-Looking Predictive Modelling

Usually, customer behaviour has been modelled by historical or backward-looking statistical regressions, which have empirically underestimated the risk due to the scarcity of extreme scenario samples and the inability to dynamically extrapolate the observed behaviour to various market conditions.

In contrast, the so-called "efficient" approach is a prudent forward-looking approach whereby policyholders behave in a way that maximizes the net present value of the future cash flows – i.e. maximize the expected discounted future accumulated wealth over the life of the product (taking lapse, mortality and annuity election modelling into account).

It is usually computed under a risk-neutral valuation framework as a prudent approach, but it can, naturally, also be computed under a real-world valuation framework.

However, for a pool of insurance policies, common sense and experience tell us that not all policyholders will behave the same way at the same time.

As a result, this efficient customer approach framework can be amended by introducing a mix between backward-looking and forward-looking approaches, considering that X% of policyholders have non-fully efficient behaviour (e.g. behaviour consistent with the backward-looking predictive methodology) while 1–X% do have efficient behaviour, which translates into the notion of "reasonable" behaviour, more consistent with empirics.

Let us illustrate an example of this "reasonable" forward-looking predictive modelling in the case of surrender behaviour: we denote by p (ti) the proportion of lapses at date ti among the contracts still active in the pool.

Inspired by the mortgage prepayment models and evaluation approaches of surrender options for other life insurance products, we assume that p is a non-decreasing piece-wise linear function of the variable F

- when the lapse rate is not zero due to policyholders' personal circumstances (including liquidity and death); these so-called "non-fully efficient" surrenders are analogues to non-economic prepayment on low-rate mortgages; while
- when $F^T(t_i) < f'(t_i -)$, some efficient surrender never occurs, and a reasonable specification of p(ti) may be illustrated as in Figure 3.78.

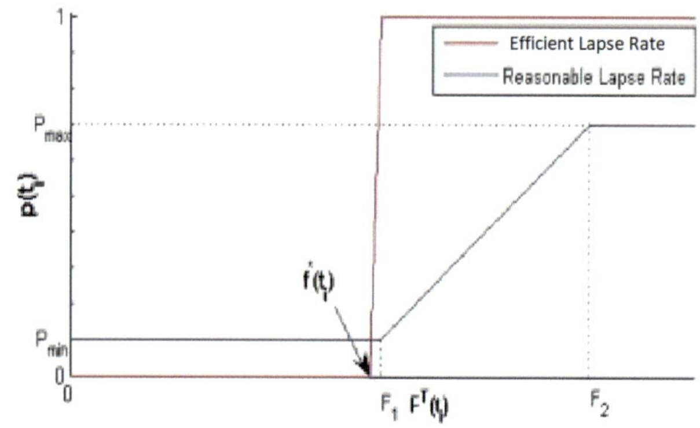

FIGURE 3.77 Efficient versus reasonable surrender predictive modelling

The four parameters F1, F2, Pmin and Pmax are determined by insurers according to some empirical tests. To be consistent with the efficient surrender assumption, F1 should be very close to f*(ti), and Pmax should be set high enough (normally ≥ 50%).

3.3.3 Predictive Modelling Case Studies

3.3.3.1 Surrender Behaviour Predictive Modelling

This complexity is found especially in the surrender issue.

The insured has the possibility to proceed with the redemption of her policy. On the day of the redemption date all warranties inherent in the contract end. No charge is applied, and the cash value is based on the value of the savings (not the value of the guarantee). The redemption can take place before or after the start of payment of a potential annuity embedded within the contract.

The dynamics of surrender and their financial consequences can be modelled in a prospective insurance model.

3.3.3.1.1 Backward-Looking Predictive Modelling To model this behaviour, a surrender rate is defined to project the surrender behaviour during the duration of the contract, with the most basic model modelled as step function as follows.

- First, a fixed rate is established that comes from historical data or an estimate specific to each country.
- A multiplicative factor is then applied, which varies according to the values of the funds and the potential financial guarantee. This provides "dynamic lapse rates", potentially taking into account the following drivers.
  - There is the moneyness of the potential financial guarantee (i.e. the ratio guarantee level/account value 1), embedded within the contract.

    In fact, if policyholder reserve is less than the minimum guaranteed amount (a so-called "out-of-the-money" financial guarantee), the value of the option will increase, making the contractor less likely to terminate the contract. In the opposite case ("in the money"), the policyholder is more likely to terminate the contract.

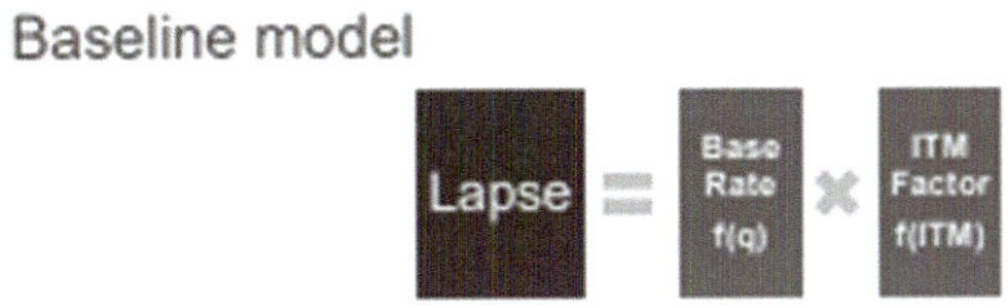

The ITM factor is computed as follows:

$$ITM_t = \frac{GuaranteeBase_t}{AccountValue_t} - 1.$$

The multiplicative factor is given by Figure 3.80.

Year	Base lapse
1	4%
2	5%
3	5%
4	7%
5	7%
6	8%
7	9%
8	15%
9+	10%

FIGURE 3.78 Base lapse

This translates into step functions such as in Figures 3.81 and 3.82.

% in-the-moneyness	Multiplicative factor
<15%	100%
15%-30%	90%
30%-50%	70%
50%-75%	50%
75%-100%	30%
>100%	20%

FIGURE 3.79 ITM factors

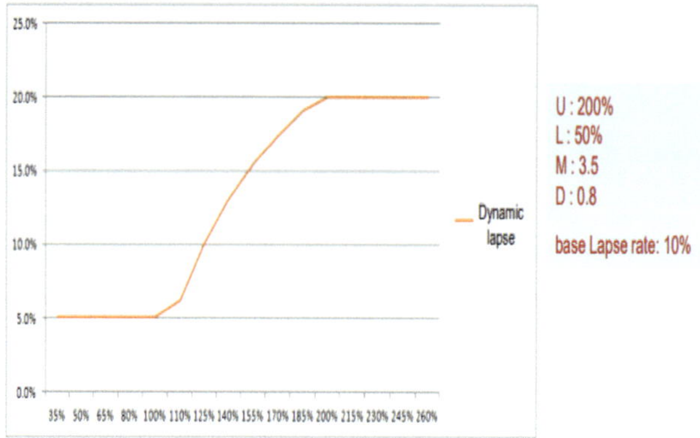

FIGURE 3.80 Dynamic lapse steps

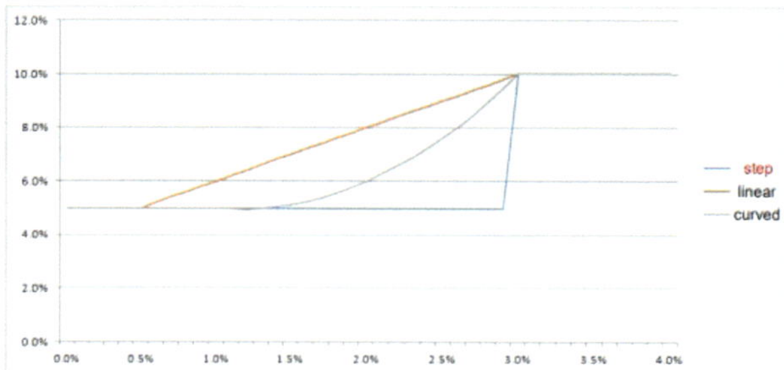

FIGURE 3.81 Curved shape exponential functions

For instance, the US principle-based reserve (PBR) standard, VM-21, applies the following dynamic lapse multiplier, λ, when calculating the cash surrender value of GMDB products:

$$\lambda = \text{MIN}\left[U, \text{MAX}\left[L, 1 - M \times \left(\frac{GV}{AV} - D\right)\right]\right]$$

where GV is the guaranteed amount, AV the policyholder account value, U, L, the upper, lower, coefficient, M the sensitivity coefficient and D an adjustment factor.

It is important to determine the type of ascending shape for this multiplier method. There are three main types of ascending shapes, and each presents different views on how policyholders react depending on the level of the dynamic variables.

For a curved shape, exponential functions may be used, such as

$$\text{Lapse rate} = \text{Base lapse rate x } e^{2\,[\text{MIN(account value/guaranteed value, 1)-1}]}.$$

- In addition, a higher future expected guarantee rate than the current guaranteed rate also provides some incentive to stay -> the ratio of the two may be a surrender driver.
- In contrast, a higher yield provided by the competition (which can be proxied by the current ten-year yield) than the one at the time of the pricing is an incentive to surrender -> the spread between the two yields may also be a surrender driver.

Then taking both moneyness and this provides the following modelling:

$$LapseRate_t = BaseLapse \cdot \frac{1}{2} \cdot \left(\frac{Account\ Value_t}{\text{Income Base}_t} \right) + \left(1 + \frac{10\_yr\_\text{rate}_t}{10r\_r\_\text{rate}_{pricing}} \right).$$

In order to model in a tractable way the general empirical phenomenon of dynamic lapse, whereby the slope of the lapse multiplier increases rapidly and then decreases near the inflection point, the arctan function may be used:

$$\text{Multiplier} = \text{f(x)} = \text{a} \times \arctan(\text{m} \times (\text{x} - \text{k})) + \text{c},$$

where k = inflection point, m = sensitivity, a & c = adjustment coefficients, $F(\infty) = U$(upper limit) and $f(-\infty) = L$(lower limit).

Then the model is adjusted through change in the c coefficient (see Figures 3.83 and 3.84).

	Upper limit (U)	Lower limit (L)	Inflection point (k)	Inflection multiplier (c)	Adjustment factor (a)	Sensitivity (m)
Coefficient	170%	70%	-43bp	112.2%	31.8%	61.7
Arctan model	f(x) = 0.318 * arctan(61.7 * (x + 0.43) + 1.122					

FIGURE 3.82 Dynamic lapse multiplier

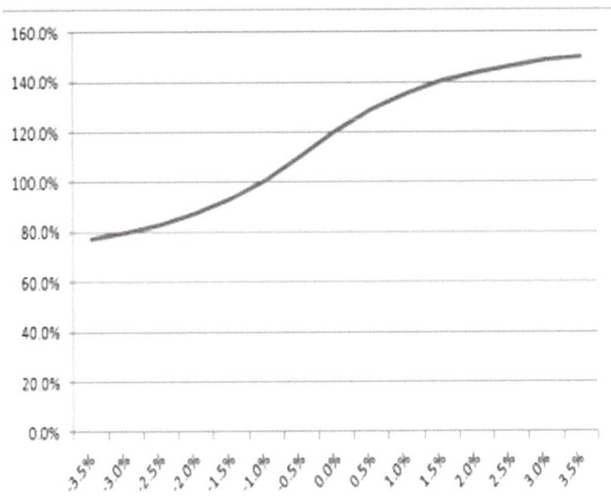

FIGURE 3.83 Dynamic lapse

- It is easy to adjust the model by changing the coefficient of the model.
  - Environmental variables used by logit and tobit time series models show that lapse depends on additional exogenous factors beyond interest rates: the unemployment rate, inflation rate, GDP growth rate, such as:

Authors	Dar and Dodds (1989)	Outreville (1990)	Kuo et al. (2003)	Kim (2005a)	Kim (2005b)	Cox and Lin (2006)	Kiesenbauer (2011)
Country	UK	US, Canada	US	South Korea	US		Germany
Lapse Data Provider	Industrial Life Offices' Association (ILOA)	American Council of Life Insurance (ACLI); Life Insurance and Market Research Association (LIMRA)	American Council of Life Insurance (ACLI)	One Korean company	Society of Actuaries' Risk Management Task Force (data on one US company)	Society of Actuaries' Risk Management Task Force (data on one US company)	German regulator BAFin and self-collected data
Type of Policy	Endowment policies	Whole-life and ordinary life insurance	All ordinary life insurance contracts*	Endowment, annuity, protection plan, education	Single premium deferred annuities	Single premium deferred annuities	Endowment, annuities including disability and long-term care, term life, group business, other
Time Period	1952-1985	1955-1979 (annually), 1966-1979 (semiannualy)	1951-1998	1997-2000	1993-2003	1993-2003	1997-2009
Methods	Standard time series regressions	Standard time series regressions	Cointegration techniques	Logit and complementary log-log model	Logit	Tobit	Logistic regression
Variables for EFH and IRH	EFH: growth in the level of unemployment and the level of actual unemployment relative to time trend IRH: internal and external rate of return	EFH: yearly unemployment rate IRH: real rate of return on alternative assets	EFH: yearly unemployment rate IRH: market interest rates	EFH: Unemployment rate IRH: reference market rate - crediting rate -surrender charges	EFH: Monthly annualized US unemployment rate IRH: Annualized five–year Treasury bond rate minus the policy credited rate	EFH: Monthly annualized US unemployment rate IRH: Annualized five–year Treasury bond rate minus the policy credited rate	EFH: Unemployment rate IRH: Bond Performance, Stock Performance
Other variables	/	Real transitory income per capita, price of insurance, anticipated inflation rate.	/	Policy age, GDP growth, surrender charge	Policy age, GDP growth, surrender charge	Policy age, GDP growth, surrender charge	Buyer confidence, GDP, company age, distribution, legal form, company size, participation rate
Main Results	EFH +; IRH -	EFH +	EFH -, IRH +	Lapse depends on additional exogenous factors beyond interest rates and unemployment rates	Logit Model closely fit the experience of the data even under extreme financial conditions	Tobit model is better than logit model; poisson and the negative binomial regression model are more appropriate to model lapse	Factors beyond interest rates and unemployment influence lapse behavior, including company characteristics

Notes: EFH: Emergency fund hypothesis; IRH: Interest rate hypothesis; *: The covered policies include universal life, variable life, variable-universal life, and traditional whole life that includes limited payment, continuous premium, joint whole life, single premium, adjustable life, monthly debit, and other fixed-premium products), term insurance (including decreasing, level and other term, term additions), and endowment insurance.

FIGURE 3.84 Lapse modelling reviews

$$\log\left(\frac{q_s}{1-q_s}\right) = -3.8414 - 0.00689 \times V_1 + 0.1908 \times V_2 + 0.0564 \times V_3,$$

where q_s is the lapse rate, V_1 the elapsed period (month), V_2 the interest rate spread of certificates of deposit and V_3 inflation.

Using the "AIC", "SC" and "–2 LogL" measures to gauge the impact of the explanatory variables on the model, we get Figure 3.86.

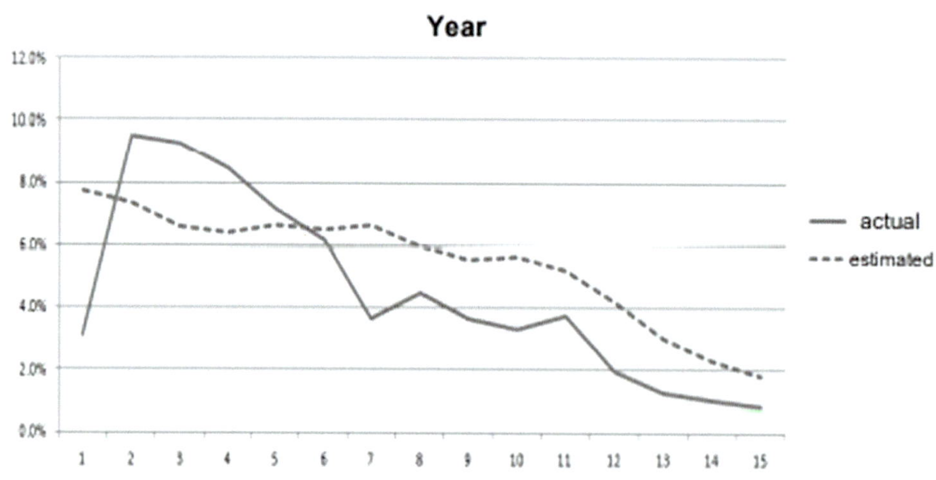

FIGURE 3.85 Savings products lapse

Remark: however, the temporary surge of the lapse rate due to the characteristics of savings products should be treated separately – i.e. the volatility of first to second year lapse rate, notably through dummy variables at the timing of the temporary surges.

- Policyholder and product characteristics (e.g. policy past and residual duration, policy year, guarantee options, policy size) are used by logistic and GLM regressions and show the importance of policy duration, policyholder age, time lapsed from contract date and contract age.

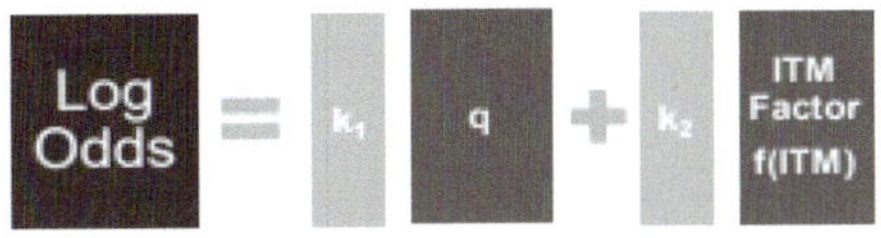

Authors	Renshaw and Haberman (1986)	Kagraoka (2005)	Cerchiara et al. (2009)	Milhaud et al. (2010)	Eling and Kiesenbauer (2011)
Country	Scotland	Japan	Italy	Spain	Germany
Data Provider	Pool of 7 companies	1 company	1 company	1 company	1 company
Type of Policy	Endowment Whole-life Temporary insurance	Annuity-type personal accident	Saving	Endowment	Endowment Annuity Term life
Time Period	1976	1993–2001	1991–2007	1999–2007	2000–2010
Methods	Logistic regression model Binomial model	Poisson model Negative binomial model	Poisson model	Logistic regression model Classification and regression tree model	Poisson model Binomial model Negative binomial model
Variables	Policyholder age and sex Contract age Product type Company	Policyholder age and sex Seasonality Unemployment rate Heterogeneity	Policyholder age Contract age Product type Calendar year	Policyholder age Contract age Product type Sum insured Risk premium Saving premium	Policyholder age and sex Contract age Product type
Main Results	four factors identified as important (age at entry, duration of policy, office, type of policy), additionally, significant interaction between policy type and duration	surrender of the insurance contracts is explained by change of unemployment rates and time elapsed from contract date	Shows the importance of policy duration, but also show the sensitivity of lapse rates to calendar year of exposure, to product class and policyholder age	Find that duration and profit benefit option are essential	Product characteristics such as product type or contract age and policyholder characteristics such as age or gender are important lapse drivers

FIGURE 3.86 Lapse modelling reviews

Among the product characteristic drivers, lapse experience after the surrender charge period is generally lower than currently assumed: policyholders seem to be most efficient at the end of their surrender charge periods.

A predictive model provides a single framework for analysing and attributing the impact to both duration and moneyness jointly.

Left tail analyses often show a large area of over-prediction by the traditional baseline model. Moving to a predictive model drops predictions to more reasonable surrender levels, notably due to the introduction of new variables as mentioned above.

3.3.3.1.2 Efficient Forward-Looking Surrender Predictive Customer-Centric Modelling Dynamic surrender behaviour is, essentially, a selection process of the policyholders against the insurer, whereby an increase leaves fewer insured to ultimately make a claim on the guarantees but reduces the fees the insurer can collect.

Whenever a financial guarantee is offered by the insurance contract, the general pattern is that more policies will lapse when the capital market is up, and fewer policies will lapse when the capital market is down.

- In an up market, the value of the minimum guarantee diminishes as the account value is likely to exceed the minimum guarantee value. As such, surrendering the policy does not create much loss to the policyholder.
- On the other hand, a down market can result in the surrender value being less than the guarantee value causing the policy to be in the money. If the policyholder surrenders at that time then she can get only the reduced surrender value, forfeiting the added value from the guarantee rider. The result is that there is strong incentive for the policyholder to keep the so-called “in-the-money” contracts in force.

Efficient forward-looking modelling reflects prudent policyholder behaviour, as experienced in the last 2008 market crash, illustrated by an initial immediate and sustained fall in surrender rates right after the crash, before an abrupt recovery consistent with interest rates. In contrast, backward-looking dynamic lapse modelling usually fails to be consistent with such empirical dynamics, with no particular dynamics after an extreme market shock.

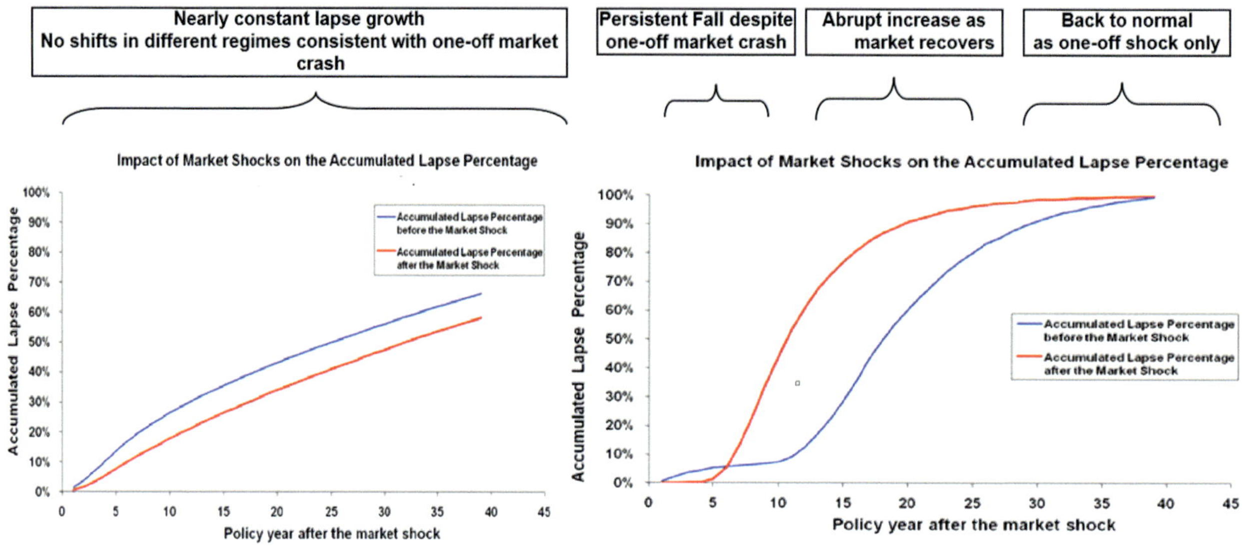

FIGURE 3.87 Backward-looking versus efficient forward-looking predictive modelling

The price evaluated by this approach can be interpreted as the fair value of the policy if all policyholders use the same efficient surrender strategy, which is similar to the optimal early-exercise strategy of Bermudan options.

The efficient forward-looking approach is illustrated below with two case studies involving financial guarantees, as a guaranteed lump sum or as periodic guaranteed withdrawals.

3.3.3.1.2.1 Guaranteed Lump Sum Case Study Let us consider a generic lump sum savings product

The guarantee can either step up and be reset to the high-water mark of the account value on the anniversary dates of the purchase (e.g. annual ratchet) when the market has performed well, or it can roll up with a fixed percentage (known as the roll-up rate – e.g. 2%) regardless of market conditions, in exchange for a periodic annual fee, denoted "c".

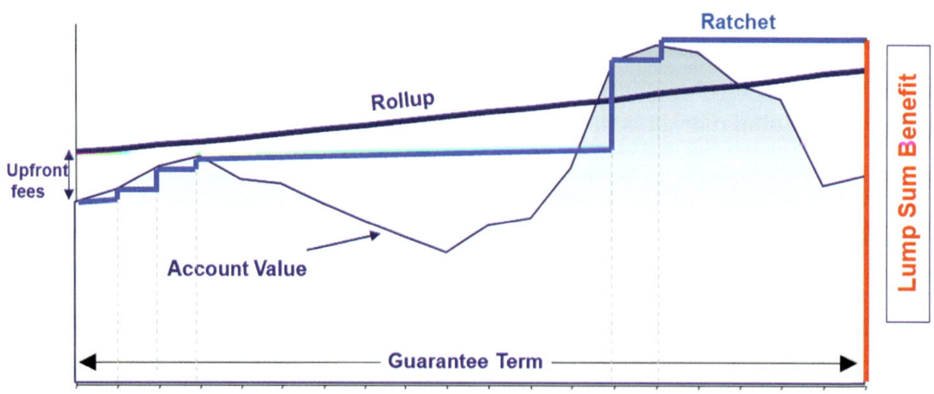

FIGURE 3.88 Guaranteed lump sum pattern

Such contracts give policyholders the ability to protect their retirement investments against downside market risks by allowing them to receive the greater of the account value and the benefit base at the maturity; so, in the event of no surrender under the risk-free neutral measure,

$$\begin{aligned} L^E(t) : \ell^E(t, r(t), \check{A}(t)) &= \mathbf{E}^{\mathbb{Q}}\left[D_t^T \max(\check{A}(T), \bar{B}) \mid \mathcal{F}_t\right] \\ &= \mathbf{E}^{\mathbb{Q}}\left[D_t^T\left(\check{A}(T) + (\bar{B} - \check{A}(T))^+\right) \mid \mathcal{F}_t\right]. \end{aligned}$$

This payoff can therefore indeed be decomposed into to the sum of the forward contract of the account value at time T Ǎ(T) and a European vanilla put (B - Ǎ(T))+ with strike equal to B.

The dynamics of the equity underlying the asset is consistent with the one described in the previous section – i.e. Black–Scholes and Hull–White:

$$\begin{cases} dS(t) = r(t)S(t)dt + \sigma S(t)dW(t) \\ dr(t) = a(\theta(t) - r(t))dt + \sigma_r dZ(t) \\ Z := \left(1 - \rho^2\right)^{\frac{1}{2}} W^{\perp} + \rho W, \end{cases}$$

where Ǎ(t) si the account value at time t with dynamics:

$$d\check{A}(t) = (r(t) - c))\check{A}(t)dt + \sigma\check{A}(t)dW(t).$$

Dij is the discount factor from tj to ti, where tj > ti,

$$D_{t_1}^{t_2} := \exp\left(-\int_{t_1}^{t_2} r(s)ds\right),$$

while B is the guarantee level at maturity, fixed at inception and equal to Ǎ(0)(1 + r)^T, where r is the roll-up rate.

To make the pricing easier, we evaluate the contract under the forward martingale measure (and not the risk-neutral martingale measure): we introduce the forward value of Ǎ(t) at T observed at date t, denoted by

$$F^T(t) = \check{A}(t) / Z_t^T,$$

where Z is the price of a zero-coupon bond maturing at T. This leads to the following dynamics:

$$\begin{aligned} \frac{dF^T(t)}{F^T(t)} &= \left(v^2 B_r^2(u) + \rho v \sigma B_r(u) - c\right)dt + \\ &\quad + \sigma dW(t) + vB_r(u)dZ(t), \end{aligned}$$

By doing the following transformations of the Brownian motions from the risk-neutral to the forward risk-neutral martingale measures,

$$\begin{aligned} dZ(t) &\to dZ^T(t) - vB_r(u)dt \\ dW(t) &\to dW^T(t) - \rho v B_r(u)dt. \end{aligned}$$

Then

$$\begin{aligned} dF^T(t) &= -cF^T(t)dt + \sigma F^T(t)dW^T(t) + \\ &+ vB_r(u)F^T(t)dZ^T(t) = \\ &= -cF^T(t)dt + \omega_u F^T(t)d\tilde{W}^T(t), \end{aligned}$$

where $\omega_u^2 = \sigma^2 + v^2 Br^2(u) + 2\rho\sigma v Br(u)$.

As a result,

$$\tilde{\ell}^E\left(t, F^T(t)\right) = e^{-cu}F^T(t) + BN\left(-d_2\right) - e^{-cu}F^T(t)N\left(-d_1\right)$$

$$d_{1,2} = \frac{\log\left(F^T(t)/B\right) - cu}{\Gamma} \pm \frac{\Gamma}{2},$$

$$\Gamma = \sqrt{\int_0^{T-t} \omega_s^2 ds}.$$

If this liability value can be easily calculated analytically under a no-surrender assumption (as it is the sum of a forward and a European put option), in contrast, once surrender optionality is introduced, we need to use some optimization numerical methods: the policyholder aims to maximize her net present value through a rational lapse boundary where PV (charges) = PV (claims), reflecting the implicit arbitrage between the roll-up rate versus the risk-free rate.

Indeed, the "efficient forward-looking" pricing including the surrender optionality involves the valuation of an optimal early-exercise strategy of a classic Bermudan option. We assume that the policyholder can surrender the contract at the end of each policy year:

$$\begin{aligned} \ell^B(T, r(T), \breve{A}(T)) &= \max(\breve{A}(T), B)\ell^B\left(t_i-, r\left(t_i-\right), \breve{A}\left(t_i-\right)\right) \\ &= ess\sup_{\tau_i \in \tau_i} \mathbf{E}^{\mathbb{Q}}\left[D_i^{\tau_i}\breve{A}\left(\tau_i\right) + 1_{\{\tau_i = T\}} D_i^T \max(\breve{A}(T), B) \# \mathcal{F}_{t_i}\right]. \end{aligned}$$

Applying the fact that r(t) and Ă(t) are all Markov processes, we have

$$\begin{aligned} &\forall i \leqslant n, \quad \ell^B\left(t_i, r\left(t_i\right), \breve{A}\left(t_i\right)\right) = \\ &= \mathbf{E}^{\mathbb{Q}}\left[D_i^{i+1}\ell^D\left(t_{i+1}-, r\left(t_{i+1}-\right), \breve{A}\left(t_{i+1}-\right)\right) \mid \mathcal{F}_{t_i}\right]. \end{aligned}$$

Similarly, with Bermudan-style options, at discrete time points ti the policyholder is supposed to compare the account value with the value of the liability to decide whether to lapse or not. If the account value is bigger than the liability, policyholders surrender the contract and get back Ă(ti). Otherwise, they continue to hold the policy. That is to say, at time ti the Bermudan-style liability should evolve as follows:

$$\begin{aligned} &\ell^B\left(t_i-, r\left(t_i-\right), \breve{A}\left(t_i-\right)\right) = \\ &= \max\left(\breve{A}\left(t_i\right), \ell^B\left(t_i, r\left(t_i\right), \breve{A}\left(t_i\right)\right)\right). \end{aligned}$$

This reflects the fact that the liability before the annuity payment is equal to the greater of the current account value and the value of continuation. To simplify the pricing process of the Bermudan-style liability, we can calculate the expectations under the forward risk-neutral measure (instead of the risk-neutral measure) – i.e.

$$\tilde{\ell}^B\left(t_i-,F^T\left(t_i-\right)\right):=\frac{\ell^B\left(t_i-,r\left(t_i-\right),\breve{A}\left(t_i-\right)\right)}{Z_i^T}=$$
$$=\max\left(F^T\left(t_i\right),\tilde{\ell}^B\left(t_i,F^T\left(t_i\right)\right)\right)=$$
$$=\max\left(F^T\left(t_i\right),\mathbf{E}^{\mathbb{Q}^T}\left[\tilde{\ell}^B\left(t_{i+1}-,F^T\left(t_{i+1}-\right)\right)F_{t_i}\right],\right.$$

where the boundary condition at maturity is

$$\tilde{\ell}^B\left(T,F^T(T)\right)=F^T(T)+\left(B-F^T(T)\right)^+.$$

As the benefit base B is fixed at inception, it is a convex, non-decreasing function in the forward value of $\breve{A}(t)$ at T. In addition, it is also a positive function; finally, for the charge fee $c > 0$, the two equations above also imply that

$$\lim_{f\to\infty}\left(f-\tilde{\ell}^B\left(t_i,f\right)\right)>0.$$

Thus, there exists a real number f* such that

$$0\le f\le f^*\left(t_i-\right)\Rightarrow\tilde{\ell}^B\left(t_i-,f\right)>f\,(\text{not lapse}),$$
$$f\ge f^*\left(t_i-\right)\ \Rightarrow\tilde{\ell}^B\left(t_i-,f\right)=f\,(\text{lapse}).$$

f* is referred to as the "critical forward account value", since the policy should be surrendered whenever the forward account is higher than the critical boundary. In contrast, within the context persistently low interest rates and equity, surrenders are expected to remain low, as experienced subsequent to the 2008 financial crisis.

Remark: thanks to the change of measure, such a critical boundary here depends only on the forward account value, rather than on both the interest rate and the account value level.

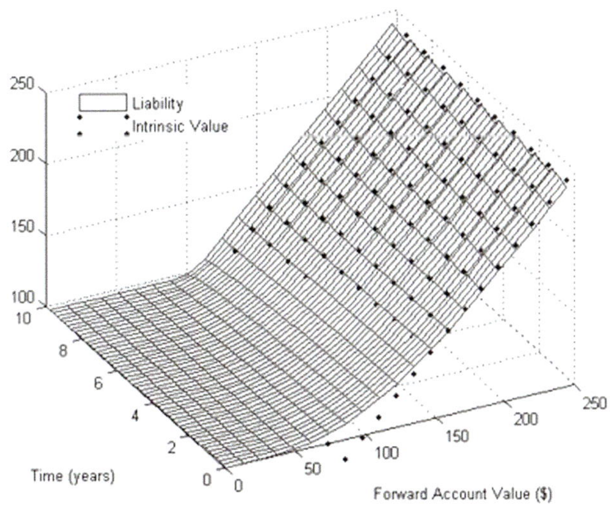

FIGURE 3.89 Critical forward account values

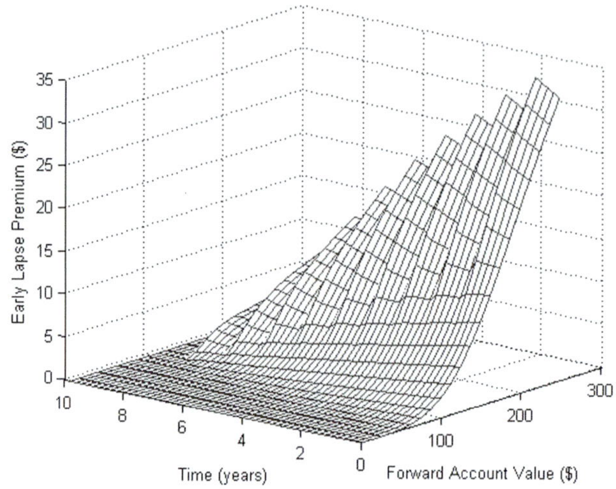

FIGURE 3.90 Early lapse premium

The difference between the surrender and no-surrender price – i.e. between l^B and l^E – is called the early surrender (i.e. contract termination) premium.

- The Bermudan liability is almost equal to the European-style liability when the account value falls to very low levels; in this case the probability that policyholders surrender the contract before the maturity is extremely small.
- However, when the forward account value increases to very high levels the early lapse premium grows almost linearly with f.
- The early surrender premium reduces gradually to zero at the expiration date.

Finally, the difference $\tilde{p}^B(t,f) := \tilde{\ell}^B(t,f) - f$ is the option value; as expected, it evolves similarly as a put. When the forward account value is significantly higher than the critical lapse boundary, the option value becomes negative between two discrete exercisable dates, due to the charge fees that policyholders are obliged to pay.

In terms of numerical implementation, the price of the liability and the critical surface f* are derived by two numerical methods: PDEs and Monte Carlo simulation schemes.

- The PDE method is faster and more precise for low-dimensional problems (< 3), especially for f nearing the critical boundary, than Monte-Carlo-based methods. In addition, it can calculate the price and risks sensitivities for different f and t at the same time.
- However, the Monte Carlo method is much more flexible and easier to be implemented. Moreover, it can be extended to other high-dimensional problems, such as path-dependent payoffs, stochastic volatility models and basket account values, and is precise.

We benchmark the PDE and Monte Carlo methods on the option value, the option value delta sensitivity and the critical boundary (see Figures 3.92 to 3.94).

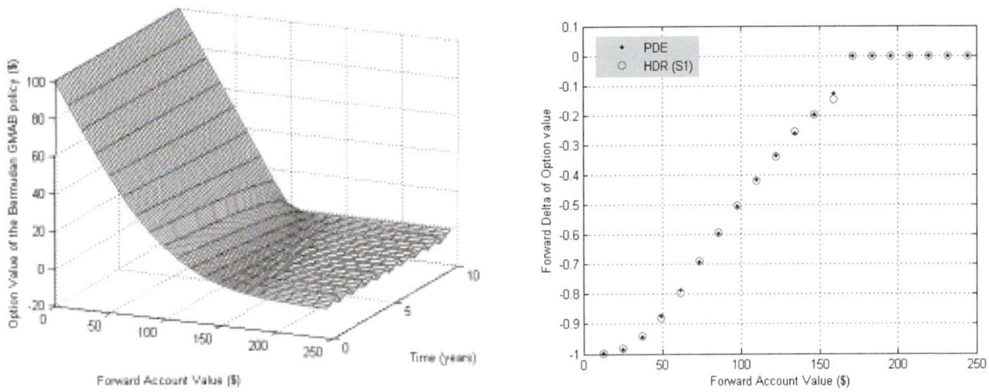

FIGURE 3.91 Option values and delta versus forward account values

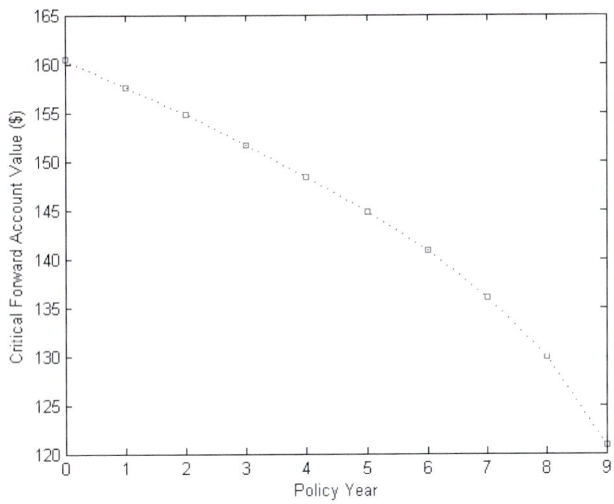

FIGURE 3.92 Critical lapse boundary

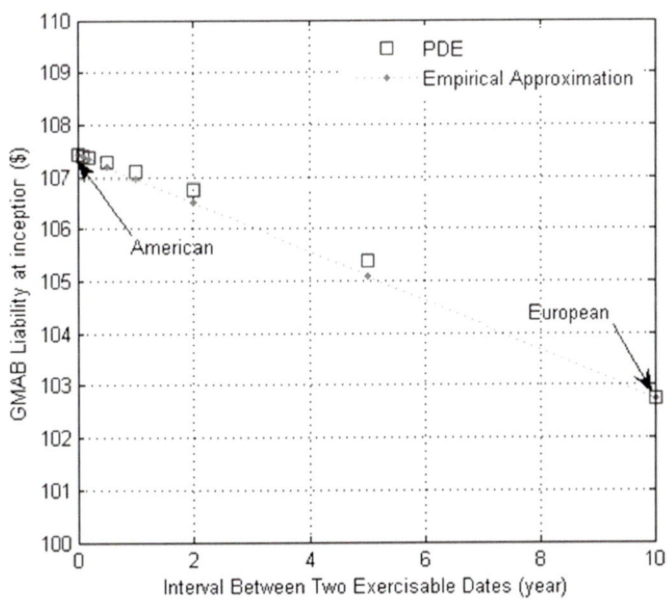

FIGURE 3.93 GMAB price by PDE versus approximation

As illustrated in these figures, the numerical results of the Monte Carlo method become increasingly unstable from the expiration to the inception of the policy, stemming from the accumulation of pricing errors caused by the linear regressions backward through time.

Regarding the PDE, an alternating-direction implicit (ADI) scheme has been used and compared to an analytical approximation based on the Barone-Adesi and Whaley (BAW) method:

Regarding the Monte Carlo simulations, the high-dimensional regression scheme (see Bouchard & Warin 2010) outperforms the traditional least squares method (see Longstaff & Schwartz 2001), for which the choice of a good global polynomial basis is typically difficult: it is more efficient to regress the value functions on an *adaptive local* linear basis rather than on some *global* polynomial basis proposed by Longstaff and Schwartz, especially for high-dimensional and long maturity problems (see Figure 3.95).

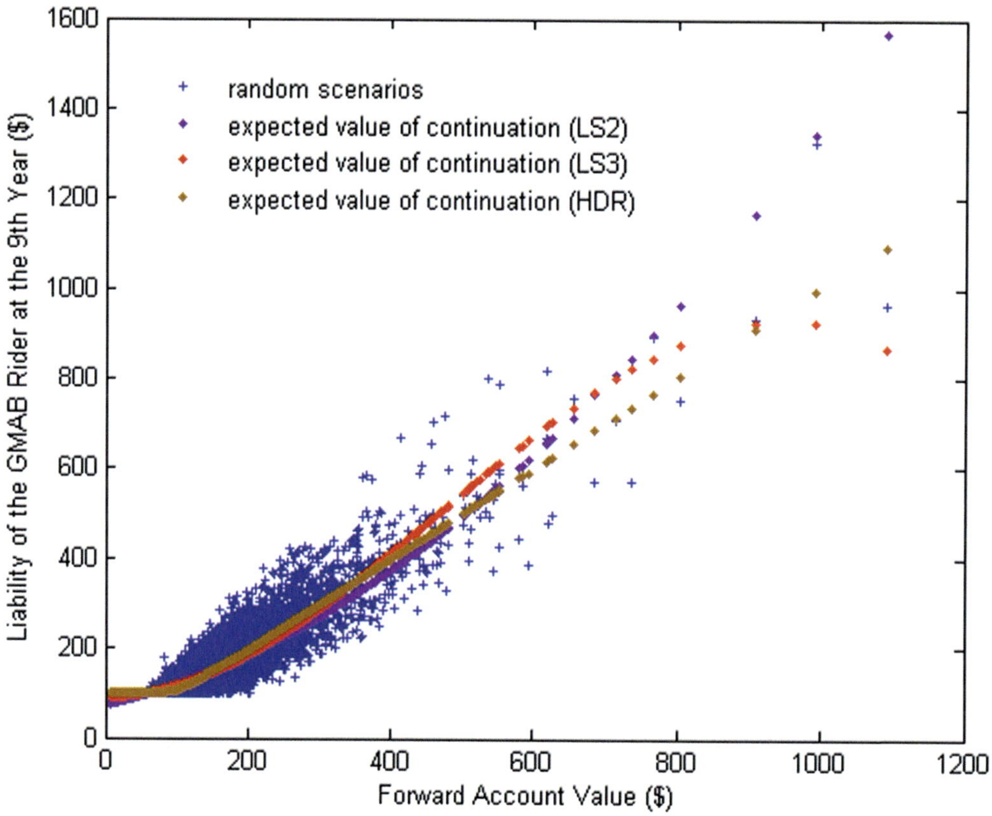

FIGURE 3.94 Liability at the ninth year versus forward account value

Notes: LS = Longstaff–Schwartz; HDR = high-dimensional regression.

Remark: more specifically, in these regressions there are two space dimensions: the interest rate r and the account value A. In the Bouchard and Warin method, the idea is to use, at each time step ti, a set of functions having local hypercube support D, which is partitioned across the two dimensions into a number of regions. On each support we define a linear function with three degrees of freedom (a constant, and the coefficients for Ã and r). At each time step the future cash flows of the liability are regressed on that function to estimate the relevant conditional expectation. The hypercubes are built so that they contain approximatively the same number of scenarios. This allows automatic adaptation to the statistical law of the underlying processes.

✓ **Quick win**: Figure 3.96 is a recap about the pros and cons of the different computational methodologies to price the surrender optionality.

	Analytical Approximation	MC Method	PDE
		LS	CN
Flexibility (Payoff)	**	***	***
Flexibility (Model)	*	***	**
Flexibility (Dimensions)	*	**	**
Stability (High Volatility)	**	**	***
Stability (Long Maturity)	**	**	***
Pricing Speed	***	*	**
Pricing Precision	**	**	***
Greek Precision	**	**	***
Surrender Boundary	**	**	***

FIGURE 3.95 Comparing pricing methodologies

Notes: LS = Longstaff–Schwartz; CN = Crank–Nicolson.

3.3.3.1.2.2 Guaranteed Withdrawals Annuity Case Study

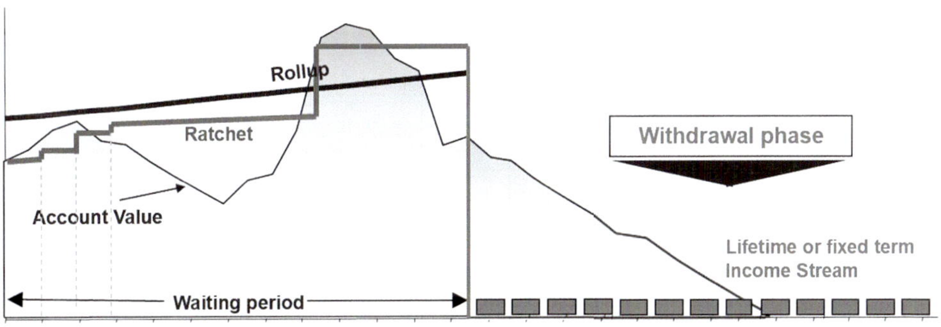

FIGURE 3.96 Guaranteed withdrawal annuity pattern

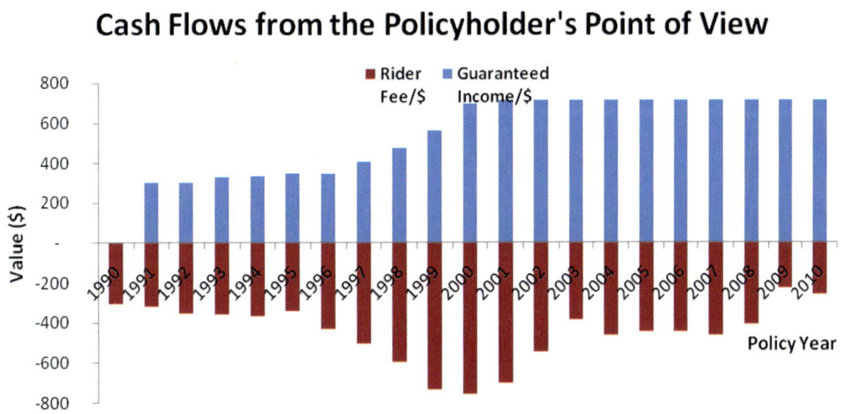

FIGURE 3.97 Cash flow patterns from the policyholder's point of view

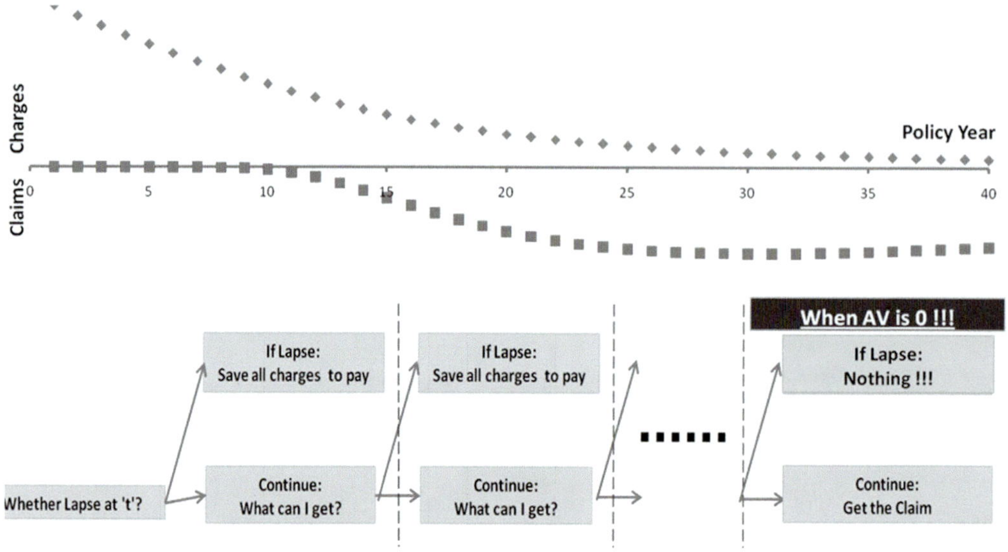

FIGURE 3.98 Lapse decision-making process

Such a guaranteed periodic withdrawal annuity gives policyholders the ability to protect their retirement investments against downside market risks by allowing the right to withdraw a fixed percentage (e.g. 4%) of the benefit base (and account value each year until death).

The benefit base can either step up and be reset to the high-water mark of the account value on the rider anniversary date (step-up policies) or can roll up with a fixed percentage annually (roll-up policies), regardless of the market conditions.

Another important feature is that the policyholder can defer the withdrawal start. For simplicity, we assume that this deferral period is zero, which means that the customers start withdrawals at the end of the first policy year.

A similar "efficient forward-looking" modelling framework as for the lump sum case is applied (see Kalife, Mouti & Tan 2015) – i.e.

$$\begin{cases} dS(t) = r(t)S(t)dt + \sigma S(t)dW(t) \\ dr(t) = a(\theta(t) - r(t))dt + \sigma_r dZ(t), Z := \left(1-\rho^2\right)^{\frac{1}{2}} W^{\perp} + \rho W. \end{cases}$$

Regarding the account value dynamics (annual fee c),

$$d\breve{A}(t) = (r(t) - c)\breve{A}(t)dt + \sigma\breve{A}(t)dW(t)$$
$$t_i \leqslant t < t_{i+1}.$$

At each anniversary date ti, the (coupon) annuity Ai is deducted from the account value:

$$\breve{A}(t) = \left(\breve{A}(t-) - A_i\right)^+ \quad \text{for } t = t_i, i \in \{1, 2, \cdots, n\}.$$

In the event of no surrender, the liability associated with no early surrender can be identified to a deterministic function:

$$L^E(t) := \sum_{j\geqslant 1}^{n} \mathbf{1}_{t<t_j} A_j \mathbf{E}^{\mathbb{Q}}\left[D_t^{t_j} \mid \mathcal{F}_t\right] + \\ +\mathbf{E}^{\mathbb{Q}}\left[D_t^{t_{n+1}} \breve{A}(t_{n+1}) \mid \mathcal{F}_t\right] = \ell^E(t, r(t), \breve{A}(t)).$$

Because of the annuity (coupon) payments at each anniversary of the policy purchase, the liability is discontinuous at the discrete time points ti:

$$L^E(t_i -) = L^E(t_i) + A_i \\ \text{and } \ell^E(t_i -, r, \breve{a}) = \ell^E\left(t_i, r, (\breve{a} - A_i)^+\right) + A_i.$$

In the event of potential early surrender, then we face a "standard" Bermudan-type option, as in the lump sum case:

$$L^B(t) := \operatorname*{ess\,sup}_{\tau \in T_t} \mathbf{E}^{\mathbb{Q}}\left[\sum_{j\geq 0} 1_{t<_j\leq\tau} D_t^{t_j} A_j + D_t^{\tau} \breve{A}(\tau) \mid \mathcal{F}_t\right].$$

As for the lump sum case, we can use the fact that $(r, \breve{A}, \bar{\breve{A}})$ is a Markov process to associate the price process to a deterministic map satisfying

$$\ell^B(t, r(t), \breve{A}(t), \bar{\breve{A}}(t)) = L^B(t) \quad \text{for } t \leq \varpi.$$

If the policyholder is not authorized to surrender the contract between two purchase anniversaries, then the liability process evolves as a martingale on each interval [ti, ti+1]. In particular,

$$\forall i \leqslant n, \quad \ell^B\left(t_i, r(t_i), \breve{A}(t_i), \bar{\breve{A}}(t_i)\right) = \mathbf{E}^{Q}\left[D_i^{i+1} \ell^B\left(t_{i+1} - r(t_{i+1} -), \breve{A}(t_{i+1} -), \bar{\breve{A}}(t_{i+1} -)\right) \mid \mathcal{F}_{t_i}\right].$$

Like the lump sum case, at discrete time points ti the policyholder is supposed to compare the account value with the value of the liability to decide whether to surrender or not. If the account value is bigger than the liability, policyholders surrender the contract and get back the account value. Otherwise, they continue to hold the policy.

The liability evolves at the anniversary date ti as follows, in the event of a roll-up guarantee:

$$\ell^B(t_i -, r, \breve{a}) = \max\left(\breve{a}, \ell^B\left(t_i, r, (\breve{a} - A_i)^+\right) + A_i\right),$$

$$\begin{aligned} \ell^B\left(t_n, r(t_n), \breve{A}(t_n)\right) &= \mathbf{E}^{Q}\left[D_n^{n+1} \ell^B\left(t_{n+1} -, r(t_{n+1} -), \breve{A}(t_{n+1} -)\right) \mid \mathcal{F}_{t_n}\right] \\ &= \mathbf{E}^{Q}\left[D_n^{n+1} \max\left(\breve{A}(t_{n+1} -), \ell^B\left(t_{n+1}, r(t_{n+1}), \breve{A}(t_{n+1})\right) + A_{n+1}\right) \mid \mathcal{F}_{t_n}\right] \\ &= \mathbf{E}^{Q}\left[D_n^{n+1} \breve{A}(t_{n+1} -) \mid \mathcal{F}_{t_i}\right] < \breve{A}(t_n). \end{aligned}$$

It follows from the properties introduced above that, at the time ti (0 < i < n + 1), there exists a function of the interest rate such that

$$\begin{aligned} 0 \leqslant \breve{a} < \breve{a}^{*}(t_i -, r) \quad &\Rightarrow \ell^{B}(t_i -, r, \breve{a}) > \breve{a} \\ \breve{a} \geqslant \breve{a}^{*}(t_i -, r) \quad &\Rightarrow \ell^{B}(t_i -, r, \breve{a}) = \breve{a}, \end{aligned}$$

referred to as the "critical account value" at time ti, since the policy should be surrendered as soon as the account value increases to this level at time ti for a given interest rate level.

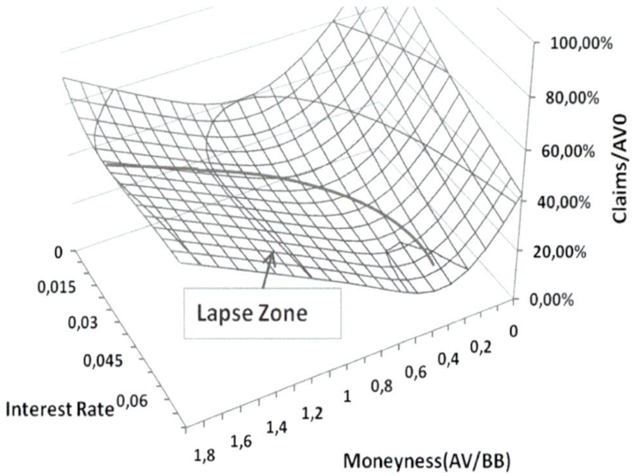

FIGURE 3.99 Lapse zone

Notes: AV = account value; BB = benefit base.

The dependence between surrender and financial market conditions is consistent with the intuition and experience since the 2008 financial crisis. Surrender behaviour again heavily depends on the account value and interest rate level.

Indeed, when the interest rate goes up, the discounted value of the insurer's annuities decreases. If the portfolio value rises at the same time, then the guarantees become worthless, and more customers should surrender their policies to get back the account value. In the opposite situation, when both interest rate and account value decrease, the contract becomes – to the contrary – much more attractive, thus leading to a reduced surrender rate.

Note that customers need not to be "efficient" per se, in the financial market sense, to adopt such behaviours. Namely, they are certainly not able to compute the optimal exercise times associated with such very complex products. However, competitors can take advantage of new market conditions to propose more attractive products, thus influencing their behaviour indirectly.

Figure 3.101 shows the liability price with efficient surrender (left) versus no lapse (right), whose difference is the early surrender premium.

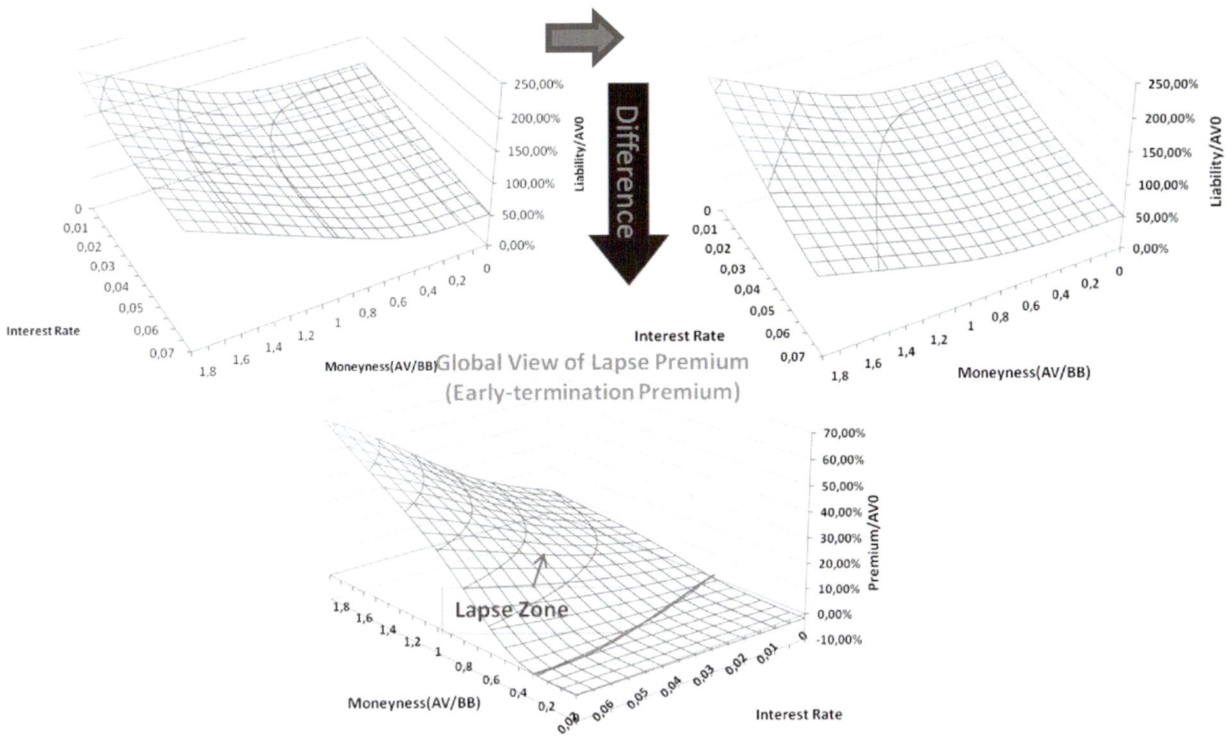

FIGURE 3.100 Global view of early surrender premium

The surrender risks sensitivities also differ significantly.

- Within the efficient forward-looking modelling framework, a jump in volatility within high equity markets (AV/BB above one) increases the probability of an equity rise, and so higher out-of-the-moneyness in terms of the embedded put, and thus a lower guarantee, implying a higher surrender rate – hence lower claims (see Figure 3.102).

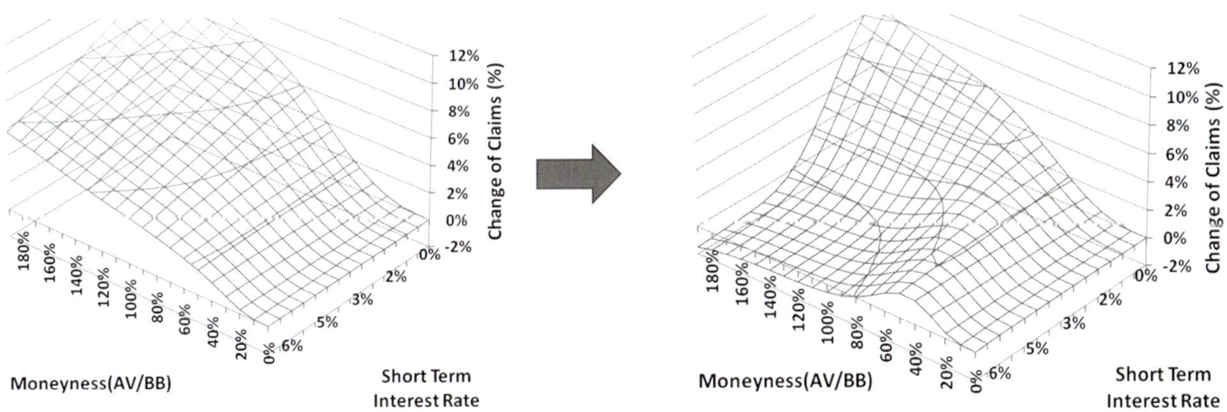

FIGURE 3.101 Sensitivity to a volatility shock

- In contrast, a drop in long-term interest rates within low equity markets (low AV/BB) tends to increase the richness of the embedded put, thus reducing surrender and increasing claims (see Figure 3.103).

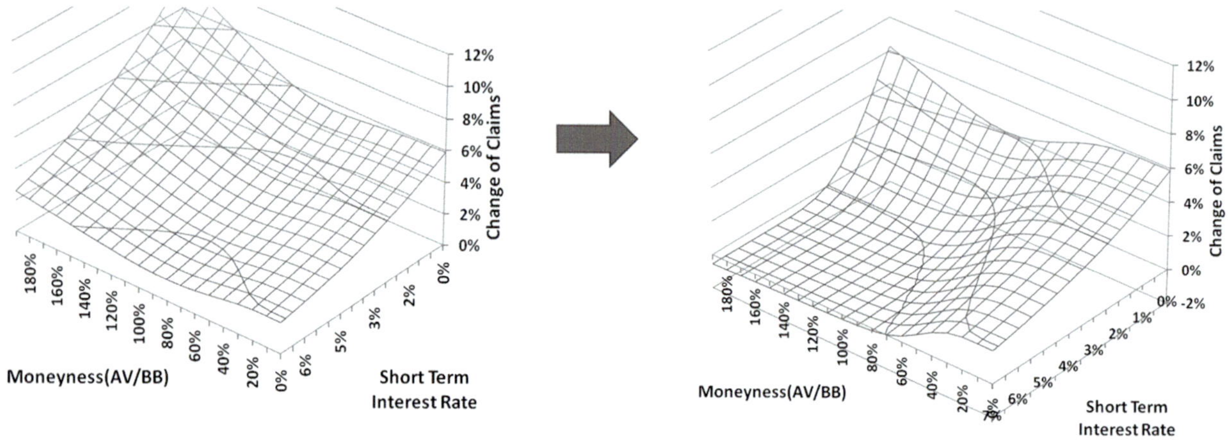

FIGURE 3.102 Sensitivity to an interest rate shock

The market risk sensitivities (delta, gamma and rho Greeks) of the embedded Bermudan put account for the difference between PV (claims) and PV (charges). A high AV/BB (and thus higher surrenders) decreases the absolute value of these three sensitivities, while a low AV/BB (and thus lower surrenders) increases them.

In terms of numerical implementation, the same methods are used as for the lump sum case, when the PDE (ADI) scheme and the Monte Carlo (HDR) method converge (see Figures 3.104 and 3.105).

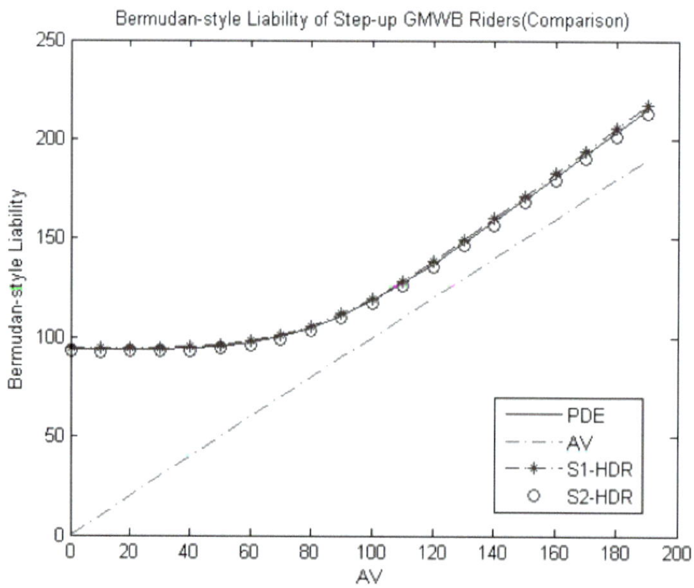

FIGURE 3.103 Liability value versus account value

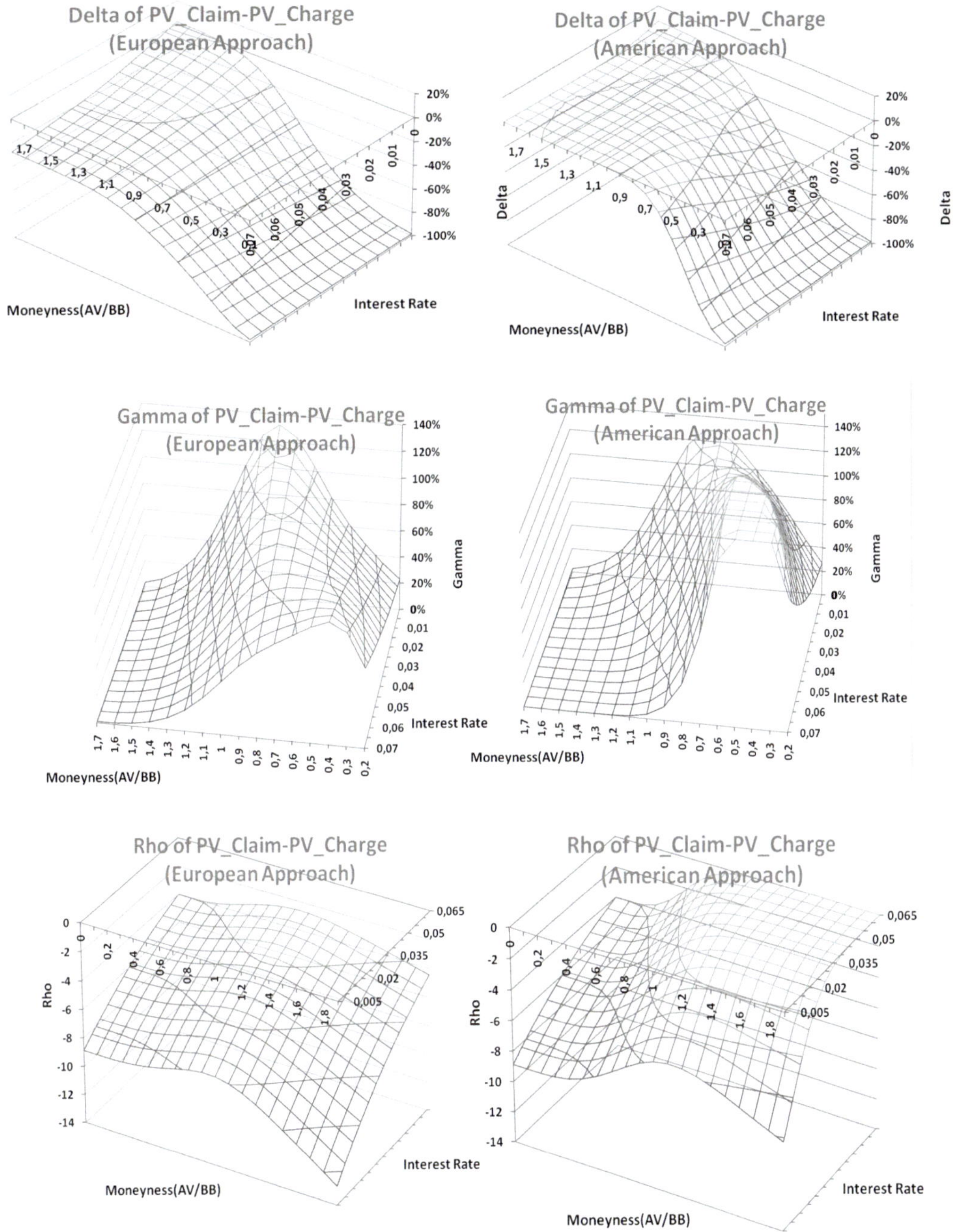

FIGURE 3.104 Greeks of PV (claims) versus PV (charges) (option values)

If we apply such an efficient customer behaviour approach to the French market on variable annuities, we indeed retrieve the empirical behaviour, notably by taking into account the French tax regime depending on the contract's past maturity, which plays a key role in accounting for the jumps in surrender rates at four- and eight-year maturities (see Figure 3.106).

Contract Maturity	Tax Rate (French)
< 4 years	35%
4 < maturity < 8 years	15%
> 8 years	8%

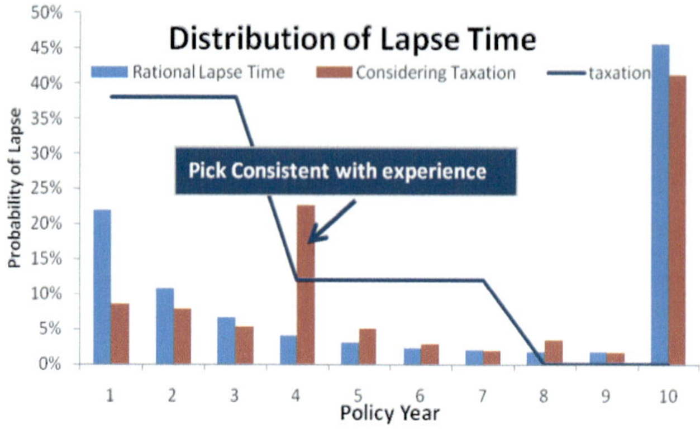

FIGURE 3.105 Distribution of lapse times versus policy year

3.3.3.2 Efficient Forward-Looking Customer-Centric Deferral Period Modelling

As illustrated in the generic, some savings and retirement contracts offer the possibility to policyholders to choose the timing of their withdrawals, and so to decide on the deferral period duration before withdrawals. In other words, policyholders hold the option of extending the deferral period.

Such an optionality can therefore be at first sight apprehended through the analyses of so-called "extendable maturity options".

3.3.3.2.1 Proxy Study through Extendable Maturity Puts Extendable maturity options are options with extendable (thus stochastic) maturities that can be exercised at their maturity date T1 but that also allow the option holder at time T1 to extend to time T2. In deriving these expressions, we also allow the strike price of the option to be adjusted from K1 to K2 at the time of the extension. Therefore, these extendable maturity puts can be considered as proxies of savings and retirement claims with potential roll-up features.

As illustrated below, the properties for these deferral period options are very different from those of conventional options with non-stochastic maturities.

3.3.3.2.1.1 Pricing Extendable Maturity Puts The current premium of an ordinary European put with strike price K and maturity T is denoted as P(X,K,T). Using this notation, the maturity condition satisfied by the extendable put at T1 is

$$\max\left(0, P\left(X, K_2, T_2 - T_1\right) - A, K_1 - X\right).$$

This payoff function can again be written as the maximum of two risky payoffs. Thus, at T1, the extendable put allows the option holder to choose between the payoff function for a call option on a put and the payoff function for an ordinary put option.

Figure 3.107 illustrates one- to two-year-maturity extendable put and one-year-maturity plain vanilla put premiums expressed in terms of notional.

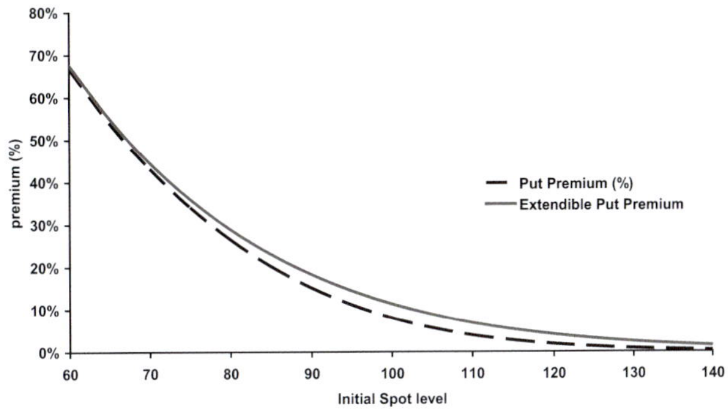

FIGURE 3.106 Extendable puts versus vanilla put premium ($K_1 = K_2 = 100, A = 0$, σ = 20%, r = 0%)

Regarding the optimal extension thresholds, the payoff function of the extendable put implies that the put is extended at T1 if and only if the underlying asset price X belongs to a range [I1, I2].

- When X < I1, the put is exercised at T1.
- When X > I2, the put is allowed to expire at time T1.

The bounds I1 and I2 can be found by solving the following two equations:

$$P(I_1, K_2, T_2 - T_1) = K_1 - I_1 + A,$$
$$P(I_2, K_2, T_2 - T_1) = A.$$

Warning: within such a modelling framework, depending on the option input parameters, the Raphson–Newton algorithms used to solve for extension parameters are sometimes unstable and give rise to spurious oscillations on graphs.

Regarding the behaviour of the bounds I1 and I2:

- I1 < K1, and I1 is an increasing function of A, K1 and r and a decreasing function of K2, T2 and σ.
- I2 > K2, and I1 is a decreasing function of A and r and an increasing function of K2, T2 and σ.

Solving the payoff coupled to the underlying dynamics assumption, subject to the above maturity condition, provides the premium of the extensible put, denoted as EP(X, K1, T1, K2, T2, A).

$$\begin{aligned} EP(X, K_1, T_1, K_2, T_2, A) &= P(X, K_1, T_1) \\ &- XN(\gamma_1, \gamma_2, -\infty, \gamma_3, \rho) \\ &+ K_2 e^{-r(T_2 - T_1)} N\left(\gamma_1 - \sqrt{\sigma^2 T_1}, \gamma_2 - \sqrt{\sigma^2 T_1}, -\infty, \gamma_3 - \sqrt{\sigma^2 T_2}, \rho\right) \\ &+ XN(\gamma_4, \gamma_2) \\ &- K_1 e^{-rT_1} N\left(\gamma_4 - \sqrt{\sigma^2 T_1}, \gamma_2 - \sqrt{\sigma^2 T_1}\right) \\ &- Ae^{-rT_1} N\left(\gamma_1 - \sqrt{\sigma^2 T_1}, \gamma_2 - \sqrt{\sigma^2 T_1}\right), \end{aligned}$$

where $N(a, b, c, d, \rho)$ is the cumulative probability of the standard bivariate with correlation coefficient ρ.

The first term is the value of an ordinary put option without the extension feature, while the sum of the remaining terms represents the value of the extension optionality.

A number of rational bounds can be placed on the value of the extendable put. For example, since the value of the extension optionality is non-negative, the extendable put is worth at least as much as the corresponding non-extendable put. In fact, the lower bound of the value of the extendable put is the maximum of the non-extendable put value $P(X,K_1,T_1)$ and the value of a call with strike price A and maturity T1 on the put value $P(X,K_2,T_2-T_1)$.

The upper bound for the extendable put is $\max\left(K_1e^{-rT_1},K_2e^{-rT_2}-Ae^{-rT_1}\right)$.

3.3.3.2.1.2 Risk Analysis of Extendable Maturity Puts The following risk metrics are used:

- the value of the extension privilege;
- the sensitivities to product design and market parameters; and
- the sensitivities to market shocks occurring at the extension period time T1.

3.3.3.2.1.2.1 The Value of the Extension Privilege This is defined as the difference between an extendable put and a plain vanilla put. This is easily shown to be an increasing function of K2, T2, and a decreasing function of K1.

We illustrate in Figures 3.108 and 109 the non-insignificant additional cost stemming from the deferral period option in the case of an extendable put, where the additional cost is 41% of a conventional put premium for an ATM extendable maturity put.

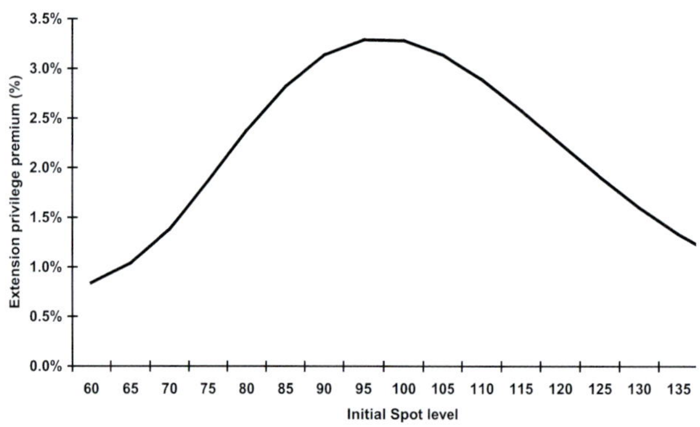

FIGURE 3.107 Extension premium ($K_1 = K_2 = 100, A = 0$, σ = 20%, r = 0%)

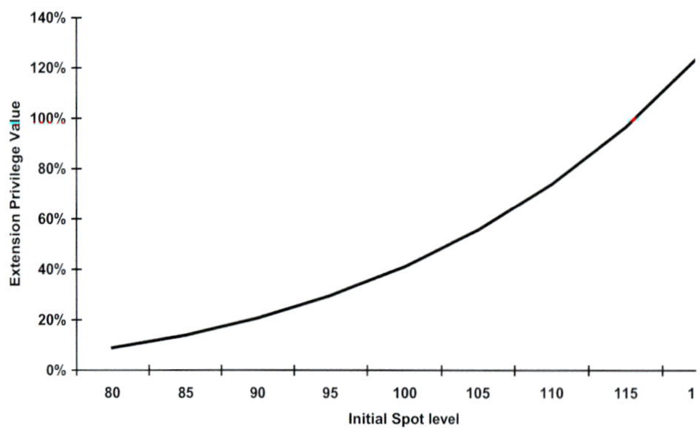

FIGURE 3.108 Extension premium as percentage of put premium ($K_1 = K_2 = 100, A = 0$, σ = 20%, r = 0%)

Furthermore, such an optionality involves complex behaviour with respect to risks; however, an increase in the underlying asset, X, has an indeterminate effect on the value of the extension optionality, as shown below for a one- to two-year-maturity extendable put. The intuition for this is that the put is more likely to be extended when the initial value of the spot X is near the optimal range [I1, I2], which increases the value and then decreases for larger and lower values of X.

Thus, a “nuanced” view is required: an option holder may sometimes prefer to have a second chance to exercise an option instead of a slightly in-the-money expiration. Consequently, an increase in the probability of an in-the-money expiration is not always favourable for option holders.

The comparative sensitivities with respect to T1, r and σ are also indeterminate, as illustrated by the crossing curves depending on the moneyness:

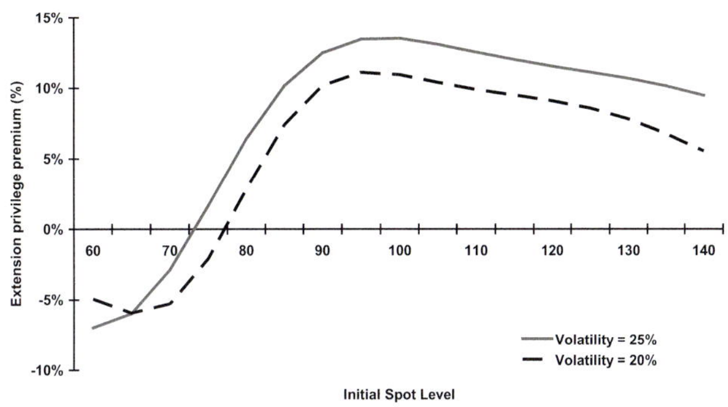

FIGURE 3.109 Impact of volatility on extension premium ($K_2 = K_2 = 100, A = 0$, r = 3%, T_1 = 4 years, T_2 = 5 years)

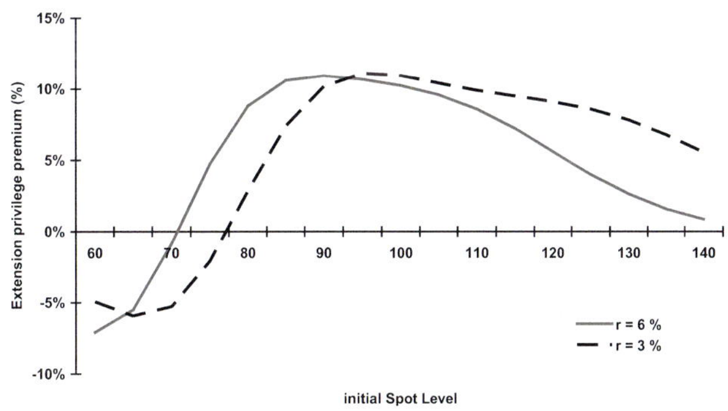

FIGURE 3.110 Impact of rate on extension premium ($K_1 = K_2 = 100, A = 0$, σ = 20%, T_1 = 4 years, T_2 = 5 years)

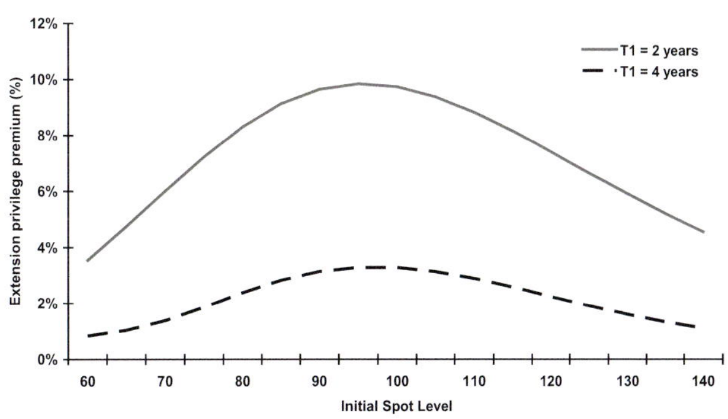

FIGURE 3.111 Impact of T1 on extension premium ($K_1 = K_2 = 100, A = 0$, σ = 20%, r = 3 %, T_2 = 5 years)

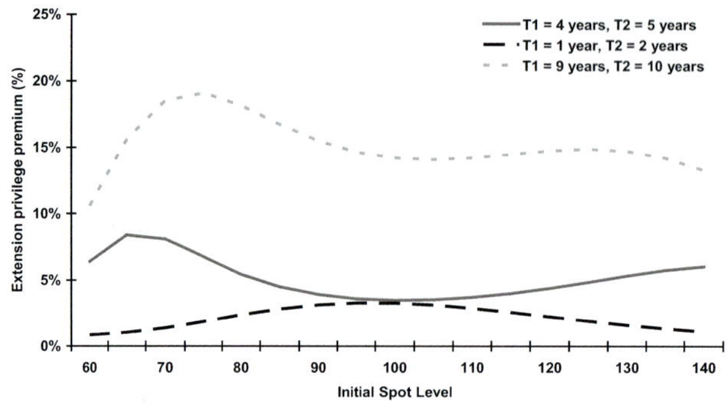

FIGURE 3.112 Impact of maturity on extension premium ($K_1 = K_2 = 100, A = 0$, r = 0%, σ = 20%)

This might be counterintuitive at first sight, because increases in r and σ and decreases in T1 not only increase the length of the optimal extension interval [I1, I2] but also increase the payoffs from an extension at every point in the interval.

Once again, careful attention through a "nuanced" view is required in order to properly assess the risks; the reason for these indeterminate sensitivities is that changes in T1, r and σ also affect the risk-neutral conditional density of S at time T1 and, therefore, the probability that an extension will occur. Thus, even though a change in T1, r and σ can increase the payoff for an extension, the probability that an extension occurs could decline sufficiently to more than offset the increased payoff and lower the overall value of the extension optionality.

Warning: optimal option exercise decisions are not monotonic in parameters but, rather, vary as a bell shape. This emphasizes the need for careful attention to the complexities of American options, in contrast to European options.

3.3.3.2.1.2.2 Sensitivities to Product Design and Market Parameters The sensitivities for the extendable put are generally similar to those for non-extendable puts. For example, the value of the extendable put is an increasing function of K1, K2, T1, T2, σ and a decreasing function of X, A, r. As illustrated in Figures 3.114 and 3.115, an extendable put is more sensitive to interest rates level than a plain vanilla put.

However, the Greeks for extendable puts can differ substantially from those for plain vanilla puts. For example, the extendable put is more sensitive to changes in the underlying asset X (delta) than a conventional put for out-of-the-money options but is less sensitive for in-the-money options.

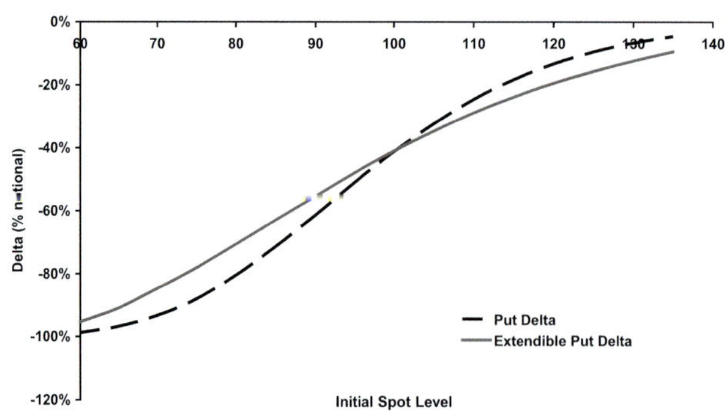

FIGURE 3.113 Extendable put delta versus conventional put delta ($K_1 = K_2 = 100, A = 0$, r = 0%, T_1 = 1 year, T_2 = 2 years, σ = 20%)

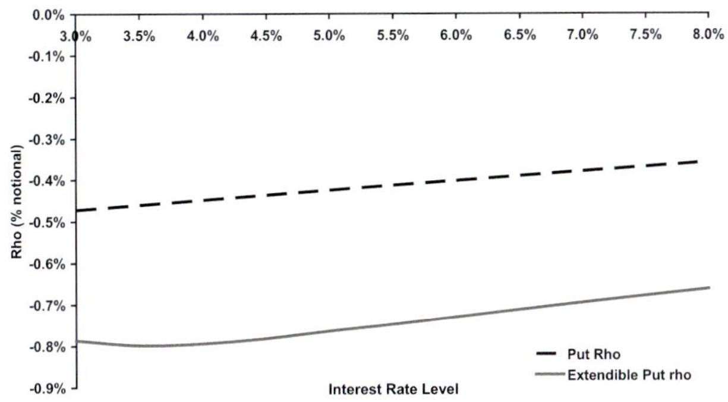

FIGURE 3.114 Extendable put rho versus conventional put rho ($K_1 = K_2 = 100, A = 0$, r = 0%, T_1 = 1 year, T_2 = 2 years, σ = 20%)

3.3.3.2.1.2.3 Sensitivities to Market Shocks Occurring at the Extension Period Time T2

If we consider an interest rate shock at the extension time T1, the sensitivity is as expected, since the extendable put is decreasing in the interest rates.

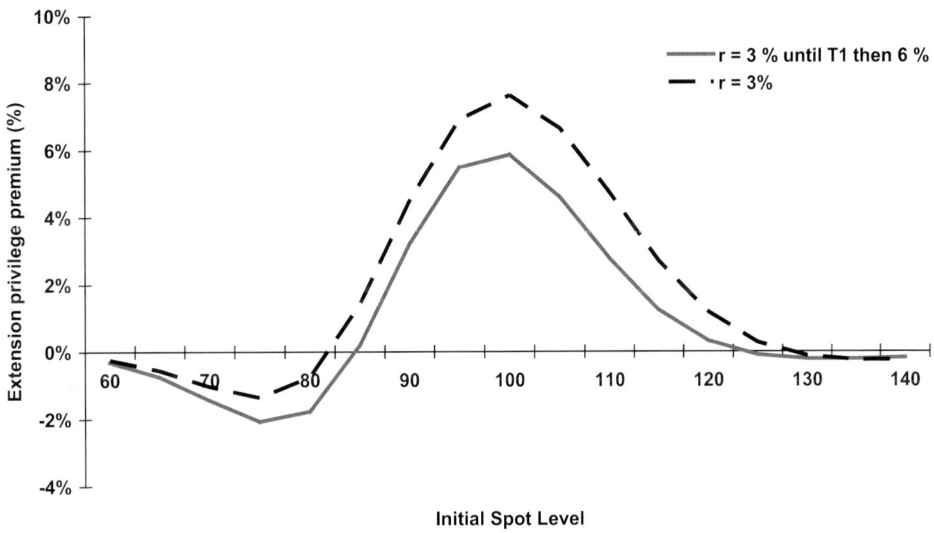

FIGURE 3.115 Impact of rates shock at the extension time ($K_1 = K_2 = 100, A = 0$, T_1 = 1 year, T_2 = 2 years, σ = 20%)

Among the key conclusions are these.

- Deferral options within accumulator products involve complex modelling issues that require careful attention. The additional cost is not insignificant.
- Optimal deferral period extension decisions are not monotonic in parameters (asset, volatility, interest rates) but, rather, vary like a bell shape, depending on the moneyness.
- In particular, deferral period options involve strong nonlinear issues with respect to key market parameters such as residual maturity, interest rates and volatility, for which the market sensitivities of the deferral option are indeterminate.

3.3.3.2.2 Pricing and Risk Analyses within a Guaranteed Withdrawal Annuity The decision to extend the deferral period lies in the balance between losing one year of withdrawals versus the increase in the guaranteed value (through a potential roll-up, ratchet feature or higher guaranteed withdrawal amount).

Usually, insurers offer a bonus in the guaranteed value (e.g. in the United States with a 200% guaranteed increase over a ten-year period, or a higher guaranteed withdrawal) as long as the deferral period is extended.

Using similar modelling to that in the surrender behaviour section illustrates that, under risk-neutral scenarios, the most efficient (i.e. maximizing the expected average discounted accumulated wealth over the life of the product) deferral period is short (between one and five years in practice), which emphasizes the interest in starting the withdrawals quite soon after the inception, even in the event of significant roll-up rates and ratchets. Naturally, under real-world scenarios such as a market bubble, it may make sense to further delay the withdrawals.

Such efficient behaviour also stems from the usual high fees of the guarantee.

The observed rising efficiency in customer behaviour has pushed most variable annuity insurers to offer a 200% minimum guarantee level in exchange for no withdrawals for ten years (thus a 8% annual increase in the guarantee level), in order to incentivize customers to delay their withdrawals.

3.3.3.3 Efficient Forward-Looking Customer-Centric Predictive Withdrawal Modelling

We consider here a guaranteed withdrawal annuity.

Such a guaranteed periodic withdrawal annuity gives policyholders the ability to protect their retirement investments against downside market risks by allowing the right to withdraw a fixed percentage (e.g. 4%) of the benefit base (and account value) each year until death.

The benefit base can either be stepped up and reset to the high-water mark of the account value on the rider anniversary date (step-up policies) or can roll up with a fixed percentage annually (roll-up policies), regardless of the market conditions.

Another important feature is that the policyholder can defer the withdrawal start. For simplicity, we assume that this deferral period is zero, which means that the customers start withdrawals at the end of the first policy year.

The withdrawal is free to be chosen by the policyholder. The withdrawals that are made at anniversary dates are deducted on a dollar-for-dollar basis from both the account value and the guaranteed base, up to a certain threshold G; then they are penalized as being deducted on a "pro-rata basis" with respect to the guaranteed base (i.e. the account value decreases more in percentage terms than the guaranteed base, since it is lower the guaranteed base):

$$f(\gamma) = \begin{cases} \gamma, & \text{if } \gamma \le G \\ \gamma - \kappa(\gamma - G), & \text{if } \gamma > G. \end{cases}$$

The optimal withdrawal size at the anniversary date can be solved through the following Bellman equation,

$$\begin{aligned} &V_t + (r-\alpha)SV_S + \frac{1}{2}\sigma^2 S^2 V_{SS} - rV = 0 \\ &V\left(t^-, S_t, B_t\right) = \max_\gamma V\left(t, S_t - \gamma, B_t - \gamma\right) + f(\gamma), \end{aligned}$$

out of which the customer-centric optimal withdrawal amount is dependent on the account value and the guaranteed base (benefit base: BB) (see Figure 3.117).

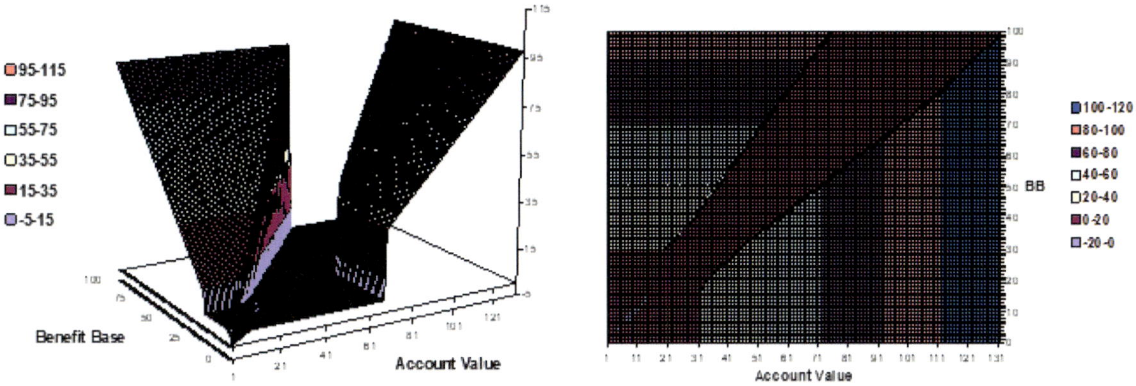

FIGURE 3.116 Optimal withdrawal strategy

3.4 COMPUTATIONAL CHALLENGES AND OPPORTUNITIES: ACCELERATING THE COMPUTATIONAL RUN-TIME BY SPEEDING UP THE CONVERGENCE OF MONTE CARLO SIMULATIONS

3.4.1 Randomized Quasi-Monte Carlo Scenario Generators

The focus here is on the potential risks associated with the number of simulations and the choice of a specific Monte Carlo (MC) simulation generator: traditional (pseudo-) Monte Carlo, quasi-Monte Carlo (QMC) or hybrid quasi-Monte Carlo. This issue requires specific convergence analyses and a nuanced view in order to assess the pros and cons of different Monte Carlo simulation generators. In fact, traditional (pseudo-) Monte Carlo simulations require a significant number of simulations not only to mitigate the seed initiation impact but also to enforce convergence and martingale tests.

We propose an approach to measure the rates of convergence of the different Monte Carlo simulation generators, depending on the complexities within products, from plain vanilla puts to exotic puts and accumulator products. In order to mitigate associated convergence issues, we introduce specific techniques: Brownian bridge technique, randomized Monte Carlo.

3.4.1.1 Context and Objectives

Here we focus on the impact of the proper choice of Monte Carlo simulation generator based on rates of convergence, which depend on:

- the number of simulations;
- the martingale tests;
- the maturity of the product to price;
- the number of underlyings; and
- the complexity of the payoff features: from plain vanilla to exotic and accumulator products.

We propose an approach to measure the pros and cons of increasingly refined Monte Carlo simulations generators. Specifically, we show that the appropriate choice strongly depends on a series of factors, but most importantly on the complexities within the option payoffs. Given the strong path dependence of accumulator products, we introduce some techniques such as Brownian bridge and randomized quasi-Monte Carlo simulations in order to improve the convergence rate of prices.

Section 2 introduces the modelling framework of Monte Carlo simulations generators, where quasi-Monte Carlo simulations are introduced. Section 3 presents a number of issues related to rates of convergence. Section 4 discusses some improvements to mitigate pitfalls of quasi-Monte Carlo simulations due to exotic features of accumulator products and introduces the randomized quasi-Monte Carlo simulation generator.

3.4.1.2 The Modelling Framework

3.4.1.2.1 Monte Carlo Simulations Amount to Integral Computation Let us start with a European call option given by the integral below:

$$C(V_t,t) = e^{-r(T-t)} E_t\left[\textit{payoff}(V_T) \right] \Rightarrow$$
$$C(V_t,t) = e^{-r(T-t)} \int_0^{\infty} \max(V_T - K, 0) \cdot p(V_T, T \mid V_t, t)\, dV_T$$
$$\Rightarrow C(V_t,t) = e^{-r(T-t)} \int_0^{\infty} f(V_T)\, dV_T.$$

- Et[.] is the payoff expectation under a risk-neutral (or martingale) measure conditional to the information at the (current) time t.
- p(VT, T | Vt, t) is the risk-neutral transition density function of V, which is log-normal for the geometric Brownian motion.
- Vt is the value of the underlying asset at the (current) instant t.
- K is the exercise price.
- T is the expiration time.
- r is the risk-free interest rate.

Because the price is an expected value, we have to evaluate an integral, and this evaluation is better if the points are uniformly scattered in the entire area of the integral above.

3.4.1.2.2 Quasi-Monte Carlo Simulations Do Not Depend on the Seed, in Contrast to Traditional Monte Carlo The traditional Monte Carlo simulation uses a pseudo-random sequence of numbers, which are in fact generated with deterministic (explaining the adjective "pseudo") algorithm. In addition, the seed (initial value of a sequence) depends on an (unpredictable) external feeder, such as the computer clock. For example, in Excel the function Rand() returns a different number from U[0, 1] every time you press the "recalculation" command (F9).

In conclusion, when using traditional Monte Carlo methodology with random numbers, the results differ depending on the seed. If you start with a different seed, you end up with a different answer. How different? It depends. But you will end up with different answers. As an illustration, the generator seed has a significant impact on 2,000 simulations sets of hybrid returns (Hull–White, Black–Scholes), in contrast to 10,000 simulations sets.

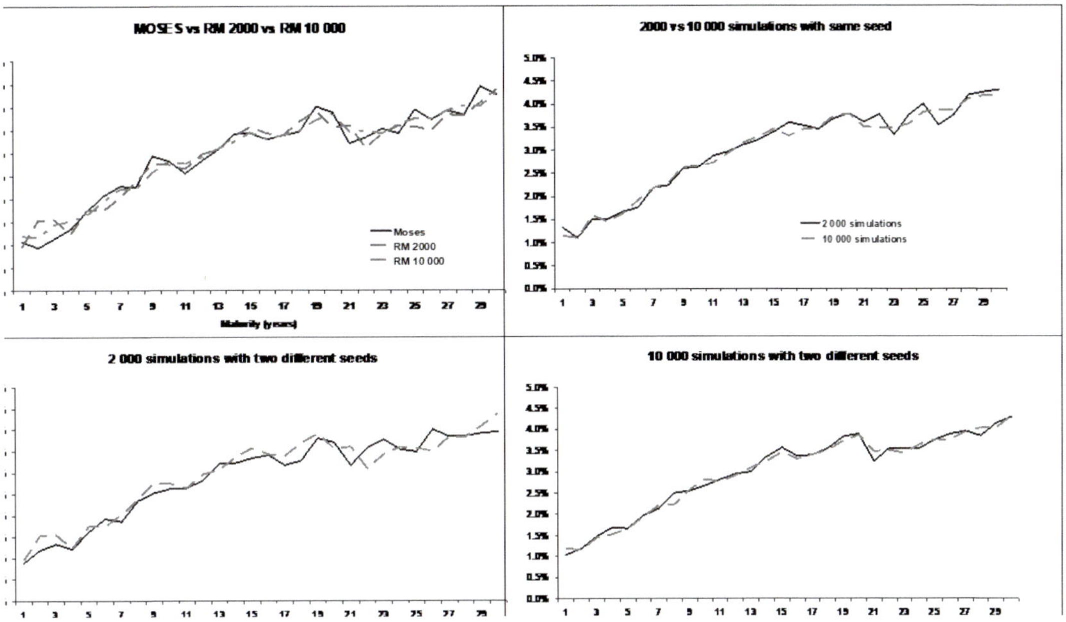

FIGURE 3.117 Impact of the generator seed

Note: RM = risk management.

3.4.1.2.3 Quasi-Monte Carlo Simulations Improve Pricing Accuracy This does not happen with quasi-Monte Carlo simulations that use quasi-random sequences, more evenly scattered throughout the region, offering shorter computational times and/or higher accuracy. Since the low discrepancy sequences are totally deterministic, there is no seed impact.

In most cases the simulation error with traditional Monte Carlo (pseudo-random numbers) is higher or much higher than the error with quasi-Monte Carlo simulation. In many simulations this error is higher by a factor of ten or more!

Figure 3.119 illustrates the nice statistical properties of a quasi-random sequence (using a van der Corput sequence in base 2; see the definition later) compared with the (theoretical) uniform [0, 1] distribution, and with two typical pseudo-random sequences (generated with Excel). Hence, a quasi-random sequence presents a better performance than a typical pseudo-random sequences.

Statistical Properties of Quasi-Random x Uniform x Pseudo Randon

N = 1,000 simulations

	Quasi-Random (b2)	Uniform [0, 1]	Pseudo-Random (2 series)	
Minimum	9.76563e-4	0	1.81917e-5	3.27922e-4
Maximum	0.998047	1	0.999004	0.999982
Mean	0.498866	0.5	0.492246	0.512323
Median	0.498047	0.5	0.484037	0.521946
Standard Deviation	0.288661	0.288675	0.2851	0.291302
Variance	0.083325	0.083333	0.081282	0.084857
Skewness	2.042816e-3	0	0.039117	-0.077605
Kurtosis	1.794803	1.8	1.833614	1.773279

FIGURE 3.118 Statistical properties of quasi-random versus uniform versus pseudo-random

Quasi-random methods, especially Sobol sequences, can achieve the given accuracy with far fewer simulations (sometimes ten to 1,000 times fewer) than pseudo-random sampling requires. Moreover, the larger number of samples used by a simple Monte Carlo to achieve this accuracy will result in a significant increase in the computational time, as illustrated in Figure 3.120 in the case of plain vanilla options using the Black–Scholes framework.

	Monte Carlo	Quasi MC	Exact
Nb Simulations	100000000	10000	
Call	13.26954569	13.26932026	**13.2696855**
Put	3.753499646	3.753352779	**3.75342733**

FIGURE 3.119 Monte Carlo versus quasi-Monte Carlo versus exact

3.4.1.2.4 Quasi-Monte Carlo Simulations' Strength Stems from Low Numbers of Discrepancies Imagine we have 1,000 points and a box that is a cube with a volume equal to one. How do we distribute the 1,000 points in the cube space, so that any (sub-) volume in the box that we sort has a number of points in it that is proportional to the respective volume?

The discrepancy is related to the worst deviation of the percentage of points in any sub-volume (or sub-rectangle, if you imagine a thin slice in the space) in relation to the exact percentage of points for a perfect evenly distributed point. The discrepancy measure is the extent to which the points are evenly dispersed. It is associated with the property that the successive numbers are added in a position as far away as possible from the others numbers, avoiding clustering (groups of numbers close to other). So the job for the numbers generated sequentially is to fill in the larger "gaps" between the previous numbers of the sequence, as illustrated in Figure 3.121.

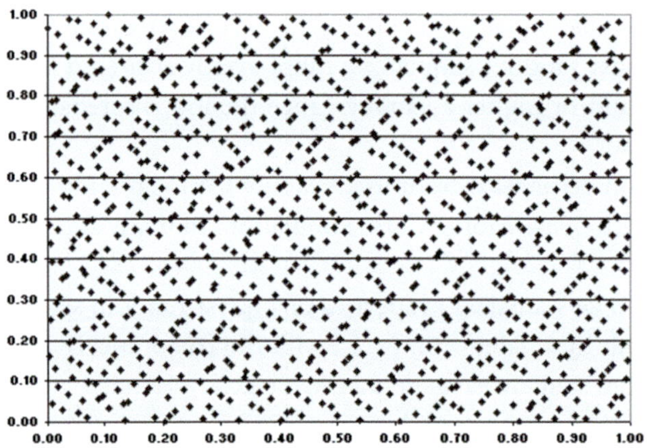

FIGURE 3.120 Dispersion of a van der Corput sequence

The picture illustrates the "filling in the gaps" dynamic from the van der Corput (1935) sequence in base 2. Notice in Figure 3.121 that there is no clustering of numbers in this interval [0, 1].

However, the van der Corput sequence exhibits a cyclicality that can be a problem in many applications: it moves upwards and falls back in cycles of two.

3.4.1.2.5 Dealing with Fat Tails in Simulations The European call option (C) over a stock that values V today (t = 0) with exercise price K is given by

$$C(t=0) = \exp(-rT)\int_0^1 \max\left\{0, V\exp\left[\left(r-\delta-1/2\sigma^2\right)T+\sigma\sqrt{T}F^{-1}(u)\right]-K\right\}du$$

where F(x) is the standard normal cumulative distribution function and u is drawn from the uniform [0, 1] distribution. The uniform distribution in the interval [0, 1] is, for practical purposes, the only distribution that we need generate for our simulations. The reason is that the samples from the other distributions are derived using the uniform distribution.

The uniform distribution can be generated with either pseudo-random numbers or quasi-random numbers, and algorithms are available to transform a uniform distribution into any other distribution. The main and direct way to do this transformation is by the cumulative distribution function inversion.

The known general way to obtain the inverse of a normal distribution is by using the Box–Muller algorithm. However, it damages the low discrepancy sequence properties (alters the order of the sequence or scrambles the sequence uniformity). In addition, it is slower than Moro's inversion, described below.

In order to deal with the tail issues, Moro's hybrid algorithm uses two algorithms, one for the central part of the normal distribution, the other for the tails of the distribution. In the following table we present the result of a limited test (only one quasi-random sample with 1,000 points) about the accuracy of Moro's inversion (using Moro's original constants) and the inversion using the Excel internal function named NORMSINV(). For this test, using a van der Corput quasi-random sequence in base 2, both inversions (Moro and Excel) presented practically the same (acceptable) accuracy. In this case, the Excel 97 inversion does not look too bad. However, weak point in the Excel inversion is the accuracy of the tails.

Comparison of normal standard inversion functions

u	Excel NORMSINV	Exact	Moro's inversion
0.001	−3.09024	−3.09023	−3.09023
0.0001	−3.71947	−3.71902	−3.71902
0.00001	−4.26546	−4.26489	−4.26489
0.000001	−4.76837	−4.75342	−4.75342
0.0000003	−7.15256	−4.99122	−4.99122
0.0000002	−5,000,000	−5.06896	−5.06896

In contrast, note that Moro's inversion gets the same result of the column "exact" to five decimal point precision. This nice performance of Moro's inversion for the tails can make a significant difference for the valuation of real options. For typical option problems, mainly for out-of-money options, the performance of the tails is important, because the option exercise is taken to occur at the upside (more extreme value), highlighting the weight of the tail in the (real) option value.

3.4.1.3 Pricing Convergence Issues

3.4.1.3.1 Rate of Convergence: a Martingale Test Approach The practitioner of simulations is always concerned with

- the computational time; and
- the numerical error of the simulation.

Both are related to the simulation rate of convergence to the true value: the standard error in the simulation can be viewed as the standard deviation of the (random) sample divided by an increasing function of N, the number of simulations or paths.

- One way to reduce the error is by using techniques of reduction of standard deviation of the sample ("variance reduction techniques"), such as antithetic variables.

- The other way to reduce the error is by increasing the number of simulations N, but the computational time increases directly with the number of simulations.

The traditional Monte Carlo method using (pseudo-) random numbers has a convergence rate of only O(N). Note that this rate is independent of the dimension d; it depends only of the number of simulations N. The convergence of MC is controlled by the total variance of the problem and the number of simulations, N. The dimension of the problem, D, does not appear.

As illustrated in Figure 3.122 in the case of hybrid equity returns used in accumulator products, the results strongly depend on time horizon and product features. The 2,000 simulations confidence interval (CI) widens significantly beyond 30 years in comparison with 40 years for 10,000 simulations, while the 2,000 simulation mean is often outside the CI beyond 30 years (depending on the seed). In contrast, sets of 10,000 simulations depend to a far lesser extent on the seed initiation. Thus, increasing the number of simulations when possible is definitely a way to mitigate the seed impact. Empirical tests suggest that 7,000 to 8,000 is a reasonable threshold, above which such concerns about the seed initiation start to significantly fade.

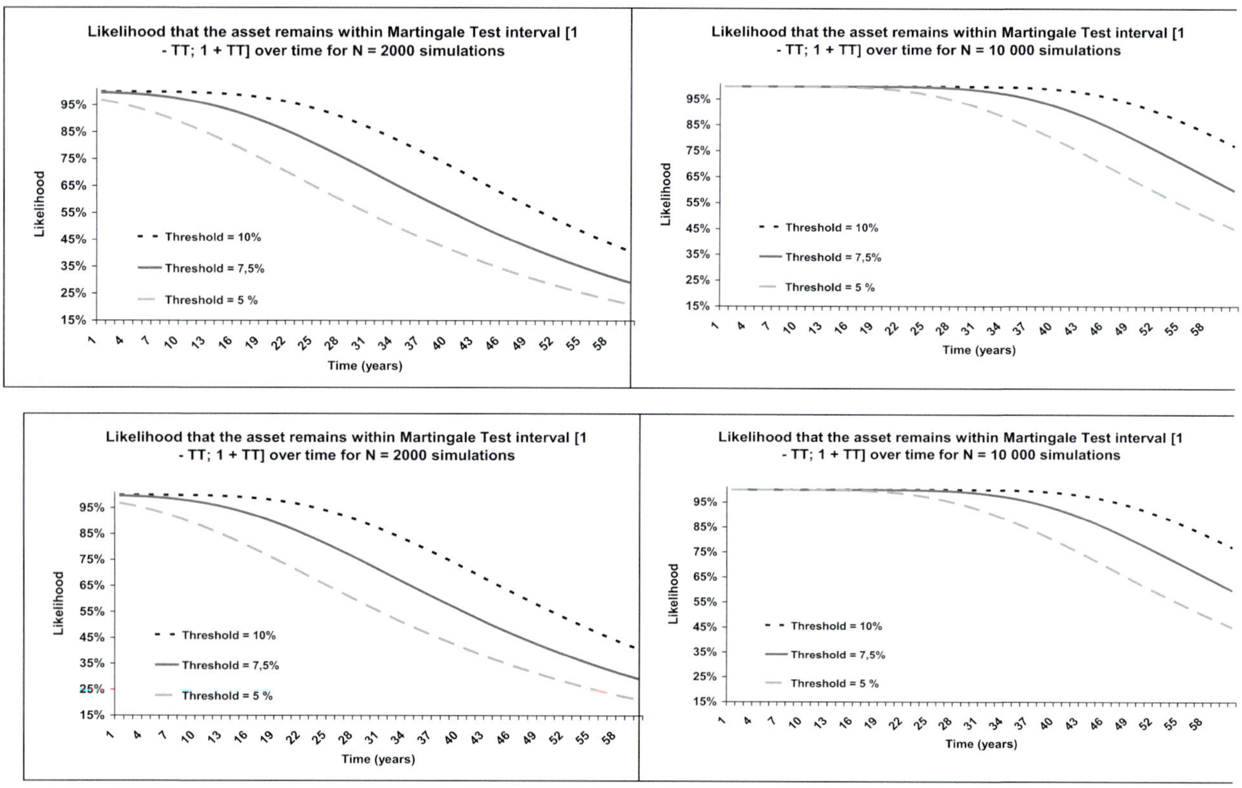

FIGURE 3.121 Martingale test intervals

Regarding Martingale tests on indices, a 2,000-simulation set passes the test over 30 years, but requires a large number of seed resets over 60 years, and so increases the run-time. Results are strongly dependent on equity volatility and correlation. In contrast, 10,000-simulation sets pass the test over 60 years at once. Obviously, these results strongly depend on the volatilities of the indices.

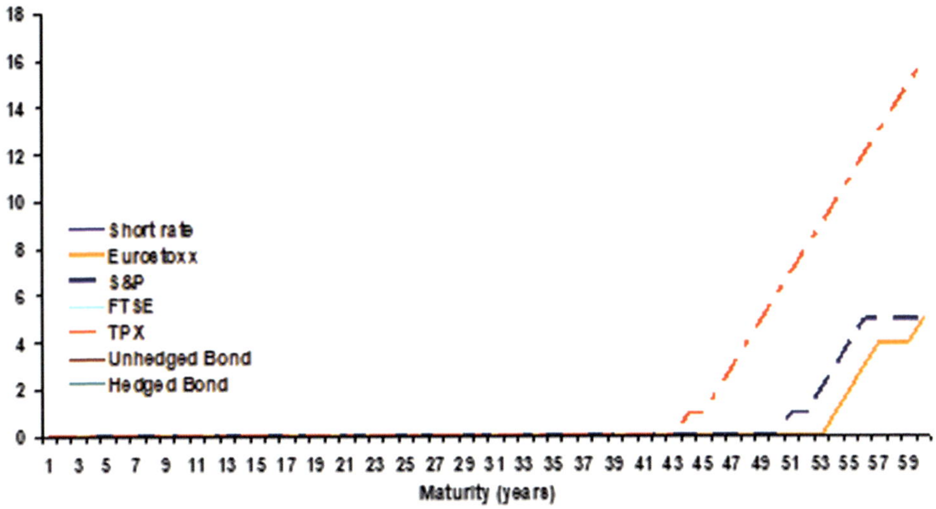

FIGURE 3.122 Cumulative number of martingale test failures over time (2,000 simulations)

3.4.1.3.2 Quasi-Monte Carlo Simulations: the Panacea regarding Convergence Issues? In contrast, the quasi-Monte Carlo rate of convergence can be much faster. One example: for MC, it is necessary to increase 100 times the number of simulations N to reduce the error by a factor of ten, whereas QMC requires in general far fewer, and often only ten times. Therefore, QMC simulations may appear to be *the* solution in order to mitigate the convergence issues as described above.

However, two kinds of limitations have to be emphasized.

- Contrary to traditional Monte Carlo simulation generators, this convergence rate can be guaranteed only for smooth integrals, and so the quasi-Monte Carlo advantage may be distorted significantly in the event of path-dependent payoffs.
- Contrary to traditional Monte Carlo simulation generators, the theoretical maximum error for the multi-dimensional low-discrepancy sequences is of O((ln N)d/N), where d is the number of dimensions. So, unlike the pseudo-random bound convergence, the quasi-random sequence performance decreases with the dimension d (the "curse of dimensionality").

		Monte Carlo	Quasi-Monte Carlo	
			Best	Worse
d	N	$1/\sqrt{N}$	1/N	$\ln(N)^d/N$
1	1,000	0.03162	0.00100	0.00691
1	100,000	0.00316	0.00001	0.00012
2	10,000	0.01000	0.00010	0.00848
5	10,000	0.01000	0.00010	6.628
10	10,000	0.01000	0.00010	439295.5
50	100,000	0.00316	0.00001	1.14626E+48

OBS: without reduction of variance techniques for both MC and QMC

FIGURE 3.123 Monte Carlo versus quasi-Monte Carlo

In practical problems, we need the increments along the path of the asset to be independent. So, the required main property along the time is the independence. The major problem for the pure quasi-random sequences is their degradation regarding independence when the dimension is large, whereby the generation process of uniformly distributed points in [0, 1]d becomes increasingly hard as d increases, because the space to fill becomes too large. The main challenge for the low-discrepancy sequences is to avoid the multidimensional clustering caused by the correlations between the dimensions.

Let us illustrate this curse of dimensionality with specific quasi-Monte Carlo sequences. The van der Corput sequence is the basic one-dimensional low-discrepancy sequence, and is also the first dimension of the Halton sequence. Faure and Sobol sequences are like the Halton sequence, but with only one base and with a reordering of elements. The main difference is related to the multidimensional construction of the sequences. The Faure and Sobol sequences change dimension with a permutation of the quasi-random vectors within each dimension.

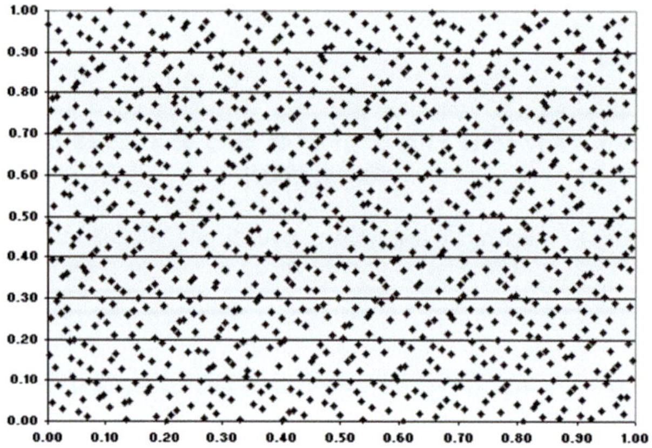

FIGURE 3.124 Halton sequence with 1,000 × 1,000

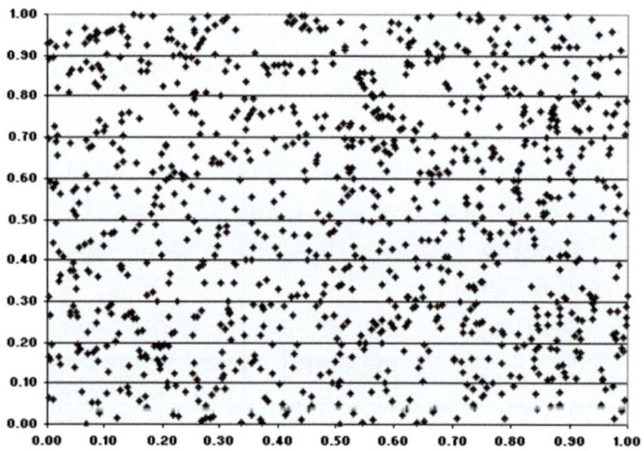

FIGURE 3.125 Two-dimensional (pseudo-) random numbers

The speed at which increasingly fine grid points are generated decreases with increasing dimension. The Halton sequence for 20 × 21 dimensions is shown in Figure 3.128, and we can see a visible degradation. In practice, due to the correlation, many people prefer to avoid the use of Halton sequence for more than about six or eight dimensions.

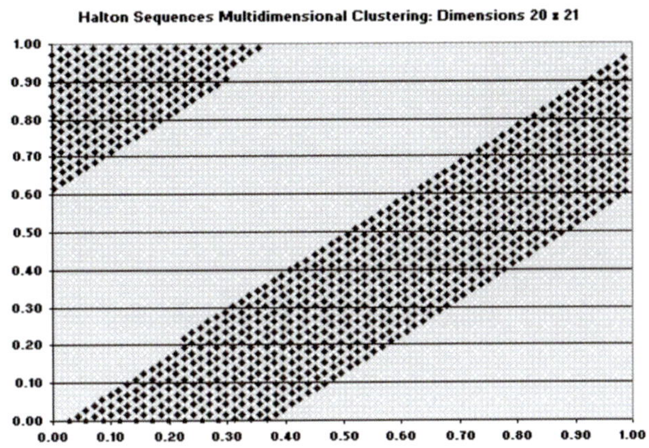

FIGURE 3.126 Halton sequence multidimensional clustering: dimensions 20 × 21

Thus quasi-random sequences in high-dimensional problems are highly correlated. The Halton sequence becomes unsatisfactory after dimension 14. The Faure sequence exhibits high-dimensional degradation at approximately the 25th dimension.

The Sobol sequence can be generated significantly more quickly than the Faure sequence, due to the shorter cycle length. Although the Sobol sequence appears to be more resistant to high-dimensional degradation, it also shows some correlation between neighbouring dimensions, but this correlation is lower than for other low-discrepancy sequences, and it starts at a higher dimension.

3.4.1.3.3 Why This "Curse of Dimensionality" Is Critical to Savings and Retirement As a matter of fact, several savings and retirement products precisely involve multidimensional and path dependence issues, whereby the number of dimensions d is the product of

- the number of funds used by the investment strategy, due to embedded options on basket indices within accumulator products; and
- the number of discrete time intervals of one sample path, specifically years or months, due to embedded exotic and path-dependent options, such as floating strike puts or floating strike lookback puts within savings and retirement contracts

3.4.1.4 Mitigating Solutions

3.4.1.4.1 Dealing with the "Curse of Dimensionality" Using the Brownian Bridge Technique We wish to generate low-discrepancy sequences

- with no correlation for every pair of dimensions (different time instants);
- across the paths at one specific time instant, we want a good uniform sample number in order to generate, for example, a good normal distribution of a basket of Brownian motions; and
- thus be able to deal with path-dependent options.

A good result for QMC in high-dimensional problems was reached using techniques such as the Brownian bridge technique to reduce the effective dimensionality of the problem. Suppose we have generated a number of points W(0), W(1), W(2), W(3), etc. of a Wiener process path by computer simulation. It is now desired to fill in additional points in the interval [0, 1] – i.e. to interpolate between the already generated points, to generate the value at intermediate time steps.

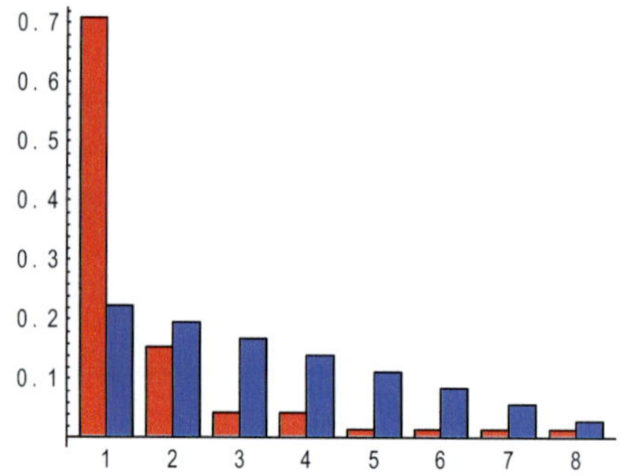

FIGURE 3.127 Comparison of effective dimensionality (variance distribution) for eight-step path for an Asian put

At the 95% level, the sequentially generated and the Brownian bridge paths have effective dimensions of seven and five respectively. This technique has been applied as bias correction of the quasi-Monte Carlo simulation of exotic option valuation. The simulation results are improved significantly by adding the bias term for the following payoffs: lookback options, Asian options, barrier options.

But the Brownian bridge is not a panacea. It has several disadvantages.

- Papageorgiou (2002) provides an example of a digital option in which the Brownian bridge performs worse than the standard discretization. Hence, the Brownian bridge does not offer a consistent advantage in quasi-Monte Carlo integration. In fact, the Brownian bridge converges more slowly whatever the QMC method that is used.
- It can demand additional computation run-time (reordering the path) for path-dependent problems.

3.4.1.4.2 Reintroducing Randomness: a Simple Randomized Quasi-Monte Carlo In order to deal with the "curse of dimensionality", we recommend in the following the use of hybrid quasi-Monte Carlo, for example using quasi-random sequences for a small number of dominant dimensions and pseudo-random sequences for the remaining dimensions. The idea is to get the accuracy of the quasi-random approach for the dimensions with higher impact on the results without the disadvantages of higher-dimensionality behaviour for these sequences.

To this end, we present a very simple hybrid quasi-random algorithm, which permits us to join

- appealing quasi-random vector properties with
- the independence property of pseudo-random numbers.

This approach uses independent (pseudo-) random permutations of quasi-random vectors, and appears to avoid all the problems of high-dimensional degradation of low-discrepancy sequences that we discuss above.

The simplest way to conjugate the appealing properties of quasi-random sequences together with the requirements of independence, avoiding the multidimensional degradation of quasi-random numbers, is by a random shift of a low-discrepancy vector, so that these points "become mutually independent after the randomization". Since the hybrid quasi-points are built as permutations of a Corput sequence with base 2, this method does not experience problems with autocorrelation in a high-dimension environment.

For a problem with D dimensions (e.g. number of periods from a time discretization: D = T/Dt) and N simulations (or paths),

- use the van der Corput sequence in the base 2 (the simplest low-discrepancy sequence) vector with N elements as the basic vector (or for the first dimension); and

- for each dimension use the van der Corput base 2 vector, but with independent random permutations for the elements of this vector.

If N >> D, independence along the path is reached as a very good approximation, because it eases the destruction of the correlation of QR sequences along the path. But, even for N = D, the sequence can be sufficiently independent for practical purposes. Fortunately, in practice it is more probable that will happen is N > D, or even N >> D.

The random permutation idea is to break the correlations (cyclicality) of the sequence of low-discrepancy vectors. The resulting vector after the permutation preserves its low-discrepancy property of an evenly distributed numbers in the interval [0, 1], because the numbers are the same in another order, and in addition the sequence in the same row from the matrix (i.e. in the same sample path) gains the property of independence – or, at least, become closer for practical purposes.

Let us see some charts showing some characteristics of these sequences in multidimensional space. Figure 3.130 (with 1,000 points) shows that the degradation of a low-discrepancy sequence in a high-dimensional case does not occur for a hybrid quasi-random sequence (van der Corput base 2).

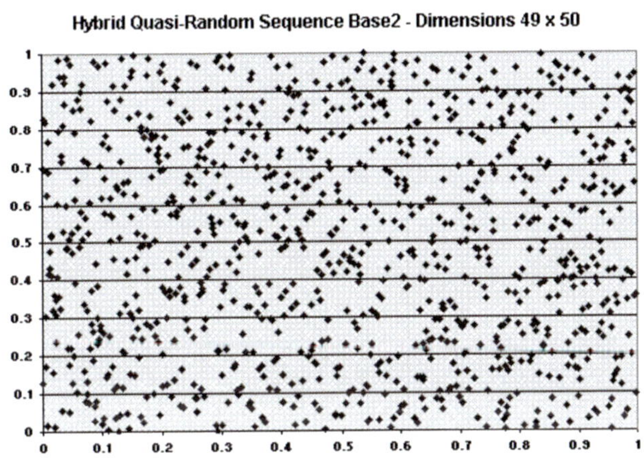

FIGURE 3.128 Hybrid quasi-random sequence base 2: dimensions 49 × 50

The hybrid quasi-random sequence seems better than pseudo-random: compared with the pseudo-random plot presented previously, it lacks the problem of degradation observed in the standard (non-hybrid) quasi-random sequences for higher dimensions. The multidimensional sequence, being more evenly dispersed in each individual dimension, has a higher potential to be evenly dispersed at higher dimensions if it is possible to eliminate the correlation between dimensions. This job is performed by the random permutation.

The only remaining important check for the quality of our very simple hybrid quasi-random numbers is the test for independence along the dimensions (along the path). The absence of clustering exhibited in the last picture above (and other plotting between high-dimensional projections, not shown) is a good indicator of no correlation between dimensions for this hybrid quasi-random sequence. For example, if successive uniform vectors exhibit positive correlation, the unit square (2D projection) will exhibit clustering around the diagonal from bottom left to top right.

However, the time to construct hybrid sequences is greater than for Halton sequences, due to the additional operations needed to calculate uniform permutations of the one-dimension sequence. Furthermore, the hybrid method gives the worst error rate among other low-discrepancy sequences.

3.4.1.4.3 Convergence Rate Comparisons: Tailoring the Monte Carlo Generator to Payoffs In the case of plain vanilla puts, even when you get close to 10,000 simulations the crude Monte Carlo method, the pseudo-random numbers, still does not give reliable estimates of the value. Even though, at the beginning, the quasi-Monte

Carlo error is high, it drops quickly and becomes very stable, and this advantage of QMC increases with maturities, as illustrated in Figure 3.131.

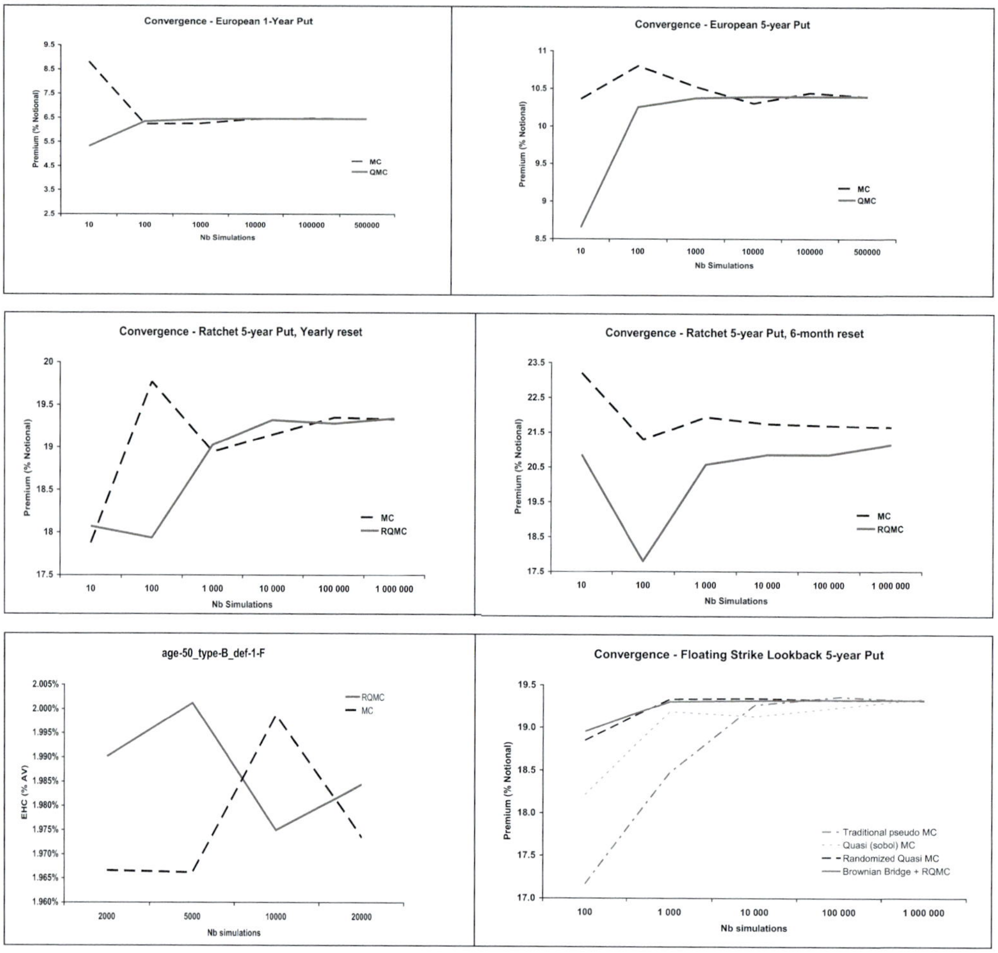

FIGURE 3.129 Convergence versus number simulations

It is even more dramatic for a put than for a call, since the value of the put is bounded; in fact, the intrinsic value of the put will never exceed the strike price of 100, and so it is going to be between zero and 100, in contrast with the call, which can go up to infinity if the price of the security goes very high.

In the case of ratchet puts, the quasi-Monte Carlo should use the Brownian bridge technique, or, preferably, be mixed with randomized (RQMC) simulations, which are required in order to mitigate spurious oscillations from wrong QMC prices.

Finally, regarding accumulators, specifically Yen VA+, where we can check that the convergence corridor is smaller in the case of randomized quasi-Monte Carlo.

Some tests have been developed using more sophisticated RQMC simulations proposed by specialists in number theory coded in C++. As illustrated in the case of a floating strike lookback five-year put, we can see how much these refined methods improve convergence by reducing the associated required number of simulations:

- a randomized quasi-Monte Carlo requires only 1,000 (antithetic) simulations, in contrast to
- traditional (pseudo-) Monte Carlo simulations, which needs between 50,000 and 100,000 simulations to reach the same level of precision.

As a consequence, run-time and hedging accuracy improvements are potentially very significant.

3.4.1.5 Conclusion

Graphic illustrations of these analyses enable us to draw a number of key conclusions.

1. In our search for general (robust) and fast simulation methods, quasi-Monte Carlo offers a good compromise. It can be easily implemented "on top" of existing Monte Carlo simulation engines and it has broad applicability.
2. However, we must be mindful of the possible pitfalls, particularly for high-dimensional problems, as illustrated by multidimensional and path-dependent features within accumulator products. Convergence tests for each (new) class of products are recommended. In addition, quasi-Monte Carlo methods offer, in principle, the ability to obtain absolute error bounds. In practice, this is very difficult to calculate and may grossly overestimate the actual error. Note that this leaves open the possibility of a "catastrophic failure" of convergence, for sample sizes of practical interest. However remote, such a possibility is worrisome, if no error estimation procedure is in place.
3. Therefore we recommend either

   a) an increase in the number of simulations, in order to mitigate the seed initiation impact and to improve the martingale test success rate;

   b) the use of Brownian bridge with quasi-Monte Carlo simulations; or

   c) most preferably, the use of randomized hybrid quasi-random for the most complex savings and retirement products (e.g. variable annuities): they offer the important guarantee of "soft failure", coupled with the ability to perform in-sample statistical error analysis, as with the standard pseudo-MC method.

3.4.2 GPU Technology

For 30 years one of the important methods for improving the performance of consumer computing devices has been to increase the speed at which the processor's clock operates.

Because of power and heat restrictions, as well as a rapidly approaching physical limit to transistor size, researchers and manufacturers have now begun to look elsewhere. Rather than solely looking to increase the performance of a single processing core, they ask: why not put more than one in a personal computer?

In recent years much has been made of the computing industry's widespread shift to parallel computing. Nearly all consumer computers over the past decade have shipped with multi-core central processors.

Among these parallel computing technologies is the graphics processing unit.

3.4.2.1 Hardware Information

3.4.2.1.1 Personal Computers ASUS Notebook G74Sx-BBK7:

- processor: Intel(R) Core(TM) i7-2630QM central processing unit (CPU) 2.00 GHz;
- installed memory (RAM): 8.00 GB;
- system type: 64-bit operating system.

3.4.2.1.2 Graphics Cards NVDIA Geforce GTX-560M 2G:

- total global memory: 2,048 Mbytes;
- four multiprocessors with 192 CUDA cores.

Memory information:

- total amount of constant memory: 65,536 bytes;
- total amount of shared memory per block: 49,152 bytes;
- total number of registers available per block: 32,768;
- warp size: 32;
- maximum number of threads per multiprocessor: 1,536;
- maximum number of threads per block: 1,024;
- maximum sizes of each dimension of a block: 1,024 × 1,024 × 64;
- maximum sizes of each dimension of a grid: 65,535 × 65,535 × 65,535;
- maximum memory pitch: 2,147,483,647 bytes.

3.4.3 Numerical Results

We test the code of Mont Carlo and compare the GPU and CPU results. We use float precision for GPUs since, although the devices support double precision, using float precision is much quicker than double precision (we have made a simple test and find that using float is five times quicker than double precision).

3.4.3.1 Black–Scholes by Closed-Form Formula

Options Count	Bandwidth	Executing Kernel	GPU Time	CPU Time	CPU T/GPU T
8000000	23.07 GB/s	512 iterations	3.47 msec	472.73 msec	136.23

FIGURE 3.130 Black–Scholes closed-form formula by GPU

3.4.3.2 Black–Scholes by Quasi-Monte Carlo Simulations

Options Count	Simu Paths	GPU Time	CPU Time	CPU T/GPU T	Error Difference
256	262144	4.59 msec	3576.57 msec	779.21	$3.08E-008$

FIGURE 3.131 Black–Scholes quasi-Monte Carlo by GPU

3.4.3.3 *Exotic Options by Quasi-Monte Carlo Simulations*

3.4.3.3.1 Lookback Options

Type	Call/Put	S	K	T	σ	r
Look-back	Call	15.0	15.0	2.0	0.1	0.06

Simu Paths	GPU Time	CPU Time	CPU T/GPU T	Price Difference of CPU and GPU
262144	14.09 msec	2895.33 msec	205.49	$1.668930E-006$

FIGURE 3.132 Lookback options by quasi-Monte Carlo by GPU

3.4.3.3.2 Asian Options

Type	Call/Put	S	K	T	σ	r
Asian	Call	12.0	10.0	1.0	0.1	0.06

Simu Paths	GPU Time	CPU Time	CPU T/GPU T	Price Difference of CPU and GPU
262144	15.28 msec	2824.11 msec	184.82	$9.775162E-006$

FIGURE 3.133 Asian options by GPU

It seems that using CUDA to price Asian options is less efficient than that for lookback options. The reason is that, to calculate the payoff of Asian options, we have to compute the average price. In CUDA, it will cost much time to use the divide operator, whereas we use only the max function to price lookback options.

CHAPTER 4

Designing Products Aligned with the Risk Appetite for Sustainable Savings and Retirement

THE COMBINATION OF DECLINING pensions and flagging state healthcare spending is forcing more Europeans to look for additional or alternative retirement funds. This confluence of circumstances produces a tremendous demand for protected saving solutions in the mass market. In general, demand for such guaranteed products has been highest among lower- and middle-income brackets, while individuals in higher-income brackets have indicated more willingness to invest in products with no protection but a higher upside potential.

Across all lines of business, there is increased demand for more affordable, transparent and customized insurance that better suits the evolving conditions and can be easily adjusted as the needs change. The new needs and expectations regarding financial wellness have translated into drastic and rapid shifts in risk appetite on the part of individuals, in terms of (i) ensuring suitable and sustainable behaviour in the use of their savings and retirement products, such as surrender, withdrawals and annuitization, becomes more "efficient" throughout a "multi-stage" life and (ii) a demand for more sustainable investments in assets, as illustrated by the significant growth in their environmental, social and governance scores.

These significant changes in customers' risk appetite and associated saving behaviour have further strengthened within the context of (i) declining and persistently low interest rates over the past decade, driven by the jump in central banks' balance sheets, and (ii) new regulatory prudential required capital (RC), accounting and consumer protection frameworks, which tend to increase not just long-term safety but also short-term costs.

As a result, life insurers have had to further expand their value propositions and product sets to address all aspects of financial wellness, centring customer experiences and streamlining their journey as a priority. Life insurers must respond to customer needs and expectations around flexibility, personalization and relevance. In this respect, life insurers have improved the sustainability of both their liabilities and their investment assets.

Regarding liabilities, section 1 illustrates how life insurers can screen sustainable saving and retirement product opportunities, balancing shareholders' versus customers' risk appetites, through capital intensity and customer-centric metrics. Section 2 then focuses on how to improve the solvency IRR through product designs, mitigate customer behaviour risk and enhance the customer experience by tailoring the product design and reconcile insurer and customer risk appetites through sustainable capital-light products (low capital requirements while offering capital protection and an upside potential).

Regarding the assets side, section 3 focuses on why life insurers are also eager to move into asset management as a capital-light business. They hope to lighten their capital burden; however, this change also puts

DOI: 10.1201/9781003218371-5

insurers in direct competition with private equity and asset managers who have a significant competitive advantage in this area and often operate at lower costs. Among the key sustainable innovations reconciling insurer and customer risk appetites are (i) the improved risk management of fund management, (ii) enhanced performance in asset selection and allocation while meeting the growing demand for ESG assets and (iii) optimized portfolio rebalancing, allowing for a lowering of transaction costs and financial risks alike.

4.1 SCREENING SUSTAINABLE OPPORTUNITIES BALANCING SHAREHOLDERS' VERSUS CUSTOMERS' INTERESTS WITH PERSISTENTLY LOW RATES AND GROWING REGULATORY CONSTRAINTS

4.1.1 The Impact of a Prolonged Low Interest Rate Environment

The interest rate environment has a significant impact on many segments of the financial sector, including the life insurance industry. Interest rates declined significantly following the 2008 global financial crisis before levelling out at historically low rates, which persisted even with monetary policy normalization efforts by the Fed beginning in 2015. Interest rates then fell precipitously further at the beginning of the Covid-19 pandemic, with the yield on the benchmark ten-year Treasury (constant maturity) dropping to an unprecedented low of 0.6%. In mid-2021 interest rates returned to near-pre-pandemic levels, but those levels were at 50-year lows. They rose further with the emergence of significant inflation, further fuelled by the war in Ukraine, not experienced since the beginning of the 1980s, although they have remained significantly below the average historical level.

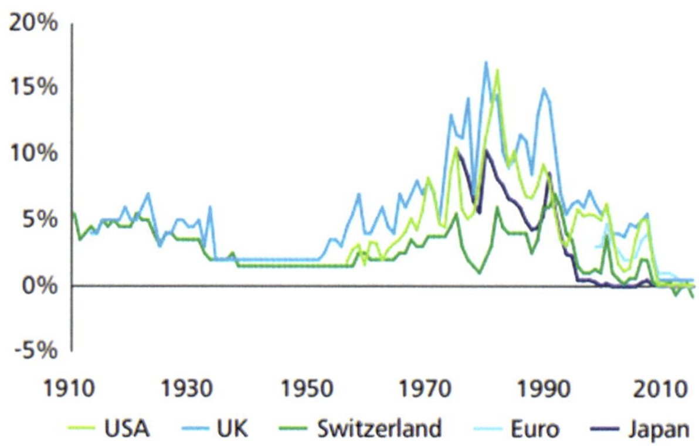

FIGURE 4.1 Overnight rates from central banks over the past century, 1910–2020

Central banks uses a variety of tools to fulfil their dual mandate of maximum employment and stable prices. Their primary policy instrument is open market operations, which is the buying and selling of short-term Treasury securities to keep the designated interest rate within its target range. Historically, central banks raised their target when the economic outlook was inflationary and lowered the target rate or range when the economy was soft or in a downturn, such as during a recession. In an effort to support the economy and financial markets during the financial crisis and 2007–2009 Great Recession, central banks dramatically lowered the target to nearly 0%, then began using unconventional monetary policy instruments, including the large-scale purchases of long-duration Treasury securities and agency mortgage-backed securities (a quantitative easing [QE] programme).

Such a persistently low interest rate environment is a key concern for life insurers, because their assets and liabilities are heavily exposed to interest rate movements, across the four dimensions of the risk appetite (earnings, value, capital, liquidity).

4.1.1.1 Earnings Impacts

The average return on equity for insurers has fallen significantly, notably in Europe, from around 11% in 2014 to about 9% in recent years.

In particular, their investments are concentrated (80%) in fixed-income securities that return interest, largely bonds, and are then repackaged into annuities and life insurance products that reflect the yields of the underlying investments. When interest rates increase, therefore, the opportunity cost of holding bonds at a lower rate over time also increases.

Moreover, their liabilities are also sensitive to interest rates. Specifically, many of their products, such as annuities, have a guaranteed rate of return, usually in the form of interest that is credited. Life insurers' earnings mostly are derived from the spread between their investment returns, which are mainly interest, and what they credit as interest on these consumer products. During times of persistently low interest rates the spread between interest earned and interest credited is compressed, which not only reduces net income for the insurers but also puts them at risk of being unable to meet contractually guaranteed obligations to policyholders.

Many insurers' earnings on in-force blocks have come under pressure, as guaranteed rates to policyholders are still as high as 150 to 400 bps in some markets, while yields on bonds have declined by 150 to 300 bps since 2010/11. Naturally, this has strained capital, as insurers have had to adjust their reserves to reflect future earnings expectations. Specifically, the spread fell from 1.8% in 2007 to 1.15% in 2009 in the United States, then remained historically low, until plummeting from 2018 to 2020 (reaching 0.63%).

Even in the event of rising rates in the coming months and years, carrier investment portfolios tend to lag behind movements in the market by several years, due to the various durations of the bonds that make up investment portfolios. Besides, most life insurance products with declared dividends or crediting rates assume that the current market environment stays constant for the life of the policy, as regulatory restrictions preclude the use of earning rates higher than what is currently being credited.

Historical analysis shows that the overall trend is for the profitability of the insurance sector to increase in an environment of rising interest rates. Overall price-to-earnings (P/E) ratios for insurance company stocks usually increase in fairly direct proportion to increases in rates.

4.1.1.2 Value Impacts

Changing interest rates also affect insurers through portfolio revaluation effects. Falling interest rates increase the value of existing interest-earning assets in the life insurers' portfolios (from about $3.6 trillion to $3.9 trillion in the 2020s, as against $2 trillion in 2006), in particular the long-term bonds massively held by insurers.

Further, life insurers typically hold bonds to maturity, because of the need to match the timing of asset returns and policyholder claims, so therefore the capital gains typically are not realized. Moreover, when a low interest rate environment persists, higher-yielding assets mature and must be replaced by low-yielding assets. An average insurance company reinvests about 12% of its assets annually, so this profitability challenge becomes increasingly acute every year.

If the first risk in a low interest rate environment is to be forced to invest at too low yields, the second risk is to face a sudden increase in interest rate; clients are likely to let their contract lapse so as to take a new opportunity with higher rates. Thus, persistently low interest rates may also affect customers' behaviour; in particular, one of the general account's complexities is predicting and managing policyholder behaviour, either lapsing earlier than scheduled or sticking longer than expected, which impacts the insurance balance sheet duration.

- In a low rate environment, benefiting from higher guaranteed rates, policyholders keep or even increase their savings -> duration increases.
- In a rising rate environment, policyholders can use their option to lapse and invest in other higher-yielding products -> duration decreases.

As a result, such changes in behaviour imply negative feedback effects on interest rate risks for the insurer.

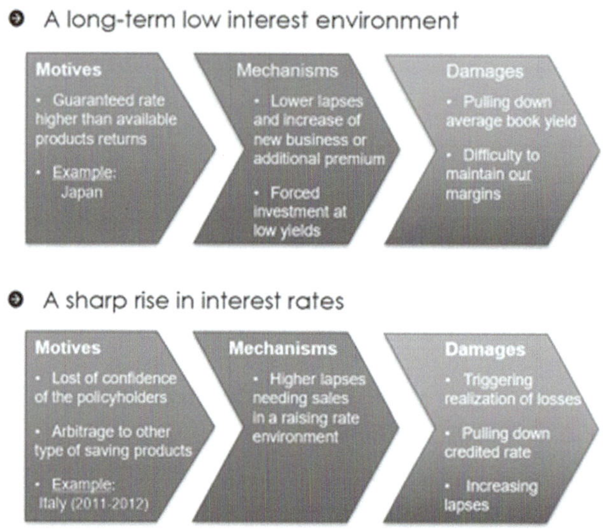

FIGURE 4.2 Impact of interest rates on lapses

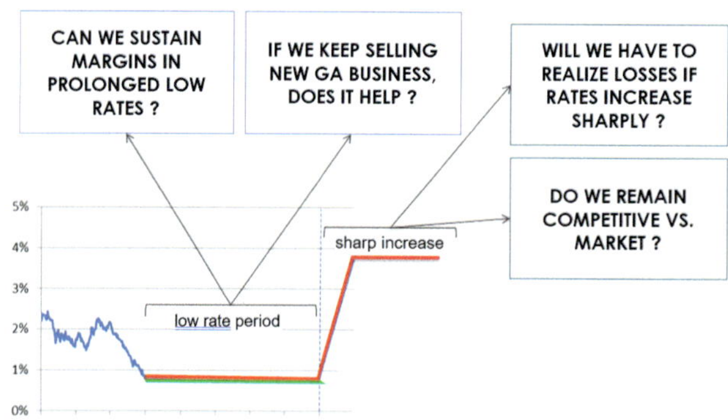

FIGURE 4.3 Rates move versus earnings

Another challenge lies in new business value: as interest rates fall the overall willingness of people to enter into long-term insurance contracts declines, since customers are less willing to potentially lock their money and their future premiums into contracts offering an expected low return. Most customers drift to alternative investment approaches, such as short-term savings vehicles (the "wait and see" approach) or directly invest in other asset classes, such as real estate. Thus, insurers' new business volumes progressively come under pressure. This is more apparent in traditional insurance markets that have historically favoured bond investments to back policyholder liabilities.

Last but not least, there is the value proposition: most markets still want insurance guarantees even though returns on traditional products are less than zero after the deduction of underlying costs. Consequently, providing insurance products that include guarantees by simultaneously offering the potential for attractive returns is impossible. The majority of European insurers are today unable to price new guaranteed investment products at competitive rates. In addition, the higher capital charges under Solvency II for investments other than government bonds add to this dilemma.

4.1.1.3 Capital Impacts

The fall in interest rates does not improve the insurers' statutory capital position, because the bonds are reported at book value: in 2014 Swiss life insurance companies had to use over 50% of the total operating result to strengthen their solvency capital.

As a matter of fact, in the past few years the cost of capital for insurers has climbed far more rapidly than it has for other industries. As a result, capital has to be managed more efficiently in order to deliver higher returns. Indeed, any recalculation of the technical provisions to reflect interest rate movements will result in a stronger drag on an insurer's capital generation as it is subsequently amortized.

However, these can reinforce one another and result in a vicious cycle: pressure on solvency ratios can force insurers to move out of higher-yielding asset classes that attract higher capital charges in order to improve the solvency ratio, and, while the "de-risking" provides a short-term improvement, ultimately it further deteriorates firms' ability to generate capital/cash. Financial revenues will decline across products, the capital requirements of guarantees will skyrocket and the value proposition of life insurers in savings products will be increasingly questioned.

Moreover, local reserving rules that include a "smoothing" element to avoid spurious volatility (e.g. by looking at historic interest rates over a long time horizon) are now fully based on a low interest rate environment.

For example, in Germany the Zinszusatzreserve (ZZR), introduced in 2011, brought in an additional reserve whenever the cost of a guarantee offered by a life insurer is higher than a "reference rate". This additional reserve is based on the average of the previous ten years of swap rates – i.e. on the past ten years of low and decreasing rates – which means that it will effectively "lock in" the effect of the current environment even after rate changes, thus forcing many firms to realize additional reserves, further compromising capital generation ability in the long term. Furthermore, German statutory reporting is likely to exacerbate the economic pressure, as life insurers have to sell the high-yielding assets in order to fund the growing ZZR requirement (introduced in 2011 and expected to reach a total of €160 billion by 2030, or more than 10% of total life insurance reserves) and reinvest them into lower-yield assets.

4.1.1.4 Liquidity Impacts

Persistently low interest rates can finally affect life insurers' liquidity. Liquidity is the ease with which assets can be converted to cash. Liquidity management is critical for life insurers. As part of asset liability management (ALM), life companies strive to match asset cash flows with cash outflows to avoid asset–liability mismatch and interest rate risk. During periods of low interest rates, cash flows from assets and liabilities can be badly mismatched, exposing insurers to losses from potentially pressured asset sales to meet current obligations to policyholders. At the same time, under conditions of persistently low interest rates liquidity demands tend to decrease, as policyholders are more likely to keep their money in life insurance investment products, such as annuities, due to the previously agreed and guaranteed return and limited availability of higher-yielding alternatives. On the flip side, as interest rates rise, interest-earning policies are increasingly likely to lapse.

Insurers thus face a dilemma. Their traditional business model no longer works in a negative rate environment. Guaranteed insurance products that provide simple retirement income are unsustainable, and the situation becomes even more acute under current solvency regulations (specifically, the European and new US NAIC frameworks, to some extent). Companies' own returns on capital and price-to-book ratios are nearly at all-time lows. At the same time, customers are clamouring for new answers as they are forced to invest in a low-yield, high-volatility environment and take on the longevity risk themselves. Insurers can resolve this dilemma for themselves and their customers, but bold action is required.

4.1.2 Growing Regulatory Concerns and Constraints

4.1.2.1 … With respect to Accounting and Capital Requirements

Recent and forthcoming regulatory changes in the industry have also encouraged entities to reassess the capital intensity and profitability for all lines of business within their portfolio.

In particular, accounting changes in the industry, such as International Financial Reporting Standards 17 and long-duration targeted improvement (LDTI), have changed the way insurers recognize profit, and certain cohorts that were profitable in the past may lead to the disclosure of losses in the future. Under the LDTI standard, it is expected that those portfolios with the following characteristics will be more adversely affected:

- older product vintages, as these are most likely have higher locked-in discount rates under current GAAP;
- long-tail liabilities for which no unlocking has occurred to date, due to adverse deviations from their pricing assumptions, such as LTC utilization and payout annuity longevity; and
- the materiality of variable annuities, especially if they are not hedged, and GMxB riders, as market risk benefits are now fairly valued and will be more volatile.

IFRS 17 will add further challenges: life insurers with large guaranteed savings books will be particularly affected, as their landscape has been shaped by a three-decade bull market for bonds. IFRS 17 will also impact the financial reporting of insurance companies. Examples of these impacts include the requirement to isolate onerous contracts from profitable ones and to disclose a loss component. In addition, insurance companies subject to IFRS 17 reporting will need to rethink their communication to investors, with new revenue and profit recognition patterns that are expected to be significantly different from those under the current reporting basis.

Finally, implementing these new regulations is more complex and expensive for entities that are exposed to a wide range of insurance products. On top of other factors, skipping this process can provide an incentive to refocus around the core business, and the increased regulatory costs have been cited as a motivation for divestiture.

The complication is that savers continue to express a strong appetite for capital protection, while the European Union's Insurance Distribution Directive (IDD), passed in 2016, and 2014 Markets in Financial Instruments Directive (MiFID) form a demanding framework for product adequacy.

Even so, most regulatory bodies have helped manage the persistently low interest rate challenge. FINMA, the Swiss Financial Market Supervisory Authority, has reduced the maximum guaranteed interest rates that life insurances may offer to their customers from 1.25% to 0.75% (0.5% for single-premium policies). The Swiss government reduced the minimum guaranteed rate of interest credit for the mandatory part of pension savings from 1.75% in 2015 to 1.25% in 2016. Under Solvency II the ultimate forward rate (UFR) was introduced, which artificially lowers the economic weight of long-term liabilities

4.1.2.1.1 The Mismatch of Assumptions of Assets versus Liabilities in Statutory Capital versus IFRS Earnings

4.1.2.1.1.1 In the United States Accounting regimes have different objectives, which tends to lead to divergences in evolution.

- On the one hand, US statutory capital requirements are designed with the protection of policyholder interests and the long-term solvency of companies in mind.
    - They are used by both companies and regulators to monitor financial strength against established benchmarks.
    - They constitute the primary metric for the evaluation of whether cash is available for dividends or investment.
    - They are more conservative than other accounting frameworks, offering faster amortization of acquisition expenses, a conservative view of deferred tax assets, a cash surrender value floor, the standard scenario floor, prudent margins, limits on asset-based fee recognition and estimates based on the worst 30% of outcomes.

- On the other hand, IFRS 17 is designed to provide a long-term view of earnings.
    - It amortizes the expense of acquiring new business (the deferred acquisition cost: DAC) over the life of the products sold.
    - It smooths other fluctuations in earnings that are caused by market movements.
    - Deferred tax assets are allowable as long as they are recoverable.
    - Annuity fee recognition is set at the inception so as to avoid any gain on sale.

4.1.2.1.1.1.1 Acquisition Cost Recognition

- Under IFRS, DAC is amortized based on the expected gross profits over the life of the policy. Amortization moves with the market.
- Under the US statutory framework, DAC is amortized on a set schedule, regardless of the commission, and it can be amortized at a faster pace only when profitability is impaired.

As a result, strong sales levels may lead to a drag in statutory operating growth relative to IFRS.

4.1.2.1.1.1.2 Guarantee Fee Recognition

- The reported statutory operating capital generation removes all guarantee fees collected and moves them to "Reserves net of hedging".
- IFRS reserving methods incorporate the value of guarantee fees that are expected to fund future projected benefit payments using the assumptions applicable for that method. The level of fees recognized is fixed at issue, and they are capped so that they are equal to benefits (i.e. any excess is ignored). The remaining fees are recognized as earned and included in operating profit.

4.1.2.1.1.1.3 Floored-Out Reserves versus the Cash Surrender Value under the Statutory Regime

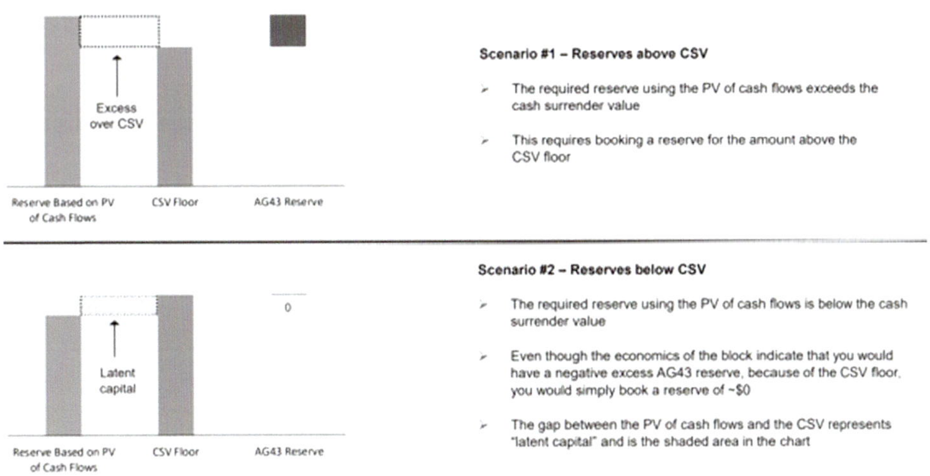

FIGURE 4.4 Floored-out reserves versus the CSV under the statutory regime

Note: AG43 = Actuarial Guideline 43, set by the NAIC.

This may lead to non-economic call option spending in upside equity.

- When reserves are floored out, further up moves in equity markets will cause a build-up of "latent capital".
- It leads to additional hedge spending, as there are no reserves to release to offset hedging losses from stronger equity markets.
- The gap as we move to the left illustrates the "release" of this latent capital as markets fall and no reserves are posted.
- In the event of little or no reserve offset for hedging losses in up equity markets -> need for an incremental call option spend to protect statutory capital in upside equity market scenarios above what the economic profile would require.

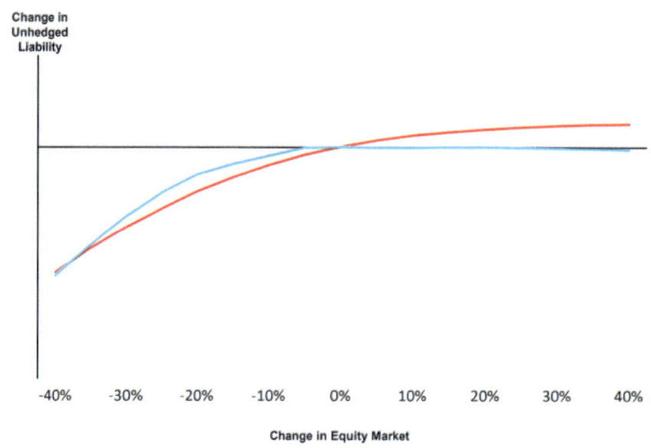

FIGURE 4.5 Local statutory framework versus economic change in unhedged liability

4.1.2.1.1.2 In Europe

4.1.2.1.1.2.1 Hedging Interest Rate Risks on a Solvency II Basis Adds Unwanted Earnings Volatility

- The Solvency II hedges do not include swaps beyond 20 years because the Solvency II rates converge deterministically after 20 years to the UFR over 40 years -> this extrapolated Solvency II yield curve mostly depends on the 20-year swap and is highly sensitive to the slope just before that point.
- -> A 50-year swap translates into a long position in a 20-year swap and a short position in a 15-year swap -> the duration reduces significantly.
- -> The Solvency II hedge is significantly more volatile than the IFRS market-consistent hedge -> hedging on the Solvency II basis adds unwanted earnings volatility.

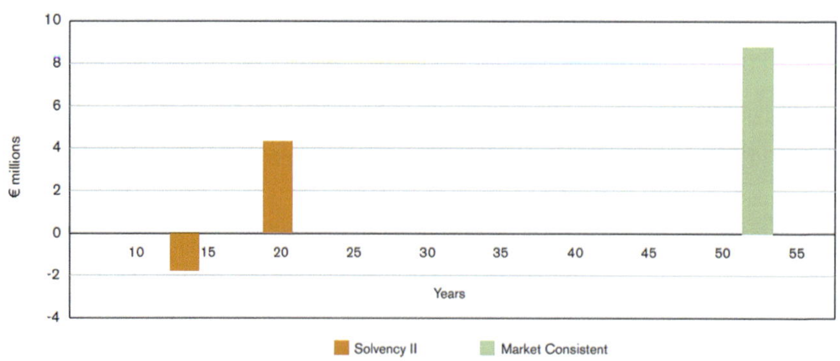

FIGURE 4.6 Solvency II hedge versus IFRS hedge

4.1.2.1.1.2.2 The Solvency II Standard Formula Recognizes Only the Current Actual Derivatives except for Automatically Rolled-Over Ones Because the standard formula is based on instantaneous shocks to the balance sheet -> an internal model would be more appropriate.

4.1.2.1.2 The Prevalence of So Many Different Embedded Value Reporting Methodologies across Asia Brings Major Challenges in Comparing EV Results Embedded value (EV) reporting started in the United Kingdom in the 1980s as a way for life insurance companies to give better guidance to analysts and shareholders on their underlying economic values. At that time accounting standards were not fully equipped to handle the unique nature of life insurance businesses, and it was very difficult to use the standard financial statements to assess a life insurer's economic value.

The methodology has since spread globally. Early EV methodologies, using deterministic approaches to value cash flows and implicitly allowing for the cost of policyholder options and guarantees, asset–liability mismatch risk, credit and other risks and the economic cost of capital through the use of a risk-adjusted discount rate (RDR), are often now named traditional embedded value (TEV).

Following some TEV-related criticism in the investment community, a group of leading European insurers, known as the European Insurance CFO Forum (CFO Forum), published more detailed agreements on principles for EV calculations and disclosures in 2004, which formed the basis for what is now referred to as European embedded value (EEV) methodology, providing greater comparability between insurers.

The latest evolution in EV reporting came in 2008, with the introduction of the market-consistent embedded value (MCEV) principles, which introduced mandatory market-consistent valuation of assets and liabilities and more specific disclosure requirements: frictional cost of capital (FCoC), which reflects the tax and investment costs that arise on the assets backing the required capital, and cost of residual non-hedgeable risks: CRNHR. The CFO Forum had originally intended introducing MCEV as the mandatory standard from 2012 onwards, but this target was withdrawn in 2011 pending the development of Solvency II and IFRS.

EV was initially introduced into Asia through the subsidiaries and joint ventures of European companies. Since then many domestic insurers have introduced EV reporting, but based on different methodologies.

- In India, almost all companies that report EV now do so on an Indian embedded value (IEV) or MCEV basis.

  In contrast, all insurers in Japan adopt MCEV or a market-consistent EEV (MC-EEV) approach.

- Domestic insurers outside India and Asian multinational corporations (MNCs) tend to report on a TEV basis. Insurers in China, South Korea and Taiwan continue to report on a TEV basis.

- The number of European MNCs reporting EV has reduced since 2016, as their parent companies have switched to using Solvency II (SII).

The prevalence of so many different EV reporting methodologies across Asia brings major challenges in comparing EV results, making a good understanding of the differences between the methodologies critical when comparing the embedded value results of insurers, including traditional embedded value, European embedded value, market-consistent embedded value and Indian embedded value.

- Regarding the risk discount rate, TEV typically use a risk-free rate plus risk margins to derive discount rates, based on two possible approaches: "top-down", with one discount curve used for all cash flows based on risks faced by the entire organization; or "bottom-up", in which each cash flow is discounted using a risk-free rate plus the risk margin subject to judgement. Under MCEV a bottom-up approach is mandatory, and the curve is typically on swap rates (with adjustments for illiquidity), while the risk margin is usually in a tight range, except for Taiwan and Hong Kong.

- Regarding the investment return assumptions, the future investment return is a key assumption for calculating VIF and VNB for TEV- and EEV-reporting companies, with a wide range being seen: typical TEV practice is to use a risk-free rate plus a risk premium approach for the main asset classes, whereby the risk premium assumptions differ by asset class; for EEV, some insurers opt to use a risk-neutral approach, while others use a risk-free rate plus a risk premium approach; for MCEV, a risk-neutral approach is typically used, whereby assets are assumed to earn returns based on a risk-free curve (swap or bond rates).

  There is also some divergence between insurers on the link between current market yields and future investment return assumptions, which some insurers derive from spot bond yields (with risk margins for other asset categories) while others position them as long-term return assumptions, with increasing divergence from spot bond yields as interest rates have fallen in recent years, which can potentially introduce some disparity (as insurers take credit for market value uplifts from falling interest rates, but only partially reduce their VIF results as investment return assumptions are not reduced to the same extent).

- Consequently, regarding the computation of the PVFP, TEV uses the projection of future profits using real-world investment return assumptions, discounted using a subjective RDR. EEV uses the projection of future profits with real-world investment return assumptions, discounted using a curve based on risk-free rates, adjusted using a risk margin, which reflects any risks not allowed for elsewhere in the valuation. Some EEV reporting firms also opt to use a market-consistent approach, which entails using risk-free rates in the certainty-equivalent approach. MCEV uses the projection of future profits using market-consistent risk-neutral investment return assumptions, discounted using a curve based on risk-free rates. Discount rates can be adjusted to include an illiquidity premium.

- Regarding the cost of guarantees, only firms reporting EEV, IEV and MCEV must calculate the time value of options and guarantees (TVOG), for which most insurers will use risk-neutral (as opposed to real-world) models for ease of calculation. Firms reporting TEV do not explicitly allow for it but, typically, include only the intrinsic value (using their deterministic investment return assumptions), but make implicit allowance for TVOG in their choices of RDR.

- Regarding the cost of capital, insurers need to make assumptions on the future level of their required solvency margin (SM) when projecting distributable earnings, which is typically based on what insurers perceive to be the minimum level that will prompt regulatory intervention.

The overall position is as follows.

- The primary advantage that EEV and MCEV approaches have over TEV is the greater standardization (and less subjectivity) of assumptions, methodologies and disclosures, leading to better comparability from an investor's viewpoint. For example, MCEV assumes that assets earn the risk-free rate of return, which avoids the use of actual risk-weighted yields or management's view of future market directions in EV calculations, as is the case with TEV (and some EEV) reporting.

- Insurers reporting on EEV or MCEV bases will typically experience greater volatility in EV results, which makes reporting complicated and is one of the reasons often cited as to why most Asian companies have not yet moved from TEV to EEV or MCEV.

- Another key reason is the increased capabilities required to fully implement EEV or MCEV reporting: the implementation of proper TVOG calculations requires the use of stochastic models to value embedded policy options and guarantees, which means using specialized economic scenario generator (ESG) software. This will add to financial reporting times. In addition, it is difficult to calibrate the ESG for Asian capital markets, which are in general not as deep or liquid as those in the United States or Europe, while long-term guarantees are so prevalent there, not explicitly allowing for TVOG, which is an obvious and significant flaw in companies' TEV financial reporting.

In particular, among recent acute challenges is the impact of changing interest rates on embedded value results, which depend upon the methodology adopted, the assumptions used and the type of business written by the company.

- Under a market-consistent EV or the Indian EV framework, a lower interest rate might lead to an increase in the MCEV or IEV if a company writes more unit-linked business (due to the uplift in starting asset values and discounting of future profits at a lower interest rate) or protection-oriented business (due to the unrealized gains on starting assets and discounting of future profits at a lower interest rate), while the opposite might be true for traditional savings business with guarantees (due to lower investment margins).
- Under a traditional EV framework, the impact of changing interest rates also depends upon whether a company alters its long-term investment return assumptions in line with the change in interest rates. If a company does not change its long-term investment return assumptions, it may lead to counterintuitive results, such as declining interest rates, resulting in a higher EV.
- A certain amount of caution must be exercised when evaluating Japanese company embedded values and their adjusted net worth (ANW: the shareholders' net assets, including free surpluses and required capital – i.e. the amount returned to shareholders should all assets be sold and liabilities settled immediately) and VIF (the discounted future distributable earnings arising from policies in force as at the valuation date) components, especially when comparisons are made across Asia.
  - Japanese companies typically report on a market-consistent basis, either MCEV or MC-EEV. In addition, many companies manage large blocks of legacy policies with relatively high investment guarantees. As a result, many companies have a very small (or even negative) VIF compared to the size of their in-force block.
  - On a percentage basis, the VIF is extremely sensitive to changes in the interest rate environment. However, due to the use of a market-consistent approach and ALM, VIF changes are usually substantially offset by changes in ANW. As a result, overall EV is far less sensitive than the individual VIF and ANW components.

EV results by their nature are typically impacted by changes in insurance regulations, among which are the following.

- In India the Insurance Act was amended to increase the foreign direct investment limit for the insurance sector from 49% to 74%.
- In Singapore the life insurance industry lowered the caps of illustrative investment returns used in policy illustrations.
- In Taiwan reductions were introduced in reserving interest rates on all policies denominated in New Taiwan dollars and US dollars by 25 bps and 50 bps respectively, effective from the second half of 2020.
- In South Korea the liability duration has been reinforced on long-term products, leading to higher required capital.

4.1.2.1.3 Changes in Capital Regulations Prompt Insurers to Revisit Their Existing Capital Management, Strategic Asset Allocation and Hedging Programmes: Focus on Asia Capital regulations for life insurance companies in Asia are complex and varied. They are also subject to change, with such changes often affecting how insurers manage their business. Most insurance markets in Asia follow some form of risk-based capital (RBC) regime, except India and Hong Kong at present, though some of them, including Hong Kong, India and Brunei, still currently use an EU Solvency I type of approach.

The assessment of required and available capital using an economic balance sheet approach has underpinned most of the recent changes in Asian capital regulations: regulators are introducing new RBC regimes or "upgrading" existing frameworks, with increasing consideration being given to consistency with the new International Financial Reporting Standard 17, International Capital Standards (ICS) and other capital regimes worldwide.

- Singapore has completed its transition to a new solvency regime (RBC 2), with changes in the valuation of liabilities, assessment of capital requirements and recognition of available capital, translating notably into a shift in product mix with fewer guarantees and less optimized asset liability management to back up long-duration liabilities.
- Hong Kong is introducing an RBC framework for 2024. The introduction of Pillar 2 (ERM, ORSA) requires the enhancement of the risk management and governance framework, business strategy and operation: identification and assessment of risks, including emerging risks such as climate change, risk measurement against risk appetite, stress and scenario testing and mitigation actions.
- The China Banking and Insurance Regulatory Commission (CBIRC) has revised its solvency regulations for insurance companies, based on a three-pillar framework, focusing on strengthening the control of insurers' solvency and protecting consumers (limiting unearned profits recognized as core capital, encouraging better interest rate risk asset liability management practices).
- In Malaysia the regulator introduced in March 2020 enhancements to improve the risk capture of the RBC framework, with revisions to the stress-testing parameters used for the interest rate risk capital charge.
- In South Korea the Financial Supervisory Service (FSS) has announced its plan to adopt a principle-based capital framework similar to ICS for 2023: insurance liabilities calculated based on market value (not historical cost basis); full and fair valuation of all assets and liabilities; and a 99.5th percentile confidence level (versus 99th today). Most insurers are expected to face a significant decrease in their solvency ratio even below the minimum requirements, requiring an increase in capital, a lower proportion of risky assets (equities and real estate) and the implementation of ALM systems.
- In Thailand the 2020 revisions included decreasing the diversification correlation factor between assets and insurance and reducing the equity risk charge when there is equity investment hedging.
- Taiwan, where the current RBC approach is based on prescribed risk factors multiplied by risk exposures, is set to move to an ICS-based regime for 2026, with three key changes: the solvency measurement will move to an economic balance sheet approach; the calculation of required capital will be on a stress-based approach to better capture the "tail risk"; and the requirements on assets qualifying as admissible capital for solvency purposes will be stricter than the current requirements. The most prominent issue of the transition is the duration and currency mismatch in assets and liabilities, with expected greater volatility in the solvency position.

In particular, for solvency purposes, an increasing number of Asian capital regimes require companies to do the following.

- Assess their assets on a market value basis (e.g. Hong Kong's proposed RBC framework, Indonesia, Singapore, Thailand, Malaysia), though some are still measuring their assets using different accounting bases (e.g. for China's C-ROSS, Japan's regulatory capital).
- Value their liabilities using a gross premium valuation (GPV) approach, as opposed to a net premium valuation (NPV) approach, allowing for an additional risk margin and, potentially, a time value of options and guarantees using a fair value approach based on "relatively market-consistent" discount factors.

Although there is a trend towards the use of an economic balance sheet framework, markets are moving at different paces, and many regulators in Asia seem to have taken a more practical approach that reflects market specifics, while ensuring a reasonable degree of conservatism (e.g. the flooring of reserves in in Japan, Indonesia, Malaysia, Singapore, South Korea and Taiwan). This leads to inconsistencies between RBC regimes across the region, notably regarding the following.

- The risk margin: while provisions for adverse deviation are mostly adopted, the approach to deriving it differs from one market to another.
- The discount rate.
  - The yield curves used to assess the best estimate of liabilities (BEL) are typically defined using a "bottom-up" approach (i.e. the discount rate = market-consistent risk-free rate + illiquidity adjustment).
  - However, the valuation of liabilities requires the use of a yield curve that extends to very long durations, reflecting both market conditions and long-term economic views, while Asian market data cover a much shorter duration.
  - Given the long-term nature of many life insurance contracts, the insurers can then potentially benefit from the risk premium, typically called an illiquidity premium adjustment (e.g. volatility adjustment, UFR), that is applied in the discount rate to reduce the short-term economic balance sheet volatility, stabilize the net asset value and better reflect the illiquid nature of liabilities.
- Regarding the capital adequacy ratio, most markets have an average regulatory solvency ratio within the range of 180 to 400%, except for Japan and Indonesia (above 450%).
- Regarding investment assets, more than a half are invested in bonds, either mostly government bonds (e.g. Thailand), or corporate bonds (e.g. Hong Kong) and alternative credit (although small). The proportion of equities varies, with markets having a material proportion of participating business (e.g. Singapore, Malaysia, Hong Kong) typically investing more in equities.
- Regarding liabilities, they also differ significantly: the unit-linked business is significant in Indonesia, India and Malaysia, while universal life business is popular in Hong Kong, Singapore and South Korea. Non-participating traditional business (e.g. whole life, credit life, term life) remains a material product category in Asia, while participating business (e.g. endowments, whole life) is popular in Japan, Hong Kong, Singapore, India and Sri Lanka.
- Regarding TVOG (which measures the in-the-moneyness of the investment guarantees), it is usually not material in Asia, notably in India (as the level of guarantees for participating products is typically low and not widespread for unit-linked business); in Indonesia (mostly unit-linked without guarantees); in Malaysia (participating products are currently out of the money); in Thailand (most products are non-participating);

in Singapore (TVOG is mostly not assessed and most investment guarantees are out of the money, although it is expected to be significant for universal life); or in Taiwan (not assessed). The exception is Hong Kong, for participating and universal life products, for which investments guarantees can be high.

A move to a more "economic" RBC regime (with a typical fair value approach used to value assets and liabilities) implies that balance sheet volatility becomes increasingly important within the currently more volatile and unpredictable economic environment. This requires the use of more sophisticated and value-risk-based techniques, to de-risk balance sheets by shifting more risks to policyholders and third-party asset managers, reducing the level and cost of guarantees while improving ALM (targeting a certain level of volatility while keeping a material exposure to achieving the upside).

In this respect, many insurers currently conduct their strategic planning with a key focus on traditional top-line revenue and bottom-line profitability growth metrics, to be accompanied by additional risk-based metrics that clearly identify the trade-off between shareholder value (e.g. measured in terms of EV or value of new business) and risk (e.g., measured in terms of RBC requirements and return on capital). For new business in particular, life insurers will need to find the right balance between maximizing the top line (by selling products with attractive returns to customers but with potentially expensive financial options and guarantees) and optimizing capital (by selling products that are more capital-efficient but that may not be so attractive to customers). Ultimately, more emphasis is likely to be placed on recognizing diversification benefits (both product and risk) for a given line of business.

Strategic asset allocations also need to be revised, with, potentially, less focus on levels of asset returns and more emphasis on risk-based metrics. Management actions need to be tailored to better reflect management decisions under stress scenarios that affect the risks faced by the company, and, ultimately, to make allowance for this within the assessment of RBC capital.

4.1.2.1.4 Tailoring the Product Design to the IFRS 17 Framework Whereas the previous standard, IFRS 4, allows insurers to use their local GAAP, IFRS 17 defines clear and consistent rules that will significantly increase the comparability of financial statements. For insurers, the transition to IFRS 17 (effective for annual reporting periods beginning on or after 1 January 2023) will have an impact on financial statements and on key performance indicators.

The main differences between the IFRS 17 and the local statutory, US GAAP or Solvency II frameworks are related to the following areas:

- measurement of insurance contracts;
- level of aggregation;
- risk sharing;
- recognition of onerous contracts;
- reinsurance;
- deferred acquisition costs;
- revenue recognition;
- accounting treatment of income on day 1;
- measurement of options and guarantees;
- separation of embedded derivatives within insurance contracts; and
- presentation and disclosure.

4.1.2.1.4.1 IFRS 17 Introduces a More Transparent Presentation of Profit and Loss Under IFRS17, the general model requires the measuring of an insurance contract at initial recognition at the total of the fulfilment cash flows (comprising the estimated future cash flows, an adjustment to reflect the time value and an explicit risk adjustment for non-financial risk) and the contractual service margin (CSM).

The CSM is a component representing the unearned profit that the entity will recognize as it provides services under the insurance contracts, while the day 1 losses are recognized immediately. Regarding a potential change in future cost assumptions, it will also be reflected in the performance profile of subsequent years. In contrast, under Solvency II or US GAAP, day 1 gains and losses are recognized immediately, and there is no deferral of acquisition costs.

For insurance contracts with direct participation features, under IFRS 17 the variable fee approach (VFA) applies as a variation on the general model. Among those contracts are unit-linked contracts (including variable annuities and certain equity-linked contracts), European 90/10 contracts and UK with-profit contracts – in contrast with term and whole life, US universal life, fixed immediate annuities and many endowment contracts (as there are often no underlying items identified), which use the general model.

Unlike many current accounting models that develop a single "best estimate", under IFRS 17 all scenarios and their associated probabilities should be considered and weighted. These are expected to be reviewed at each reporting date. This is in contrast to the US GAAP, for which the original assumptions used for future policy benefits are locked at contract inception and held constant over the term of the contract (except for long-duration contracts).

The most common methods for measuring financial options and guarantees are the stochastic "risk-neutral" and "real-world/deflator" methods, consistent with the cost of hedging. There are alternative "real-world" stochastic methods, used today under local statutory accountings or US GAAP, whereby some asset classes (such as equity instruments and real estate) are assumed, based on historical market averages, to outperform fixed income asset classes. These "real-world" methods are not permitted under IFRS 17, because financial options and guarantees would not then be measured consistently with observable current market prices.

Under IFRS17, investment returns are not included in the cash flows used in measuring the insurance liability. In contrast, under US GAAP, for long-duration contracts, the interest assumptions are still based on estimates of investment yields (using low credit risk fixed-income instrument yield) expected at the time the insurance contracts are issued.

While under US GAAP most insurers use discount rates based on the assets held by the insurer, under IFRS17 the discount rate can be determined using either of the following two approaches.

- In the bottom-up approach, it is computed by adding a liquidity risk premium to the risk-free rate (based on a reference portfolio). The illiquidity adjustment reflects the fact that policyholders often either cannot terminate insurance contracts at all or can terminate them only subject to surrender penalties. It is different from Solvency II, which relies on the "matching adjustment" and the "volatility adjustment", while IFRS 17 requires the maintenance of duration matching with liabilities.
- Or there is the top-down approach, whereby it is computed by removing credit risk from the yield to maturity (based on a reference portfolio).

In practice, the two discount rates will often be different. An entity is required to use one of the two approaches, and is not required either to calculate or to reconcile the result to the approach not used.

Regarding the risk adjustment for non-financial risks, IFRS 17 does not require entities to use any specific technique (confidence level, conditional tail expectation [CTE] or cost of capital). Expected changes in revenue are recognized, thereby impacting the pattern and emergence of profit. It is distinct from and lower in size and sensitivity than the risk margin under Solvency II, the market value margin (MVM), as it captures only non-financial risks and does not reflect the risks that do not arise from the insurance contract (e.g.

operational risk). Under US GAAP, such provision for the risk of adverse deviation is applied only to traditional life long-duration contracts.

Regarding the level of aggregation of contracts, IFRS 17 subdivides each portfolio into groups of (i) contracts onerous (if the total of insurance acquisition cash flows, cash flows occurring on initial recognition of the insurance contracts and fulfilment cash flows result in a net cash outflow) at initial recognition, if any, (ii) contracts at initial recognition having no significant possibility of becoming onerous subsequently, if any, and (iii) the remaining contracts in the portfolio, if any. Contracts issued more than one year apart shall not be included in the same group. Insurance contracts can become onerous when adjustments to the contractual service margin exceed the amount of the contractual service margin, which must be recognized immediately in profit or loss. In contrast, US GAAP indicates that insurance contracts shall be grouped consistent with the entity's manner of acquiring, servicing and measuring the profitability of its insurance contracts. Moreover, US GAAP does not prohibit accrual of a net loss (i.e. a loss in excess of deferred premiums) that probably will be incurred on insurance policies that are in force, provided that the loss can be reasonably estimated.

Regarding deferred acquisition costs, IFRS 17 defines insurance acquisition cash flows as cash flows arising from the costs of selling, underwriting and starting a group of insurance contracts that are directly attributable to the portfolio of insurance contracts to which the group belongs, including cash flows that are not directly attributable to individual contracts. And they are reflected as an immediate cash outflow at the start of the contract boundary. In contrast, US GAAP defines acquisition costs as those that are related directly to the successful acquisition of new or renewal insurance contracts. And such costs are deferred and amortized.

Regarding revenue recognition, under IFRS 17 entities are required to report as insurance revenue the consideration for services on an earned basis, while under US GAAP long-duration contract premiums are recognized as revenue over the premium-paying periods of the contracts when due from policyholders.

Overall, liabilities are expected to be less volatile under IFRS than under Solvency II. This also means that companies that manage risks under IFRS 17 are likely to see more volatile Solvency II results. Even so, these various impacts may profoundly change the presentation of profits to shareholders. As importantly, CSM is a key driver in the timing of profit recognition under IFRS 17 and the reason for more granular tracking of liabilities' movements over time, and it is expected to drastically influence the product design, as illustrated in the next section.

4.1.2.1.4.2 The Impact of IFRS 17 on Product Design as opposed to Solvency II

4.1.2.1.4.2.1 The Distribution Pattern of Profits under IFRS 17 against Those Implied by Solvency II IFRS 17 represents a fundamental change in accounting rules, and insurers may be unsure of how, or even whether, their product pricing and design may change as a result of potential changes in the new business value, here computed as the probability-weighted discounted value of "real-world" distributable profits, with the "distributable profit" computed as the change in distributable NAV (which is the excess of available capital – own funds from the Solvency II balance sheet – over required capital – i.e. target solvency ratio × SCR under Solvency II).

By "real-world" distributable profits, we mean that the projection of distributable profits will use real-world (as opposed to risk-neutral) investment returns. This is a key point, as, given that the shareholders' required return is used as the discount rate, the difference between projected investment returns and discount rate will mean that the NBV is sensitive to the timing of the emergence of profit. This is an important point for participating savings businesses in particular, whose viability often depends on real-world investment excess profits over the risk-free rate (so-called "uplifts", ignored under risk-neutral approaches). Hence, in this case, it is not just the timing of the projected future profit stream that is different but also the assumed size of the stream.

This is particularly so under the VFA approach, in which these real-world uplifts may not be recognized fully in one shot but spread out over the current and future service period. Thus the impact of real-world uplifts may be deferred compared with the impact under Solvency II. However, the picture may be complicated, as distributable profits may initially emerge more slowly under SII (e.g. because the SCR for market risk is higher than the IFRS 17 CSM), but then gain pace over time. However, at the same time, IFRS 17 VFA profits will start to be held back by gathering the unamortized real-world uplifts whose recognition is spread over future service.

Nonetheless, if the required capital is much higher than the CSM (which itself could be due to investment in risky assets), IFRS 17 may not necessarily defer profits overall compared with Solvency II. In contrast, the CSM will decelerate the emergence of distributable profit whenever CSM > required capital, because SII (before required capital) and IFRS 17 (before CSM) are assumed to recognize a similar level of future margins at the outset of the contract.

If we make the further simplifying assumption that the CSM and the required capital run off at the same rate, then this is equivalent to saying: initial CSM > initial Solvency II required capital. Given that both Solvency II and the IFRS 17 CSM align with market-consistent valuation methodology, initial CSM is close to initial Solvency II available capital. So the condition above for IFRS 17 to act to delay the emergence of distributable profit becomes: initial Solvency II available capital > initial Solvency II required capital. In other words: if a contract generates a surplus on a Solvency II basis at the outset, then the introduction of IFRS 17 may broadly be expected to constrain distributable profits.

This then leads to the following high-level conclusions: for products with low profitability (market-consistent new business value: MCNBV) and/or high capital consumption (Solvency II), IFRS 17 will not add any constraint beyond limits on distribution due to Solvency II; for products with high profitability (MCNBV) and/or low capital consumption (Solvency II), IFRS 17 could lead to a material constraint on the emergence of profits compared to the distribution pattern implied by Solvency II.

Overall, the distributable profit stream for new business with either low profitability – assuming that the contracts are not (or do not become) onerous under IFRS 17 – and/or high capital requirements is unlikely to change due to the introduction of IFRS 17. Nevertheless, IFRS 17 could act as a constraint on the timing of profit distribution in the case of products that have high profitability and/or low capital requirements. However, when real-world investment profits are an important contributor to profitability, in some cases this may make IFRS 17 more inclined to slow down distributable profits than would otherwise be the case.

Given this impact on the emergence of profits compared to the distribution pattern implied by Solvency II, the main question is the extent to which IFRS 17 will alter the expected pattern of profit distribution to shareholders, and hence impact the product design.

4.1.2.1.4.2.2 The Impact of IFRS 17 on Product Design for Insurers under the Solvency II Capital Requirement Framework For traditional savings products with profit sharing, interest guarantees imply low margins on a market-consistent basis in current financial conditions (i.e. low initial CSM), but in some cases may generate a significant capital requirement (i.e. high SCR). In this case, limiting dividends to IFRS 17 profits would not normally be expected to create a constraint compared with a Solvency II basis.

However, a key component of value under Solvency II come from investing in "risky" assets, as shareholders' share of real-world "uplifts" translates into increased own funds each year (as assets increase by more than technical provisions, particularly if guarantees are in the money). However, this effect will be partly offset by the cost of additional capital requirements. Considering a single ten-year premium with annual profit sharing equal to the investment return less the financial margin, subject to minimum interest guarantee, the VFA of IFRS 17 acts to hold back profits generated by the real-world uplifts in each year, and so constrains the NBV.

For a unit-linked product without investment guarantees, a good level of profitability implies a high initial CSM. The SCR will depend on the asset mix (e.g. level of equities) and the lapse risk, but, when required capital is comparatively low and profitability high, IFRS 17 may constrain distributable profits compared with

Solvency II, given that the high CSM will act to hold back the emergence of profit to a greater extent than the RC will.

- Increasing the initial premium charge increases the profitability but the required capital is relatively unaffected and remains low, which does not therefore act to hold back distributable profits under Solvency II. However, we now have a high CSM, and this will act to hold back IFRS 17 profits, and so may constrain distributable profits materially.
- Increasing the annual management charge increases the profitability and also the required capital, because (unlike an up-front charge) an annual management charge is affected by lapse and market shocks. With the same high CSM as previously, we can expect IFRS 17 profits to be held back over the projection. However, with RC now higher, it will also act to hold back distributable profits on the Solvency II basis, and hence, while we can expect IFRS 17 to constrain distributable profits compared with Solvency II, this will be to a lesser extent than observed for the adjustment to the initial charge.
- Increasing the surrender penalty increases the profitability but the required capital increases even more than the previous case (due to the loss of very high surrender penalties arising from lower lapses throughout). With RC now approaching a similar magnitude to the CSM, we can expect profits to be held back to a similar extent over the projection under Solvency II compared with IFRS 17. IFRS 17 does have an effect on NBV (3,206/3,348 = 96%), but this is not nearly as pronounced as that given by the other forms of adjustment.

4.1.2.2 … With respect to Consumer Protection

A shift in focus in financial regulation can be observed within the European Union in recent years. Regulators are turning their attention towards market conduct regulation with a focus on retail investor protection. The objective is to strengthen the rules applying before, during and after the sale of financial products within the European Union, building on the initiatives of the three European supervisory authorities (European Banking Authority: EBA; European Securities and Markets Authority: ESMA; and European Insurance and Occupational Pensions Authority: EIOPA) that have made consumer protection their key priority.

In addition, within the context of the persistently low interest rate environment, the return of investment on savings and retirement has drastically decreased, which puts into question the sustainability and credibility of these businesses from a customer appetite perspective.

This has raised increasingly strong attention from regulators, who have reached the following notable conclusions.

- In order to facilitate retail investors' access to investment products, the interests of retail clients should be promoted and their needs met, in particular through a product governance scheme, by which investment firms must analyse the products they need to design or distribute, especially by assessing their target markets and the typology of their customers.
- For European citizens to be encouraged to invest, they must benefit from information that is relevant, fair and proportionate; in this respect, work should be done to adapt obligations regarding existing data in order to make them more useful and adapted to the needs of investors, and therefore more understandable.
- To be attractive to retail investors, investment products must also be profitable, which calls for the introduction of the concept of "value for money".

Two directives and one regulation proposed by the European Commission aim to reinforce consumer protection in the insurance sector and ensure that the entire insurance product life cycle targets the consumer's best interests.

1. Revision of the Insurance Mediation Directive (IMD 2) was proposed by the Commission to address the fragmentation of EU insurance markets. It carries the biggest impact for the sector, as it focuses on insurance exclusively and reinforces the rules applying to the development and sale of insurance products in order to protect policyholders.
2. The IDD (Insurance Distribution Directive) focuses on the means and channels of sale, mainly. It increases expectations and requirements, including guarantees or sanctions, for those selling insurance products. The IDD sets as well the basis for POG (product oversight and governance) requirements, which will demand that insurers build a product approval process, including target markets, for the approval of each insurance product, ensuring that the right product is sold to the right person.
3. The revision of the Market in Financial Instruments Directive (MiFID 2) was undertaken after the crisis to make financial markets more efficient, resilient and transparent, and to strengthen investor protection for a wide range of financial products and instruments. Insurance was not in the original scope, but, given the delay in the legislative process for IMD 2, European legislators decided to use MiFID 2 to set up a first set of requirements for insurance-based investment products, essentially covering conflicts of interest.
4. The Regulation on Key Information Documents for Packaged Retail and Insurance-based Investment Products (PRIIPs) is a new initiative taken after the crisis to protect retail investors against excessive complexity on the part of financial products. It sets up requirements to disclose key information about financial products to investors before the sale, standardizing the way information has to be shared and communicated, so that the consumer will be best positioned to understand what she buys, its main features, the costs and expected return and the underlying risk.

The three texts cover two main categories of issues.

- First are rules on the development and design of insurance products, under the denomination of product oversight and governance.
  - Product approval processes are to be developed by firms, which have to specify an identified target market of customers, identify the risks associated with the product and check whether the intended distribution strategy is consistent with the target market.

Remark: however, it is essential to ensure that the POG requirements stay at a high level and do not include overly detailed requirements, as supervisors should not be in the driving seat for the product launch. There is indeed a risk that an over-prescriptive POG stance could kill any innovation, which would negatively impact customers

  - There must be an ongoing product review, which has to ensure that the product continues to meets its initial objectives and that it is actually distributed to the identified target market, to ensure that consumers' interests are taken into consideration throughout the life cycle of a product.
  - The EIOPA has published 13 policy proposals, among which are:
    - governance with sign-offs (by CRO, CFO and local executive committee);
    - target market awareness: identification of groups of customers for whom the product is not suited is not considered today;

- product testing: a "customer-centricity" document and testing the adequacy of the product with a quantitative criterion;
- in-force product monitoring; and
- distribution channels: processes/controls to be put in place to prevent misselling of the product and ensure the adequate alignment of distribution with marketing goals and customer protection.

- The other main category concerns rules on the distribution of insurance products:
  - requirements concerning intermediaries' qualification, notably including ongoing training requirements;
  - rules regarding the sales process (a duty to inform and advise clients as to the suitability of the products to their needs);
  - measures to avoid conflicts of interest between clients, intermediaries and undertakings; and
  - rules on intermediaries' remuneration, including enhanced transparency and restrictions made on certain commissions.

Two main issues have been identified.

1. Enhanced transparency on the distribution costs of the products, including on the remuneration of intermediaries. The aggregated costs solution has been adopted in the final outcome of PRIIPs, but the European Parliament advocates full transparency.

Remark: the customer needs to be able to understand the cost of the "advice" she is receiving, so that an assessment can be made as to whether that advice offers value for money. This entails full transparency around the actual commission being received.

Remark: intermediaries should definitely provide information on their relationship with the product provider (tied, independent) in order to avoid conflicts of interest. Such requirements should be given a large scope to cover the diverse forms of insurance intermediation in Europe, including direct sales and price comparison websites on the internet. This would allow levelling the playing field between channels and thus reinforcing fair market competition in the client's favour. However, rules to avoid conflicts of interest should not be network-blind. In particular, brokers and tied agents should not be submitted to the same rules, because, whereas a broker acts on behalf of his client, a tied agent represents the interests of the insurance company.

2. Restrictions on the potential for third parties to remunerate intermediaries.

Remark: disclosures of intermediaries' remuneration and bans on commissions should be carefully weighed and their impact assessed beforehand, notably in view of avoiding the unintended consequences of the Retail Distribution Review (RDR) in the United Kingdom, which banned commissions from third parties for independent intermediaries. This caused the number of independent advisers in the United Kingdom to plummet (there was a 23% drop in the number of intermediaries in 2013, with 60% of the remaining advisers expecting greater use of restrictions) and the cost of advice to rise. The rise created an "advice gap" between well-off investors, who know the value of advice and are ready to pay an increased price for it, and the majority of others, who have given up on advised sales. Therefore, policy-makers should avoid a "one size fits all" approach on distribution, and distinguish in particular between tied agents and brokers (which is currently not always the case). As a consequence, restrictions on commissions should be left as a choice.

Overall, there are multiple opportunities and risks within enhanced consumer protection (see Figure 4.7).

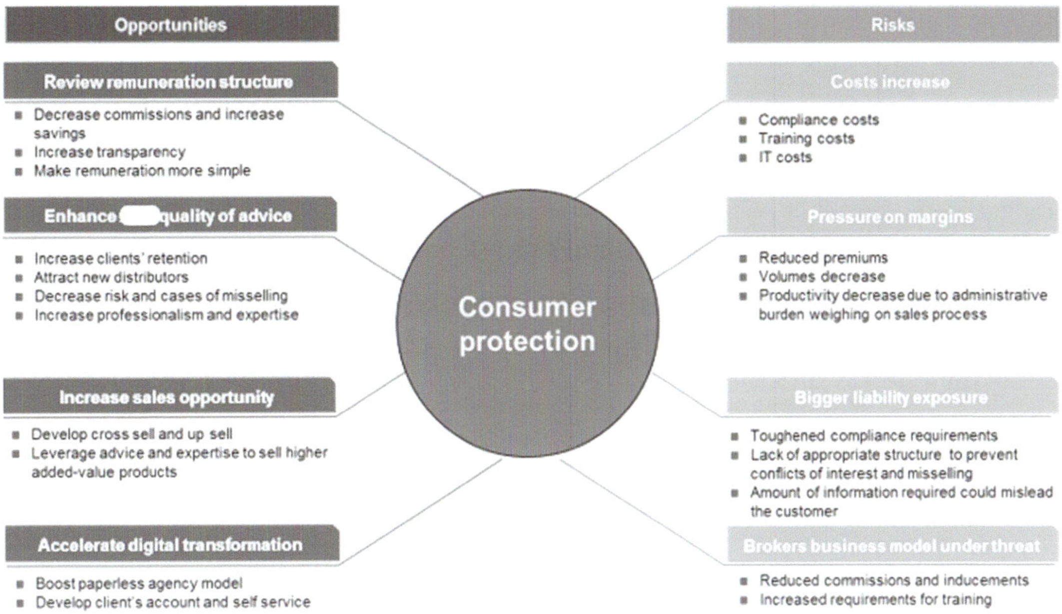

FIGURE 4.7 Opportunities and risks in enhanced consumer protection

4.1.3 Renewing the Life Insurance Proposition

Life insurance companies should review their strategies. As a prerequisite, though, it is critical that management teams reject the usual incremental approach to strategic planning and consider steps such as the following.

- Work with scenarios that extend beyond the current capital market consensus (such as spikes in rates or lapses, normalization of credit spreads, or exits from the Monetary Union).
- Perform a structured review of modelling choices and actuarial assumptions in the SII balance sheet to better reflect risk profile and release capital when current assumptions are inappropriate or overly prudent (e.g. management actions embedded in BEL, calibration of internal model).
- Put all options for improving the economics and balance sheet of the in-force business on the table. These include buying out legacy contracts and up- or cross-selling capital-light products, deferred annuities, pure savings and loans. Consider closing the general account for new business to lock in earnings and protect solvency.
- Reducing interest rate exposure through hedging will reduce solvency sensitivity to changes in rates and free up capital, which can be used to bolster capital generation through higher-yielding investments. Firms that rely on derivatives to manage their interest rate exposure will also have more discretion in terms of the duration profile.
- Strengthen distribution, service and data capabilities. Make a step change in productivity, reviewing all options to outsource pieces of the value chain, address overheads and simplify the IT architecture and legacy tariffs. Insurers must accelerate their ability to leverage technology to deliver new propositions at low cost so as not to erode already thin returns and maximize value to customers.
- Upgrade efforts to improve customer relevance and engagement, including better capturing and converting of leads, providing personalized, pre-underwritten offers and giving digitally enabled advice and guidance. Companies first need to get closer to their customers and provide guidance and advice to help them navigate the complexity.

Explore opportunities for diversification into health, wealth and beyond, towards a lower exposure to interest rate swings, such as the following.

- Offer unit-linked with-protection policies, where bundled risk coverages can be an attractive source of differentiation for insurers, and riders related to term life or disability risks can satisfy customer demand for capital protection, while offering much lower capital intensity. More sophisticated solutions tailored to specific customer needs, such as educational benefits, can constitute additional value-added services.
- But the product volume is, by nature, restricted, and margins are shrinking due to heightened competition. For example, consider the 20 to 25% decline of premiums in Germany, Switzerland and the United Kingdom from 2010 to 2015.
- Offer accumulation products with decumulation benefits: residential and healthcare benefits have grown – for example, for Chinese insurers. Policyholders can opt in on personal needs and health conditions during the decumulation phase.
- Insurers can lower the terms of new policies (e.g. by lowering guaranteed rates), thereby progressively lowering future payouts and mitigating compression of the credit spread, by, for example, decreasing crediting rates on in-force universal life policies, decreasing dividend rates on whole life policies or increasing the costs of insurance (monthly cost of insurance protection based on age, gender, health benefit and death benefit).

From a new product point of view, life insurers have been responding to the challenges of persistently low interest rates squeezing already tight margins and making it difficult to cover rising capital requirements from prudential regulation by rethinking their product offerings to less capital-intensive products. New (capital-light) investment solutions that provide yield in accumulation and income and low volatility in retirement are essential. These require new investment capabilities in multi-assets, real assets and novel solutions that smooth volatility and generate income in retirement.

They are gradually shifting towards “capital-light” products, such as unit-linked (UL) products or products with lower or no guarantees, or hybrids as a combination of traditional savings and unit-linked features, thus transferring part of the investment risk to policyholders, in an attempt to consume less in the way of capital requirements.

In this respect, life insurers should consider several tactical levers:

- applying additional criteria to regulate the underwriting of new guaranteed products;
- reviewing pricing policies to better account for capital guarantee options (e.g. lapse risk is usually mitigated by surrender penalties, market value adjustments and tax advantages);
- revisiting investment possibilities and asset allocation policy;
- improving the mechanisms of non-guaranteed bonuses: reductions in the volatility thanks to managed-volatility funds (see later); cuts in policyholders’ dividends; changes in profit-sharing discretionary rules; and
- reducing savings guarantees: reduced guaranteed insurability rider (GIR) on the in-force portfolio; lower implied guaranteed rates; lower technical interest rates.

Selling more sustainable new business can improve the margin if the guaranteed rate for the new business decreases more than the decrease in investment yield. In fact, since the third quarter of 2014 low interest rates, combined with the imminent implementation of Solvency II, have generated steep drops in the capital ratios of European life insurers, often reaching levels just above 100%. However, insurers that pulled these tactical levers have often regained 40 to 70% of such drops – even though their impact depends on interest rate fluctuations, portfolio characteristics and insurers' effectiveness in executing such levers.

However, while this may help reduce product risk, there are obstacles to the transition to a less capital-intensive product model.

- Market dynamics, as well as policyholder needs and expectations, may slow the uptake of capital-light products.
- The shift towards capital-light products is slow to implement, and legacy guarantees still make up most of the insurers' technical liabilities. Heavy legacy books, which are embedded with guarantees, will continue to weigh insurers down. It will probably be years, or even decades, before insurers see product moves translate into technical provisions with reduced guarantees in their legacy books and stronger risk profiles. Beyond a certain timing, book yields become significantly lower than market yields, and unrealized gains would be too low to bridge this gap.

  Let us highlight the case of Germany, with by far the highest average guarantee rate, with only a slight decrease in recent years (about 0.05 to 0.15% per year), in contrast to the strong decline in interest rates: the guarantee rate has been below the zero-year rate since 2010.

  Moreover, guarantees, which account for a half of business, are usually lifelong and do not have repricing options in the accumulation phase, and have to be provided on a yearly basis (and not only at the maturity of the contract, as is the case in the Spanish life insurance market, for instance). This, combined with the long maturity of insurance contracts (19-year durations, double the European average) relative to the maturity of available fixed-income instruments, implies a very significant and persistent duration gap, and so interest rate risk.

  Finally, the nonlinear nature of guarantees means that the value of liabilities is highly sensitive to yields as the investment return approaches the level of the guarantee.
- Reducing exposure to earnings volatility stemming from inherent product features is likely to be a challenging long-term game.

Moreover, by leaving customer demands unanswered, distribution channels in this environment run the risk of losing relevance. In addition, the value proposition for pure unit-linked products can seem weaker by measures such as customer costs and payoffs compared with asset management products. In this respect, there is the need to screen sustainable opportunities balancing shareholders' against customers' interests.

4.1.4 Screening Sustainable Opportunities Balancing Shareholders' versus Customers' Risk Appetites

Many life insurers are at a strategic crossroads, facing new options and decisions about where to invest and grow the portfolio, where to harvest and which areas to exit, and thus how to achieve the optimal product mix; and where to participate along the value chain, and which customer segments and countries to target. To successfully execute these choices, they must identify which capabilities are critical to develop and master, and which can be "good enough".

Whatever their choices, insurers must do more than get the business basics right. They also must clearly communicate a compelling story to investors, so that investors understand why and how the business is performing and can gauge the model's sustainability. In fact, since the financial crisis, investors have increasingly baulked at the opacity of life insurance financials, because much of what gets booked as earnings is subject to assumptions. Investors are concerned about some insurers' legacy exposure to soured businesses such as long-term-care insurance and variable annuities. The low interest rate environment will make it even harder to create acceptable levels of distributable cash and returns on capital. As a result, choosing the right metrics to manage and communicate will give investors confidence that an insurer has made the right strategic and operational decisions.

In order to sustainably update their strategies by meeting both shareholders' and customers' risk appetites within the persistently low interest rate environment while complying with the growing need for transparency from investors and regulations, life insurers need to introduce screening for sustainable opportunities balancing shareholder and customer interests, through dedicated metrics focusing not only on capital risk but also on a customer value basis, and this at a granular – i.e. product – level.

As such, the measurement of return on capital and customer experience becomes a crucial part of governance, driving decisions to move capital away from underperforming to high-achieving products or business sectors – and, when possible, giving back to shareholders and lenders.

4.1.4.1 New Business and In-Force Capital-Based Metrics at Product Level

Life insurers need to succinctly communicate how they will create value for shareholders and customers alike.

More investors want to understand a company's approach to intrinsic value and non-traditional metrics such as free cash flow, not just the traditional metrics of revenue growth, return on investment and dividend growth. This explains why some insurers now incorporate cash metrics into their internal management and external communications. Prudential UK, for example, has made a string of structural changes to grow in a capital-efficient manner, and has spent considerable time educating investors on differing capital flows for new and in-force business.

Life insurers have made capital management a fundamental part of their presentation to investors, in order to illustrate how well they manage their increasingly expensive capital much more effectively. Most provide regular updates on the progress achieved year on year, emphasizing the implementation of risk-adjusted measures and new ways of managing, for example, life in-force portfolios, or optimize portfolio initiatives, such as exiting riskier businesses, investing in capital-light opportunities, thinning out product portfolios and reviewing new business.

How to measure the productivity of capital involves three perspectives whether we look at new business or in-force portfolios:

- the return on risk capital, to measure profitability against the volatility of expected results;
- the capital efficiency/intensity ratio, at the finest possible granular level (business unit or product level); and
- the wealth consumption measure, to track profitability against actual invested capital at the finest possible granular level (business unit or product level).

The two perspectives are complementary, and equally important for disaggregating and assessing business unit performance.

4.1.4.1.1 Solvency Internal Rate of Return

4.1.4.1.1.1 Definition The internal rate of return is an indicator used to measure the return on an investment, and it can be used to assess the viability of an investment by comparing it with the return on similar investments, or the cost of capital of the investment itself.

It is defined as the discount rate at which the net present value of all cash flows from the investment is equal to zero. For an investment incurring a cash flow in each period t, the IRR is the solution of the following equation:

$$\sum_{t=0}^{T} \frac{\text{Cash Flow}_t}{(1+\ \text{IRR}\ )^t} = 0.$$

For a typical investment with negative cash flow in the initial period representing the initial investment, the IRR can be understood as the discount rate at which the present value of future cash flows equals the initial investment:

$$\text{Initial investment} = -\text{Cash flow}_0 = \sum_{t=1}^{T} \frac{\text{Cash flow}_t}{(1+\ \text{IRR}\ )^t}.$$

The definition of the IRR in insurance follows the same logic, with the cash flows reflecting the specificities of the life and savings business.

- The initial cash flow corresponds to the initial investment made to sell new business. It consists mainly of expenses related to acquiring the new business (known as "strain"), plus an initial amount of required capital.
- The future cash flows (known as distributable earnings or free cash flows) are composed of the statutory earnings generated from the business, the income on required capital and the change in required capital.

The solvency internal rate of return, therefore, is the discount rate at which the present value of distributable earnings is equal to the strain and initial required capital invested to support the business underwritten. It is thus defined as the solution to the following equation:

$$\sum_{t=1}^{n} \frac{\textit{Distributable earnings}}{(1+IRR)^t} = \textit{Strain} + \textit{Initial required capital}.$$

- The strain represent the first year's acquisition cost of the new business (i.e. the strain of acquiring the new business, including the acquisition and administration costs, plus the initial required capital);
- Distributable earning_0 = Statutory earning_0 - Required capital_0
  Distributable earning_t = Statutory earning_t - Required capital_t + Required capital_{t-1},

  including the interests and charges on the required capital;
- The statutory earnings are, therefore, the profits net of taxes, compliant with the local accounting standards.
- The required capital is the so-called hard capital = max (target SCR – soft capital, floor) – i.e. the amount of capital on top of the soft capital (which represents the unrealized profits) in order to reach the target SCR.

  The so-called "soft capital" is $\mathbf{VIF_t - MVM_t}$

  Thus

$$\textbf{Required capital}_t = \textit{Hard capital} = \textbf{Max}\left(\textit{Target SCR} - \left(\mathbf{VIF_t - MVM_t}\right), 0\right).$$

✓ **Quick win**: the MVM is to be compared with VIF (thus looking at the difference VIF – MVM, which makes a significant portion of the AFR). MVM is significant for general account savings, translating into potential negative AFR. Even so, a significant MVM can be compensated for by high VIF, resulting in a positive AFR, as in protection and health and UL policies.

✓ **Quick win**: in terms of capital leverage, the MVM is to be positioned against the SCR: a high MVM relates to a high SCR, maturity of business run-off (more than 15 years) and projection of future premiums (e.g. Switzerland, Hong Kong, Germany, Japan).

The amount and timing of distributable earnings are based on four key components:

- the first year's investment (i.e. strain to acquire new business, including acquisition and administration costs, plus initial required capital);
- statutory earnings (SE, including investment return on capital), which are usually negative in the first year due to the high investment then;
- the release of required capital (RC), which is also negative due to the initial capital put up to meet regulatory requirements; and
- the financial and non-financial assumptions underlying the projection.

The solvency IRR may be computed on a base case scenario or on a stress scenario (e.g. market crash or economic deflation).

The impact of a change in the regulatory capital requirement framework may be significant, depending on the magnitude of risks: for example, due to the low-risk nature of pure unit-linked products, the AFR is sufficient to cover the majority of the SCR, implying a lower hard capital requirement under Solvency II, while the Solvency II IRR is higher than the Solvency I IRR.

Another indicator that complements IRR is the payback period, defined as the length of time required for the accumulated cash flows to become positive (i.e. pay back the initial investment). The payback period ignores the time value of money. The payback period is hence the smallest period t∗ satisfying these conditions:

$$\sum_{t=0}^{t^*} \text{Cash flow}_t \geq 0. \quad \sum_{t=0}^{t^*} \text{Distributable earnings}_t \geq 0.$$

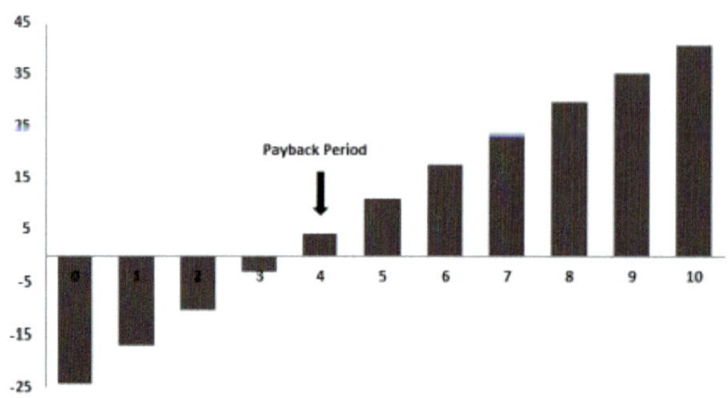

FIGURE 4.8 Payback period

✓ **Quick win**: loading on payments makes it possible to significantly increase the IRR without limit from the point of view of the insurer. However, it must not be too important for the product to be attractive to the customer. It is possible to present a large IRR with characteristics such as loading on payments that impact only the first cash flow without the product being really profitable thereafter. It may then be interesting to look with a large IRR that the payback return period is not too low.

4.1.4.1.1.2 Application The solvency IRR can be used to assess the viability of an investment by comparing it with the return on similar investments, or the cost of capital of the investment itself.

The solvency IRR can be improved by changing product features. For example, regarding unit-linked products, these are possible options:

- increasing the insurance policy amount;
- changing the charge structure; and
- introducing surrender charges.

Let us illustrate on a basic UL product with the following features (see Figure 4.9).

Basic Product Feature

Premium Type	Single Premium
Policy term	30 years
Benefits on Death	Account Value
Benefits on Surrender	Account Value
Asset Management Charge	1% p.a. of account balance
Premium Charge	5%
Surrender Charge	N/A
Partial withdrawal charge	N/A
Basic Commission	5%
Fund allocation	70% Equity : 30% Bond

FIGURE 4.9 Basic product features

We consider the following product design changes:

- tweak 1: increase policy amount by 50%;
- tweak 2: + AMC (asset management charge), charged as 1% of initial premium;
- tweak 3: + include surrender charge (SC) of 5% in first five years; and
- tweak 4: + decrease the lapse rate in first five years (due to SC);

resulting in the following impacts on the Solvency II IRR (see Figure 4.10).

Baseline	Tweak 1	Tweak 2	Tweak 3	Tweak 4
-1.6%	-1.4%	-1.3%	-1.0%	-1.1%
33.8%	43.0%	38.0%	88.1%	68.1%
4	4	4	3	3

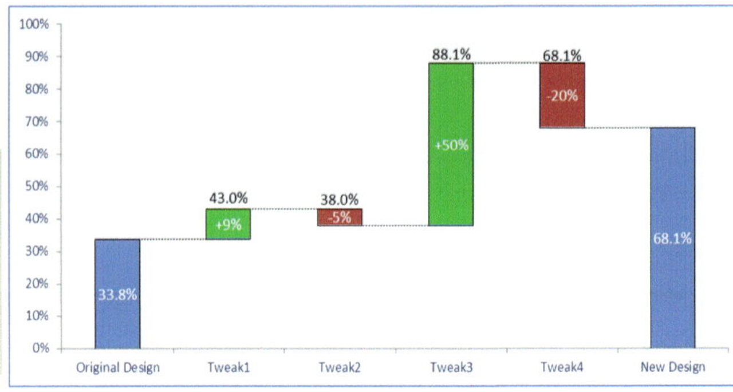

FIGURE 4.10 Impact of design changes on Solvency II IRR

Due to the low-risk nature of the pure UL product, the Solvency IIIRR is usually higher than the Solvency I IRR.

The AFR (VIF – MVM) is sufficient to cover majority of STEC, leading to very small Solvency II hard capital requirement (lower than under Solvency I).

Beyond the sole use of solvency IRR and payback period, screening and benchmarking the quality of businesses are then possible through the chart of IRR versus coverage ratio (= AFR / SCR) (see Figure 4.11).

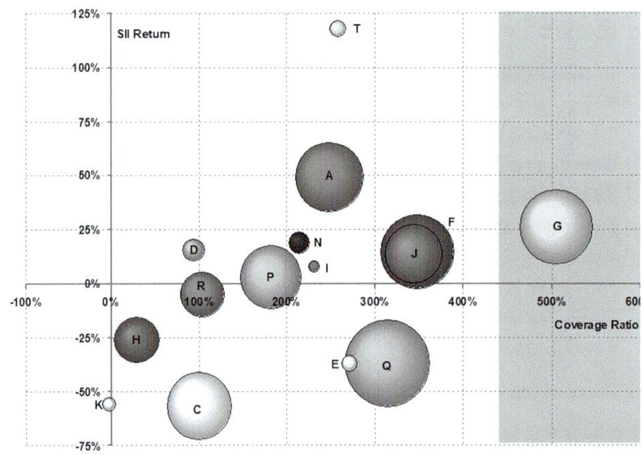

FIGURE 4.11 IRR versus coverage ratio

4.1.4.1.1.3 Limitations Figure 4.12 shows the limitations and advantages of IRR versus payback period metrics.

	Advantages	Limitations
IRR	▪ Simple to communicate externally, understood by analysts ▪ Meaningful to assess the return on invested capital	▪ Can have multiple solutions or be infinite/undefined ▪ The size of investment is ignored ▪ Mechanically assumes that cash flows coming out of the project are reinvested at the IRR ▪ Very volatile for a product entailing a high degree of path-dependent optionality
Payback Period	▪ Simple to communicate externally, and well understood ▪ Meaningful to assess the time length needed to recover the initial investment.	▪ Future cash flows after the payback period are ignored. This could be misleading when they are negative

FIGURE 4.12 Limitations of IRR and payback period metrics

4.1.4.1.2 Product Capital Efficiency/Intensity Among profitability/capital efficiency combined metrics are the following.

1. Short-term return = earnings / SCR.
2. Long-term return = (VIF – MVM – CoC × PVSCR) / PV(SCR); this reflects the long-term margin adjusted for the cost of capital required to finance the SCR. PVSCR provides the average view of the capital used over the maturity of the product.

   This accounts for the product profitability net of non-hedgeable risks, expressed as a percentage of capital requirements.

   The MVM deduction is justified by the fact that market risks may be mitigated by financial hedge assets, in contrast with non-hedgeable risks, which require reinsurance or product design changes.

These two KPIs are used to rank new business products across entities.

4.1.4.1.3 Product Financial Wealth Consumption Following the historical decline over the past ten years (in line with central banks' balance sheets) and the expected risk exposure to rising interest rates, the so-called "financial wealth", defined as "accumulated reserves + unrealized capital gains", is a strategic competitive advantage for savings and retirement businesses.

The wealth is also a decisive driver of the SCR that insurers try to decrease. Although financial wealth consumption increases future market risks, the financial wealth consumption metric amounts to the crediting rate + loadings – Solvency II interest rate, which is the sum of BEL + VIF (gross of tax) minus the cash value.

Two ratios can be tracked:

- financial wealth consumption (FWC) / present value of economic profit (PVEP) (to be compared to NBV / PVEP to make sure that the wealth is consumed by the most profitable products); and
- FWC / NBV, to identify the effectiveness with regard to value creation: close to or above 100% causes the value created to be questioned.

This latter ratio illustrates the amount of current wealth distributed to the new product as compared to the value it generates.

- The main challenge is that some products rely too heavily on the wealth consumption, such as pure general account products, whole life products or endowment with premium indexation, and some traditional life with high guaranteed rates. Among ways to mitigate are the following.
  - Lowering the financial guarantee.
  - Reducing the dependence on the investment income -> lower FWC -> lower FWC / NBV.
  - Promoting products with profits driven by the technical margin -> lower FWC & higher NBV -> lower FWC / NBV.
- In contrast, a pure UL product will consume zero current wealth and will have a ratio equal to zero.
- Regarding guaranteed products, a high wealth consumption / (BEL + VIF) ratio usually illustrates "deeply in-the-money options", and therefore a vulnerability to the financial scenarios in the portfolio.
- Using standalone pricing is also a way not to consume the financial wealth.

- Favouring bundles of savings + protection and health products, as the the wealth consumption of savings products can be more than offset by the wealth creation of disability products, as observed in Germany.

Remark: financial wealth consumption / NBV = (FWC / statutory reserve) * (statutory reserve / NBV) -> FWC / NBV is driven by the (statutory reserve / NBV) ratio, which makes sense from an economic perspective.

Regarding the in-force policies, the financial wealth consumption / VIF ratio is usually high if the profitability is low or, at best, if there is little possibility of an upside from a favourable financial scenario or if its profitability relies on financial margins, and consequently on the unrealized capital gains, meaning a high and increasing financial SCR, notably if rates remain low.

Even so, such financial wealth consumption metrics embed some limitations.

- Regarding assets, it measures the unrealized capital gains of the assets, so a decrease in the interest rate environment implies an increase in financial wealth consumption, even if the structure of liabilities does not change during the period. The link with liability management is therefore not so direct and clear.
- A short-term product could consume less financial wealth only because its maturity is short, so the financial wealth consumption / product maturity ratio needs to be analysed.
- Potential underestimates of the statutory reserves would artificially reduce financial wealth consumption.
- Financial wealth consumption / VIF may be biased, because VIF takes only future cash flows into account – i.e. does not consider margins realized in the past.
- A high financial wealth consumption / VIF ratio does not always mean low profitability.
- The link between financial wealth consumption and the sustainability of the credited rates is not straightforward.

✓ **Quick win**: in order *not* to be too conservative, the financial wealth consumption approach should consider a risk premium for risky assets, and thus come in line with an earnings approach and not be limited only by the risk-neutral approach of Solvency II, for example.

✓ **Quick win**: instead of defining thresholds for the ratios (as can be done for the capital intensity ratios), it is far better to monitor the ratios over time. The pace with which the ratio changes over time provides insights into recommendations for changes in the credited rate.

4.1.4.2 Introducing Customer Value Metrics for Consumer Protection

From a customer's risk appetite perspective, "client fairness" is a principle to ensure that distributed products provide sufficient "value for money" to the client. It is a key issue with respect to customer-centricity. Offering products with good "value for money" strengthens clients' trust in brands.

From an insurer's risk perspective, consumer protection is increasingly raised by regulations (MiFID), while it also serves to limit the potential for lawsuits, as illustrated by the spread of "class actions" over the world. In the context of digital activity, it is also a strong measure for reducing reputational risk. For example,

- for unit-linked products with guarantees, the upside potential should be sufficient, while
- A product whose gross margin is almost fully allocated to the insurer and the distribution would be hardly defensible in front of a judge

For savings and retirement products, a strong component of "value for money" from a client perspective constitutes return on investment. As a result, the core methodology relies on a "return on investment from a client perspective" (based on all the cash flows the customer receives). As products could combine death and life benefits, cash flows paid to the client should be weighted by the corresponding probability. Surrender rates should also be considered to value the "availability" of savings. As the client can receive payment in case of life (annuity, withdrawals, lump sum, surrender, etc.) and death, cash flows have to be weighted by their probability of occurrence:

$$CF_t \cdot = q \cdot CF_t^{death} + l \cdot CF_t^{lapse} + (1-q-l) \cdot CF_t^{life},$$

where q is the probability of death, l the probability of lapse and (1 – q – l) the probability of still being in the contract at the end of the period [t; t + 1]. The time 0 cash flow is negative and equal to the gross premium paid by the client (denoted P).

The return on investment (RoI) is then defined by solving the equation

$$0 = -P + \sum_{t=1}^{N} \frac{CF_t}{(1+RoI)^t}.$$

In the case of a standard one-year-horizon life insurance contract with a gross premium P that pays an amount K in case of the insured's death, the RoI is computed as

$$P = \frac{q \cdot K}{1+RoI} \Leftrightarrow RoI = \frac{q \cdot K}{P} - 1.$$

4.2 SUSTAINABLE LIABILITY DESIGN ENHANCEMENTS

4.2.1 Improving the Solvency IRR through Liability Designs: Focus on General Accounts under Solvency II

4.2.1.1 Solvency II versus Solvency I IRR

The internal rate of return is a metric that measures the profitability rate of an investment (acquisition expenses, capital requirement, etc.). The IRR is used in the decision whether to launch a product if it reaches a target defined by entities.

It corresponds to the discount rate at which the present value of distributable earnings is equal to the strain and initial required capital invested to support the business underwritten. It is thus defined as the solution to the following equation:

$$\sum_{t=0}^{n} \frac{Distributable\ earnings_t}{(1+IRR)^t} = 0,$$

$$\sum_{t=1}^{n} \frac{Distributable\ earnings_t}{(1+IRR)^t} = Strain + Initial\ required\ capital$$

where

- the strain represent the first year's acquisition cost of the new business;
- Distributable earning$_0$ = Statutory earning$_0$ − Required capital$_0$
  Distributable earning$_t$ = Statutory earning$_t$ − Required capital$_t$ + Required capital$_{t-1}$,

  including the interest and charges on the required capital;
- the statutory earnings are the profits net of taxes, compliant with the local accounting standards; and
- the required capital is the so-called hard capital = max (target SCR – soft capital, threshold) – i.e. the amount of capital on top of the soft capital (which represents the unrealized profits) in order to reach the target SCR.

The amount and timing of distributable earnings is based on four key components.

- The first year's investment (i.e. strain to acquire new business, including acquisition and administration costs, plus initial required capital).
- Statutory earnings (SE, including investment return on capital), which are usually negative in the first year due to the high first-year investment.
- Release of required capital (RC), which is also negative due to the initial capital put up to meet regulatory requirements.
- The financial and non-financial assumptions underlying the projection.

Moving from Solvency I to Solvency II implies a change in the capital requirement (IRR component) calculation: whereas the SI capital requirement depends only on the account value, the SII capital requirement depends on each risk.

In order to better apprehend the difference between the two solvency IRRs, let us first remind ourselves of the pattern of cash flow distribution for a general account product (see Figures 4.13 and 4.14).

FIGURE 4.13 Cash flow pattern under Solvency I versus Solvency II

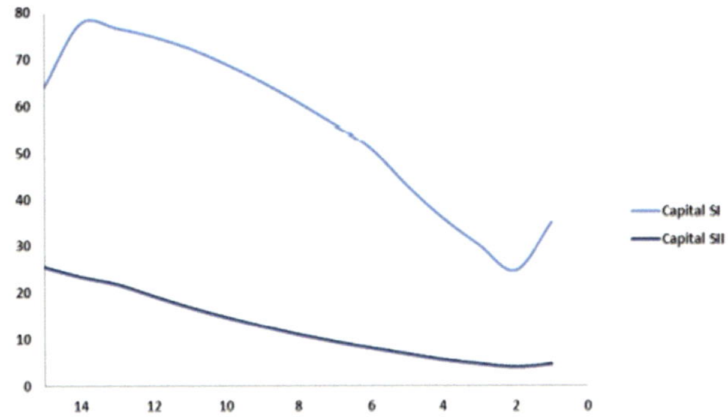

FIGURE 4.14 Capital versus residual maturity

Notes: The life table used is the French TH-00 02 and the redemption function used is constant by 5% each year.

The IRR is very sensitive to the first cash flows, since they are not very much discounted. Thus, the IRR is more important under SII, as the SII cash flows are much more favourable in the first two years than under SI, while the differences in cash flows in the following years are small. Despite the fact that the last cash flow is much more favourable under SI, it has little weight on the IRR because it is updated over 15 years.

To understand these differences in cash flows, we must look at the only component that varies between SI and SII: the required capital. Here is the evolution over time (years) of the capital under SI and SII.

- Under Solvency I, the capital requirement is very important at the beginning of the contract, then decreases sharply since the account value decreases following the various contract exits (redemptions and death). The two irregularities between the first two years and between the last two years are due to unrealized capital gains or losses (soft capital). The irregularity between the last two years makes it possible to justify the penultimate negative cash flow since it is necessary to reinvest capital and not to release it.
- In contrast, under Solvency II the capital requirement is much less important, and decreases, since the risks are greater at the beginning of the contract (because they impact all the years) than at the end of the contract (when they impact only the few remaining years).

The difference between SI and SII cash flows over the first two years is therefore explained by the fact that the SI capital is much larger than the SII capital. The following cash flows are then slightly more favourable to SI, since the SI capital decreases a little more quickly over time.

Given the strong dependence of SII SCR on risks, let us mention the risk components.

- Life risk losses make up just under three-quarters of the SII SCR. We can consider two types of life risk for this fund in euros.
    - There are risks involving a loss of customers (increase in redemptions, mortality and disasters), and therefore a decrease in profits; these comprise about 90% of a life SCR in the first year. Indeed, the insurer's profits for a fund in euros come mainly from the levying of management fees, which becomes impossible for customers out of the contract.
    - There is also the risk of an increase in maintenance costs: about 10% of a life SCR in the first year.
- Market risk losses amount to 22% of the SII SCR. These losses are due to a decrease in the financial margin on the return on assets.

- Interest rate and spread risks have the same consequences and reduce the income of the bond portfolio (state and company), which corresponds to 85% of the total portfolio. The sum of the SCRs of these two risks thus represents more than a half of the market SCR.
- Equity and real estate risks represent 22% of the market SCR while equities and real estate represent only 10% of the asset portfolio, since they are risky assets compared to bonds and cash.
- The risk of illiquidity represents 22% of the market SCR and mainly impacts bond yields

- Losses related to the risk of bond market default constitute 6% of the SII SCR.

4.2.1.2 Solvency IRR Sensitivities to General Account Design Features

Computing the IRR sensitivities to general account features will enable improvements in the product design in order to maximize the solvency IRR.

We consider here the following characteristics of a single premium general account (see Figure 4.15).

Maturity	15 years
Age	44
Single Premium	€1000
Acquisition costs	€50
Maintenance fees	2,5
Premium loadings	4,5%
Annual Management Fees	0,9% (of l'Account Value)
Minimum Guaranteed rate	0% (Annual)

FIGURE 4.15 Product features

4.2.1.2.1 Impact of the Loadings on Premiums on the Solvency IRR, Which Has a Strong and Major Impact on the First Year's Cash Flows

The SI and SII SCRs are generally stable with the loadings on premiums: under SI, the SCR decreases very slightly, since the level of the account value decreases with the loading on payments; however, under SII, it is generally stable, since loading occurs only in the first year, so there are no increasing risks with loading on payments.

But, since the first cash flow as it is is not discounted in the IRR equation, the IRR is very sensitive to it, and therefore very sensitive to the loading on premiums and increases sharply with the latter, even though the sensitivity is lower under SII.

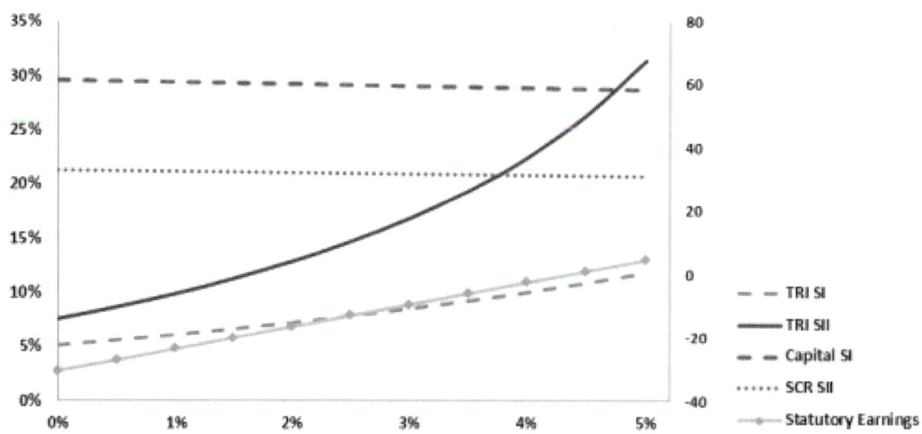

FIGURE 4.16 IRR versus loadings

4.2.1.2.2 Impact of the Minimum Guaranteed Rate (Expressed as a Percentage of the Account Value) The increase in the guaranteed rate reduces the insurer's profits, in two steps: a decrease in the financial margin when the guaranteed amount and management fees are lower than the financial gains; then an increase in financial losses when the return of the portfolio is in line with the guaranteed amount and management fees

A higher guaranteed interest to the customer implies an increase in the SCR: under SI, because increasing the guaranteed rate increases the level of the account value; and under SII, because it increases the risks of not having a sufficient return to serve it.

There is therefore a decrease in statutory earnings and an increase in SCR when the guaranteed rate is increased. This implies a decrease of the SI and SII IRR (see Figures 4.17 and 4.18) (Bouhris 2016).

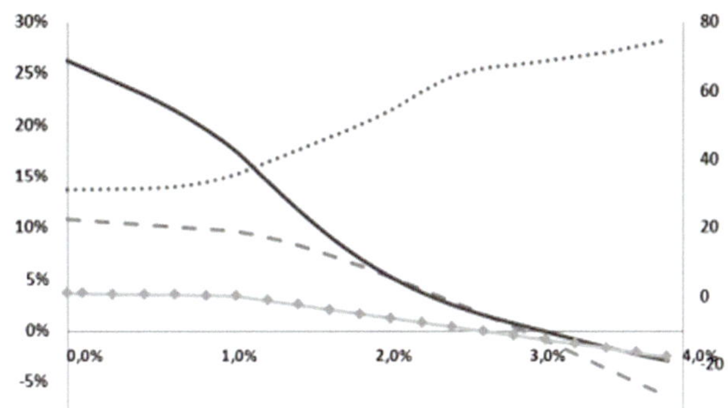

FIGURE 4.17 IRR versus minimum guaranteed rate

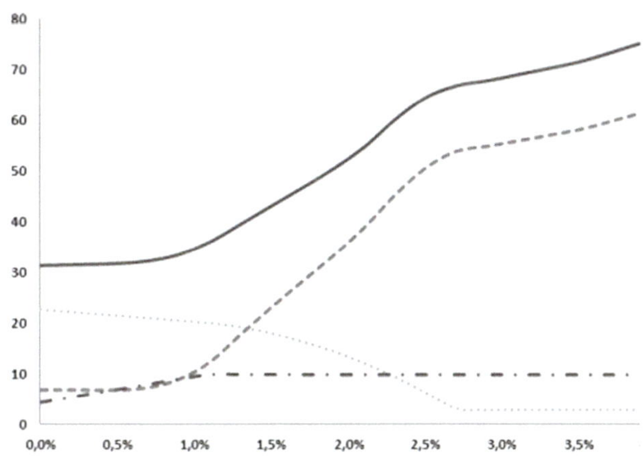

FIGURE 4.18 SCR split by risks (total, life, market, default) versus guaranteed rate

However, the solvency IRR does not decrease in the same way under SI and SII.

Under SI, the evolution of the IRR is led by the evolution of statutory earnings and the decline is done in two steps:

- a small decrease for a guaranteed rate between 0 and 1%, which corresponds to a decrease in the financial margin;
- then a faster decrease beyond 1% of the guaranteed rate, which corresponds to an increase in financial losses.

On the other hand, under SII, the decline is a little more complex. The IRR curve is first concave and then convex. The evolution of the IRR seems to be inversely led by that of the SII SCR. Indeed, the SII IRR and the SII SCR have a correlation coefficient of –95% and are inversely related by 97%.

The evolution of the SII IRR is therefore mainly explained by the evolution of the SII SCR. To better understand this evolution, here is the breakdown of the SCR according to the different risk classes.
The SCR is driven by the market SCR, which evolves across three phases of the guaranteed rate.

- It is stable for a guaranteed rate between 0 and 0.8%. The return on assets is large enough to pay the guaranteed rate and deduct management fees, and risk-free assets are sufficient to provide this return.
- For a guaranteed rate between 0.8 and 2.7%, the market SCR increases. Risk-free assets are no longer sufficient to pay the guaranteed rate and deduct management fees, and the return on risky assets is becoming significant, which explains the growth of the market SCR, since the corresponding risks have increased.
- Finally, when the guaranteed rate is too high (above 2.7%) and becomes higher than the return on assets, the financial losses are greater than the management costs, so there are no longer any sources of profit and the product is no longer profitable. The market SCR continues to increase because the insurer promises a higher guaranteed rate to the customer, but to a lesser extent because the insurer no longer makes a profit.

On the other hand, the life SCR decreases when the guaranteed rate increases.

- The product becomes more and more profitable for the customer when the guaranteed rate increases, and therefore the losses in case of increased mortality or redemptions become lower.
- The life SCR is almost zero when the product is no longer profitable for the insurer (beyond 2.7% of guaranteed rate). The insurer is no longer sensitive to an increase in redemptions or mortality. The life SCR is not completely zero, since the cost risk does not depend on the guaranteed rate and is not zero. We also observe the appearance of a very slight SCR linked to the risk of longevity. This quasi-cancellation of the life SCR corresponds to the slowdown in the increase in the market SCR.

Finally, the default risk SCR slightly increases with the guaranteed rate and then becomes stable. However, the amount of the default in the event of a shock hardly depends on the guaranteed rate.

- The SCR increases slightly when the return on assets is sufficient to pay the guaranteed rate and deduct the management fees, because the default decreases the bonus interest of the client and the financial margin of the insurer. The defect impacts the customer and the insurer.
- When the guaranteed rate and management fees are higher than the financial gains, the client no longer has a bonus interest. The default shock impacts only the insurer, so the SCR becomes stable.

The position of the inflection points under SI as under SII depends on the management case used, since it is linked to the return on the asset.

4.2.1.2.3 Impact of the Management Fees Management fees are an amount deducted from financial gains each year, calculated as a percentage of the account value and included in the insurer's profits. They therefore impact all cash flows as follows.

- When management fees are lower than financial gains (2% in the first year), increasing the percentage of management fees increases statutory earnings.
- When management fees are greater than financial gains (above 2% in the first year), statutory earnings become constant, since the portion of management fees greater than financial gains is recognized as financial losses.

Specifically, the Solvency I SCR and Solvency II SCR move in opposite directions with respect to management fees.

- Under Solvency I, the capital of the first year decreases only slightly with the management fees, so the IRR increases continuously with the management fees.
- In contrast, under Solvency II, the first-year capital significantly increases with the management fees. The IRR does not increase continuously with management fees but presents an optimum when the SCR exceeds a certain threshold (Bouhris 2016).

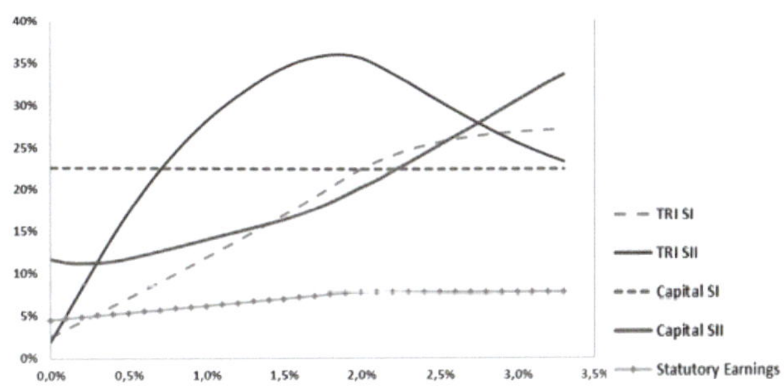

FIGURE 4.19 IRR versus management fees

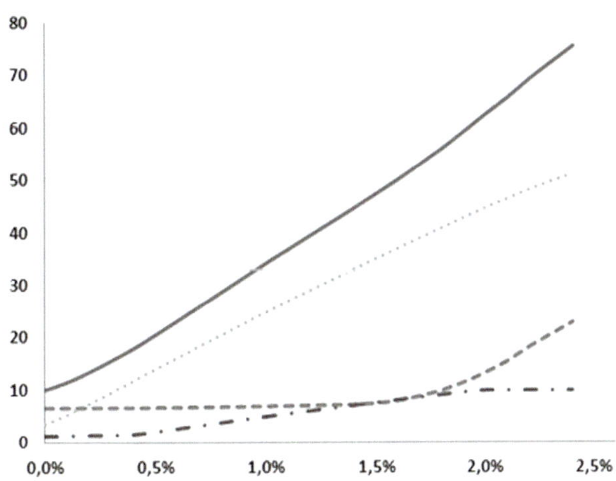

FIGURE 4.20 SCR split by risks (total, life, market, default) versus guaranteed rate

To understand this difference, the evolution of the capital under SI and SII over time for different levels of management fees is shown in Figure 4.21.

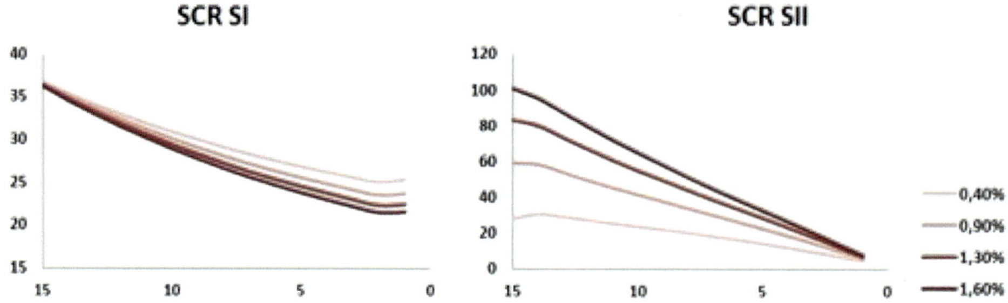

FIGURE 4.21 SCR versus residual maturity for different management fees

Under Solvency I, the SCR decreases with the management fees, since they reduce the level of the account value and therefore the SCR. This decrease is especially visible at the end of the contract.

- In the first year the different account values have a very similar value, since the management fees are levied only once.
- At the end of the contract the management fees are deducted several times, which increases the difference between the account value for the different levels of management fees.

In contrast, under Solvency II, the SCR increases with the management fees. The difference in SCR between the different management fees decreases over time. Here we give the breakdown of the SII capital according to the different risk classes (life, market and default) to better our understanding.

The life SCR increases significantly with management fees: this increase comes mainly from the risk of buybacks. An upward shock of redemptions reduces profits, because the insurer cannot deduct management fees from the last few years once the client has left the contract. This loss of profits in the event of an increase in redemptions is therefore all the more important as the management fees are important, which explains the increase in the life SCR with the management fees.

Regarding the market SCR, it is generally stable for low management fees, and increases beyond a certain threshold of management fees: as long as the management fees are not too high (less than 1.7%), risk-free assets are sufficient to provide the necessary return for the deductions of management fees. Beyond 1.7% of management fees, risk-free assets are no longer sufficient and the return on risky assets becomes significant, hence an increase in market risk and therefore in its capital.

Finally, the capital linked to the risk of default increases with the management fees the first time, and then becomes stable. However, the amount of the default in the event of a shock hardly depends on the management fees: when management fees are lower than financial gains, the default reduces the interest served to the customer and the financial margin, and so there is an increase in capital related to the risk of default since the default does not impact only the insurer. On the other hand, when the management costs are higher than the financial gains, the insurer presents financial losses, and the default does therefore impact only the insurer, leading to a stabilization of the default SCR.

The sensitivity of the IRR to the management fees levied above was made in the absence of a guaranteed minimum rate. However, the evolution of the IRR (in particular SII) according to the management fees depends, among other things, on whether or not the sum of the minimum amount paid to the customer and the fees levied exceed the financial gains. Promising a guaranteed rate reduces the financial margin of the insurer, and therefore reduces profits and increases risks – and thus the SCR too. Therefore, as can be seen below, the guaranteed rate decreases the IRR, and, moreover, the optimum and stabilization of the SII IRR according to the management fees are moved to a lower management fee rate.

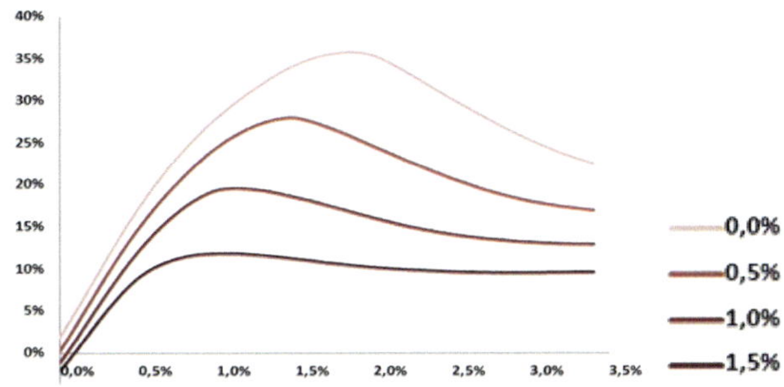

FIGURE 4.22 Solvency II IRR versus management fees depending on minimum guaranteed rate

4.2.1.2.4 Impact of the Contract Maturity The greater the maturity, the more the insurer can make a profit. Indeed, each year of the contract allows the making of profits, and a large maturity allows the placing of the client's money at longer-term rates, generating more interest. Hence, there is an increase in VIF with maturity.

However, a large maturity implies the tying up of more capital, for longer, which represents a significant cost. Thus, the capital under SI and under SII increases with the maturity. However, there is an irregularity in the evolution of the capital for low maturities under SI and SII due to the management case used.

Therefore, as in the case of management fees for SII, the IRR under both SI and SII show an optimum when varying maturity.

However, it is noticeable that the SII IRR is much more sensitive than the SI IRR. On the one hand, the SI capital is larger than that under SII, and, on the other hand, the capital under SII increases more strongly than under SI.

- The capital under SI increases since a larger maturity makes it possible to place the money at longer-term, more interesting rates. This implies a better return, leading to an increase in account value and therefore in capital.

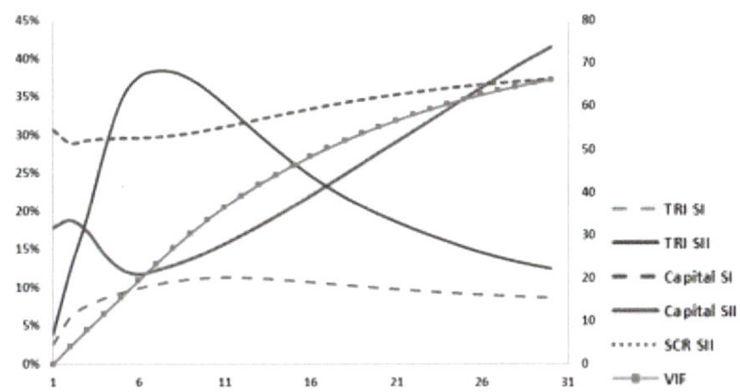

FIGURE 4.23 IRR versus maturity

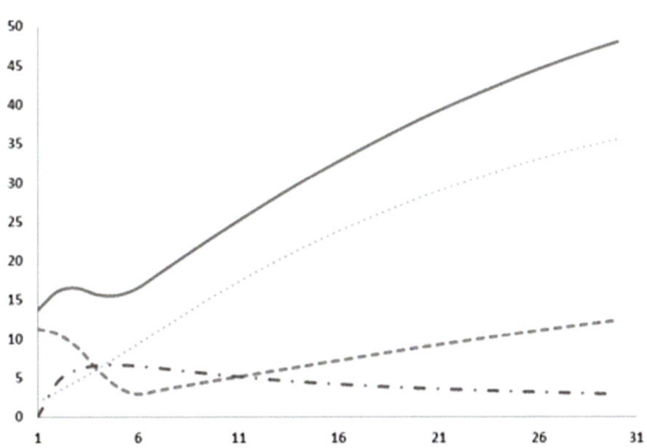

FIGURE 4.24 SCR split by risks (total, life, market, default) versus guaranteed rate

- Under SII, the capital increase is due to a longer exposure to different risks. Figure 4.24 shows the breakdown of the SII SCR by risk classes for a better understanding.

As for the sensitivity to management fees, it is mainly the life SCR that explains the total capital, explained chiefly by the lapse SCR. The greater the maturity, the more the number of deductions from management fees increases (one levy each year). Consequently, an upward shock in buybacks has a greater impact on a long-term maturity contract than a short-term contract, since it prevents more management fees for a long-term contract than a short-term contract.

Regarding the market SCR, it decreases for low maturities and then increases beyond a maturity of six years. The fall is due to the management case, since the return on equities in neutral risk is negative in the first years (as we are interested here in the SCR of the first year, the date of calculation is the first year and the projections of the following years are made in neutral risk). The growth of the market SCR after six years is explained by the increasing risk of investing in the longer term.

4.2.1.2.5 Impact of the Potential Guaranteed Minimum Death Benefit The floor guarantee provides the beneficiary with a minimum amount in the event of the death of the insured. The cost of this guarantee is deducted from the account value each year. The mortality margin taken by the insurer is 50%.

Increasing the amount of the floor guarantee increases statutory earnings and the VIF. Indeed, the mortality margin makes it possible to get a slightly higher guarantee price than the actual value of the guarantee. The account value must be large enough to pay the cost of the guarantee – i.e. the sum insured in the event of death must be reasonable in relation to the payment made, and the mortality margin must not be too large either.

- Below an insured sum of €1,000, statutory earnings are stable, since the sum insured is less than the initial payment.
- Beyond €1,000, statutory earnings increase in proportion to the sum insured.

Under SI, the capital of the first year decreases slightly with the sum insured. As a result, the IRR continuously increases with the sum insured. In contrast, under SII, the SCR and the capital increase with the sum insured. The IRR increases thanks to an increase in profits (linked to the mortality margin) but presents an optimum when the capital exceeds a certain threshold. However, it should be noted that the optimum is positioned for an insured sum disproportionate to the initial payment. The product looks more like a death guarantee than a savings product.

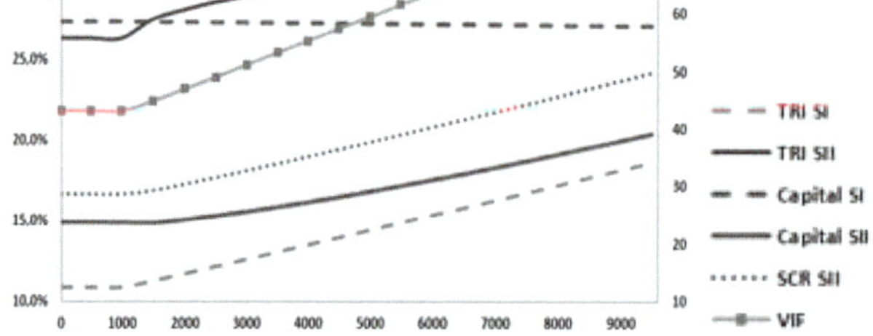

FIGURE 4.25 IRR versus maturity

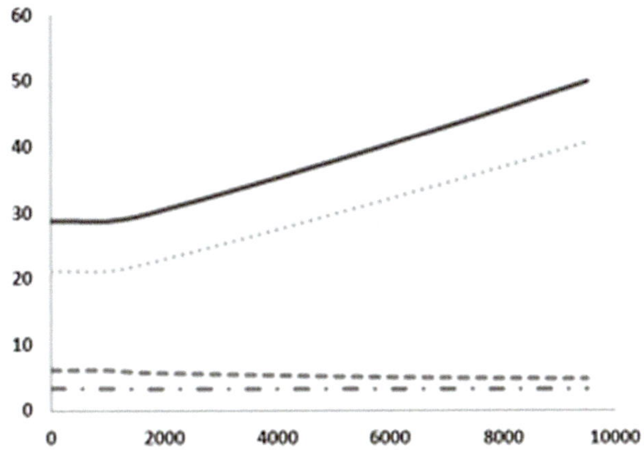

FIGURE 4.26 SCR split by risks (total, life, market, default) versus guaranteed rate

As when we varied the management fees, we see that the capital under SI and under SII does not evolve in the same way with the sum insured. The evolution of the SCR SI and SII over time for different sums insured is depicted in Figure 4.27.

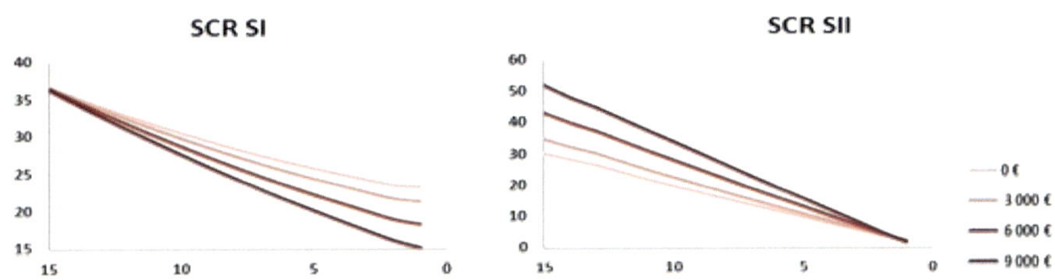

FIGURE 4.27 SCR versus residual maturity depending on the sum assured

The SI SCR decreases with the sum insured because it increases the cost of the guarantee and thus decreases the level of the account value. As in the case of management fees, this difference is mainly seen at the end of the contract, because the price of the guarantee has been deducted several times. In contrast, the SII SCR increases with the sum insured; here is the decomposition of the SII SCR according to the different risk classes.

The life SCR increases with the sum insured. This increase is related to two types of losses.

- First is a loss related to the impossibility of deducting the cost of the guarantee, and therefore benefiting, from the mortality margin following the exits of contracts (redemptions and death). This loss is all the more important as the sum insured is large, since the amount of the margin on the cost of the guarantee increases with the sum insured.

- The second type of loss relates to benefits to be paid that are larger than expected following a greater morbidity (mortality risks and life disaster). Similarly, this loss is all the more important as the sum insured is large, because the amount of benefits to be paid increases with the sum insured.

The market SCR is generally constant, since market returns do not depend on the sum insured and the cost of the guarantee. The default SCR, similarly, is generally stable, since the amount of the default does not depend on the sum insured.

Remark: these studies above were carried out with the assumption of a constant surrender rate. Even so, if there is a consequence on the numerical results, the shape of the curves does not change compared to the assumption of a constant lapse rate. Therefore, the results and preliminary analyses on the evolution of the IRR are in this respect consistent and reliable.

4.2.1.3 Improving the Solvency IRR through Product Design

4.2.1.3.1 A Static Lapse Rate Loading on payments makes it possible to significantly increase the IRR without limit from the point of view of the insurer. However, it must not be too important for the product to be attractive to the customer. It is possible to present a large IRR with characteristics such as loading on payments that impact only the first cash flow without the product being really profitable thereafter. It may then be interesting to see if, with a large IRR, the Payback return period is not too low.

Since the IRR is strictly decreasing with the guaranteed minimum rate, the only way to operate the IRR is a guaranteed rate of 0%. The rate paid to the customer therefore corresponds only to profit sharing. The risk of a guaranteed rate is that it is determined at the beginning of the contract according to the current context. With the latter evolving during the life of the contract, the insurer is not sure of being able to pay this guaranteed rate in a few years. This is a problem encountered in the current period of low interest rates. It is therefore preferable to offer a guaranteed rate of 0% even if it means reducing management fees.

Contrary to what one might have thought with a profit-only vision, the SII IRR does not increase continuously in line with management fees; the IRR admits an optimum according to these fees. In our case of a 0% guaranteed rate, this optimum is reached for 1.8% of management fees. This rate is beyond what is generally applied. This optimum depends in particular on the guaranteed rate: with a guaranteed rate of 1%, the optimum IRR is reached for 1% of management fees – a more reasonable rate.

Depending on the maturity of the contract, the IRR has an optimum obtained for a maturity of seven years. This corresponds to a short duration for a life insurance contract, and in contrast to the tax regulations, which encourage the customer to keep her contract for more than eight years.

If the insurer decides to offer a floor guarantee in the event of death, it can determine the amount insured in order to optimize the IRR. This sum is €6,000 in this application. This corresponds to six times the initial payment, so this sum is disproportionate. However, offering a floor guarantee between €1,000 and €6,000 does not reduce the IRR calculated in the absence of a guarantee.

4.2.1.3.2 A Dynamic Lapse Rate The studies above were carried out with the assumption of a constant surrender rate. It is possible to increase the IRR to the detriment of customer profitability, but in practice the redemptions will not be constant. For example, if we consider a guaranteed minimum rate of 0% and significant management fees, what remains of the financial gains after the deduction of the management fees is low, and therefore the rate is as well. Some customers will want to buy back their contract because it pays little. The example is extreme, but it does reveal that the assumption of a constant redemption rate is unrealistic.

Surrender risk is one of the main risks in connection with a savings product. The hypothesis of a dynamic function of buybacks, which depends, among other things, on market rates, therefore has a significant impact on the IRR.

The lapse function considered here is a traditional dynamic lapse function, within which the lapse rate can be split into a sum of two lapse rates, the structural lapse rate and the cyclical lapse rate: $\lambda_t = \lambda_t^S + \lambda_t^C$.

- Structural lapse corresponds to buybacks made regardless of the economic context (rate of return etc.). The insured has an immediate need for money and therefore buys back her savings. We consider the constant rate of structural buybacks 2%.

- Cyclical lapse corresponds to lapse depending on the economic context. These redemptions therefore depend, among other things, on customer satisfaction on the rate served by their savings contract compared to a market reference rate. For instance, we can consider that half the customer's satisfaction depends on the satisfaction of the previous year (to take account of a memory effect) and the other half on the difference between the rate served and the market reference rate of the previous year. Here, for a fixed reference rate, satisfaction at a given date is an affine function of the rate served.

We now consider this cyclical buyback function (see Figure 4.28).

$$\lambda_t^C = \begin{cases} \lambda_{max} & \text{si } S_t < S_{inf} \\ \dfrac{\lambda_{min} - \lambda_{max}}{S_{sup} - S_{inf}} S_t + \dfrac{\lambda_{max} S_{sup} - \lambda_{min} S_{inf}}{S_{sup} - S_{inf}} & \text{si } S_{inf} \leqslant S_t < S_{sup} \\ \lambda_{min} & \text{si } S_{sup} \leqslant S_t \end{cases}$$

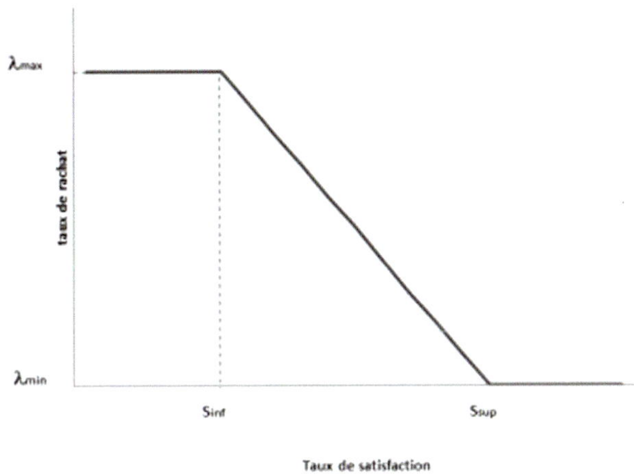

FIGURE 4.28 The dynamic lapse function

As illustrated in Figure 4.29, the shape of the curves does not change compared to the assumption of a constant lapse rate. Therefore, the results and preliminary analyses on the evolution of the IRR are in this point consistent and reliable.

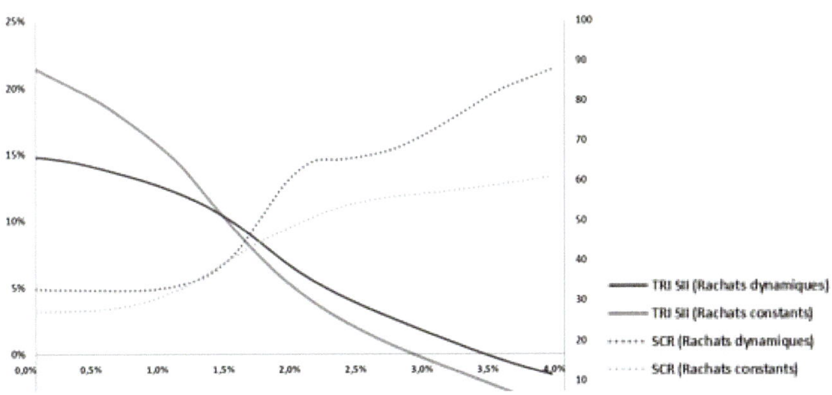

FIGURE 4.29 IRR and SCR versus minimum guaranteed rates

Taking into account such a dynamic lapse function, we get the results shown in Figure 4.30.

Maturity	15 years
Age	44
Single Premium	€1000
Premium loadings	4,5%
Annual Management Fees	1,5% (of l'Account Value)
Minimum Guaranteed rate	1% (Annual)

FIGURE 4.30 Product features

The corresponding IRR is 10% and the client's annualized rate of return is 1.1%.
Let us now consider three changes in product design in order to improve the Solvency II IRR.

- A decrease in management fees to 0.8%: as explained above, reducing management fees has a limited impact on the IRR, but it does impact on customer satisfaction, and therefore on redemptions.
- A decrease in the guaranteed rate to 0%, to reduce the risk related to the uncertainty of future rates.
- An increase in the load on payments to 5%, to compensate for the decrease in management fees.

The subsequent impacts are as follows (see Figure 4.31).

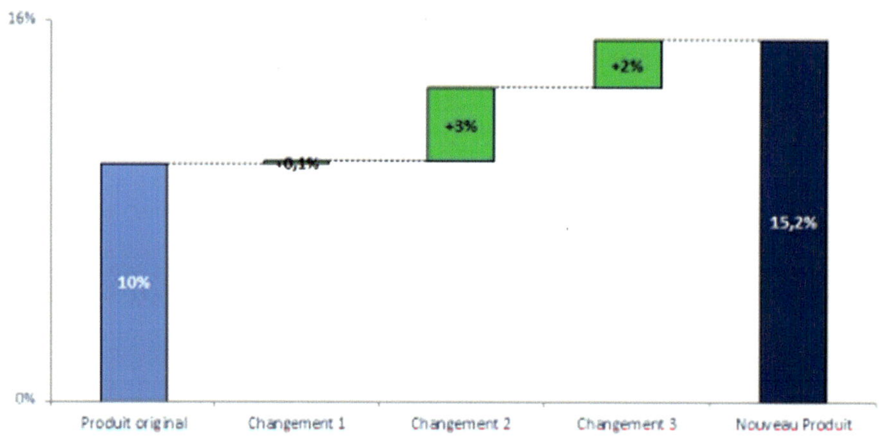

FIGURE 4.31 Impact of product design on Solvency II IRR

- The first change in product design provides a new IRR of 10.1%. One would have thought that the IRR would decrease, but reducing the management fees makes it possible to reduce the SCR and to invest more, and therefore to increase the level of the account value.
- The second change increases the IRR to 13.2%. This improvement is not a surprise, since it reduces the risk related to the uncertainty of future market conditions – i.e. the SCR.
- The final change increases the IRR to 15.2%. Loading on premiums occurring only in the first year has no impact on the SCR and makes it possible to increase profits and therefore the IRR without risk. However, the assumptions do not take into account the attractiveness of the product for the customer.

4.2.2 Mitigating Product Risks through Liability Design: Focus on Policyholder Behaviour Risks

4.2.2.1 Mitigating Customers' Behaviour Risks Embedded within the Deferral Period

Following the modelling approach on the extendable puts in Chapter 3, two potential product design enhancements can significantly decrease the risks for the insurer.

4.2.2.1.1 Shifting Down the Extended Strike K2 – i.e. Reducing Roll-Up Rates Roll-up rates enable us to increase the extended strike, which implies a higher extension optionality value, as illustrated in Figure 4.32. Therefore, it might be appropriate to limit the roll-up features in order to reduce the risk exposure.

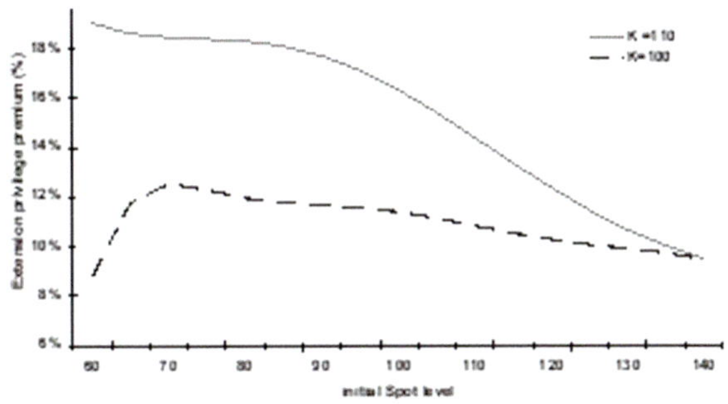

FIGURE 4.32 Impact of K2 on extension premium, $K_1 = 100, K_2 = 110, A = 0$, r = 0, T1 = 4 years, T_2 = 5 years

4.2.2.1.2 Introducing a Penalty Payment Introducing a penalty payment should insurers enable to decrease the extension privilege value, as illustrated in Figure 4.33. To the extreme, the delta also comes closer to the conventional put delta: choosing a = 10 (% notional) in the case in Figure 4.34 makes the extendable put delta equivalent to the plain put delta.

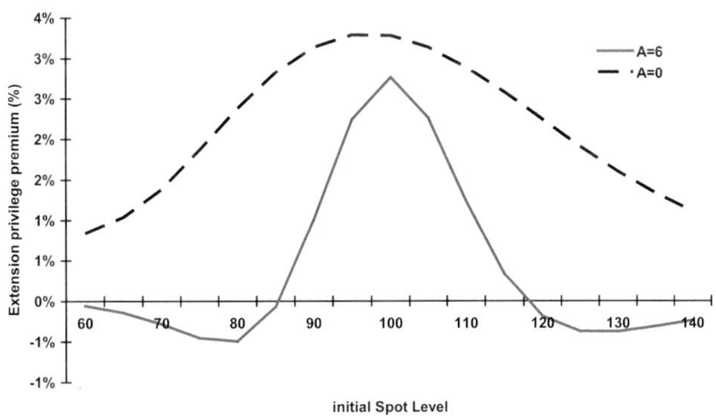

FIGURE 4.33 Impact of increasing additional payment A to the option writer on extension privilege value, $K_1 = 100, K_2 = 110, A = 10$, r = 0, T1 = 1 year, T2 = 2 years

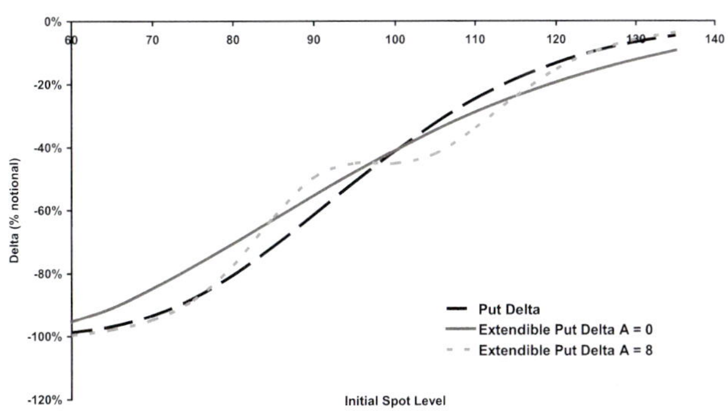

FIGURE 4.34 Extendable put delta versus vanilla put delta, $K_1 = 100, K_2 = 110, A = 10$, r = 0, T1 = 1 year, T2 = 2 years

Warning: the link between the penalty A and the extension privilege value is nonlinear. Therefore, product redesign regarding the additional payment A requires some subtle analyses in order to properly reduce risks.

4.2.2.2 Reconciling Insurer and Customer Lapse Risk Appetites through Fee Product Design

Consistent with the modelling approach developed in Chapter 3, behavioural simulations as presented here combine individual decision-making rules and artificial intelligence agent-based modelling to model policyholder behaviour.

Such agent-based modelling can be used for a variety of business applications, such as product management and asset retention, pricing and product design, distribution strategy, capital and risk management, and strategic analysis.

We consider here the guaranteed withdrawal annuity product and the modelling of efficient (or "rational") customers surrender behaviour using agent-based modelling, which can be used in product designs to stir policyholder behaviour while also fitting client appetite.

As expected, the efficient ("rational") lapse modelling usually implies a higher economic cost, as it reflects more extreme possible/theoretical policyholder behaviour, illustrated by a no-arbitrage charge annual charge fees rate c, which puts the balance sheet in equilibrium at inception between assets and liabilities.

Different Lapse Strategies	Annual Charge
0-Lapse	2,78%
Dynamic Lapse	2,60%
Tax-Based Efficient Lapse	3,78%

FIGURE 4.35 Annual charge depending on lapse pricing

However, it is usually very difficult to find the no-arbitrage charge fees in the "efficient" (or so-called "rational") lapse assumption. To overcome this difficulty, we introduce the up-front charge fees measured by: PV (claims) – PV (charges).

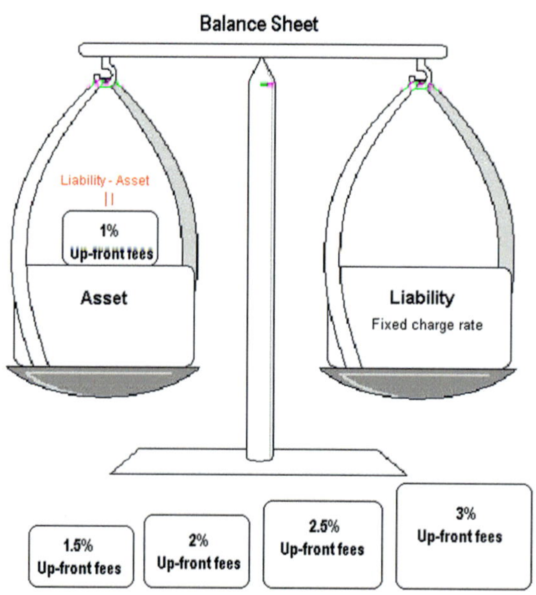

FIGURE 4.36 Up-front charge

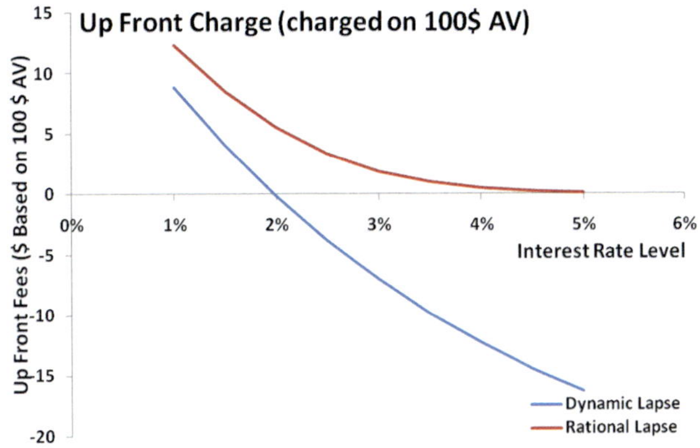

FIGURE 4.36 (Continued)

Such a charge fee may be based on one or several different references: account value, net premium, guarantee base, sum of premiums. It is itself significantly dependent on the interest rate environment:

Given the usual higher efficient behaviour cost, such an efficient behaviour charge fee may be mixed with a proportion X% with the standard "dynamic lapse" charge fee based on past observations, in order to improve risk management while staying competitive.

The proportion of efficient surrender behaviour may also be extracted from past observations, and thus the dynamic lapse charge, considering that α% customers behave efficiently while (1 – α%) behave statically:

$$\alpha\% \times \text{efficient behaviour charge fee} + (1 - \alpha\%) \times \text{static behaviour charge fee} = \text{dynamic behaviour charge fee.}$$

Finally, different mixes of charging structures are possible, offering multiple trade-offs in terms of charging structures, depending on the client's appetite (see Figure 4.37).

- **Upfront Fee + Surrender Charge**
- **Upfront Fee + trailed Charge**
- **Upfront Fee + trailed Charge + Surrender Charge**

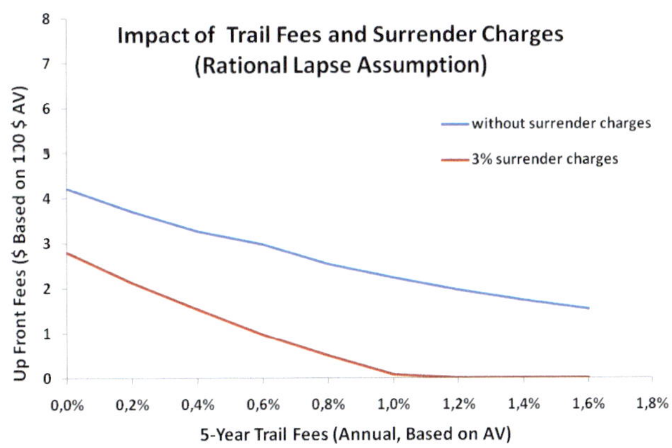

FIGURE 4.37 Impact of different fees designs on up-front fees

4.2.3 Enhancing the Customer Experience of Savings and Retirement Products through Sustainable Customer-Centric Product Design

4.2.3.1 Enhancing the Sustainability of Indexed Annuities from the Customer's Risk Appetite Perspective

The expected utility framework is used here, based on the classic Merton portfolio selection problem: an investor who wants to maximize the expected utility of terminal wealth is considered and, in addition, to have a payoff at least as good as that of an EIA.

The contract therefore maximizes the investor's expected utility subject to the payoff being at least as good as that of the selected EIA.

4.2.3.1.1 The Merton Solution

We first consider the general problem of how an investor selects an optimal portfolio. This is a classical problem in financial economics and it was solved by Robert Merton (see Merton 1971).

We show how the solution is modified when there is a guarantee. We focus on the solution to the optimal portfolio problem and the structure of the payoff, assuming that the investor makes the optimal decision.

We consider an investor with initial wealth x0 and a utility function given by u(.), where u is strictly increasing, strictly concave and twice differentiable. The available assets consist of a risky asset S and the risk-free asset. We assume no arbitrage, no friction and that the market is complete. Given these assumptions, there exists a unique state-price process {ξt}. We assume that the investor wishes to maximize the expected utility of terminal wealth, denoted by X(T), over time horizon T. The investor's optimal terminal wealth is given by

$$\mathrm{X}_T^* = I\left(\lambda^m \xi_T\right),$$

where $\mathrm{E}\left[\xi_T I\left(\lambda^m \xi_T\right)\right] = x_0$.

I(.) is the inverse of the investor's marginal utility function u'(.). We refer to the payoff in the previous formula as the Merton solution.

Now we modify the investor's objective function by including a guarantee at maturity. In this case the investor wishes to maximize the expected utility of terminal wealth subject to the constraint

$$\mathrm{X}_T \geq x_0 e^{gT},$$

where g is the guaranteed rate. This is a natural constraint given how often it occurs in practice. We can also find the solution to this problem. In this case

$$\mathrm{X}_T^* = \max\left[I\left(\lambda^g \xi_T\right); x_0 e^{gT}\right],$$

$$\text{where } \mathbf{E}\left[\xi_T \max\left\{I\left(\lambda^g \xi_T\right); x_0 e^{gT}\right\}\right] = x_0.$$

This last solution gives the optimal payoff when there is a guarantee and the new Lagrange multiplier, λg, will differ from the Merton one, λm. In fact, λg > λm. It turns out that the investor's expected utility is reduced when we impose the constraint. This result is evident from inspection of the two payoffs. However, it is noteworthy given the prevalence of such guarantees that their inclusion serves to reduce the investor's expected utility relative to the unconstrained problem.

We can examine the difference between these two solutions in the case in which the risky asset has a log-normal distribution, since this case is analytically very tractable. In this case ξ also has a log-normal distribution, and is given by

$$\xi_T = a\left(\frac{S_T}{S_0}\right)^{-b},$$

$$\text{where } a = e^{\left(\frac{\theta}{\sigma}\left(\mu-\frac{1}{2}\sigma^2\right)T-\left(r+\frac{1}{2}\theta^2\right)T\right)}, b=\frac{\theta}{\sigma}, \theta=\frac{\mu-r}{\sigma}.$$

We assume that the investor has a log utility function and initial wealth x0. In this case, the optimal wealth under the Merton solution is

$$X_T^* = I\left(\lambda^m \xi_T\right) = I\left(\frac{\xi_T}{x_0}\right) = \frac{x_0}{a}\left(\frac{S_T}{S_0}\right)^b.$$

The corresponding optimal terminal wealth when there is a guarantee is

$$X_T^* = \max\left\{I\left(\lambda^g \xi_T\right), x_0 e^{gT}\right\} = \max\left\{\frac{1}{\lambda^g a}\left(\frac{S_T}{S_0}\right)^b; x_0 e^{gT}\right\}.$$

4.2.3.1.2 Optimal Design of a "Point-to-Point" EIA Consider a point-to-point EIA with a fixed guarantee of x(0) * exp(gt) at maturity and participation rate α. We assume that the investor wishes to maximize the expected utility of terminal wealth with the constraint that the payoff has to be at least as good as that of the point-to-point EIA.

This analysis is similar to that of the last subsection except that the investor's optimal payoff will now contain the payoff of the point-to-point EIA as the guaranteed payoff.

We assume that the investor has a utility function u(.) and that the terminal wealth under the EIA is Γ,

$$\text{where } \Gamma = \max\left\{x_0 e^{gT}, x_0\left(\frac{S_T}{S_0}\right)^\alpha\right\}.$$

Hence the optimal terminal wealth of this investor is equivalent to the solution of the following maximization problem, max E[u(XT)], subject to the constraint that the investor's terminal wealth XT satisfies

$$X_T \geq \max\left\{x_0 e^{gT}, x_0\left(\frac{S_T}{S_0}\right)^\alpha\right\}.$$

The optimal form of XT can be derived using the general theory of portfolio selection in the Merton framework. It is similar to the optimal design we considered in the last subsection. The optimal payoff at time T is given by

$$X_T^* = \max\left\{I\left(\lambda^g \xi_T\right), x_0 e^{gT}, x_0\left(\frac{S_T}{S_0}\right)^\alpha\right\}, \quad E\left[\xi_T \max\left\{I\left(\lambda^g \xi_T\right), x_0 e^{gT}, x_0\left(\frac{S_T}{S_0}\right)^\alpha\right\}\right] = x_0,$$

where the claim $\max\left\{I\left(\lambda^g \xi_T\right), x_0 e^{e^{gT}}, x_0\left(\frac{S_T}{S_0}\right)^\alpha\right\}$

is the optimal payoff for an investor with utility function u(.). This equation means that the no-arbitrage value of the optimal payoff is equal to x0. Of course, this is the ideal contract from the customer's viewpoint, and there is no allowance for transaction costs.

In practice, the premium x0 will exceed the no-arbitrage value of the contract.

We now consider a numerical example when Γ is the payoff under a specific point-to-point EIA. We fix the participation rate α = 45%. The guaranteed contract payoff, Γ, provides the greater of a guaranteed return 2%, and participation in the market with α = 45%. The optimal terminal wealth for this case is

$$X_T^* = max\left\{ I_0 e^{S^T}, I_0 \left(\frac{S_T}{S_0}\right)^a, \frac{1}{\lambda^S a}\left(\frac{S_T}{S_0}\right)^b \right\} \qquad \Gamma = max\left\{ I_0 e^{S^T}, I_0 \left(\frac{S_T}{S_0}\right)^a \right\}.$$

$$= \max\left\{1.1052; S_T^{0.45}; 1.0541\sqrt{S_T}\right\} \qquad = \max\left\{1.1052; S_T^{0.45}\right\}$$

The payoff under this optimal contract is greater than Γ for a wider range of values of the market index S. It is clear that the contract payoff Γ is not the optimal terminal wealth.

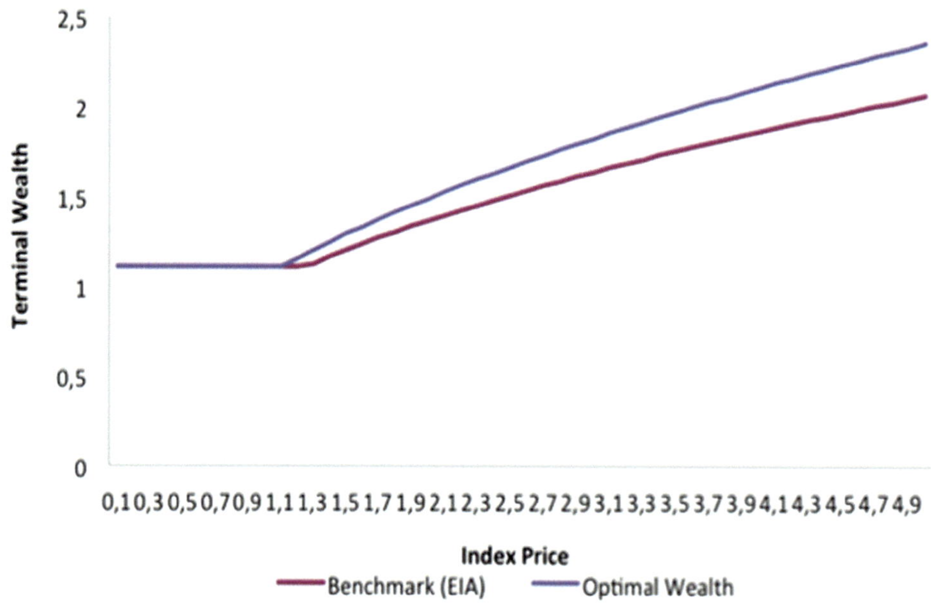

FIGURE 4.38 Optimal wealth versus EIA benchmark for a 45% participation rate

4.2.3.2 Tailoring the Product Design to the Customer's Needs, as illustrated by an Increasingly Differentiated Savings Behaviour Pattern

Life insurers must respond to customer needs and expectations around flexibility, personalization and relevance.

Savings behaviour has undergone significant changes towards more efficient usage over the last decade, as illustrated by the emergence of increasingly differentiated retirement savings patterns, in line with new needs and the subsequent life phases.

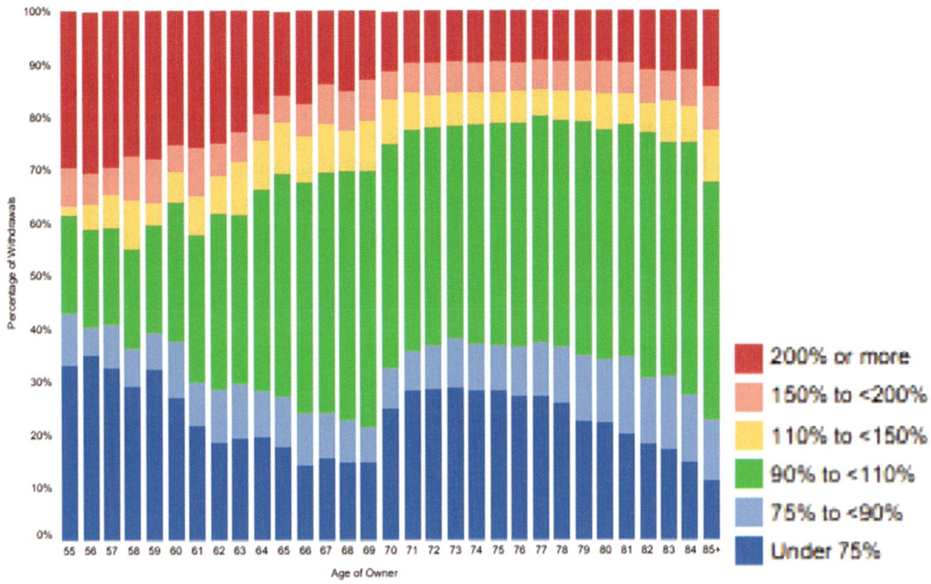

FIGURE 4.39 Withdrawal patterns across ages

To this end, there are three types of savings use behaviour patterns, illustrated by the frequency and amounts of withdrawals, and strongly influenced by age.

- Before the age of 60: one-off withdrawals of heterogeneous sizes, most often massive (over 200% of the annual guaranteed withdrawal), illustrative of the consumption of expensive durable goods, real estate purchases, financing of education and vocational training, children's studies and management of periods of unemployment.
- From 60 to 70 years of age: regular withdrawals, of moderate and homogeneous sizes during retirement, illustrative of the entry into annuity.
- Over 70 years of age: mostly the entry into annuity, with regular moderate and homogeneous withdrawals, but also excess withdrawals (for tax reasons or donation).

This segmentation of customer profiles reflects a strong differentiation of savings behaviours. Meeting the expectations of the insured then requires an optimization of the product design, through product features tailored to the savings profiles, for an enhanced customer experience.

The savings contract considered here entitles the policyholder to receive an income stream at maturity for the rest of her life, after a waiting period during which savings are invested in financial assets ("account value"), and benefit from an increasing guaranteed level at a minimum "roll-up" rate. The insured is also allowed to withdraw a certain amount on a yearly basis during the waiting period. Moreover, if the contract contains a death benefit, then a certain amount is paid to the beneficiaries in case the policyholder dies during the term of the contract.

4.2.3.2.1 Formulation and Notations We consider an x-year old policyholder possessing a savings and retirement contract. At inception, an initial endowment is invested in a risky asset St. The specifications of the contract include a set of dates $0 = t_0 < t_1 < t_n < t_N = T$, where $t_0 = 0$ is the contact inception and $t_N = T$ its maturity. These so-called contract anniversaries are the dates on which events can take place – i.e. withdrawals, payments, etc. An analysis of the design of general equity-indexed annuities from the investor's perspective and a generalization of the conventional design is proposed in Boyle and Tian (2008).

4.2.3.2.1.1 The Contract Assumptions **The Financial Market**
The pricing is based on the common pricing literature, which assumes the existence of a risk-neutral measure $\mathbb{Q}$ under which future cash flows can be valued as their expected discounted values. The existence of such measure implies an arbitrage-free financial market.

We assume that the risky asset S_t, which serves as an underlying mutual fund for the variable annuity, follows a geometric Brownian motion with constant coefficients under $\mathbb{Q}$:

$$\frac{dS_t}{S_t} = rS_t dt + \sigma_t dW_t,$$

where σ is the volatility or the risky asset, r the risk-free rate and W a standard Brownian motion under Q. The money market evolves with risk-free interest rate, and the numéraire process Bt is given by

$$dB_t = rB_t dt.$$

Under the risk-neutral probability measure, the discounted asset process $B_t^{-1}\,St$ is a martingale.

The Mortality Assumption
It is common practice among insurers to use a deterministic mortality rate to evaluate and replicate their policy pool. The common assumption is also made that financial markets and biometric events are independent.

Let us introduce the mortality notations as:

- $x0$: the policyholder's age at the contract inception;
- qn: the probability that the policyholder, aged $x0$ at inception, dies between time $tn-1$ and tn;
- pn: the probability that the policyholder, aged $x0$ at inception, is alive at time tn; and
- ω: the limiting age beyond which survival is impossible; according to the definition, we have $pn = (1 - qn)pn-1$, where $n \in \{1,2,\ldots, N\}$.

From the insurer's perspective, the percentage of active contracts in a large policy pool of policyholders aged $x = x_0 + t_n$ at a given time t_n is thus given by p_n.

4.2.3.2.1.2 The Contract State Variables At a given anniversary date t_n, the value of the contract, purchased by an x_0-year-old policyholder at inception, is determined by three main state variables: the account value, the benefit base and the two-states variable determining if she is alive or dead at time t_n

- Account value A_t: the value of the investment account, this is indexed on the asset value S_t, and reduced by withdrawals and fees.
- Benefit base G_t: also referred as the guaranteed account, this is an "imaginary" wealth upon which annuities, guaranteed withdrawals and benefits are calculated. However, if the insured wants to lapse the contract, she will not be able to get this wealth.
- Death process I_n: a two-states variable in $\{0,1\}$ informing us if the policyholder died during (t_{n-1}, t_n), or is still alive at t_n. The death probability in the interval (t_{n-1}, t_n) is given by $q_n = \mathbb{P}(I_n = 0 \mid I_{n-1} = 1)$, which depends on the policyholder's age at inception.

We restrict our analysis to single-premium contracts $A_0 = G_0$ – i.e. one premium at inception with no additional contributions. The policyholder can either withdraw money or exercise the income benefit. Withdrawals include "zero" withdrawals and complete surrender of the contract – i.e. lapse.

4.2.3.2.1.3 The Income and Death Benefits At maturity, the holder of the contract can select to take a lump sum of the account value A_{t_n} or annuitize the benefit base at a pre-specified guaranteed annuitization rate. Annuity factors, which give the annuitization rates, denoted by $\ddot{a}^{act}_{tN}$ for the actual and $\ddot{a}^{gua}_{tN}$ for the guaranteed, are defined as the price of an annuity paying \$1 each year, either at the market's rates curve or at an internal guaranteed rate defined by the insurer. The calculations of the annuity factors take into account the probability that the insurer is alive in the future with probability p. They are given by

$$\ddot{a}^{(.)}_{t_N} = \sum_{t_i=t_N}^{\omega-x_0} p_i e^{-r^{(.)}_{t_i}(t_i-t_N)}.$$

These factors are increasing, since an older insurer will probably have less in the way of annuities than a younger one. Therefore, annuitizing the account value is equivalent to a lump sum, and annuitizing a benefit base G is equivalent to the amount I_{t_n}, equal to

$$G\frac{\ddot{a}^{gua}_{t_N}}{\ddot{a}^{act}_{t_N}}.$$

If the contract also contains a death guarantee, then the D amount is paid to the beneficiaries in the event of the death of the insured during the term of the contract and the cash flow ends, with $\delta=(\delta_1,\delta_2,\ldots,\delta_N)$ the death indicator variable and the probability of death $q_n = P(\delta_n = 1) = (1 - p_n)$.

Cash flows are recognized at discrete annual interim dates t_n between the signing and expiry of the contract: $0 = t_0 < t_1 < \cdots < t_n < \cdots < t_N = T$.

For the sake of simplicity, we assume that the policyholder can take withdrawals each policy anniversary t_n and denote by γ_n the withdrawals amount. The income benefit also starts at anniversary dates, and, in the event of a death benefit, the latter is paid out at these dates as well. Thus, the state variables described above may have discontinuities at times $t_1, \ldots t_n$. Therefore, for a state variable Y, we distinguish between its value "Yt_n–" before and "Yt_n+" after events take place at the anniversary date.

4.2.3.2.2 Contract State Variables Dynamics

4.2.3.2.2.1 Development between Two Policy Years (tn-1, tn) Assuming that an annual guarantee fee α is continuously charged by the issuer, the value of the account value A_{tn} evolves as

$$A_{t_n^-} = A_{t_{n-1}^+} \times \frac{S_{t_n}}{S_{t_{n-1}}} exp(-\alpha\Delta t), \quad n = 1,2,\ldots,N,$$

where $\Delta t = t_n - t_{n-1}$.

In practice, the guaranteed fee is charged discretely and proportionally to the account value, which can easily be incorporated into the wealth process. Denoting the discretely charged fee with the annual basis as $\bar{\alpha}$, the wealth process becomes

$$A_{t_n^-} = A_{t_{n-1}^+} \frac{S_{t_{n-1}}}{S_{t_n}}(1-\bar{\alpha})\Delta t.$$

The guarantee offered by the contract is denoted G, increasing at a guaranteed annual rate η during the waiting period, and serving as the basis for converting the contract into a life annuity I, associated with predefined annuity factors at the signing of the contract. The benefit base remains constant between two policy years – i.e.

$$G_{t+1} = (1+\eta)G_t,$$

$$G_{t_n^-} = G_{t_{n-1}^+}.$$

4.2.3.2.2.2 Transition at a Policy Year t_n The contract events take place at the discrete policy years. On each date t_i the future cash flow f_n can therefore take three forms:

- with probability $q_n = (1-p_n)$ death insurance amount D_n;
- with probability p_n:
  - withdrawal $\gamma_n \in \left[0; A_{t_n}\right]$ (which reduces A_{t_n} and the guarantee G_{t_n}, in an equal or even way); or
  - exercise the income benefit I_{t_n}.

4.2.3.2.3 Contract Valuation Traditionally, customers' behaviour has been modelled by historical or backward-looking statistical regressions, which have empirically been not enough to anticipate or dynamically extrapolate the observed shifts in behaviour and meet customers' risk appetites. In contrast, an "efficient" behaviour strategy valuation is a prudent forward-looking approach whereby policyholders behave in a way that maximizes the net present value of future cash flows, depending on key drivers such as market conditions and product design features (see Wang et al. 2015 and Kalife et al. 2018).

4.2.3.2.3.1 The Stochastic Control Problem The customer experience is measured by the value of the retirement savings contract (i.e. the average consumption of accumulated savings discounted), and is therefore the average of the sum of the future cash flows f_n discounted over the life of the retirement savings contract:

$$V(t_0, A_0, I_0) = \mathbb{E}_{t_0}^{\mathbb{Q}} \mathbb{E}_{t_0}^{\delta}\left[\sum_{n=1}^{N} e^{-rt_n} f_n\right],$$

where $\mathbb{E}_{t_0}^{\mathbb{Q}} \mathbb{E}_{t_0}^{\delta}$ denotes that the expectation is conditional on information available at time t_0 – i.e. with respect to both the financial risky asset process under $\mathbb{Q}$ and the mortality process under the real probability measure δ.

Noting $f = (f_1, f_2, \ldots, f_N)$, the savings strategy, it is considered that the empirical profiles correspond to an optimal strategy that maximizes customer satisfaction:

$$f^* = \text{argsup}_f V(t_0, A_0, I_0)$$

This falls within the framework of standard optimal stochastic control problems for a controlled Markov process, whose numerical resolution is carried out in a retrograde manner starting from the terminal condition of date N. Finding the contract value $V(t_n, A_n, I_n)$ at time t_n is done via a backward Bellman equation.

Since the account value A evolves between two anniversary dates, whereas the benefit base is a constant piecewise function (i.e. changes at anniversary dates only), the value is driven by a partial differential equation with jump conditions at each withdrawing date to link the prices at the adjacent periods; the required backward recursion is thus written between t_{n+1}^- and t_n^+ as

$$\begin{aligned} V(t_n^+, A, G) = \mathbb{E}^{\mathbb{I}}\Big[B_{n,n+1}\Big(1_{I_{n+1}=1} \times \mathbb{E}_{t_n^+}^{\mathbb{Q}}\left[V\left(t_{n+1}^-, A_{t_{n+1}^-}, G_{t_{n+1}^-}\right) \mid A, G\right] \\ +1_{I_{n+1}=0} \times \mathbb{E}_{t_n^+}^{\mathbb{Q}}\left[D\left(t_{n+1}^-, A_{t_{n+1}^-}, G_{t_{n+1}^-}\right) \mid A, G\right]\Big)\Big] \\ = (1-q_{n+1})\mathbb{E}_{t_n^+}^{\mathbb{Q}}\left[V\left(t_{n+1}^-, A_{t_{n+1}^-}, G_{t_{n+1}^-}\right) \mid A, G\right] + q_{n+1}\mathbb{E}_{t_n^+}^{\mathbb{Q}}\left[D\left(t_{n+1}^-, A_{t_{n+1}^-}, G_{t_{n+1}^-}\right) \mid A, G\right] \end{aligned}$$

with jump condition

$$V\left(t_n^-, A, G\right) = \max \quad \gamma_n \in \mathbf{A}_n \left(\bar{f}_n\left(A, G, \gamma_n\right) + V\left(t_n^+, h^A\left(A, G, \gamma_n\right), h^G\left(A, G, \gamma_n\right)\right)\right).$$

4.2.3.2.3.2 Numerical Scheme The algorithm starts from a final condition for the contract value at t_N^-. Subsequently, solving the PDE gives the solution for the contract value at t_{n-1}^+.

The PDE used to calculate the expected value under the assumed risk-neutral process for the risky asset St is easily derived using the Feynman–Kac theorem.

When the risky asset follows a geometric Brownian motion process, the governing PDE right after a withdrawal decision t_n^+ to right before the following one t_{n+1}^- for $n = N - 1, N - 2, \ldots, 0$ is expressed as follows,

$$\partial_t \Phi + \frac{1}{2}\sigma^2 A^2 \partial_{AA} \Phi + \left(r - \alpha^A\right) A_t \partial_A \Phi - \alpha^G G_{t_n^+} \partial_A \Phi - r\Phi = 0,$$

to which we add boundary conditions (defined in the next section). It is solved using the Crank–Nicolson finite difference method (see Dai, Kwok & Zong 2008 and Huang, Forsyth & Labahn 2012).

Note again that the benefit base changes only at the anniversary dates and is a constant parameter between two anniversary dates. Applying the jump condition to the solution at t_{N-1}^+, we obtain the solution at t_{N-1}^-, from which further backward time stepping gives us a solution at t_{N-2}^+, and so on. The numerical algorithm takes the following key steps.

1. Generate a finite grid for the account value A and benefit base G – i.e. $A_0 < A_1 < \ldots < A_J$ and $0 = G_0 < G_1 < \ldots < G_K$.
2. At t_N, define the final condition for each note point (A_j, G_K), j = 1, 2,..., J and k = 1, 2,..., K to get Φ (t_n^-, A, G) and the boundary conditions for Amin and Amax for each potential $G_{k\in\{1,2,\ldots,K\}}$.
3. For each potential benefit G_k, k = 1, 2,..., K, solve the PDE using the Crank–Nicolson finite difference scheme to obtain $\Phi\left(t_{N-1}^+, \cdot A, G\right)$.
4. Apply the jump condition to obtain Φ (t_{N-1}^-) for all the values of γ_{N-1} and find the withdrawal amount γ_{N-1}^* that minimizes $\Phi(t_{N-1}^-, A, G)$. In general, this involves a two-dimensional interpolation in (A,G).
5. Repeat (3) and (4) for $t = t_{N-2}, t_{N-3}, \ldots, t_1$.
6. Evaluate the PDE for the backward time step t_1 to t_0 to obtain solution $\Phi(t_0, A, G)$ at A_0 and G_0.

4.2.3.2.3.3 Localization and Boundary Conditions Within each time interval (t_{n-1}, t_n), only the account value varies, since all the benefit bases, death and life, remain constant. Thus, for $t \in (t_n^{\mp}, t_{n+1}^-)$, the annuity value Φ(t, A,G) solves the following linear PDE for each fixed value of the benefit base G:

$$\partial_t \Phi + \mathcal{L}\Phi = 0, \mathcal{L}'\Phi = \frac{1}{2}\sigma^2 \, A^2 \partial_{AA} \Phi + \left(r - \alpha^A\right) A \partial_A \Phi - \alpha^G G \partial_A \Phi - r\Phi.$$

This equation is originally posed on the domain (t, A) ∈ [0,T] × [0,∞). For computational purposes, and because asset prices are finite, and so is the account value, one needs to localize this domain to [0,T] × [0, A_{max}], where A_{max} is large enough not to be attained by the account value during the lifetime of the annuity. Thus, we need to add complementary boundary conditions. We consider that we are between two anniversary dates $t_n^{\mp}$ and t_{n+1}^- backwards.

- When A = 0, the policyholder no longer has the possibility of making any withdrawals from her account. However, if the income benefit election is possible, then the income period begins, given that the policyholder is alive, and the death benefit is activated if she is dead at t_{n+1}. Since the account value is equal to zero, then the annuitization will be indexed on the benefit base. Therefore, we have

$$\Phi(t,0,G) = e^{-r(t_{n+1}-t)}\left(p_{n+1}\frac{\ddot{a}^{gua}_{t_{n+1}}}{\ddot{a}^{act}_{t_{n+1}}}G + p_n q_{n+1} D\left(t^-_{n+1},0,G\right)\right).$$

- When $A = A_{max}$, we consider retrieving all the cash more interesting than any other strategy if the policyholder is alive. If she dies, the death benefit will be activated. Therefore, the Dirichlet boundary condition for this case is

$$\Phi\left(t,A_{max},G\right) = e^{-r(t_{n+1}-t)}\left(p_{n+1}\,A_{max} + p_n q_{n+1}\,A_{max}\right) = e^{-r(t_{n+1}-t)} p_n\,A_{max}.$$

Let us define the solution domains:

$$\bar{\Omega}_n = \left[t^+_{n-1}, t^-_n\right]\times\left[0, A_{max}\right],$$
$$\bar{\Omega} = \bigcup_{t_n}\left[t^+_{n-1}, t^-_n\right]\times\left[0, A_{max}\right].$$

Construction of the Scheme

Let $(A_0, A_1,..., A_J)$ be the equally spaced grid in the direction of the account value with $A_0 = 0$ and $A_J = A_{max}$. Analogously $(G_0,..., G_K)$ is an equally spaced grid for the benefit base with $G_0 = 0$ and $G_K = G_{max} = A_{max}$. The spatial steps for both variables are considered to be equal:

$$\Delta A = \Delta G_j, \text{where } \Delta A = \frac{A_{max} - A_0}{J} \text{ and } \Delta G = \frac{G_{max} - G_0}{K}.$$

Hence, $A_j = j\Delta A$ and $G_k = k\Delta G$, $\forall j,k$. The discrete time steps are denoted by $n\Delta t$ for $n = 1,..., N$, where $T = N\Delta t$. Since, in our analysis, we consider that events occur only at anniversary dates that are yearly, $\Delta t = 1$ and each time t_n coincides with the discrete time step $t_n = n$.

The numerical procedure to solve the approximation is the standard finite difference approach, using the general theta scheme (the Crank–Nicolson scheme). We employ the two-level implicit finite difference scheme to discretize.

Recall that changes in the benefit base occur only at withdrawal dates. After withdrawing the amount γ_n at time t_n, the account value changes from At^-_n to At^+_n, and the benefit base drops from Gt^-_n to Gt^+_n.

The application of the jump condition decreases the account value and benefit base. For each G_j, a continuous solution from the PDE is associated. We can restrict the possible values for the withdrawal amount to multiples of ΔA. This implies, for a given account value A_j at time t^-_n, that the withdrawal amount γ takes j possible values: $\gamma = A_j - A_i$, $i = 1, 2,..., j$. However, numerical tests show that a finer grid is preferable for the withdrawal amount. Therefore, it is not guaranteed that the account value, nor the benefit base after the withdrawal, At^-_n and Gt^+_n, fall within their respective grid nodes. To solve this issue, a two-dimensional interpolation is required.

4.2.3.2.4 Enhancing Customer Experience through Product Features that Meet Customer Expectations The introduction of certain product features leads to savings consumption behaviours in line with the observed patterns, for an enhanced customer experience.

Behaviours take the form of (i) withdrawals $\frac{\gamma^*}{G} > 0$ (expressed as a percentage of the guarantee G) or (ii) the activation of the life annuity $\left(\frac{\gamma^*}{G} = -1\right)$ in Figure 4.40, depending on the past duration t and the ratio A/G (investment account A divided by guarantee G).

The choice of less risky S investment assets (illustrated by a lower volatility: 10% versus 30% in the figure) favours earlier withdrawals, thus moving the behaviour towards the empirical pattern observed for those below 60 years.

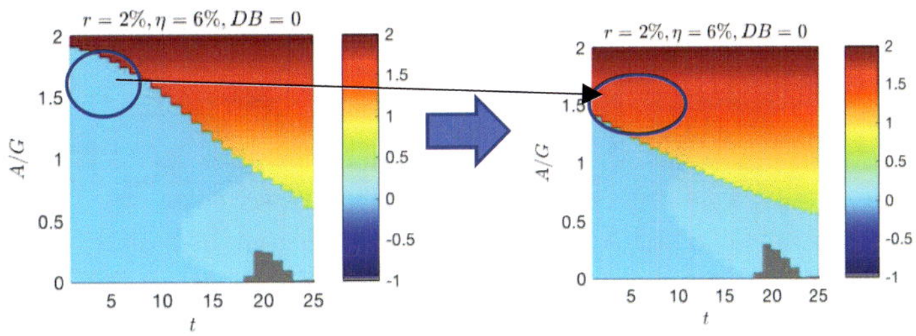

FIGURE 4.40 Impact of asset volatility

The introduction of a dynamic roll-up rate η for the guarantee G (i.e. decreasing as interest rates fall), on the one hand, reduces the disparity in withdrawal sizes but, on the other hand, promotes a more frequent activation of the life annuity, thus moving the behaviour towards the empirical pattern of retirees aged from 60 to 70 years.

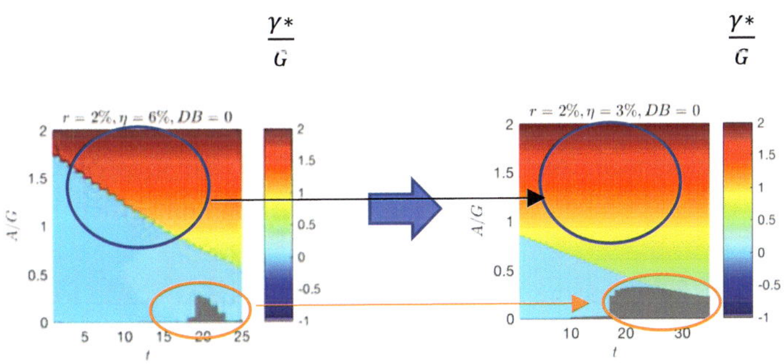

FIGURE 4.41 Impact of a variable roll-up guarantee rate

The introduction of a death guarantee, on the one hand, promotes the smaller size of withdrawals but, on the other hand, postpones the withdrawals and the activation of the life annuity, thus moving the behaviour towards the empirical pattern of retirees over 70 years.

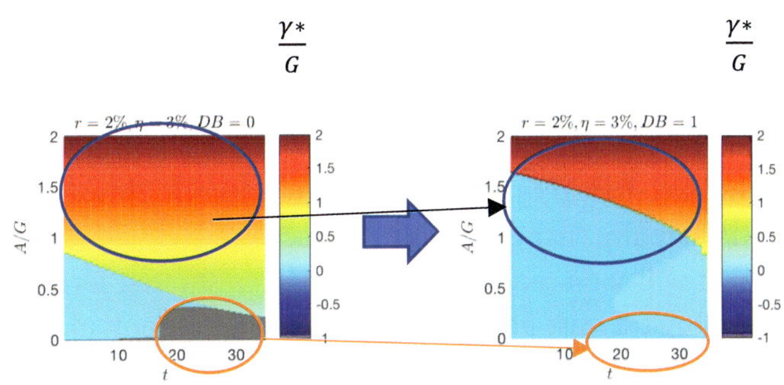

FIGURE 4.42 Impact of a death guarantee

Savings behaviours depend, as expected, on the economic environment: a low investment account and/or a high guarantee provide an incentive to reduce the frequency and size of withdrawals, and to activate the life annuity earlier.

Similarly, the effectiveness of product design changes is influenced by the economic environment: withdrawals increase in frequency and size with interest rates and the level of investment assets.

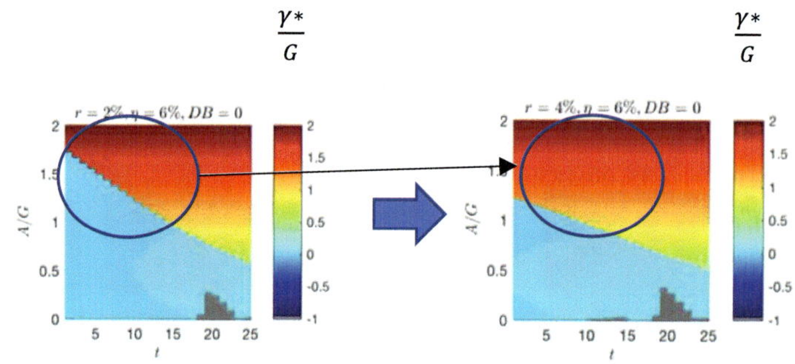

FIGURE 4.43 Impact of rising interest rates

Conclusion

Longer life expectancy, increasing lifestyle and health care costs and the persistence of low interest rates have contributed to the emergence of increasingly differentiated savings behaviour patterns. The explicit quantitative modelling of efficient savings behaviour makes it possible to select the product features that meet customers' needs and expectations, in line with their empirical savings behaviour patterns, for an enhanced customer experience.

4.2.4 Reconciling Insurer and Customer Risk Appetites through Sustainable Capital-Light Products: Reducing Capital Requirements while Increasing the Upside Potential

Traditional life insurance products with long-term guarantees have the greatest exposure to investment risk. They are very costly unless insurers can reduce the duration mismatch, which usually entails high hedging costs and higher fees for the customer.

Therefore, this product type has become unattractive over the last decade for both the issuer (given the regulatory requirements combined with the outlook of a long-lasting market environment with low or negative interest rates) and the customer (given that the level of guarantees has been steadily declining and the potential upside returns are very limited).

Initially life insurance companies reduced the cost of guarantee through various ways, such as offering a guarantee for only a proportion (typically 60 to 90%) of the premium instead of 100% or charging fees higher than the "nominal" guarantee rates.

Then life insurers introduced two alternative product design initiatives.

- One approach has tried to undertake an "incremental optimization" on a slow basis (the preferred choice for the majority of insurers), which takes a market perspective based on the idea that observed low interest rates are a temporary phenomenon, combined with the argument that long-term insurance liabilities are uniquely illiquid, and so can be marked down for an illiquidity premium. Therefore, these companies are not fundamentally changing their product strategy but relying on the expectation that the economic environment will soon recover with rising interest rates. As a result, they continue to offer insurance products including (at least partial) guarantees while simultaneously focusing on the reduction of their cost base in order to reduce the guarantee product's capital intensity.

- The alternative approach targets a "fundamental strategic shift" of the insurers' business (as in the United Kingdom, Sweden and Norway), which takes the perspective that companies are currently not rewarded for taking the investment risk of offering guarantees and low interest rates might persist for a longer period. These insurers move away from products with "unmanageable" long-dated guarantees and focus on unit-linked products with lower capital intensity through lower guarantee richness or lower equity exposure. Some (e.g. Standard Life, Legal & General, Ameriprise) have also expanded their own asset management capabilities in order to generate at least some investment management margins in house.

Given these two alternative approaches, insurers are trying to develop their product offerings accordingly, with hybrid products mixing unit-linked and traditional product features. The challenging trend towards innovative investment-linked products offering certain guarantees while simultaneously requiring less risk capital has been approached from two emerging product categories.

- The first category can be described as the pure unit-linked products, whereby the policyholder takes the investment risk, and so the insurer bears no investment risk. However, from an insurer's perspective there is no value added compared to a pure asset manager, and thus it is less attractive for the customer due to the usual large expense loadings. However, many European policyholders are not comfortable with taking the full investment risk and are seeking some kind of guarantee. This pure UL product is then enriched by adding guarantees in such a way that one portion of the customer's money is invested in funds directly (and the policyholder takes all the investment risk) while another portion is treated as a traditional product with a certain minimum guaranteed sum assured. These developments result in so called "static" or "dynamic" hybrid products, such as dynamic UL products (e.g. CPPI policies), variable annuities or UL products with guarantees (see Chapter 2 for details of their respective product designs).
- The second product category consists of the traditional life insurance products with fixed guarantees. However, fixed guarantees are very capital-intensive to interest rate risk. These products are thus gradually being modified towards "traditional products with profits", or so-called "new traditionals", including an optional index participation or smaller but more flexible guarantees. Capturing the potential short-time upsides of the market while retaining some floor guarantee level makes these products more attractive for the customer, while holding less capital intensity for the insurer.

Life insurers have thus introduced a variety of new so-called "capital-light" products aimed at consuming fewer capital requirements: structured annuities, index-linked products, unit-linked products with guarantee mechanism, constant proportion portfolio insurance (CPPI) in underlying funds, individual CPPI policies, static and dynamic hybrids, and "select" products (index-linked surplus) are only a few examples, as categorized in Figure 4.44 (see Chapter 2 for details).

Product Innovations	Product Features
Hybrids	Combination of traditional and UL product Static or dynamic allocation 2 or 3 baskets
UL iCCPI	UL product with dynamic asset allocation Guarantee given by third party
New Traditionals	Traditional products with reduced guarantees in return for higher upside Reduced guarantees possible, e.g., in form of lower guaranteed rate (0%) only
Traditional with Index Linkage	Traditional with profits product and lower guarantee Usage of levered certificates in order to increase exposure level Bonus invested into certificate linked to index

FIGURE 4.44 Product innovations

All these innovations in product design acknowledge that life insurance companies need to handle the complications arising from the conflict between low interest rates (giving rise to high required capital requirements) and the customer's pursuit of a certain security level, while simultaneously limiting the level of complexity as part of an effective distribution strategy.

In terms of business volumes, such hybrid unit-linked premiums have significantly increased in Europe since the introduction of the Solvency II capital requirement regulation, though the business sometimes developed more on the banking side than on the insurance side.

- In Germany, although it is characterized by high guarantees and long liability duration, and thus high interest rate risk, the hybrid UL products as described before, which have contributed merely about 8% to overall technical provision, have recently reached 25 to 30% of new business premiums.

- The Dutch life insurance market looks similar to the German one, in that traditional savings products represent nearly two-thirds of gross premium income and of technical provisions, while the average guarantee level is as high at about 3%. To mitigate exposure to interest rate risk, companies have been massively shifting their business towards banking savings and capital-light products (75% of new individual savings products) and away from individual life products (including UL). Furthermore, some Dutch insurers, such as Aegon and NN, are diversifying into asset management in order to acquire additional fee income and reduce their capital requirements, though their margins are usually low.

- French insurers' technical liabilities are still dominated by traditional products, and have only gradually increased the UL part (less than 20% of overall technical provision) since 2017, with 30 to 40% of new premiums now, versus 20% in 2016. That said, the average guarantee rate in the book has continued to decrease to an average of about 0.8%, while French insurers have taken measures to avoid new premiums being invested in old policies with high guarantees. Consequently, they face less significant interest rate risk than their German or Nordic peers, despite the relatively long duration of insurance contracts (11 years) and having increasingly invested in shorter-term corporate bonds.

- Italian life insurers' back book of guaranteed interest policies (75% of technical reserves) are generally well matched in the duration of assets and liabilities. The market average guaranteed interest rate on participating life reserves has declined by about 20 to 30 basis points to around 1%, though it is sufficiently covered by returns on segregated funds. The average effective maturity is five to seven years, with combined surrenders and lapses of about 10% (as most savings products' guarantees offer a guaranteed surrender value if redeemed early), exposing Italian life insurers to the risk of losses on their bond portfolios when yields and surrenders spike. Solvency II and low interest rates push insurers towards capital-light products through hybrid life policies, accounting for 15 to 20% of new life insurance business.

- Spanish life insurers' guarantees, though reducing, remain relatively high at between 2.5 and 3.0% for those in force, as against less than 1% for the new business. However, life insurance products in Spain have a shorter duration than in the rest of Europe, and Spanish insurers have historically mitigated much of the reinvestment risk by closely matching assets and liabilities, as well as applying market value adjusters in the event of early redemptions, thus passing the risk of negative movements in the market onto the policyholders.

- Swiss life insurers started to steer their product portfolios and asset allocation towards less capital-intensive products earlier than most European peers, motivated by Switzerland's ultra-low interest rate environment and by a very strict regulatory regime. Nevertheless, insurers' back books are dominated by traditional products with guarantees (about 95% of total reserves), despite guarantee products making up 10 to 20% of new business premiums. It will take time for the positive effects of the capital-light products to be visible in the back book.

- The Nordic life insurance markets (Denmark, Finland, Norway and Sweden) are exposed to material interest rate risk due to significant guarantees in the back book, despite the amount of guarantees (2.5 to 3.5%). Like in the rest of Europe, we are seeing a clear move towards capital-light products, such as UL products or products with lower or no guarantees. In Sweden, for example, it is becoming more popular to have only about 80% of paid premiums guaranteed. Nordic insurers are restructuring their portfolios more rapidly than in other European markets. In Norway the amount of UL business in terms of premiums increased to 35% in 2016 from 21% in 2012, and to about 18% from 12% in terms of reserves, in part because local regulators support the conversion of life insurers' traditional in-force business into UL products.

4.3 SUSTAINABLE INVESTMENT ASSET DESIGNS

Many life insurers are eager to move into asset management as a capital-light business. They hope to lighten their capital burden; however, this change also puts insurers in direct competition with asset management companies, which have a significant competitive advantage in this area and often operate at lower costs. Revenues have been growing more quickly for asset management than for life insurance, generally with a higher return on equity as well. Investors find it easier to assess earnings projections in asset management, so, for a given growth rate, asset managers tend to be rewarded with a higher multiple than straight life insurers.

But, while the grass might seem greener in asset management, that industry comes with its own substantial challenges. Fees have steadily eroded as low-cost passive strategies continue to gain market share at the expense of active strategies. In some countries investor protection regulations have cut commission revenues, while stricter compliance requirements have created significant costs for client advisers.

To create and sustain value in asset management, insurers will have to develop and master some new capabilities. Publicly listed insurers subject to quarterly earnings pressures have faced competitors with more visibility in underlying cash flows and returns, as well as with longer-term investment horizons.

The competition driven by private acquisitions of in-force books is growing. In mature markets, publicly listed life insurers also face competition from mutual companies, due to the mutuals' access to lower-cost capital.

4.3.1 The Growing Competition of Private Equity Firms with Capital-Intensive Offloadings by Insurers

Private equity and pension fund investors such as Guggenheim, Blackstone and Athene Holding have crowded into life insurance through buying legacy books.

On the one hand, the traditional life insurance business has become less attractive: many insurers' earnings on in-force blocks have come under pressure, as guarantee rates to policyholders are still as high as 150 to 400 bps in some markets, while yields on bonds have declined by 150 to 300 bps since 2010/11. Naturally, this has strained capital, as insurers have had to adjust their reserves to reflect future earnings expectations. An average insurance company reinvests about 12% of its assets annually, so this profitability challenge becomes increasingly acute every year.

With the expectation that rates will persist at low levels for a significant period of time, there has been continued pressure on short-term returns and restrictions to the flexibility in pricing of new and existing products. Therefore, the focus has shifted to exiting sub-optimal lines of businesses. In addition, operational and IT issues have continued to challenge profitability and require significant investment and management attention to address – e.g. migrating legacy policy administration systems and investing in automation can be attractive in the medium term but require careful management to prevent technical or servicing issues.

On the other hand, reinsuring or selling legacy capital intensive or non-profitable books is an opportunity to shift strategy and move into more attractive capital-light, fee-based business model businesses (e.g. investment management in defined-contribution pensions, in which assets are growing at 6 to 8% annually across Europe and the United States), as a way to either stabilize their capital ratio, pay out dividends or deploy capital for growth. As a result of such strong fee income in their revenue mix, these insurers will typically trade at higher valuations (often nine to 12 times P/E ratio, or 1.1 to 1.7 times book value) than those in capital-intensive businesses (usually five to eight times, or less than 1.1 times book value).

In particular, several US player have divested their closed block of variable annuities or fixed index annuities to reduce the volatility of earnings and refocus on capital-light businesses (e.g. Equitable Holdings, Voya, Prudential, Allianz). Investors responded well: over the subsequent three years their stocks outperformed the life index by several percentage points, with one carrier's price-to-book (PB) ratio rising from 1.2 to 1.6 times, at a time when the broader industry's PB ratio fell from 1.3 to 1.0 times. One recent reinsurance transaction is Equitable Holdings reinsuring its $12 billion block of annuities to Venerable Holdings in 2021 (140,000 fixed-rate variable annuities, $14 billion, 13% of total variable annuity in force, 6% guaranteed rates sold between 2006 and 2008), which reduced Equitable's CTE 98 required capital by some 64%, while halving the net amount at risk, creating $1.2 billion of value, releasing $800 million capital and increasing the RBC ratio by 60 points. Among other examples are Blackstone's purchase of Allstate Life Insurance Company, Sixth Street's purchase of Talcott Resolution Life Insurance Company, and KKR's strategic transaction with Global Atlantic. One notable example in which geographic and product expansion was used as justification for divestiture is the spinoff of Jackson Financials from Prudential plc, when the company will use excess capital to fund its growth in the Asian market. Another example is American Financial Group's decision to sell its annuity business to Massachusetts Mutual Life Insurance Company for $3.5 billion.

Several companies have also entered the group life, disability and voluntary benefits businesses through acquisitions in order to get to the customers at the top of the value chain. Some recent transactions include MetLife's $1.7 billion acquisition of Versant Health in 2020 and New York Life's $6.3 billion acquisition of Cigna Group's life and disability insurance business in 2019. Following these transactions, New York Life and Metlife became the second largest market participants in group life and voluntary benefits respectively.

In 2021 private equity (PE) firms announced deals to acquire or reinsure more than $200 billion of liabilities in the United States. Such investors now own over $900 billion of life and annuity assets in western Europe and North America. Assuming the pending deals close successfully, private investors will own 12% of life and annuity assets in the United States, totalling $620 billion, and represent more than a third of US net written premiums of indexed annuities. All five of the largest PE firms by assets have holdings in life insurance, representing 15 to 50% of their total assets under management.

The core attraction is straightforward: closed life insurance books are interesting to investors primarily because the cash flows that result from them are regular and can be easily forecast, even for decades to come. While the balance sheets of life and annuities companies are well stocked with assets (to match the liabilities of future payouts and indemnities), until payout these assets need to be invested so as to generate returns. And, in many cases, the cost of servicing the liabilities is significantly lower than the potential investment return, so the spread in between represents an attractive margin.

The most common way for PE firms to capture the spread is to set up an insurer that they control through an equity investment and then acquire or reinsure books from other insurers. To ensure they earn the required returns on acquired books, they typically influence the strategic asset allocation (SAA) and apply their investment management capabilities to earn alpha on some of the asset classes.

The benefits for the private equity firms are threefold.

- First, PE firms provide expertise and greater tolerance for offshore and unique capital structures than some insurance companies, including partnering with pension funds requiring low rate of returns. The investment management focus on finding the optimal capital efficiency and desire to up-risk the portfolio will help these insurance products stay profitable in low interest rate environments. As such, private equity firms serve as enablers for insurance companies seeking to increase yields on capital. PE firms' investment strategies can generate internal rates of return of 10 to 14%, largely by rotating the asset allocation into classes that are higher risk and higher return (e.g. asset-backed securities such as collateralized loan obligations [CLOs], and private loans) and achieving higher alpha (a 62 basis point higher investment yield than the industry average), while the approximately ten times asset-to-equity ratio typical of insurers amplifies the impact of strong investment performance.

- Second, by acquiring insurance companies, private equity firms can increase their assets under management considerably with low risk, which allows them to maintain strong credit quality due to most insurance companies having heavy fixed-income portfolios and a strong capital base. PE firms can then leverage the steady stream of reliable assets offered by insurance books, providing a stable base to rapidly build their alternative credit capabilities. Indeed, acquiring a life book immediately provides long-term assets for the firm's credit arm to invest, and so is a much faster way to reach scale and far less onerous than raising several credit funds, providing a significant source of fee-related earnings, which are more resilient to market fluctuations and more stable than carried interest. In particular, PE-backed insurers typically use structured credit products as core assets in their life insurance books. Origination of these asset classes is reaching record levels (e.g. $760 billion CLOs). Besides, such loan risk is less important because most liabilities will not be paid until their long-term maturity. For example, thanks to such massive amounts of low-cost capital supply by Athene, Apollo can originate massive loans directly at a time of growing demand while the US Dodd–Frank regulations have limited banks' lending ability, which in addition yield more than usual because there are no intermediaries (brokers, investment banks) and hold higher yields for tailored credit solutions (middle-market direct lending, CLOs, equipment leasing, structured products).
- Third, life insurance offers the potential for scale. Traditional life liabilities in Europe total €4.5 trillion; in the United States, life and annuity insurers carry $4.5 trillion of assets on the general account, with an additional $1.5 trillion in separate variable-annuity liabilities, and there are $3 trillion of private sector defined-benefit liabilities. Even after many large PE acquisitions, a huge supply continues to be available.

In order to assess the quality of the potential acquisition of life insurance books, PE firms may use the following metrics not only for the aggregated book but also across the business lines of the book (to the most granular way).

- Traditional financial ratios: ROE (e.g. 14 to 16%); earnings per share (EPS) growth rate (e.g. 5% p.a.); P/E ratio, debt-to-equity ratio (e.g. 25 to 28%), earnings coverage, dividend yield + share buyback (4 to 6%); WACC to be compared to ROE or return on invested capital (ROIC) (using the Sharpe econometric model for the cost of equity based on long-term estimates – e.g. risk-free rate ~2.5%, beta ~1.2, equity risk premium ~6%; cost of debt 5 to 6% -> WACC ~6 to 7%). The CoC is the indicator to meet the bottom line of shareholders' expectation, whereas ROE is to show the target for shareholders.
- Insurance industry specific metrics: net premium / equity (over two times); (yield – guaranteed rates) gap, potential VIF monetization opportunities.
- Investment equity allocation (less than 10%) and ALM sensitivities.
- Capital-based metrics: solvency ratio (e.g. 170 to 230%); solvency IRR (e.g. 15%); share price / unrestricted tier 1 (or SII equity shock delta) (~1); capital generation / share price (10%); SCR sensitivities to market and behaviour risks; quality and structure of capital (percentage of subordinated debts); solvency ratio triggers for dividend increase (160%); percentage of capital-light policies among life products, capital intensity and wealth consumption metrics.

Once they have acquired a book, PE firms usually use six levers to improve ROE by up to four to seven percentage points.

- Investment performance: optimization of the strategic asset allocation and delivery of alpha within the SAA. Most PE firms view this as the main way to create value, and typically have deep skills in core investment management areas, such as strategic asset allocation, asset/liability management, risk management and reporting, as well as access to leading investment teams that have delivered alpha. For example, Athene asset yield tends to be +30 bps, and 30 bps more efficient in operating expenses than average US life insurance companies. As a result, Athene's implied ROE outperforms by ~300 bps the average US life insurance company.
- Capital efficiency: optimization of balance sheet exposures – e.g. active management of duration gaps. This lever presents a great opportunity, given their different treatment under generally accepted accounting principles, enabling PE firms to apply a longer-term lens and reduce the cost of hedging.
- Operations/IT improvement: reduction of operational costs through simplification and modernization. Cost cutting is a paradox here: on the one hand, insurers have generally not cut costs as rapidly as other industries, and the books in question are often high-cost operations; on the other hand, acquirers sometimes underestimate the complexity that drives these costs, given the complicated nature of multiple legacy systems and nuances across policy vintages – to say nothing of new costs for post-merger integration.
- New entrants have a particular advantage here, as they can adopt a digital-first approach to data and technology, unencumbered by legacy system issues. Empirically, in the United States PE firms enjoy costs 20 to 40% lower than general life insurers, while in Europe they are burdened with costs 30 to 60% higher, in part due to the more complex books they have acquired. Process streamlining, changes to the operating location and efforts to reduce overhead costs and scale benefits are the usual levers. However, unlocking scale benefits often requires action to reduce the complexity of the book, by offloading legacy products or decommissioning legacy IT systems, which may reduce dividends in the short term but offer an attractive return over the longer term. New AI techniques, including machine learning and data virtualization applied to system migration and data extraction, allow insurers to bring down part of the costs before executing an outsourcing contract, and therefore to retain more value.
- Technical excellence: improvement of profitability through price adjustments, such as reduced surplus sharing. Here again, AI can enhance an insurer's understanding of customer blocks and enable it to develop a segmented approach with targeted interventions.
- Commercial uplift: cross-selling and up-selling higher-margin products. Once more, AI offers additional benefits, such as avoiding lapsing through a better understanding of customers and identifying opportunities to cross-sell or up-sell.
- Franchise growth: acquiring new blocks or new distribution channels. Again, advanced analytics can unlock new opportunities here as well: applying machine learning to model policyholder behaviour in a target book, for example, can be 20 to 50% more accurate than traditional actuarial methodology.

However, there are a few risks that PE firms should be aware of.

- The asset side of the balance sheet, including illiquidity and credit risk, introduced by rotating the portfolio into higher-risk credit assets. Managing the credit risk of the underlying assets, maintaining sufficient liquidity as needed for policyholders and managing the mark-to-market volatility on the credit portfolio during a credit downturn to maintain regulatory and rating stability are all critical. This risk has been latent, given the relatively benign credit environment in the past decade, but a future emergence could put stress on balance sheets.

- On the liability side, calculating the value for the technical provisions – i.e. the present value of future benefits to customers, including future profit participation – as well as the costs associated with servicing the contracts is complex. Should the new management plan to significantly change its profit-sharing strategy, this would be directly reflected in the technical provisions. And the cost situation in a run-off portfolio might conceivably get worse expenses associated with servicing the insurance contracts; for example, the outsourcing of particular activities also typically changes with the separation from the previous insurance group, with the result that the future costs will be different.
- The solvency capital requirements may also significantly increase as a result of the outsourcing of contract management, since the loss absorption capacity that had existed beforehand can no longer be used.
  - These elements specifically are increasingly scrutinized by regulators to protect the interests of the policyholders and ensure that they are at all times backed up by sufficient capital and that the commitments made can be fulfilled, both in the base case and in the event of a 200-year occurrence. A key concern is therefore regulatory uncertainty. Although regulators are getting used to the idea of PE ownership of life carriers, approval can take time, as PE firms are entering a highly regulated industry for the first time.
- The first question is that of the ability to appropriately manage the portfolio, which depends on whether existing systems and employees of the ceding undertaking are also transferred. If the safety level is expected to drop, the supervisory authority will demand safeguards. One example is that the acquiring company must make a commitment that for a certain period of time its capital resources will not fall below a level significantly higher than the legal minimum required, to prove that they are backed up by sufficient funds. Another possible safeguard can consist of a preclusion (by way of contractual relationships within the new group) of customers being burdened with higher administrative costs than previously.
- Public opinion can be another obstacle. In two recent western European deals, concerns about the impact of a PE owner meant that late-stage negotiations did not succeed.

4.3.2 The Insurance Company Hunt for Yield amid Tightening Regulations

Insurers face a challenge: in this period of historically low rates, where to find yield? In fact, since 2006 the net investment yield of insurance portfolios has massively declined (from 6 to 4% in the United States).

Since its creation in 1986, the Bloomberg US Aggregate Bond Index ("the Agg") has been the most widely used proxy for the US bond market. However, the broad fixed-income universe has become more diversified in structure and quality over the past 36 years with the growth of sectors such as asset-backed securities (ABSs), non-agency residential mortgage-backed securities (RMBSs), high-yield corporate bonds, leveraged loans and tax-exempt municipal bonds.

In contrast, the composition of the Agg has not kept pace with these changes: Treasuries today comprise 49% of the Agg, which, when combined with agency securities, amounts to more than 75% of the Agg. In fact, the Bloomberg US Aggregate Bond Index represents less than a half of the fixed-income universe in the United States ($24 trillion out of $52 trillion). Thus, investors reluctant to stray far from the Agg's composition are disadvantaged by sector concentration in low-yielding securities.

As a result, over the last decade declining insurance portfolio yields have led companies to seek out higher-yielding assets in which they had not historically participated, such as structured credit and private credit.

Others looked to introduce equity-like returns into their bond portfolio by using hybrid investment vehicles, some of which mirror to a degree the policy structures that insurers sell to policyholders, such as the fixed indexed annuity, sold by practically every life insurer in recent years.

However, easy financial conditions may be masking the risks associated with the potentially damaging long-term effects of reflating the economy through debt accumulation. Thus, over the past several years the regulators have become increasingly suspect of certain asset classes, and therefore predisposed to increase capital requirements on those investments. For example, in the United States insurers used to rely on the fact that, once a bond was rated by a nationally recognized statistical ratings organization, the bond was deemed exempt from filing. As of year-end 2021, principal-protected securities and combo notes no longer have filing exemptions, which means that insurers can no longer forecast capital requirements on a rating alone but must now file the security and wait for the organization's conclusion in hindsight, limiting insurers' ability to know with certainty the cost of allocating capital to an investment. For private letter ratings, insurers now are required to file a credit package for review before being able to utilize the rating. Moreover, the NAIC is looking at the potential of requiring two ratings on private securities.

This increased level of oversight is having a detrimental effect on insurers. In addition to the lack of clarity and consistency, and increased time needed to transact, this is having a direct economic impact on the business. Every asset on an insurance company's balance sheet has an associated designation and capital charge. While asset risk is only one of several risks for which insurers need to maintain capital, for life insurers – and particularly for annuity-focused writers – it is the largest, as the resulting capital factor seeking to capture the underlying credit risk is exponential. For example, a downgrade of a security from a 1.G designation to a 4.C designation would lead to an approximate 12 times increase in the capital required.

At a time when it is vital for insurers to find differentiated investments, these requirements are, essentially, forcing them to become index investors. It steers insurers into lower-yielding core asset classes, further pressuring investment yields. Indirectly, policyholders (and shareholders) are potentially compromised by the narrowing investment opportunities to which insurers have access, which may lead to lower crediting rates and higher insurance costs.

4.3.3 Neither Active nor Passive Asset Management Styles Seem to Have Significantly and Sustainably Managed Recurrent Choppy Markets during the Last Decade

The asset management industry has been facing four main challenges since the 2008 financial crisis, regarding not only alpha generation but also risk mitigation.

- First, there is the rise in valuations over the past few years, which is significantly due to falling interest rates rather than higher growth rates, driven by a $20 trillion stimulus by central banks (the Fed's $7 trillion balance sheet has nearly doubled since August 2019, growing by some $120 billion per month).
- Second, there have been drastic changes in volatility regimes patterns, as illustrated by a drag on volatility driven by persistently low interest rates and significant share buybacks (stirring 40% of the S&P 500 growth since 2012), combined with recurrent unexpected and short-lived but nonetheless severe volatility spikes (June 2010, August 2011, June 2012, August 2015, February 2016, June 2016, February 2018, October–December 2018, March–May 2020).
- Third, there is the increased competition between asset managers, driven by demand for uncorrelated returns.
- Fourth, the industry has been tentative and lacking in strong conviction, as illustrated by the frequent small gains/losses.

4.3.3.1 These Four Challenges Have Significantly Penalized Active Management over the Past Decade …

This is true whether analysis is based on fundamentals or on technical indicators. Regarding active management based on fundamentals, such as the sector rotation strategies, diversification is dying: there is a 75% cross-correlation of stocks within the S&P 500, as it is dominated by a few stocks (five techs account for 25% of its performance, and 50% when combined with the consumer and health industries' stocks), and value strategies have been systematically destroyed by debt-induced corporate shares buybacks and central banks.

4.3.3.2 … In Favour of Passive Managed-Volatility Fund Strategies, Massively Used by the Insurance Industry over the Past Decade

Such limitations have initially favoured low-fee passive management (e.g. managed-volatility funds), which aim to deliver a portfolio with a stable level of volatility in all market environments, by systematically reducing an allocation to securities with lower expected risk-adjusted return in exchange for securities with a higher level of expected risk-adjusted return.

This relies on the ability to forecast the near-term risk, by taking advantage of the three features usually observed over the past two decades.

- There has been a highly negative correlation between equities and bonds during times of exceedingly equity volatility, as noted by traditional investment companies, which used to increase the fixed-income allocation to offer some protection during equity bear markets.
- Volatility has persisted, as illustrated by a 55% correlation between adjacent 60-trading-day periods' standard deviation, in contrast to nearly zero for returns; past returns do not predict near-future returns, but past volatility predicts to some extent future near-term volatility.
- The third feature is translating in frequently adjusting the portfolio risky exposure inversely to predicted future near-term volatility (usually extracted from a simple extrapolation of recent experience, stemming from the significant correlation called "volatility clustering" – i.e. large changes tend to be followed by more large changes while small changes tend to be followed by further small changes). These strategies would indeed make no sense if risky asset returns followed an i.i.d. normal distribution. Volatility clustering and fat tails in return distributions are indeed the two major drivers explaining the success of managed-volatility strategies.

Regarding the mechanism of such a managed-volatility fund, let us illustrate with a so-called "target volatility fund" (TVF) or "capped volatility fund" (CVF), dynamically rebalanced between a risky and a safe basket to control realized volatility (RV): the exposure to the most volatile "risky" assets (i.e. equity) is shifted to less volatile "safe" assets (bonds, cash) whenever the basket volatility rises beyond the predefined target.

It offers a flexible and adaptive approach to asset allocation that aims to reduce portfolio volatility without sacrificing long-term returns while improving risk-adjusted returns.

At each weekly rebalancing date, the CVF algorithm calculates the new weights (risky/safe) to reflect the target volatility constraint (target volatility as limit), and triggers a rebalance if it exceeds the rebalancing threshold:

$$equity\,exposure = \min\left(\frac{\sigma_{\text{limit}}}{\hat{\sigma}_{\text{equity}}}, 100\%\right),\ \alpha^2\sigma_R^2 + 2\alpha\beta\sigma_R\sigma_S\rho_{R,S} + \beta^2\sigma_S^2 = \sigma_{\text{Target.}}^2$$

The volatility metric is usually an exponentially weighted moving average (EWMA) starting from the first 21 days' historical annualized rolling volatility:

$$\sigma_t^2 = \lambda \sigma_{t-1}{}^2 + (1-\lambda) r_t^2, \text{with } \lambda = 0.94.$$

The volatility of the fund is "capped" because the fund would never increase its target equity exposure above the maximum (90% in the case of CVF 90) even if the volatility of the equity is lower than the target volatility.

The ATM algorithm adjusts/reduces the level of equity exposure any time the realized volatility reaches a certain limit.

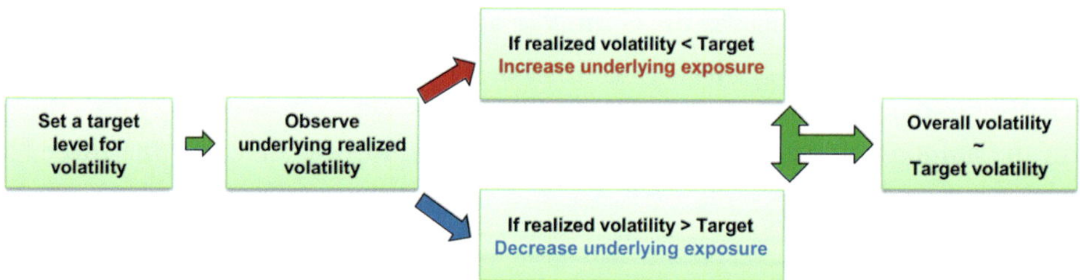

FIGURE 4.45 Managed-volatility funds mechanism

Empirically, over the past 90 years passive managed-volatility fund strategies have tended to normalize distribution tails of short-term market trends at low costs. In fact, they consistently contained volatility within a much tighter range while reducing outsized drawdowns in extreme market conditions, improving risk-adjusted returns by a tangible 20 to 30% for holding periods up to a quarter, enhancing portfolio skewness and achieving robust tail risk reduction.

Overall, managed-volatility strategies allow insurers to:

- reduce the fees – not only the management fees (~25 bps versus 100 to -150 bps for mutual funds) but, more importantly, usually half the guarantee fees;
- reduce the product repricing frequency;
- pay only for realized volatility, saving the volatility risk premium;
- have a guarantee value insensitive to the guarantee price; and
- remove the need for any more volatility hedging.

As a result, the performance has proved superior to the static allocation: it reduces the underperformance during periods of market decline (also leveraging from the usual corresponding lower interest rates through the higher bond allocation), while giving up the limited upside potential during market recovery (characterized by high volatility and market growth, with very low underperformance – e.g. –0.03% during 2012–2014). As a result, the equity participation pattern is similar up to the 90th percentile of the return distribution, and so far higher than within CPPI mechanisms:

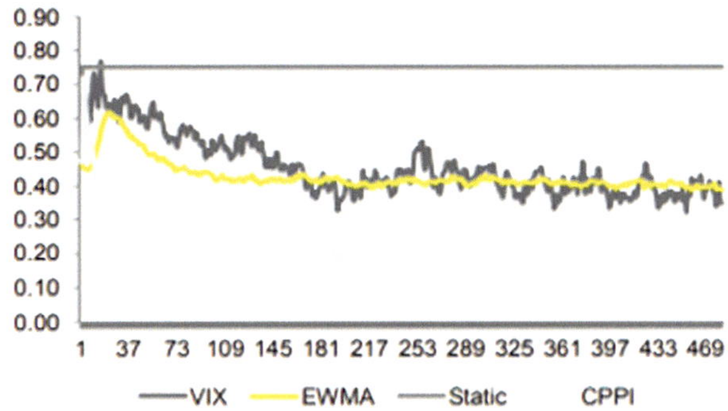

FIGURE 4.46 Average equity allocation

	CVF 90 - Vol 16%	Static 80/20
average	7.3%	6.0%
stdev	11.4%	13.6%
Min	-26.9%	-40.5%
0.5%	-24.2%	-36.6%
1%	-23.0%	-34.4%
5%	-18.5%	-28.9%
10%	-10.6%	-13.1%
25%	2.3%	1.2%
50%	9.9%	8.7%
75%	14.8%	12.1%
90%	19.4%	18.5%
95%	22.3%	21.6%
99%	28.1%	34.6%
99.50%	29.2%	38.7%
Max	33.3%	52.4%
Sharpe Ratio	64.1%	43.8%

Better Downside protection with Volatility management Strategies

CVF has better performance with higher equity exposure (90%) during calm periods than Fixed Allocation Funds (80%)

CVF gives up limited potential upside

FIGURE 4.47 Distribution of returns

Static 80/20			CVF 90 - Vol 16%
12.8%	average		11.7%
9.0%	stdev		4.2%
4.7%	Percentiles	0.5%	5.3%
4.9%		1%	5.5%
5.9%		5%	6.7%
6.4%		10%	7.2%
7.3%		25%	8.5%
10.2%		50%	11.1%
13.0%		75%	13.9%
22.3%		90%	16.6%
29.8%		95%	18.8%
57.9%		99%	26.1%
63.0%		99.5%	33.2%

FIGURE 4.48 Distribution of volatility

A capped volatility fund is very efficient in controlling the volatility risk exposure.

- The 1-year realized volatility of the CVF is, more than 90% of the time, below its target.
- The 1-year realized volatility 99.5th percentile is very close to the target.
- The vol of vol risk is significantly reduced (the standard deviation is significantly lower).

Remark: the short-term volatility management (21-day volatility) of the CVF can show inefficiency due to the volatility indicator being based on the last 21 days of historical data (a delay in rebalancing), but it still give important downside protection compared to the static allocation strategy.

There are several investment applications for which this volatility-stabilizing feature is valuable, and it has emerged as an increasingly popular asset class.

1. The primary example is the variable annuity industry, which can significantly reduce the hedging costs to provide the capital guarantee versus the use of put options (roughly by a percentage equal to the ratio of the realized managed volatility to the put implied volatility), implying lower capital requirements. The variable annuity AuM invested in such funds rose from about $55 billion in 2011 to some $360 billion by 2019, providing much cheaper funds (0.25% versus 1.5 to 3% for mutual funds annual fees).
2. Stabilizing volatility can also complement the long-term horizons of traditional valuation advisers: valuation strategies are good at forecasting returns for very long-time horizons, while managed-volatility strategies tend to react dynamically to shorter-term trends in volatility by lowering the equity exposure immediately after a sell-off. In fact, since 1926 there have been 37 drawdowns of 15% or more, which illustrate that, on average, volatility remains highly elevated throughout a ten-month period following the drawdown, while returns are disappointing relative to the "median".

There are various mechanisms for reducing the equity exposure, such as:

- the "volatility target", whereby %AuM equity allocation = volatility target level/today's volatility estimate, which theoretically makes the volatility of the fund over the rebalancing period exactly equal to the target volatility, assuming we can perfectly estimate equity volatility;
- the "volatility cap", whereby %AuM equity allocation = min (X%; volatility cap level/today's volatility); and
- the "variable volatility cap", whereby the volatility cap level above decreases as the market falls, in order to further limit the risk of loss.

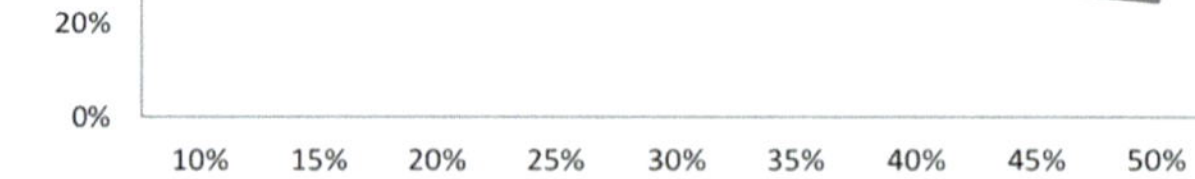

FIGURE 4.49 Managed-volatility equity allocation depending on volatility level

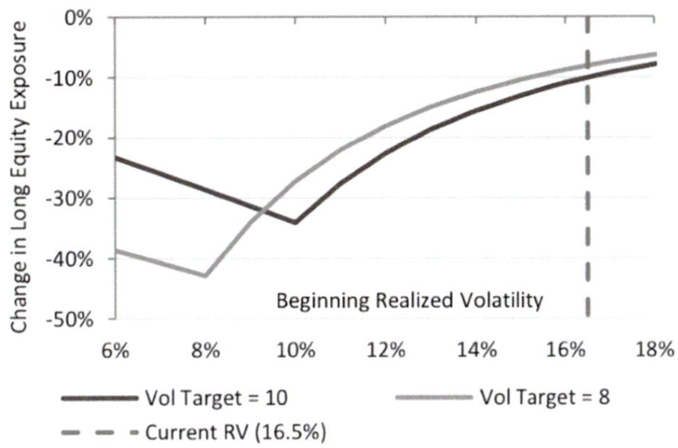

FIGURE 4.50 Managed-volatility equity de-allocation mechanism

In addition to reducing the equity exposure immediately after a sell-off, managed-volatility strategies will also employ leverage (i.e. higher than 100% equity exposure) during low-volatility episodes (i.e. whenever the short-term volatility forecast falls below the long-term target) that get along with positive equity returns (such as during equity bubbles), to compensate for opportunities lost by de-risking in crisis periods. Symmetrically, managed-volatility strategies will short futures contracts whenever the RV becomes higher than the target or cap.

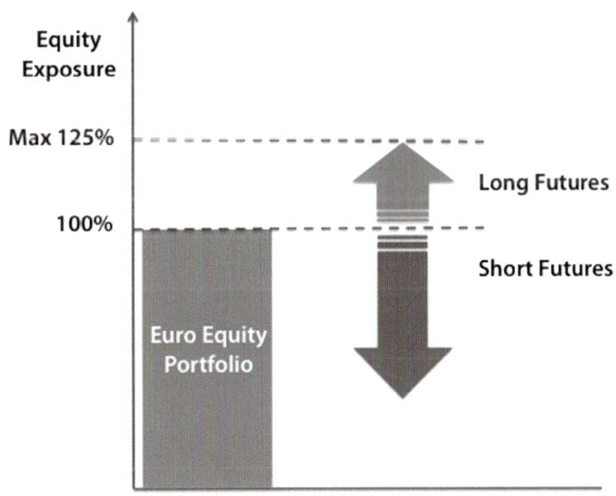

FIGURE 4.51 Managing equity allocation using futures

The robustness of managed-volatility strategies for short-term trends can be assessed across sample periods, with 20 to 30% improvements in Sharpe ratios across holding periods up to a quarter (i.e. the length of the window over which the investor evaluates performance), and decreasing marginal improvements beyond. For holding periods beyond one year, the long-term growth potential and mean reversion tendencies of equity markets take over as driving factor in favour of market valuation approaches (see Moreira & Muir 2017).

Investors naturally prefer high skewness and low kurtosis, which managed-volatility strategies consistently provide, as illustrated by kurtosis close to Gaussian distribution and right and left conditional value at risk tail metrics, as they adjust the equity exposure to the conditional expectation of volatility; the managed-volatility algorithm normalizes the left tail by reducing the equity exposure during periods of high volatility, but also normalizes the right tail by increasing the equity leverage beyond 100% during low-volatility regimes in order to compensate for lower exposure during equity bubbles, which usually accompany high volatility.

4.3.3.3 A Few Equity Volatility Spikes combined with Increasingly Correlated Markets Have Made Most Passive Management Inefficient with the Emergence of a New Volatility Regime Pattern since 2013

Managed-volatility funds' mechanisms suffer from multiple statistical weaknesses and the lack of fundamental catalysts embedded within their modelling; thus, for them to be effective requires not only expertise and experience, in order to fine-tune their dynamic adaptive allocation, but also a sound interpretation of the economic fundamentals and market activity. In addition, their protection levels are on a "best efforts" basis, with no "hard" guarantee.

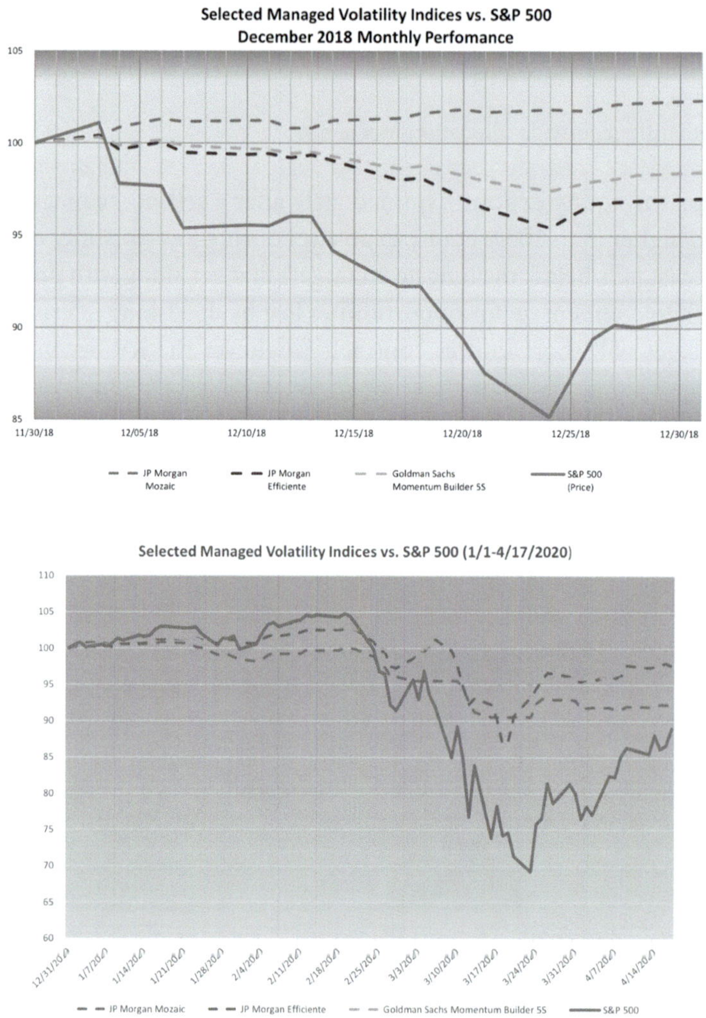

FIGURE 4.52 Selected managed-volatility indices versus S&P 500, (a) December 2018, (b) 1 January–17 April 2020

Note: Managed-volatility strategies cushioned market sell-offs over the last decade, with small negative returns and halved maximum drawdowns (e.g. 2018, 2020) compared to major stock indices.

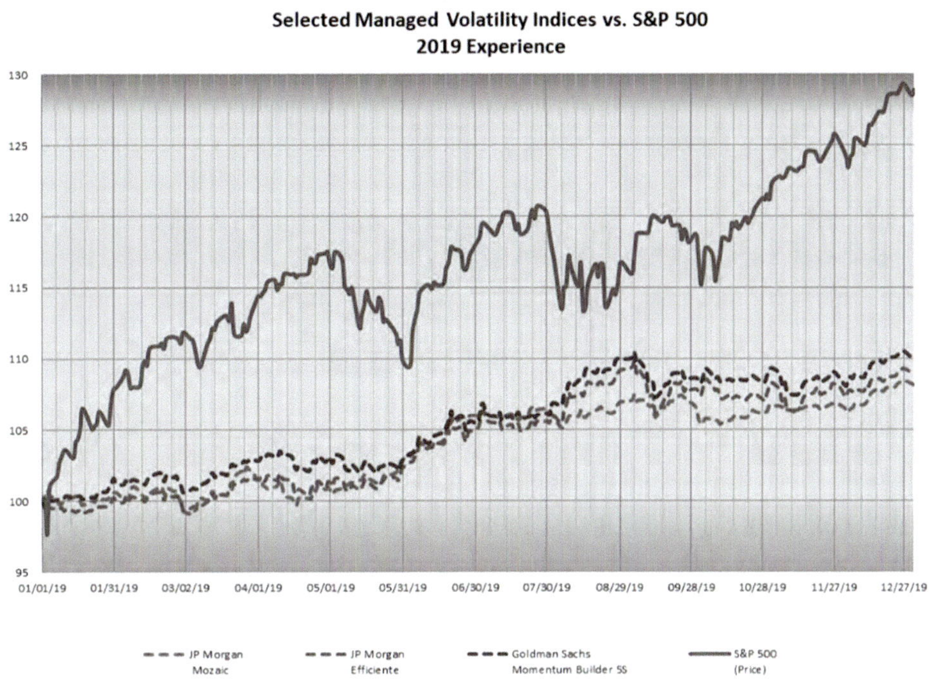

FIGURE 4.53 Selected managed-volatility indices versus S&P 500, 2019

Note: Managed-volatility strategies significantly lagged behind market recoveries (e.g. 2019).

This is illustrated by the poor performance of many passive management funds during the Covid-19 market crash in 2020 – e.g. −40% for equity alternative risk premiums. In fact, they suffer from three key empirical weaknesses, driven by technical and fundamental drivers.

1. Managed-volatility strategies do not meet all investors' risk appetite and may incur significant transaction costs, due to their dynamic adjustment of equity exposures and high tracking error to a market benchmark for some periods of time, as empirically observed for flexible funds embedded within life insurance products.
2. Managed-volatility strategies experience under-risking and over-risking patterns, as they assume that future volatility can be estimated with accuracy, which is not true, as it is typically estimated using an exponentially weighted moving average of historical equity volatility: even though the long-term realized volatility tracks the target volatility (thanks to the autocorrelation of daily returns), the short-term volatility may oscillate widely and be significantly away from the target. Exponential smoothing almost completely removes long-term vega, but short-term vega remains, which implies potential under-risking during uptrends and over-risking during downtrends.
3. Large-scale short-term volatility is also caused by dynamic surrender, ratchets/clicks – which are not mitigated by a target volatility fund mechanism.

The target volatility fund mechanism removes the volatility term premium (the spread long-term implied volatility minus short-term implied volatility) but does not remove the short-term implied–realized volatility spread. As a result, target volatility funds need to be hedged by rolling over short-term (one- to three-month) hedge instruments, such as puts or variance swaps, on the TVF.

Even though the long-term realized volatility tracks the target volatility, short-term volatility can oscillate widely, implying potential under-risking/over-risking. This stems from the lack of accuracy of the widely used

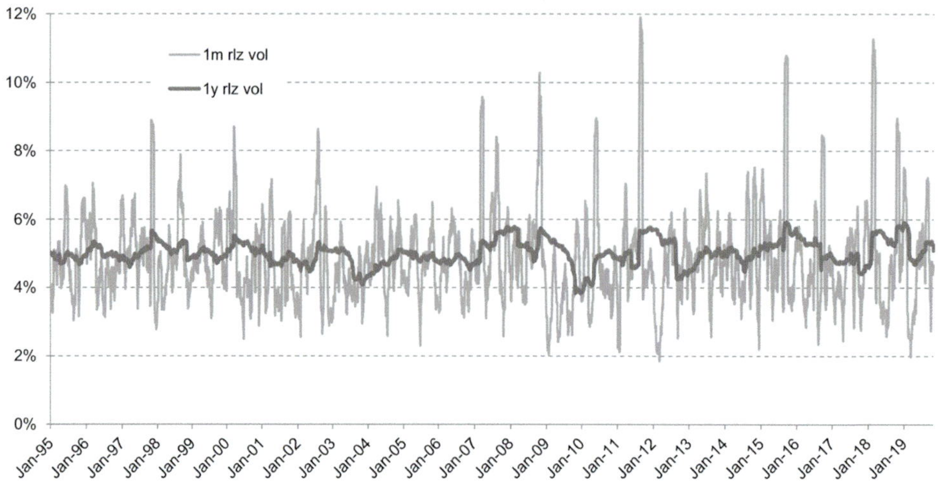

FIGURE 4.54 One-month versus one-year realized volatility, 1995–2019

EWMA volatility estimator (based on i.i.d. returns with zero-mean assumptions), which, notably, has the following features.

- It lacks a robust inverse relationship with equity drawdowns beyond daily time horizons.
  - Managed-volatility strategies assume that heightened volatility results in negative returns, while the relationship between historic volatility and future expected returns is not clear, not only in terms of direction but also in terms of magnitude.
  - In fact, although there is mostly an average statistical one-to-one negative relationship between the "daily" volatility and "daily" equity return (an increase of the VIX by 1% is consistent with a decrease of equity by 1%), the statistical metrics R^2 is only 56% on a weekly time horizon.

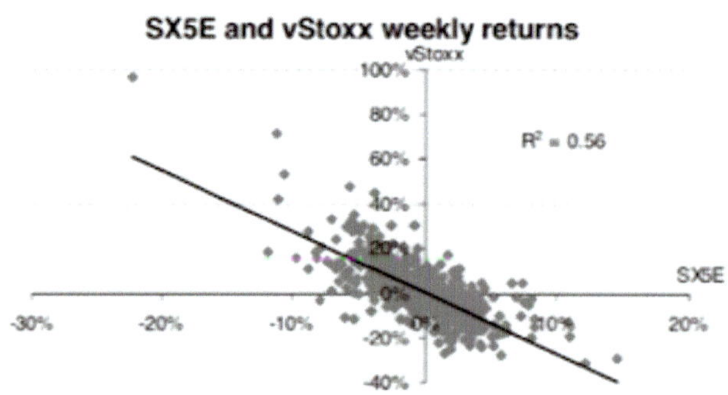

FIGURE 4.55 SX5E and VSTOXX weekly returns

Note: Implied volatility is negatively correlated to equity markets, but is a poor hedge.

  - The relationship may even be positive during equity bubbles.
  - The one-day expected move in equity √((E(|r|))^2) is not, as expected, equal to the one-day volatility √(E(|r|^2)), but larger than it (by Jensen's inequality).
- It is too much sensitive to drawdown outliers.

Although such estimates hold interesting statistical features consistent with empirical observations over the medium term (e.g. autocorrelation of daily returns), while avoiding key pitfalls compared to the traditional equally weighted moving average or standard deviation and providing better smoothness and robustness, they still suffer from several statistical weaknesses.

- $\sigma_t^2 = (1 - \lambda) \Sigma \lambda^{i-1} r_{t-i}^2$ – i.e. $\sigma_t^2 = (1 - \lambda) r_{t-1}^2 + \lambda^{i-1} \sigma_{t-1}^2$ – where r is the return, σ is the volatility and where a low λ provides highly reactive volatilities that quickly die away. As this EWMA is based on past volatility, it is not forward-looking enough compared to market-implied volatility, and thus works only retroactively.
- The EWMA can be roughly estimated as falling between the one-month and two-month realized volatility levels. As a result, an equity downward jump (i.e. one extreme return – a statistical outlier) implies a massive increase in the EWMA estimate (more than in the equally weighted moving average, albeit less than within a standard deviation), and so drives a correspondingly drastic equity de-allocation on a handful of trading days, which drives the overall fund volatility significantly below the volatility target.

  If the starting volatility level is just enough for a fund to be at its maximum equity allocation, as happened during the Brexit vote in the United Kingdom in June 2016 (–45% equity de-allocation) and in August 2015 (-55% equity, from 90% down to 40%), these massive equity de-allocations are all the more required to reduce the overall fund volatility than equities, which contribute ~90% to the overall fund realized volatility in practice (most portfolios hold 60% 20%-volatility equities versus 40% 5%-volatility bonds).

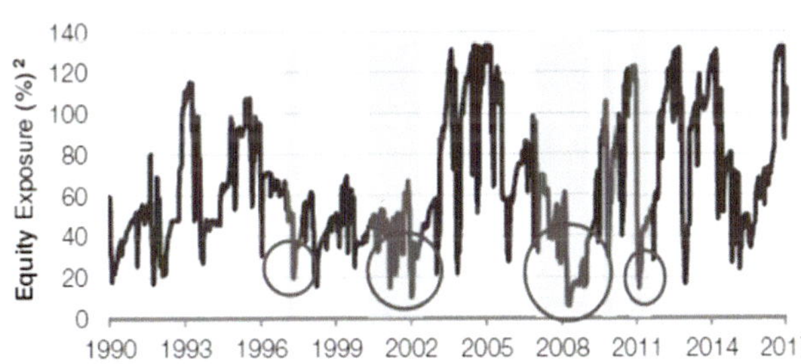

FIGURE 4.56 Significant de-equitization during major equity crashes, 1990–2017

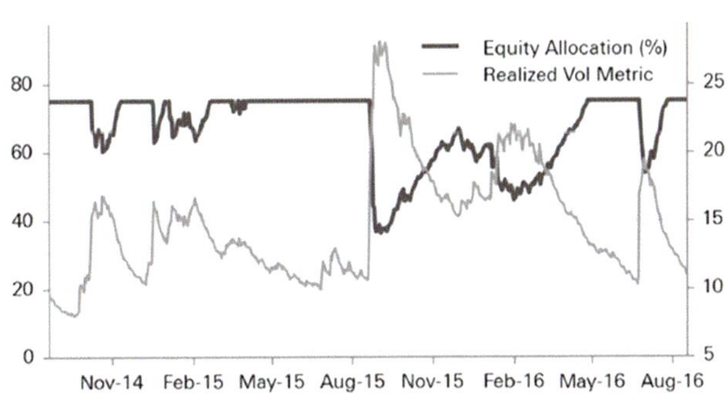

FIGURE 4.57 Aggressive equity de-allocation on a handful of trading days, September 2014–September 2016

- Although the EWMA does not suffer from the so-called "ghost effect" embedded within the equal weighted moving average (thanks to the low weighting factor λ providing a fast market downward shock and recovery), it is still a moving average estimate, and therefore misses most of the sudden and drastic shifts in volatility regimes while taking weeks to bring equity allocations back to pre-crisis levels (as observed during the 2015 and 2016 mini-crashes).

This often translates into too aggressive equity de-allocation (August 2011; June 2012; August 2015, with $50 billion equities sold; June 2016, with $25 billion sold; February 2018, with $150 to 200 billion sold; March 2020) and too slow re-equitization (taking many weeks, if not months). For example, after the worst December loss since the 1929 Great Depression, the S&P 500 achieved its biggest quarterly gain (13%) in a decade in the first quarter of 2019, as opposed to no more than a single-digit rebound by most managed-volatility strategies. The same applies for the whole year 2019, with 29% versus a single-digit outcome.

This also applies to volatility modelling, whether constant, stochastic or jumps. In fact, the model choice for the fund equity volatility has an impact on the quality of fit of the target volatility embedded within the insurance guarantee.

- Although a constant volatility makes the EWMA estimator fit the model volatility and retrieve the target volatility, it is not a realistic assumption.
- Using the more realistic "stochastic volatility model, the EWMA estimator still tends to lag behind the model volatility, although it retrieves, on average, the target volatility with some recurrent deviations.
- Using a much more realistic jump modelling assumption provides lower volatility than the target after the crash, but higher volatility before the crash, which results in a significantly higher guarantee cost compared to the constant volatility assumption (~+30%).

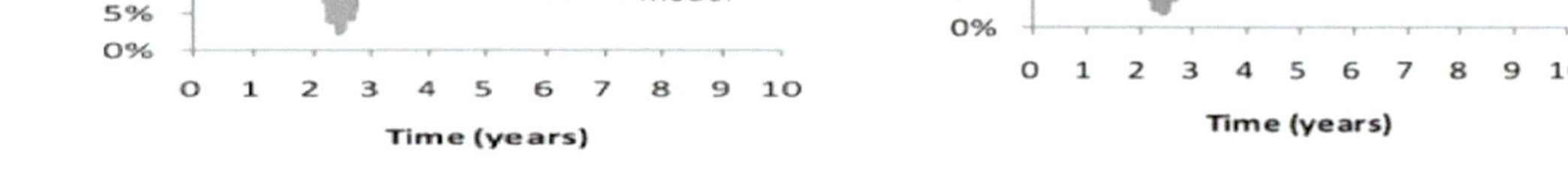

FIGURE 4.58 Stochastic volatility assumption fits the target volatility with some deviations

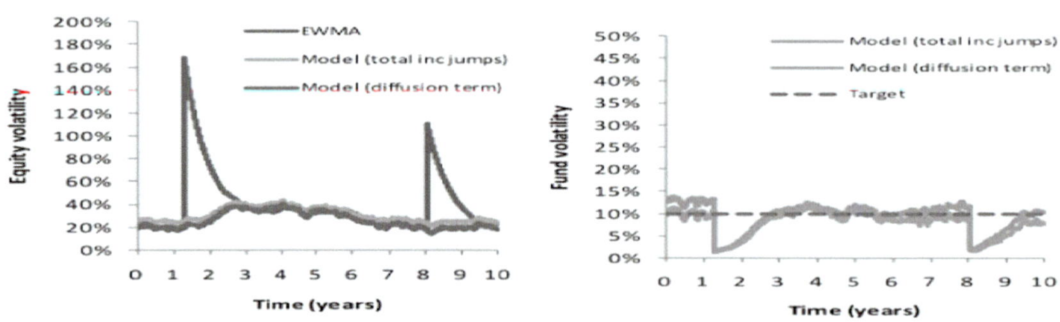

FIGURE 4.59 Jumps model gives lower volatility than target after the crash but higher beforehand

- As it is based on an i.i.d. returns model with mean zero, a EWMA will forecast the same average volatility whether the forecast is over the next ten days or over the next year. Such an i.i.d. assumption is indeed very unrealistic: the volatility of financial asset returns changes over time, with periods when volatility is exceptionally high interspersed with periods when volatility is unusually low. Although the use of GARCH models of volatility capture this volatility clustering (when volatility forecasts can be higher or lower than average over the short term), as the forecast horizon increases the GARCH forecasts still converge to the constant long-term volatility.

As a matter of fact, such a lack of volatility control effectiveness has increased even more with the emergence of a new volatility regime pattern since 2013, illustrated by short-lived but sharp transitions from low volatility to high volatility ("volatility spikes") – as many as in the prior two decades – while equally many episodes of large-scale volatility collapsed between 2010 and 2016 as for several decades beforehand. In contrast, before 1998 the S&P 500 could suffer a multi-sigma sell-off but still have low volatility (with realized volatility *not* jumping much at all – even for a short period).

FIGURE 4.60 Smaller but more frequent volatility spikes over the past decade, January 1990–January 2018

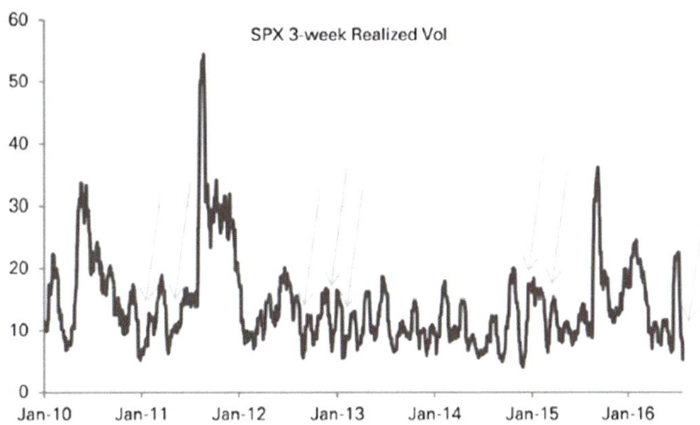

FIGURE 4.61 SPX three-week realized volatility, January 2010–June 2016

Note: Sudden drops in SPX realized volatility from high levels to very low levels have become increasingly common.

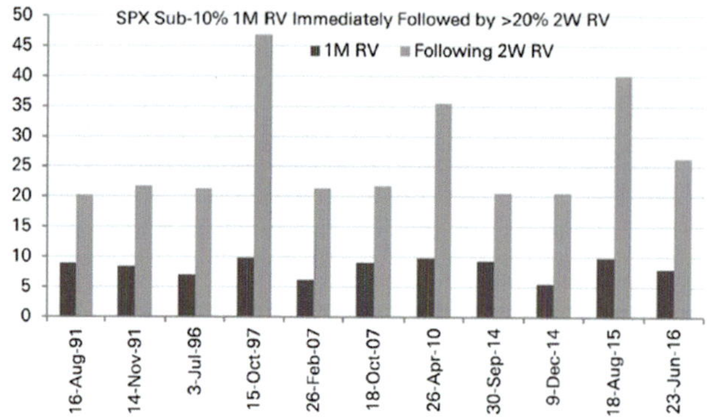

FIGURE 4.62 SPX sub-10% one-month realized volatility immediately followed by >20% two-week realized volatility, August 1991–June 2016

Note: There were almost as many transitions from lows to highs between 2009 and 2016 as in the previous 20 years.

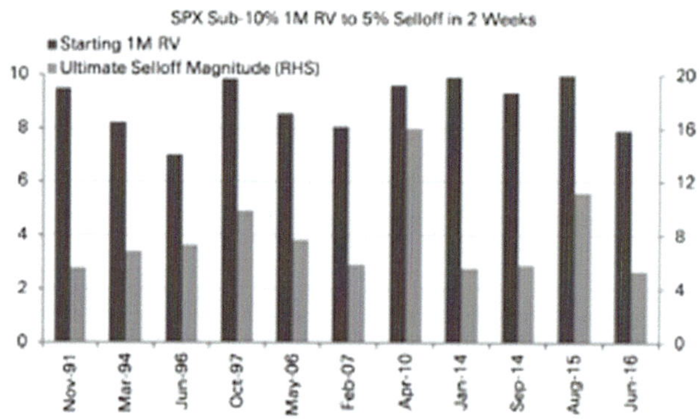

FIGURE 4.63 SPX sub-10% one-month realized volatility to 5% sell-off within two weeks, November 1991–June 2016

Note: The pattern is one of short-lived and severe sell-offs.

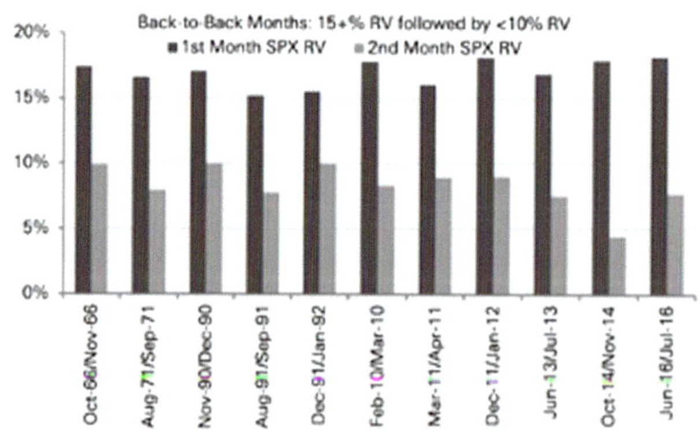

FIGURE 4.64 Back-to-back month volatility spikes: SPX 15+% realized volatility followed by <10% realized volatility, October 1966–June 2016

This translates more often than expected into breaches of the volatility target/cap levels, potentially implying higher "fair" fees (than in CPPIs during market crashes) or calling for puts on target volatility indices.

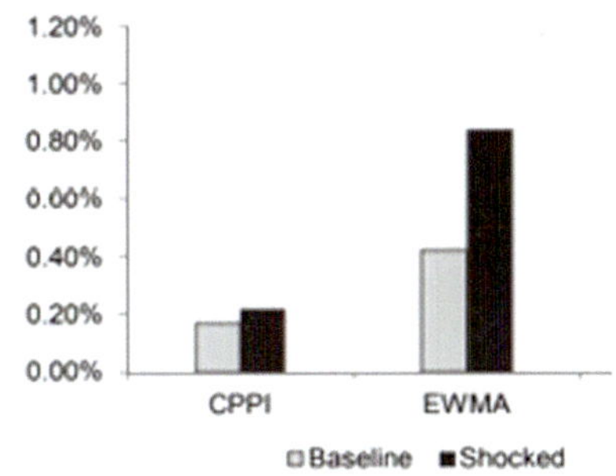

FIGURE 4.65 Fair rider fees

Managed-volatility equity de-allocation into bond mechanisms is efficient as long as there is a sufficiently high negative correlation between equities and bonds during equity market downturns, which provides profitable shifts of investments into fixed-income instruments, as experienced during the past two decades.

This is in contrast to the positive correlation between equities and bonds between 1995 and 2002, or 2006/2007, 2013, 2015, 2016, 2018 or March 2020. In fact, since 1885 bonds and equities have moved in opposite directions only 11% of the time, as opposed to 30% of the time in tandem.

Moreover, even in cases of negative correlation between equities and bonds, the protection offered by bonds during market sell-offs is limited within the usual 60% equity/40% bond portfolios, as the contribution of equities to realized volatility is 90%, which requires a massive de-allocation in equities in order to reduce the volatility of the overall basket.

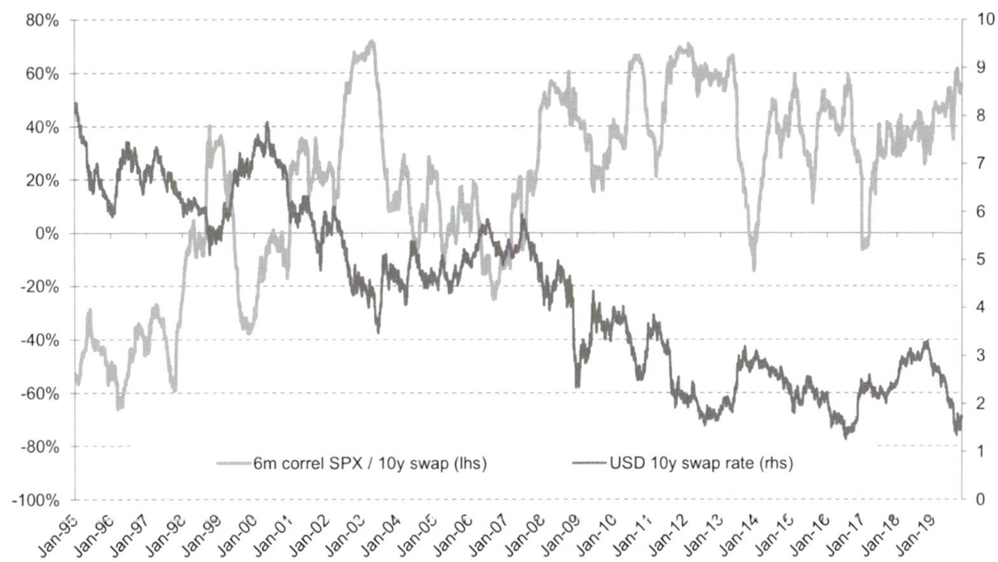

FIGURE 4.66 Correlation between equity and rates is not always positive, January 1995–December 2019

Most managed-volatility strategies lack analyses of the volatility regime, which usually provides anticipative signals regarding the behaviour of future near-term volatility, through the behaviour of fixed strike implied volatility, skew, vol of vol and volatility convexity.

- For instance, a much more volatile long-dated than short-dated implied volatility is indicative of a shift from low- to high-volatility regimes.

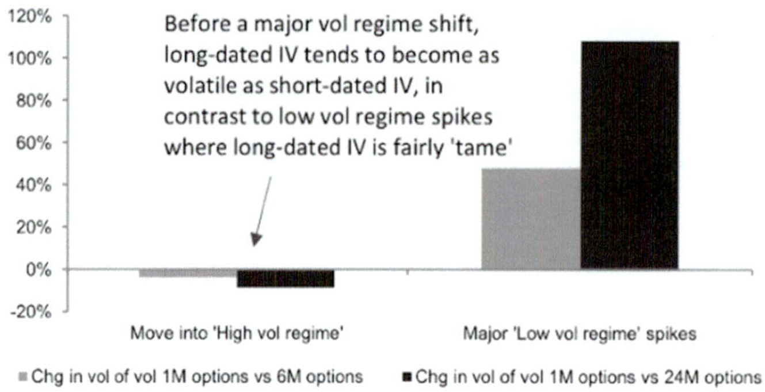

FIGURE 4.67 Long-dated implied volatility is as volatile as short-dated before a volatility regime shift

- A high volatility risk premium encourage investors to look for volatility short-selling opportunities, thus leading to lower volatility.

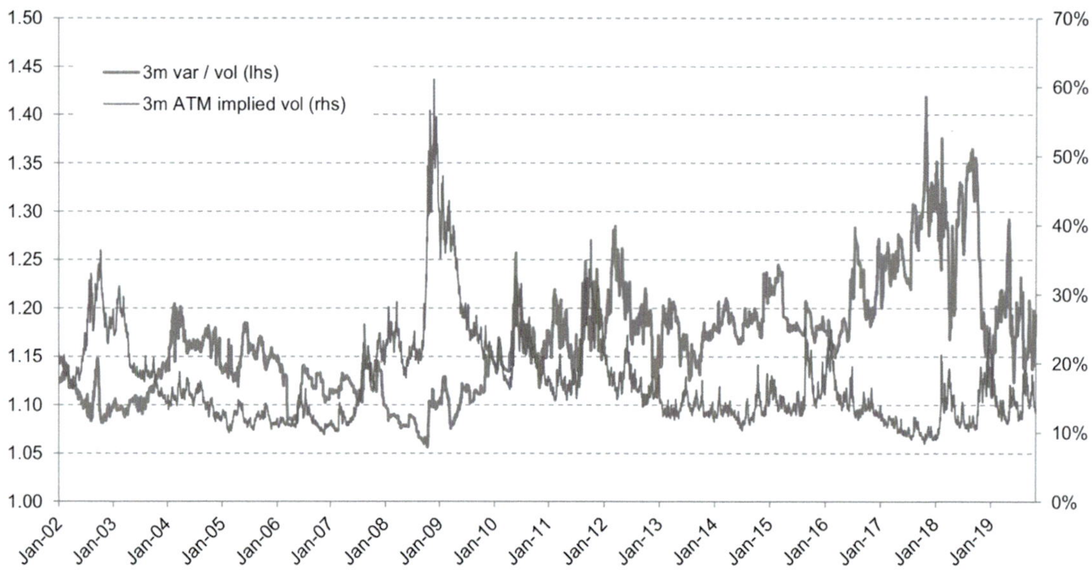

FIGURE 4.68 High variance/volatility ratios illustrative of remaining vol of vol risk, January 2002–December 2019

- Positive net long equity gamma positions dampen volatility (as dealers sell equities in rising markets and buy in falling markets), while net short equity gamma positions tend to exacerbate volatility (as dealers buy equities in rising markets and sell in falling markets).

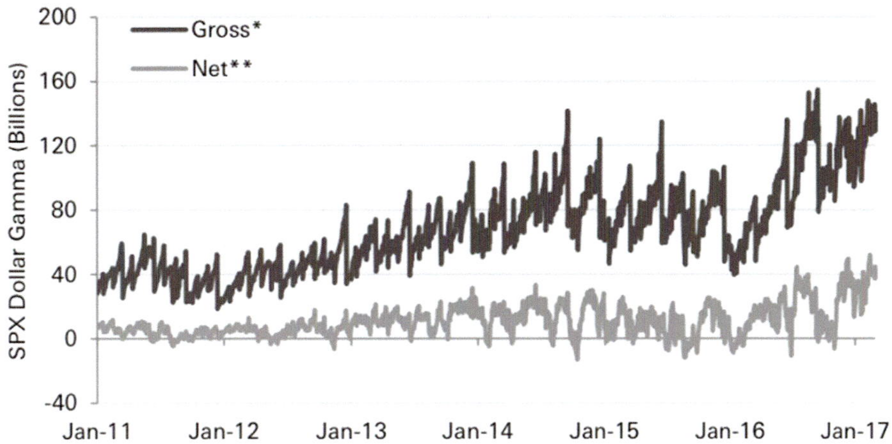

** Gross dollar gamma calculation assumes dealers are long all options*
*** Net dollar gamma calculation assumes that dealers are long 75% of calls outstanding and short 50% of puts. Dealer positioning has been 'flatter' on the downside as a result of tighter risk controls.*

FIGURE 4.69 Estimated gross and net equity gamma positioning, January 2011–June 2017

Notes: Gross dollar gamma calculation assumes that dealers are long all options. Net dollar gamma calculation assumes that dealers are long 75% of calls outstanding and short 50% of puts. Dealer positioning has been "flatter" on the downside as a result of tighter risk controls.

Managed volatility is characterized by significant trading activity, selling within a day of the emergence of elevated volatility, which can, ironically, increase market volatility as so much money moves out of equities – all the more so as there has been a growing concentration within the industry (for Vanguard it is 25% of passive management).

	Overall market share (percent)*		Passive fund AUM, March 2018 ($bill.)
	March 1999	March 2018	
Vanguard	11	23	3,404
BlackRock	0	8	1,410
State Street	0	3	613
Fidelity	14	9	422
Charles Schwab	0	1	174

FIGURE 4.70 Market share and passive fund AUM, March 1999 & March 2018

- In fact, they sell equities when volatility rises and buy when volatility falls, thus exacerbating sell-offs and rallies, becoming a key driver of the repeated "feedback loop" pattern of volatility spikes (sharp sell-offs then rebounds) observed during the ten years from 2010 to 2020 (June 2010, August 2011, June 2012, August 2015, February 2016, June 2016, February 2018, October–December 2018, March–May 2020).

- In particular, by serving as marginal equity sellers on initial market sell-offs, from a high allocation within falling market/high volatility/high put open interest and put–call open interest, or vega ratios/ short equity future positioning/negative equity put gamma positioning and excess over equity call gamma/concentration of vega over the 80 to 100% strike range. Specifically, near the market close (in particular if concentrated on ATM short-dated options, e.g. 30% listed SP 500 options volume within 2% of spot and less than two weeks to expiry in 2017).

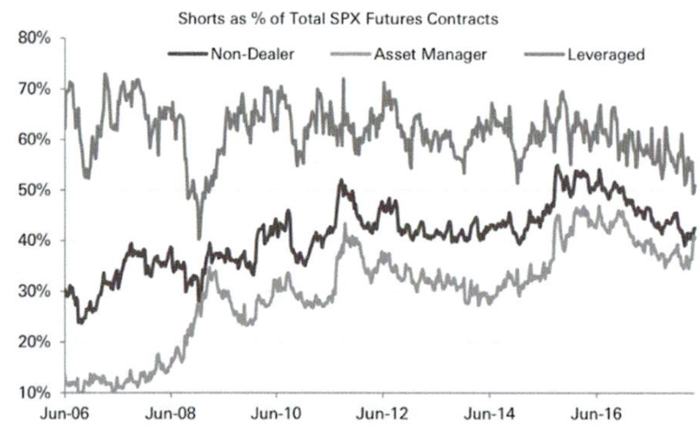

FIGURE 4.71 Shorts as a percentage of total SPX futures contracts, June 2006–December 2017

Note: Futures positions by asset managers point to hedging activity.

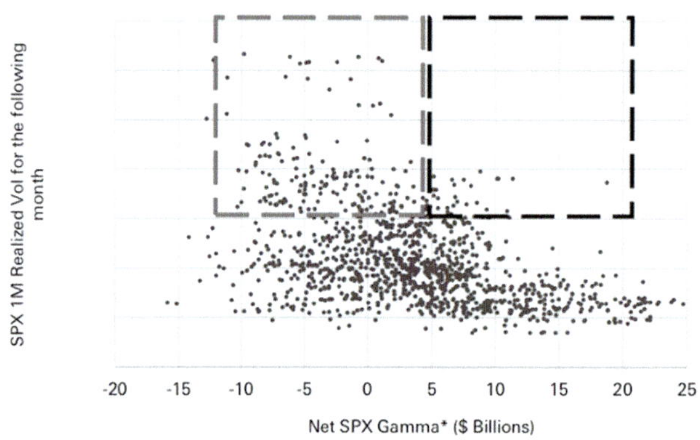

FIGURE 4.72 Increases in one-month realized volatility occur in the next month when dealers net short gamma

- In this situation, combined with low volatility and low vol of vol, even a small movement in terms of volatility represents a large percentage change and can cause managed-volatility strategies to exacerbate a volatility event; a sudden 4% equity sell-off would drive $40 billion of selling by the $400 billion AuM of managed-volatility strategies. In August 2015 managed-volatility strategies sold $50 billion in equities, and $25 billion in June 2016. In February 2018 their selling pressure doubled in a week from ~10% to ~20%, so these strategies reduced equity exposure by up to ~50%, leading to $150 to 200 billion of selling. There is often a delay of one to three days between when a signal is triggered and trade implementation, and positions are often reduced over several days.

Managed-volatility strategies' efficiency does not only depend on the tracking of the target volatility of the assets, as their rebalancing is also driven by the asset–liability mismatch, and thus by the volatility risk (equity gamma) of the liability – which itself is very volatile and path-dependent. In fact, hedging P&L will be driven by the product of the asset variance gap times the net asset–liability gamma.

The profit or loss of a dynamically hedged liability depends on the implied volatility used to price the option and implement the hedges:

$$P\&L(\sigma) = \frac{1}{2}\sum_{i=0}^{N-1} e^{r(T-t_i)}\Gamma_{t_i} S_{t_i}^2 \left(\sigma^2 \Delta t_i - \left(\frac{\Delta S_{t_i}}{S_{t_i}}\right)^2\right).$$

Break-even volatility is a method of solving for the volatility that yields no profit or loss based on replicating the delta-hedging procedure of a liability based on a historical share price time series – i.e. such that the average theta–gamma P&L above the liability is zero over the time horizon:

$$\sigma^2 \delta t = \sum_{i=0}^{N-1} w_i \left(\frac{\Delta S_{t_i}}{S_{t_i}}\right)^2, \text{where } w_i = \frac{e^{r(T-t_i)}\Gamma_{t_i} S_{t_i}^2}{\sum_{j=0}^{N-1} e^{r(T-t_j)}\Gamma_{t_j} S_{t_j}^2}.$$

The larger the gap between the volatility target and the break-even volatility, the larger the loss (as the insurer is the seller of the liability). This is why low-volatility targets tend to perform better than high-volatility target funds in practice.

Remark: a strength of the break-even methodology is that it is non-parametric in nature; it makes no assumptions as to the true underlying stock price process.

Remark: the break-even volatility is a weighted average of squared returns or realized variances. For a specific volatility – i.e. specific strike and term – the average will be dominated by the returns where the Γt values are highest. This happens for price changes close to the strike price and towards maturity of the option; these price changes have arguably the greatest effect. Thus, points on a break-even skew are averages of the same returns with different weights due to Γ's dependence on the strike. The downside of this is that an average of break-even volatilities for a specific option term leads back to the historical volatility for that period. Constructing the skew is then susceptible to sample variation because it depends on the chosen period.

✓ **Quick win**: given the strong path dependence of the P&L above, the mean P&L needs to be resilient, specifically if the short gamma exposure of liabilities is significant.

a) The break-even volatility is then recommended to be the average over a large range of stressed-volatility and rangy asset P&L scenarios; then the pricing and hedging vol = MAX (average break-even stressed-volatility; 85th historical percentile).
b) The P&L is also driven by the delta-hedging rebalancing frequency, and so a prudent approach lies in the specification of a minimum delta-hedging rebalancing to minimize the P&L volatility. With reference to Derman and Kani (1994), a proxy for a zero-mean P&L volatility close to ATM is

$$\text{Volatility}(PnL) = \sqrt{\frac{\pi}{4}}\frac{\partial P}{\partial \sigma}\frac{\sigma}{\sqrt{n}},$$

$$n^* = \frac{\pi}{4}\left(\frac{\partial C}{\partial \sigma}\frac{\sigma}{\text{Volatility }(PnL)}\right)^2,$$

where n is the number of rebalancing and C the liability price (proxied as a put option).

The empirical volatility (PnL) could therefore be compared to the theoretical weekly delta-hedging volatility (PnL) as above, all the more so in that capped volatility fund returns are different from a normal distribution.

Such a proxy for a minimum number n* of delta hedgings could also be used as an indicator of PnL volatility, but only as a "weak" indicator, since, given the short gamma exposure of liabilities, too many frequent hedgings could be damageable to the mean (PnL) if close to ATM.

c) Make sure that extreme realized volatility percentiles (e.g. 8th, 95th, 99.5th percentiles) of the break-even volatility distribution are close to the volatility target.

Managed-volatility strategies' protection levels are on a "best efforts basis"; they do not constitute a guarantee. Managed-volatility strategies may therefore fail to be suitable for investors seeking explicit "hard" downside risk protection, or if performance is to be measured against a market benchmark such as an index.

As a result, neither active nor passive styles seem to have significantly and sustainably captured the velocity of changes over the past decade. Going beyond the "passive versus active" asset management paradigm through a mix of fundamentals and systematic quantitative investment strategies may then be appropriate to help capture the velocity of changes.

4.3.4 Sound Mixes between Fundamentals and Technicals Improve Sustainability in Choppy Markets

Mixing the two approaches is likely to improve the P&L, whereby fundamental approaches notably select assets based on KPIs and macro drivers, while systematic ones detect technical patterns and industrialize mitigating actions. Let us detail three value propositions.

4.3.4.1 Embedding Specific Fundamentals within Systematic Algorithmic Passive Asset Management to Mitigate Part of Their Structural Weakness

This is done so that they do not overreact following an initial market fall but only steadily on the basis of sound economic concerns, while enabling a recovery consistent with the market, by introducing fundamentals of the following four types.

i. First, there are fundamental economic catalysts, such as economic growth (e.g. persistently low real interest rates and inflation, indicative of low growth, and thus of a low cost of capital and low volatility), M&A activity, share buybacks (placing downward pressure on volatility) and central bank policies (e.g. quantitative easing and increasing balance sheets, thus depressing volatility).

ii. Second, there are the fundamentals of the volatility regime, which provides anticipative signals regarding the behaviour of future near-term volatility. Examples include

- a much more volatile long-dated than short-dated implied volatility, which is indicative of a shift from low- to high-volatility regimes; and

- a high volatility risk premium, which encourage investors to seek volatility short-selling opportunities, thus lowering volatility.

iii. Third, there are supply and demand fundamentals, such as market dealers' hedging gamma positions (as positive net long equity gamma positions dampen volatility, while net short equity gamma positions tend to exacerbate volatility), all the more so given their increasingly massive use of weekly options (30 to 50% of all options versus less than 5% in 2011).

iv. Lastly, there are liability features, which mostly drive asset rebalancing and are volatile and path-dependent. Technically, the P&L is the product of the spread between the expected and realized volatility times the net assets versus liability gamma.

4.3.4.2 Selecting Resilient Assets Supported Both by Fundamentals and by Systematic Market Technical Indicators

Contrary to conventional wisdom, less volatile stocks empirically tend to perform more strongly over the long term, as they lose significantly less during drawdowns. In contrast, volatile stocks have to work much harder, first to restore the value lost during periods of decline and then to grow.

This historical performance of low-risk stocks defies the central paradigm of traditional finance theory, which states that lower risk goes with lower returns (a "negative mean-variance relationship"), stemming from three main drivers:

- a lottery mentality, driving most investors to consistently overpay for the small chances of winning big in riskier stocks;
- an inclination to avoid low-beta stocks; and
- the general use of log-Gaussian modelling assumption in returns distributions (characterized by a left-skewed skewness in the distribution of returns – i.e. a long tail to the left of the returns distribution).

As a result, a sound mix of fundamentals and systematic market technicals within the asset selection and the asset allocation processes, which combines statistical returns distribution patterns with the selection of stocks of high fundamental quality (healthy and stable profitability, strong free cash flows, low debt and shareholder-friendly practices, above-average dividend payouts, low net equity issuance), can produce a stronger performance. This has proved to be a more efficient way since the 2008 crisis, not only to mitigate significant declines but also to generate significantly higher returns at similar levels of risks.

In particular, the price/earnings to growth (PEG) ratio offers an additional way to screen resilient candidates. This ratio is calculated by dividing the current price to earnings (PE) ratio by the growth in earnings per share over the previous 12 months. Stocks with a PEG ratio of one or less are considered as undervalued. For example, a corporation with a PE ratio of 25 and a 25% EPS growth rate will have a PEG ratio of one. In addition to looking at the PEG ratio, you should also consider if the current growth pattern will continue in the coming quarters. This will result in a forward PEG ratio calculation, which is a more meaningful metric than the trailing PEG ratio. Forward PEG ratios will be appealing if forward growth is expected to be strong. A PEG ratio that is higher than one indicates that the stock has been overrated and is unfavourable in the market. On the other hand, PEG ratio values less than one show that the stock has been underpriced.

Leveraging on the lower regulatory capital requirements for long-term assets is also a resilient solution across choppy markets, as illustrated within Solvency II by the "Long Term Equity Investments" category, whose cost of capital is divided by two compared to traditional equity investments.

4.3.5 Sustainable Investment Strategy Enhancements

4.3.5.1 Reconciling Sustainability, Market Performance and Customers' Risk Appetite through an ESG Risk-Based Optimal Asset Allocation

4.3.5.1.1 ESG and Insurance Company Investments An asset with a high ESG score means good sustainability practices within the company: among the "E", "S" and "G" factors,

- the environment factors refer to the behaviour on environmental issues, such as resource depletions, climate change, pollution;
- the social factors are related to the treatment regarding people, workers and local communities, including health and safety; and
- the governance factors refer to corporate policies and governance, including tax strategy, corruption, structure and remuneration.

Such ESG scoring has been devised by the major ratings agencies (see Figure 4.73). A positive relationship between highly rated ESG stocks and financial performance may be assumed because companies engage with stakeholders for value-enhancing purposes, sometimes referred to as "doing well by doing good" (Benjamin Franklin).

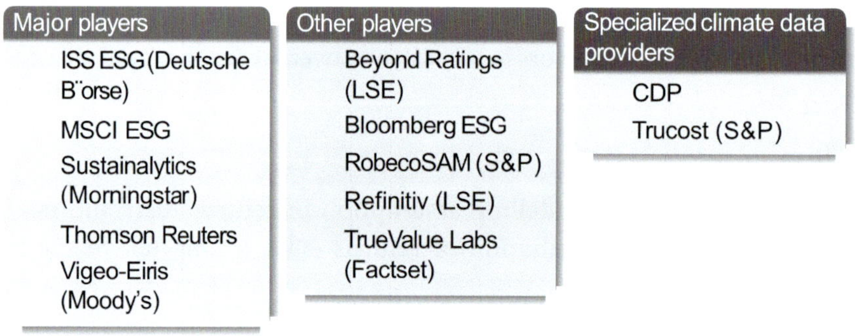

FIGURE 4.73 The main ESG scoring agencies

Under this value-enhancing view of CSR, managers engage with stakeholders simply because such projects are deemed to have a positive net present value. Alternatively, it has been recognized that companies employing ESG practices are better positioned to deal with future challenges in a rapidly changing global landscape in which climate risk is investment risk, while some companies seen as embracing fewer ESG issues are facing bans, such as companies producing single-use plastics, which are being forced to adapt their production or see their performance worsen. Indeed, academic research conducted since the 1970s has shown that there is a positive correlation between return and sustainably managed companies.

These companies may first experience less exposure to ESG-related risks, which will have a positive influence on their stock performance. Adherence to ESG criteria should help minimize the downside risk in a firm's financial performance (and thereby its stock price), based on the premise that ESG-compliant firms are likely to be well managed overall. Focusing on ESG is also expected to make it easier for companies to obtain debt and equity financing in the future. Conversely, companies that do not focus on ESG are expected to experience an increase in their cost of debt and equity.

A high ESG score is thus likely to help a company to (i) attract more investors, (ii) lower the cost of capital and (iii) produce a better operational performance, and thus stock performance.

Therefore, the ESG score is becoming an increasingly key factor in portfolio asset allocation. ESG AuM is expected to reach $50 trillion by 2025 (a third of global AuM) within the context of growing customer demand and burgeoning regulatory requirements, specifically in the insurance industry.

ESG assets by geography and year in $ millions

	2016		2018		
	Amount	%	Amount	%	CAGR
Europe	12,040	53%	14,075	46%	8.1%
United States	8,723	38%	11,995	39%	17.3%
Japan	474	2%	2,180	7%	114.5%
Canada	1,086	5%	1,699	6%	25.1%
Australia	516	2%	734	2%	19.3%
Total	**22,839**	**100%**	**30,683**	**100%**	**15.9%**

FIGURE 4.74 ESG assets under management by geography and year, 2016 & 2018

ESG assets as % of Managed Assets

	2014	2016	2018
Europe	58.8%	52.6%	48.8%
United States	17.9%	21.6%	25.7%
Canada	31.3%	37.8%	50.6%
Australia	16.6%	50.6%	63.2%
Japan		3.4%	18.3%

FIGURE 4.75 ESG assets under management as a percentage of all managed assets, 2014–2018

In fact, ESG factors are of most relevance to long-term products with an investment component, primarily unit-linked and pension savings products and, to a lesser extent, with-profits products.

Asset management interest in ESG considerations is being driven by greater client demand (mostly institutional). The vast majority of millennials have expressed an interest in sustainable investing. To the extent that younger generations are likely to be investing for the long term (e.g. for pensions), ESG investments could be an appropriate choice, given that green technologies, for example, will take some time to become mainstream.

Given the increasing popularity of ESG funds, and the fact that unit-linked investors bear the market risk associated with their investments, it seems more likely that unit-linked providers will be required to offer ESG fund options within product offerings.

More importantly, since 2015 regulations have been putting increasing pressure on the asset management and insurance industries to incorporate ESG scoring into their investment policy, while the credit rating agencies are including ESG factors in their ratings to better evaluate companies' risk of default.

Even so, it is only a limited number of providers that currently offer funds that are specifically marketed and named as ESG funds. However, some insurers are setting standards for their investment managers: Standard Life, AXA, Allianz, Zurich and Generali, for example, have published online their policy on how they incorporate ESG factors into their unit-linked investments, such as:

- having policies on aspects such as how they engage, report and vote on ESG matters, and on any sectors or activities they specifically exclude from funds;
- being aware of how legislation changes will impact exposure to investments;
- being able to demonstrate how they incorporate ESG considerations into their investment processes; and

- proactively engaging with companies – e.g., using voting rights to encourage positive practices and reporting on their voting and engagements.

ESG factors are primarily used in the initial screening process at present but they are expected to become a bigger driver of investment decisions over time. There are various approaches to ESG investing, ranging from simple exclusionary screening to the integration of ESG considerations into fundamental analysis of various firms.

ESG Investing Strategies	Description
Negative/exclusionary screening	Excluding companies or sectors based on ESG specific criteria
Norms-based screening	Screening of investments that comply with certain norms and standards (OECD, ILO, UN, UNICEF); May also be exclusionary
Positive/best-in-class screening	Investing in companies or sectors that demonstrate positive ESG performance versus peers
Sustainability-themed investing	Investment in themes or assets that address specific issues like climate change, food, water
ESG integration	Explicit overlay of ESG factors on financial analysis
Active ownership	Direct corporate engagement following comprehensive ESG guidelines
Impact investing	Targeted investments meant to solve social or environmental problems

FIGURE 4.76 The main ESG investing strategies

Notes: OECD = Organisation for Economic Co-operation and Development; ILO = International Labour Organization; UN = United Nations; UNICEF = United Nations Children's Fund.

ESG is primarily a gating factor for active investors currently: a high ESG score alone is unlikely to help a company attract more investors and attain a higher multiple, but a low score could hurt its ability to attract capital and hurt its multiple.

ESG factors can be used as a source of valuable economic information on downside risk, rather than upside.

- Nearly 80% of respondents say ESG factors have caused them to limit exposure to or divest entirely from particular assets: tobacco companies, arms manufacturers and companies in the extraction, power and real estate sectors due to climate change.
- Investors express some concern about the overuse of ESG information. This reluctance is not as much about ESG as it is an economic credit decision. However, if ESG criteria and negative screening are too broad, they put too much restriction on the typical portfolio. On the upside, if a company can perform well economically, and at the same time have a positive social or environmental impact, then it is typically going to be more successful.

4.3.5.1.2 ESG Factors and Market Performance In terms of market performance, market timing usually plays a key role for ESG assets.

- They offer downside protection during a sustained crisis (2008, 2020).
- They drive performance in the long run (Europe, North America).
- They are used as beta in Europe and alpha strategy in the United States.
- However, they have volatile performance trends: they underperformed in 2010–13, then outperformed other market sectors.

Even so, in terms of a literature review, reconciling market performance and sustainability is still an open issue, and, with no conclusive statistics, different researchers' findings and conclusions are highly conflicted and have no clear answer.

- Until the early 1990s ESG integration suggested wasteful costs and competitive disadvantage, adding constraints: Rudd (1981) talks of "a damaging permanent bias". Screening asset universes (with ESG constraints) restricts the number of stocks available, which would limit the diversification effect.
- The second assumption is a neutral one: the decision to increase a company's costs should lead to an increase in the benefits by matching funds (Revelli & Viviani 2015; Chong, Her & Phillips 2006).
- The third axis: there is a positive correlation between return and sustainably managed companies. They may, first, experience less exposure to ESG-related risks, which will have a positive influence on their stock performance (less specific risk: Hoepner 2010; better risk management). Second, sustainable managed companies may have better operational performance by, for instance, reducing waste or reducing their greenhouse gas emissions or improving their energy efficiency, which can lead to long-term cost savings and put them in the front line of providing innovation. Lastly, sustainable managed companies may be linked to the ability to better attract talent and create customer loyalty due to the company's good brand image and reputation.
  - Statman (2004), Verheyden, Eccles and Feiner (2016): there is a positive relationship between ESG-performing firms and higher financial performance, and ESG screening improves risk-adjusted returns. Positive net selectivity: the loss in diversification is not compensated for enough by sufficient extra α. Moreover, the lower diversification is quite stable over time. However, the 25% screened universe displays a negative net selectivity, which implies that the gain from screening is not monotonous: it is valid only for the best of the best ESG-friendly firms.
  - Friede, Busch & Bassen (2015) looked at more than 2,000 empirical studies on the link between ESG and financial performance: 90% of studies found a relationship between ESG and financial performance – positive for a large majority.
  - Clark, Feiner and Viehs (2015) looked at more than 200 academic studies, industry reports, newspaper articles and books: 80% showed that strong sustainability practices have a positive influence on investment performance.
  - Heinkel, Kraus and Zechner (2001) find that the cost of capital of companies excluded by green investors increased. Chava (2014) finds in his comparison that investors demand higher expected returns on stocks excluded by environmental screens. Lenders also charge a higher interest rate on bank loans issued to firms with these environmental worries.
  - Hoepner et al. (2018) look at how engagement on ESG issues can benefit shareholders by reducing firms' downside risk. Even so, Hvidkjaer (2017) in his literature review identifies several studies with different conclusions around the performance of "sin stocks"; sectors such as tobacco can have attractive characteristics, including their defensive characteristics in difficult market conditions.
  - Kaiser and Welters (2019) find that absolute returns are significantly lower when investing in companies rated highly in terms of ESG factors.
  - Agarwal and Ouaknine (2019) find that high ESG ratings are correlated with outperformance.
  - Ashwin Kumar et al. (2016) and Verheyden, Eccles and Feiner (2016) find better Sharpe and Treynor ratios.

- Different time horizons can also be studied and show conflicting results: Derwall, Koedijk and Ter Horst (2011) find that positive screening shows abnormal returns, whereas the performance of these socially responsible stocks is not maintained in the long run.

Overall, there is no clear answer as to the performance of the best-in-class ESG stocks compared to passively investing in an index stock market. Both positive and negative returns can be found in the literature, which can be due to a lack of supervision around ESG activities, and a large number of studies could not conclude on either positive or negative performance because of statistically insignificant results.

4.3.5.1.3 Reconciling Sustainability, Market Performance and Customers' Risk Appetites Here we focus on one-year performance using weekly returns. This study covers a period from 4 January 2010 to 27 December 2019 across Europe. The studied companies are selected from the supranational index Stoxx600, which gathers the biggest European market caps from all industries. We use two types of index, the Stoxx600 and the Stoxx600 ESG Index (which follows standardized ESG exclusion screens and thus excludes companies dealing with controversial weapons, tobacco or coal), as benchmarks. Starting from this benchmark and to construct our portfolios, only the firms with ten years of ESG scores and close values were kept in the dataset – i.e. 410. The weekly Euribor rates have been used as a proxy for the risk-free rate.

The ESG scores are provided by Refinitiv for the whole period from 2010 to 2019, computed on a yearly basis until 2019. These scores are available for more than 9,000 companies worldwide, covering 2,100 firms in Europe alone. The Refinitiv framework divides "ESG" into ten categories, with each one of them depending on the firm's industry among a classification of 54 industries in total initiated by Thomson Reuters. Each industry has around 186 specific meaningful indicators to evaluate a company on a total of over 500 initial measures. The ten categories eventually aggregate into the well-known three pillars with a specific weighting scheme for each indicator. All the metrics used are publicly available.

4.3.5.1.3.1 Passive ESG Portfolio Asset Allocations Underperform the Usual Benchmark Indices Even so, although a naive diversification based on an equally weighted best ESG stock portfolio provides over time a positive trend in performance relative to the benchmark, and there is some outstanding performance during specific periods (e.g. the 2020 market crash) or in the long term, it still most often underperforms the worst ESG stock portfolios over shorter time horizons (e.g. one year) in terms of overall returns, risk and cost of capital.

- Lower Sharpe, Sortino (same logic as the Sharpe ratio but replacing the total volatility by a downside deviation) and Treynor ratios (same logic as the Sharpe ratio but replacing the total volatility by the beta, considering the specific risk as fully diversifiable), and lower information (excess return to benchmark divided by its tracking error volatility).
- Higher volatility, beta, VaR and max drawdown for the best ESG portfolios relative to a market benchmark, or even the worst ESG portfolios.

FIGURE 4.77 Absolute returns, 2010–2019

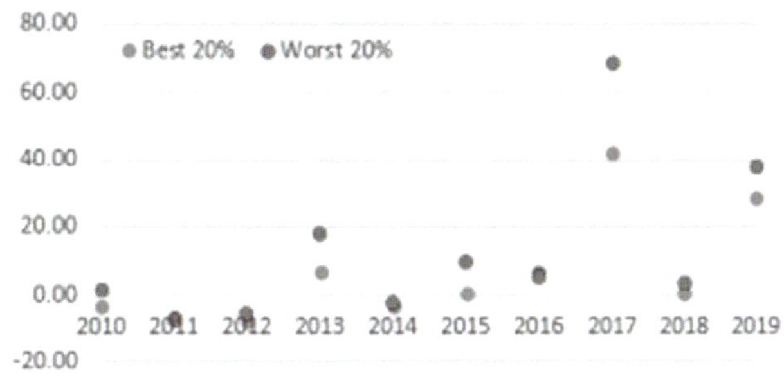

FIGURE 4.78 Sharpe ratios, 2010–2019

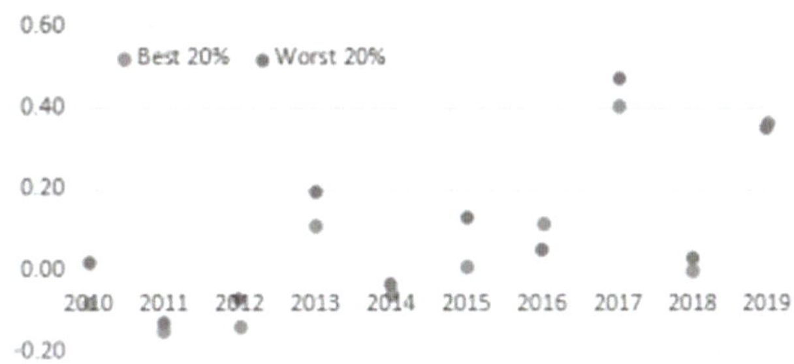

FIGURE 4.79 Sortino ratios, 2010–2019

Potentially, the poorer performance of ESG stock portfolios stems from the ESG screening, which tends to reduce diversification, as measured by the following.

- The diversification ratio:

$$DR = \frac{\sum_{i=1}^{n} (x_i \sigma_i)^2}{\sigma_p^2}.$$

The lower the ratio is, the less powerful the diversification effect is; or, equivalently, the higher the diversification ratio is, the more diversified the portfolio is. In other words, a portfolio with a high diversification ratio maximizes the number of non-correlated sources of risk.

An immediate property is that the screened universe (such as the ESG-screened portfolio) cannot have a higher diversification ratio than the unscreened universe, simply because it is a subset of this whole universe, so the sources of non-correlated risk must be smaller.

- The diversification measure:

$$D = \left(\frac{\sigma_p}{\sigma_m} - \beta_p \right) * (r_m - r_f).$$

The lower diversification is *not* more than compensated for by an extra alpha, as illustrated by the negative net selectivity (= alpha – diversification measure) for the best ESG portfolios, and positive for the worst.

Such results are consistent with major studies on passive ESG portfolio asset allocations. For example, Bennani et al. (2018), in particular, do not find evidence of outperformance of ESG-focused ETFs compared to benchmark indices; and this applies even during the pandemic (Pavlova & de Boyrie 2022), although their performance seems to be better compared to the pre-pandemic period. Several studies have attributed the greater returns of ESG-focused portfolios during the pandemic to spillover effects – i.e., investors redirecting capital to ESG-focused portfolios because of their risk aversion and, as a result, attention to corporate fundamentals (to avoid downside risk) increasing in response to the market shock created by the pandemic (see, e.g., Singh 2020).

As a result, a modified asset allocation has been devised by integrating the ESG constraints directly into the Markowitz approach.

4.3.5.1.3.2 Constrained ESG Active "Optimal" Portfolio Asset Allocations Tend to Outperform the Usual Benchmark Indices

4.3.5.1.3.2.1 The ESG-CAPM Theoretical Model (Pedersen, Fitzgibbons & Pomorski 2021) Within such a model, equilibrium asset returns satisfy an ESG-adjusted capital asset pricing model (CAPM), showing when higher ESG assets have lower or higher equilibrium expected returns.

Three types of investors are considered here.

- "Type U" (ESG-unaware) investors are unaware of ESG scores and simply seek to maximize their unconditional mean-variance utility. They take expected excess returns to be E(r), with risk given by the variance-covariance matrix, var(r).
- "Type A" (ESG-aware) investors also have mean-variance preferences, but they use assets' ESG scores to update their views on risk and expected return. They use assets' expected excess return, $\mu = E(r|s)$, conditional on the ESG information, s, and the conditional variance-covariance matrix of excess returns, var(r|s).
- "Type M" (ESG-motivated) investors use ESG information and also have preferences for high ESG scores. They seek a portfolio with an optimal trade-off between a high expected return, low risk and a high average ESG score.

While optimizing across three characteristics (risk, return, ESG) can seem challenging, the investor's problem can be reduced to a trade-off between ESG and Sharpe ratio. In other words, risk and return can be summarized by the Sharpe ratio. Specifically, for each level of ESG, we compute the highest attainable Sharpe ratio (SR), translated into the ESG–SR frontier (illustrated in Figure 4.80), which is independent of investor preferences: an investment expert can first mechanically compute the frontier, and then the investment board can choose a point on the frontier based on preferences. In other words, investors with the same information should agree on the frontier even if they prefer different portfolios on the frontier. This separation property resembles that of the standard mean-variance frontier, which also depends only on security characteristics.

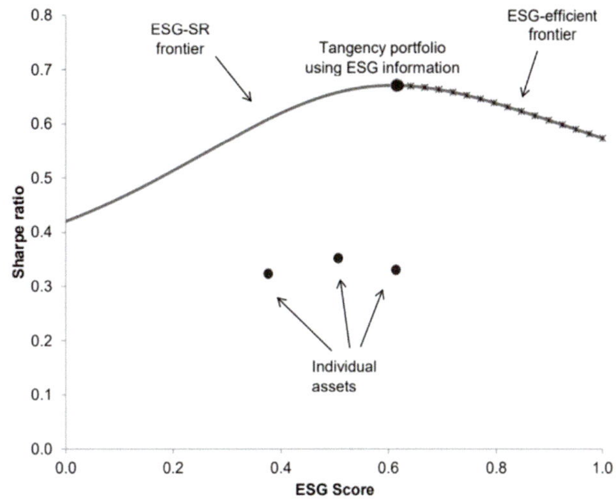

FIGURE 4.80 The ESG–SR frontier

To understand why the ESG–SR frontier is hump-shaped, consider first the tangency portfolio known from the standard mean-variance frontier, which has the highest SR among all portfolios, so its ESG score and SR define the peak in the ESG–SR frontier.

Further, the ESG–SR frontier is hump-shaped because restricting portfolios to have any ESG score other than that of the tangency portfolio must yield a lower maximum SR.

Type A investors choose the portfolio with the highest SR – that is, the tangency portfolio using ESG information. Type M investors have a preference for higher ESG, so they choose portfolios to the right of the tangency portfolio, on the ESG-efficient frontier. Type M investors with a small preference for ESG choose a portfolio just to the right of peak with nearly the maximum SR (higher than the SR achieved by type U), and type M investors with strong preferences for ESG choose portfolios on the far right of the ESG-efficient frontier (possibly with lower Sharpe ratios than type U investors). Finally, type U investors may choose a portfolio below the frontier, because they compute the tangency portfolio while ignoring the security information contained in ESG scores (they condition on less information).

The portfolio problem for types U and A investors has the standard Markowitz type of solution (albeit with potential different estimates of risk and expected returns), so we focus here on the solution for type M investors, who seek to maximize their utility U over final wealth W and average ESG score s, given the extended mean-variance framework

$$U = E(\widehat{W} \mid s) - \frac{\overline{\gamma}}{2}\mathrm{Var}(\widehat{W} \mid s) + Wf(\overline{s}), \widehat{W} = W\left(1 + r^f + w'r\right),$$

where W is the initial wealth, W^ is the final wealth, r is return of portfolio, rf is risk-free return, f is a function f: R → R of the average portfolio ESG score and γ¯ is the absolute risk aversion parameter.

Thus, with μ = E[r|s],

$$U = W\left(1 + r^f + w'\mu\right) - \frac{\overline{\gamma}}{2}W^2 w' \Sigma w + Wf(\overline{s})$$
$$= W\left(1 + r^f + w'\mu - \frac{\gamma}{2}w' \Sigma w + f(\overline{s})\right).$$

Considering rf is constant, we consider only

$$\max_{\forall w}\left(w'\mu - \frac{\gamma}{2} w'\Sigma w + f(\bar{s}) \right).$$

In line with the standard Markowitz mean-variance analysis, given an ESG level, the maximum SR that can be achieved (denoted the ESG–SR frontier) is

$$SR(\bar{s})_{\text{s.t. } \bar{s}=\frac{w's}{w'1}} = \max_{w}\left(\frac{w'\mu}{\sigma} \right).$$

In order to be able to use this definition of the highest Sharpe for each ESG level, we first rewrite the utility maximization programme as

$$\begin{aligned} &\max_{\bar{s}}\left[\max_{\sigma}\left(w'\mu - \frac{\gamma}{2} w'\Sigma w + f(\bar{s}) \right) \right] \\ &SR(\bar{s})\sigma = w'\mu \\ &\sigma^2 = w'\Sigma w, \end{aligned}$$

so that it becomes

$$\max_{\bar{s}}\left[\max_{\sigma}\left\{ SR(\bar{s})\sigma - \frac{\gamma}{2}\sigma^2 + f(\bar{s}) \right\} \right].$$

This means that the investor's problem can be thought of as first choosing the portfolio with the highest SR given a level of risk σ and an ESG score s¯, and then maximizing over σ and s¯.

The first step is solved by optimizing against volatility σ. After solving for σ, the problem becomes

$$\max_{\bar{s}}\left[(SR(\bar{s}))^2 + 2\gamma f(\bar{s}) \right].$$

Not surprisingly, ESG affects the optimal portfolio choice, given that ESG is in the utility function. Such a result illustrates this trade-off using a part that depends only on securities (the ESG–SR frontier) and another part that depends only on preferences (2γf(s)); just like the standard Markowitz theory with the mean-variance frontier, the ESG–SR frontier can be computed independent of preferences, and then the investor can decide where on the frontier to place herself based on risk aversion.

Remark: this method works because the investor is assumed to care about the average ESG score only, which does not change when the investor chooses the risk level in the second step by choosing her cash holding. If investors care about total ESG instead of average ESG, then the investor's problem cannot be summarized as the ESG–SR frontier, which also shows that our frontier results are not trivial.

Understanding the ESG–SR frontier shows how differences in risk aversion and in ESG preferences can be distinguished.

- If an investor has no direct preferences for ESG (f ≡ 0) but differs in her risk aversion γ, then she should invest in the portfolio of risky assets with the corresponding Sharpe ratio and average ESG score, but the more risk-tolerant should put a larger fraction of their wealth in this portfolio (i.e. own fewer cash instruments).

- If an investor has the same risk aversion but differs in terms of having stronger ESG preferences, she should buy a portfolio with lower SR but a higher average ESG score.
- SR matters less when an investor is more risk-averse (because she knows that she will take less risk anyway), so, in relative terms, ESG becomes more important for her.

The steady-state equilibrium security prices and returns are derived below, where expected returns are given by an ESG-adjusted CAPM, depending on the relative portions of types of investors, starting with the simplest cases, in which all investors are of the same type.

- If all investors are type U (i.e. ignore ESG), then we are back to a standard CAPM equilibrium: all investors hold the unconditional tangency portfolio that equals the market portfolio, and each security's expected excess return is driven by its unconditional market beta, $E\left(r_t^i\right)=\beta^i E\left(r_t^m\right)$.

  What is new here is that a (small) investor who understands that ESG scores are informative can exploit this insight: from the perspective of an investor who uses ESG scores, the equilibrium conditional expected excess return now depends linearly on ESG,

$$E\left(r_t^i \mid s\right)=\beta^i E\left(r_t^m\right)+\lambda \frac{s^i-s^m}{p^i},$$

  where s^i is the stock I ESG score while s^m is the weighted average ESG score of the market portfolio.

  If a high ESG score is indicative of a high future profit – i.e. if $\lambda > 0$ – then stocks with ESG scores above average have higher conditional expected returns than those with below-average ESG scores. This is in line with empirical findings, such as those of Gompers, Ishii and Metrick (2003).

- If all investors are type A (all investors use ESG signals, but without ESG preferences), then we get a conditional CAPM equilibrium, and investors can no longer profit from using the informational value of ESG scores because this information is already incorporated into prices. This theoretical prediction is in line with the empirical finding of Bebchuck, Cohen and Ferrell (2009), who show that the measures from Gompers, Ishii and Metrick (2003) do not predict abnormal returns out of sample.
- Finally, suppose that all investors use ESG in their signals and in their identical ESG preferences (i.e. are all type M); then the equilibrium conditional expected excess return is further modified, given by the standard conditional CAPM expression (for type A investors), which is here adjusted for whether the ESG score is above or below that of the market:

$$E\left(r_t^i \mid s\right)=\bar{\beta}^i E\left(r_t^m \mid s\right)-\pi\left(s^i-s^m\right),$$

  where π is the parameter catching the strength of the preference for ESG ($\pi = 0$ for type A investors).

In other words, the firm's cost of capital is lower if its ESG score is higher, or, equivalently, the firm can issue shares at higher prices. This low cost of capital encourages high-ESG firms to make real investments, because, using this low discount rate, more projects will have a positive net present value.

If all types of investors exist, then several things can happen, depending on the relative importance of each investor type, leading to a relation between ESG and expected returns that is positive, negative or neutral.

- When there are many type U investors and when high ESG predicts high future profits, high-ESG stocks deliver high expected returns: high-ESG stocks are profitable, yet their prices are not bid up by type U investors, leading to high future returns.
- When the economy has many type A investors, then these investors bid up the prices of high-ESG stocks to reflect their expected profits, thus eliminating the connection between ESG and expected returns.
- Further, if the economy has many type M investors, most investors are willing to accept a lower return for a portfolio with higher ESG levels, then high-ESG stocks actually deliver low expected returns.

Whether favourable ESG characteristics signal good profitability or low profitability (the sign of λ) is an empirical question. For instance, Pedersen, Fitzgibbons and Pomorski (2021) find E, S and overall ESG as weaker predictors of future profits.

4.3.5.1.3.2 From Theory to Practice: Trailblazing a Hard Way to Outperformance versus the Benchmark Indices As a result, given that it is in practice impossible to measure the relative importance of each investor type, and extremely hazardous to estimate both λ and the π parameter (preference for ESG), in the following the optimal asset allocation is restricted to the assumption $\lambda = \pi = 0$.

First we make sure that all required modelling assumptions of the CAPM are passed by econometrics tests (regarding the non-autocorrelation, the normality and the non-heteroscedasticity of returns), as illustrated in Figure 4.81.

	Best 20% Portfolio		Worst 20% Portfolio		
Assumptions	Y/N	p-value.s	Y/N	p-value.s	Test
1- Normality	✓	0.83848	✓	0.65974	χ^2
		0.77462		0.83069	JB
2- Homoskedasticity	✓	0.57165	✓	0.27558	BP (as.)
		0.53233		0.24539	White (as.)
3- Non Autocorrelation	✓	0.243	✓	0.0997	BG (as.)

FIGURE 4.81 Econometric tests

4.3.5.1.3.2.1 Using Too Many Restrictive ESG Constraints Improves the Trend but Does Not Yet Outperform Why don't they outperform? Because the "best-in-class" constraint on the ESG score is too restrictive: numerous analyses on single names illustrate that best-in-class ESG stocks (e.g. best 10%) do not provide higher market performance (neither improved excess return nor risk-adjusted – namely Sharpe, Sortino or Treynor ratios), while illustrating a larger VaR versus the benchmark index (S&P 500), in contrast with lower-rated ESG assets of companies, which progressively integrate ESG concerns into their activities, and thus have rising ESG scores. Further profit-driven investors should even consider bad ESG ratings stocks from a diversification point of view.

In this respect, MSCI's ESG research study "Can ESG add alpha?" (Nagy, Kassam & Lee 2015) shows that companies that managed to improve their ESG score saw an increase in their share price during the following 12- to 18-month period. Overall, the ESG Momentum Index strategy outperformed significantly the MSCI World Index between 2008 and 2014.

Even so, such a 10% best-in-class ESG active "optimal" portfolio asset allocation improves diversification, net selectivity (positive now and four times better than above) and returns (both absolute and risk-adjusted ones), although the metrics remain weaker versus a 10% worst-in-class ESG active "optimal" portfolio asset allocation.

4.3.5.1.3.2.2 Using an ESG Constraint on the Average Portfolio Score Enables Outperformance of the Benchmark Indices The average ESG criterion is as follows: the average ESG score of the optimal portfolio is required to be higher than the benchmark index ESG score.

Remark: requiring the average ESG of the optimized portfolio to be significantly higher than the average ESG score of the benchmark index adds significant constraints on its ability to outperform in terms of returns and risk, as illustrated by De Spiegeleer et al. (2021), which imposes as the average ESG score constraint

$$\mathbf{ESG}^t\mathbf{w} \geq \mathbf{ESG}_{\text{MV}} + 0.1 \cdot (\max\{\mathbf{ESG}\} - \min\{\mathbf{ESG}\}).$$

In order to comply with the customer's risk appetite, another risk constraint is added to the average ESG criterion.

As a result, an optimal asset allocation can be built based on a nonlinear Markowitz programme upgraded by ESG information (the average ESG score of the optimal portfolio is required to be higher than the benchmark index ESG score) and risk criteria (volatility, value at risk, conditional value at risk), which deals with the Solvency II capital requirement constraints, as follows:

$$\begin{cases} Max & r_p \\ \text{s.c.} & V^{opt} \leq V^n \\ & \overline{s}_p^{opt} \geq \overline{s}_p^{n}, \end{cases}$$

specifically maximizing the absolute returns under two constraints:

- a minimum ESG score; and
- a maximum value at risk.

Even if the scope of stocks is restricted to the best-in-class ESG stocks (20%), it turns out that it delivers a clear and resilient outperformance in terms of overall returns, risk and cost of capital, to the benefit of customers and insurance companies alike. More specifically,

- there have been higher excess returns and a downside Sharpe ratio (Sortino ratio) since 2016;
- the net selectivity is now positive and four times better for the best ESG compared with the worst ESG assets; and
- there has been a lower downside deviation (considering only returns that fall below a minimum threshold) since 2013.

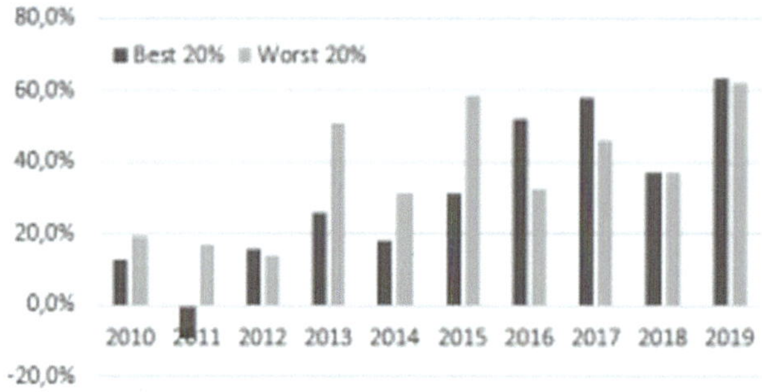

FIGURE 4.82 Excess returns, 2010–2019

FIGURE 4.83 Sharpe ratios, 2010–2019

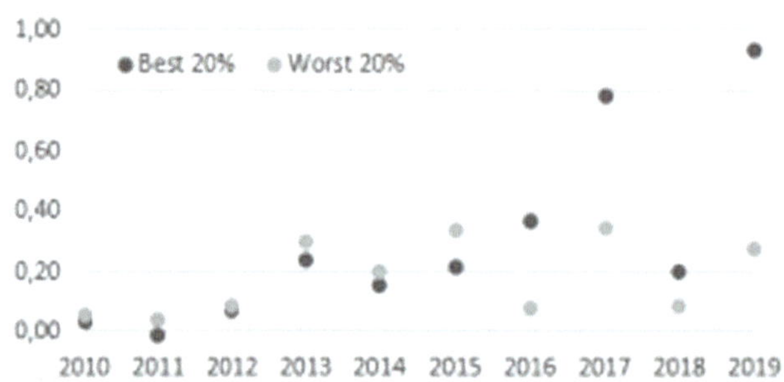

FIGURE 4.84 Sortino ratios, 2010–2019

Remark: such results are consistent with the findings of Roncalli et al. (2021) and Bennani et al. (2018).

The results can be further improved, in both magnitude and duration, if we extend the scope of stocks to, for example, the 80% best-rated ESG stocks (by excluding the 20% worst ESG assets). Moreover, the constraint on ESG portfolios can be further alleviated by introducing "hybrid" ESG scoring metrics, instead of plain ESG scoring metrics, which combine both ESG scoring *and* corporates KPIs (e.g. debt or EBITDA [earnings before interest, taxes, depreciation and amortization] metrics). As a result, the outperformance has been confirmed since the availability of ESG data (2009), and this across all KPIs.

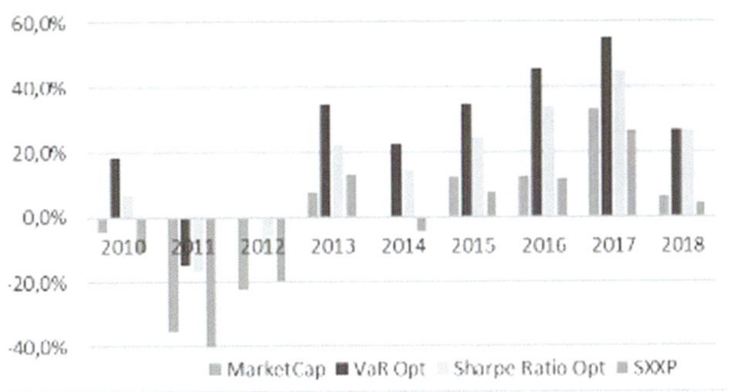

FIGURE 4.85 Excess returns, 2010–2018

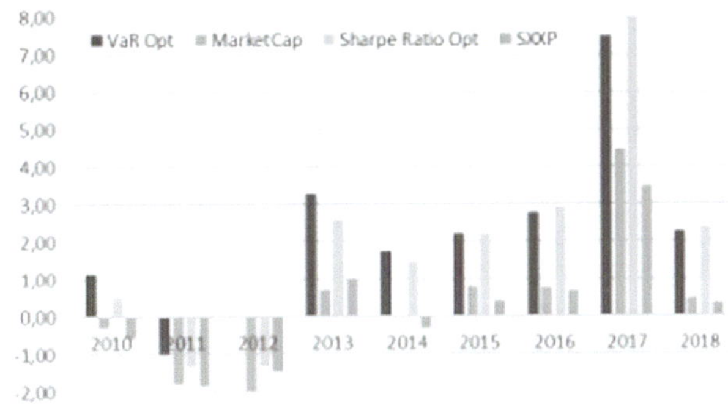

FIGURE 4.86 Sharpe ratios, 2010–2018

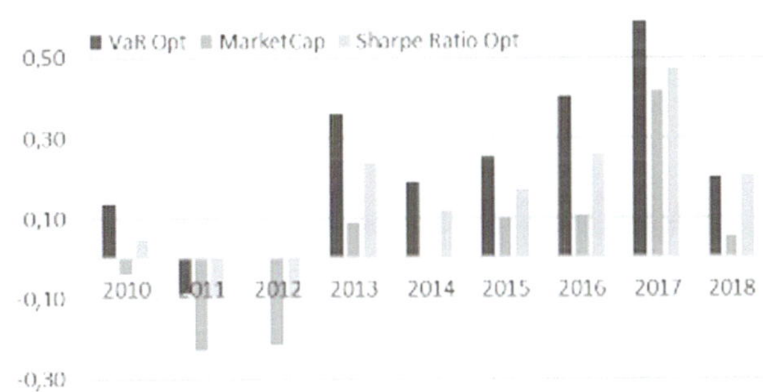

FIGURE 4.87 Sortino ratios, 2010–2018

Such hybrid ESG scoring metrics have indeed evolved in parallel with the Sharpe ratios (see Figure 4.88).

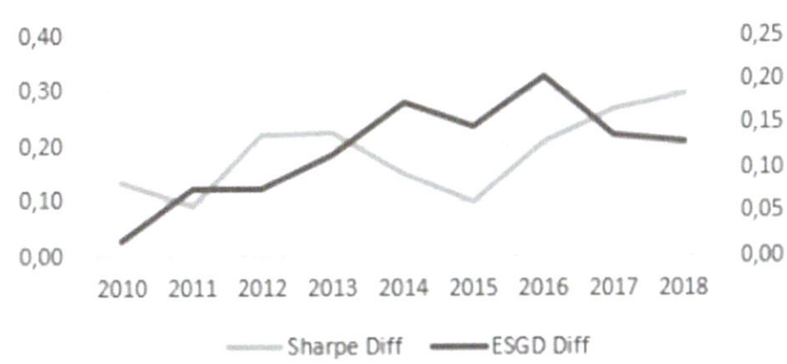

FIGURE 4.88 Hybrid ESG change versus Sharpe ratio change

Given the multiple constraints, the optimal asset allocation involves a nonconvex optimization, translating into multiple feasible regions with multiple local optimums as well, so the usual optimization methods cannot be used. In this respect, the combination of a global stochastic evolutionary algorithm with a local deterministic nonlinear algorithm is needed when functions are non-smooth, including non-differentiable and discontinuous functions in the programme, such as the value at risk computation in our case.

- The global stochastic evolutionary algorithm is a stochastic process that is used to find a global optimum (or near-optimum). It uses a genetic approach: the algorithm generates a population of solutions that will be exposed to several random natural selections; the "best" solutions will be "kept alive" and the others will be "killed".

  The population is a solution set going through an evolution process in the hope that the fittest individuals – solutions – will survive all the steps of selection. To put it simply, the solutions that fulfil the conditions for survival are selected and subject to reproduction, with a low probability of mutation in the process. The mutation event occurs to simulate the random modification of genes, so that solutions do not get "stuck" in a local optimum and can relatively easily access the range of other solutions to naturally reach the global optimum. Mutations come in addition to crossovers and recombinations of several existing and fitting points. In short, select → reproduce → mutate & repeat to create increasingly better solutions.

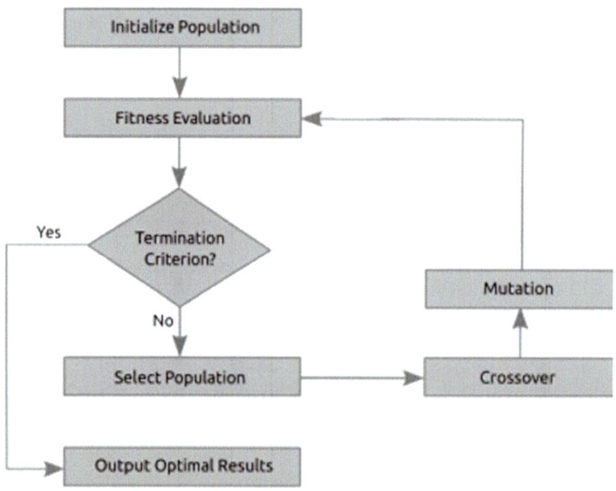

FIGURE 4.89 An evolutionary algorithm

  In the end, the remaining solutions are likely to be the optimums. The main drawback of this algorithm is that it is limited by time; we could perfectly find a better solution if we could let the algorithm run for longer.

  In short, the situation at one possible solution set gives very little information about where to look for a better solution. This greatly impedes the efficiency of genetic algorithms, as one generation may not lead to a better one when "reproducing" – mixing individuals to find better matches in the next generation. To counter this effect, we use the randomly generated mutation of individuals. However, the probability of being stuck in a local optimum remains, and many solution sets exist, so this method relies on a random sampling of possible solutions such that it determines a good but not perfect solution set. This Excel Solver type is thus stochastic and can yield different results on each run depending on the randomly sampled points.

- The generalized reduced gradient (GRG) nonlinear algorithm is a deterministic process that is used to find a local optimum. By making small changes in the variables' initial values, it tries to improve the objective. However, it stops when it finds a peak; a remedy is to try to start the algorithm with different initial values.

The starting point of the optimizations is the weights of an equally weighted portfolio. Then a two-step process is launched: first we run the evolutionary algorithm in order to find a global optimal point. Then we launch the GRG nonlinear algorithm to make sure that we did find an optimum. If the GRG nonlinear results in the same solution, we can be sure that we have found the global optimum. In contrast, if it gives us another one, we now have an (at least) local optimum.

4.3.5.1.4 Conclusion Within the context of growing customer appetite for ESG assets and increasing regulations on sustainability issues, reconciling sustainability and market performance is still an open issue for the academics, as illustrated by the poor performance of passive ESG portfolios, which usually underperform versus benchmark indices.

Trailblazing a way to reconcile sustainability, market performance and customer risk appetites is still possible through a balanced optimal ESG-constrained and risk-based active portfolio management that also meets customers' risk appetites.

Beyond performance and risk metrics, the asset manager needs to make sure that the stock components of the resulting optimized portfolio meet his targets in terms of portfolio diversification, market capitalization, liquidity and sector diversity, which can also be included in the optimization programme.

4.3.5.2 Decreasing the Transaction Costs of Investment Strategies through Sustainable Risk-Appetite-Based Optimal Asset Rebalancing Strategies

4.3.5.2.1 Why? Over time, as a portfolio's different investments produce different returns, the portfolio drifts from its target asset allocation, acquiring risk and return characteristics that may be inconsistent with a customer's goals and preferences (e.g. a 60/40 equity/bonds target).

But, as asset values move randomly, asset ratios diverge from the desired target ratio. A portfolio's return relative to its target asset allocation is the appropriate framework for evaluating a rebalancing strategy's effectiveness in risk control. Funds "rebalance" by trading infrequently to restore the asset proportions to their target levels and recapture the portfolio's original risk and return characteristics. Rebalancing typically occurs at a regular time interval – e.g. every month or every quarter.

- In asset allocation, risk and return are absolutes: for instance, an expected annual return of 10% might be associated with annualized volatility of 15%, while a return of 5% might be associated with volatility of 7%.
- In a rebalancing strategy, by contrast, risk and return are measured relative to the performance of the target asset allocation. However, like the selection of a portfolio's target asset allocation, a rebalancing strategy involves a trade-off between risk and return.
- Fortunately, in most market environments, where returns follow a random walk pattern, successful relative risk control also reduces absolute risk

To ensure that a portfolio's risk and return characteristics remain consistent over time, a portfolio needs to be rebalanced.

A rebalancing strategy precisely addresses this risk by formalizing guidelines about how frequently the portfolio should be monitored, how far an asset allocation can deviate from its target before it is rebalanced and whether periodic rebalancing should restore a portfolio to its target or to some intermediate allocation.

4.3.5.2.2 The "How Often?", "How Far?" and "How Much?" Paradigm We take as given the target asset ratio and focus only on keeping close to it; in other words, maximizing the fund value is not the primary objective, but keeping tight control on the tracking error while minimizing transaction costs is.

Three questions therefore need to be addressed.

- How frequently the portfolio should be monitored?
- How wide should the drift bands around the target allocation be? In other words, what degree of deviation from the target allocation triggers the rebalancing?
- And how far back towards the target should a portfolio be rebalanced? Should the portfolio be rebalanced to its target or to an intermediate allocation?

4.3.5.2.3 Risk Appetite Considerations Just as there is no universally optimal asset allocation, there is no universally optimal rebalancing strategy. An institution selects a rebalancing strategy based on its tolerance for risk relative to a target allocation, meaning some specific risk appetite combinations.

If a portfolio is never rebalanced, it will gradually drift from its target asset allocation to higher-return, higher-risk assets.

- Consider a portfolio with a target asset allocation of 60% stocks and 40% bonds that is never rebalanced.
- Because stocks have historically outperformed bonds, the portfolio's asset allocation gradually drifts to 90% stocks and 10% bonds.
- As the portfolio's equity exposure increases, the portfolio moves toward the upper-right end of the rebalancing frontier: higher risk and higher return.
- The rebalancing frontier can be adapted to account for the costs of rebalancing – transaction costs, taxes – which diminish the portfolio's return. As the portfolio is rebalanced more frequently, transaction costs become a bigger drag on performance

As illustrated in Figure 4.90, the return distributions of unrebalanced and rebalanced portfolios are both bell-shaped, but the rebalanced bell curve (solid line) is thinner and taller than the unrebalanced portfolio (dotted line), and slightly to its right.

The bell curve of the rebalanced portfolio is thinner and taller because it is less likely to experience extreme returns and more likely to experience moderate returns. Most investors would be happy to give up some potential for extremely high returns in exchange for a reduced likelihood of extremely low returns.

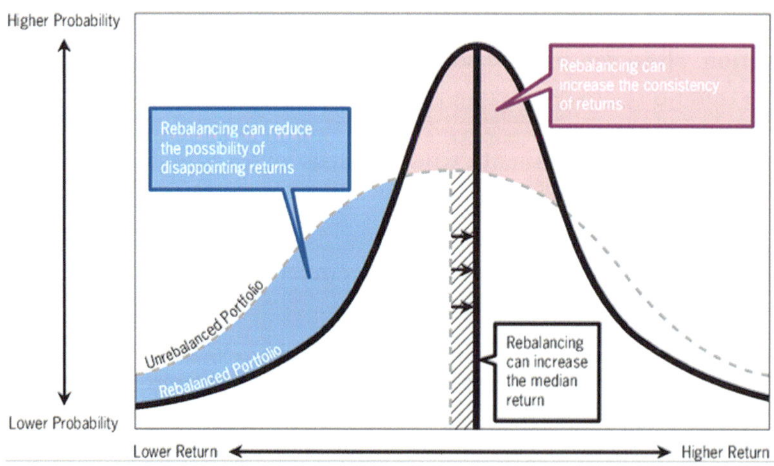

FIGURE 4.90 The advantages of rebalancing

Due to differing risk tolerances, two institutions with identical asset allocations may prefer different rebalancing strategies.

- Real-world experience indicates that investors often view risk in absolute terms.
- The risk of not adhering to the asset allocation target as expressed by the loss in a relative risk control, which is measured by the degree of tracking error, weighted by the utility cost per unit of tracking error λ:

$$L = \lambda\left(w(t) - w^{*}\right)^{2} dt.$$

4.3.5.2.4 Key Drivers of Rebalancing Strategies The appropriate rebalancing strategy depends on a number of drivers, such as the market environment and asset-class characteristics, including the following.

- The average returns: a high expected return for a particular asset causes a portfolio's allocation to drift towards this asset class more quickly. Significant differences between the expected returns of portfolio holdings increase the risk of significant deviation from the target allocation, and thus increase the need to rebalance.

  The impact of a rebalancing strategy on a portfolio depends not only on the return level but also on the return patterns over time.

  - If security returns approximately follow a random-walk pattern, then rebalancing more frequently or within tighter bands reduces a portfolio's downside risk (absolute risk as well as relative to the target asset allocation).
  - In a trending or mean-reverting market, the impact of rebalancing may be somewhat different when viewed on an absolute or relative-to-target basis.
- The volatilities: high return volatility increases the fluctuation of the asset class weightings around the target allocation and increases the risk of significant deviation from the target. Greater volatility implies a greater need to rebalance. In the presence of time-varying volatility, rebalancing occurs more often when volatility rises.
- The correlations: high correlation between the returns of asset classes means that they tend to move together. When the returns of all the assets in the portfolio move in the same direction, the asset allocation weightings tend to remain unchanged, reducing both the risk of significant deviation from the target allocation and the need to rebalance.

Overall, the rebalancing increases with the dispersion of returns across the portfolio assets and with the volatility implying more fluctuation around the target allocation, while higher correlation between the returns means that they tend to move together, thus reducing the need to rebalance.

The other key factors to consider when implementing a rebalancing strategy are the costs: the transaction costs, which will reduce the return of the portfolio; and the risk of not adhering to the asset allocation target, as expressed by the loss in a relative risk control L, as mentioned above. The total cost is the sum of the two:

$$V\left(w(t); w_{max}, w_{min}\right) = E\left\{\left[\int_{t=t}^{\infty} e^{-r(t-t)} \lambda\left(w(\tau) - w^{*}\right)^{2} d\tau + PV\{transactions\ costs\}\right] \text{given } w(t)\right\}.$$

We can then optimize the λ through minimization of tracking error.

4.3.5.2.5 Contributions from Theory: the Liquidity Premium and the Trigger Point

If transaction costs are proportional to the dollar amounts traded, the conventional strategy of periodic rebalancing to the target ratio will not be optimal.

Theoretical developments (see Constantinides 1986 and Dixit 1991) show that, in the presence of proportional transaction costs, the optimal policy is characterized by a "no trade" interval about the target asset ratio. These drift bands are tolerance ranges set around allocation targets for each asset.

- As the ratio of asset values moves randomly within this interval, the investor refrains from transacting and no trading is needed.
- When the asset ratio wanders outside the no-trade interval, asset proportions should be adjusted and the portfolio is rebalanced.

The region of no transactions is found to be wide, and, therefore, an investor's demand for the assets is sensitive to the current composition of the portfolio. Furthermore, the average demand for one of the two assets, which is subject to transaction costs, is substantially reduced.

Therefore, transaction costs have a first-order effect on demand, whereby investors accommodate large transaction costs by drastically reducing the frequency and volume of trade.

This also means that investors are insensitive to deviations of the asset proportions from those that are optimal in the absence of transaction costs. Therefore, transaction costs have only a second-order effect on asset returns, and a small "liquidity premium" is sufficient to compensate for deviating significantly from the target portfolio proportions.

In other words, transaction costs may play a significant role in preventing investors from purchasing an underpriced asset with the intention of selling it a few months later and realizing a profit net of transaction costs. However, transaction costs do not explain systematic and significant deviations of assets' expected returns adjusted for risk premiums.

Among the conclusions of theoretical developments are the following.

- The width of the region of no transactions is insensitive to the variance of the asset's rate of return.
- An increase in the risky asset's variance increases the frequency and amount of transactions.
- An increase in transaction costs broadens the region of no transactions, although it is due to the assumption of Brownian motion (BM) for assets' dynamics. Because the BM is an infinite variation process, the costs accompanying frequent rebalancing, even levied at a low rate, would quickly outweigh the benefits of precisely optimal diversification.

To determine when rebalancing is efficient, we weigh the benefits against the costs.

- The costs of rebalancing transaction fees grow steadily as the portfolio drifts further out of alignment (Figure 4.91, red area), because selling twice as much of a security costs about twice as much to trade.
- The principal benefit of rebalancing is the reduction of tracking error risk. This benefit increases as the portfolio drifts further from its target allocations. And, because risk is quadratic, the tracking error quadruples as the portfolio drifts twice as far off target (blue area).

Since rebalancing costs are linear while rebalancing benefits are quadratic, at some point the benefits of rebalancing will begin to outweigh the costs, and the net benefit of rebalancing (bottom panel, white line) will become positive. At that point, referred to as the "trigger point", rebalancing becomes efficient.

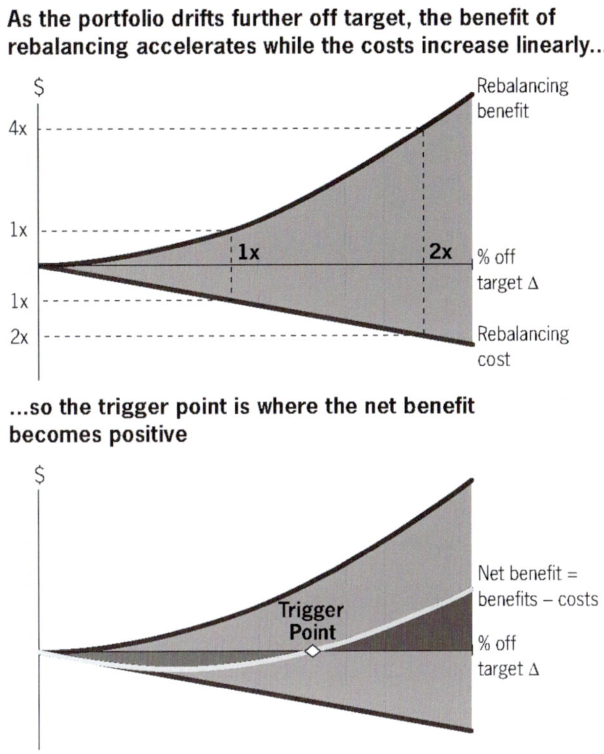

FIGURE 4.91 Linear costs versus quadratic gains

Here, minimizing transaction costs while controlling portfolio risk consists of returning the "out of balance" asset class halfway back to its initial percentage.

The optimal rebalancing also depends on whether transaction costs are fixed or proportional.

- When trading costs are mainly fixed, independent of the size of the trade, rebalancing to the target asset allocation is optimal in order to avoid the need for further transactions (Zakamouline 2006).
- On the other hand, when trading costs are mainly proportional to the size of the trade, rebalancing to the closest rebalancing boundary, which minimizes the size of the transaction, is optimal (Zakamouline 2006).
- If trading costs are largely proportional to the trade size, albeit lower for large trade sizes, then portfolios may be better off rebalancing closer to the target asset allocation.

4.3.5.2.6 From Theory to Practice

4.3.5.2.6.1 The Impact of Market Conditions We have conducted simulations to identify returns patterns in which rebalancing strategies and the decision not to rebalance produce the greatest discrepancies.

- Trending markets.
    - If equity prices rise in every period, regular rebalancing implies continually selling the strongly performing asset and investing in the weaker performer. The result is a lower return compared with a less frequently rebalanced portfolio. A good example is the US stock market's steady upward surge during the mid- to late 1990s.
    - If stock prices fall in every period, then the portfolio is continually buying equities as their prices decline, experiencing returns below those of a portfolio that is never rebalanced. The two and a half

years following the US stock market's March 2000 peak were an example of a downward-trending market – again, an environment that made rebalancing unattractive.

- The more frequently a portfolio is rebalanced, the tighter its risk control relative to the target asset allocation, even during periods of strong price momentum. This tighter risk control is accompanied by the higher potential cost of rebalancing, as indicated by the greater number of rebalancing events and the higher turnover rate.

- Mean-reverting markets.
  - In a mean-reverting market, a portfolio's returns can be enhanced by rebalancing, buying an asset after it has decreased in value and selling it after it has appreciated. For example, in 1987 and 1988 the US stock market followed a pattern of mean reversion.
  - Although a portfolio's buy and sell decisions are generally well timed when rebalancing in a mean-reverting market, our simulation indicated that as a portfolio was rebalanced more frequently, or within tighter bands, its absolute average return decreased.
  - The additional return produced by well-timed purchases and sales was less than the additional return produced by the higher equity allocation in a less frequently rebalanced portfolio.
  - Risk differences between the different portfolios were relatively modest.
    - The small scale of the differences in this simulation was partly due to the low standard deviation of all portfolio returns in mean-reverting markets.
    - When each upward or downward price movement was followed by its opposite, a portfolio's fluctuation around its average return tended to be modest.
- Random-walk environment.
  - Although the equity markets may seem to experience periods of trending and mean reversion, a more realistic model of discounted equity price movements is a random walk.
  - A portfolio that was rebalanced more frequently, either because it was monitored more frequently or because it had tighter rebalancing thresholds, tracked the target asset allocation more closely.
  - Although a more frequently rebalanced portfolio will have less risk than a less frequently rebalanced portfolio, the cost of rebalancing may place upper limits on the optimal number of rebalancing events. Transaction costs detract from the portfolio's return, potentially undermining the risk control benefits of some rebalancing strategies
  - The decision depends on the type of rebalancing cost, as described thereafter.

4.3.5.2.6.2 Optimal Rebalancing Metrics The drift bands' width constitutes an important metric, with these factors to consider.

- Large versus thin drift bands.
  - Rebalancing too soon (i.e. with very narrow drift bands) or allowing excessively large drift tolerances are both less than optimal from a return perspective: returns tend to increase initially as the bands are widened, but eventually reach a peak and begin to decline.
- Fixed versus percentage bands.
  - For small allocations, the percentage approach produces relatively tight bands, while the fixed absolute approach produces relatively loose bands. The reverse is true with large allocations.

- Tight bands can trigger excessive rebalancing and, in turn, higher trading expenses. Less frequent rebalancing caused by relatively loose bands may increase portfolio volatility and decrease returns.

The turnover metrics are also important.

- Expressed in annual terms, the turnover is defined as the marginal present value (PV) of the transaction costs alone.
- The expected turnover is inversely proportional to the width of the no-trade interval.
- The turnover varies inversely with the cube root of the transaction costs.
- When parameters other than λ remain fixed, to a close approximation turnover is proportional to the cube root of λ.

The size of the optimal no-trade interval ($w_{max} - w_{min}$) must be taken into consideration.

- When parameters other than λ remain fixed, the no-trade interval is inversely proportional to the cube root of λ.
- It varies directly as the cube root of transaction costs (TC).
- It is proportional to the cube root of the variance, or, equivalently, is proportional to the standard deviation raised to the 2/3 power.

Tracking error (TE) metrics also need to be addressed.

- Tracking error is measured by the annualized standard deviation of w.
- The standard deviation of the TE is inversely proportional to the expected turnover, while proportional to the width of the no-trade interval.

Risk/reward metrics are a further area to consider.

- Fund value metrics: mean and variance.
- Transaction costs versus tracking error need to be traded off against each other.
- Transaction costs versus the number of rebalancing events are also subject to a trade-off.

Risk appetite metrics are the final set to look at.

- Maximizing portfolio fund value & minimizing transaction costs × tracking error
- Maximizing portfolio fund value & minimizing VaR (P&L) × transaction costs × tracking error.

4.3.5.2.6.3 Results based on Risk Appetites

4.3.5.2.6.3.1 Forward-Looking Approaches Various market conditions are considered, including different equity risk premiums in relation to bond and equity returns:

- low- versus high-volatility regimes;

- low versus high interest rate levels; and
- high versus low transaction costs.

Overall, the optimized strategies incur far less in the way of transaction costs for any particular magnitude of transaction costs. They rebalance more frequently. However, this increase is only marginal, and they incur lower transaction costs per rebalancing.

As expected, in most market environments, when returns follow a random-walk pattern, a successful relative risk control also reduces the absolute risk.

- Strategic rebalancing is not detrimental to portfolio fund value, as can be seen in Figure 4.92.

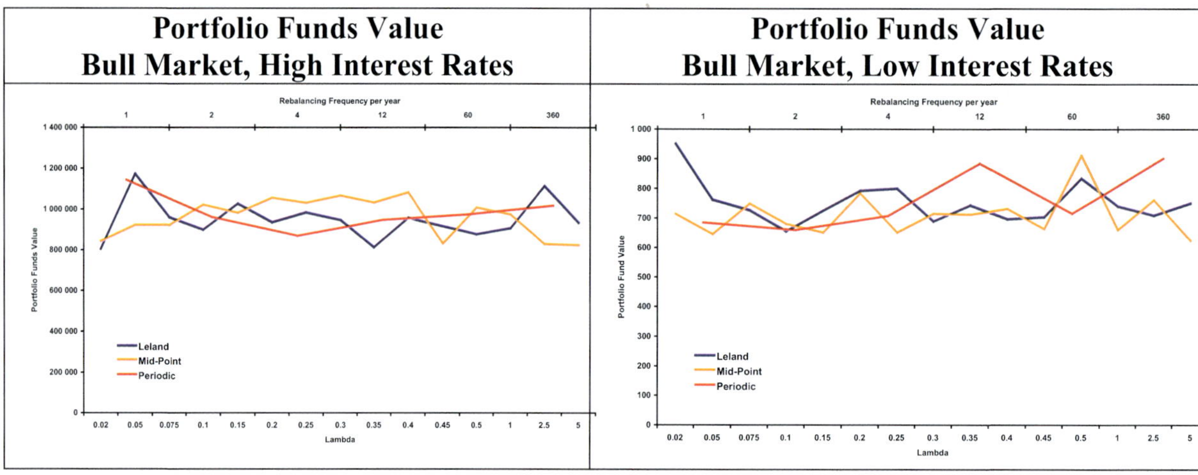

FIGURE 4.92 Portfolio value across various market regimes

- A comparison between the halfway method and the Leland method in terms of rebalancing frequency and transaction costs is shown in Figure 4.93.

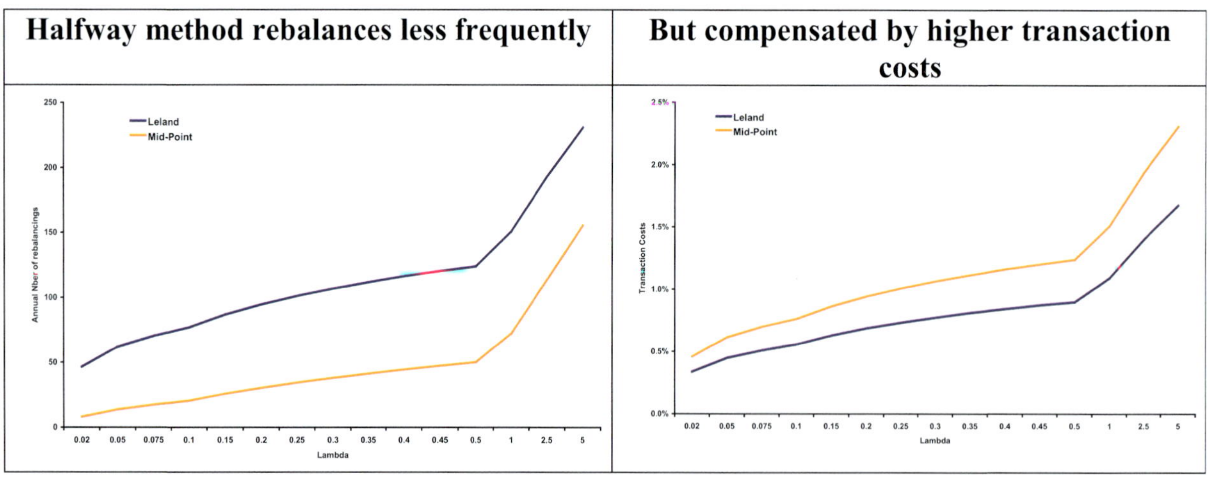

FIGURE 4.93 Annual number of rebalancings versus the lambda parameter

However, when considering both aspects, the two strategic rebalancing methods are close.

- Both outperform the periodic rebalancing significantly.
- The lower the risk tolerance, the lower the tracking error, but the higher the transaction costs.
- Whatever the risk tolerance, the product of TE × TC is constant, illustrating the relevance of such metrics.

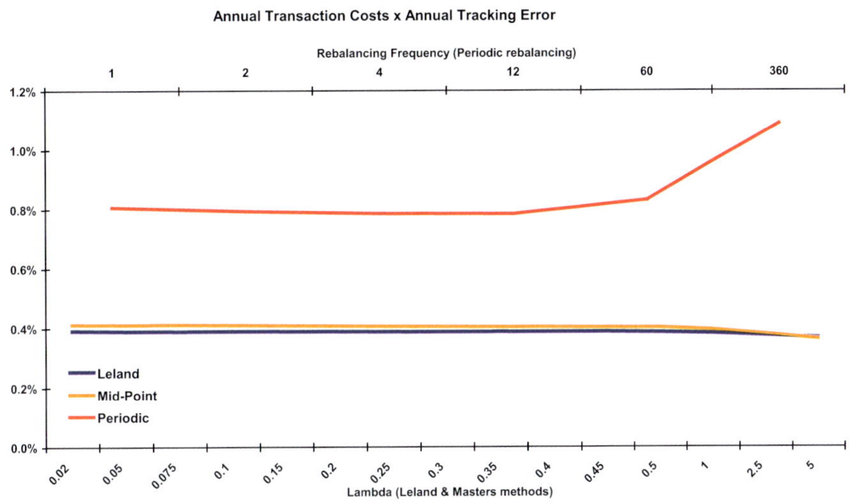

FIGURE 4.94 Strategic rebalancing versus periodic rebalancing in terms of annual transaction costs times annual tracking error

- Whatever the market conditions, when dealing with significantly different assets within the basket, in terms of both volatility (20% versus 10%) and correlation (30%), trigger points are significantly improved by using either the Leland or the mid-point methods.
- The Leland and mid-point methods outperform equally, whatever the transaction costs.

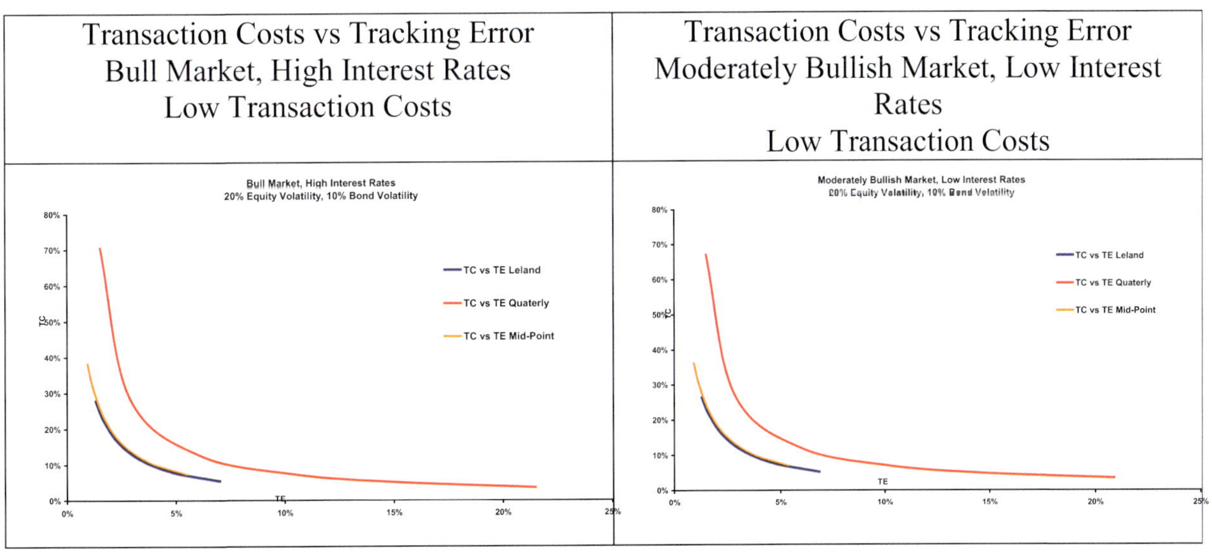

FIGURE 4.95 Transaction costs versus tracking error

- The risk appetite metrics illustrate that the methods improve by a factor of two.
- The higher the number of rebalancing events, the larger the advantage of the Leland method.

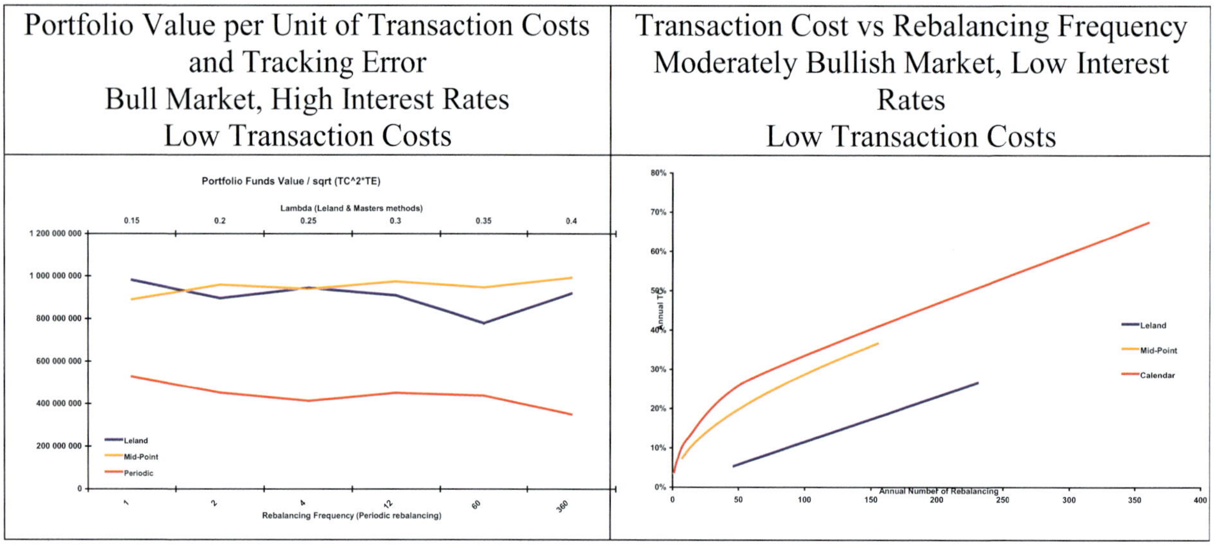

FIGURE 4.96 Portfolio value per unit of transaction costs and tracking error

If we consider the variability of the fund value, the methods improve the strategy by four to nine times

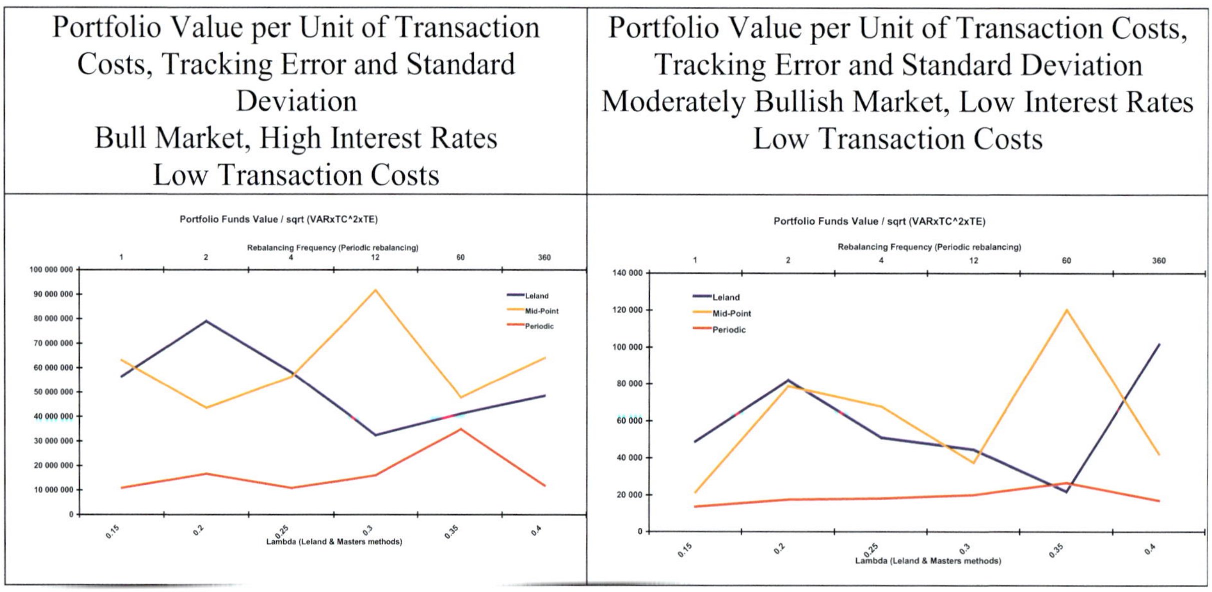

FIGURE 4.97 Portfolio value per unit of transaction costs and tracking error and standard deviation

4.3.5.2.6.3.2 Back-Testing Approach

Overall, the following points can be made.

- The optimized methods incur far fewer transaction costs for any particular magnitude of transaction costs. The optimal size of the no-trading zone then depends on risk appetite, represented by the parameter λ.

- As expected, the break-even between TE and TC (i.e. the "trigger point") is lower; but the risk/return outperformance is still significant.

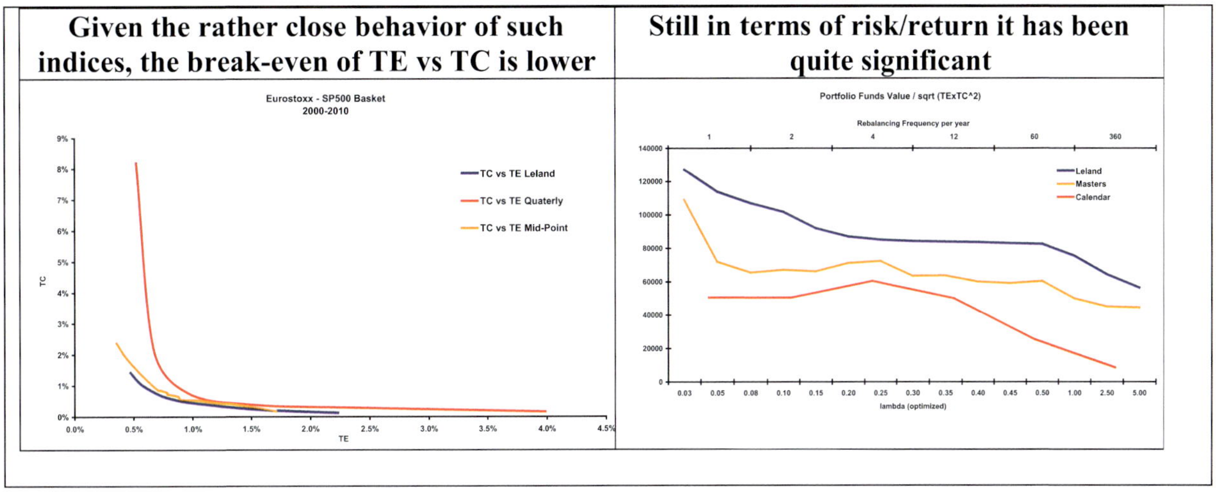

FIGURE 4.98 Transaction costs versus tracking error

4.3.5.2.6.3.3 *Extension to a Three-Asset Portfolio*

As there is no extension of the theoretical framework to more than two assets, a specific algorithm has been built in order to deal with multi-asset portfolios as an extension of the Leland strategic rebalancing strategy. Then the forward-looking approach illustrates that the efficiency of a two-asset framework is maintained regarding the three-asset framework, by a factor of two.

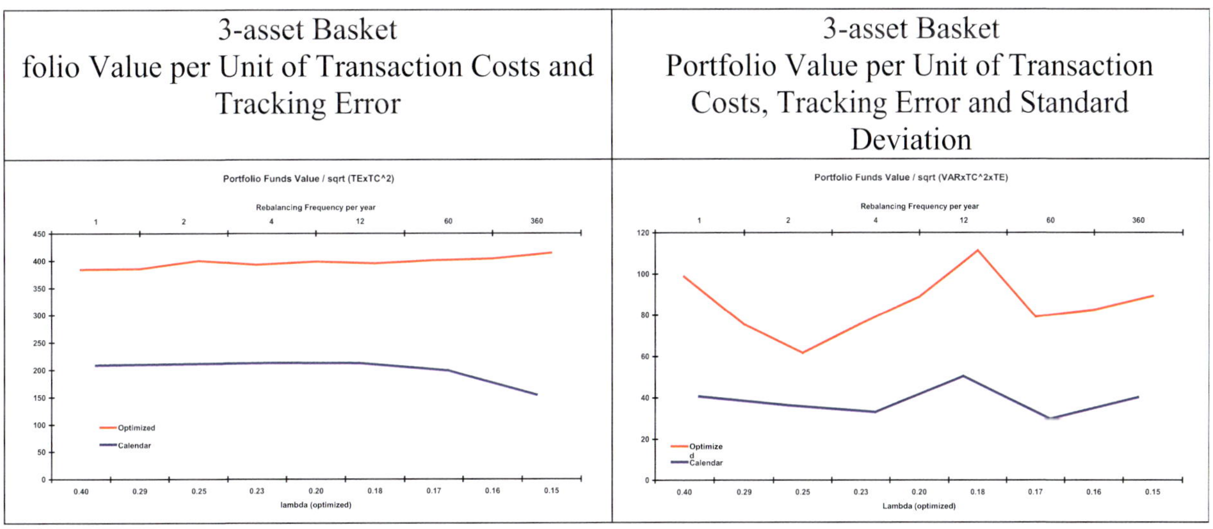

FIGURE 4.99 Three-asset basket portfolio value

4.3.5.2.6.3.4 *Conclusions*

Let us first look at strategic rebalancing versus periodic rebalancing.

- Rebalancing only when a target allocation band is violated has resulted in less frequent rebalancing than an annual programme, with generally lower transaction costs.

- The times when rebalancing-to-target costs exceeded those of rebalancing annually were during volatile markets, when it was also the most rewarding.

In terms of strategic rebalancing outperformance, the following can be said.

- Across a range of returns patterns, market conditions and transaction cost levels, illustrated through both forward- and backward-looking approaches, the two strategic rebalancing approaches have proved to be consistently efficient, as expressed by significantly improved risk appetite metrics.

In connection with the "Leland" versus the "halfway" rebalancing strategies in terms of rebalancing frequency, we can conclude as follows.

- As expected, the "halfway rebalancing" method rebalances less frequently, given that it rebalances only halfway back to the target allocation. However, it is surprising that the increase in rebalancing frequency is only marginal at most tolerances.
- The relative outperformance can be explained by its use of smaller rebalances, resulting in significantly lower transaction costs per rebalance, only partly offset by the marginal increase in rebalancing frequency.

For the "Leland" versus the "halfway" rebalancing strategy in terms of relative performance in risk appetite metrics, the following conclusions can be drawn.

- In terms of the trade-off between tracking error and transaction costs for each of the rebalancing methods, the Leland and halfway methods provide similar significant outperformance.
- In terms of risk appetite metrics, the Leland and halfway methods provide significant outperformance, with the difference between them depending on the tolerance with respect to the tracking error (parameter λ).

  The methodology involves a multi-step automatized process using back-testing and stress tests exploring a wide range of patterns, and selects a rebalancing strategy based on its risk tolerance relative to a target allocation, after transaction costs have been allowed for. It is a customer-centric approach, as the tracking error is weighted by the customer's risk tolerance.

We can conclude that the optimized rebalancing strategy significantly outperforms the periodic rebalancing approach.

- The rebalancing has to be carried out more frequently, but this is more than offset by the far lower transaction costs.
- The optimized "rebalancing halfway" strategy rebalances less frequently than the optimized "to the target boundary", which is compensated for by the higher transaction costs per rebalancing.

FIGURE 4.100 Portfolio value

4.3.5.3 Decreasing the Cost of Capital through Sustainable Asset Selection

4.3.5.3.1 Long-Term Equity Investment The prudential treatment of equity investments was reviewed by the European Union's regulator in March 2019. Among the measures that have come into force, the European Commission has introduced a new class of long-term equity, long-term equity investment (LTEI), that benefits from a reduced equity shock under the Solvency II capital requirement framework. However, in order to benefit from this favourable regulatory treatment, the regulation (article 171bis) stipulates a number of eligibility conditions.

LTEI shares benefit from a 22% shock to the market value instead of the generic shocks of 39%/49% adjusted for symmetric adjustment, subject to conditions. This measure is advantageous in terms of capital savings, for two reasons.

- On the one hand, it places the equity investments thus identified at the lower limit of the interval of the capital cost of the shares, regardless of the market configuration.
- On the other hand, it makes it possible to have the regulatory cost comparable, without taking into account the effects of diversification, to that of a BBB fixed-rate bond with a credit duration of 12 years.

This reduced shock applies to type 1 shares (listed in the European Economic Area: EEA) and to the subset of type 2 shares (unlisted shares of companies with their registered office in one of the EEA member states). In this context, an insurance or reinsurance undertaking may therefore benefit from this regulatory treatment, provided it meets a number of requirements.

- First, the asset portfolio must be assigned to an identified liability. None of the identified portfolios, nor the sum of those portfolios, may represent more than 50% of the total size of the insurance undertaking's balance sheet. This excludes shares held in front of equity or unit-linked contracts in so far as their holding period depends, with some exceptions, on the insured and not on the insurer.
- Second, like a portfolio of assets already ring-fenced for contractual or management reasons, the portfolio of assets and its counterpart to the liabilities thus identified must be strictly managed and organized

separately from the other activities of the undertaking. In short, do not cover losses from other activities outside the identified liability and keep separate accounts.

- Third, within the asset portfolio, a subset of equity investments must be designated. The holding period of all the investments that are part of it must be evaluated. The average holding time of the subset must remain above five years. Favourable treatment may apply as soon as the subset is identified, even if the five-year holding threshold has not yet been reached.

  The calculation of the five-year holding period can still be open to interpretation: the holding periods can be weighted by the cost price or by the market value, knowing that the first method of calculation allows better stability in terms of the average duration.

  Another point of interpretation is the inclusion or not of accounting management operations, knowing that the effect of these operations has no impact on the average holding period from an economic point of view.

- Fourth, the company's policies must reflect the intention of long-term ownership and ensure that, in a situation of tension, the company will be able to avoid for at least ten years the forced sale of one of the securities in the portfolio. However, the text does not define these situations of tension.

  It therefore seems possible to consider referencing stressed situations already defined in the ORSA and to extrapolate the projection period to ten years.

- Finally, an internal control system will have to accompany the establishment of long-term equity portfolios. The use of this device is not subject to prior approval by the supervisory authority but will require ex post documentation of compliance with the various conditions. This is consistent with investment policies, asset/liability management and management mandates, which will need to be adapted when necessary.

The application of this measure requires a significant operational and documentary investment, but, in return, it opens the way to significant capital savings. At current equity market levels, it theoretically allows a division of the cost of capital close to two for type 1 shares. In addition, the stocks that will benefit from this system will be aggregated within the market SCR and will therefore benefit from the diversification of the standard formula. This theoretically allows insurers to significantly improve the Solvency II ratio (by ten to 20 basis points, depending on the insurer's equity allocation), or to double the equity allocation (equivalent to the situation prior to the introduction of Solvency II).

However, the feedback from insurers highlights the fact that the portion of eligible equity seems to be significantly limited in practice by the five restrictions mentioned above, which reduces the hypothetical Solvency II ratio gain. In particular, there are these three stipulations. The average duration of the assets must be greater than five years – which is far stricter than a principle of intention to hold for a minimum of five years; the possibility of avoiding for at least ten years the forced sale of any investment in shares of the subset has to be justified; and the portfolio of assets and the portfolio of liabilities are to be managed separately.

Thus, few institutions today use this measure. The gap between the operational cost and the associated gain has been larger than expected in many cases. This is why the European Commission put pressure on the EIOPA at the end of 2021 to review most of the strict conditions mentioned above, which are expected to be significantly lightened from an operational perspective in the coming years. For instance, the ten-year ALM stress test of condition 4 is expected to be replaced by a far less cumbersome instantaneous stress test, while being validated by 90% of the French general account in force.

4.3.5.3.2 Long-Term Real Assets (Infrastructure) Insurance companies with long-dated liabilities that have traditionally been financed using investment-grade corporate bonds have faced a collapse in yields over recent years, which has encouraged them to look further down the credit spectrum (to capture additional credit risk premium) or to give up liquidity (to capture an illiquidity premium).

The hunt for extra yield this encourages is a concern: no one wants to repeat the mistakes of the financial crisis, when many insurance investors found themselves with apparently high-rated bond-like assets that did not end up delivering the cash flows required.

Infrastructure debt is fundamentally different, as it may be considered to have superior credit fundamentals to the corporate bond sector. Infrastructure debt is, essentially, a fixed-income instrument that allows investors to capture an illiquidity premium (typically 70 to 100 basis points of extra spread when compared to an equivalently rated publicly traded corporate bond).

Infrastructure projects and companies with a similar risk profile are essential for the growth and prosperity of an economy. Typical examples include roads, power distribution facilities, airports and communication networks. These tangible assets generate long-term cash flows, and these flows secure the coupon principal payments for investors.

Historically, European infrastructure companies have been financed by banks. With the growing interest from institutional investors for the asset class, infrastructure funding is now more often achieved via the combination of short-term loans (seven years or less), underwritten by banks, and long-dated fixed-income instruments in the form of a loan or a bond (ten years or more) privately placed with investing institutions. This trend is also encouraged by borrowers seeking to diversify their source of financing and by their shareholders seeking to reduce their exposure to bank financing cycles. Finally, regulatory and capital deleveraging pressure on the banks' balance sheets has reduced their enthusiasm for long-term infrastructure investment, creating an opportunity for institutional investors to access these investments.

Institutional debt accounts for some 20% of infrastructure financing today, and this figure is expected to continue to grow on the back of increasing demand from institutional investors, including pension funds and, in particular, insurance companies.

The key reasons why insurance investors ought to consider infrastructure debt as a component of their overall portfolios are as follows.

- There is a prospect of long-term, stable, contractual cash flows: the assets available are, essentially, the same as those received from corporate bonds, and they are straightforward for an insurance company to use for liability-matching purposes
- Infrastructure debt has, historically, had superior risk/return characteristics compared to the corporate bond market.
- It also has superior credit fundamentals and a stable credit spread: insurers are often operating in a near-monopolistic environment, and this may be factored into the credit assessment. Frequently, they have contractually guaranteed cash flows with explicit or implicit government backing. It is also common to see relatively low operational costs.
- Low default rates are a characteristic of infrastructure debt, which has consistently generated very stable cash flows and experienced lower default rates (and higher recovery rates) than comparably rated corporate bonds. In addition, it is usually privately placed, and, as such, investors have access to management and a wide range of private information, including due diligence reports and the company's financial models, to assist in the accurate assessment of the investment. Financing documentation is also more protected, allowing investors to monitor metrics such as cash flow coverage for debt service and financial leverage.
- Infrastructure debt has a wider credit universe: the current liquid corporate credit universe, composed of around 75 issuers, accounts for only around 20% of the entire infrastructure financing market. Unlisted infrastructure debt allows investors to access a wider array of sectors, geographies and risk profiles.

- Favourable capital treatment also characterizes infrastructure debt. On 1 April 2016 the European Commission recognized the particular features of infrastructure debt we described above and approved a reduction in the Solvency II SCR spread capital charge on qualifying rated infrastructure debt of roughly 30% compared to similarly rated corporate bonds.

As a result, infrastructure debt ought to be able to provide investors with very good excess return on capital compared to corporate bonds – i.e. the risk-adjusted return in excess of that at which liabilities grow, once the cost of capital has been taken into account:

$$E \approx \cdot S \cdot - \{[SCR \times D] \times COC\} \cdot - \lambda,$$

where E= excess return, S = spread over swap, SCR = solvency capital requirement, D = allowance for reduction in SCR by diversification against other risk factors (set to 0.9), COC = cost of capital of insurer (set at 6%) and λ = annualized cumulative loss over ten years

In conclusion, the capital charge reduction for infrastructure debt is likely to sharpen the insurance industry's focus on the asset class. Infrastructure debt can offer insurance investors:

- a significant capital advantage, broadly 30% lower than that for similarly rated corporate bonds;
- the ability to improve return due to the pick-up of an illiquidity premium;
- superior credit fundamentals compared to corporate bonds, translating into a low loss rate; and
- contractual, stable, long-dated coupons and maturity cash flows with which to match liabilities.

CHAPTER 5

Hedging Risks Aligned with the Risk Appetite for Sustainable Savings and Retirement

THE INSURER'S BALANCE SHEET is actually composed of the following elements.

- The insurer's liabilities in relation to the commitments to the policyholders.
- The mathematical provisions – i.e. the savings, made up of customers, for life contracts and capitalization contracts, but also for temporary deaths and lifetimes: including premiums, commissions, claims, lapses and taxes.
- The capitalization reserve, which requires insurance undertakings to set aside capital gains in the event of a fall in rates. Symmetrically, it allows them to take them back in times of rising rates. The higher an insurer's capitalization reserve, the better it will be able to cope with a rise in interest rates.
- The provision for profit sharing, which corresponds to the revaluation not distributed during a year and set aside. It must be redistributed within a few years following its settings: in France it stands at 90% of the technical positive margin (100% in the event of a loss); 85% financial margins (0% in the event of a financial loss).
- The insurer's assets are made up of:
  - assets representing the core equity; and
  - assets representing the mathematical provisions – i.e. savings contracts, made of bonds coupons and principals, equity dividends.
- Net asset value = asset – liabilities.

In order to meet their obligations for guarantees embedded within the liabilities (e.g. in annuities), life insurance companies have developed hedging programmes to manage market risks, within which they use investment derivatives because of the flexibility and liquidity of the derivatives markets. The use of derivatives by life insurance companies has been massively expanding since the 1990s, albeit at various pace across countries, in line with economic and regulatory changes.

DOI: 10.1201/9781003218371-6

Hedging is actually a technique derived from the option management techniques used by investment banks, which consist of building a portfolio of assets (hedge assets) that will vary like the guarantee.

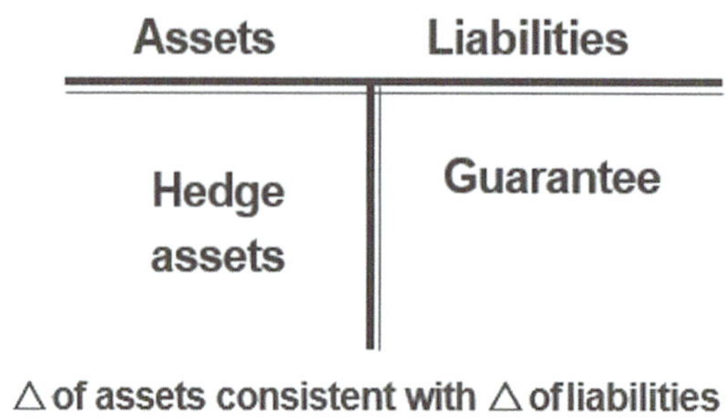

FIGURE 5.1 Assets versus liabilities

Hedging is designed to move in the opposite direction to insurance liabilities for the guarantees, which helps offset the increase in reserve liabilities, and allows the insurance company to maintain the required statutory surplus for the life insurance product.

Naturally, not all risks are hedgeable using financial derivatives, such as policyholder behaviour or longevity risks, while others are not hedgeable but are still measurable (e.g. basis risk, mortality, very long-term rates and volatility).

Even so, there are many different types of derivatives used to manage the risks with respect to life insurance liabilities, including but not limited to equity options, futures or forward contracts, interest rate swaps and total return swaps. The derivative contracts may be either traded on a regulated exchange (e.g. S&P Futures), or traded over the counter (OTC) and negotiated between the insurance company and an investment bank. Derivatives also vary in their maturities, ranging from three-month futures contracts to five-year through 30-year durations for OTC options and interest rate swaps.

In contrast with the usual simplistic theoretical approaches based on perfect market assumptions (e.g. Black–Scholes model), using derivatives is anything but trivial, as it involves several practical issues that make the hedging most often imperfect and partial: structural short convexity position held by insurers, uncertain volatility, transaction costs and P&L dispersion, lack of long-dated and liquid derivatives, hedge effectiveness versus the cost of carry, performance drag versus tail risk protection, recurrent gap risks.

Hedging sustainability is apprehended through the decomposition of the hedging P&L into the sensitivities to the market risk drivers ("Greeks"), the potential control of which is illustrated through sustainable mitigating techniques that leverage on both structuring expertise and hedging experience: risk measurements (duration/convexity, break-even and par volatilities, hedge effectiveness and cost of carry, performance drag versus tail risk protection measures), mathematical modelling (e.g. optimal control theory, Lévy processes, nonlinear PDEs and associated BSDEs), product design innovations and the use of specific hedge instruments, mostly aligned with the customer's risk appetite.

The chapter's first section is dedicated to hedging principles, processes and P&L attribution, including the impact of accounting and tax treatment. The different hedging methodologies are then developed in the second section, from dynamic to semi-static strategies, from theory to practical implementation, applied to directional then volatility hedging (convexity, vega and equity rate hedging). Finally, the third section focuses on the growing hedging liquidity constraints and their mitigating techniques.

5.1 HEDGING PRINCIPLES, PROCESSES AND P&L ATTRIBUTION

5.1.1 Hedging Principles

5.1.1.1 Why Hedging? Illustration with a Financial Guarantee Embedded within a Life Insurance Product

As part of their liabilities, insurers provide customers with investment opportunities that can be coupled with financial protections depending on the client's risk appetite.

- The investor keeps a significant upside exposure but pays a premium to ensure downside protection: asset + protective put.
- The investor prefers lower but stable returns (through bonds) but pays a premium to get the potential upside (through call options): CPPI, general account, Euro growth fund.
- Such downside protection/stable return product values are subject to market uncertainties -> in case of financial market turmoil, the insurer will suffer a loss instead of the client.

Given that the downside protection is subject to financial market downturns, the insurance company may suffer a loss in the event of financial market turmoil, as illustrated in the example (see Figure 5.2) of a product characterized by a partial guaranteed lump sum amount after a waiting period: a policyholder invests €100 in a given equity fund and the insurer guarantees her a minimum of €80 at a one-year horizon.

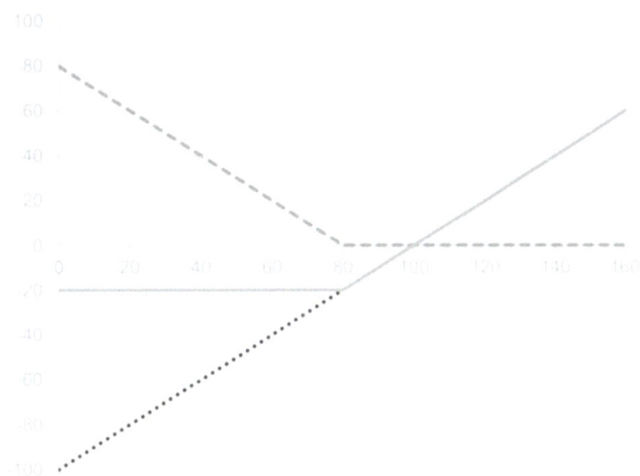

FIGURE 5.2 Downside protection illustration

The payoff for the client is given by the blue line (which can be decomposed between the fund – dotted blue – and a put option – dotted red).

- If the equity fund terminal value is above €100, then the client makes a profit.
- If the equity fund terminal value is between €80 and €100, then the client suffers some loss (between €0 and €20).
- If the equity fund terminal value is below the guarantee level, €80, then the loss for the client is capped at €20, while the insurer pays the difference between the terminal value and €80, which can be significant, which is called the put option payoff. This risk of loss to the insurer must be taken into account in the

insurance premium at contract's inception, as any future potential loss cannot be passed on to the customer at a later date.

From then on, the customer's gain, which must be included in the insurance premium, is broken down into stock + put, thus putting the insurer at risk.

Before the emergence of financial derivatives, life insurance companies managed risks through assets and customer's diversification and reinsurance, which reached limits with rising volatility that could mostly be mitigated by financial derivatives at low cost.

In fact, in order to mitigate those financial risks borne by the insurer, this downside protection has to be "hedged" by the insurer, by investing in "hedge assets" (e.g. futures, options) whose variations with respect to economic variables offset the variation of the downside protection embedded within the savings and retirement product – and whose expected cost has to be integrated into the insurance premium. Indeed, insurance companies are major users of "hedge" assets to protect the value of their large AuM total, as illustrated by the significant increase in their notionals from \$786 billion as of FY 2010 to \$2,600 billion as of FY 2018, out of which \$1.1 trillion are invested in equity hedges.

5.1.1.2 Some Historical Background about the Growing Hedging Needs

The use of derivatives by life insurance companies has been expanding since the 1990s, albeit at various paces across different countries, in line with economic and regulatory changes.

- Basically, there are three features of life insurance liabilities that are driving the need for hedging: their long-term nature, the large volumes and the significant market risk. Among US insurers, 97% derivatives are used by life insurance, and 95% for hedging purposes (49% swaps, 45% options, 7% futures and forwards)
- The first use of derivatives started in the 1990s, along with the structural fall in interest rates, and so there was a need to protect the value and earnings embedded within guaranteed pension and general account products, as illustrated by the bankruptcies of major insurance companies.
- Their use for earnings protection then massively increased after the Internet bubble crash in 2003, notably in the United States, with the disappearance of reinsurance for the variable annuity business embedding very rich guaranteed living benefits.
- Derivatives use then diversified into cost of capital mitigation after the 2008 crisis, within a persistently low interest rate environment combined with recurrent market mini-crashes and the emergence of more stringent capital requirement regulations (both in the United States and Europe).
- Finally, derivatives have recently been directly introduced into product design innovations in the United States, highlighting the need to strengthen the relationships between insurance carriers, investment banks and regulators for better customer experience and customer protection.

Major shifts in the offering of derivatives within the rapidly changing economic and regulatory landscape have massively increased their use over the past decade.

- There has been a sharp increase in the use of mark to market (MTM) approaches, involving adjusting the value of an asset to reflect its value as determined by current market conditions, and the costs of the regulatory environment have also risen. The new regulatory regimes (Solvency II, Dodd–Frank, EMIR,

DOL) have increased the cost of capital in favour of hedging using derivatives: more market-consistent valuations with a risk-sensitive approach -> credit given to advanced risk mitigation techniques.

- The limits of reinsurance, in terms of capacity and cost, have been reached: popular in the United States in the 1990s, most reinsurance disappeared after the equity market falls of 2000–03, in particular as it was expensive (there was a need to give up M&E, in addition to rider fees) in the low interest rate environment and subject to regulatory review/approval, while also being limited in terms of capacity – without even talking about the counterparty risk. Moreover, it is impossible and/or uneconomical to fully hedge market risk in all scenarios, especially as it relates to varying policyholder behaviour, while extreme market conditions can also lead to ineffective hedging, which leads to hedge breakage
- Another factor was the persistence of these historically low interest rates (triggered by the 2008 credit crisis and further fuelled by the European sovereign bond crisis) and recurrent equity market crashes.
    - Such crashes occurred in 2012, 2014, 2016, 2018, 2020 and 2022.
    - The persistently low rates drastically reduce the yields of general account products, which need hedging to lock in their business value.
    - The recurrent equity crashes and volatility spikes put at risk the business development of unit-linked policies in Asia and Europe, which also need to be hedged.
    - These all reduce the effectiveness of earnings diversification and encourage ALM and asset management fee hedging.
- A further factor has been new customers' needs and experiences -> more flexible designs with optionalities.
    - The recurrent equity market crashes strongly increase the need for guarantees.
    - Policyholders' behaviour becomes increasing efficient, which also has the effect of working in favour of more hedging.

As a result, developing the liquidity of derivatives is the key for the sustainable development of hedging.

- The "one-way market" nature and magnitude of mortgages, structured products and variable annuity hedging activities -> potential market impact independently of fundamentals but related to the growth of the options market.
- "Liquidity is the new leverage."

Overall, the combined effect of these driving factors has been to make a major contribution to pushing up the use of derivatives by life insurance companies, from $786 billion as of FY 2010 to $2,600 billion as of FY 2018.

5.1.1.3 Hedging Mechanism Basics

A hedging strategy consists in investing in assets whose variations with respect to economic variables offset the variation of the liability.

In this example (see Figure 5.3), we make the assumption that the guarantee liability variation is +0.2 for –1% (delta = –20%) in asset variation.

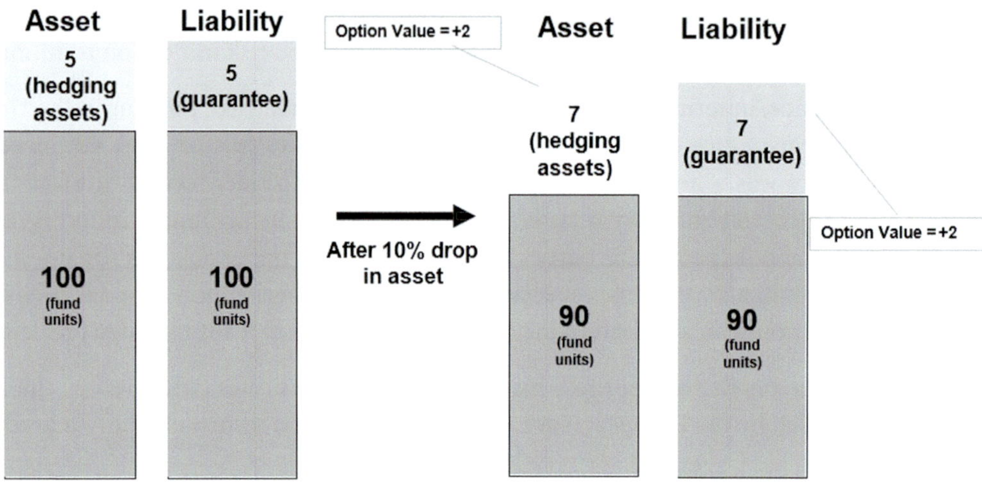

FIGURE 5.3 Hedging mechanism basics

After a 10% drop, assets and liabilities are matched -> the change in hedging assets is equal to the change in the option value, while the insurance company P&L = 0 (−8 for the asset versus −8 for the liability).

5.1.1.4 Hedging and Risk Appetite Frameworks (Earnings, Value, Capital, Liquidity)

Hedging protects the earnings, value, capital and liquidity needs of the life insurance company, based on commitments made to policyholders and the beneficiaries of contracts.

The uses of hedging are indeed multiple across the four dimensions of the risk appetite framework regarding life insurance liabilities.

- Hedging decreases the earnings sensitivities to market risks.
- Hedging enables insurers to lock in significant part of the business value.
- Hedging decreases the capital requirements.
- Hedging reduces the cash liquidity needs.

In order to do this, it is necessary to identify, measure and control the risks of mismatch between asset and liabilities, within each specific framework of the overall risk appetite framework.

There are two main methods: either static or dynamic.

- Static approaches
    - take only the current assets and liabilities held in the balance sheet; and
    - use cash flow matching only for short term analyses, and duration matching for longer term analyses.
- Dynamic approaches include future contracts/add-ons, and the impact of rates on contract yields and on policyholders' behaviour.

Given the significant convexity of life insurance liabilities, hedging for large and non-parallel changes needs options (45% of derivatives notionals versus 49% for swaps and 7% for futures and forwards by insurers in the United States).

The majority of insurers use static hedging with a dynamic overlay as their core hedging strategies (through a mix between futures, swaps and options), where the need for dynamic feature stems from (i) the lack of liquidity of many derivatives (equity options and variance swaps [varswaps] in particular), (ii) the difference in Greek behaviour between the liabilities and the hedge assets and (iii) the strong budget constraints on hedge fees (the transaction costs of futures and swaps being lower than those of options).

All participants leveraged OTC derivatives as part of their hedging programmes; participants with a large dynamic component as part of their strategy mentioned that they utilize listed options to manage their near-term exposure. Equity futures are favoured for dynamic hedgers to manage residual exposures.

A common rationale for leaving a particular cohort unhedged was usually the cost and low liquidity associated with placing a hedge. In this respect, a company's macro-hedge programme would eventually account for any residual risk left by the outstanding cohort.

The practice of using derivatives is further complicated by the mismatch of assumptions of assets versus liabilities across the value, earnings, capital and liquidity dimensions of the risk appetite.

In addition to the liabilities, derivatives are also used to protect the earnings of the life insurance fees – notably the asset management fees, which are volatile, as they mostly fluctuate in response to unit-linked movements. As a result, securing asset management fees, reducing earning volatility and the capital requirements, makes sense. However, this practice has to take into account transaction costs, and also remove any upside from rising markets. Moreover, the impact is not completely straightforward, due to the different accounting treatments between assets and liabilities: hedge asset loss recognized immediately versus an asset management charge (AMC) gain only over time under IFRS.

5.1.2 Hedging Processes

Strengthening risk management is everyone's job. The "three lines of defence" strategy reinforces overall attention to risk and control issues.

- Hedging operations management is the first line of defence, whose objectives/responsibilities are as follows.
  - Better understanding, measuring and managing:
    - monitoring and reduction of operational risks; helping implement management's action plans and results from the external review;
    - monitoring of model and hedging risk;
    - stress-testing and analysis;
    - developing independent risk measurement/analysis tools; and
    - contributing to P&L attribution analysis.
  - Contributing to providing risk-related information for decision-making:
    - involvement in product design/development; and
    - involvement in refinement of the hedging strategy.

This translates into the following.

- Dissecting the hedging process into logical processes and sub-processes, based on an information flow:
  - access to data and personnel as well as ample time to understand, test and analyse is required;
  - continuous communications (no games) are called for; and
  - "flag"/escalation has to be agreed.

- Creating a matrix that captures operational controls and the risk management environment. For each sub-process the following are required:
  - documenting operational controls;
  - cataloguing the known risks embedded in hedging activities as well as operations; and
  - brainstorming on emerging/unknown risks.
- Creating an inventory of existing and a desired risk measures and monitoring techniques (from both business and risk management perspectives).
- Prioritizing, establishing thresholds and agreeing on responsibilities (who runs which analytics) and the escalation process.
- Cataloguing the tools/technology used to conduct the operations and measure and monitor risks.
- Documenting the requirements and rationale for additional (or replacement) tools/technology.

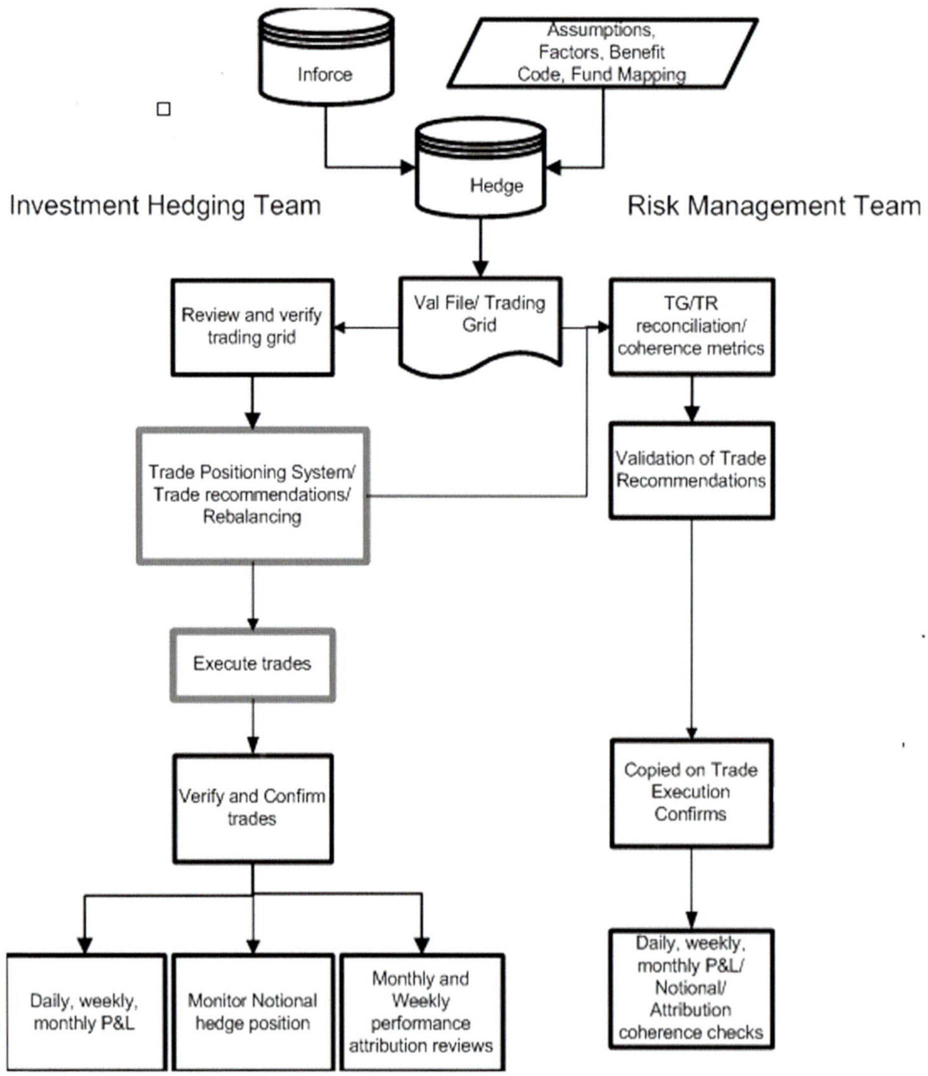

FIGURE 5.4 Information flow

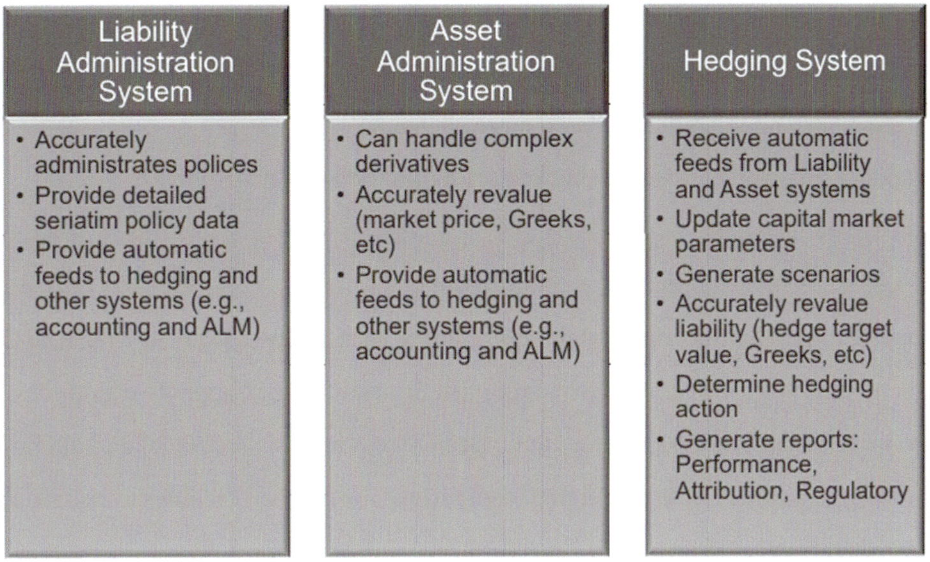

FIGURE 5.5 Infrastructure administration system

- Risk management constitutes the second line of defence, which notably does as follows.
  - Proceeds to regular modelling reviews:
    - analysis on modelling issues.
    - Asset side: impact of random dynamics via independent risk management hybrid ESG.
    - Liability side: systematic "second eye" check on models developed/used for hedging services
    - Progressive ability to analyse product design risks via an independent risk management ALM pricer: given a high-level concept of the types of guarantees that are desired, it is necessary to investigate the feasibility of various guarantee levels, asset mixes and product features, such as ratchets. This is done through the pricing process, which calculates the market-consistent cost of hedging particular combinations for the central model points. Comparison of the cost of hedging versus the acceptable price that can be charged in the market puts important constraints on the design features that can be offered in a given market environment.
    - The use of nested stochastic simulations enables dynamic hedging strategies to be more accurately modelled and residual risks to be assessed. As these investigations are computationally very intensive, they require the use of distributed grid computing resources in order to keep run-times down to manageable levels.
    - As well as the market-consistent cost of hedgeable risks, careful consideration also needs to be given to the risks that cannot be hedged. Such risks either need to be minimized through appropriate product design or have a cost of capital allocated (by reference to the market-consistent cost of the hedgeable risks):
  - systematic review of sensitivity results.
    - Risk management from an actuarial/hedging angle based on information provided by client services.
    - Risk classification: "product design", "market", "behaviour" and "fund allocation" risks.
    - Progressive ability to analyse hedging risks via an independent risk management ALM sensitivity tool.
  - Pre-launch hedge design review.

- Monitoring of the hedge programme, in particular:
  - the selection of risks to hedge;
  - the frequency and tolerances of hedging rebalancing;
  - the adequacy of and consistency between risks and hedging instruments;
  - the hedge budget; and
  - periodic adjustments to the hedge programme.
- Develop the internal model for capital requirements, such as Solvency II or RBC (see Chapter 1).
- Design and monitor the operational hedging risk limits of the risk appetite framework:
  - review of risks and risk management practices, across financial/actuarial and operational risks;
  - defining operational limits of hedging, comparing "normalized earnings at risk" to P&L, based on
    - risk drivers and
    - stressed scenarios,
  - and an allocation key to combine the risk factors (via the Greeks) into one adverse or extreme event,
  - across the four dimensions of the risk appetite framework – earnings, value, capital and liquidity – using their corresponding KPI and KRIs (see Chapter 1),
  - with corrective actions to improve the risk profile of businesses.

- The internal audit is the third line of defence, and it includes the points of attention shown in Figure 5.6.

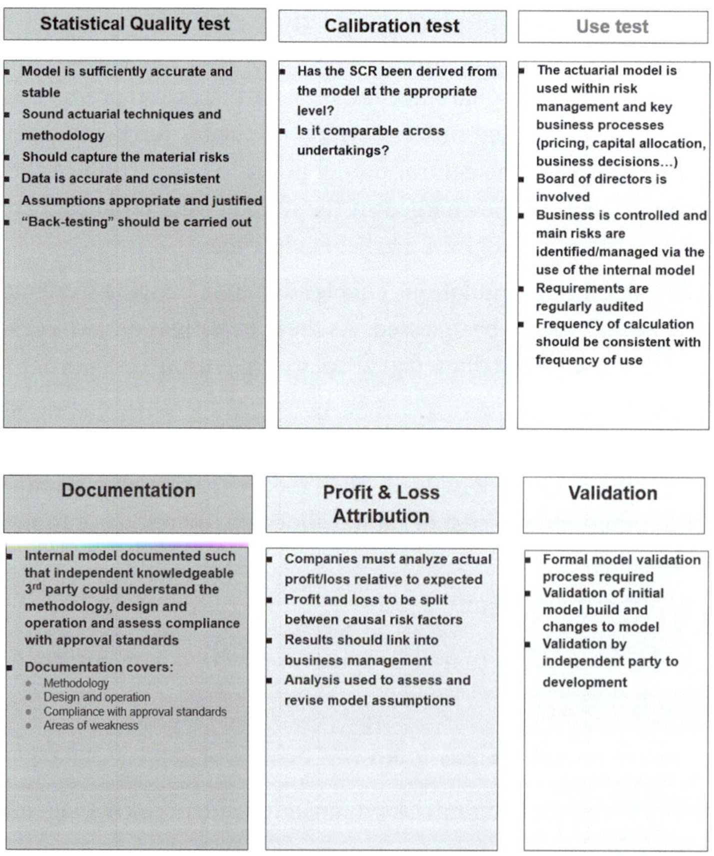

FIGURE 5.6 Internal audit points of attention

5.1.3 From P&L Attribution to Hedging Implementation

5.1.3.1 From Greeks to Hedging P&L Attribution

The most common risk metrics observed are those involving the Greek sensitivity mismatch of liabilities versus hedges.

Hedging leverages on the following P&L attribution, using a proxy Taylor approximation based on the different Greek sensitivities, where f may be the liability or the asset:

$$\Delta f(S,t,r,\sigma) = \underbrace{\frac{\partial f}{\partial S}}_{\text{delta}}\Delta S + \underbrace{\frac{\partial f}{\partial t}}_{\text{theta}}\Delta t + \underbrace{\frac{\partial f}{\partial S}}_{\text{rho}}\Delta r + \underbrace{\frac{\partial f}{\partial \sigma}}_{\text{vega}}\Delta\sigma + \frac{1}{2}\underbrace{\frac{\partial^2 f}{\partial S^2}}_{\text{gamma}}(\Delta S)^2 + \ldots$$

Depending on which terms are hedged, the approaches incorporating the different so-called "Greeks" can be classified into the following:

- delta hedge;
- delta, gamma hedge;
- delta, gamma, rho hedge;
- delta, gamma, rho and vega hedge;
- and beyond.

The idea is to use a hedging portfolio for which the partial derivatives (Greeks) with respect to the key financial risk drivers are the opposite of the liability portfolio.

Usually the three key financial risk drivers are as follows.

- Low interest rates, which call for rho hedging.
  - On an economic basis, low interest rates are strongly increasing the value of guarantees, especially in the long term (rho risk).
  - Low interest rates also impact the fixed-income investments.
  - Low interest rates are all the more damaging than roll-up rates if guarantees are high, although it will probably take several years for economic claims to come through.
- A low equity level, which makes guarantees more in the money, and which calls for delta and gamma hedging.
- High volatility, which increases the value of the guarantees on an economic basis, and the hedge costs, which calls for vega hedging.

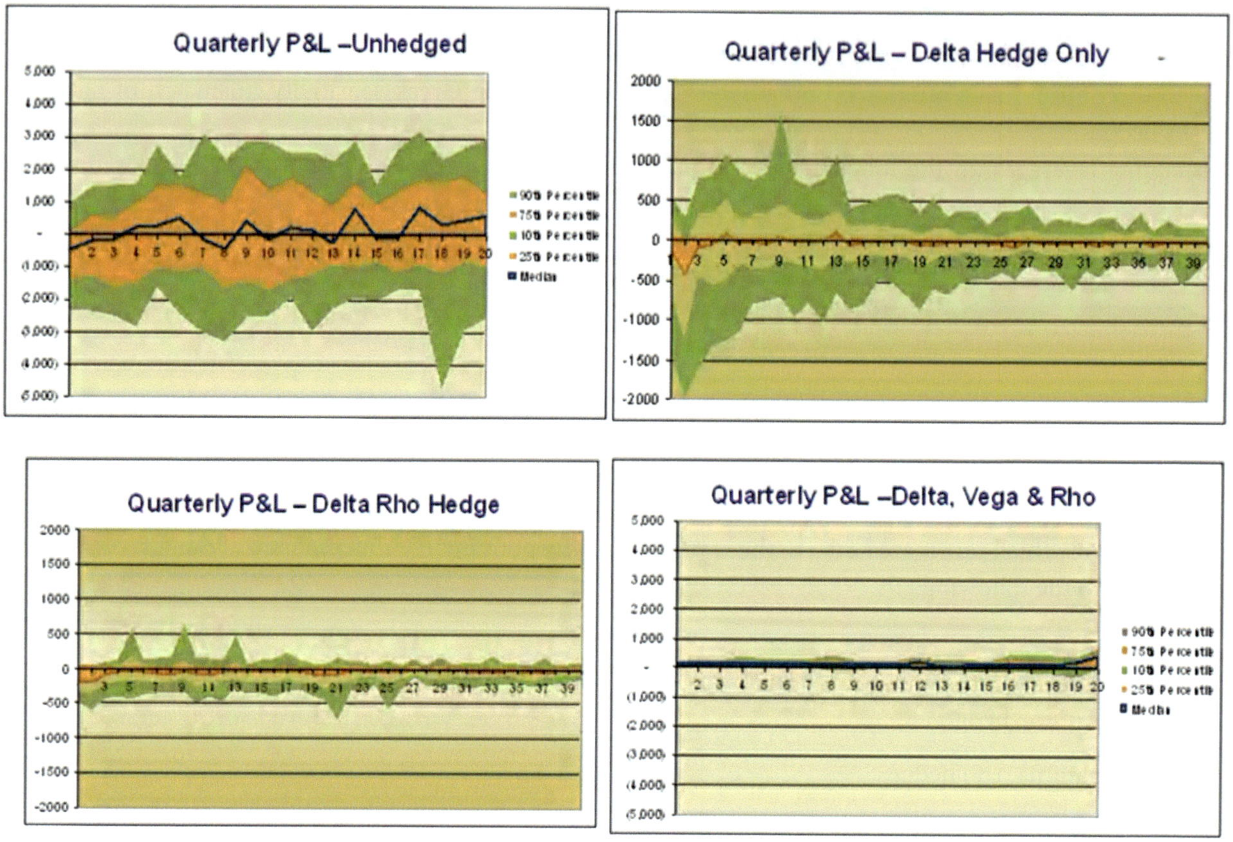

FIGURE 5.7 Hedging P&L depending on Greeks

As a result, the "three-Greek" hedge approach is widespread among life insurers, with the impacts in terms of hedge effectiveness as follows.

- The hedge transactions in place mainly consist of short-term (less than five-year) puts and futures.
- Some insurers under-hedge rho risk (assuming that interest rates will rise) to avoid locking in interest rates at historical lows.
- Hedging the tail risk offers a higher risk-based return

5.1.3.2 Interest Rate (Rho) Hedging

On the one hand, lowering interest rates causes an increase of the present value of life insurance liabilities (economic loss), through the use of lower discount rates. On the other hand, fund performance depends on both equity and fixed-income investments.

- If no correlation is assumed between equities and interest rates (which is a tough assumption), the equity fund return is not affected by low interest rates.
- Conversely, the fixed-income investments generally have to be rolled over. Hence, lower interest rates imply an actual loss due to the lower cash flows received in the future.

Hence, hedging rho is not a trivial choice. Rho hedging lies in offsetting the first-order change in the liability reserve due to interest rates movement (i.e. liability rho PnL) with first-order P&L from the hedges (i.e. asset rho PnL), using instruments such as bond futures, swaps, etc.

Using swaps as a hedge instruments for rho hedging, the overall change in rates impact can be expressed with the use of swaps:

$$\textit{Rho hedging mismatch} = \textit{dR impact on OV} - \textit{Swap change}$$

$$\textit{where} \begin{cases} \textit{dR impact on OV} = \sum \textit{Rho liability}_i \times \textit{Change in rate}_i \\ \textit{Swap change} = \textit{P\&L swaps}_t - \textit{P\&L swaps}_{t-1} \end{cases}.$$

5.1.3.3 Delta Hedging

In parallel with the rho hedging, the equity delta hedging lies in offsetting the first-order change in the liability reserve due to equity market movement (i.e. liability delta PnL) with first-order P&L from the hedges (i.e. asset delta PnL), using instruments such as equity index futures, equity total return swaps, etc.:

$$\textit{Delta hedging mismatch} = \textit{dS impact on OV} - \textit{Indices change}$$

$$\textit{where} \begin{cases} \textit{dS impact on OV} = \sum \textit{Delta}_i \times \textit{Change in index}_i \\ \textit{Indices change} = P\&L_t - P\&L_{t-1} \\ P\&L_t = P\&L_t \textit{of futures} + P\&L_t \textit{of bonds} + P\&L_t \textit{of FX} \end{cases}$$

5.1.3.4 Vega Hedging

So far implied volatility has been assumed constant over the period, where higher volatility causes an increase of the present value of the liabilities (economic loss).

As a result, the P&L holds an additional explicit term, which is the vega hedge P&L, which is equal to the vega times the change in implied volatility between two rebalancings of the hedge options.

Regarding vega hedging, the hedging options can be short- or long-dated.

- Long-term options usually match life insurance liabilities better and reduce roll risk. However, they are expensive and illiquid, implying higher transaction cost (1 versus 0.6 volatility points for equity options).
- Short-term options are cheaper and liquid. However, they have to be rolled over. In this case, higher volatility implies a higher hedge cost – i.e. an actual loss due to the higher premiums paid to roll over the hedge.

Hence, hedging vega is not a trivial choice.

5.1.3.5 Hedging Strategy Implementation

The hedging strategy needs to consider what to protect (statutory capital, IFRS earnings, value?), how much to protect (mean P&L, CTE 70 or CTE 90?) and how to fund the protection (using swaps or futures for interest rates risks, index options or VIX for equity volatility risks?).

Depending on the hedging strategy goals, its implementation may be one of the following.

- Static: under this strategy, the company will attempt to match the Greeks of issued liabilities (adjusted for expected decrements) with a derivative of equivalent terms.
- Static with a dynamic overlay: under this strategy, the company will match all issued liabilities (adjusted for expected decrements) up to a certain percentage of the account value. The portfolio of liabilities and hedges is then rebalanced dynamically according to predefined Greeks.
- Dynamic: under this strategy, the company will attempt to match the Greek profile of outstanding liabilities with offsetting hedge transactions.

Most participants tend to move onto a more dynamic hedging approach, to accommodate evolving product designs and to improve hedging efficiency when prevailing dealer rates are deemed to be high.

5.1.4 Hedge Accounting and Tax Treatment: Focus on US Variable Annuities (GMxBs) since the 2008 Crisis

In the autumn of 2010 the life insurance industry and the IRS began the Industry Issue Resolution (IIR) process regarding the recognition of hedging gains and losses for guaranteed benefits on variable annuity contracts, and in 2014 it published section 446 of the Internal Revenue Code (IRC), the Directive Related to Hedging of Variable Annuity Guaranteed Minimum Benefits by Insurance Companies, the intent of which was to provide "an efficient and uniform method of accounting" for certain GMxB hedges.

Most GMxB hedges do not qualify for GAAP hedge accounting under FAS 133, because FAS 133 requires detailed documentation and "hedge effectiveness" testing. Although most companies can design an FAS 133 test that measures the high degree of correlation between the GMxB hedge and capital market risks, they often do not meet the FAS 133 standard of a "highly effective" test that measures actuarial assumptions and policy holder behaviour. The GMxB liability may also be treated as an "embedded" derivative for GAAP.

Since FAS 133 does not allow hedge accounting if you are hedging a derivative (i.e. an embedded derivative) with another derivative, GMxB hedge gains and losses are required to be mark to market through the income statement. The GMxB liability for the embedded derivative is also MTM through the income statement, under FAS 157. The GMxB unrealized hedge gains and losses are also recognized on an MTM basis through the statutory surplus.

Tax hedge treatment is important to insurance companies because hedge gains and losses are taxed as ordinary income, and not as capital gains and losses. Ordinary treatment is preferred, since capital losses can only offset capital gains and can be carried back only three years or carried forward five years, whereas ordinary losses can offset either capital or ordinary gains and can be carried back three years or carried forward 15 years for a life insurance company before expiring. Tax hedge accounting also allows the company to match the timing of the GMxB hedge gain or loss with the timing of the item being hedged (i.e. the liability for the GMxB).

Historically there has been a diversity of practice in how insurance companies accounted for VA hedge gains and losses pursuant to US Internal Revenue Service (IRS) regulation § 1.446-4. For example, some companies used one method for VA hedge gains and another method for VA hedge losses, while other companies spread the net hedge gain or loss over different amortization periods.

Under the 2014 directive, in order to qualify as a hedging transaction for tax purposes, a hedge must be (1) entered into in the ordinary course of business, (2) used to manage the risk of price changes with respect to ordinary property and (3) clearly identified in the taxpayer's books and records on the day the hedge is created (IRC § 1221(b)(2)).

Pursuant to the directive, all the GMxB hedges are aggregated. All hedge gains and losses are then netted for the current year and allocated between VA contracts issued before 31 December 2009 and those issued on or after 31 December 2009. Before 31 December 2009 the following applies.

- Net hedge gains for contracts issued are recognized up to the amount of the net tax deduction for the year, the net tax deduction being the amount of GMxB accrued plus (or minus) the increase (or decrease) in tax reserves held for the GMxB.
- Net hedge losses for those contracts, on the other hand, are deducted on an MTM basis, except to the extent of tax reserve increases for GMxBs in the current year.
- In either case, excess hedge gains or losses for the current year are then carried forward to the subsequent year.
- One reason for the different tax treatment of hedge gains and losses for those contracts is the asymmetric relationship between hedging and tax reserves. For example, in a market crash, similar to 2008, there were significant hedge gains that were partially offset by increases in GMxB tax reserves. But in a rising market, such as from 2010 to 2013, companies' VA hedges may generate significant hedge losses each period, but the corresponding decreases in tax reserves may begin to taper off because the account values are much higher than the guaranteed amounts – i.e. the guarantees are out of the money. Tax reserves can never decrease below zero and are always capped by statutory reserves

For the VA contracts issued after 31 December 2009 the directive does not provide detailed guidance for accounting for hedge gains and losses, other than that they "should be accounted for using a method consistent with the matching requirements in Treas. Reg. § 1.446-4(e)(1)," which requires "the timing of income, deduction, gain, or loss from the hedging transaction must be matched with the timing of the aggregate income, deduction, gain, or loss from the items being hedged".

5.2 FROM DYNAMIC HEDGING TO SEMI-STATIC HEDGING

5.2.1 Sustainable Directional Hedging

5.2.1.1 Interest Rate (Rho) Hedging

5.2.1.1.1 From Cash Flow Matching to Duration Gap Management (e.g. for General Account and Traditional Life Insurance Products)

5.2.1.1.1.1 The Cash Flow Matching Methodology The cash flow matching methodology lies in building a portfolio of bond investments such that its future cash flows are equal, on each date, to the cash flows generated by the company's commitments to policyholders, using bonds and swaps as hedge instruments.

This bootstrap methodology is as follows.

- Choose a fixed-rate bond whose maturity coincides with the latest date of outflow of funds.
- Invest in this bond an amount such that the flow generated by this bond at maturity is equal to the flows generated by the liabilities.
- Deduct from the other flows resulting from the liabilities (corresponding to earlier dates) the amounts of coupons detached by the previous bond.
- Start again for the penultimate date furthest from the release of funds, and so on ...

Although the cash flow matching technique may be effective in the short term, it does not give satisfactory results in the long term, since the flows of assets and liabilities are themselves dependent on interest rates.

$$\alpha_n = \frac{F_{tn}^{P\prime}}{A_{tn}^{On}}$$

$$\alpha_{n-1} = \frac{F_{tn-1}^{P\prime} - \alpha_n F_{tn-1}^{On}}{F_{tn-1}^{On-1}}$$

etc

$$\alpha_{n-k} = \frac{F_{tn-k}^{P\prime} - \sum_{l=0}^{k-1} \alpha_{n-1} F_{tn-k}^{On-1}}{F_{tn-k}^{On-k}}$$

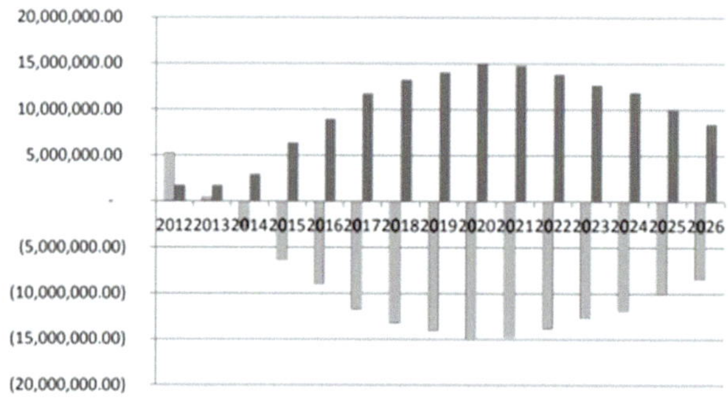

FIGURE 5.8 Cash flow matching approach

This leads to the duration gap management approach, based on the modified duration interest rate sensitivity.

5.2.1.1.1.2 The Modified Duration This measures the percentage change of a price after a change in yields of 1%.

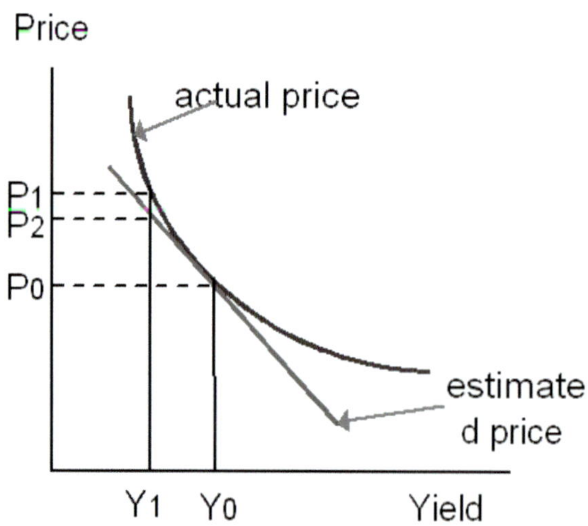

FIGURE 5.9 Price versus yield

- The Modified Duration provides a linear price-to-yield relationship as it

$$\Delta P = \left(\frac{dP}{dS}\right)\Delta Y + \frac{1}{2}\left(\frac{d^2P}{dY^2}\right)\Delta Y^2 \Rightarrow \frac{\Delta P}{P} = \frac{1}{P}\left(\frac{dP}{dS}\right)\Delta Y + \frac{1}{2P}\left(\frac{d^2P}{dY^2}\right)\Delta Y^2$$

$$= (\,Modified\ duration\,)\Delta Y + \frac{1}{2}(\,Convexity\,)\Delta Y^2$$

So, if we consider small changes in the yield curve, we can neglect the convexity second-order term and compute the modified duration in a simple "proxy" way as follows:

$$\frac{\Delta \text{P}}{\text{P}} \times 100 = -D_{mod} \times D_{i,}$$

$-\text{D}_{\text{mod}}$ = the modified duration for the bond

Di = yield change in basis points divided by 100

P = beginning price for the bond

and so

$$ApproxModDur = \frac{(PV_-)-(PV_+)}{2\times\Delta Yield\ \times PV_0}.$$

Such interest rate sensitivity holds several useful properties for interest rate risk management:

- it is additive (portfolio duration = SUM (WiDi) for each security i); and
- it increases with maturity and decreases with yield and coupon, and is lower than maturity.

Given that the aim of hedging is to offset as much on the hedge asset as on the liability to be hedged for the same shift in rates, the notional of hedge asset to offset the first-order interest rate risk of the liability can be easily computed:

Notional of liabilities × DV01 (liab) = notional of hedge asset × DV01 (hedge)

→ Notional of hedge asset = notional of liabilities × DV01 (liab) / DV01 (hedge)

Note: DV01 is the change in price ($) for 1 basis point (0.01%) change in yield = modified duration × price.

However, the duration metrics hold several limitations: as a first-order interest rate sensitivity, it is a reliable proxy for interest rate sensitivity only for small and parallel changes in the yield curve. Fortunately, as the second-order convexity is always positive, in the event of large yield movements the duration is a "prudent" interest rate sensitivity – i.e. the actual price of a liability is always higher than that predicted by the duration, by an amount equal to the always positive convexity (which reduces losses in falling interest rates while increasing gains in rising interest rates).

In terms of interest rate risk management, the concept of duration can be extended to the duration gap approach.

5.2.1.1.1.3 The Duration Gap Management The duration gap measures the gap in terms of duration between the hedge assets and the liabilities (weighted by their respective notionals):

If the duration gap is positive, a rise in interest rates hurts earnings while a decrease in interest rates helps earnings.

Example: assets of $1,000 with a duration of 4.2 years. Liabilities of $800 with a duration of 4.4 years (note: shareholders' equity must be $200). Duration gap = 4.2 – ($800/$1,000) × 4.4 = 0.68 years. When interest rates go up, this company's assets will drop more rapidly than its liabilities, eating into its equity.

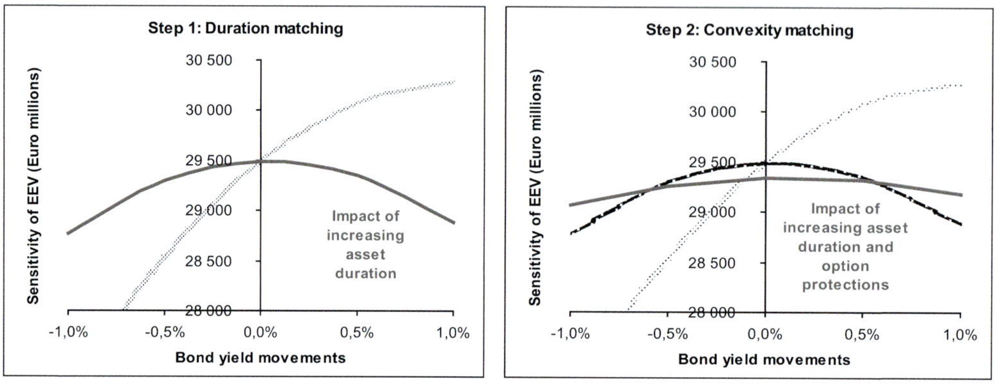

FIGURE 5.10 Duration and convexity hedging

For example, in order to offset a rise in the liability duration, the assets' duration can be increased through buying bond futures, receiver swaps or floors: the duration of a receiver swap is duration fixed leg – duration floating leg ~ duration fixed leg. Indeed, the fixed side of the swap has the standard positive duration of a fixed coupon bond, whereas the floating side of the swap has only a slightly positive duration, because the interest rate is reset frequently (if the interest rate were continuously reset, the floating side would have zero duration) -> the net duration is then almost that of the fixed-rate side, which is like holding a long fixed-rate bond.

Risk managers will set a maximum positive and negative limit on the exposure of a portfolio to a 1 bp shift in interest rates at any single point on the par curve. Whenever the net exposure is in breach of this limit, the portfolio manager must rectify this by reducing its value through buying or selling new swaps.

However, although the duration metrics hold the advantage of being very synthetic, they consequently suffer from a lack of coverage accuracy during non-parallel movements in the yield curve. In the event of non-parallel shifts in the yield curve, a vector of key rate duration metrics (the sensitivity in price to a 1% change in yield for a specific maturity) should be used and offset through an optimal allocation of a portfolio of interest rate hedge assets across the different maturities, which would minimize the gap between the duration vectors.

Example: assume bond X has a one-year key rate duration of 0.5 and a five-year key rate duration of 0.9. Bond Y 1.2 and 0.3, respectively -> bond X is half as sensitive as bond Y on the short-term end of the curve, while bond Y is one-third as sensitive to interest rate changes on the intermediate part of the curve.

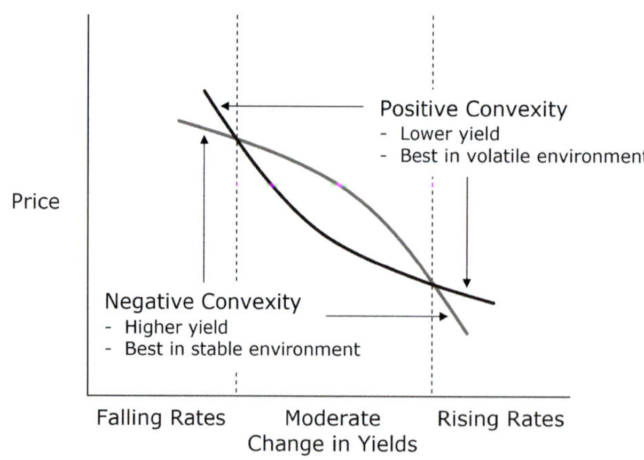

FIGURE 5.11 Convexity

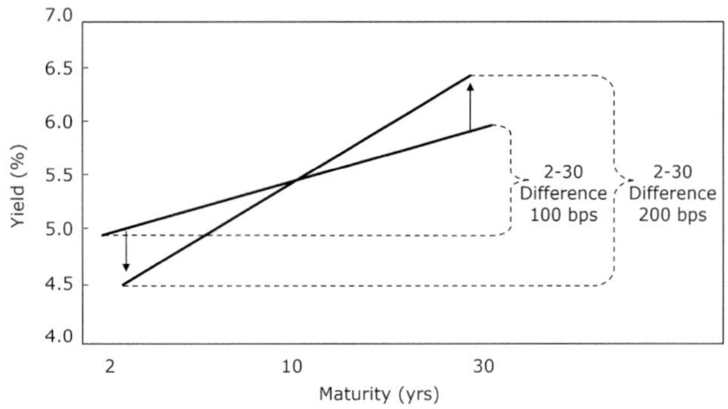

FIGURE 5.12 Rotation around the ten-year maturity

An example of a duration gap strategy usually implemented by life insurers is the stop loss rebalancing derivatives strategy, which intends to keep the short-term duration gap below a specific threshold expressed in terms of DV01 $m o number of years.

Life insurers make duration gap management decisions depending on the impact of expected scenarios across the full spectrum of their risk appetite, as measured by earnings (IFRS) / value (EEV) / capital (RBC) / liquidity (cash needs) metrics.

5.2.1.1.1.4 The Carry/Roll-Down Strategy As interest rates fell from 2008 to 2012 to persistently low levels while the yield curve was steeply upward-sloping, rho hedges carried positively as time passed, driven by the spread in yields between hedges (e.g. constant-maturity 20-year Treasuries) versus the funding rate (three-month repo). This holds as long as the yield curve does not rise back too significantly; in fact, such carry gains may be put at risk by potential roll-down losses whenever the yield curve rises and steepens more than expected (roll-down depends on the local steepness of the curve: if the yield curve is stationary, a bond's yields roll down the curve as time passes, the bond thereby appreciating in value), in order to build a sustainable carry/roll-down strategy. Another application may be to remove hedges if the yield curve is expected to flatten or invert (thus switching from positive to negative carry) or to rise sufficiently (implying some negative roll-down potentially more than compensating for the positive carry).

Life insurers assess their risk appetite for such a strategy by comparing their views to the break-even rise in interest rates, as illustrated in the example below.

Let us say we buy a three-year bond and keep it for one year – i.e. finance the holding for a year. We will benefit from the "carry" – the difference between the yield on the three-year bond and the one-year funding cost.

We will also hope to benefit from the "static roll". In other words, if the curve stays where it is (static), then the three-year bond will move down the curve to a lower yield when it becomes a two-year bond in one year's time. As the yield goes down, the price of course rises, giving a capital gain. The static roll in this case can be calculated as (yield on three-year bond – yield on current two-year bond) × modified duration of two-year bond.

The risk, of course, is that the curve is not static: we are worried that the X-year rate might have risen in one year's time, giving a capital loss. So the question becomes: does the carry from a steep yield curve more than offset the roll-down risk – i.e. the risk of rates rising quickly?

The break-even yield rise can be calculated as follows:

$$\left(\frac{\text{Carry} + \text{Roll}}{\text{Modified duration two-year bond}}\right)$$

This break-even yield rise tells us by how much the two-year rate can rise in one year's time before we start to make losses from the trade. But how volatile the two-year rate is would also give us some idea of the possibility of the two-year rate rising above the break-even rate. Thus, a common way of measuring this degree of risk and return is to use a Sharpe ratio:

$$\text{Sharpe ratio} = \frac{\text{Break-even yield rise}}{\text{Implied volatility}}$$

Example: if we think that the curve will steepen, we would receive the two-year rate and pay the ten-year rate, over, let us say, a one-year horizon. If the two-year price value of a basis point (PVBP) is 1.8898 and the ten-year PVBP is 7.7442 (each found by marking to market the respective swaps after a 1 bp shift in the relevant rate), then the correct weighting for the two legs of the trade is 7.7442/1.8898 = 4.1. In other words, we need 4.1 times the notional on the two-year swap versus the ten-year swap.

- The carry is: ((weighting × 2Y) – 10Y) – 3.1 × Euribor = ((4.1 × 4.09) – 5.2) – 3.1 × 3.4 = 1.03.
- If the 1Y into 9Y rate is 5.47% and the 1Y into 1Y rate is 4.8%, then the roll is: ((2Y – "1Y into 1Y") × MD1 × weighting) + ("1Y into 9Y" – 10Y) × MD9) = ((4.09 – 4.8) × 0.96 × 4.1) + (5.47 – 5.2) × 7) = –0.9.
- Thus the carry + roll = 0.13.

5.2.1.1.2 Case Study: Mitigating Rho Hedging for Fixed Indexed Annuities through Product Design

Here we embed the hedging risk management into the product design as a clear way to reduce the P&L uncertainty, which is applied here to a FIA-structured product.

From a client point of view, FIAs present a clear advantage as a savings product. Indeed, the policyholder is certain to get back her capital at maturity T, and get an upside coming from the participation rate times the equity call option. Moreover, the fees paid by the client are a half of what is paid by the client in the case of variable annuities (1 to 2% versus 3 to 4%).

Let us remind ourselves of the generic price of a FIA (or EIA):

$$P = K \times \exp(-r * T) + (1-\alpha) \times \sum_{i=1}^{T} r_i + \alpha \times Callprice,$$

with r being the risk-free interest rate, α the so-called participation rate, r_i the ten-year US Treasury rate and $Callprice$ the price of the call, giving the upside of the product (the strike of the call is K). K is the guaranteed capital, so therefore $K \times \exp(-r * T)$ is a zero-coupon bond ensuring that the client gets back her capital at maturity T. Usually the maturity T is ten years.

The innovation in this product comes from the fact that we have calibrated the α as a market adjuster.

What we wanted to do is integrate the hedging into the participation rate, in order to mitigate the lapse rate occurring in the event of rising interest rates.

The two innovations in our product are the following:

- the usual fixed rate is changed and is variable, indexed on the ten-year US Treasury rate; and
- the participation rate is variable and indexed on the market, via an adapted design.

Indeed, the biggest risk for this kind of product is the lapse risk, coming from high-rising interest rates. When interest rates rise high, the client can be tempted to lapse because she considers that the guaranteed rate she obtains via this product is much lower than what she could expect from another product in the market.

This is why the roll-up rate is designed as variable, with control of this rate via the variable participation rate α. From an insurer's point of view, this product is interesting, since it is clearly advantageous in the context of rising interest rates, as the variable guaranteed rate is controlled by the participation rate, which is slightly lower than what is proposed in a classical FIA.

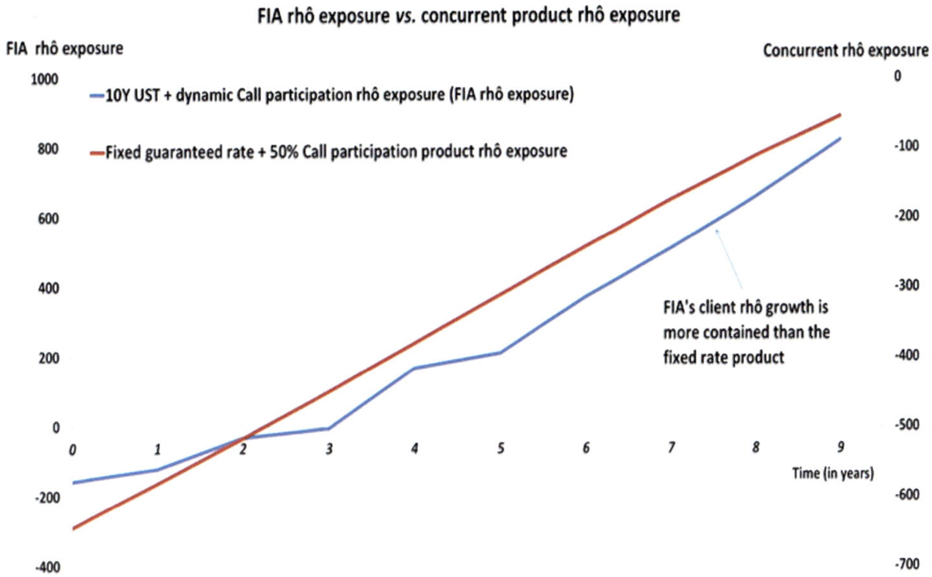

FIGURE 5.13 Change of risk profile via adequate design

In Figure 5.13 we compare the rhos of a classical FIA product versus our product, and we can see that we go from a negative rho profile to a positive rho profile, which is of a great interest in the case of high-rising interest rates, mitigating the lapse risk. Moreover, the upside risk is under control, since the variable participation rate is here to "cap" the client value, reducing the cost of the product.

The rho exposure seems to be irregular from years 0 to 5. This comes only from the number of simulations; since we have a path-dependent product, we must launch a high number of Monte Carlo simulations to generate a regular risk profile.

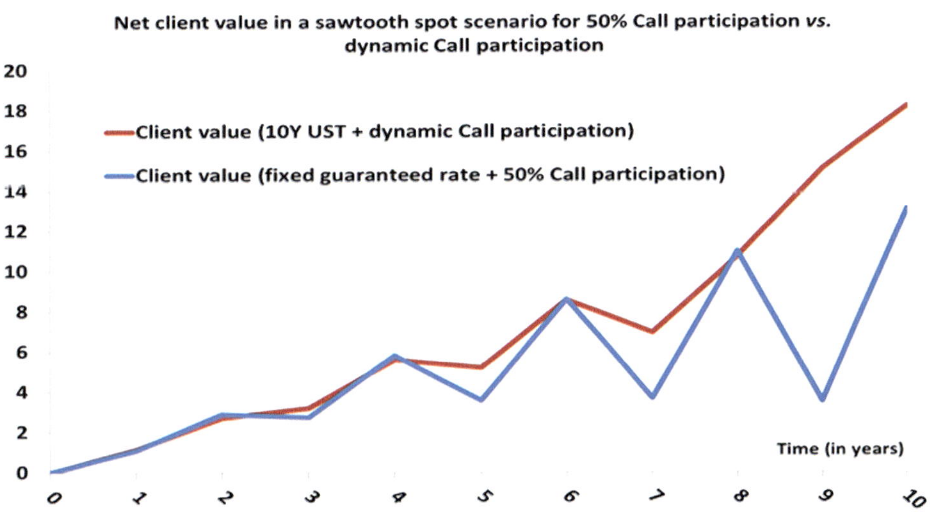

FIGURE 5.14 How the product is robust versus a classical FIA under a sawtooth equity spot scenario

In Figure 5.14, we can observe that the product is clearly more stable in a sawtooth equity spot scenario, since the client value is less volatile. This is mainly due to the variable participation rate, which "rolls with the punches" more easily than a classical FIA product.

Reducing the volatility is the key to hedging in general. And here we can see the efficiency of the calibration of α, because it reduces the volatility of the client P&L profile, which can cause high hedging fees.

Moreover, in Figure 5.15, we can see that adjusted alpha is maintained under control during the lifetime of the product, even when placed in a high equity spot scenario.

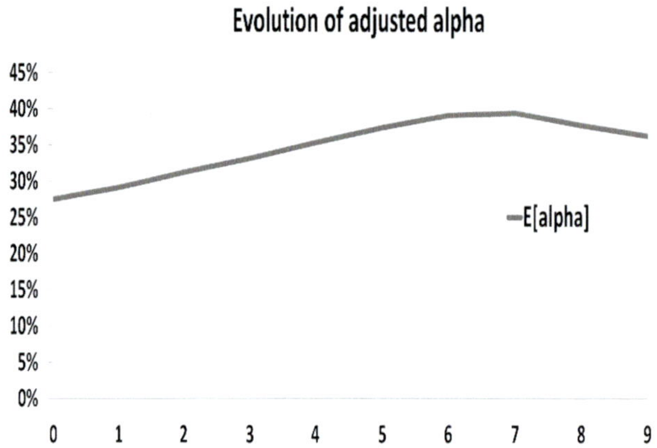

FIGURE 5.15 The evolution of adjusted alpha

The other key fact with this product design is that, when equity markets tumble, it clearly performs better than a classical FIA.

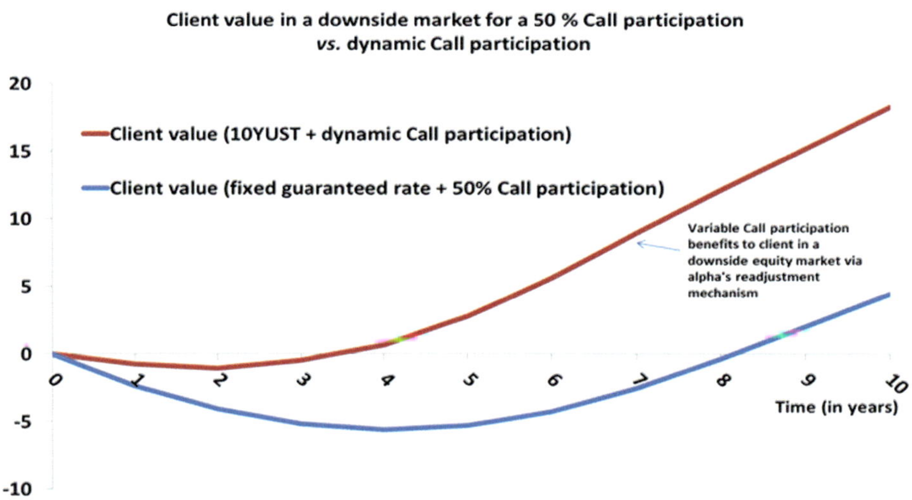

FIGURE 5.16 Better performance of the product versus the classical FIA

As illustrated in Figure 5.16, we observe an outperformance from the product in the event of a downside equity market, and the client's value is approximately three times larger than the FIA.

5.2.1.2 *Equity Delta Hedging*

5.2.1.2.1 Delta-Hedging Effectiveness As long as transaction costs can be considered as negligible, the Black–Scholes framework, which assumes no market frictions ("perfect financial markets"), is relevant to design a hedging strategy with the three following properties: self-financing, no risk and no arbitrage.

$$\Delta c = c_t \Delta t + c_S \Delta S + \frac{1}{2} c_{tt} (\Delta t)^2 + \frac{1}{2} c_{SS} (\Delta S)^2 + c_{tS} \Delta t \Delta S + \ldots$$

Now consider the hedged portfolio:

$$\begin{aligned} P &= c + \alpha S \\ \Delta P &= \Delta c + \alpha \Delta S \\ &= c_t \Delta t + (c_s + \alpha)\Delta S + \frac{1}{2} c_{tt} (\Delta t)^2 + \frac{1}{2} c_{ss} (\Delta S)^2 + c_{ts} \Delta t \Delta S + \ldots \\ &\Downarrow \\ &\alpha = -c_S \quad \text{eliminates the risk of } \Delta S. \end{aligned}$$

Note: the second equality derives from the self-financing assumption of Black–Scholes, while the last equality stems from the no risk assumption, the other components of the development being either deterministic (in particular, the squared change in S as a consequence of the Itô lemma, which holds for infinitesimal change of S consistent with the no-transaction-costs assumption of Black–Scholes) – or with higher-order negligible risk.

In this context, the delta represents the number of stocks to buy or sell in order to hedge.

The effectiveness of the delta-hedging strategy can, essentially, be defined as delta $_{\text{asset}}$ × return $_{\text{asset}}$ – delta $_{\text{liability}}$ × return $_{\text{roll-forward liability}}$; using notations:

$$\sum (\Delta_A \times \mathrm{R}_A) - \sum (\Delta_L \times \mathrm{R}_L)$$

+ P&L from intraday trades.

Equivalently:

$$\underbrace{\sum (\Delta_A - \Delta_L)\mathrm{R}_L}_{\text{Equity rebalancing efficiency}} + \underbrace{\sum \Delta_A (\mathrm{R}_A - \mathrm{R}_L)}_{\text{MVA}}$$

+ P&L from intraday trades.

This can also split into three categories.

1) Equity rebalancing efficiency: measures the P&L impact associated with the investment decision to rebalance or not the hedge portfolio to achieve delta neutrality daily:

$$(\Delta_{asset} - \Delta_{liability})R_{liability}.$$

2) MVA (market value adjustment – spot versus futures): measures the timing impact of cash versus futures markets closing at different times within the month, but at the month's end it measures the P&L breakage/slippage due to the return differential between the hedge portfolio and the liability roll-forward returns:

$$\Delta_{asset}(R_{asset} - R_{liability}).$$

3) PnL due to liability changes (in-force or model-change-related): PnL associated with not achieving a delta-neutral portfolio due to the fact that residual liability Greeks (equity delta) after the in-force change over or after the model change are not always fully hedged due to the timing of information received from the liability team to the rebalancing team

Remark: in a market environment in which deltas are smooth, we might expect the cumulative year-to-date market value adjustment to be relatively small, since by construction it would be associated with "timing"-related discrepancies between the index and futures price levels.

The delta hedging provides a significant reduction not only in the P&L variance but also in the tail risks, as illustrated in Figures 5.17 to 5.19.

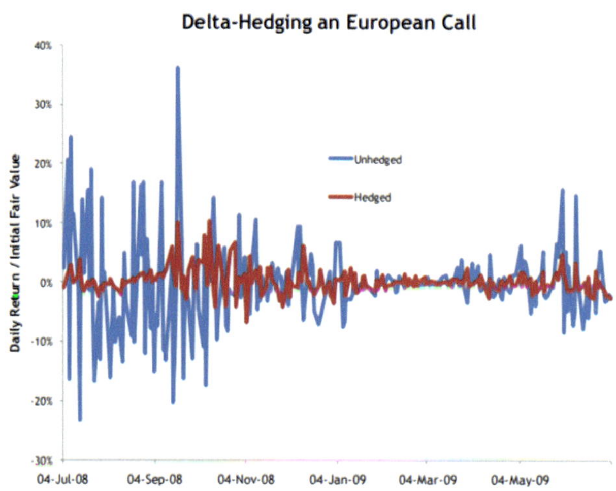

FIGURE 5.17 Delta-hedged versus unhedged ATM European calls, 1 January 2008–30 June 2009

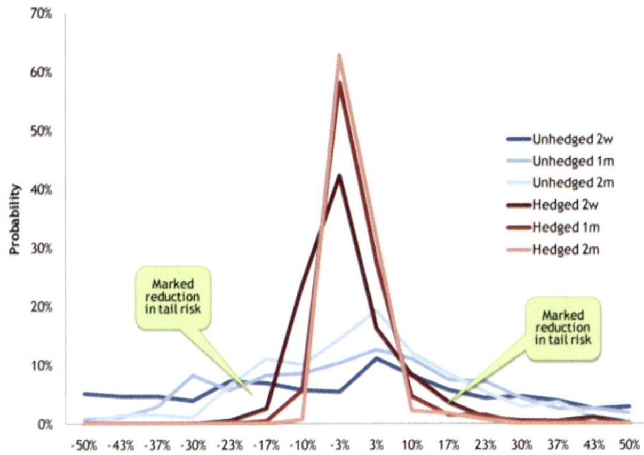

FIGURE 5.18 Delta hedging's impact on the return distribution

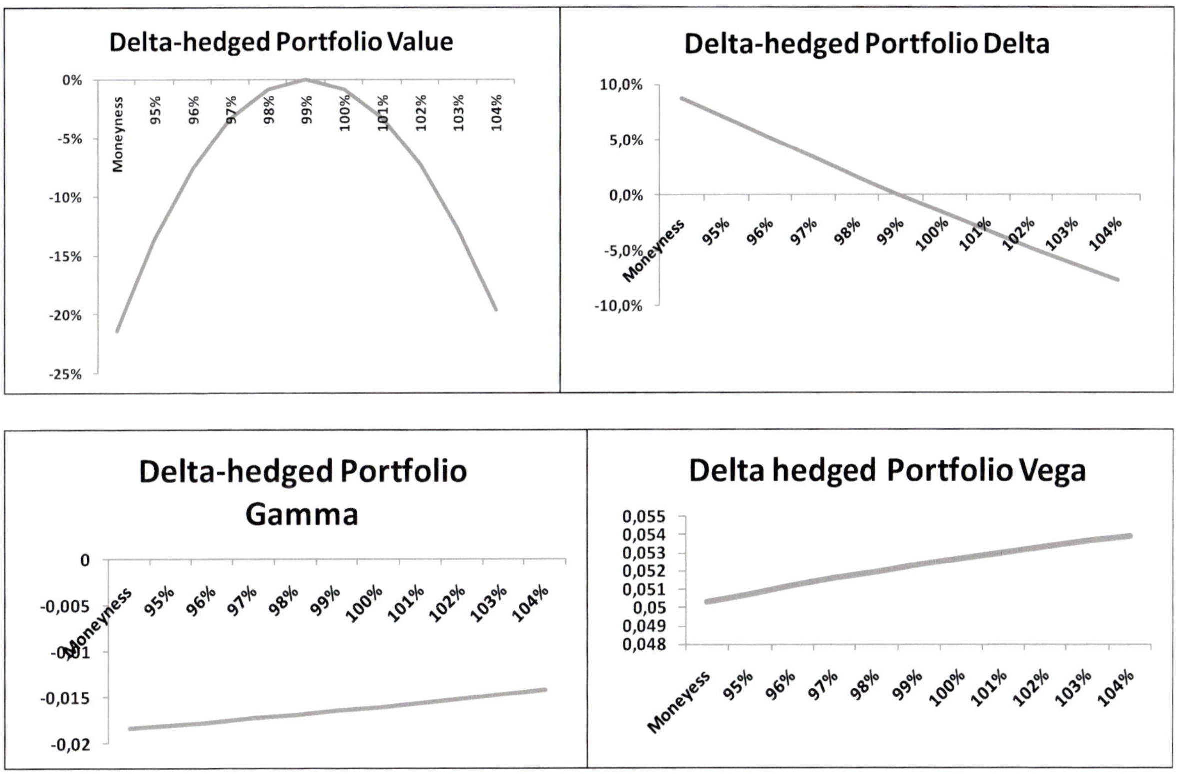

FIGURE 5.19 Value and Greeks of a delta-hedged portfolio

Once delta-hedged, the portfolio value and Greeks have the shapes shown in Figure 5.19.

5.2.1.2.2 Delta-Hedging Short Convex Life Insurance Liabilities in the context of Transaction Costs For small moves in the underlying, the replicating hedge is accurate. For larger moves in the underlying (which are potential whenever transaction costs are non-zero, making the Itô lemma non-applicable, and so the squared change in S is no longer deterministic in time), the long options position always outperforms the replicating hedge in both directions. For a delta-hedged option, the gamma P&L will be the outperformance of the option over the replicating hedge. The more the delta changes, the more the replicating delta hedge will underperform the long options position.

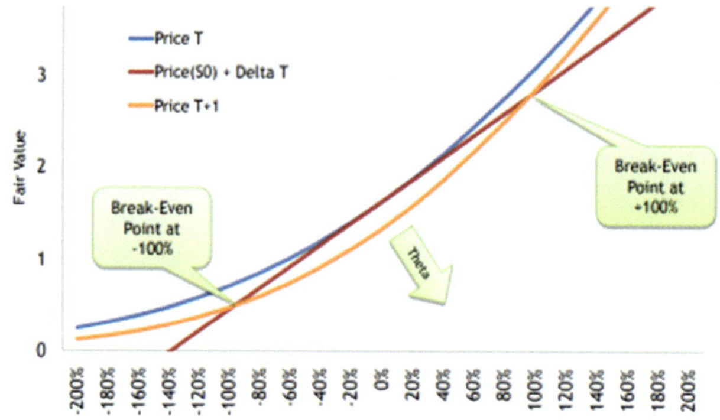

FIGURE 5.20 Hedging a European call

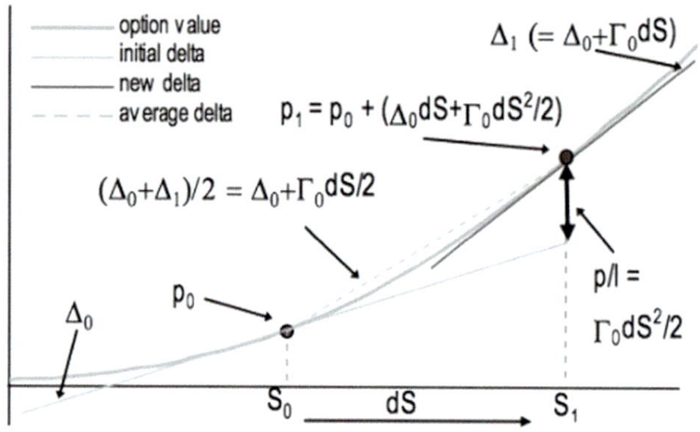

FIGURE 5.21 Gamma P&L from delta-hedged options

How will much the delta hedge underperform? The actual amount depends on the difference between the initial and final delta of the option, which is gamma times the change in the spot price. Let us illustrate with an example:

1) Spot is initially $S_{0,}$ option is qorth $p_{0,}$ delta and gamma are initially Δ_0 and Γ_0 respectively.
2) Sopt increases, by dS, to $S_{1.}$
3) Delata increases, to $\Delta_1 = \Delta_0 + \Gamma_0 dS$.
4) Average delta is therefore $\Delta_1 = \Delta_0 + \Gamma_0 dS/2$.
5) New premium $p_1 = p_0 + \Delta_{Avg}\, dS$.
6) $p_1 = p_0 + \Delta_0\, dS + \Gamma_0\, dS^2/2$ (*by 4 and 5*).
7) p/1 on long option is $\Delta_0 dS + \Gamma_0 dS^2/2$ (*by 6*).
8) p/1 on short futures in $- \Delta_0$ dS.
9) Total p/1 is $\Gamma_0\, dS^2/2$ (*by 7 and 8*).

Whenever life insurers sell embedded guarantees to policyholders that are characterized by convex payoffs (e.g. embedded put options), the risk management of such short convex liability's exposure implies negative convexity (gamma) P&L exposure whenever the financial market moves.

Whenever transaction costs cannot be considered as negligible given the large size of life insurers books (multibillion dollar exposure), the continuous-time Black–Scholes framework cannot be considered as realistic, since the usual life insurance convex liability payoffs (due to embedded implicit put options) cannot be replicated by an infinite number of piecewise linear hedge instruments (such as equity or bond futures), which would then imply an infinite transaction cost of hedging and ultimately bankruptcy. As a result, the hedging rebalancing can be made only at discrete time intervals, thus missing some potentially non-insignificant convexity component, as measured by the so-called gamma P&L exposure (see Figure 5.22).

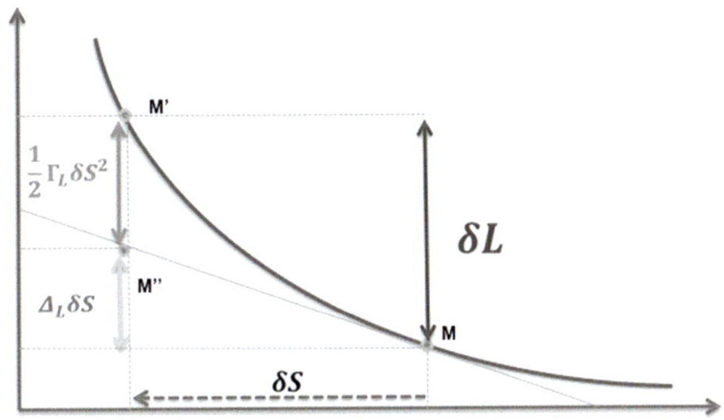

FIGURE 5.22 Decomposition of a hedging P&L

As we can see in the figure, a long position in option is convex and there is a positive gamma. To delta-hedge, the trader will need to sell stocks if the stock price goes up and buy stocks if the stock price goes down (sell high, buy low). The trader can sit on the bid and offer.

A short position in option is negative gamma. In this case, the trader will need to sell stocks if the stock price goes down and buy stocks if the stock price goes up to be delta-hedged (sell low – buy high). Given the combined negative convex payoff positions and discrete hedging rebalancing (due to non-zero transaction costs), the risk management implies some negative gamma P&L whenever the market moves up or down, as illustrated by the following example of delta-hedging a short at-the-money-put option position, thus with an absolute delta close to 50%.

- If the market moved up, the put would be more out of the money with a lower absolute delta position, which would require the buying of additional stocks at a higher than expected price, since the rebalancing occurs discretely, and so after the stocks move up.
- IF the market moved down, the put would be more in the money with a higher absolute delta position, which would require the selling of additional stocks at a lower-than-expected price, since the rebalancing occurs discretely, and so after the stocks move down.

As a result, the hedging rebalancing strategy would follow the pattern "buy high – sell low", which means systematic potential losses whenever the market moves sufficiently to compensate for the initial option premium received at time 0. This further emphasizes the need to consider taking gamma hedging into account.

To make the gamma correction we need to adjust our estimate of delta and plug in a revised delta (and hopefully improved delta) in our simulation model:

$$C(S_{t+h}) = C(S_t) + \varepsilon\Delta(S_t) + \frac{1}{2}\varepsilon^2\Gamma(S_t)$$

- > the results with gamma correction are more tightly centred around zero, as illustrated in Figures 5.23

and 5.24.

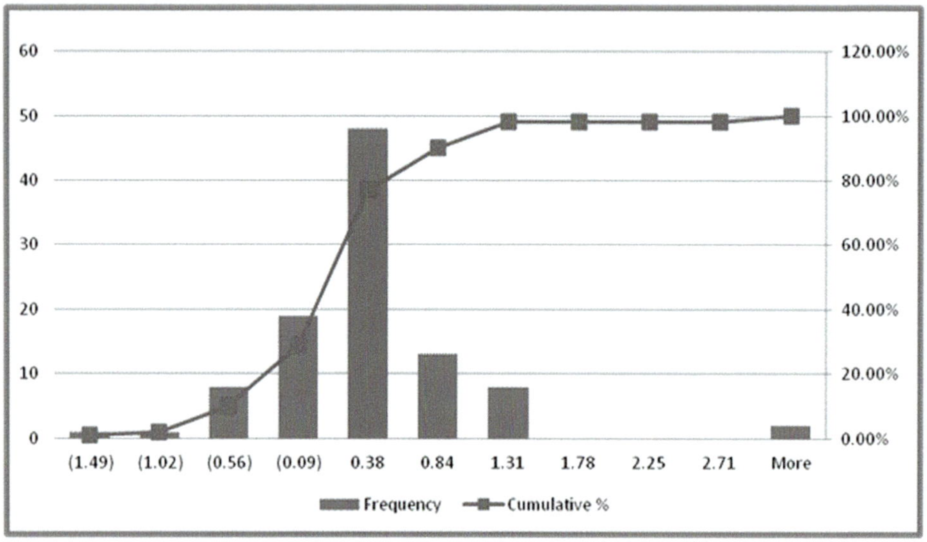

FIGURE 5.23 Call delta-hedging P&L

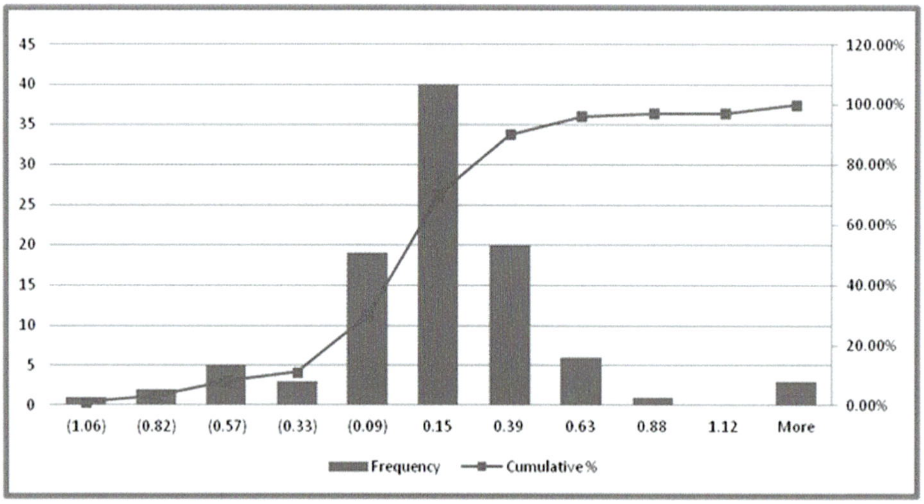

FIGURE 5.24 Call delta with gamma-adjusted hedging P&L

Once delta-gamma-hedged, the portfolio value and Greeks have the shapes shown in Figure 5.25.

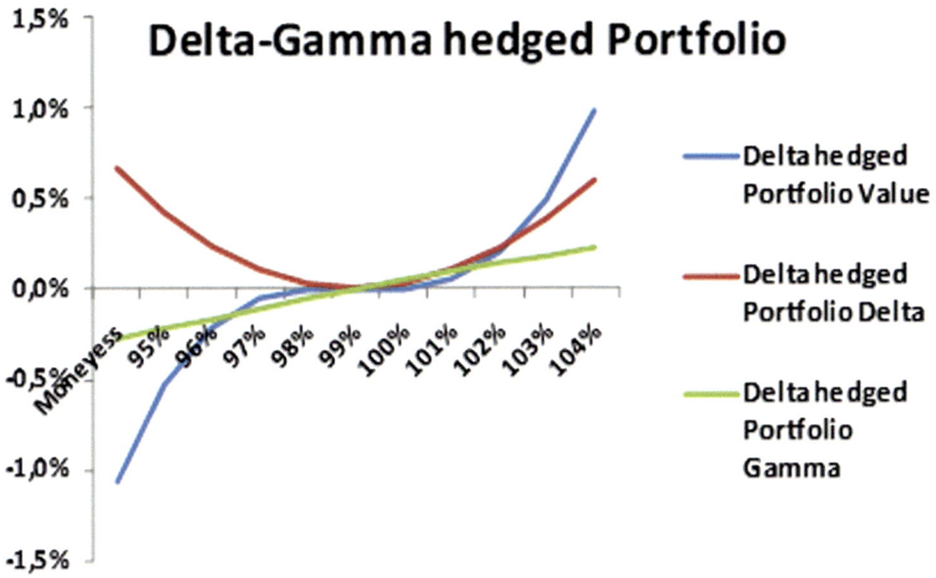

FIGURE 5.25 Delta-hedged value, delta and gamma

5.2.1.2.3 Sustainable Delta Hedging within an Environment of Uncertain Volatility Besides the structural short gamma position held by the life insurer, the delta-hedging P&L is also impacted by the change in future volatility.

- If the stock market does not move significantly before the updating of the replicating portfolio, the gamma impact is much smaller than the delta effect. Hence, in a low-volatility market, the delta hedge is not much less effective than a delta gamma hedge.
- Conversely, the delta hedge is considerably less effective when the equity volatility is high (see Figure 5.26).

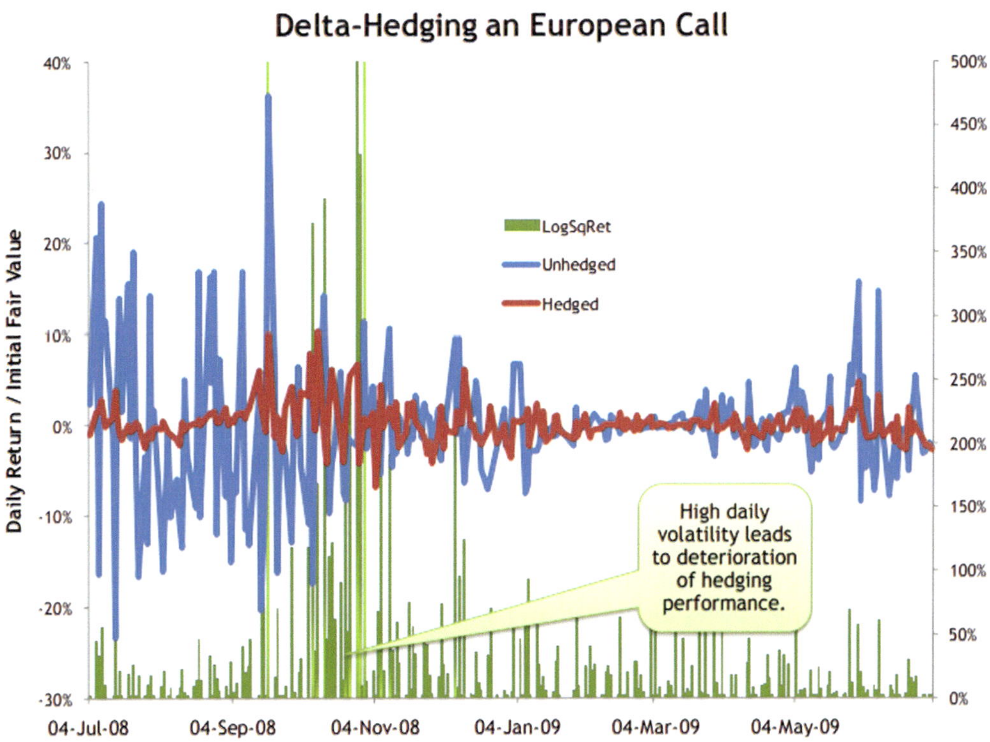

FIGURE 5.26 Delta-hedged P&L of a European call, 1 January 2008–30 June 2009

This is to the profit of the delta-gamma-hedging strategy, which is exposed to stochastic movements in the realized volatility, which can give rise to profits or losses depending on how the realized compares with the pricing implied volatility.

- In fact, if a position is continuously delta-hedged with the correct delta (calculated from the known future volatility over the life of the option), then the payout is not path-dependent. Even though one path is always ATM while the other has most volatility OTM, delta hedging gives the same profit for both options, as long as they hold equal volatility and the same start and end points. The cause of this relationship is the fact that, while the ATM option earns more due to delta hedging, the total theta cost is also higher (and exactly cancels the delta-hedging profit). This result is due to the fact that a European option payout depends only on the stock price at expiry. Therefore, the volatility over the life of the option is the only input to pricing. The timing of this volatility is irrelevant.

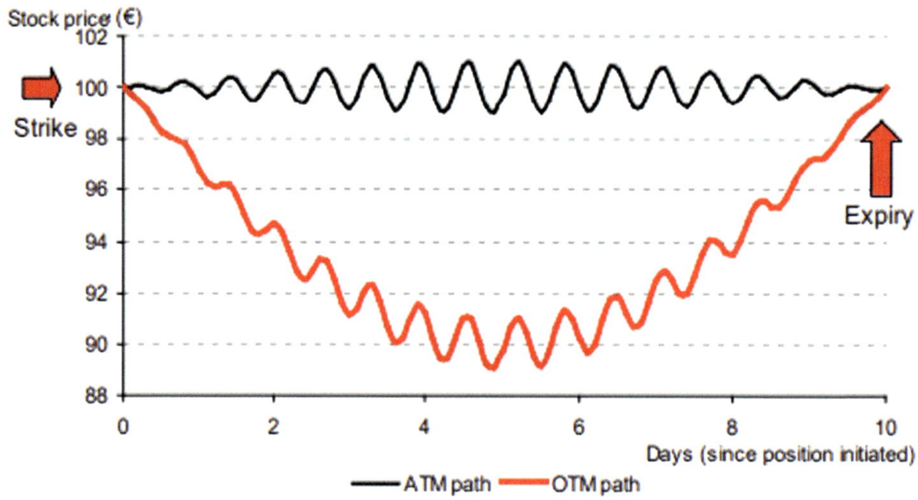

FIGURE 5.27 Stock prices over time

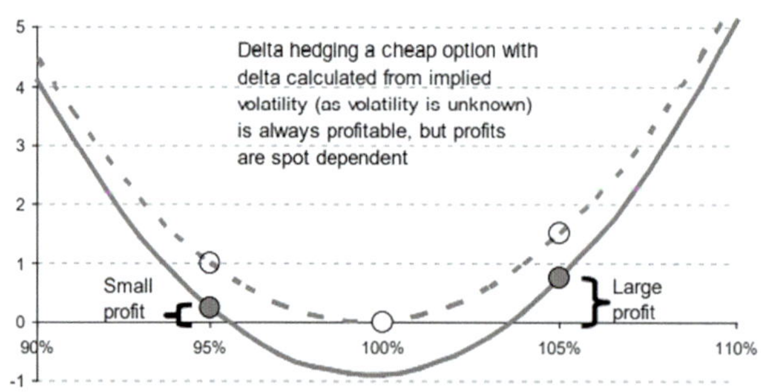

FIGURE 5.28 Delta-hedging profits are spot-dependent

- But, if volatility is unknown, the correct delta cannot be calculated. The implied volatility is often used to calculate deltas, which causes the position to have equity market risk and, hence, it becomes path-dependent (although the average or expected profit remains unchanged). The P&L from delta hedging is no longer independent of the direction in which the underlying moves.

As a result, although European-style options are path-independent, once we start to evaluate the impact of dynamic hedging, dynamic hedging makes the options become path-dependent, as a direct consequence of the limitations of the Black–Scholes assumptions, since continuous hedging at a known constant volatility would result in a definite value with no variation (hence, you would achieve not just path independence, but independence of the final underlying asset value as well). Sporadic dynamic hedging and stochastic volatility make the realized value of a dynamic hedging strategy dependent on the full price history of the underlying asset.

The fact that there is a difference between the correct delta (calculated using the remaining volatility to be realized over the life of the option) and the delta calculated using the implied volatility means that returns are dependent on the direction of equity markets. Delta hedging is a cheap option with the delta calculated from implied volatility (as volatility is unknown), and is always profitable, but the profits are now spot-dependent. If there is a difference between the actual delta and the estimated delta, there is market risk, but not enough to make a cheap option unprofitable (or an expensive option profitable). This is because in each infinitesimally small amount of time a cheap option will always reveal a profit from delta hedging (net of theta), although the magnitude of this profit is uncertain. The greater the difference between implied and realized, the greater the market risk and the larger the potential variation in profit.

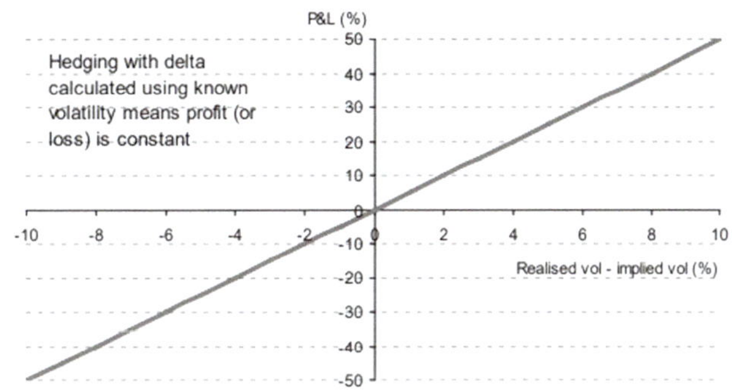

FIGURE 5.29 P&L versus realized volatility minus implied volatility

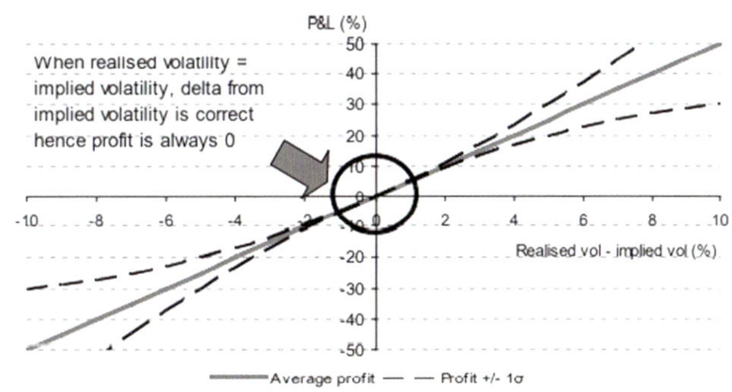

FIGURE 5.30 P&L versus realized volatility minus implied volatility

Only if the implied volatility is equal to the realized volatility will the estimated delta calculated from the implied be equal to the actual delta (calculated from the realized), and so profits from hedging will exactly match the theta cost for all paths, so it is path-independent.

This is accounted for in the P&L decomposition through the gamma–theta P&L component once the liability is assumed to be delta-hedged (under the assumption of zero interest rates) shown in Figure 5.31.

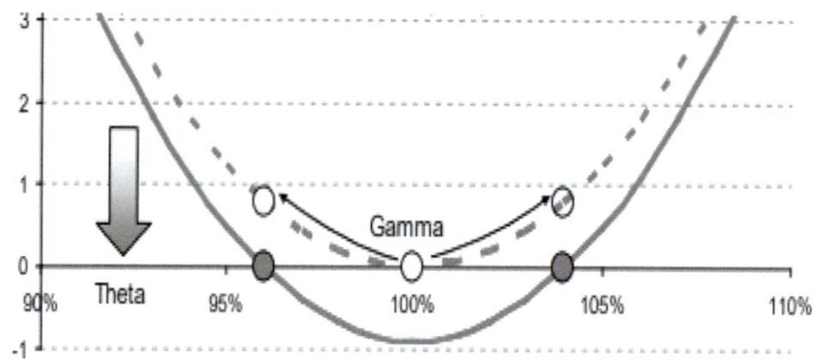

FIGURE 5.31 Daily P&L

Typically, gamma and theta have opposite signs: for a long call or put position, gamma is positive and theta is negative – i.e. the trader is long shocks/volatility (he makes money as the stock price moves) and short time (he loses money as maturity approaches.). For a short call or put position, the situation is reversed.

In each unit of time, the stock can either go up or down. As the position is initially delta-neutral (i.e. delta is zero), the gamma (or convexity) of the position gives it a profit for both downward and upward movements. While this effect is always profitable, the position does lose time value (due to theta). If an option is priced using the actual fixed constant volatility of the stock, the two effects cancel each other, and the position does not earn an abnormal profit or loss as the return is equal to the risk-free rate: theta pays for gamma, and the amount earned from gamma is equal to the loss of time (theta).

Moreover, based on the Black–Scholes (delta-hedging) equation (assuming zero interest rates and constant implied volatility over the period),

$$\Theta \approx -\frac{1}{2}\Gamma S^2 \sigma^2_{Implied}.$$

Plugging this equality into the daily P&L equation then provides

$$\begin{aligned} Daily\,P\,\&\,L &\approx \frac{1}{2}\Gamma\left[(\Delta S)^2 - S^2\sigma^2_{Implied} \times \Delta t\right], \\ &\approx \frac{1}{2}\Gamma S^2\left[\left(\frac{\Delta S}{S}\right)^2 - \left(\sigma_{Implied}\sqrt{\Delta t}\right)^2\right], \end{aligned}$$

$$Daily\,P\,\&\,L \approx \frac{1}{2}\Gamma S^2\left[\left(\frac{\Delta S}{S}\right)^2 - \left(\sigma_{Implied}\sqrt{\Delta t}\right)^2\right].$$

This equation tells us that the P&L on a delta-hedging strategy assuming constant implied volatility over the period is driven by two factors:

- the variance spread (squared realized versus squared implied), which determines the sign of the P&L; and
- the so-called "dollar gamma", which has the role of a scaling factor and does not determine the sign of the P&L.

Thus, a life insurer who is short dollar gamma will make money if the subsequent realized variance is lower than the pricing implied volatility embedded within the liability, break even if they are the same and lose money if realized is above implied. In addition, as the final P&L is the sum of the daily variance spread weighted by the dollar gamma, days when dollar gamma is high will tend to dominate the final P&L.

However, volatility is not constant. Furthermore, the gamma – and thus the magnitude of the P&L – also depends on where that volatility is realized in relation to the liability strike and to the liability maturity. This emphasizes the strong stochasticity and path -dependence of the hedging P&L. If the volatility realizes close to maturity when the underlying trades near the strike, then the gamma will be very high and the absolute value of the P&L will be larger. As a result, for non-constant volatility, it is possible to buy and delta-hedge an option at an implied volatility smaller than the subsequent realized volatility, and still lose money from delta-hedging!

Thus, the "perfect storm" scenario lies in both high realized volatility and high gamma.

Remark: for example, life insurance liabilities are usually out-of-the-money long-term options. Hence, generally the delta hedge is a reasonable approach, at least at inception. However, the hedge has to be monitored carefully, because the gamma impact can surge in fast downward-moving markets close to at-the-money guarantees.

Remark: the use of a constant volatility assumption may be misleading, in particular in the event that the average realized volatility consists of very high volatility during a short period when the market plunges sharply and is preceded and followed by periods of much lower volatility, when the higher-than-average volatility during the period when prices fall sharply can cost much more than saved from the lower-than-average volatility during the other periods. This is what happened during 2011 with the downgrade of the US debt ratings by the major ratings agencies.

✓ **Quick win**: a random process of significant price jumps, in which no rehedging is permitted until after the jump is completed, can be used as a simulation of periods of illiquidity.

Such refinements in the hedging P&L call for a "delta–gamma–theta" approximation:

$$C(S_{t+h}, T-t-h)$$
$$= C(S_t, T-t) + \varepsilon\Delta(S_t, T-t) + \frac{1}{2}\varepsilon^2\Gamma(S_t, T-t) + h\theta(S_t, T-t).$$

	Starting Price	$\epsilon\Delta$	$\frac{1}{2}\epsilon^2\Gamma$	θh	Day Later (h = 1 day) Predicted	Actual
S_{t+h} = \$40.75	\$2.7804	0.4368	0.0183	−0.0173	\$3.2182	\$3.2176
S_{t+h} = \$39.25	\$2.7804	−0.4368	0.0183	−0.0173	\$2.3446	\$2.3452

FIGURE 5.32 Delta–gamma–theta P&L approximation

Overall, a significant "risk" that we run with a (discrete) delta hedge in practice is that we might not be using the correct assumption for underlying volatility: as illustrated in the gamma–theta P&L formula above, it is expected to have losses if the realized volatility is higher than the pricing level and to have profits if it is lower.

This will generally be estimated from historic data, or maybe from the implied volatility of derivative instruments in the market, but, clearly, we are at risk of having this key assumption in our hedging formula wrong. Note that this is the risk of having the wrong delta, which is impacted by the assumed distribution of the realized volatility (see Figures 5.33 and 5.34).

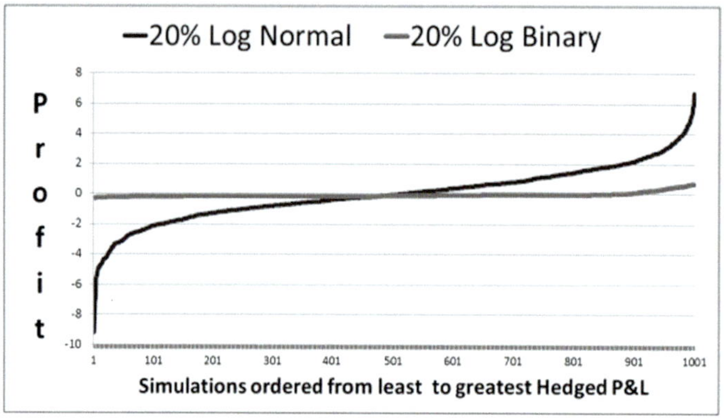

FIGURE 5.33 Simulations ordered from least to greatest hedging P&L

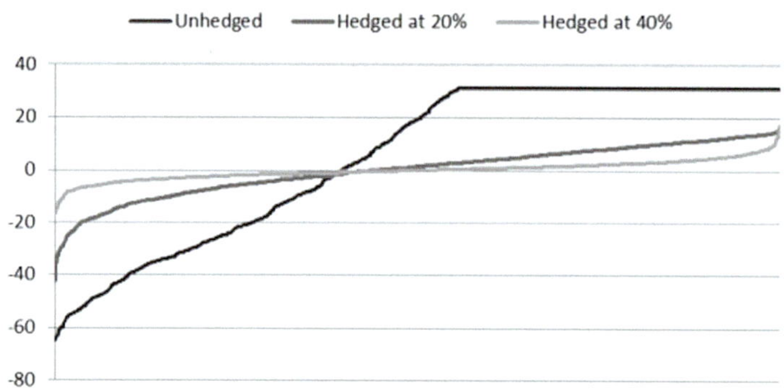

FIGURE 5.34 Simulations ordered from least to greatest hedging P&L

However, regarding the impact on the P&L, if we have underestimated the underlying volatility, by hedging with a 20% volatility assumption as opposed to an actual 40% realized volatility, then it can certainly be said that we have got it fairly badly wrong – and yet the hedging is still quite effective, and not too far off the effectiveness of a hedging programme that had the correct 40% volatility assumption, at least in a broad middle range of outcomes.

This is not saying that the final outcome is not sensitive to the underlying volatility; indeed, the initial value of the put option using 40% volatility is more than double the value using 20% volatility. What it is saying is that the effectiveness of hedging in reducing the dispersion of final outcomes is far less sensitive to having chosen the "correct" volatility assumption.

The bad news is that, if the option is fundamentally mispriced, there is no way to hedge your way out of that situation.

✓ **Quick win**: in practice, life insurers use a prudent estimate of underlying volatility, such as the 70th or 80th percentile of historic realized volatility, although the appropriate reference should be the historic implied volatility, because the liability volatility risks are hedged with options whose price is based on implied volatility and not realized volatility.

Once delta-gamma-vega-hedged, the portfolio value and Greeks have the shapes shown in Figure 5.35.

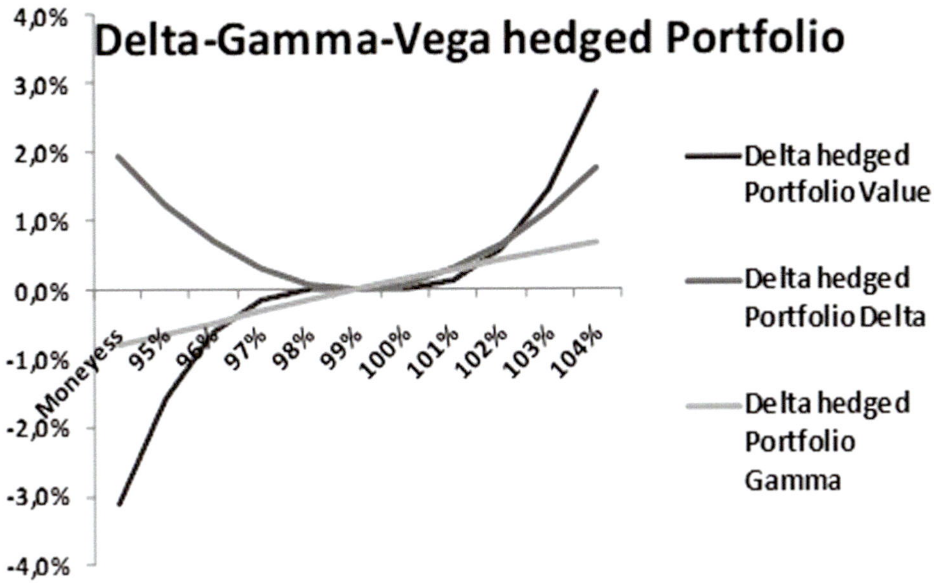

FIGURE 5.35 Value, delta and gamma of a delta-hedged portfolio

5.2.1.3 Sustainable Delta-Hedging Rebalancing (e.g. for FIAs, Vas and EIAs)

5.2.1.3.1 Delta-Hedging Rebalancing and P&L Dispersion If the realized volatility is equal to the pricing volatility, then the P&L volatility depends only on the hedging portfolio rebalancing frequency.

The realized volatility risk is also a function of the size of the interval for delta rebalancing.

With continuous delta hedging (with known or unknown volatility), it is impossible to lose money on a cheap option (an option whose implied volatility is less than the realized volatility over its life). However, as the error from discrete hedging is independent from the profitability of the trade, it is possible to lose money on a cheap option (and make money on an expensive option).

The size of the hedging error can be reduced by increasing the frequency of hedging. An approximation (shown in Figure 5.36) is that, if the frequency of hedging is increased by a factor of four, the hedging error term halves.

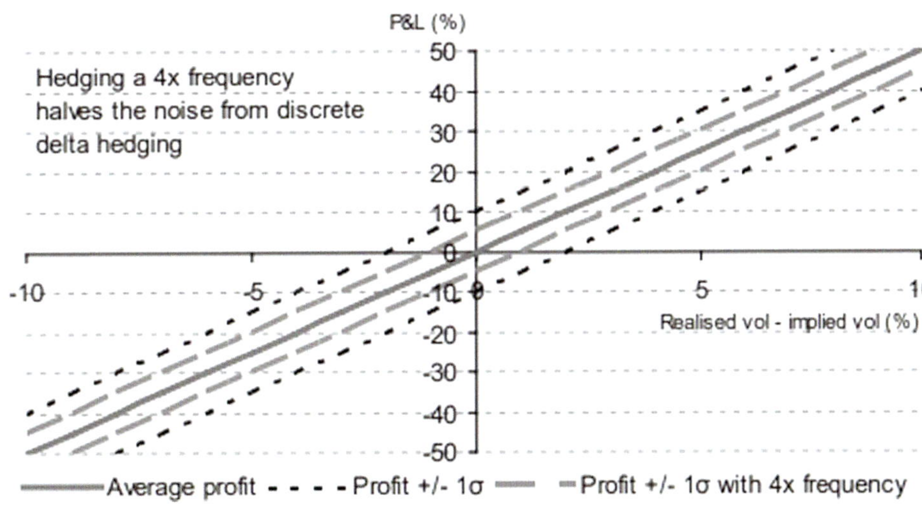

FIGURE 5.36 Hedging P&L versus realized minus implied volatility

The underlying model is that realized volatility – i.e. the realized square of the price changes – is of the order of the time interval, and its variance is of the order of the square of the time interval. So, if we split the time interval into n sub-intervals, each sub-interval will display a variance in the square of the price change of an order $1/n^2$ the variance for the whole interval.

Summing these independent variables over the sub-intervals, we find that the variance of the realized volatility over the whole interval is 1/n times the variance over the whole interval without subdivision. Hence, the realized volatility risk reduces in proportion to the square root of the size of the rebalancing interval. For example, the realized volatility risk for weekly rebalancing might be only 50% of that for monthly rebalancing.

Consequently, it can be further proved that the realized volatility – i.e. dispersion – of the P&L reduces in proportion to the square root of the size of the rebalancing interval, as long as the ex post realized volatility remains close to the ex ante implied pricing volatility.

Proof: see Derman (1999).

- $\frac{\Delta S}{S} = \mu\Delta t + \sigma\varepsilon\sqrt{\Delta t}$ and assuming that realized volatility = implied volatility.
- We build a portfolio made of long one call and short a delta quantity of stocks:

$$\pi = C - \left(\frac{\partial C}{\partial S}\right)S.$$

- Then the hedging error (P&L) accumulated over time due to the mismatch between a continuous versus discrete time step hedge is

$$\begin{aligned} HE &= \pi + d\pi - \pi e^{r\Delta t} \\ &\approx -r\Delta t\left[C - \left(\frac{\partial C}{\partial S}\right)S\right] + \left[C_t\Delta t + C_S dS + C_{SS}\frac{\sigma^2 S^2 \varepsilon^2}{2}\Delta t - C_S dS\right]. \\ &\approx \left(C_t + C_{SS}\frac{\sigma^2 S^2 \varepsilon^2}{2} - r\left[C - \left(\frac{\partial C}{\partial S}\right)S\right]\right)\Delta t \end{aligned}$$

- Plugging the Black–Scholes PDE:

$$r\left[C - \left(\frac{\partial C}{\partial S}\right)S\right] = C_t + C_{SS}\frac{\sigma^2 S^2}{2},$$

$$HE = \frac{1}{2}C_{ss}\sigma^2 S^2\left(\varepsilon^2 - 1\right)\Delta t.$$

- Thus, over n steps,

$$HE = \sum_{i=1}^{n}\frac{1}{2}\Gamma_i S_i^2 \sigma_i^2\left(\varepsilon_i^2 - 1\right)\Delta t \quad \text{and} \quad \sigma_{HE}^2 = E\sum_{i=1}^{n}\frac{1}{2}\left[\Gamma_i S_i^2\right]^2\left(\sigma_i^2\Delta t\right)^2.$$

Considering a close to ATM option from Black–Scholes: $E\left[\Gamma_i S_i^2\right]^2 = S_0^4\Gamma_0^2\sqrt{\frac{T^2}{T^2-t_i^2}}$ and $S_0^2\Gamma_0 = \frac{1}{\sigma(T-t)}\frac{\partial C}{\partial\sigma}$ and $(T-t)/(\Delta t) = n$.

- Thus

$$\begin{aligned}\sigma_{HE}^2 &= \frac{1}{2}\sum_{i=1}^{n} S_0^4\Gamma_0^2\sqrt{\frac{T^2}{T^2-t_i^2}}\left(\sigma^2\Delta t\right)^2\\ &= \left(S_0^4\Gamma_0^2\right)\left(\sigma^2\Delta t\right)^2\frac{1}{2\Delta t}\int_t^T\sqrt{\frac{T^2}{T^2-\tau^2}}d\tau\\ &= \left(S_0^4\Gamma_0^2\right)\left(\sigma^2\Delta t\right)^2\frac{\pi T}{4\Delta t}\\ &= \frac{\pi}{4}n\left(S_0^2\Gamma_0\sigma^2\Delta t\right)^2,\end{aligned}$$

$$\sigma_{HE}^2 = \frac{\pi}{4}n\left(\frac{1}{\sigma(T-t)}\frac{\partial C}{\partial\sigma}\sigma^2\Delta t\right)^2 = \frac{\pi}{4}n\left(\frac{1}{n}\frac{\partial C}{\partial\sigma}\sigma\right)^2 = \frac{\pi}{4n}\left(\frac{\partial C}{\partial\sigma}\sigma\right)^2,$$

$$\sigma_{HE} \approx \sqrt{\frac{\pi}{4}}\frac{\partial C}{\partial\sigma}\frac{\sigma}{\sqrt{n}},$$

$$\sigma_{P\&L} \approx \sigma\times vega\times\sqrt{\frac{\pi}{4N}},$$

where N is the number of times the position is hedged in a year.

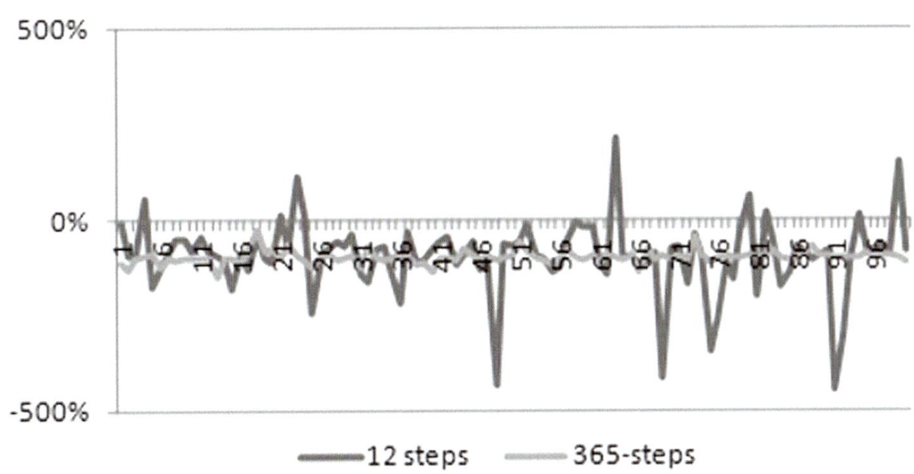

FIGURE 5.37 Delta-hedging P&L

Although it is a case of the higher the rebalancing frequency, the lower the replication cost, such cost is also proportional in the volatility (see Figures 5.38 and 5.39).

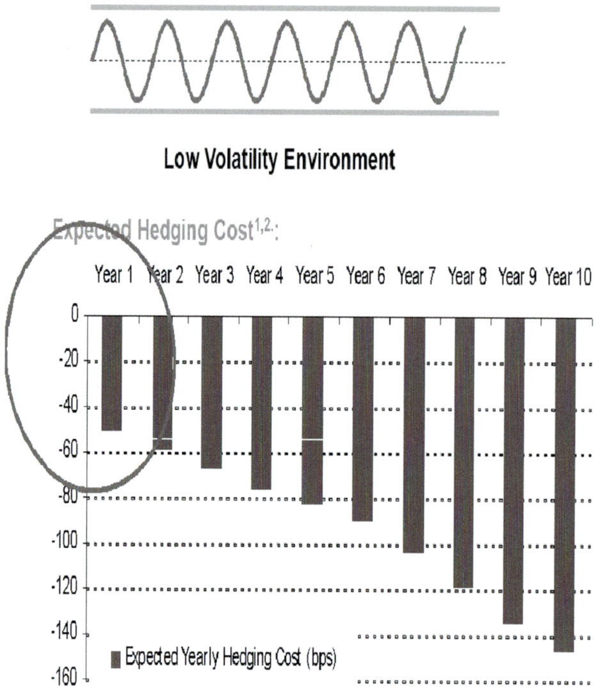

FIGURE 5.38 Expected hedge cost in low-volatility regime

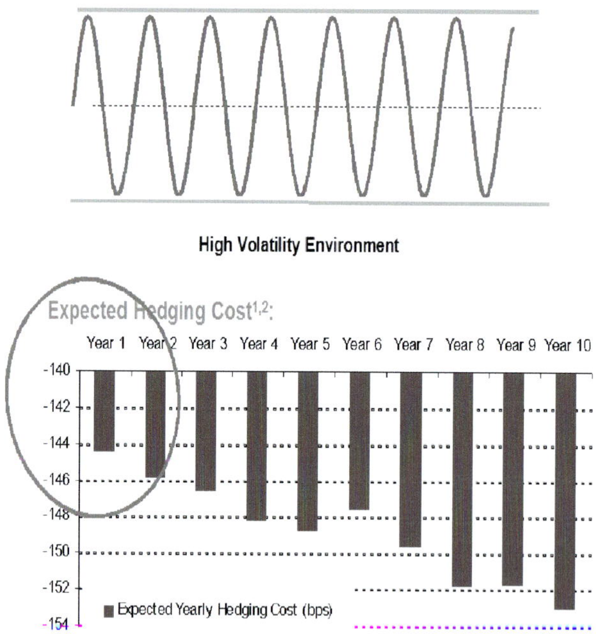

FIGURE 5.39 Expected hedge cost in high-volatility regime

And the impact of such volatility on the delta-hedging cost may increase for different business time units depending on the market environment, and thus flows: intra-daily, daily, weekly, etc. Specifically, when the daily realized volatility increases, large moves towards the end of day may occur. The market often "bounced back" the following day, leading to mean reversion, and therefore costly portfolio rebalancing, if based on intraday or daily time units (as opposed to weekly).

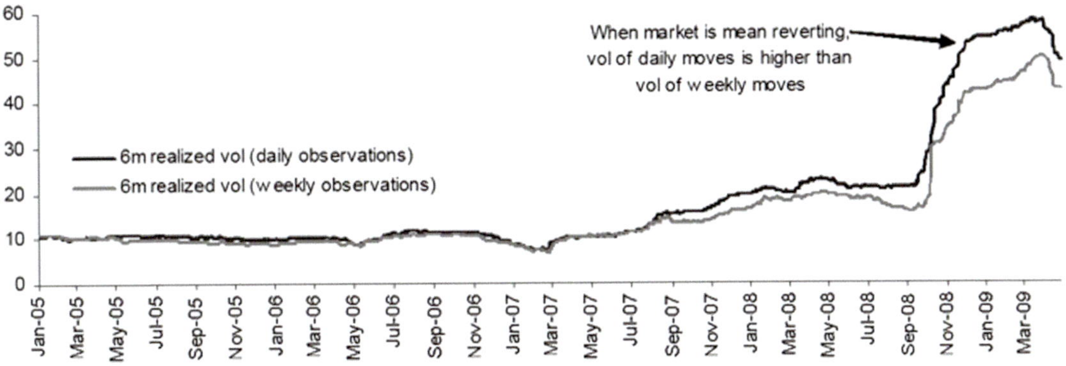

FIGURE 5.40 P&L depending on rebalancing frequency, January 2005–May 2009

- Overall, the variation in the profit (or loss) is caused by the variation due to discrete hedging and the inaccuracy of the delta (as volatility is unknown). This combination of discrete delta hedging and unknown volatility makes it possible for a delta-hedged cheap option to reveal a loss.

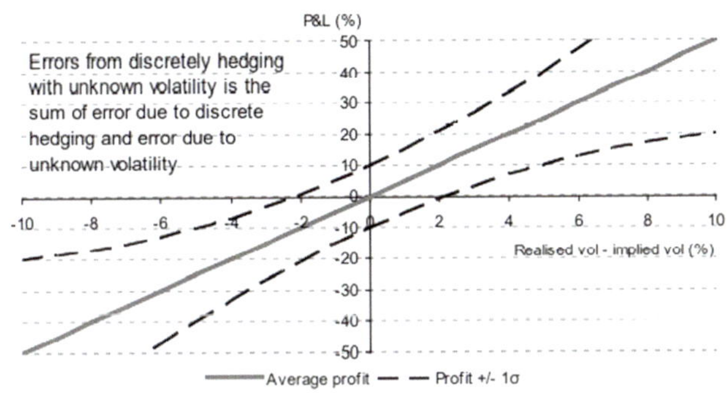

FIGURE 5.41 P&L versus realized minus implied volatility

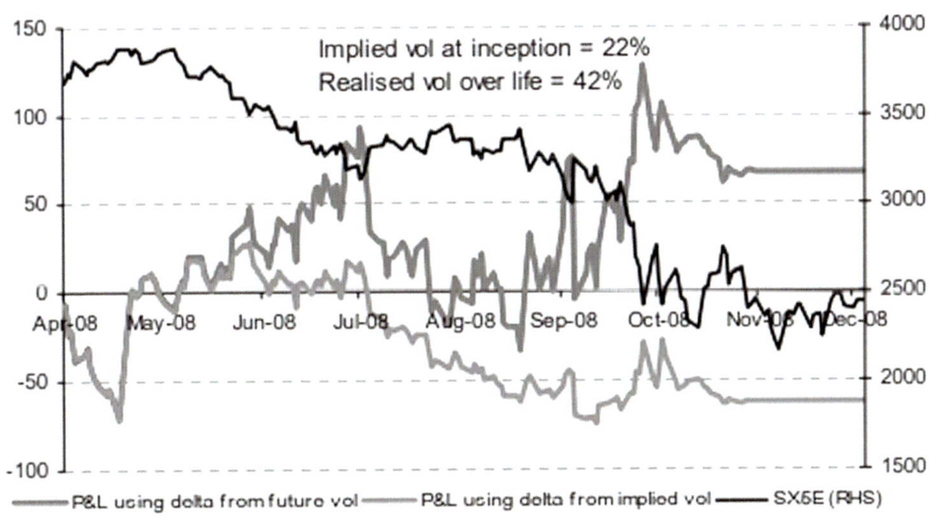

FIGURE 5.42 P&L versus realized minus implied volatility

As a result, the payout of a delta-hedged option cannot be known (unlike in the Black–Scholes model), so there is the possibility of losing money under discrete delta-hedging, as illustrated in this example.

- If you had bought an ATM option in April 2008 with 22% implied volatility, you would have lost money despite realized volatility being almost twice as much (42%).
- The loss was due to delta, which was up to 24% different from that expected by the implied volatility (i.e. when using future realized volatility instead of implied volatility, 60% versus 84%); as the market declined before volatility spiked, the delta using implied was near to 100%, but should have been less. Hence, traders bought too many futures to delta-hedge, and suffered a loss.

✓ **Quick win**: hedging with delta using implied volatility is bad for long volatility strategies. Typically, when volatility rises there is often a decline in the markets. The strikes of the option are therefore likely to be above spot when actual volatility is above implied. This reduces the profits of the delta-hedged position, as the position is actually long delta when it appears to be delta flat.

Alternatively, the fact that the position hedged with the realized volatility over the life of the option is profitable can be thought of as due to the fact that it is properly gamma-hedged, as it has more time value than is being priced into the market. Hence, if a trader buys an option when the implied looks five points too cheap, then the hedge using delta should be calculated from a volatility 5 pts above the current implied volatility. Using the proper volatility means that the profit is approximately the difference between the theoretical value of the option at inception (i.e. using actual realized volatility in pricing) and the price of the option (i.e. using implied volatility in pricing). As a result, use expected, not implied, volatility for delta computations.

See Section 5.2.2.1.2.1 (Risk Appetite Framework for a Sustainable Liability Equity Volatility) regarding the methodology for a sustainable expected volatility.

5.2.1.3.2 Optimizing Delta-Hedging Rebalancing Transaction Costs versus Tracking Error One approach to hedging equity volatility risk is to purchase equity volatility protection via options or varswaps in the capital markets, which may be expensive; an alternative solution to explore is to review/revisit the effectiveness of a full daily equity delta rebalancing strategy, net of transactions costs, to see if this is still optimal and compare this strategy to alternative rules-based rebalancing strategies.

The choice of the rebalancing frequency then becomes a trade-off between hedge effectiveness and transaction costs.

5.2.1.3.2.1 Analysis of Transaction Costs The elements of transactions costs to be explored are as follows.

5.2.1.3.2.1.1 Volatility Adjustments The first adjustment is to adjust the measurement of volatility to reflect a) the nature of stock price processes as different from geometric Brownian motion (which is implicit in Black–Scholes) and b) to introduce the transaction costs of dynamic replication.

5.2.1.3.2.1.2 Statistical Adjustments The implied volatility measurement is a recovered parameter from observed option prices using the Black–Scholes formula, and it includes a combination of pricing adjustments and statistical adjustments.

The statistical adjustments include amounts for a) heteroscedasticity, being the non-constant volatility parameter; b) allowances for specific jumps (whose impact is short-lived); c) the structural relationships between interest rate levels and volatility levels; and d) adjustments for trading days and other conventions.

5.2.1.3.2.1.3 Bid–Ask of Replication In the presence of transaction costs, continuous (infinite) rebalancing leads to the transaction costs becoming infinite, and thus risk-free replication through continuous rebalancing breaks down. As a result, hedging occurs in discrete time, and there is an optimization of residual hedging error and hedging costs to determine the fair cost of hedging.

Thus, we can identify that the optimal position is related to hedge costs, volatility and hedge rebalancing frequency.

An example of such an optimization or formula is that developed by Leland (and expanded further by Wilmott into a generalized formula depending on whether we are short or long gamma). This approach seeks to adjust the volatility parameter, on the basis that the need for trading is proportionate to the volatility of the underlying, and, in the case of our short put position, it leads to an increase in the statistical volatility as follows:

$$\sigma_{tc} = \sigma^* \left\{1 + (2/\pi)^{1/2} * k / \left(\sigma^* \sqrt{dt}\right)\right\}^{1/2},$$

where

- σ = historic/expected volatility of the underlying unit;
- σtc = our volatility parameter adjusted for hedge costs;
- k = the round trip transaction cost; and
- dt is the discrete time step, and 1/√dt is a function of the frequency of trading.

Thus (frequency of trading) * (round trip transaction cost) drives the adjustment.

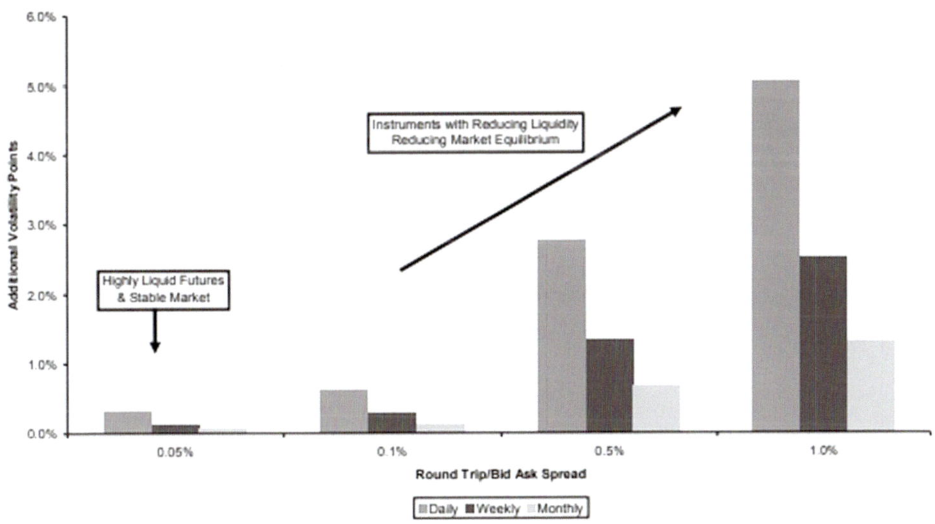

FIGURE 5.43 Impact of liquidity and trade frequency on hedging cost

While markets are stable and liquid, therefore, the bid–ask cost can be relatively benign, as a proportion or amount of volatility; however, as markets dislocate, the costs escalate rapidly and materially. Moreover, how sensitive the adjustment is to the underlying level of volatility is very much a second-order effect.

Thus, the liquidity of the underlying instrument, as manifested in bid–ask prices, is a direct input into the option replication price, through the volatility parameter. This leads to the simple, but important, result that options on liquid instruments are cheaper than options on illiquid instruments, all other elements equal.

Indirect consequences of this are a) in the limit, options on illiquid underlying units become unhedgeable/not replicable due to price considerations and b) variation in market implied volatility (vega risk) countenances changes not just in expected volatility but also in the liquidity of the instruments used to replicate them.

5.2.1.3.2.2 Optimized Delta-Hedging Strategy for Higher Hedging Accuracy and Lower Transaction Costs

5.2.1.3.2.2.1 The Impact of Transaction Costs on Option Price and Volatility In the presence of transaction costs, it is necessary to use optimization to determine a delta-hedging strategy. A trade-off exists between achieving a lower standard deviation of results utilizing more frequent hedging and achieving a higher expected return utilizing less frequent hedging, leading to lower transaction costs.

The usual pricing models are derived on the basis of the assumptions of frictionless markets and continuous trading. However, in a practical environment, we face transaction costs when hedging an option, and trades occur only at discrete time intervals. Since transaction costs can rise significantly during adverse and extreme market conditions, those hedging strategies that work well under normal market conditions may deteriorate in performance during crisis periods.

The introduction of transaction costs could lead not only to a change of return for options but also to a change in the volatility, as illustrated below. Let us consider the option price dynamics (see Leland 1985):

$$dP = P_t * dt + P_z * dz.$$

From Taylor expansion, the log-normal dynamics of the asset S, and the Sharpe ratio, we get

$$\frac{\dfrac{\partial P}{\partial t} + \mu S \dfrac{\partial P}{\partial S} + \dfrac{1}{2}\sigma^2 S^2 \dfrac{\partial^2 P}{\partial S^2} - rP}{\sigma S \dfrac{\partial P}{\partial S}} = \frac{\mu - r}{\sigma}.$$

Let us introduce transaction costs L:

$$dP = dP_0 - Ldt + dP_l,$$

following which the previous equation becomes

$$\frac{\dfrac{\partial P}{\partial t} + \mu S \dfrac{\partial P}{\partial S} + \dfrac{1}{2}\sigma^2 S^2 \dfrac{\partial^2 P}{\partial S^2} - L - rP}{\sqrt{\left(\left(\sigma S \dfrac{\partial P}{\partial S}\right)^2 + \left(\dfrac{1}{2} k \sigma S^2 \dfrac{\partial^2 P}{\partial S^2}\right)^2 \left(1 - \dfrac{2}{\pi}\right)\right)}} = \frac{\mu - r}{\sigma}$$

Close to expiry, a proxy for such a price is then

$$P = SN(d_1) - Xe^{-(r-q)t} N(d_2), \quad q = -(\mu - r) \quad \sigma = \sqrt{\left(\sigma^2 - (\mu - r)\left(1 - \frac{2}{\pi}\right)^{\frac{1}{2}} k - k\sigma\sqrt{\frac{2}{\pi \delta t}}\right)}$$

This equation confirms that the value of the option price changed not only because transactions costs led to a changed return but because transactions costs could also influence volatility.

5.2.1.3.2.2.2 The "How Often?", "How Far?" and "How Much?" Questions Violating the Black–Scholes assumption of no transaction costs and applying proportional transaction costs on the Black–Scholes model would imply the need for hedging in discrete time to avoid infinitely large trading costs.

Rebalancing the portfolio at discrete time intervals will reduce the transaction costs in total, and the risk will increase, leading to a hedging error as the hedge will no longer be perfect.

In this respect, a rebalancing delta-hedging strategy precisely formalizes guidelines about how often the delta should be rebalanced, how far the delta can deviate from its target and how much delta rebalancing should restore its target.

We seek here the best rebalancing delta strategy regarding both hedging accuracy (measured by the hedging error mean and variance) and transaction costs.

The different delta rebalancing strategies are ranked based on a mean-variance framework.

The comparison is illustrated here on a short put position usually embedded within life insurance guarantees. This study explores on a put how transaction costs affect the optimality of hedging when using the Black–Scholes option pricing model.

Within this context, the P&L of such a delta-hedged hedged short put position becomes

$$PnL_{\text{moy}} = -(\mu - r)^2 \sum_{i=0}^{N-1} \int_{t_i}^{t_{i+1}} \mathbb{E}\left[S_t e^{r(T-t)} \int_{t_i}^{t} S_v \Gamma_v dv \right] dt,$$

which illustrates that:

- on average, the discreteness of hedging has a negative impact on the P&L of a hedger who is short of some liabilities;
- this negative P&L impact is proportional to the time step of the hedging strategy;
- this negative P&L impact is proportional to the square of the equity risk premium; and
- the higher the gamma (i.e. the more the option ATM), the higher the negative impact.

A number of models developed to improve the hedging results, when accounting for transaction costs, are analysed and compared. To numerically evaluate these strategies, extensive Monte-Carlo simulations are generated and the results are computed in risk appetite frameworks. Back-testing analyses are also performed in order to confirm/inform these results.

Remark: the mean-variance rule is subject to criticisms that its criterion is appropriate only under restrictive assumptions. These assumptions include that the investor has a quadratic utility function and that asset returns are normally distributed. Investors with quadratic utility functions display increasing risk aversion behaviour. Even so, Levy and Markowitz (1979) and Kroll, Levy and Markowitz (1984) show, for various utility functions and empirical return distributions, that an expected utility maximizer will be indifferent in making choices between using direct utility maximization and the mean-variance rule.

5.2.1.3.2.2.3 The Different Delta Rebalancing Strategy Alternatives These can be divided into "time-based" and "move-based" strategies.

- In a time-based strategy, the rebalancing occurs at regular time intervals without looking separately at the optimality of each rehedging. It is rather comparable with the discretization approach of the Black–Scholes model. Examples: discrete Black–Scholes; Leland approach.
- In a move-based strategy, the portfolio rebalance is dependent on the movements of the underlying asset. A bandwidth around the Black–Scholes delta is then computed, forming a so-called "no-transaction" region. Rehedging occurs only when the hedge ratio lies outside the boundaries of this region. Examples: delta tolerance; asset tolerance; fixed-bandwidth delta; variable-bandwidth delta; asymptotic utility-maximization-based strategies.

Let us explore further the different examples mentioned above, starting with the two time-based delta rebalancing strategies.

- Black–Scholes hedging at fixed regular intervals: this is the simplest hedging strategy, as the hedging is repeated in the same manner for all subsequent time intervals. In fact, when n is large, the variance of hedging errors is small and transaction costs are large because of the frequent hedging.

  The choice of n is related to the hedger's risk aversion.

  - On the one hand, a less risk-averse hedger will choose a small n because he is willing to accept larger hedging errors and save on transaction costs.
  - On the other hand, when a hedger is risk-averse he will choose a large n so that he can rebalance his hedge position frequently and minimize hedging errors at maturity. However, this will involve large transaction costs.

- Leland's hedging: a modified Black–Scholes hedging strategy to ensure that no infinite transaction costs are incurred in hedging an option, no matter how small the rehedging interval is.

  Instead of perfect replication at every time interval, the hedger focuses on matching the option payoff with a modified hedging volatility that depends on the proportional transaction costs and the constant hedging frequency. Thus, supposedly proportional transaction costs are accumulated in the Black–Scholes volatility:

$$\sigma_m^2 = \sigma^2 \left(1 + \frac{\sqrt{\frac{2}{\pi}}k}{\sigma\sqrt{\delta t}} \right).$$

  In incorporating the transaction costs of trading in the price movements of the underlying asset, the volatility will increase. As a result, this modified volatility improves the risk–return trade-offs of the hedging portfolio, by allowing the hedging errors to be negatively correlated with the transaction costs.

  For example, when a hedger is in a short gamma position, the option is hedged with an increased hedging volatility. This increase in volatility will decrease the absolute value of the gamma in a high-gamma region, and therefore reduce transaction costs because of the reduced rehedging activity.

  The strategy is optimized with respect to transaction costs and the length of the time interval, without considering if it is optimal to hedge at every time interval.

  As with the Black–Scholes discrete hedge, the choice of transaction frequency δt reflects the hedger's risk aversion level. A very risk-averse hedger will choose to rebalance as frequently as possible and select a small δt, while a relatively risk-tolerant hedger will select a large δt.

Now let us switch our focus to the four alternative, move-based delta rebalancing strategies.

- Asset tolerance strategy: the hedger monitors the market continuously and rebalances by hedging the portfolio back to the perfect hedging position when the percentage change in the price of the underlying asset exceeds a predetermined amount:

$$\omega_1 = t, \quad \omega_{i+1} = \inf\left\{ \omega_i < \omega < T : \left| \frac{S_\omega - S_{\omega_i}}{S_{\omega_i}} \right| > h \right\}, \quad i = 1, 2, \ldots,$$

where h is a given constant percentage.

The choice of h depends on the hedger's risk aversion level. A risk-averse hedger will select a small h in order to obtain small hedging errors at maturity, but at the cost of large transaction costs. On the other hand, a more risk-tolerant hedger is willing to face larger hedging errors, but enjoys low transaction costs.

- Delta tolerance strategy: the hedger will rebalance his portfolio to the Black–Scholes delta when the hedging ratio moves outside the tolerance level H. The series of stopping time, ω, is recursively given by

$$\omega_1 = t, \quad \omega_{i+1} = \inf\left\{\omega_i < \omega < T : \left|\Delta - \frac{\partial C}{\partial S}\right| > H\right\}, \quad i = 1,2,\ldots$$

H is related to the hedger's risk aversion level: a risk-averse hedger will choose a small H while a more risk-tolerant hedger will choose a large H.

Such a delta-hedging rebalancing threshold can be expressed as an absolute net delta reference, as for Germany, the Mediterranean region and Switzerland, or as a relative one, as for France and Belgium.

- Hedging to a fixed bandwidth around delta: the boundary of the no-transaction region for this simpler hedging strategy is as follows:

$$\Delta = \frac{\partial C}{\partial S} \pm H.$$

As with the previous case, H also represents the hedger's risk aversion level. For this reason, a risk-averse individual will choose a small no-transaction region while a relatively risk-tolerant agent will choose a relatively large no-transaction region.

This strategy is closely related to the delta tolerance hedging strategy. The major difference between this strategy and the delta tolerance strategy is that the hedger will rebalance the hedge ratio to the nearest boundary of the hedging width. On the other hand, the delta tolerance strategy will have a hedge ratio that is equal to the perfect Black–Scholes hedge position.

These four move-based strategies generate fixed boundaries for a no-transaction region around the Black–Scholes delta. While they evidently reduce transaction costs, as they are rehedged depending on the movement of the underlying asset, they are nevertheless not exceptionally dynamic.

Davis and Norman (1990) show it to be more optimal to hedge only to the nearest boundary in order to save in transaction costs. This would thus logically imply that it is more optimal to hedge using the fixed-bandwidth strategy than the delta tolerance, and supposedly also the asset tolerance strategy. Under certain conditions, these three move-based strategies are superior to simple time-based strategies – that is, when the underlying asset is volatile, transaction costs are small and the hedger is less risk-averse.

The drawback with these three strategies is that they cannot account for certain risks that evolve with the value of the underlying asset.

- The asset tolerance strategy simply uses the price of the underlying asset to rehedge without using any sort of risk measure.

- In contrast, the delta tolerance and the fixed-bandwidth delta strategy take into consideration the delta measure of the hedged option.

- Other sensitivity measures (Greeks), such as gamma, are not a part of the strategies. In a high-gamma area a small change in the underlying asset will have a large effect on the delta. Thus fixed bandwidths cannot adjust to this and reduce the rebalancing.
- Hedging to a variable bandwidth around delta: the utility-based hedging. Such approaches derive an option hedging strategy by explicitly taking into consideration the hedger's risk preference specified by a utility function. A no-transaction region is then formed where the boundaries define where the investor is indifferent between the utility of rehedging and the utility of not making any changes to the portfolio.

While all utility-based strategies seem to outperform other strategies, they have restricted usability, as they do not provide a closed-form solution, and so are computationally time-consuming and need to be monitored and adjusted continuously. There the asymptotic approach of Whalley and Wilmott's variable-bandwidth strategy is faster and simpler to compute. The optimal hedging strategy derived from this approach is to rehedge the portfolio to the nearest boundary of the no-transaction region (or hedging bandwidth) when the hedge ratio moves out of this region. Otherwise, do nothing.

As the no-transaction region is not fixed, as in the three move-based strategies analysed above, it better adjusts for changes affecting the strategy in reality: the no-transaction region is adjusted by both delta and gamma. A high gamma would generally imply more frequent rehedging, the no-transaction region is positively related to gamma, and thus, by increasing the bandwidth of the no-transaction region when gamma is high, the transaction costs can be reduced. In these strategies, the boundaries are simply computed depending on the risk aversion with respect to the movements of the underlying asset.

Whalley and Wilmott (1999) propose another way to alleviate the computationally time-consuming problem in obtaining the optimal hedging strategy. They provide an asymptotic solution to the problem posed by Hodges and Neuberger (1989) by assuming that transaction costs are small. They show that the boundary of no-transaction region is of the following form:

$$\Delta = \frac{\partial C}{\partial S} \pm \left(\frac{3}{2} \frac{e^{-r(T-t)} kS\Gamma^2}{\gamma} \right)^{\frac{1}{3}}.$$

Γ is the Black–Scholes gamma, γ is the hedger's absolute risk aversion and k is the proportional transaction costs. As k approaches zero, the optimal hedge will become the Black–Scholes hedge. The above formula can be further simplified as follows:

$$\Delta = \frac{\partial C}{\partial S} \pm h\left(e^{-r(T-t)} S\Gamma^2\right)^{\frac{1}{3}},$$

where h is a given constant tolerance level that reflects the hedger's risk aversion.

Instead of hedging to a fixed bandwidth around delta, the hedger will rebalance the hedge ratio to the nearest boundary of the variable hedging width once the delta movement is beyond the non-hedging bandwidth. The bandwidth is determined by option gamma, spot underlying asset price, risk-free rate and time to maturity. The bandwidth thus changes over time.

As we expect there will be more rehedging activities in regions with high gamma, the optimal control region is adjusted to be positively related to the gamma value so as to reduce the transaction costs that occur in frequent rehedging.

Another feature of the optimal strategy is that the no-transaction region becomes larger as the option gets close to maturity. This implies that a hedger tends to trade less towards the end of the option position and that most hedging transactions have been done in the early stages.

The role of the spot underlying asset price in determining the no-transaction region is linked to the fact that there is no one-to-one correspondence between the change in the underlying price and the change in the delta. The inclusion of the underlying asset price would allow us to adjust the delta of the hedge only for the amount that is required to keep within the no-transaction region and to reduce some transaction costs.

5.2.1.3.2.2.4 From Theory to Practice: Relative Performance of Delta Rebalancing Strategies through Simulations and Back-Testing The transaction cost literature concludes that, in the presence of transaction costs, the hedging strategy chosen by a hedger depends on his risk aversion. This is because a hedger is facing a trade-off between hedging accuracy and transaction cost expenditure. As a result, a simple comparison of hedging strategies by using hedging errors as a benchmark for performance measure is inappropriate for assessing the effectiveness of option hedging with transaction costs.

Thus we generate a range of risk aversion parameters for each hedging strategy and test how hedging errors and transaction costs vary according to risk aversion. We provide a simulation of the performance of different hedging strategies to benchmark the hedging errors of different strategies.

Our empirical results show that all hedging strategies perform badly during crisis periods. In our context, hedging errors represent the profit and loss of a hedging portfolio, and therefore they are chosen to be the performance measure in this study. The empirical hedging errors presented in the result section are the hedging errors net of transaction costs.

The resulting cash position after transaction costs for the hedge is

$$B_t = C_t - \Delta_t S_t - k \times \left|\Delta_t - \Delta_{t-1}\right| \times S_t - const.$$

The hedging errors net of transaction costs, the difference between the current value of the hedging portfolio built at time t and the market value of the put option are as follows:

$$\text{HE}(t + \delta t) = \Delta_t S\left(t + \delta_t\right) + B_t e^{r\delta_t} - C\left(t + \delta_t, \tau - \delta_t\right).$$

The hedging error is chosen to quantify hedging performance because it represents the profits and losses that a hedger faces. We evaluate hedging performance under two criteria: mean hedging error and root mean squared error, which is the square root of mean square hedging error.

Since positive and negative hedging errors tend to offset each other, mean hedging error measure does not appear to be a good performance measure. However, it does provide indicative information on whether a particular hedging strategy systematically over- or under-hedges the target position. On the other hand, RMSE is widely used in the literature, and it puts more weight on dollar-value error.

We focus on hedging short positions because these positions involve the possibility of negative cash flows at maturity. Consider a situation faced by a put option writer. Suppose that the put option matures in six months and is written at the money. The writer intends to hold the position until the option has expired, or the position is closed. The writer can hedge his position by using the underlying asset during the holding period.

For the class of time-based strategies, we learn that the Black–Scholes strategy performs better for all periods except during crisis periods. On the other hand, unlike the theoretical prediction of the existence of a negative relationship between hedging errors and risk aversion, the move-based strategies tend to have a flat relationship between hedging errors and risk aversion.

The empirical analysis also demonstrates that all hedging strategies consistently under-hedge the target put option position.

The Monte Carlo simulations framework is constituted as follows:

- 100,000 simulations using constant volatility; and
- various market conditions taking into account
  - different equity risk premiums between bond and equity returns;
  - low- and high-volatility regimes;
  - high and low transaction costs; and
  - several delta-hedging rebalancing thresholds.

(i) Results of Time-Based Delta Rebalancing Strategies

We observe that the mean hedging errors for the time-based Black–Scholes and Leland hedging strategies increase when the level of risk aversion decreases. This is consistent with the model prediction that the more risk-tolerant agent will accept a greater amount of hedging errors.

Further, the hedging errors of both the Black–Scholes and Leland hedging strategies are initially negative, and they become less negative when the level of risk aversion decreases, as the portfolio is rebalanced less frequently.

The Leland hedging strategy systematically produces smaller hedging errors than the Black–Scholes and the gap of the hedging errors is larger when the risk aversion level increases (meaning fewer transactions). In contrast, the difference of the average cumulative transaction costs between the Black–Scholes and the Leland strategies is smaller when the agent becomes less risk-averse.

This observation is partly due to the fact that the Leland strategy requires smaller amounts of transaction costs than the Black–Scholes strategy when the hedging portfolio is rebalanced at a short (weekly) time interval.

As expected, as unit transaction costs become insignificant, the two strategies present a similar performance.

(ii) Results of Move-Based Delta Rebalancing Strategies

The volatility level has no influence on the ranking of strategies.

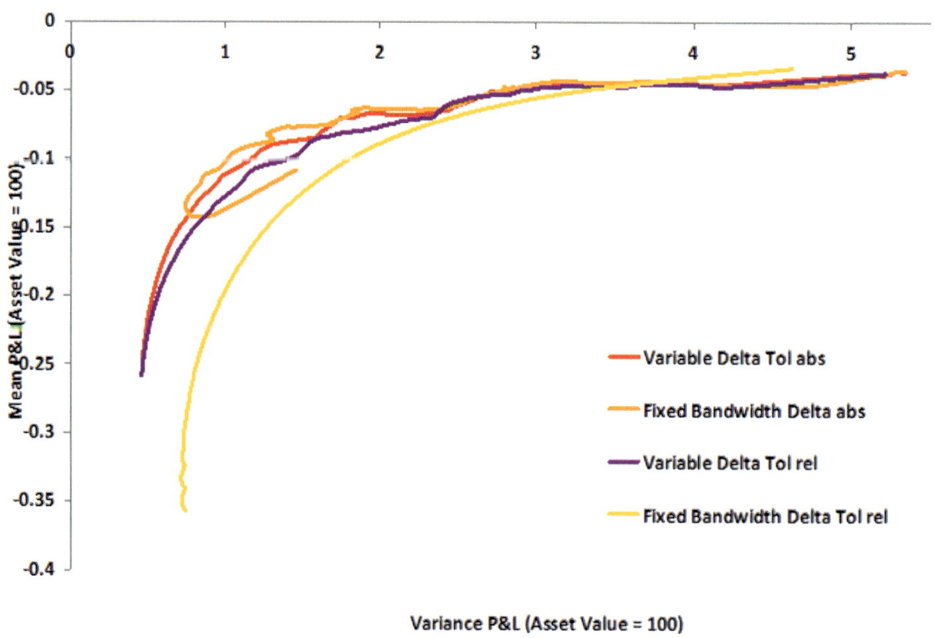

FIGURE 5.44 P&L variance versus mean P&L

If now we switch to back-testing for 1987 to 2010, we get the results seen in Figure 5.45.

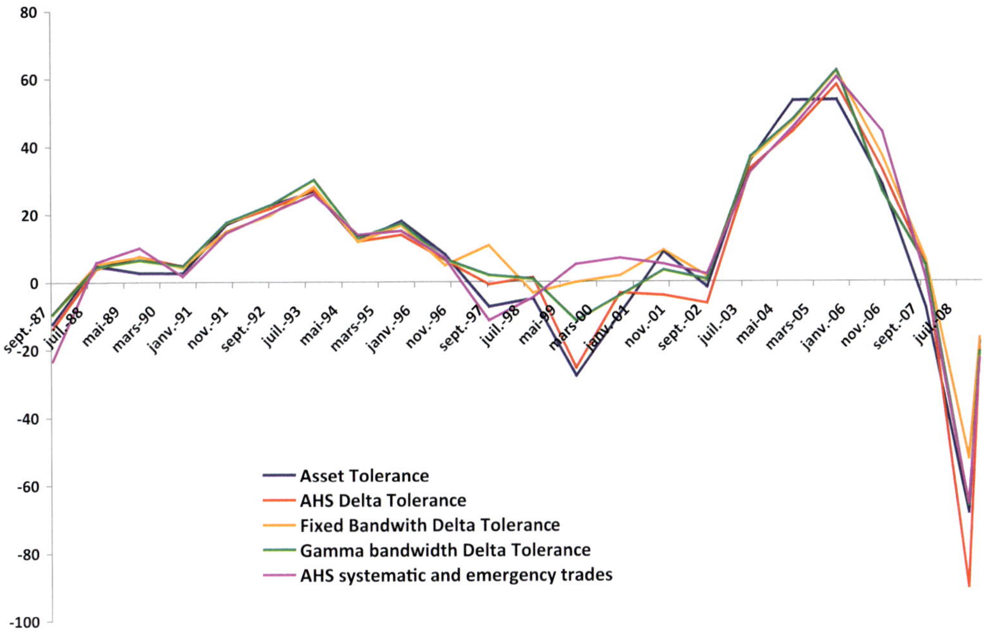

FIGURE 5.45 P&L back-testing, 1987–2010

Overall, from a mean-variance perspective, the optimized delta-hedging rebalancing tends to outperform the "periodic rebalancing" significantly, notably by making use of adequate hedge instruments.

- As long as unit transaction costs are small (e.g. bond futures), systematic weekly delta-hedging rebalancing coupled with daily emergency thresholds (based on variable-bandwidth delta metrics), aka variable-bandwidth delta tolerance, proves to be the most efficient in most market conditions.
- As unit transaction costs rise (such as for swaps in adverse market conditions), some alternatives may prove more efficient, such as the so-called gamma bandwidth delta tolerance or asset tolerance rebalancing. In extreme market conditions, a fixed-bandwidth delta rebalancing may also prove more efficient.

✓ **Quick win**: the savings in transaction costs are about 20 to 30%.

Remark: these results are in line with Sepp (2015).

Remark: our study reveals that the delta-hedging forms by all hedging strategies tend to under-hedge the target option position. If non-constant volatility and the corresponding vega hedge are added to the framework, then the delta-vega-hedging tends to over-hedge the target position: the mean and standard deviation of the net hedging error in the delta-vega-neutral hedge are larger than those in the delta-neutral hedge (see Chow 2017). In fact, given the randomness of transaction costs, hedging an extra risk dimension may have increased net hedging error risk through the adjustment of the delta position due to the use of the additional hedge instrumental option.

5.2.2 Sustainable Volatility Hedging

Life insurance hedging started with delta-hedging programmes as illustrated above, by, for example, shorting futures to neutralize the effect of the long position in the segregated fund assets.

However, as highlighted above, restricting the hedging programme to delta hedging only (which used to be the case for most life insurers until the early 2010s) provides the following limitations.

- Dynamic hedging programmes closely monitor the current position and thus require constant adjustments, as markets fluctuate and policyholder behaviour develops.
- Delta hedging avoids paying up-front option premium and may be less expensive than using options if realized volatility remains below implied option volatility; delta hedging is seen as a way to save the volatility risk premium. However, it may be difficult to place at low costs if volatility increases pricing and remains at a constant level, as regularly experienced during the last decade.
- The cost of hedging is unknown and can drastically jump during a market crash with a sell-off (during which delta hedging may create increasing downside pressure on prices and generate liquidity holes). Moreover, delta hedging implies frequent rebalancing of the hedging assets, which can be quite expensive due to the transaction costs.
- There are more onerous requirements for getting capital relief (the conditional tail expectation calculation under C-3 Phase II), because rolling strategies find it difficult to model the projected impact (e.g. complex nested stochastic models), while the standard scenario does not give credit for future hedging.
- There is higher earnings volatility if reserves uses implied volatility, given its significant changes.
- There is less up-front investment in hedges but more up-front investment in infrastructure.
- It is operationally intensive to maintain proper positions, given any deviations from original assumptions on markets and policyholder behaviour.
- Model risk: given the very long maturities of the products under consideration, hedging strategies must take into account a high degree of uncertainty about the true dynamics of the underlying asset.
- It is more complex to implement than static hedging.

As a result, hedging volatility risk (notably through options or varswaps) makes sense given the embedded optionalities when the liability exhibits some convexity.

If decrements could be ignored, then the short put (here a one-year 5% in the money, so 105) in the annuity contract could be hedged exactly by writing (going long) a corresponding one-year put spread (90 – 100) option (all hedges are calculated based on a 20% implied volatility versus a 33% realized volatility; transaction costs are based on a bid–ask spread of 25bps).

	Standard Deviation of P&L		Transaction costs	
Number of Rebalancing	Delta hedge only	Combined Static + Delta Hedge	Delta hedge only	Combined Static + Delta Hedge
10	50,6%	6,3%	1,5%	0,1%
20	41,5%	6,7%	2,2%	0,2%
50	40,9%	5,5%	3,5%	0,4%
100	42,6%	4,9%	5,0%	0,6%
200	41,6%	4,8%	7,1%	0,9%
300	39,9%	3,8%	8,5%	1,1%
400	40,1%	4,1%	9,9%	1,2%
500	38,4%	3,9%	11,2%	1,4%
600	35,9%	3,4%	12,0%	1,5%
700	41,0%	3,5%	13,3%	1,6%
800	39,2%	3,7%	14,4%	1,7%
900	40,0%	3,8%	15,0%	1,9%

FIGURE 5.46 Standard deviation and transaction costs versus number of rebalancings

The lowest level for the standard deviation of results that can be achieved by the mixed strategy is about one-tenth of what can be achieved by the pure strategy, roughly 4% compared to roughly 40%. The inability to reduce the standard deviation of results below a lower bound is due to both the uncertainty of volatility and the use of incorrect volatility inputs in forming hedge ratios.

Regarding transaction costs, they are also divided by a large factor (around eight) thanks to the introduction of static hedging. It is important to recognize the distinction between the two aspects of dynamic hedging costs: transaction costs that arise from bid–ask spreads, and gamma hedging costs from buying high and selling low that would be present even if all trades were at midmarket. Transaction costs are a direct function of the frequency of rehedging, and a trade-off occurs between the higher transaction costs and the lower variability of P&L with less frequent rehedging.

By contrast, there is no a priori reason to believe that the level of gamma hedging costs will vary in any systematic way with the frequency of rehedging. A good way to see this latter point is to look at how P&L is related to the gap between actual hedges held and the theoretical hedge called for by the Black–Scholes formula. The expected value of this P&L under the standard Black–Scholes assumption is given by this formula (as the theoretical delta will lead to an expected return of zero, so any holdings above or below the theoretical delta can be regarded as proprietary positions that will lead to the same expected return as an outright position in the underlying forward):

$$\sum_{\substack{small\\ time\\ periods}} \left(\Delta_{actually\ held} - \Delta_{theoretical}\right) \times \text{expected price change of underlying forward.}$$

The consequence of this formula for the relationship between gamma-hedging costs and the frequency of rehedging is that, as rehedging becomes less frequent, it widens the gap between Δactually held and Δtheoretical. However, unless a correlation between the sign of this gap and the sign of the expected price change in the underlying forward is expected for some reason, the expected value of the incremental P&L should be zero. (Although this formula is strictly correct only in the case when the Black–Scholes assumptions hold, Monte Carlo simulations with stochastic volatility show similar results.)

Extending the programme to include other Greeks increases the cost of the programme, including trading costs, but this must be balanced against the increased effectiveness, which permits the holding of less capital, both under internal models and under the RBC C-3 Phase II risk-based capital requirements.

5.2.2.1 Sustainable Convexity Hedging

Some life insurers have remained partially hedged by shorting futures contracts, then tracking the market daily and adjusting periodically to rebalance the hedge, with the belief that extreme downside risks would be remote, temporary and thus manageable.

However, in addition to the level of the interest rates and the underlying market, the embedded guarantees and optionality values within life insurance liabilities will vary based on other underlying forces, especially volatility.

In fact, dynamic hedging should work well for small market movements only. The expression "It works well until it doesn't" has some truth to it, as illiquidity and price jumps may occur during large market swings. In an increasingly mark-to-market regime, insurers will benefit from products that act to offset the volatility in their liabilities, all the more if they are short both gamma and vega and thus suffer from rising volatility. Static hedging using options may therefore be used to hedge the volatility tail risk (vega) and curvature (gamma and rho convexity) of market movements more effectively.

With this in mind, the most popular approaches have been to delta hedge by shorting futures and to purchase options from institutional dealers in order to hedge the curvature arising from larger and longer-dated market moves.

In fact, insurers have been using options for hedging annuity liability convexity risks since the mid-1990s, starting with the guaranteed annuity and equity indexed annuity markets, and then the variable annuity market since 2004, growing notably with the diminished availability of reinsurance alternatives and increased scrutiny from rating agencies regarding unhedged insurance risks.

Beyond the use of options, managing volatility risk also involves setting some sustainable liability volatility assumption, in line with the risk appetite framework, since volatility P&L is at risk whenever the hedge volatility rises above the liability volatility.

Unfortunately, life insurance contract maturities may be estimated at 30 to 40 years, with an average life of approximately 20 years, while OTC options beyond ten years are generally highly illiquid. As a result, shorter-dated OTC options and variance swaps have become popular, as they enjoy high liquidity in the major US indices, where put options are usually purchased with varying maturities and strikes to reflect the term structure of liability exposure based on actuarial analysis.

To reduce the cost of purchasing puts, some insurers have defined their exposures more precisely and have chosen to purchase put spreads or collars or other, cheaper customized OTC options. Even so, volatility increases as markets sell off, due to higher perceived risk, which is often answered with more put option buying. The increased demand for puts usually drives prices and implied volatility rates higher, which is the source of increasingly damaging negative feedback effects.

5.2.2.1.1 Interest Rate Convexity ("Rho Convexity") Hedging

5.2.2.1.1.1 Key Interest Rate Duration and Convexity Matching for Non-Parallel Shifts of the Yield Curve In the first place, it is necessary to find the vector of the durations and convexities of the liabilities while having a series of cash flows Fi, where Fi is the liability flow corresponding to the date ti. Letting dFi be the duration relative to the flow Fi, we construct using the yield curve the vector of duration (dF1, dF2,……, dFP) and (CF1, CF2,……,CFP), which represents the duration vector of the liabilities.

Our objective is to build an asset portfolio composed of N interest rate products with a duration vector as close as possible to (dF1, dF2, ……, dFP) and (CF1, CF2, ……, CFP).

The objective is to calculate, for each title, its weight wl, where wl is the optimal weight of the title l for l from 1 to N. This will be done by numerical optimization by the method of least squares aimed at approaching as close as possible the two duration vectors (*and* the convexity vectors, if combined):

$$\delta(\mathrm{w}) = \sqrt{\sum_{i=1}^{P}\left(d_i^w - d_i^F\right)^2} = \sqrt{\sum_{l-1}^{P}\left(\sum_{l-1}^{N} w_l d_i^l - d_i^F\right)^2} \qquad w^\star = \arg\min \delta(\omega).$$

5.2.2.1.1.2 Application to the Investment Margins Using the investment margins schedule over several set of scenarios shows:

- the scenarios that need to be protected;
- the profile of margins over time; and
- the right period of time for hedge derivatives to reduce the drop of the investment margins.

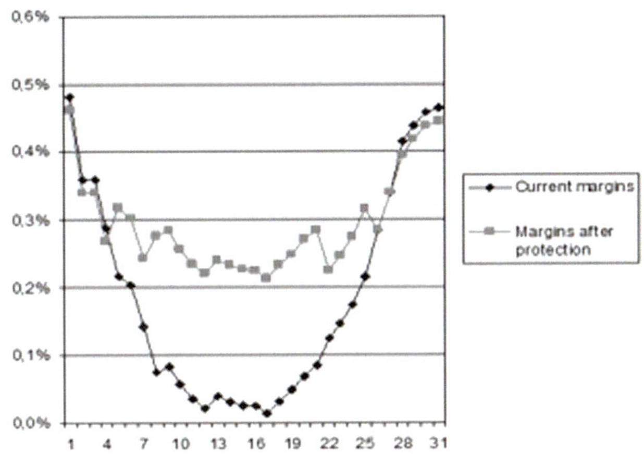

FIGURE 5.47 Margins

5.2.2.1.1.3 Using options for Interest Rates Convexity hedging

5.2.2.1.1.3.1 Interest Rate Options Pension and insurance companies are exposed to negative convexity within their minimum guaranteed rates annuities and are, therefore, buyers of long-dated volatility as hedges.

In order to mitigate convexity risk, in practice a combination of swaps, caps/floors and swaptions are used. In fact, long-term swaps do hold interest rate convexity sensitivity, which in practice can manage up to a half of the interest rate convexity risks of life insurance liabilities (e.g. minimum guaranteed rate annuities). Bond options are usually not very much traded, due to their short maturity (up to two years).

Interest rate options (caps/floors and swaptions) are derivatives on forward rates given the dependence of minimum guaranteed rate annuities on such underlying, since the annuity payout will be dependent on rates starting at the time of the payout period – i.e. on forward rates viewed from today.

There are two types of options: exchange traded options on the Chicago Mercantile Exchange (CME); and OTC options.

- Listed options on the CME:
  - maturity available in sizes only up to three months out; and
  - limited liquidity for six months out on the screen.
- OTC options on US Treasuries.
  - OTC options on US Treasuries are available on either the bond futures or specific underlying bond. As with listed options, liquidity for six months out is limited.
  - Products of a variance swap type on interest rates are illiquid and seldom traded.
  - The US Treasury option market does not appear as an option for implementing interest rate gamma/volatility hedging.

Liquid swaption choices are available at long and short maturities, and wide bid–ask spreads are mostly a function of moneyness (see Figure 5.48).

b/a spread in bps vol		Strike: ATMF, Plus:				
Mat.	Tenor	-100	-50	0	50	100
6 mo	10 yr	2	1.5	1	1.5	2
1 yr	10 yr	1.75	1.375	0.75	1.375	1.75
2 yr	10 yr	1	0.75	0.5	0.75	1
5 yr	10 yr	0.75	0.625	0.5	0.625	0.75
10 yr	10 yr	0.75	0.625	0.5	0.625	0.75
6 mo	20 yr	2	1.5	1	1.5	2
1 yr	20 yr	1.75	1.375	0.75	1.375	1.75
2 yr	20 yr	1	0.75	0.5	0.75	1
5 yr	20 yr	0.75	0.625	0.5	0.625	0.75
10 yr	20 yr	0.75	0.625	0.5	0.625	0.75
6 mo	30 yr	2	1.5	1	1.5	2
1 yr	30 yr	1.75	1.375	0.75	1.375	1.75
2 yr	30 yr	1	0.75	0.5	0.75	1
5 yr	30 yr	0.75	0.625	0.5	0.625	0.75
10 yr	30 yr	0.75	0.625	0.5	0.625	0.75

FIGURE 5.48 Bid–ask spread versus tenor and strikes

Note: ATMF = at-the-money forward.

A comparison across key criteria between interest rates hedges is drawn in Figure 5.49.

	Transaction cost	Liquidity	Counterparty risk	Convexity richness	Customization	Bleeding time value	Net income
Futures	Very low	< 2 years	No	None	Low	No	Small
Swaps	low	2 – 20 years	Yes	Medium if long term	High	No	Medium
Caps/floors	Significant	2 x 5; 2 x 10; 5 x 10	Yes	Significant	High	Yes	High
Swaptions	Significant	10x10	Yes	Very significant	High	Yes	High

FIGURE 5.49 Comparison of interest hedges

Using interest rate options enables us to reduce the immediate IFRS underlying earnings exposure to interest rate volatility.

- Either long- or short-dated positions can be used to reduce the overall net interest rate gamma.
- This reduces the future IFRS underlying earnings exposure to interest rate volatility and cross-rate volatility.
- Shorter-dated positions result in greater uncertainty for future gamma hedging costs (roll risk), while longer-dated positions can provide gamma over a longer timeframe, although their effectiveness is path-dependent.
- Exposure to the term structure created by short-dated positions is less complex than that for longer-dated positions.

However, long option positions can create adverse exposure to falling and volatile implied volatility, and so introduces vega P&L and greater instability in net income, as regularly observed for variable annuity carriers during the past decade

Using interest rate options also protects value and significantly reduces capital requirements.

5.2.2.1.1.3.2 Swaption Hedging Strategy Implementation

(i) Two Main Approaches

There are two illustrative approaches to adding rho convexity (the dollar value of a basis point: DV01/bp) through options (e.g. swaptions): the short-term approach spreads maturities across six to 12 months, and the mixed-maturity (recommended) approach using longer-term options, implying different behaviour on the part of their respective average implied volatilities.

The short-term approach spreads short-dated options (across six to 12 months), which have convexity highly concentrated around the strike.

- They are effective only as a gamma hedge in a narrow interval around the strike.
- Concentration increases as expiry approaches.
- In order to minimize this impact, strikes should be spread above and below current forward rate levels.
- The concentration of convexity at strike also increases at lower implied volatility.
- Spreading strikes to higher interest rates should improve fit at increased cost per period.

Short-term portfolio

				Implied volatility:		All following in $ million:		Swaption vega:	Daily standard deviation:		
Maturity	Tenor	Strike range	Net gamma	bp/day	bp/yr	Notional	Premium	per bp/day	per bp/yr	bp/yr volatility	Vega P/L
6m	15y – 30y	ATM +/- 100	3,667	5.8	93	825	7	3	0.2	3.4	0.6
9m	15y – 30y	ATM +/- 100	3,667	6.0	95	796	13	4	0.3	2.9	0.7
1y	15y – 30y	ATM +/- 100	3,667	6.1	97	829	18	5	0.3	2.5	0.9
Total/*average*			11,000	6.0	95	2,450	38	12	0.8	2.8	2.1

FIGURE 5.50 Short-term portfolio

The mixed-maturity (recommended) approach seeks a semi-static fit to liability exposure with one- to ten-year maturities (see Figure 5.51).

Liability-Fitting Portfolio

				Implied Volatility:		All Following in $ million:		Swaption Vega:		Daily Std Deviation:	
Maturity	Tenor	Strike Range	Net Gamma	bp/day	bp/yr	Notional	Premium	per bp/day	per bp/yr	bp/yr Vol	Vega P/L
1y	15y-30y	ATM +/- 100	3,300	6.1	97	746	17	5	0.3	2.5	0.8
2y	15y-30y	ATM +/- 100	2,681	6.3	100	755	32	8	0.5	2.3	1.2
3y	15y-30y	ATM +/- 100	2,269	6.2	99	745	40	9	0.6	2.1	1.2
5y	10y-20y	ATM	2,200	6.2	99	1,169	93	15	0.9	1.6	1.5
10y	15y-20y	AIM	550	5.2	02	361	31	6	0.4	1.7	0.6
TOTAL/*Average*			11,000	6.2	98	3,776	213	43	2.7	2.0	5.3

FIGURE 5.51 Mixed-maturity portfolio

- The swaptions selected to match convexity are to some extent on a key-rate-specific basis – e.g. 63% of net convexity added from short-dated one-year options.
- The range of maturities is selected to limit the rebalancing required, and conscious of estimated liability vega – i.e. involves larger vega (given the longer-term options), but on a less volatile part of the volatility surface (compare both the vega and the standard deviation of the vega P&L).

- The portfolio may require adjustments for market conditions, and the .spreading of maturities and strikes
- OTM options are better suited to improve fit than to reduce costs:
    - lower premium per $ convexity for out of the money than at the money; and
    - the premium in excess of the assumed volatility is a different story: it may be higher if the implied volatility is higher.
- Higher volatility results in higher theta, and a higher hedge cost.

Swaption Issued ATM w/ Net Convexity 1,000

Swaption Premium

	Up 100bps	At Issue	Down 100bps
6Mx20Y	17,931,975	5,857,645	693,195
5Mx20Y	17,769,208	5,364,329	480,799
4Mx20Y	17,625,456	4,818,717	290,698
3Mx20Y	16,449,862	3,537,963	94,222
2Mx20Y	16,345,212	2,796,093	21,076

Premium at 77 bp

	Up 100bps	At Issue	Down 100bps
6Mx20Y	17,511,947	4,206,445	137,453
5Mx20Y	17,469,454	3,857,152	78,613
4Mx20Y	17,439,339	3,470,865	35,808
3Mx20Y	16,316,941	2,360,126	5,457
2Mx20Y	16,304,828	1,838,454	432

Premium above 77 bp

	Up 100bps	At Issue	Down 100bps
6Mx20Y	420,028	1,651,199	555,742
5Mx20Y	299,754	1,507,177	402,186
4Mx20Y	186,117	1,347,853	254,890
3Mx20Y	132,921	1,177,837	88,765
2Mx20Y	40,384	957,638	20,644

Swaption Issued 100 bp OTM w/ Net Convexity 1,000

Swaption Premium

	Up 100bps	At Issue	Down 100bps
6Mx20Y	13,026,196	2,146,366	121,787
5Mx20Y	11,950,529	1,565,527	56,006
4Mx20Y	10,760,987	1,018,464	18,473
3Mx20Y	8,542,868	436,823	2,271
2Mx20Y	7,308,575	144,210	84

Premium at 77bp

	Up 100bps	At Issue	Down 100bps
6Mx20Y	8,493,267	300,930	605
5Mx20Y	7,813,728	175,375	119
4Mx20Y	7,062,498	82,126	11
3Mx20Y	5,177,346	15,572	0
2Mx20Y	4,484,013	1,750	0

Premium above 77 bp

	Up 100bps	At Issue	Down 100bps
6Mx20Y	4,532,928	1,845,436	121,183
5Mx20Y	4,136,801	1,390,153	55,886
4Mx20Y	3,698,489	936,339	18,461
3Mx20Y	3,365,522	421,251	2,271
2Mx20Y	2,824,562	142,460	84

Maturity	Gross Conv	Net Conv	(% Total)	Implied Vol	Premium	Notional	Vega
1	2,829	2,407	63%	102	20	605	0.2
5	1,342	823	22%	88	25	350	0.3
10	1,240	570	15%	81	29	399	0.4

FIGURE 5.52 Swaption portfolio characteristics

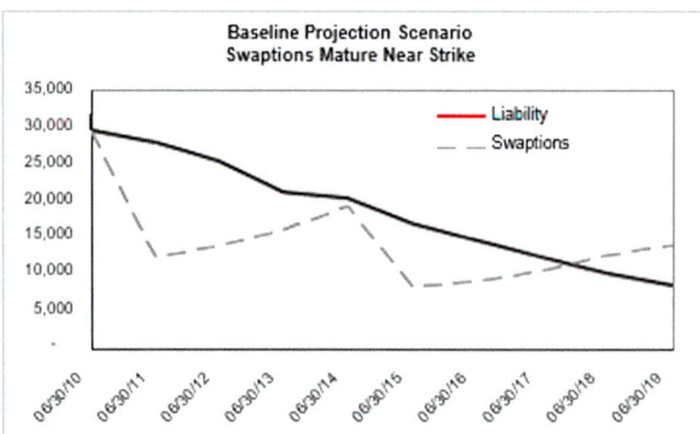

FIGURE 5.53 Baseline projection scenario: swaptions mature near strike

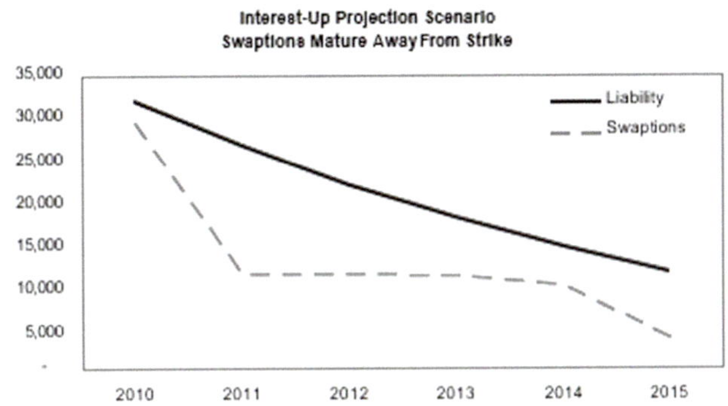

FIGURE 5.54 Interest-up projection scenario: swaptions mature away from strike

(ii) Key Considerations

- Short-term and liability-fitting solutions have similar cost levels, but with the liability-fitting option costs extend more into future years, so entering large swaption position with implied volatility in excess of the reserving assumption may require the reserving assumption to be revisited.
- Swaption gamma evolution is predictable but highly path-dependent: Gamma increases rapidly if the option matures near the strike; inversely, it fades to zero if the option matures far from the strike. In contrast, the evolution of the liability interest rates gamma (rho convexity) is different:
  - there is no real "strike" level
    - The swaption portfolio net gamma is dependent on the rate level;
  - gamma decreases as rates move away from the average swaption strike;
  - neither proposed portfolio is a static hedge: although the longer-dated "liability-fitting" portfolio has more stable gamma, liability gamma is somewhat more stable and the future gamma fit of either hedge is path-dependent; and
  - the loss of gamma is not equivalent to the loss of value: swaption is delta-hedged, so the initial premium not necessarily "lost" if gamma coverage deteriorates due to rate movements, though additional future swaption trades, if required, will be subject to an unknown future implied volatility level:
    - future gamma is much lower in high-interest-rate scenarios, and
    - generally more stable across rate levels.

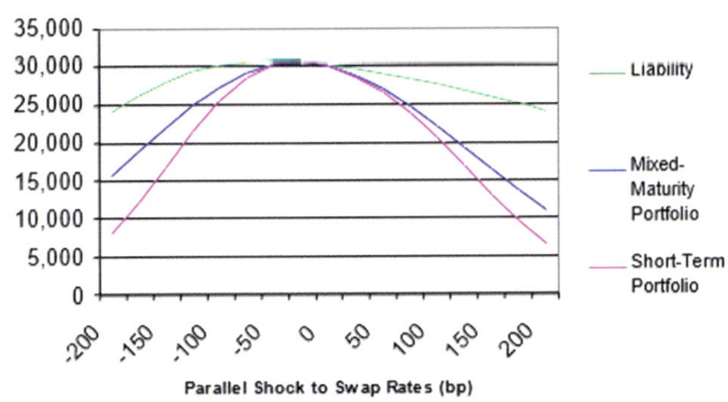

FIGURE 5.55 Net gamma sensitivity to rate level

- Liability rho convexity (or gamma) has a specific term structure.
  - The main exposure is usually to the long-term rates, while there is little gamma exposure to the shorter-term rates.
  - A closer fit is possible to the one- and two-dimensional key rate profiles using short-dated portfolios (which avoids building up large negative gamma in short-term buckets).
  - The gamma term structure matching impacts the key rate make-up of DV01 gained or loss due to rate movements. Unlike with an overall gamma mismatch (P&L always negative if short gamma), the P&L impact of key rate mismatch is directionally dependent.
  - On balance, longer-dated swaptions represent a better fit to liability portfolio and a more stable solution to gamma exposure.
    - Although higher vega increases net income volatility, it is a necessary result of matching liability, and is advantageous for some market-consistent value measures.
    - It provides protection for future interest rate gamma, beyond a short-term horizon; high-volatility regimes may last significantly longer than one year.

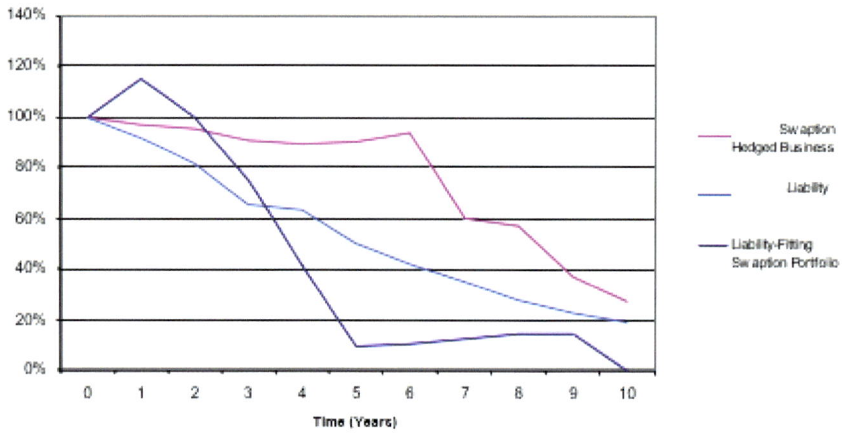

FIGURE 5.56 Projected net gamma as a percentage of initial net gamma

  - It is strongly dependent on the modelling assumptions.
    - Model changes and other realities may alter gamma unpredictably.
    - If persistence is higher than currently modelled, all future Greeks should be higher.
    - If the volatility assumptions are unlocked, gamma will move accordingly.
- Option purchases decrease the volatility of underlying earnings and increase the volatility of net income.
  - There is an impact through the dollar vega impact per the normal bp of implied volatility, where substantial mark-to-market gains or losses are possible depending on the volatility movement.
  - We would see offsetting sensitivity to volatility in liability if we were using market-consistent valuations.
- Implied volatility is not strictly a net income item, as market implied volatility has vega and theta implications.

- The vega impact immediately flows to net income.
- The theta impact flows to underlying earnings over the life of the trade.

(iii) The Choice of the Specific Interest Rates Option

The choice of interest rate option depends on the level of interest rates and convexity, combined with liquidity

For high-rate protection, the preferred choice is often a payer swaption.

- Payer swaptions: the preferred choice from a convexity and liquidity perspective.
  - The decision on maturity of payer depends on whether we should wait to buy the option or we should buy longer-dated (three-year versus two-year) options.
  - Buying three-year options is the preferred alternative:
    - it prevents us from waiting too long and missing out on protection for unexpected, sudden increases in rates;
    - it does not decay expensively in the next year, and does not give up on convexity compared to the two-year option; and
    - waiting before buying options could mean higher volatility on the option after the waiting period.
- LIBOR caps: less liquid compared to payers, less convex and more vega if we want to hedge ten-year and 30-year buckets in the curve.
- Constant-maturity swap caps: less liquid compared to LIBOR caps and payers, and cheaper but with lesser convexity than comparable vanilla payer swaptions under certain scenarios.
- Bermudan payer swaptions: less liquid, with no natural source of supply, such as Bermudan payers; in addition, we need more convexity protection when rates go up, so vanilla payers suit us.

For low-rate protection, the preferred choice is often a Bermudan receiver swaption, because these retain value when rates go up and there is a natural source of supply in the market.

Regarding the choice of tails in connection with the payer swaption, the liability key rate duration is estimated to be significantly distributed across ten-year, 20-year and 30-year key rates, and so it would be worthwhile to consider hedges equally split between ten-year and 30-year key rates.

However, the ten-year tenor may look rich from a roll-down and volatility perspective:

- the ten-year versus 30-year curve may be steep;
- ten-year tail volatility may be higher than 30-year tails; and
- the ten-year tenor skew may be rich.

If so, then it means that, for the same amount of premium, we are able to buy comparable notionals of the two-year–ten-year spread and the two-year–30-year spread. This means that, for the same premium, we get significantly lesser DV01 and convexity, meaning that, when rates go up 200 bps, the P&L delivered from 2y10y is going to be lesser (for a constant-premium strategy).

5.2.2.1.2 Sustainable Equity Convexity (Gamma) Hedging For example, assume that a product guarantees a policyholder, after ten years, the policy on S&P 500 total return. Thus, one can conclude that this GMAB payoff equates to an ATM put. Sometimes promises made in policies reflect similar payout profiles to options traded in the capital markets.

A simple guaranteed life insurance product will pay its accumulated value (based on crediting rates and S&P 500 total return performance), but at no time pay less than the principal amount, while preserving the upside potential, which can be decomposed into a combination of a fund and a long-term (more than ten-year) put option: the client invests $100 in a given fund and the insurance company guarantees her a minimum of $100 at a ten-year horizon.

- If the final fund value is above $100, the client makes a profit.
- If the final fund value is below $100, the embedded put option is activated and the loss from a client perspective is limited to $0 -> the insurer pays the difference between the final fund value and $100, which can be significant in the event of a crisis.

As a result, the risk is asymmetric and nonlinear, which cannot be statically fitted with futures instruments that are linear instruments with respect to the fund. In contrast, put options resolves such nonlinear asymmetric risk profiles of liabilities through their similar nonlinear asymmetric payoff structure. Using put options offers many advantages: the ability to hedge against a loss if the asset falls below a certain level; profits from rising levels of volatility in time of crisis; the ability on the part of the investor to adjust the degree of exposure without having to adjust the composition of the insurance product. As life insurance guaranteed products have long-term maturities, they are dynamically hedged using short- to medium-term put options in combinations with futures.

However, the problem is not solved so readily, since the ATM put in some cases may be exercised against the issuing insurer at unexpected times due to surrender, withdrawals or annuity election. The maturity may thus be uncertain. A typical GMDB may have one, two or all of the following features: 1) principal: this protects beneficiaries upon the insured's death, against a possible drop in value of underlying, and also guarantees a pre-specified minimum equal to the total value of the premium paid; 2) a rising floor: the initial investment grows at a minimally guaranteed rate (e.g. 5%); 3) a ratchet (look-back option): upon the insured's death, this guarantee ensures the highest anniversary account value. For the issuer, the risks are primarily those of mortality, surrender and market movements (both embedded optionality and reinvestment rate risk).

As insurers ponder the best method for hedging these complex liabilities, the trend has begun to favour using a combination of approaches and products to isolate key risks and hedge individual components of the overall exposure to limit tail risk. In addition, as more regulators and rating agencies utilize stochastic models, insurance companies that do not manage the risk of these policies may be penalized compared to those that hedge.

5.2.2.1.2.1 Risk Appetite Framework for a Sustainable Liability Equity Volatility The realized volatility of the underlying assets of the liability tends to be significantly asymmetric, between 2008 and 2013 standing 20% of the time more than 25 points over the average for that period, for instance.

As a result, the reserving assumption of the usual 16 to 18% for realized volatility at most life insurance annuity carriers (except Jackson, at 25%) is pretty low, though it is consistent with the long-term behaviour of realized volatility.

Such structural empirical low liability volatility assumptions contribute to some extent to the gap between the operating earnings and the net income, which is often lower by 25% to 50% for variable annuity carriers. The other source of the gap lies in the exposure of hedge assets to changes in volatility (through vega P&L), in contrast to a zero-vega P&L for the liabilities due to their constant assumption.

While the cost of protection embedded within the hedge assets (summarized by the implied volatility) is typically a few points higher (e.g. for a one-year maturity the spread is two points on average, but may reach ten to 15 points under extreme market conditions; the spread is lower for shorter-dated maturities) than actual realized volatility: ex post realized volatility tends to be lower than the ex ante cost. Consequently, the market cost of long-term protection is high relative to reserving levels and historical realized levels: it increases with the protection maturity, although longer maturity also reduces the roll-over risk of hedge assets.

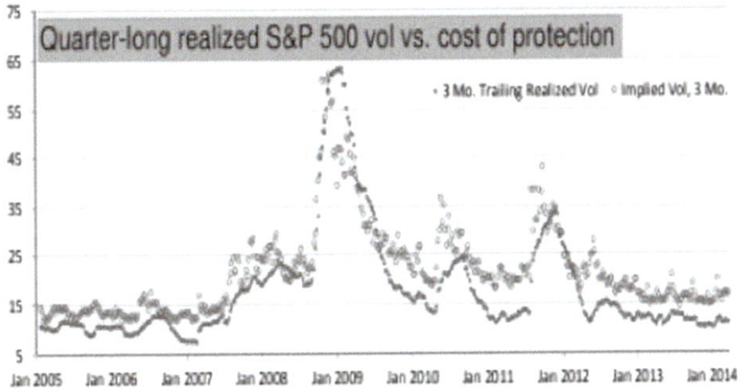

FIGURE 5.57 Three-month realized S&P 500 volatility versus cost of protection (implied volatility), January 2005–June 2014

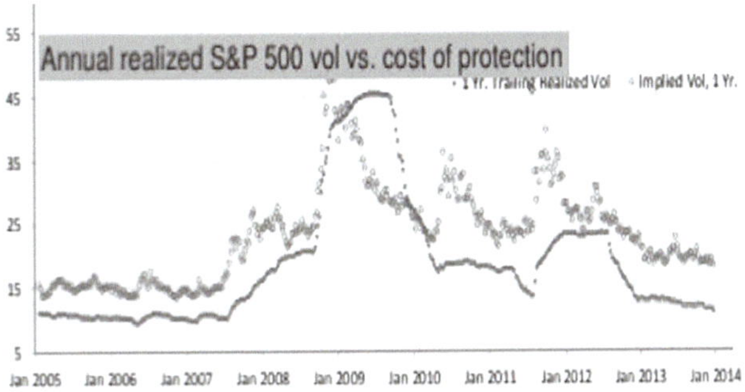

FIGURE 5.58 12-month realized S&P 500 volatility versus cost of protection (implied volatility), January 2005–June 2014

This section focuses on a potential optimal liability volatility for hedging in practice.

Setting a liability volatility assumption depends on risk appetite – i.e. determining how much volatility risk to retain/mitigate, as volatility risk can be costly, especially when equity gamma is high and interest rates are low.

5.2.2.1.2.1.1 Break-Even Volatility Each day a P&L is generated. The hedging P&L minimum-variance volatility is simply that volatility which cancels the daily P&L of the hedged portfolio over the hedging period. In other words, hedging with break-even volatility in the Black–Scholes framework cancels the hedging P&L of the hedged portfolio –i.e. assuming zero interest rates,

$$\text{Daily } P\,\&L = \frac{1}{2}\Gamma S^2 [\underbrace{\left(\frac{\Delta S}{S}\right)^2}_{\substack{\text{realized} \\ \text{one-day} \\ \text{variance}}} - \underbrace{\sigma^2 \Delta t}_{\substack{\text{implied} \\ \text{one-day} \\ \text{variance}}}], \qquad \text{with } \Gamma = \frac{\phi(d_1)}{S\sigma\sqrt{t}}$$

$$d_1 = \frac{\log(S/K) + \left(\sigma^2/2\right)}{\sigma\sqrt{t}}$$

$$Daily\ P\&L \approx \frac{1}{2}\Gamma S^2\left[\left(\frac{\Delta S}{S}\right)^2 - \left(\sigma_{implied}\sqrt{\Delta t}\right)^2\right].$$

The gamma P&L for one strike is linked to the quadratic average of the returns (vertical peaks) weighted by the gamma of the option (surface with grid) corresponding to that strike (see Dupire 2006 for more details).

The hedging daily P&L break-even volatility is therefore defined as

$$\sigma^2_{\text{break even}} := -\frac{\Theta}{2\Gamma^{\$}dt}.$$

Given that we are looking for a break-even volatility over the residual maturity, over a sum of hedging daily P&L, the break-even volatility is thus the unique strictly positive volatility that cancels daily P&L summed until maturity:

$$\sigma_{BEV}{}^2 = \frac{\sum_t \Gamma_t\left(\sigma_{BEV}\right)\times \Delta S_t{}^2}{\sum_t \Gamma_t\left(\sigma_{BEV}\right)\times S_t{}^2\times \Delta t}.$$

Break-even volatility can therefore be viewed as a quadratic average of returns weighted by the gamma of the option. The weighting scheme is centred at the strike level and adds up to one at each date, concentrated close to maturity into the strike. The break-even volatilities for different strikes are the average of the same returns but weighted with different schemes

To compute the break-even volatility, we take a spot path, a strike and a maturity as inputs and compute the corresponding break-even volatility. This function uses a numerical procedure to calculate the solution. We can compute the implied break-even volatility using Monte Carlo simulations.

Given that the computation can be unstable, especially for low and high strikes, tractability can be improved through the following modified version of break-even volatility, which does not require any numerical inversion to be computed while remaining very similar to the original one and generating approximately the same smile: the only difference is that now the gamma terms are computed at a constant level of volatility (i.e. the implied volatility of the underlying at strike k):

$$\tilde{\sigma}_k^{BEV^2} = \frac{\sum_i \Gamma_i\left(\sigma_k^{\text{implied}}\right)\Delta S_i^2}{\sum \Gamma_i\left(\sigma_k^{\text{implied}}\right)S_i^2\Delta t}$$

5.2.2.1.2.1.2 Break-Even Volatility Features The implied skew is actually bigger than the break-even volatility skew. In fact, the break-even volatility should "record" the realized volatility of the option at the strike k considered:

$$\tilde{\sigma}_k^{BEV^2} = \frac{\sum_i \Gamma_i\left(\sigma_k^{implied}\right)S_i^2\left(\frac{\Delta S_i}{S_i}\right)^2}{\sum_i \Gamma_i\left(\sigma_k^{implied}\right)S_i^2\Delta t^2} \quad \text{-----Realized volatility term}$$

.

In the definition of break-even volatility, the realized volatility is weighted by the gamma of the option.

- When the spot is close to the strike k, the realized volatility is weighted by a large gamma.
- When the spot is far from the money, the realized volatility of the spot is "considered less", because it is multiplied by a small gamma.

Hence, if we have a spot path close to the strike, it produces a break-even volatility that is close to the realized volatility spot around the strike: this volatility, in a Monte Carlo simulation, is the implied volatility at the strike k, while the resulting break-even volatility is very close to the spot implied volatility.

Consider now a spot path where the price of the underlying (around 100% for the entire life of the option) is always far from the strike (80%). The realized volatility in each period is weighted by a small gamma. Then, the break-even volatility generated by this path is close to the realized volatility of the path. If this path is generated by a Monte Carlo simulation, the break-even volatility turns out to be very close to the implied volatility at 100%

In a Monte Carlo procedure we simulate both paths that "cross" the strike and paths that remain far from it.

- Then we consider the break-even volatilities generated by all the simulations and we take their average.
- On the right wing of the smile we have the opposite effect: break-even volatility is higher than implied volatility.
- On the left wing of the smile the break-even volatility is smaller than the implied volatility; this happens because all the "bullish" paths contribute to the final break-even volatility with a low value, which provides a reduced skew exposure.

The break-even volatility dampens the realized skew because it gives the same importance to all the paths. The break-even volatility of a path that is always far from the strike has the same weight as that of a path that crosses the strike k (it is an arithmetic average).

A possible solution to go around this problem is to consider a different definition of volatility: the par volatility.

5.2.2.1.2.1.3 The Par Volatility as Sustainable Liability Volatility Assumption This is the only volatility such that the expectation of the gamma P&L is equal to zero – i.e. you break even in expectation:

$$E\left[Gamma\ P\&L\left(\sigma_k^{PAR}\right)\right] = E\left[\sum_i \Gamma_i\left(\sigma_k^{PAR}\right)S_i^2\left(\left(\frac{\Delta S_i}{S_i}\right)^2 - \sigma_k^{PAR^2}\Delta t\right)\right] = 0.$$

In the par volatility we are taking the expectation of P&Ls, not of volatilities. Hence, we are weighting fewer paths that are far from the strike k. A product based on par volatility is very closely related to the gamma P&L of a delta-hedged options. For risk management purposes, we need to make the mean hedging PnL equal to zero among adverse scenario patterns, so that we can estimate the corresponding break-even volatility using a relatively simple minimum solver applied to the standard deviation of the daily P&L function.

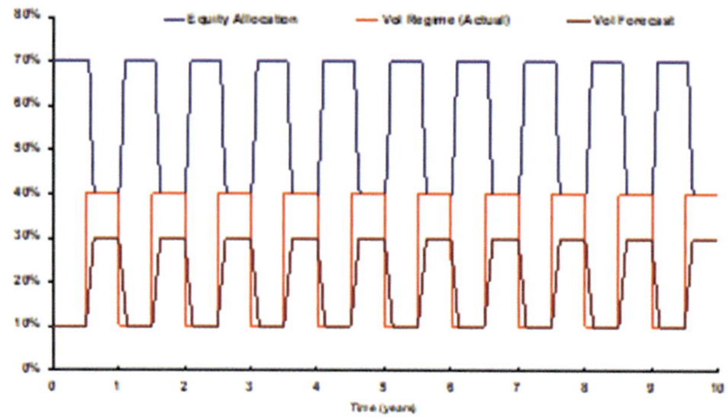

FIGURE 5.59 Volatility regime stress test

5.2.2.1.2.1.4 The Minimum P&L Variance Volatility as Sustainable Liability Volatility Assumption However, the hedging of a contingent claim is also intended to reduce the hedging P&L variance: dealers want to be able to sell options for more than they are worth, or buy them for less, and then lock in those profits by hedging. In the Black–Scholes setting, hedge ratios depend on the volatility used, σ, while s, the performance of Black–Scholes delta hedging, materially depends on the volatility used. Therefore, an interesting question in the Black–Scholes framework is: which volatility produces hedge ratios that minimize the hedging P&L variance?

Each day a P&L is generated. The minimum-variance hedging P&L volatility is simply that volatility which minimizes the variance of the hedging P&L. In other words, hedging with the minimum-variance hedging P&L volatility in the Black–Scholes framework minimizes the variance of the hedging P&L of the hedged portfolio, and hence is a reasonable candidate for the optimal Black–Scholes volatility. The standard deviation of daily P&L shows a consistent, roughly quadratic dependence on hedge volatility.

There again, we can estimate the corresponding minimum-variance hedging P&L volatility using a relatively simple minimum solver applied to the standard deviation of the daily P&L function. Empirically, the minimum-variance hedging P&L volatility is empirically close to realized for quiet market conditions (such as pertained from mid-2003 to mid-2005) and for the market turbulence induced by the 2008 credit crunch.

On average realized volatility – if we knew what it was going to be – it would be a good volatility to hedge at. But there have indeed been particular paths where hedging at realized can fail to minimize the P&L variance. In fact, whenever realized volatility significantly changes (i.e. jumps), and the option's gamma in one volatility regime is significantly different from that in the other, hedging at a volatility close to the realized is fine as long as the gamma is of reasonable size, though it is not when the gamma becomes small (so the gamma-weighted delivered volatility is not large and the P&L is relatively insensitive to the hedge volatility).

This highlights the practical importance of gamma-weighted realized volatility.

In contrast, the implied volatility differs from it for the whole period, indicating that it is, at best, an imperfect estimate of the right hedge volatility.

Option pricing using the minimum-variance hedging P&L volatility produces a fairly flat P&L with low variance. Implied volatility, in contrast, typically produces a profit, but at the cost of significant earnings variance. Moreover, using implied during the crunch period was loss-making.

5.2.2.1.2.2 From Options to Variance Swaps for Equity Convexity (Gamma) Hedging As mentioned above, volatility has a direct impact on hedging effectiveness. For this very reason, it is difficult to measure the overall P&L of a dynamic hedging strategy, partly due to the uncertainty of volatility.

Hedging volatility usually comprises a mix of variance swaps (spot and forward) and option tranches, for two main reasons.

- They can be sized to track the gross liability volatility risk profile. When using a variance swap, we get a flat gamma profile with respect to the equity market level, while, when using an option tranche, we size up the options to target the liability profile.
- Option tranches tend to cost less than varswaps for maturities longer than six months, but this does not always hold for shorter maturities, especially for those within three months, due in part to the larger volatility skew for short maturities. Even so, in practice, the volatility cost will be generally two points lower than with a variance swap.

However, there are several risks and challenges associated with option tranches.

- They are trading options with a sufficient range/number of strikes.
- The liability gamma profile will change if rate levels change differently from expected.
- The liability gamma profile might change due to model change – the same issue as with variance swaps.
- Option tranches will become less stable when they get very close to maturity; maturities can be staggered to minimize the effect.
- Options will have higher-order terms that do not have an offset on the liability side; P&L attribution is not as clean as a variance swap.

5.2.2.1.2.2.1 Why Variance Swaps for Life Insurance Hedging Partly due to its use in Black–Scholes, volatility has historically been used as the measure of deviation for financial assets. However, the correct measure of deviation is variance (or volatility squared).

- Variance takes into account the implied volatility of all strikes with the same expiry, while ATM implied will change with spot, even if the volatility surface does not change. For example, if an option with a fixed strike that is initially ATM is bought, then as soon as spot moves it is no longer ATM. However, the traded level of variance is applicable no matter what the level of spot and is not path-dependent, which makes it the ideal measure for deviation.
- Deviations need to be squared to avoid cancelling themselves out. Mathematically, if deviations were simply summed, then positive and negative deviations would cancel out. This is why the sum of squared deviations is taken (variance) to prevent the deviations from cancelling out. Taking the square root of this sum (volatility) should be considered a derivative of this pure measure of deviation (variance).
- P&L from a delta-hedged option depends on the square of the return (see above). Due to the convexity of an option, if the volatility is doubled, the profits from delta hedging are multiplied by a factor of four. For this reason, variance (which looks at squared returns) is a better measure of deviation than volatility.

Buyers and sellers of volatility have typically traded straddles and range accrual notes, which are very sensitive to volatility. However, the payoff profiles of both products are not ideal, and the products are both still very dependent on the absolute level of the underlying.

In contrast, variance swaps have no dependence on the absolute level of the underlying, and are a pure volatility play, expressing a view on realized equity market volatility; just enter into a variance swap to get long volatility. Moreover, the hedge for a variance swap is relatively simple – in contrast with volatility swaps, for which there is, as a result, a lot less liquid market in volatility swaps, and margins on volatility swaps are a great deal higher than they are for variance swaps.

Activity in variance swaps died down after volatility spiked in late 2008, causing many dealers to experience hefty losses, particularly in single stocks. The revival of variance swaps came from the need for a pure volatility product – i.e. one with close to no delta hedging – in markets with reduced liquidity, where variance is not an option, such as Asian single-stock markets.

5.2.2.1.2.2.2 Variance Swaps as Constant Dollar-Gamma Derivatives Variance swaps pay out an amount equal to the difference between realized variance (volatility squared) and a pre-agreed strike level, multiplied by the vega notional (vega is the sensitivity of an option's price to changes in volatility). On the trade date, no cash flows are exchanged, but the mark to market on the trade will begin the day after the trade date and will flow through the insurer's earnings statement, just like any other trading asset.

Variance swap is an exotic derivative instrument whereby one party agrees to receive at maturity the squared realized volatility converted into dollars for a pre-agreed price:

$$\textit{Variance swap payoff} = \$1 \times \left[\frac{252}{N} \sum_{t=1}^{N} \underbrace{\left(\ln \frac{S_t}{S_{t-1}} \right)^2}_{\text{squared log-return}} - K_{\text{var}}^2 \right],$$

where K_{var} is the variance strike expressed in volatility percentage points (e.g. 30%) and is not to be confused with the strike of a vanilla option, which is a stock price level.

Because variance is additive in time dimension, the mark to market of a variance swap can be linearly decomposed at any point in time between realized and implied variance:

$$MTM_t = e^{-r(T-t)} \left[\frac{t}{T} Realized_t^2 + \frac{T-t}{T} Implied_t^2 - K_{\text{var}}^2 \right],$$

$$Varswap_t = \text{National} \times PV_t(T) \times \left[\frac{t}{T} \times \left(\text{Realized vol}(0,t) \right)^2 + \frac{T-t}{T} \left(\text{Implied vol}(t,T) \right)^2 - Strike^2 \right],$$

where PVt(T) is the present value at time t of \$1 received at maturity T.

In order to characterize the specificity of variance swap equity gamma hedging purposes, let us highlight the connection between the gamma–theta P&L and the payoff of a variance swap:

$$\textit{Variance swap payoff} = \frac{252}{N} \sum_{t=1}^{N} \left(\ln \frac{S_t}{S_{t-1}} \right)^2 - K_{\text{var}}^2$$

$$\textit{Final option } P\&L = \sum_{t=1}^{N} \gamma_t \left[r_t^2 - \left(\sigma_{Implied} \sqrt{\Delta t} \right)^2 \right]$$

The final option P&L above is a weighted sum of squared realized returns minus a constant that has the role of the strike, and so is very similar to the payoff of the variance swap above except for the weighting known as dollar gamma, which makes it highly path-dependent.

The difference between the two equations above therefore lies in the weighting of the squared daily log-returns: variance swaps are constant equally weighted, and thus could be hedged by an option with constant dollar gamma: the log-contract. By contrast, the final P&L of a delta-hedged vanilla option position is dollar-gamma-weighted, which changes over time.

✓ **Quick win**: the huge advantage of variance is the fact that the gamma coverage is constant, day in, day out, across the life of the contract. There is no "decay" of the coverage. As a result, if the long-dated life insurance liability model generates a gamma for the portfolio that is fairly "constant" across both time (say the next year) and price shocks, then a variance contract may be a suitable tool to mitigate risk.

Thus, its key success in hedging lies in its efficient source of volatility exposure (dollar gamma and vega) despite movements in the stock (versus straddle positions that then require continuous rehedging of both gamma and vega), which can then match the also stable dollar-gamma positions during stable and rising markets for the guarantees embedded within many guaranteed annuities, not only return of premium but even enhanced watermark optionalities usually embedded within variable annuities.

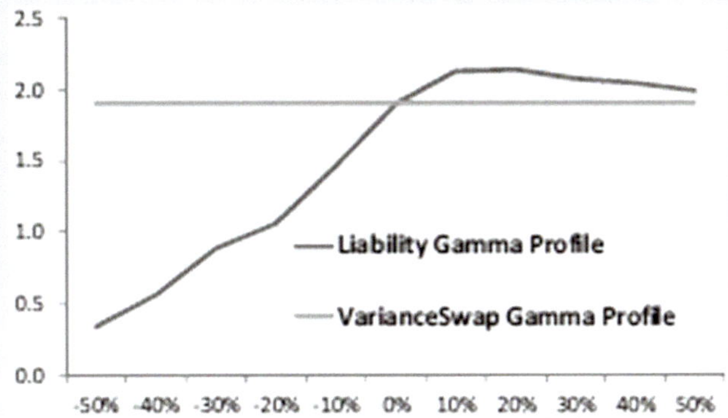

FIGURE 5.60 Liability versus varswap gamma

5.2.2.1.2.2.3 Replication of Variance Swap Contracts For the final P&L to be path-independent, the dollar gamma must be constant, which means that, for some constant a (Bossu 2006),

$$\gamma_t = c \rightarrow \Gamma = \frac{c}{S^2} \rightarrow \frac{\partial^2 f}{\partial S^2} = \frac{c}{S^2} \rightarrow f = c\ln S + bS + a.$$

In other words, the perfect static hedge for a variance swap would be a combination of the log-asset (a derivative that pays off the log-price of the underlying stock), the underlying stock and cash. As a result, the link between the variance swap and the log-contract comes from the following analytic formula for the value of a log-contract:

$$\ln F - \frac{1}{2}\int_0^T \sigma^2 dT.$$

We now have a hedging strategy for a variance swap. Buy a replicating set of vanilla options for twice the volume of log-contracts as the volume of variance swaps sold (twice the volume in order to counteract the "½" in front of the integral in the formula). Delta-hedge these vanilla options. Since the log-contract is losing value at exactly the rate of $1/2\int_0^T \sigma^2 dT$, the delta hedging should be producing profits at exactly the rate needed to cover payments on the variance swap.

Unfortunately, log-contracts are not traded. But can't we find a combination of vanilla calls and puts with the same maturity as the variance swap such that the aggregate dollar gamma is constant?

Formally: find quantities ("weights") such that

$$\gamma_{\text{Aggregate}} = \sum_{i=1}^{N^{Put}} w_i^{Put} \gamma_i^{Put} + \sum_{i=1}^{N^{Call}} w_i^{Call} \gamma_i^{Call} = c.$$

Using the Black–Scholes closed-form formula for gamma, if rates are zero, the dollar gamma actually has a bell-shaped curve, which peaks slightly after the 100 strike, when S is equal to

$$\Gamma^S(S,K) = \frac{S}{\sigma\sqrt{2\pi T}} \exp\left(-\frac{\left(\ln(S/K) + 0.5\sigma^2 T\right)^2}{2\sigma^2 T} \right), \; S^* = Ke^{\sigma\sqrt{T} - \sigma^{2\gamma}/2}.$$

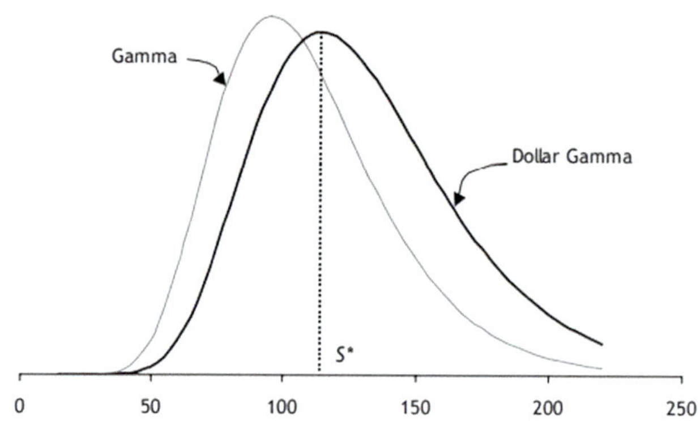

FIGURE 5.61 Gamma versus dollar gamma

So, in order to produce a constant dollar gamma across the broadest range of stock prices, the weights should be inversely proportional to the strike (i.e. w(K) = c / K, where c is a constant) (Bossu 2006).

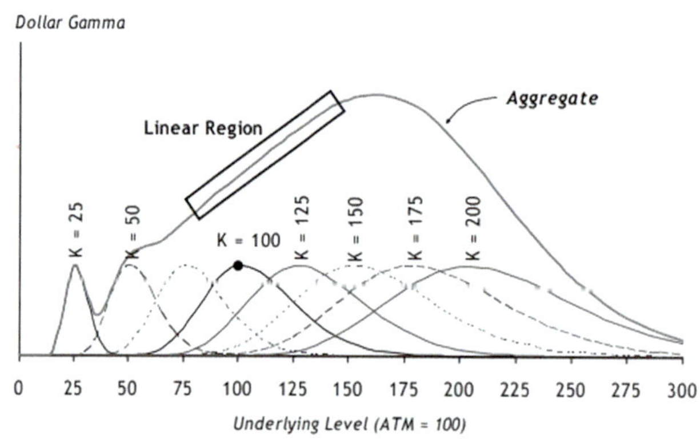

FIGURE 5.62 Dollar gamma versus spot level

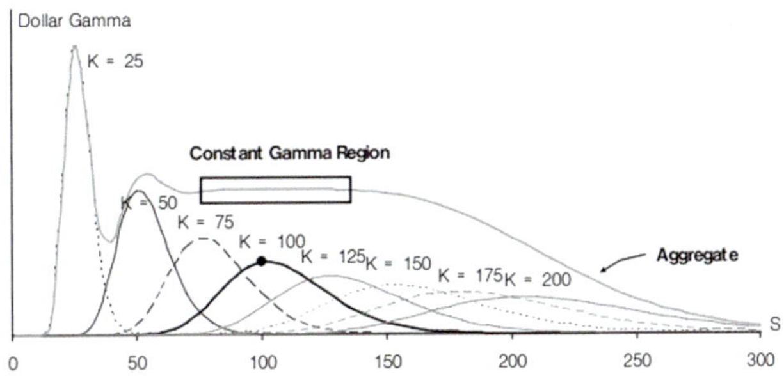

FIGURE 5.63 Dollar gamma versus spot level

However, the aggregate gamma is still non-constant, although we also notice the existence of a linear region when the underlying level is in the range 75 to 135.

This observation is crucial: if we can regionally obtain a linear aggregate gamma with a certain weighting scheme w(K), then the modified weights w'(K) = w(K) / K will produce a constant aggregate gamma. Since the naïve weights are inversely proportional to the strike K, the correct weights should be chosen to be inversely proportional to the squared strike.

Such an optimal weight can also be deduced from the fact that a variance swap has a flat vega profile that doubles in height and width if the strike is doubled.

An equal-weighted portfolio is not suitable, as the vega profile of an option increases in size and width as the strike increases (i.e. an option of strike 2K has a peak vega double the peak vega of an option of strike K and is also twice the width):

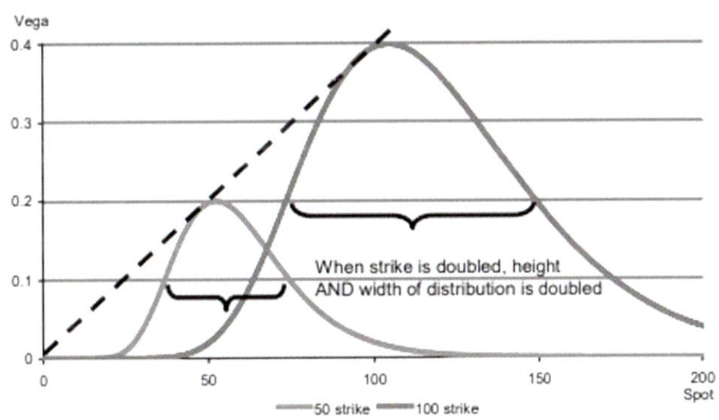

FIGURE 5.64 Vega versus spot

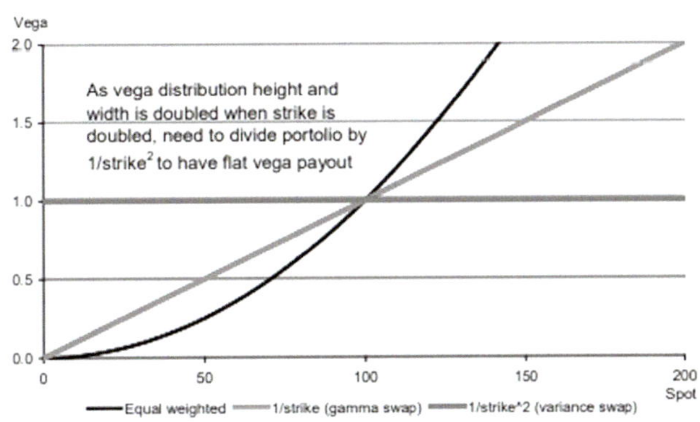

FIGURE 5.65 Vega versus spot

In other words, as the vega profile doubles in height and width if the strike is doubled, the portfolio has to be divided by strike K once, to correct for the increase in height, and again, to compensate for the increase in width (for a combined weight of $1/K^2$).

Proof: vega of option = $\tau \times S \times f(x, v)$,
where $x = K / S$ (strike a ratio of spot), τ = time to maturity, $v = \sigma^2\tau$ (total variance), while

$$f(x,v) = \frac{1}{\sqrt{2\pi}} \times e^{\frac{-d_1^2}{2}}$$

$$d_1 = \frac{\text{Ln}\left(\frac{1}{x}\right) + \frac{v}{2}}{\sqrt{v}}.$$

If we have a portfolio of options where the weight of each option is w(K), then the vega of the portfolio of options V(S) is

$$V(S) = \tau \int_{K=0}^{\infty} w(K) \times S \times f(x,v) dK.$$

As $K = xS$, this means $dK / dx = S$; hence $dK = S \times dx$, and we can change variable K for x:

$$V(S) = \tau \int_{x=0}^{\infty} w(xS) \times S^2 \times f(x,v) dx.$$

In order for the portfolio of options to have a constant vega no matter what, the level of spot dV(S)/dS has to be equal to zero:

$$\frac{dV}{dS} = \tau \int_{x=0}^{\infty} \frac{d}{dS}\left[S^2 w(xS)\right] \times f(x,v) dx = 0.$$

And, by the chain rule,

$$\Rightarrow \tau \int_{x=0}^{\infty} \left[2Sw(xS) + S^2 \frac{d}{dS} w(xS)\right] \times f(x,v) dx = 0$$

$$\Rightarrow \tau \int_{x=0}^{\infty} \left[2Sw(xS) + S^2 \frac{d}{dS} w(xS)\right] \times f(x,v) dx = 0.$$

As d/dS = (d/dK) × (dK/dS), and dK/dS = x,

$$\Rightarrow \tau \int_{x=0}^{\infty} S\left[2w(xS) + xS \frac{d}{dK} w(xS)\right] \times f(x,v) dx = 0.$$

As K = Xs,

$$\Rightarrow \tau \int_{x=0}^{\infty} S\left[2w(xS) + K\frac{d}{dK}w(K)\right] \times f(x,v)dx = 0$$

$$\Rightarrow 2w + K\frac{d}{dK}w(K) = 0 \text{ for all values of S}$$

$$\Rightarrow w(K) = \frac{\text{constant}}{K^2}.$$

Overall, the portfolio of options with weights inversely proportional to $1/K^2$ produces a vega that is virtually independent of stock price S, as long as S lies inside the range of available strikes and far from the edge of the range, and provided the strikes are distributed evenly and closely. You will therefore obtain an exposure to variance that is independent of stock price – just what is needed to trade a variance swap.

Variance swaps are thus typically hedged by replicating the payoff using a portfolio of vanilla options, a set of out-of-the money puts and calls on the underlying equities:

$$\Pi(S, \sigma\sqrt{\tau}) = \sum_{K>S.} \frac{1}{K^2} C(S, K, \sigma\sqrt{\tau}) + \sum_{K<S.} \frac{1}{K^2} P(S, K, \sigma\sqrt{\tau}).$$

The $1/K^2$ weighting means that a larger amount of OTM puts are traded than OTM calls (approximately 60% is made up of puts). This causes a log-contract (portfolio of options weighted $1/K^2$) to be long skew. The curved nature of the weighting means the wings (very out-of-the-money options) have a greater weighting than the body (near-ATM options), which means that a log-contract is long volatility surface curvature.

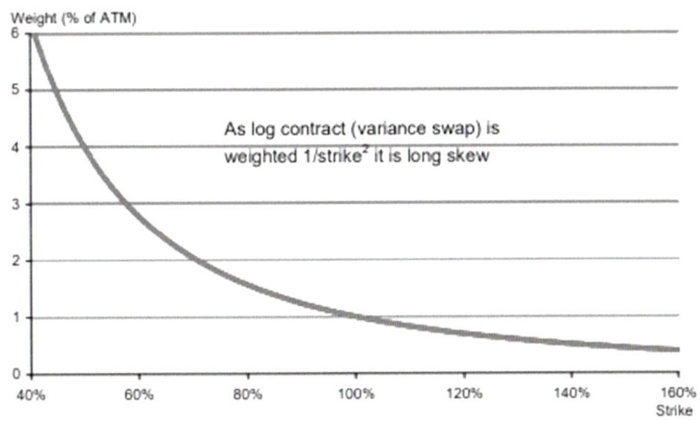

FIGURE 5.66 Weight (% ATM) versus strike

The inclusion of OTM (and hence convex) options means that the log-contract is also long volga (or vega convexity), but they are not the same thing. Long OTM (wing) options are long vega convexity, but not volatility surface curvature (unless they are shorting the ATM or body at the same time). The curvature of the volatility surface can be defined as the difference between 90–100 skew and 100–110 skew (i.e. the value of 90% + 110% – 2 × 100% implied volatilities).

At time t you can sum the options values to show that the total portfolio value is

$$\Pi(S, \sigma\sqrt{\tau}) = \frac{S - S_*}{S_*} - \log\left(\frac{S}{S_*}\right) + \frac{\sigma^2 \tau}{2}.$$

Clearly, the variance exposure of Π is

$$V = \frac{\tau}{2}.$$

To obtain an initial exposure of \$1 per volatility point squared, you need to hold (2/T) units of the portfolio Π:

$$\Pi(S, \sigma\sqrt{\tau}) = \frac{2}{T}\left[\frac{S - S_*}{S_*} - \log\left(\frac{S}{S_*}\right)\right] + \frac{\tau}{T}\sigma^2.$$

- The first term in the payoff describes 1/S* forward contracts on the stock with delivery price S*, whose value represents a long position in the stock (value S) and a short position in a bond (value S*), which can be statically replicated, once and for all, without any knowledge of the stock's volatility.
- The second term describes a short position in a log-contract with reference value S*, a so-called exotic option, whose payoff is proportional to the log of the stock at expiration, and whose correct hedging depends on the volatility of the stock. All the volatility sensitivity of the weighted portfolio of options we have created is contained in the log-contract.

For now, assume that we are in a Black–Scholes world where the implied volatility σ_I is the estimate of future realized volatility. If you take a position in the portfolio Π, the fair value you should pay at time 0 is

$$\Pi_0 = \frac{2}{T}\left[\frac{S_0 - S_*}{S_*} - \log\left(\frac{S_0}{S_*}\right)\right] + \sigma_I^2.$$

At expiration, if the realized volatility turns out to have been σ_R, the initial fair value of the position captured by delta hedging would have been

$$\Pi_0 = \frac{2}{T}\left[\frac{S_0 - S_*}{S_*} - \log\left(\frac{S_0}{S_*}\right)\right] + \sigma_R^2.$$

So the net payoff on the position, hedged to expiration, will be

$$\text{payoff} = \left(\sigma_R^2 - \sigma_I^2\right).$$

By rehedging the position in log-contracts, therefore, you have, in effect, been the owner of a position in a variance swap with fair strike Kvar = σ^2_I and face value \$1. You will have profited (or lost) if realized volatility has exceeded (or been exceeded by) implied volatility.

Thus, the fair strike of a variance swap (i.e. the level of Kvar such that the swap has zero initial value) is then given as

$$K_{\text{var}}^* = \sqrt{\frac{2e^{rT}}{T}\left[\int_0^1 \frac{1}{k^2} Put(k)dk + \int_1^{+\infty} \frac{1}{k^2} Call(k)dk\right]}.$$

In theory, a dealer is required to buy a continuum of strikes down to zero and up to infinity, appropriately weighted, to replicate a variance swap. In fact, a finite number of strikes fails to produce a uniform vega, as the stock price moves outside the range of the available strikes. As long as the stock price remains within the strike range, trading the imperfectly replicated log-contract will allow variance to accrue at the correct rate. Whenever the stock price moves outside, the reduced vega of the imperfectly replicated log-contract will make it less responsive than a true variance swap.

For each strike, a dealer would be required to buy an amount of options inversely proportional to the square of the strike.

$$Varswap_0 \approx \frac{2}{T}\left[\sum_{i=1}^{N^{puts}} \frac{p_0\left(k_i^{put}\right)}{\left(k_i^{put}\right)^2}\left(k_i^{put}-k_{i-1}^{put}\right)+\sum_{i=1}^{N^{calls}} \frac{c_0\left(k_i^{call}\right)}{\left(k_i^{call}\right)^2}\left(k_i^{call}-k_{i-1}^{call}\right)\right]$$
$$-PV_0(T)\times\left(K^{VS}\right)^2.$$

Usually it is near-impossible to find deep out-of-the-money puts or calls on a stock or index, and so the far end of the range – corresponding to very improbable high levels of volatility – are left unhedged.

The problem for dealers arises when less deeply out-of-the-money puts (30 to 60%), representing relatively probable high-volatility scenarios, become illiquid. The out-of-the-money strike puts hold a significant weight in the hedging portfolio, and, if they cannot be found, hedging the swaps in advance becomes exceptionally difficult, with dealers having to resort to use at-the-money volatility, which results in a more cost-efficient product with tighter bid–ask spreads.

Such a hedging strategy is cheaper and doable for volatility swaps – and more so than for variance swaps: if the spot drops and volatility climbs, dealers are not required to buy as many deep out-of-the-money puts as they would to hedge a variance swap, because the vega does not move as dramatically with volatility as it does in the case of variance swaps. Usually the long volatility position outperforms the long variance position in profit and loss terms for realized volatilities between 20 and 60, as the dynamic behaviour of volatility swaps is less aggressive, in that dealers will be less exposed to volatility convexity and skew than they would trading a variance swap. But a volatility swap remains an exotic product, so maintaining a competent hedge in all market conditions will take a good deal of effort. The difficulty in hedging volatility swaps drove liquidity towards the variance swap market, particularly during the 2002 equity collapse. As variance swaps can be replicated by delta-hedging a static portfolio of options, it is not necessary to buy or sell options during the life of the contract.

Variance swaps are traded for an amount of vega (sensitivity to a 1% increase in volatility):

$$Variance\,notional = \frac{Vega\ notional}{2\times\ Strike}.$$

To price a variance swap, the traders must compute the weighted fair values of the options required to hedge the swap. For any given variance swap quote, then, the strike price is set to reflect the aggregate cost of the hedge portfolio. So, when pricing a variance swap, the major factor is the cost of the strip of options. Because a whole strip of options is used, the volatility skew has a major impact on pricing, usually resulting in the variance swap strike being higher than the at-the-money implied strike. The payoff of a variance swap is convex in volatility, which causes the "fair" strike to be around the 90% put implied volatility, which is slightly higher than "fair" volatility.

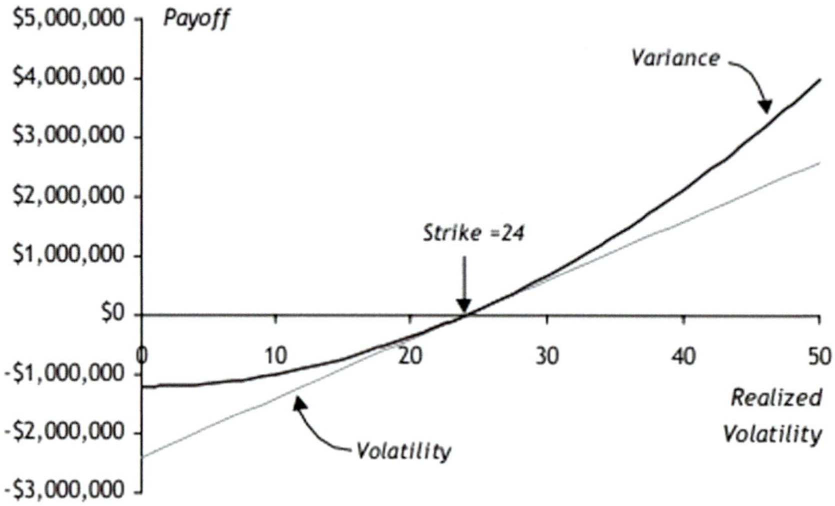

FIGURE 5.67 Varswap payoff versus realized volatility

However, varswaps also hold some weaknesses, such as potential over-hedging whenever OTM the liability dollar gamma is flat or slightly increasing under falling equity markets. As a result, the gamma hedge ratio would be lower than in the ITM case, implying the need to buy larger amounts of varswaps for hedging.

Remark: in the previous section, we explained how to replicate a variance swap by means of a portfolio of options whose payoffs approximate a log-contract. Although our explanation depended on the validity of the Black–Scholes model, the result – that the dynamic hedging of a log-contract captures realized volatility – holds true more generally. The only assumption we will make about the future underlier evolution is that it is diffusive, or continuous; this means that no jumps are allowed:

$$\frac{dS_t}{S_t} = \mu(t,\ldots)dt + \sigma(t,\ldots)dZ_t, \quad V = \frac{1}{T}\int_0^T \sigma^2(t,\ldots)dt, \quad F = E\left[e^{-rT}(V-K)\right].$$

Thus the fair delivery value of future realized variance is the strike for which the contract has zero present value:

$$d(\log S_t) = \left(\mu - \frac{1}{2}\sigma^2\right)dt + \sigma dZ_t \ldots \frac{dS_t}{S_t} - d(\log S_t) = \frac{1}{2}\sigma^2 dt$$

$$\begin{aligned} V &\equiv \frac{1}{T}\int_0^T \sigma^2 dt \\ &= \frac{2}{T}\left[\int_0^T \frac{dS_t}{S_t} - \log\frac{S_T}{S_0}\right], \quad K_{var} = \frac{2}{T}E\left[\int_0^T \frac{dS_t}{S_t} - \log\frac{S_T}{S_0}\right] \end{aligned}$$

As there are no actively traded log-contracts for the second term, we must duplicate the log payoff, at all stock price levels at expiration, by decomposing its shape into linear and curved components, and then duplicating each of these separately. The linear component can be duplicated with a forward contract on the stock with delivery time T; the remaining curved component, representing the quadratic and higher-order contributions, can be duplicated using standard options with all possible strike levels and the same expiration time T.

For practical reasons, we want to duplicate the log payoff with liquid options – that is, with a combination of out-of-the-money calls for high stock values and out-of-the-money puts for low stock values. We introduce a new arbitrary parameter S* to define the boundary between calls and puts:

$$\log\frac{S_T}{S_0} = \log\frac{S_T}{S_*} + \log\frac{S_*}{S_0}$$

The second term is constant, independent of the final stock price, so only the first term has to be replicated:

$$\begin{aligned} -\log\frac{S_T}{S_*} = \; & -\frac{S_T - S_*}{S_*} && \text{(forward contract)} \\ & +\int_0^{S_*} \frac{1}{K^2}\operatorname{Max}\left(K - S_T, 0\right) dK && \text{(put options)} \\ & +\int_{S_*}^{\infty} \frac{1}{K^2}\operatorname{Max}\left(S_T - K, 0\right) dK && \text{(call options).} \end{aligned}$$

This represents the decomposition of a log-payoff into a portfolio consisting of: a short position in forward contracts struck at S*; a long position in put options struck at K, for all strikes from 0 to S*; and a long position in call options struck at K, for all strikes from S* to ∞; all contracts expire at time T:

$$\begin{aligned} K_{var} = \frac{2}{T}\Bigg(& rT - \left(\frac{S_0}{S_*}e^{rT} - 1\right) - \log\frac{S_*}{S_0} \\ & + e^{rT}\int_0^{S_*}\frac{1}{K^2}P(K)dK \\ & + e^{rT}\int_{S_*}^{\infty}\frac{1}{K^2}C(K)dK \Bigg) \end{aligned}$$

5.2.2.1.2.2.4 Practical Issues with Variance Swaps A variance swap is, theoretically, equivalent to a dynamically adjusted constant-dollar exposure to the stock, together with a static long position in a portfolio of options and a forward that, together, replicate the payoff of a log-contract. This portfolio strategy captures variance exactly, provided the portfolio of options contains all strikes between zero and infinity in the appropriate weight to match the log-payoff, and provided the stock price evolves continuously.

As a result, the replication of variance swaps face two main practical problems. First, there is imperfect replication due to limited strike range: because these strikes fail to duplicate the log-contract exactly, they will capture less than the true realized variance (the vega and gamma of a limited strike range both fall to zero when the index moves outside the strike range). Therefore, they have a lower value than that of a true log-contract, and so produce an inaccurate, lower estimate of the fair variance, and even more dramatically so for longer expirations (as it is more likely that the stock price will evolve outside the strike range).

More importantly, typical discrete replications (see Demeterfi et al. 1999; as opposed to piecewise linear or trapezoidal functions, as in Jiang and Tian 2007) significantly underestimate tail events, by, for instance, approximating the varswap payoff with a piecewise linear function:

$$\ln\frac{F(T,T)}{F(0,T)} = \ln\frac{F(T,T)}{K_0} + \ln\frac{K_0}{F(0,T)}, \quad f(x) = \frac{x}{K_0} - 1 - \ln\left(\frac{x}{K_0}\right).$$

The linear term can be omitted, as it can also directly be replicated by shares, leading to the exact same price by linearity, and providing the following weights for the calls and puts:

$$w^c(K_0) = \frac{f(K_1^c) - f(K_0)}{K_1^c - K_0},$$

$$w^c(K_i^c) = \frac{f(K_{i+1}^c) - f(K_i^c)}{K_{i+1}^c - K_i^c} - \sum_{j=0}^{i-1} w^c(K_i^c),$$

$$w^p(K_0) = -\frac{f(K_1^p) - f(K_0)}{K_1^p - K_0},$$

$$w^p(K_i^p) = -\frac{f(K_{i+1}^p) - f(K_i^p)}{K_{i+1}^p - K_i^p} - \sum_{j=0}^{i-1} w^p(K_i^c),$$

$$V(0,T) = u^2 \frac{2}{T}\left(1 - \frac{F(0,T)}{K_0} + \ln\frac{F(0,T)}{K_0}\right) + u^2 \frac{2}{T} DF(T)\left(\sum_{i=0}^{n} w^c(K_i^c) C(K_i^c, T) + \sum_{i=0}^{n} w^p(K_i^p) P(K_i^p, T)\right).$$

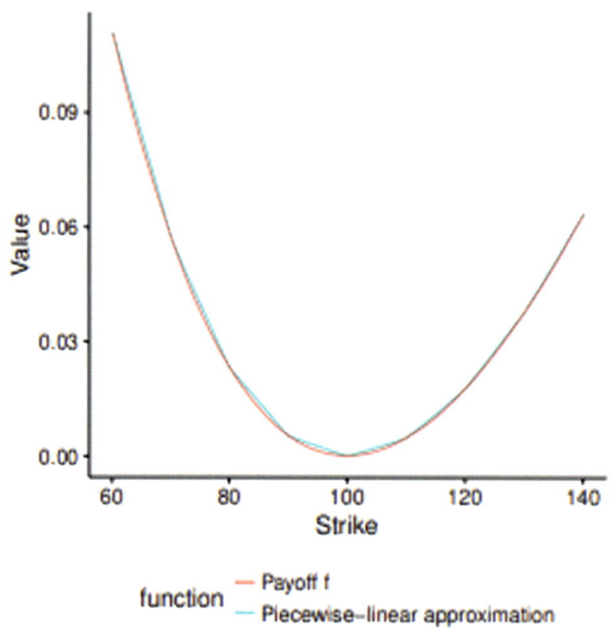

FIGURE 5.68 Payoff versus piecewise linear proxy

This leads to an artificially low variance swap price and invalidates the simple discrete hedging strategy effectiveness in practice.

When the stock price jumps, the log-contract may no longer capture realized volatility, for two reasons. First, if the log contract has been approximately replicated by only a finite range of strikes, a large jump may take the stock price into a region in which variance does not accrue at the right rate. Second, even with perfect replication, a discontinuous stock price jump causes the variance capture strategy to capture an amount not equal to the true realized variance. In reality, both these effects contribute to the replication error:

$$V = \frac{2}{T}\left[\sum_{i=1}^{N}\frac{\Delta S_i}{S_{i-1}} - \log\frac{S_T}{S_0}\right], \quad V = \frac{2}{T}\sum_{i=1}^{N}\left[\frac{\Delta S_i}{S_{i-1}} - \log\frac{S_i}{S_{i-1}}\right].$$

Suppose that all but one of the daily price changes are well behaved – that is, all changes are diffusive, except for a single jump event, characterized by the parameter J, the percentage jump downwards, from S -> S (1 – J):

$$V = \frac{1}{T}\sum\left(\frac{\Delta S_i}{S_{i-1}}\right)^2 = \frac{1}{T}\sum_{\text{no jumps}}\left(\frac{\Delta S_i}{S_{i-1}}\right)^2 + \frac{1}{T}\left(\frac{\Delta S}{S}\right)^2_{\text{jump}}.$$

The contribution of the jump to the realized total variance is given by

$$\frac{1}{T}\left(\frac{\Delta S}{S}\right)^2_{\text{jump}} = \frac{J^2}{T},$$

while the impact of the jump on the quantity is captured by our variance replication strategy:

$$\frac{2}{T}\left(\frac{\Delta S_i}{S_{i-1}} - \log\frac{S_i}{S_{i-1}}\right)_{\text{jump}} = \frac{2}{T}[-J - \log(1-J)].$$

In the limit that the jump size J is small enough to be regarded as part of a continuous stock evolution process, those last two are equal. It is only because J is not small that the variance capture strategy is inaccurate. Therefore, the replication error, or the P&L due to the jump for a short position in a variance swap hedged by a long position in a variance capture strategy, is

$$\text{P\&L due to jump} = \frac{2}{T}[-J - \log(1-J)] - \frac{J^2}{T}, \quad -\log(1-J) = J + \frac{J^2}{2} + \frac{J^3}{3} + \ldots,$$

$$\text{P\&L due to jump} = \frac{2}{3}\frac{J^3}{T} + \ldots.$$

The leading correction is cubic in the jump size J and has a different sign for upward or downward jumps. A large move downwards (J > 0) leads to a profit for the (short variance swap) – (long variance capture strategy), while a large move upwards (J < 0) leads to a loss. Furthermore, a large move one day, followed by a large move in the opposite direction the next day, would tend to offset each other.

Consider now the combination of these two effects within a short position in a variance contract hedged as above while a downward jump is large enough to move the stock price outside the range of option strikes. If the log-payoff were replicated perfectly, the constant-dollar exposure would cancel the linear part of the stock price change, and lead to a convexity gain. Although the log-payoff is not being replicated perfectly, there is still a convexity gain from the jump, but it is smaller in size. However, after this jump, with the stock price now outside the strike range, the vega and gamma of the replicating portfolio are now too low, so the gain from the jump has to be balanced against the subsequent failure of the hedge to capture the smooth variance. In addition, a large move upwards will be doubly damaging: there will be convexity loss due to the jump and the hedge will not capture variance if the jump takes the index outside the strike range.

5.2.2.1.2.2.5 The Delta of a Variance Swap Contract Regarding the sensitivities of variance swaps (assuming a zero interest rate), it is worth remembering that a variance swap is a contract whose payoff is the realized variance of an asset minus a variance "strike" agreed upon at contract initiation. Its payoff at maturity is

$$
\begin{aligned}
Payoff &= N\left(\frac{1}{T-t_0}\sum_{i=1}^{M}\left(\frac{S_t - S_{t-1}}{S_{t-1}}\right) - K_{\text{var}}\right),\\
&= N\left(\frac{1}{T-t_0}\sum_{i=1}^{M}\left(R_t^2\right) - K_{\text{var}}\right),\\
&= N\left(V\left(t_0, T\right) - K_{\text{var}}\right)
\end{aligned}
$$

where V is the realized variance.

In theory, a variance swap has zero delta, but in practice it has a small "shadow delta" due to the correlation between spot and implied volatility (skew): as the spot price goes down, the at-the-money implied volatility will grow, and hence increase the variance swap price. For this reason, the delta of a variance swap is known as the skew delta. In the case of a flat volatility surface, the delta must be zero.

$$
\begin{aligned}
\Delta &= \frac{2}{S_0} e^{-rT}\left\{1 - F\int_0^{\infty}\frac{N\left[-d_1(K)\right]}{K^2}\mathrm{d}K\right\},\\
&= \frac{2}{S_0}\int_0^{\infty}\frac{vega(K)}{K}\frac{\partial\sigma(K)}{\partial K}\mathrm{d}K,
\end{aligned}
$$

where the second formula clearly exhibits the dependence of the variance swap delta on the skew of implied volatility. In the case of a flat volatility surface (i.e. no skew), the delta of the variance swap is equal to zero exactly. In addition, as the skew is generally negative in the equity market, the delta will also be negative.

Proof: the standard variance swap price is given by the formula

$$
V = e^{-rT}\left\{2\left[\int_0^{F}\frac{1}{K^2}PUT(K)\mathrm{d}K + \int_F^{\infty}\frac{1}{K^2}CALL(K)\mathrm{d}K\right] - K_{\text{var}}^2\right\}.
$$

Differentiating the equation with respect to spot price S_0, while bearing in mind that both under-integral functions PUT(K) and CALL(K), as well as the integral limit F, depend on the spot price S_0, the Leibniz integral rule should be applied:

$$
\Delta = 2e^{-rT}\left\{\left.\frac{PUT(K)-CALL(K)}{K^2}\right|_{K=F}\times\frac{\partial F}{\partial S_o} + \int_0^{F}\frac{\partial PUT(K)/\partial S_0}{K^2}dK + \int_F^{\infty}\frac{\partial CALL(K)/\partial S_0}{K^2}dK\right\}.
$$

Taking into account the put–call parity relation and evaluating it for the strike equal to the forward price (K = F), we note that the first term in the equation above is equal to zero.

Plugging in the Black–Scholes expressions for the undiscounted delta of the put and calls,

$$\Delta = 2e^{(\mu-r)T}\left\{-\int_0^F \frac{N(-d_1)}{K^2}dK + \int_F^\infty \frac{1-N(-d_1)}{K^2}dK + \right\}$$

$$= 2e^{(\mu-r)T}\left\{-\int_F^\infty \frac{1}{K^2}dK - \int_0^\infty \frac{N(-d_1)}{K^2}dK\right\},$$

$$= 2e^{-rT}\frac{e^{\mu T}}{}\left\{1 - F\int_0^\infty \frac{N(-d_1)}{K^2}dK\right\}$$

$$\Delta = \frac{2}{S_o}e^{-rT}\left\{1 - F\int_0^\infty \frac{N[-d_1(K)]}{K^2}dK\right\},$$

we note that the volatility smile is present implicitly in the last formula via the function $d_1(K)$. However, it is not yet clear under which condition the delta will be equal to zero or whether it will have a positive or a negative sign. We start by integrating by parts the integral,

$$I \equiv F\int_0^\infty \frac{N[-d_1(K)]}{K^2}\,dK = F\int_0^\infty N[-d_1(K)]d\left(-\frac{1}{K}\right) = F\int_0^\infty \frac{1}{K}N'[-d_1(K)]\frac{d[-d_1(K)]}{dK}dK$$

$FN'(d_1) = KN'(d_2)$, by direct substitution of the expressions for d_1 and d_2 into the definition of the normal distribution function: N ′(x) – exp(– x? / 2) / 2. Substituting gets

$$I = \int_0^\infty N'[-d_2(K)]\frac{d[-d_1(K)]}{dK}dK.$$

We replace the derivative of d_1 by the derivative of d_2 using the relation

$$d_1(K) = d_2(K) + \sqrt{T}\sigma(K),$$

$$I = \int_0^\infty N'[-d_2(K)]\frac{d[-d_2(K)]}{dK}dK - \sqrt{T}\int_0^\infty N'[-d_2(K)]\frac{\partial\sigma(K)}{\partial K}dK,$$

$$= \int_{-d_2=-\infty}^{-d_2=\infty} N'[-d_2]d(-d_2) - \sqrt{T}\int_0^\infty N'[-d_2(K)]\frac{\partial\sigma(K)}{\partial K}dK,$$

$$= 1 - \sqrt{T}\int_0^\infty N'[-d_2(K)]\frac{\partial\sigma(K)}{\partial K}dK.$$

$$\Delta = \frac{2}{S_0}e^{-r\gamma}\sqrt{T}\int_0^\infty N'[-d_2(K)]\frac{\partial\sigma(K)}{\partial K}dK, \quad vega(K) = e^{-rT}K\sqrt{T}N'[-d_2(K)],$$

$$\Delta = \frac{2}{S_0}\int_0^\infty \frac{vega(K)}{K}\frac{\partial\sigma(K)}{\partial K}dK.$$

Similarly, theta is not necessarily as constant as it should be in theory, as movements of the volatility surface can cause it to change.

5.2.2.1.2.3 Sustainable Equity Convexity Risk (Gamma) Semi-Static Hedging: from Theory to Practice

5.2.2.1.2.3.1 Minimizing a Quadratic Deviation as Theoretical Framework In practice, the idea is to minimize a quadratic deviation: such strategies, to be interpreted as orthogonal projections on the subspace of portfolios admitting hedging, are often explicit and, more importantly, linear. This is a definite advantage over hedging techniques based on utility functions.

There are two ways to define coverage of this type.

1. The first is to minimize the quadratic error of hedging at maturity for a self-funded strategy:

$$\inf_{V_0,\phi} \mathbb{E}\left[\left|V_T(\phi)-H\right|^2\right] \text{où } V_T(\phi)=V_0+\int_0^T \phi_t^0 \, dS_t^0+\int_0^T \phi_t \, dS_t\$,$$

with S^0 the risk-free asset and H the payoff. If the interest rates are constant, we can choose the zero-coupon maturity T for the risk-free asset, and the problem is then rewritten

$$\inf_{\check{V}_0,\phi} \mathbb{E}\left[\left|V_T(\phi)-H\right|^2\right] \text{où } V_T(\phi)=\check{V}_0+\int_0^T \phi_t \, d\check{S}_t.$$

2. The second approach is to allow policies that are not self-funded but that require replication to be accurate (i.e. VT (φ) = H). In an incomplete market, this means that the option seller will have to continually inject money into the portfolio. This injection is identified as the cost of the strategy. It is written

$$C_t(\phi)=V_t(\phi)-G_t(\phi),$$
$$V_t(\phi)=\phi_t^0 S_t^0+\phi_t S_t,$$
$$G_t=\int_0^t \phi_8^0 \, dS_8^0+\int_0^t \phi_8 \, dS_8.$$

This minimization strategy, introduced by Föllmer and Sondermann (1986), replicates the payoff and has a strategy cost that should be as small as possible. We therefore seek to minimize at each moment t a residual cost, given by

$$\mathbb{E}\left[\left(\check{C}_T-\check{C}_t\right)^2 \mid \mathcal{F}_t\right],$$

among all eligible strategies resulting from subsequent temporal strategies. When the price of the discounted stock is a martingale, such a strategy still exists, but in the general case it is not always possible to exhibit such a strategy (even for simple cases).

This difficulty led Föllmer and Schweizer (1991) to introduce the concept of risk minimization locally. We then seek to find the extrema of the strategy vis-à-vis small disruptions of the strategy.

The risk of the strategy is then minimized locally in time. Before detailing the hedging strategies, it is necessary to specify the probability measure that makes it possible to calculate the expectation in the previous formulas. In particular, you can choose from the following.

1. P, which is the historical probability. It makes sense but can be difficult to identify.
2. Q, which is the "risk-neutral" probability. It can be obtained by calibration on market prices. Suppose, therefore, that we have chosen the probability risk-neutral Q under which the discounted asset S^ is a martingale. It is also assumed that the payoff H ∈ L 2 (Ω, F, Q) and that S^ is of integrable square. If we consider the portfolios of the form

$$\mathcal{S} = \left\{ \phi \text{càdlàg predictable} \, \mathbb{E}\left| \int_0^T \phi_t \; d\check{S}_t \right|^2 < +\infty \right\}$$

then the set A of the attainable payoffs is a closed subspace of L 2 (Ω, F, Q), and the quadratic hedging problem becomes an orthogonal projection

$$\inf_{\check{V}_{0,\phi}} \mathbb{E}\left[\left| V_T(\phi) - H \right|^2 \right] = \inf_{A\in^-} || H - A ||^2_{L^2(Q)} \, .$$

We can then use the Galchuk–Kunita–Watanabe decomposition, which states that H can be represented by

$$H = \mathbb{E}[H] + \int_0^T \phi_t^H \; \mathrm{d}\check{S}_t + N_T^H$$

where N H T is an integrable square martingale orthogonal to S^. The optimal hedging strategy is then given by φ H and the initial cost of the portfolio is V0 = exp (−r(T −t)E[H]. If we now introduce the martingale H^ t = E[H| Ft], we get

$$\check{H} = \mathbb{E}[H] + \int_0^t \phi_s^H \; \mathrm{d}\check{S}_s + N_t^H .$$

Orthogonality implies

$$\left\langle \check{H} - \int_0 \phi_s^H \; \mathrm{d}\check{S}_s , \check{S} \right\rangle = 0,$$

which means that the optimal hedging can be expressed using a covariance between the option price and the stock price:

$$\phi_t^H = \frac{\mathrm{d}\langle \check{H}, \check{S} \rangle_t}{\mathrm{d}\langle \check{S}, \check{S} \rangle_t} .$$

We then notice that, by optimizing the overall error, we obtain a strategy to minimize the risk of quadratic hedging. Indeed, for any strategy φ, we have

$$\mathbb{E}\left[\left(\check{C}_T-\check{C}_t\right)^2 \mid \mathcal{F}_t\right] =\left(\check{H}_t-\check{V}_t\right)^2+\mathbb{E}\left[\left(H-\check{H}_t-\int_t^T \phi_s \, d\check{S}_s\right)^2 \mid \mathcal{F}_t\right]$$
$$=\left(\check{H}_t-\check{V}_t\right)^2+\mathbb{E}\left[\left(N_T-N_t\right)^2 \mid \mathcal{F}_t\right]+\mathbb{E}\left[\left(\int_t^T\left(\phi_s-\phi_s^H\right) d\check{S}_s\right)^2 \mid \mathcal{F}_t\right],$$

so, to minimize this expression, we must take $\varphi = \varphi^H$ and choose φ^0 such that V˄t = H˄t for all t. And, in this case, the cost of the strategy becomes

$$\check{C}_t=\check{V}_t-\int_0^t \phi_8^H \, d\check{S}_8=\mathbb{E}[H]+N_t^H .$$

However, it is necessary to take this result with tweezers, because the choice of a minimization of a moment of order two under the probability "risk-neutral" is not trivial. Q represents a "pricing" rule and does not correspond to a statistical description of the market, so the P&L of a portfolio can have a greater variance than its "risk-neutral variance". Nevertheless, it is often difficult to distinguish a stock from a martingale, because it would require data over a very long period (sometimes much longer than the option's own life). In addition, hedging is often a "local" activity, in the sense that rebalancing is essentially daily and calculated according to localized temporal data – time during which the stock actually behaves almost like a martingale.

5.2.2.1.2.3.2 Equity Convexity Semi-Static Hedging of European Options In the insurance market, products are kept for the long term. For this and other technical reasons, it is necessary to try to minimize the number of rebalances during the life of the product.

We therefore no longer use a dynamic hedging but, rather, a static, or even semi-static, hedging. Here we detail the characteristics of such a replication. We consider an asset S and a sufficiently regular function h such that the payoff is written h(S). We can then obtain a development of the payoff around any point κ in the form

$$h(S)=h(\kappa)+h'(\kappa)(S-\kappa)+\int_\kappa^{+\infty} h''(K)(S-K)^+ dK+\int_0^\kappa h''(K)(K-S)^+ dK.$$

The first two terms correspond to a tangential part of the payoff at point κ while the last two are a correction related to the nonlinearity of the payoff. We can also see a decomposition following h(κ) zero coupons, h'(κ) forward contracts of strike κ, h"(κ)dK strike calls greater than κ and h"(κ)dK strike puts less than κ. In general, we can write the initial value $V_0[h]$ as a function of the initial prices of zero coupons B_0, calls $C_0(K)$ and puts $P_0(K)$ in the form

$$V_0[h]= h(\kappa)B_0+h'(\kappa)\left[C_0(\kappa)-P_0(\kappa)\right]+\int_0^\kappa h''(K)P_0(K)dK+\int_\kappa^{+\infty} h''(K)C_0(K)dK.$$

Take, for example, a European payoff in the currency – i.e. h(S) = (S – Kc) + with $K_c < F_0$. Using the previous decomposition, we find

$$C_0\left(K_c\right)=\left(F_0-K_c\right)B_0+P_0\left(K_c\right)\$,$$

which is precisely the call–put parity. In fact, the function h does not really need to be regular, and we can consider more general functions. If we write V (t, S; T) the value of an option whose payoff is written h(ST), we have

$$V\left(t_0,S;T\right)=\int_0^{\kappa} V(t,K;T)\frac{\partial^2}{\partial K^2}P\left(t_0,K;t,K\right)\mathrm{d}K+\int_{\kappa}^{+\infty} V(t,K;T)\frac{\partial^2}{\partial K^2}C\left(t_0,K;t,K\right)\mathrm{d}K.$$

By integrating twice per part, we get

$$\begin{aligned}V\left(t_0,S;T\right)=&V(t,\kappa;T)B\left(t_0;t\right)\\&+\frac{\partial}{\partial S}V(t,\kappa;T)\left[C\left(t_0,S;t,\kappa\right)-P\left(t_0,S;t,\kappa\right)\right]\\&+\int_0^{\kappa}\frac{\partial^2}{\partial S^2}V(t,K;T)P\left(t_0,S;t,K\right)\mathrm{d}K\\&+\int_{\kappa}^{+\infty}\frac{\partial^2}{\partial S^2}V(t,K;T)C\left(t_0,S;t,K\right)\mathrm{d}K.\end{aligned}$$

If we grow t to T, we find a result similar to that obtained previously in general:

$$\begin{aligned}V\left(t_0,S;T\right)=&h(\kappa)B\left(t_0;t\right)+h'(\kappa)\left[C\left(t_0,S;T,\kappa\right)-P\left(t_0,S;T,\kappa\right)\right]\\&+\int_0^{\kappa}h''(K)P\left(t_0,S;T,K\right)\mathrm{d}K+\int_{\kappa}^{+\infty}h''(K)C\left(t_0,S;T,K\right)\mathrm{d}K.\end{aligned}$$

And, if we choose κ again as the forward F price, we come across a simplified result:

$$\begin{aligned}V\left(t_0,S;T\right)=&V(t,F;T)B\left(t_0;t\right)\\&+\int_0^{F}\frac{\partial^2}{\partial S^2}V(t,K;T)P\left(t_0,S;t,K\right)\mathrm{d}K\\&+\int_{F}^{+\infty}\frac{\partial^2}{\partial S^2}V(t,K;T)C\left(t_0,S;t,K\right)\mathrm{d}K.\end{aligned}$$

Let us take the example of a strike call K_c and maturity T. Decomposition is rewritten

$$\begin{aligned}C\left(t_0,S;T,K_c\right)=&C\left(t,F;T,K_c\right)B\left(t_0;t\right)\\&+\int_0^{F}\frac{\partial^2}{\partial S^2}C\left(t,K;T,K_c\right)P\left(t_0,S;t,K\right)\mathrm{d}K\\&+\int_{F}^{+\infty}\frac{\partial^2}{\partial S^2}C\left(t,K;T,K_c\right)C\left(t_0,S;t,K\right)\mathrm{d}K.\end{aligned}$$

If we assume that the gammas are positive, then the value of the call at the present moment is higher than the present value it will have at time t if the asset is at the expected price F. The difference consists of a weighted sum of calls and puts outside the maturity currency t. In this case, one can buy the zero coupon of maturity t close to t_0 as well as the options, calls and puts, lead them to maturity t and continue without additional cost until maturity T.

Illustrating on a single European put option, we can compare the distribution histograms and see that the standard deviation is greatly improved (see Figure 5.69).

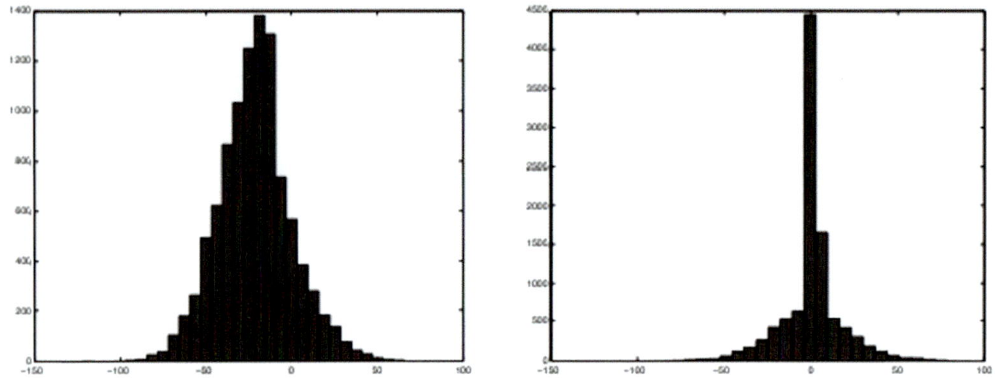

FIGURE 5.69 Delta-hedging P&L versus semi-static hedging P&L

5.2.2.1.2.3.3 Equity Convexity Semi-Static Hedging of Path-Dependent Exotic Options In general, this formula can also be used to replicate "path-dependent" payoffs. Indeed, suppose that the payoff at time T depends on both S_T and S_t. If we condition this payoff by S_t, then the payoff is now a function of S_T only and can therefore be replicated using T maturity options. If we now note U(t, S_t; T) the random future cost of this replicated portfolio, then it can itself be replicated using zero coupons and t-maturity options. Let us take the example of a ratchet option. It is a European maturity option T whose strike is a function $k(S_t)$ with $t \in [t_0, T]$. The formula gives

$$V(t_0,S;T) = C(t,F;T,k(F))B(t_0;t) + \int_0^F \frac{\partial^2}{\partial S^2} C(t,K;T,k(K))P(t_0,S;t,K)\,\mathrm{d}K + \int_F^{+\infty} \frac{\partial^2}{\partial S^2} C(t,K;T,k(K))C(t_0,S;t,K)\,\mathrm{d}K.$$

We now consider a wider class of options, the one whose payoff at time T depends on the values of an asset in n distinct dates. We can then write this payoff in the form

$$V_T = f\left(S_{t_0}, S_{t_1}, \ldots, S_{t_n}\right)$$

for $n \in N$, $t_0 = 0$ and $t_n = T$. It will be assumed that we can define a recursive representation of the payoff as follows: there is a random variable H_T and a function φ such that

$$V_T = \phi(H_T)$$
$$H_{t_i} = g_i\left(H_{t_{i-1}}, S_{t_{i-1}}, S_{t_i}\right) \text{for } i = 1, \ldots, n,$$

where g_i is an explicit function and H_0 and S_0 are given constants.

This hypothesis of recursion is not very restrictive, since it allows us to treat the case of options (dependent on time in a discrete way) of Asian type, barrier, lookback, Bermuda, ratchets, etc. However, it is not possible to deal with options dependent on a continuous time interval or American options.

In a Markovian framework, we can exhibit at each moment (ti, i = 0, 1, . . ., n − 1) functions giving the value of calls and puts. Note $C_{i,i+1}$(S, K; θ) (respectively Pi,i+1(S, K; θ)) such a function giving the price of a call (respectively a put) and linking the price of the asset S_{ti}, the strike K and the vector parameter θ (which represents the market of maturity options t_{i+1}). The parameter θ may depend on the initial market data, but it is obviously not allowed to vary randomly. It is assumed that at every discrete moment we have access to a continuum of European options of maturity t_{i+1} for all possible strikes. The semi-static replication algorithm is then defined recursively from the date t_{n-1}.

- At time t_{n-1}, conditional on the history and the value of the asset $S(t_{n-1})$, the final payoff is only an explicit function of the final value of the asset:

$$V_T = \phi\left(g_n\left(H_{t_{n-1}}, S_{t_{n-1}}, S_T\right)\right) = f_n\left(S_T; H_{t_{n-1}}, S_{t_{n-1}}\right).$$

- Using T-maturity options, the final payoff can be replicated:

$$\begin{aligned} f_n\left(S_T; H_{t_{n-1}}, S_{t_{n-1}}\right) &= f_n\left(\kappa_n, H_{t_{n-1}}, S_{t_{n-1}}\right) \\ &\quad + f_n'\left(\kappa_n; H_{t_{n-1}}, S_{t_{n-1}}\right)\left[\left(S_T - \kappa_n\right)^+ - \left(\kappa_n - S_T\right)\right] \\ &\quad + \int_0^{\kappa_n} f_n''\left(K; H_{t_{n-1}}, S_{t_{n-1}}\right)\left(K - S_T\right)^+ \mathrm{d}K \\ &\quad + \int_{\kappa_n}^{+\infty} f_n''\left(K; H_{t_{n-1}}, S_{t_{n-1}}\right)\left(S_T - K\right)^+ \mathrm{d}K. \end{aligned}$$

- The value of the exotic option can then be expressed as a function of the risk-free rate and the options at time t_{n-1}:

$$\begin{aligned} V_{t_{n-1}}^{f_n} &= f_n\left(\kappa_n, H_{t_{n-1}}, S_{t_{n-1}}\right)e^{-r(T-t_{n-1})} \\ &\quad + f_n'\left(\kappa_n; H_{t_{n-1}}, S_{t_{n-1}}\right)\left[C_{t_{n-1}}\left(\kappa_n, T\right) - P_{t_{n-1}}\left(\kappa_n, T\right)\right] \\ &\quad + \int_0^{\kappa_n} f_n''\left(K; H_{t_{n-1}}, S_{t_{n-1}}\right)P_{t_{n-1}}\left(\kappa_n, T\right)\mathrm{d}K \\ &\quad + \int_{\kappa_n}^{+\infty} f_n''\left(K; H_{t_{n-1}}, S_{t_{n-1}}\right)C_{t_{n-1}}\left(\kappa_n, T\right)\mathrm{d}K. \end{aligned}$$

$$\begin{aligned} V_{t_{n-1}}^{f_n} &= f_n\left(\kappa_n, H_{t_{n-1}}, S_{t_{n-1}}\right)e^{-r(T-t_{n-1})} \\ &\quad + f_n'\left(\kappa_n; H_{t_{n-1}}, S_{t_{n-1}}\right)\left[C_{n-1,n}\left(S_{t_{n-1}}, \kappa_n; \theta\right) - P_{n-1,n}\left(S_{t_{n-1}}, \kappa_n; \theta\right)\right] \\ &\quad + \int_0^{\kappa_n} f_n''\left(K; H_{t_{n-1}}, S_{t_{n-1}}\right)P_{n-1,n}\left(S_{t_{n-1}}, \kappa_n; \theta\right)\mathrm{d}K \\ &\quad + \int_{\kappa_n}^{+\infty} f_n''\left(K; H_{t_{n-1}}, S_{t_{n-1}}\right)C_{n-1,n}\left(S_{t_{n-1}}, \kappa_n; \theta\right)\mathrm{d}K \\ &= V\left(H_{t_{n-1}}, S_{t_{n-1}}, t_{n-1}\right). \end{aligned}$$

Just do the calculation for all possible values of $H(t_{n-1})$, and $S(t_{n-1})$.

At time t_{n-2}, the value at time tn−1 of the exotic option is a known function of the value of the stock at time t_{n-2} – i.e.

$$\begin{aligned}V_{t_{n-1}}^{f_n} &= V\left(H_{t_{n-1}}, S_{t_{n-1}}, t_{n-1}\right)\\ &= V\left(g_{n-1}\left(H_{t_{n-2}}, S_{t_{n-2}}, S_{t_{n-1}}\right), S_{t_{n-1}}, t_{n-1}\right)\\ &= f_{n-1}\left(S_{t_{n-1}}; H_{t_{n-2}}, S_{t_{n-2}}\right).\end{aligned}$$

Using maturity options tn−1, it is possible to replicate V, and the algorithm continues until time t_0.

- At time t_0, we know H_0 and S_0, so the conditioning is no longer carried out on all possible values. This algorithm is possible if the number of intermediate prices involved in the payoff is not too large.
- This algorithm can be adapted by considering the "trading" times as the exit times (barrier options). It depends heavily on the value functions of the prices of calls and puts. In short:
  - one can replicate a path-dependent payoff by a semi-static hedging;
  - then we use options for hedging; the more precise we are, the fewer updates we will have to make and the less dependent we will be on future price models;
  - we can use the replication formula by playing on the "strike–maturity" entanglement; we can choose, for example, a maturity that minimizes a certain criterion (number of calls very far from the currency, number of puts available on the market, etc.).

5.2.2.1.2.3.4 Improving the Equity Convexity Semi-Static Hedging P&L by Using Lévy Processes Lévy processes offer the advantages of modelling the asset returns as an infinite series of pure jump processes of various size, something that fits nicely with our intuition. Naturally, they hold some restrictive assumptions, such as independence and the same distribution law of returns increment; however, some of these be alleviated, in particular the independence of increments (as illustrated in Chapter 3).

Lévy process exponential asset models generally correspond to incomplete markets, making exact replication impossible. The hedge should then be interpreted as an approximation of the payoff by an eligible portfolio.

$$\Delta = \frac{\exp(-\alpha k)}{\pi}\int_0^{+\infty}\exp(-ivk)\frac{\exp(-rT)\dfrac{\partial\phi(T, v-(\alpha+1)i)}{\partial S_0}}{\alpha^2+\alpha-v^2+i(2\alpha+1)v}dv.$$

Now assume that the (risk-neutral) model for the price of the underlying at time T is of the form

$$S_T = S_0\overline{S}_T,$$

$$\begin{aligned}\frac{\partial\phi(T, v-(\alpha+1)i)}{\partial S_0} &= \frac{\partial}{\partial S_0}\mathbb{E}\left[\exp\left(i(v-(\alpha+1)i)\left(\log\left(S_0\right)+\log\left(\overline{S}_T\right)\right]\right.\right.\\ &= \frac{\phi(T, v-(\alpha+1)i)(\alpha+1+iv)}{S_0}.\end{aligned}$$

$$\Delta = \frac{\exp(-\alpha k)}{\pi}\int_0^{+\infty}\exp(-ivk)\frac{\exp(-rT)\phi(T, v-(\alpha+1)i)}{S_0(\alpha+iv)}dv.$$

$$\Gamma = \frac{\exp(-\alpha k)}{\pi}\int_0^{+\infty}\exp(-ivk)\frac{\exp(-rT)\phi(T, v-(\alpha+1)i)}{S_0^2}dv.$$

Let us now consider a put that we delta–gamma semi-statically hedge.

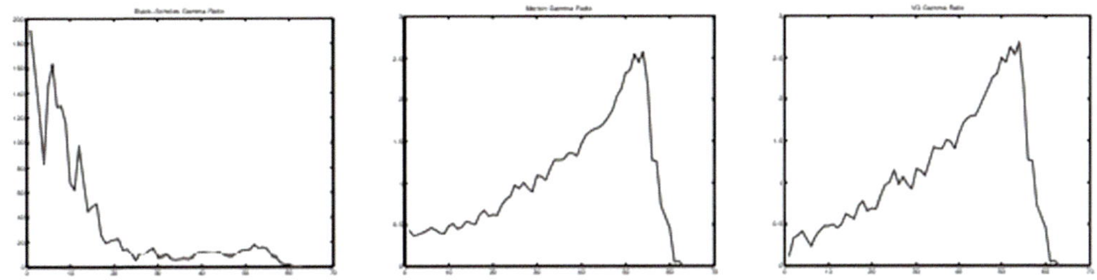

(a) Gamma Ratio - Black-Scholes (b) Gamma Ratio - Merton (c) Gamma Ratio - VG

FIGURE 5.70 Gamma ratio depending on asset modelling

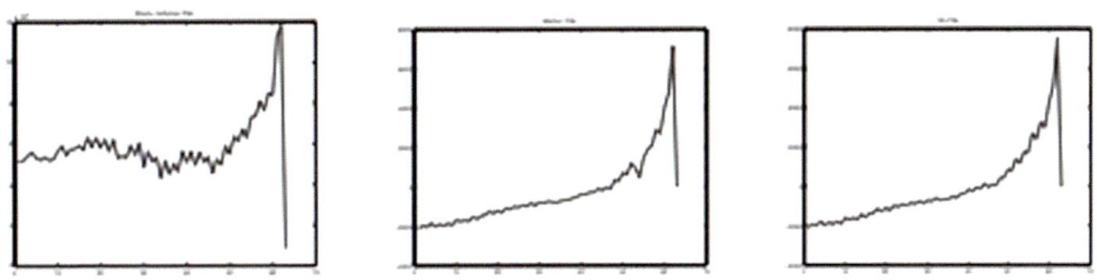

(d) P&L - Black-Scholes (e) P&L - Merton (f) P&L - VG

FIGURE 5.71 Hedging P&L depending on asset modelling

The stability of the gamma ratio is greatly improved, so the rebalancing is less, and allows us to limit not just the large variations of P&L but also the transaction costs: there is an improvement by a factor of five for the ratios, and by a factor of 15 for the hedging P&L!

5.2.2.2 Sustainable Vega Hedging

5.2.2.2.1 Gamma and Vega Hedging Implementation Since a spot, forward or future position is linear in its payoff, it has no second-order derivative, in contrast with life insurance liabilities, which usually embed put-like options that hold nonlinear (asymmetric) payoffs. While we can get away with hedging delta with a linear position because delta is a linear approximation (a tangent used to estimate the rate of change), we cannot hedge higher-order Greek (such as gamma) exposure by buying a position in the underlying.

Vega brings additional challenges. We do not directly observe volatility, and the impact of changes in volatility are not linear and cannot be modelled or hedged using linear instruments. They vary based on moneyness and the time to maturity of the option. To hedge these nonlinear changes represented by gamma and vega we have to try a nonlinear recipe. We have to buy similar (ideally, cheaper) options.

Neutralizing delta, gamma and vega requires trading in two different hedging options, and we can neutralize both gamma and vega at the same time by solving two equations (see Figure 5.72).

$$\begin{aligned}\Gamma_{\text{target}} + \beta_{\text{hedge1}}\Gamma_{\text{hedge1}} + \beta_{\text{hedge2}}\Gamma_{\text{hedge2}} &= 0\\ V_{\text{target}} + \beta_{\text{hedge1}}V_{\text{hedge1}} + \beta_{\text{hedge2}}V_{\text{hedge2}} &= 0\end{aligned} \quad \text{->} \quad \begin{aligned}\beta_{\text{hedge1}} &= \frac{\Gamma_{\text{hedge2}}V_{\text{target}} - \Gamma_{\text{target}}V_{\text{hedge2}}}{\Gamma_{\text{hedge1}}V_{\text{hedge2}} - \Gamma_{\text{hedge2}}V_{\text{hedge1}}}\\ \beta_{\text{hedge2}} &= \frac{\Gamma_{\text{target}}V_{\text{hedge1}} - \Gamma_{\text{hedge1}}V_{\text{target}}}{\Gamma_{\text{hedge1}}V_{\text{hedge2}} - \Gamma_{\text{hedge2}}V_{\text{hedge1}}}\end{aligned}$$

FIGURE 5.72 Portfolio weight of hedge assets within gamma–vega hedging

Should you leave some residual gamma and vega exposure or should our hedging portfolio neutralize it in its entirety?

From the fixed-income and ALM world comes an interesting parallel. In the ALM world, to preserve surplus or immunize capital from interest rate changes, we match the weighted average of the first-order sensitivity (modified duration) but we also require that the second-order sensitivity (convexity) of assets be greater than the convexity of liabilities.

Similarly, in the options world, while we may match delta completely, we would aim to create a hedging portfolio with excess gamma and vega exposure compared to our original short position. If that is not possible we will use exposure limits to track our sensitivity to these two factors. Our hedge is going to be an imperfect semi-hedge, which will leave some room for benefiting from unanticipated changes in market volatility as well as large unexpected jumps in the price of the underlying.

We begin by identifying the universe of options available for hedging, for which the Greeks are computed. These individual option Greeks are used to calculate weighted average Greeks for the overall hedge portfolio. Unhedged Greek exposure is the difference between the Greek for the short position and the hedge position portfolios.

For our objective function we set up the objective function cell as gamma–vega-unhedged, which we will aim to minimize, while ensuring non-negative values, and using a quadratic model for estimates and a conjugate method for searching for optimal solutions.

Next we define the constraints with respect to expiry and premium as mentioned earlier: ensure that our average hedge portfolio expiry is less than the target maturity, and that the hedge portfolio gamma and vega figures are at least the same as, but ideally greater than, the gamma and vega estimates for our target. Note that there are a number of additional constraints on the number of option contracts the hedge portfolio can include, as well as the proportion of the original premium available for hedging.

		Unhedged Limit	Unhedged Exposure
Portfolio Gamma	0.076	0%	0.000
Portfolio Vega	0.203	19%	0.047
Portfolio Delta	0.269	52%	0.29
Portfolio Rho	0.038	53%	0.04
Portfolio Theta	- 0.988	8%	- 0.08
Portfolio Price	0.670	47%	1.43
Portfolio Expiry	**1.000**	1	0.00
Gamma-Vega-Unhedged	**0.047**		

Allocation	Date	Expiry	Expiry (Years)	Strike	Money-ness	C
71%	05/14/14	360	1.00	23.53	30	C
24%	05/14/14	150	0.42	19.91	10	C
2%	05/14/14	720	2.00	20.815	15	C
2%	05/14/14	720	2.00	21.72	20	C
1%	05/14/14	1080	3.00	23.53	30	C
1%	05/14/14	1080	3.00	24.435	35	C
1%	05/14/14	720	2.00	25.34	40	C
0%	05/14/14	720	2.00	26.245	45	C
0%	05/14/14	720	2.00	27.15	50	C
0%	05/14/14	90	0.25	24.435	35	C

FIGURE 5.73 Hedging portfolio positions

Here the gamma–vega-unhedged objective function has been minimized to 0.047%, with gamma completely neutralized but some residual vega exposure left. While we did not focus on the minor Greeks (delta, rho, theta), these exposures have also been reduced.

The solution picked two major positions and five minor positions; the majors were OTM with expiries within a year, with the minors all deep OTM with longer-dated expiries. The model is, essentially, calculating the cheapest ratio of gamma to premium while complying with the constraints, whereas the vega solution is possibly more intricate.

Is there a better solution possible? Let us add two new conditions to refine the model: hedge portfolio gamma and vega greater than the target position, forcing us to buy multiple options (> 400% allocation versus 100% only beforehand) to help in this respect.

As a result, Excel's Solver failed to find a feasible solution, but at the point it gave up it had arrived at a partial solution (see Figure 5.74).

		Unhedged Limit	Unhedged Exposure
Portfolio Gamma	0.076	0%	- 0.000
Portfolio Vega	0.075	70%	0.174
Portfolio Delta	0.113	80%	0.45
Portfolio Rho	0.009	89%	0.07
Portfolio Theta	- 1.031	4%	- 0.04
Portfolio Price	0.202	11%	1.89
Portfolio Expiry	**1.000**	1	- 0.00
Gamma-Vega-Unhedged	**0.174**		

FIGURE 5.74 Hedging portfolio positions: Solver's solution

We now have ten major and seven minor positions, with all maturities under one year and deep out-of-the-money options.

In order to improve the vega position, we focus on our attention solely on the vega difference between the hedge portfolio and the original short position (versus gamma and vega), while the allocation has also been increased to force multiple purchases of 900% or more in the positions.

As Solver again fails to find a solution, we drop the multiple position force constraint and remove an additional vega limit condition, which enables to find a solution; however, the cost of the hedge portfolio has gone up. Positions are down to two with shorter maturities. Indeed, the constraint that forces multiple position purchases pushes Solver to focus on deep out-of-the-money options and it allows us to keep the hedging cost to a minimum.

Rather than limiting ourselves to a single short position, we have now sold multiple options. Given the fact that we have a range of maturities in our short portfolio and the fact that volatility varies across maturity buckets, we will need to revise our approach by tracking vega and gamma exposures across maturity buckets.

From an optimization point of view, though, given the complexity of this problem and Solver's inherent limitations, in the first instance we calculate the cumulative bucket gap and use that as the objective function. The first two constraints specify that the hedge portfolio Greeks (vega and gamma) should be at least as high as the short position Greeks. The next two constraints require that Greek differences in each maturity bucket amount to zero. The last two constraints ensure that we do not spend more than what we have (premium available) and forces hedge portfolio allocation across multiple positions.

The solution does a reasonable job of hedging gamma and vega exposure (the maturity buckets are also perfectly offset) but appears to be a bit pricey. However, when we review the portfolio allocation we run into issues. The first is the number of options used to hedge the short portfolio. There are 29 individual positions to hedge our 13 short positions across nine maturity buckets.

We therefore change the objective function. Previously we minimized the cumulative bucket gap. In this iteration we minimize the cell representing hedge portfolio cost. We also remove the additional constraint around total portfolio allocation, to give Solver the freedom to move around our hedge universe and pick as many positions as it needs to come up with a more cost-effective solution.

As before, the maturity bucket grid is a match. The hedge portfolio cost has also declined. While gamma and vega remain perfectly hedged, the delta, rho and theta unhedged exposures have increased. However, the premium saving adequately compensates us for this change.

Do we really need the four column constraints relating to the maturity buckets (i.e. maturity buckets constraints and difference by maturity bucket constraints)? Can we just get by with two rather than all four of these constraints?

- We go ahead and drop both (vega and gamma) maturity bucket constraints, assuming that the difference by maturity bucket constraints should be sufficient to reach a solution. The answer, unfortunately, is "No". Solver cannot find an optimal solution without these constraints, so they are not redundant.
- Rather than using the cumulative bucket gap that adds both gamma and vega maturity gap buckets together, we use the cumulative maturity gap for gamma only. When we run Solver we reach a solution quite close to our second pass. This results in a marginally lower premium for our hedge portfolio over our previous solution. Thus, we do not need the vega bucket constraint to solve for the optimal solution.
- However, a bigger issue now is the number of positions required to hedge the portfolio. While our new and improved solution is certainly cheaper than our first pass, there are now 86 positions required to hedge our 13 short call positions. We add a new constraint that restricts the number of positions first to 30 and then to 50. However, in both instances a feasible solution is not found. Nonetheless, when we increase the limit to 55 a feasible solution is identified, and it comes with an added bonus of lower cost. In the last iteration the number of positions is down to 52 and the premium is down to \$7.94, from \$9.62 earlier.

Let us now look at the solution from a slightly different point of view. While our gamma and vega exposures stay perfectly hedged, the hedge premium cost and left-over delta exposure goes up and down in an inverse manner. Premium cost comes down, delta goes up; delta goes down, premium cost goes up. Rather than minimize the premium, we would like to see if we could find an optimal solution that would also balance the left-over delta amount side by side with our gamma and vega hedging solution. So we create a new objective function that focuses on delta.

Interestingly enough, in this solution it is not just delta that comes down, as the number of positions required to hedge our exposure also drops, by 50%, with about 11 major positions and 15 minor exposures. However, premium jumps by 50%.

The two solutions should point out an obvious lesson. Shorter-dated options are more sensitive to changes in the price of the underlying (gamma at work) and less sensitive to changes in implied volatility (vega at work). Longer-dated options are more sensitive to changes in implied volatility. To find the optimal solution, you need to find the right balance between the two, which, within our limited setting, Solver does reasonably well. Any left-over delta exposure can then be hedged using the delta-hedging models we have discussed earlier.

5.2.2.2.2 Vega and Skew of Variance Swaps

One critical source of credibility for the life insurance business is the potentially large instability of its net income (in particular regarding variable annuities), which notably suffers from the high vega of derivatives assets, in contrast to the usual long-term constant volatility assumption implying zero-vega P&L of the liabilities under US GAAP or IFRS.

(i) Vega Sensitivity of a Variance Swap

The variance swap's vega is

$$Vega = \frac{\partial Varswap_t}{\partial \sigma_{implied}} = Notional \times \left(2\sigma_{implied}\right) \times \frac{T-t}{T},$$

which is independent of the stock level. As variance is additive, the vega decays linearly with time. For example, 100K vega of a one-year variance swap at inception will have 75K vega after three months, 50K after six months and 25K after nine months.

Variance swaps "accrue" variance based on each day's realized return innovation. Mark-to-market pricing at time t is therefore a weighted average of the "accrued variance" and the expectation of variance looking ahead. As such, the value of the contract can change dramatically based on the change in expected future variance. The mark-to-market value of a variance swap is

$$MTM = Ne^{-r(T-t)}\left(\lambda\left(V\left(t_0,t\right)-K_{\text{var}}\right)+(1-\lambda)\left(K_t-K_{\text{var}}\right)\right),$$

where λ is the proportion of time elapsed between t_0 and t:

$$\lambda(t) = \frac{t-t_0}{T-t_0} = 1-\frac{\tau}{T-t_0}.$$

Thus, the exposure of the variance swap to changes in fair-market variance, so-called "variance vega" and denoted Υ, is positively related to the proportion of time remaining to the expiration of the swap:

The change in variance vega Υ for a change in time to maturity is therefore related to the inverse of the total period covered by the contract, so that the closer we move to expiry, the lower the variance vega. If, for example, we have a two-year variance swap, its variance vega will fall by a half after one year, illustrating that an investor rolling into a new variance swap would experience varying levels of variance vega over time.

While variance swaps allow investors to trade an at-the-money delta-hedged position without delta hedging and without adjusting for the moneyness of the options, they do not – as often purported – represent "pure" volatility exposure, in that variance vega varies dramatically over the life of a variance swap.

(ii) Skew Sensitivity of Variance Swaps

The steeper the skew, the higher the fair value. Unfortunately, there is no straightforward formula to measure skew sensitivity, but we can have a rough idea using the rule of thumb for linear skew by Demeterfi et al. (1999):

$$K_{vat}^2 \approx \sigma_{ATMF}^2\left(1+3T\times Skew^2\right),$$

$$Skew\,sensitivity \approx 6\times Notional\times \sigma_{ATMF}^2 \times \frac{T-t}{T}\times Skew.$$

Example: consider a one-year variance swap on a vega notional of \$200,000, struck at 15. At-the-money-forward volatility is 14, and the 90–100 skew is 2.5 vegas. According to the rule of thumb, the fair strike is approximately 14 × (1 + 3 × (2.5 / 10)2) = 16.62. If the 90–100 skew steepens to three vegas, the change in mark to market will be

$$\Delta MTM \approx \underbrace{6\times\frac{200{,}000}{2\times 15}\times 14^2\times\frac{2.5}{10}}_{sensitivity}\times\underbrace{\left(\frac{3-2.5}{10}\right)}_{\Delta skew}\approx 100{,}000.$$

5.2.2.2.3 Forward-Starting Variance Swaps With the commoditization of variance swaps, variance is becoming an asset class of its own. A number of volatility indices have been launched or adjusted to follow the weighting methodology of the replicating portfolio, in particular the new Chicago Board Options Exchange's SPX Volatility Index (VIX) and the Deutsche Börse's VSTOXX Volatility Index. Due to the realization that variance, not volatility, was the correct measure of deviation, on Monday 22 September 2003 the VIX index moved away from using ATM implied towards a variance-based calculation.

Two popular alternative approaches have been for insurers to enter into a forward-starting variance swap or to purchase an option on a variance swap. The forward-starting variance swap allows insurers to hedge against future increases in volatility while locking in at today's relatively low strike levels. An option on a variance swap presents the advantage of limiting the downside costs if volatility moves against the purchaser yet providing upside as necessary to hedge the liability. The payout profile would resemble the "hockey stick" shape of a call option, but would possess the additional benefit of a "curved handle" due to the positive convexity of the long variance.

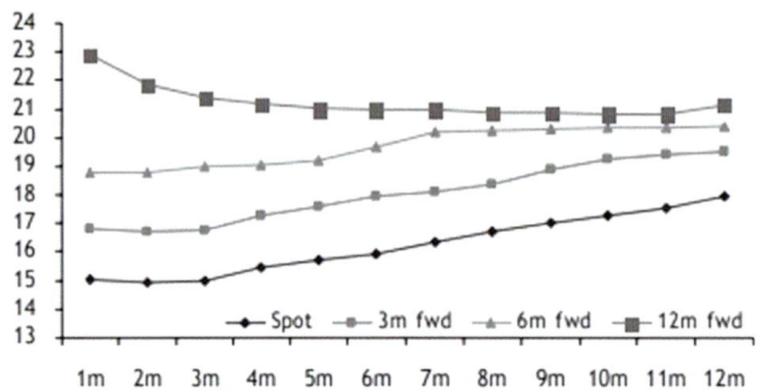

FIGURE 5.75 Forward-starting variance versus maturity

From a hedging perspective, the main attraction of forward-starting products is that they provide investors with long-term volatility (or vega) exposure, without having exposure to short-term volatility (or gamma). Forward-starting products are low in cost, but also lower in payout.

As there is zero gamma until the forward-starting product starts, the product does not have to pay any theta. Forward-starting products can therefore be seen as a low-cost, lower-payout method of trading volatility. However, if there is a rise in volatility surfaces before the forward-starting period is over, they are likely to benefit less than vanilla options (this is because the front end of volatility surfaces tends to move the most, and this is the area to which forward start has no sensitivity). While forward-starting products have zero mathematical theta, they do suffer from the fact that volatility and variance term structure are usually expensive and upward-sloping. The average implied volatility of a forward-starting product is likely to be higher than a vanilla product, which will cause the long forward-starting position to suffer carry as the volatility is re-marked lower.

The easiest forward-starting product to trade is a variance swap, as it can be hedged with two static variance swap positions (one long, one short): a forward-starting variance swap between T1 and T2, to be replicated via a long variance swap to T2, and short variance swap to T1. We define N1 and N2 to be the notionals of the variance swaps to T1 and T2 respectively:

$$\text{Payout long variance to } T_2 = \text{Payout long variance to } T_1 + \text{Payout long variance } T_1 \text{ to } T_2$$

$$\Rightarrow N_2 \frac{\sum_{i=1}^{T_2}\left[\text{Ln}(\text{return}_i)\right]^2}{T_2} = N_2 \frac{\sum_{i=1}^{T_1}\left[\text{Ln}(\text{return}_i)\right]^2}{T_2} + N_2 \frac{\sum_{i=T_1+1}^{T_2}\left[\text{Ln}(\text{return}_i)\right]^2}{T_2}$$

$$\Rightarrow \text{Payout short variance} = N_2 \frac{\sum_{i=1}^{T_1}\left[\text{Ln}(\text{return}_i)\right]^2}{T_2} (= \text{Payout long variance to } T_1)$$

$$\Rightarrow N_1 \frac{\sum_{i=1}^{T_1}\left[\text{Ln}(\text{return}_i)\right]^2}{T_1} = N_2 \left[\frac{\sum_{i=1}^{T_1}\left[\text{Ln}(\text{return})_i\right]}{T_2}\right]^2$$

$$\Rightarrow \frac{N_1}{T_1} = \frac{N_2}{T_2}$$

$$\Rightarrow N_1 = \frac{T_1}{T_2} N_2 = (\text{Notional of near-dated variance as factor of far-dated variance notional}).$$

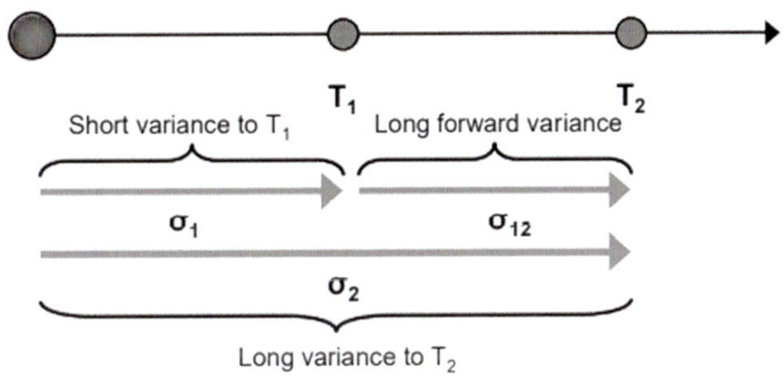

FIGURE 5.76 Forward-starting varswap components

Intuitively, this makes sense, as the near-dated variance swap to T_1 needs only to cancel the overlapping period of the longer-dated variance swap to T_2. The notional of the near-dated variance swap to T_1 therefore has to be scaled down, depending on its relative maturity to T_2.

The presence of skew causes a correlation between volatility and spot. This correlation produces a negative shadow delta for all forward-starting products. The rationale is similar to the argument that variance swaps have negative shadow delta.

5.2.2.2.4 Exotic OTC Options Various other exotic options have been structured in the past, with differing levels of success as considered from a hedge potency and cost-effectiveness standpoint. The ideal OTC products should be easy to understand, related to the liability hedging need at hand and pertinent to generating cash flows when needed, whether directly or on a marked-to-market basis. Other preferred attributes include liquidity and model-friendly structures that provide a robust framework for back-testing and scenario analysis.

Barrier options are the most popular type of light exotic product, as they are used within structured products or to provide cheap protection. The payout of a barrier option knocks in or out depending on whether a barrier is hit. Barrier options (like all light exotics) are always European (if they were American, the price would be virtually the same as a vanilla option, as the options could be exercised just before the barrier was hit).

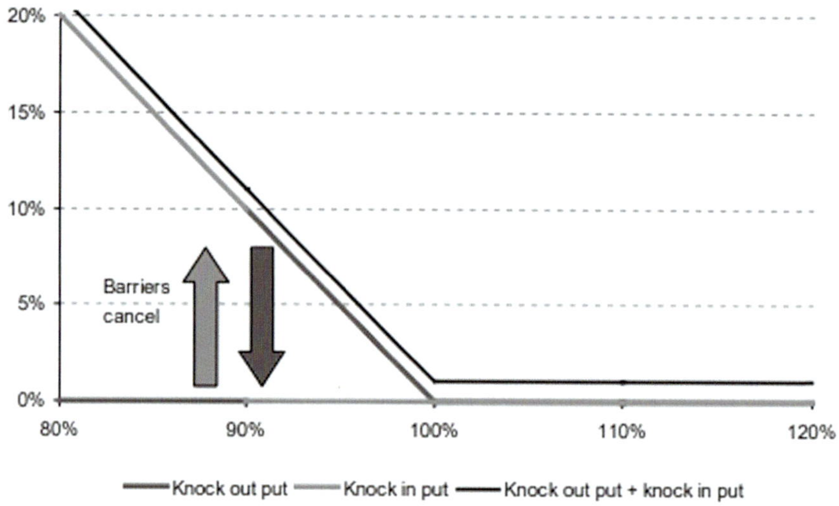

FIGURE 5.77 Knock-in versus knock-out puts

The payout of a knock-out put is equal to a "shark fin" until the barrier is reached. A "shark fin" is equal to a short digital position (at the barrier) plus a put spread (long put at strike of knock-out put, short put at barrier of knock-out put). The price of a knock-out put can therefore be considered to be the cost of a put spread, less a digital and less the value of the knock-out. As pricing digitals and barriers is not trivial, comparing the price of a knock-out put as a percentage of the appropriate put spread can be a quick way to evaluate value (the knock-out will have a lower value, as it offers less payout to the downside). For reasonable barriers between 10 and 30% below the strike, the price of the knock-out option should be between around 15 and 25% of the cost of the put spread.

5.2.2.2.5 Conclusion Overall, volatility hedging through option- and varswap-based hedging offers these pros.

- It more closely replicates underlying liability risks.
- It can lock into low volatilities (i.e. low prices of options).
- Hedges are placed for long-term risk management; they can be designed to cover a wide range of policyholder behaviour and do not necessarily have to be adjusted.
- Long-dated hedges address vega risk and protect against wide movements in the market that may not be accounted for in dynamic hedging.
- It provides vega, gamma and rho protection as well as delta.
- It pays implied volatility, which may be higher than actual volatility.
- There is significant capital relief under C-3 Phase II that is easy to evaluate, and a capital relief "haircut" for unexpected deviations in policyholder behaviour.
- It introduces net income volatilities, as the liability volatility assumption is fixed versus marked-to-market hedge asset implied volatility, which is expected to significantly decrease with the introduction of the LDTI in US GAAP.
- The cost of hedging is known up-front, specifically during market declines.
- It is efficient in terms of administration: there is minimal operational risk and it is easy to implement.

- It automatically adjusts and offsets to a large extent spot, rate and volatility risks.
- Customer behaviour can be managed to some extent through the use of static hedges with a conservative persistency estimate, and/or overlaid with dynamic hedging and/or adjustment of notional amounts.
  - Dynamic lapse reduces the lapse rate when the option is in the money, and hence can substantially increase the vega.
  - As the market rose over the last few years, the vega of the position increased substantially as policy benefits reset to higher levels.
  - As the market has declined recently, the vega has declined; while the "effective maturity" of the guaranteed minimum withdrawal benefit (GMWB) option has shrunk, which is reflected in the substantial decrease in vega-weighted maturity -> dynamic lapse assumptions currently result in higher absolute levels but a shorter duration of vega.

However, option- and varswap-based hedging holds some cons too.

- Long-dated options are far less liquid (up to five to six years only in the United States).
- Long-dated volatility data are not as readily available.
- More up-front investment in hedges (can be managed through financing hedges) and less investment in infrastructure.
- Liquidity may be limited.
- There is potential for credit exposure.
- Vega sensitivity to the volatility level.
  - As with most short option positions, vega increases with increases in absolute volatility levels. As volatility goes up, a dynamic hedger will have to purchase more vega to stay vega-neutral.
  - Vol of vol has spiked recently, requiring frequent trading of vega.
  - Long-dated volatility convexity products are still not available, and recent market performance will make this product set even more expensive when they are offered.
- Potential for costly roll of options.
  - Insurers attempting to match liability vega duration closely have been forced to roll hedge maturities down in the face of historically expensive market conditions.
  - Further, bids and offers for long-dated volatility products have widened in line with limited liquidity in the shorter end and for many derivative products.

5.2.2.3 Sustainable Hedging of Equity–Interest Rate Correlation Risk

Equity–interest rate correlation has experienced several changes in regime over the past three decades. As illustrated in Figure 5.78, it was elevated in late 2008 and remained high for a long time, which (along with higher volatility and higher deltas) contributed to additional hedging costs.

Such high correlation increased hedging costs, as most life insurance guarantee writers are short equity–interest rate correlation. Moreover, as interest rates decline, a dynamic hedger has to buy more vega to stay neutral.

Over a third of the time period since 1960 has seen the market in a low-volatility regime, in which equities rally and long-term rates trend mildly upwards, while daily correlation is highly negative.

- Over 50% of the period has experienced a medium-volatility regime, when equities tend to sell off mildly and rates stay relatively flat, with daily rates and equities returns staying negative.
- Some 7% of the time the volatility/correlation regime has shifted aggressively to a high-volatility regime, in which both equities and rates volatility realize negative trends and positive correlation.

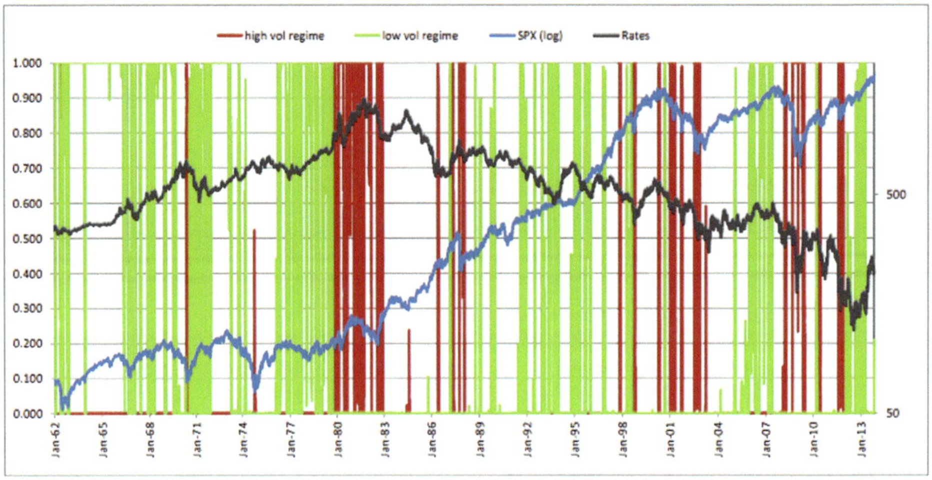

FIGURE 5.78 One-year equity–interest rate correlation, 1962–2013

During a crisis all asset classes and markets tend to crash similarly.

- Correlations between asset classes approach 100%.
- Correlations between different markets approach 100%.

These are the implication of higher correlations in a crisis.

- Higher correlations of asset classes mean higher volatility (both implied and expected).
- This makes diversification and asset allocation much less effective.
- Diversification is lacking when you really need it.
- The implication for product pricing: the correlation matrix in the assumption book probably underestimated the cost of hedging and overestimated the profitability.
- Many hedging programmes are under-hedged in a crisis (and could be over-hedged in boom markets). It is smart to purchase basket options when implied correlations are lower than crisis correlations.

Life insurance guarantee writers are short equity–interest rate correlation (in particular with GMWB for life and GMIB). As interest rates decline, a dynamic hedger has to buy more vega to stay vega-neutral. One of the key risks is that both equity and interest rates fall.

Equity–interest rate hedges are being more widely and frequently traded.

- Fixed strike put options with notional leverage based on rates (low rates = high leverage) are increasing as interest rates fall, along with rising insurance liabilities. This allows the buying of additional vega when interest rates decrease, preventing any readjustment.
- There are also more equity put options with strike price floating based on rates (low rates = high strike).
- Similarly, there are more fixed strike put options with deferred payout at maturity (low rates = high present value).

However, these OTC exotic puts are rather illiquid, so the use of long-dated (e.g. ten-year) puts (on an ATMF total return index) for equity–rates correlation risk mitigation represents a growing alternative.

- It mitigates convexity exposures: interest rate gamma, equity gamma and very significant portion of cross gamma exposures (for a relatively small equity gamma notional).
- It provides some equity delta and significant ten-year DV01 relief, thereby saving transaction costs associated with these instruments.
- ATM forward options have lower implied volatility (due to market skew), and thus cost than usual hedging put options away from the money.
- A small gamma coverage usually amounts to a near-full reduction in ten-year interest hedges, thereby saving total return swap (TRS) roll transaction costs.
- A long-dated equity put has similar Greek profile characteristics to variable annuities.
- It fixes the long-term cost of hedging, in line with the ten-year implied volatility.

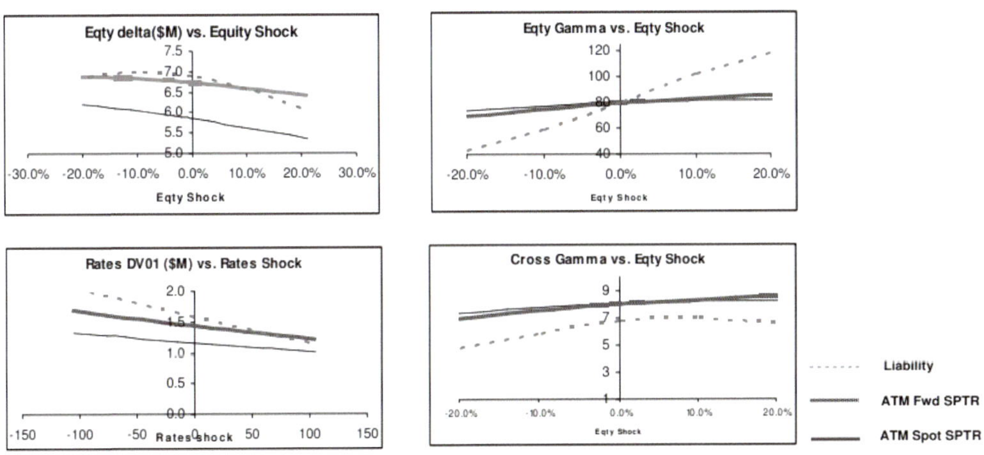

FIGURE 5.79 Liability versus ATM forward SPTR hedge asset Greeks

Remark: option theta can be dominated by the positive carry effect given the long maturity. Theta does indeed have two parts: due to optionality (negative) and due to carry (positive); when rate and time to maturity are relatively high, the carry effect dominates; if rates are low and/or time to maturity short, then optionality decay dominates.

Among the considerations are the following.

- Vega exposure: it introduces non-negligible vega exposure, because it is a long-dated option.
- Liquidity: 15-year (or even longer-maturing) puts provide a better hedge for variable annuities from a key rate bucket perspective, but are illiquid compared to ten-year puts, which are the longest-dated maturity with a reasonable amount of liquidity and market transparency.
- Delta, DV01 and gammas (equity, interest and cross) are similar regardless of model choice (for models that impose a normal distribution for rates, such as Hull–White, G2++): there is an indicative calibration result for options struck at the money forward.

	Current Market B-S Implied Vol	Equity Vol	Equity/Rate Covariance	Rates Vol	Vol Spread = Total Vol, less Equity Vol
7yr	30%	28%	5%	9%	2%
10yr	31%	27%	7%	13%	4%
15yr	33%	25%	10%	18%	8%
20yr	34%	22%	13%	22%	12%

FIGURE 5.80 Volatility and covariance

However, the hedging P&L attribution is still model-dependent, as vega and theta decay definitions depend on the model used.

- Interest rate exposure: this will offset the liability DV01, but it rolls down buckets as time passes; puts' actual interest rate exposure is to the ten-year funding curve (and not to the Treasury curve), so this will monitor/manage the ten-year overnight indexed swap/UST basis thereby introduced.
- There are significant higher-order Greeks.

Notional : 50B, ATM Spot, SPTR puts								
Term	Delta ($M)	Vega convexity ($M)	DdeltaDvol ($M)	DgammaDvol ($K)	DvegaDrho ($K)	DdeltaDTheta ($K)	DVegaDTheta ($K)	Speed ($K)
10y	(121)	(4)	2	(56)	(330)	(18)	(57)	(25)

FIGURE 5.81 Higher-order Greeks

- Empirical gamma P&L breakdown: equity gamma 50%, rates gamma 20%, cross 30%.

5.2.3 Sustainable Hedge Effectiveness versus Cost

5.2.3.1 Trade-off between Sustainable Hedge Effectiveness and Hedge Cost

High hedge effectiveness can increase the hedging cost and reduce profit. For instance, dynamic hedging strategies with high hedge effectiveness can have higher losses for a given scenario. It may be better than being naked at a given point in time, but not over a longer period.

Given the long maturity of life insurance guarantees, the hedge portfolio "gains" or "losses" need to be sustainable over the full lifetime of the hedge and not at a single point in time, during which a typical hedging strategy has a "saw-tooth" pattern, with a bleed during normal times and upward jumps during crises.

There are two axes to potentially improve such a trade-off.

- Select some specific optimized hedging strategy depending on the Greeks at stake and the rebalancing frequency.
- Select some specific optimal hedging asset.

In fact, the costs vary widely across the most usual hedging strategies:

Strategy	Execution	Description	Cost / Risk Profile
Static Hedging	OTC, exchange traded options and exotic options	•Match embedded option in liabilities •Monitor / hedge basis risk •Premium paid for options •Long volatility (hedge portfolio) •No exposure to implied vol	•Hedge cost known •Highest expected hedge cost
Semi-Static Hedging (active / passive)	Portfolio of options	•Approximate match embedded option in liabilities •Monitor / hedge basis risk •Premium paid for options •Long volatility •Exposure to implied vol	•Hedge cost not known until maturity •High expected hedge cost

Strategy	Execution	Description	Cost / Risk Profile
First Order Dynamic Hedging*	Delta, Rho, Vega* hedging with futures and swaps	•Replicate 1st order sensitivity of embedded option in liabilities •No initial cash outlay •Short volatility •Exposure to realized vol	•Hedge cost not known until maturity •Lowest expected hedge cost (next to naked) •Highest tail exposure (next to naked)
Second Order Dynamic Hedging	Gamma, Vega, Vanna, Vomma hedging w/ options, vol swaps	•Replicate 1st and 2nd order sensitivities •Premium paid for options •Long volatility (hedge portfolio) •Rollover risk = exposure to implied vol	•Hedge cost not known until maturity •High expected hedge cost •Greatest potential for losses if not executed properly

FIGURE 5.82 Cost–risk profile of hedging strategies

While most of the popular ways to reduce the bleed (cost reduction) have some risks associated with them, there is indeed most often a trade-off between higher expected profit and tail exposure (and cost of capital), as illustrated in Figure 5.83 depending on the Greek order of the hedging

Strategy	Average Payoff	Tail Ranking	Hedge Effectiveness
Naked	Best	4 (worst)	N/A
Delta only	Lower	3	~ 50%
Delta / Rho	Lowest	2	~ 70%
Delta / Rho / Vega	Lower (same as delta)	1 (best)	~ 90%

FIGURE 5.83 Risk–reward analysis

This translates into the trade-off chart shown as Figure 5.84.

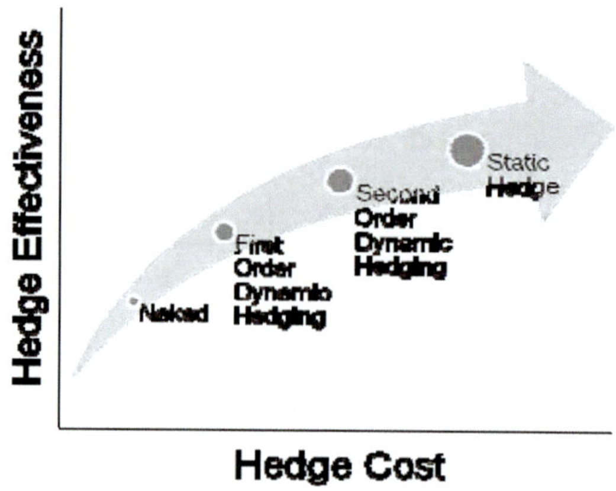

FIGURE 5.84 Vega versus spot

If we look at variable annuities (GMAB), the replication cost of a static hedging may be higher on average, albeit significantly lower in extreme market scenarios.

Regarding the potential usual candidates for the optimal hedge asset, the trade-off between hedge effectiveness (residual risk) and cost is usually as shown in Figure 5.85.

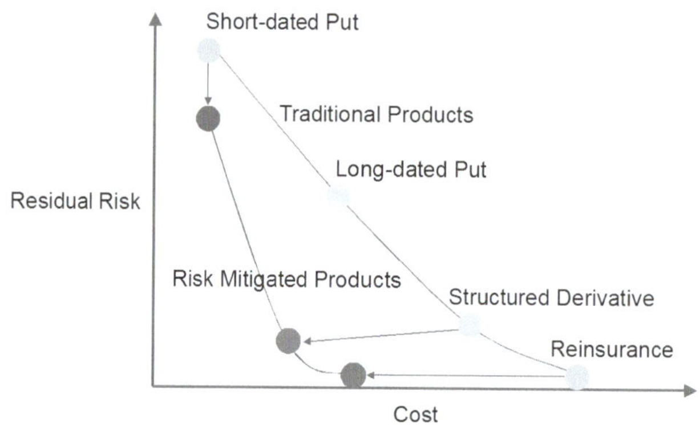

FIGURE 5.85 Risk versus cost

5.2.3.1.1 Hedge Effectiveness and Hedge Cost for Short-Dated versus Long-Dated Puts Indeed, in order to mitigate the delta-hedging issues, some life companies hedge with an array of short-dated options with varying strike levels through a "rolling short-term options strategy".

- This can be considered as an "in-between" choice between pure static and pure dynamic, since
  - options embed delta hedging, so they are likely to offer some protection against gap risk if strikes are carefully chosen; and
  - transaction costs can be mitigated compared to a daily delta hedging.
- How can one set the weights/strikes/notionals/tenors so that the short-dated option strategy appropriately spans the set of possible liability values?

The semi-static hedging methodology provides portfolio weights that enable investors to span a portfolio of short-term options in order to replicate a longer-dated option, for which the weighting factors are given by the gamma of the long-dated option at the maturity date of the short-dated option. The principle behind gamma hedging is as follows: the T-year put option can be viewed as the present value of the (T-t)-year option (T-year option in t years' time) probability gamma-weighted over all possible outcomes of the underlying asset. The value of a European put of strike K and of maturity T is related to the prices of a spectrum of European puts at a shorter maturity $t < u < T$ by the following integral equation:

$$P(S,t;K,T) = \int_0^{\infty} \omega(E)P(S,t;E,u)dE$$

$$\omega(E) = \frac{\partial^2}{\partial S^2} P(S,u;K,T).$$

The weight on the call of maturity u and strike E is proportional to the gamma that the target call would have at time u should the underlying be at price E then.

As the function relating a call's gamma to the underlying spot is a bell-shaped curve in most models, the strike weight is centred on the strike K of the target call of maturity T. As u tends to T, the gamma becomes more concentrated about K, and hence the relative weight on these calls increases. In the limit when $u = T$, all the weight is on the call of strike K, and the integral equation reduces to a tautology. The integral equation is generally approximated with a weighted sum over a finite number of call values; the determination of strikes and weights follows some standard quadrature rule.

As an illustration, we focus on a ten-year × 30-year receiver swaption after deterministic shocks on the yield curve (parallel shift) and the implied volatility, for which the replicating strategy performs rather well with the long-term option (there is a clear reduction in vega risk sensitivity).

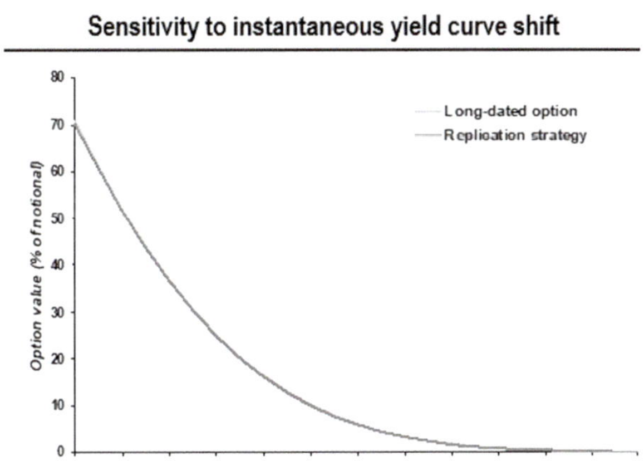

FIGURE 5.86 Sensitivity to an instantaneous yield curve shift

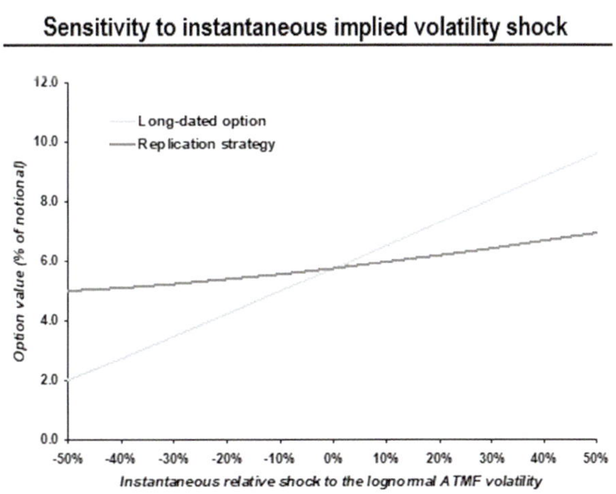

FIGURE 5.87 Sensitivity to an instantaneous implied volatility shock

However, this rolling short-term option strategy results in higher convexity in market sell-offs, and comes at the expense of a higher carry cost during benign market conditions.

In addition, shorter-dated options usually have an extra insurance premium (spread between the implied and realized volatility) due to their higher demand.

Moreover, rolling one-year puts annually is not optimal due to a significant specific strike on the roll date, while convexity varies considerably throughout the year.

As a result, systematically hedging with one-year put options (which have less insurance premium and lower convexity but also a lower cost of carry) empirically significantly outperforms systematic hedging with one-month put options.

An efficient alternative lies in buying 1/12th notional one-year puts and holding them to maturity, which holds a basket of puts with different maturities (one to 12 months) and strikes, benefiting from lower transaction costs and lower path dependence.

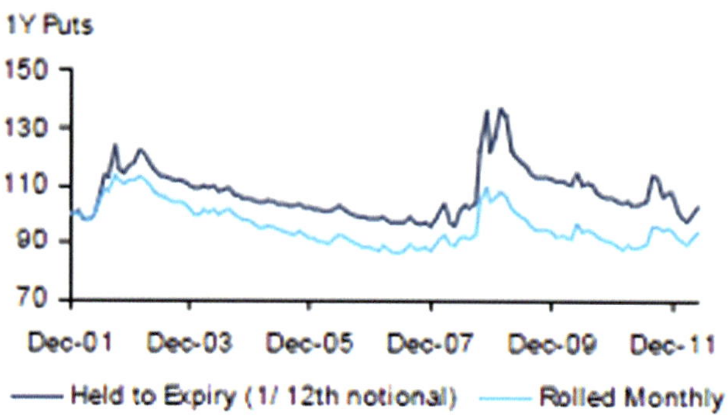

FIGURE 5.88 Overlapping rolling puts outperform, December 2001–June 2012

5.2.3.1.2 Hedge Effectiveness and Hedge Cost of Call Overwriting A way to help reduce hedging costs can be selling OTM (e.g. by 100 bps) equity call (or payer swaptions) to buy OTM (e.g. 25%) put options (or receiver swaptions), or through low-cost (if not zero-premium) put spread collars with rolling-down strikes to manage potential sharp declines, while taking advantage of volatile range-bound stocks (since the position sells more options than it buys), as regularly observed during the post-crash market recovery.

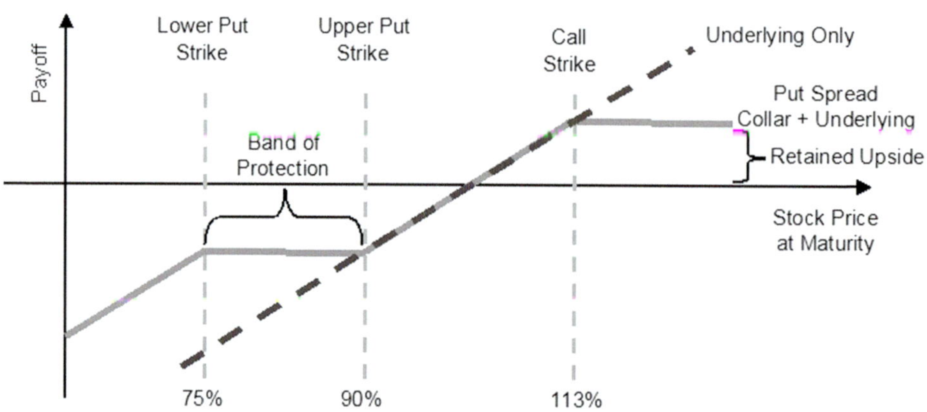

FIGURE 5.89 Put spread collar decomposition

Shorter-dated call overwriting performs better due to the higher insurance premium baked into them, while decreasing the risk of being called away in grinding rallies.

As a result, call overwriting has worked well as a yield enhancement strategy monetizing the high implied/realized premium. In line with the rationales above, using shorter-dated calla is better here, as they hold larger insurance premiums.

5.2.3.1.3 Benchmarking Hedge Effectiveness versus Hedge Cost for Downside Protection Vanilla Overlays Downside protection overlays as developed above refer to strategies in which risk mitigation steps are taken to reduce the downside risk of the specific equity exposures held, for which the underlying investments are usually based on the same benchmark exposure. It has a very big component of the hedge return coming from actual market movements of the underlying exposure (known as delta exposure). It delivers protection from bad outcomes, which do not have to be of the severity of the far-left tail event but might deliver less of a hedging gain than a tail risk hedge in the event of a sudden and extreme correction.

Downside protection seeks to complement equity market exposure such that total exposure outperforms in a down market. The performance of a downside protection overlay is therefore considered in combination with the equity exposure it is hedging, with set floors, drawdown limits or target volatility (TV) levels. Examples of downside protection overlay structures include puts and put spreads (with their low initial premium and protection in less extreme markets), and in some circumstances a portion of the market upside may be sold to finance the protection.

These downside-protected equity exposures usually outperform in a down market with more modest corrections, as more modest corrections happen more frequently than the severe, violent corrections a tail risk hedge offers protection against, such as a 1–2 put spread that entails selling an ATM or near-ATM put and buying two OTM puts while delta hedging. Such a simple strategy finances tail risk protection by selling ATM gamma, and provides positive carry, long skew, long convexity and deep tail risk protection. In return for the above benefits, the seller is exposed to high implied or realized volatility without a significant sell-off in spot.

Let us consider the projected return distribution for a put struck at 80% and a similar put spread 90–78 with a similar up-front premium of 2%, which reduces the expected return from its starting value of 5% to 3.47%. By making the protection weaker through the put spread with the same up-front premium (2%), the drag on expected return is reduced (i.e. there is an increased expected return of 4.23% versus 3.47% with the put), but with the caveat of a potential very negative annual return in the event of a major correction (i.e. greater than –22%).

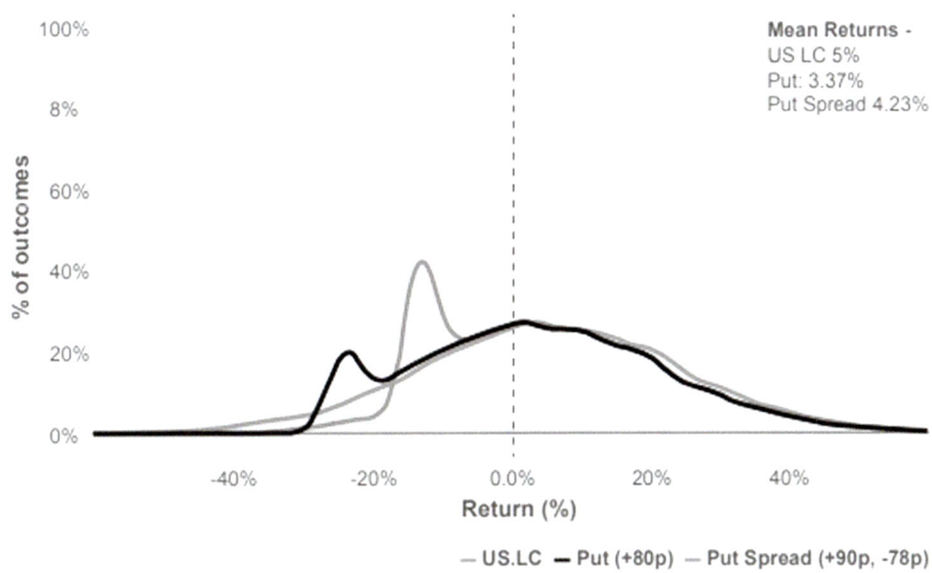

FIGURE 5.90 Return distribution

The put hedges the tail, while the put spread protects the left shoulder of the distribution at a much lower drag on expected return. There is a much higher probability of the market expiring in the zone where the put spread outperforms the put (16.6% probability) than in the zone where the put outperforms (1.5% probability). However, if this were to occur, the severity of it (1.5%) could be quite significant, and so could be complemented by a tail risk strategy.

As an extension, let us compare the four usual basic hedges – a six-month 95% put (P), six-month 80–95% put spread (PS), six-month 90–110% collar (C) and 80–95–110% put spread collar (PSC) – in terms of hedge effectiveness versus cost. Here the cost is measured by the net hedging P&L, while the hedge effectiveness is apprehended by the following four KPIs linked to the risk appetite: absolute returns; downside volatility; maximum drawdown of hedge + index; maximum underperformance of hedge.

- Put spread collars tend to provide the best protection (i.e. higher net returns) on a hold-to-maturity basis, but do not provide the same down volatility reduction as a collar or a put, especially in bear markets. In bull periods the strong outperformance of put spreads and put spread collars is due to the lower premium.
- The maximum underperformance (versus the index) is often very severe for puts and collars. For puts, this is usually due to the premium spent and periods of very ITM puts having a delta close to negative one (so limiting participation on market rebounds). For collars, this is usually due to the capped upside during some of the sharp market rebounds, again highlighting the mark-to-market risks. However, the maximum hedges and best maximum drawdowns are also provided by puts and collars.
- A comparison of returns in bull markets to returns in bear markets – similar to a gain–loss ratio – makes put spread collars look very compelling versus other hedges, particularly versus puts.
- The relative performance of different strategies mostly remains the same across regimes (e.g. put spread collars and collars often give good returns during all bear markets). However, between regimes the best and worst performers change dramatically: in bear markets collars provide good returns and reduce volatility strongly, but in bull markets they cap the upside and underperform other strategies.

Even so, most back-tests suffer from the starting date: a put spread collar strategy implemented the day before the market spikes higher is exposed to the risk of heavily capping the upside of the underlying. Over the past decade we have seen violent market trends both up and down, and so a few "unlucky days" could have conceivably hurt the performance of a hedge, particularly any that are "short the wings", such as collars, put spreads or put spread collars.

- In this respect, puts look like a dramatically underperforming strategy, as at no starting date were they the "best" hedge.
- Put spreads also flip regularly from being the best strategy to the worst strategy: in severe down markets they provide less protection for the same premium than other hedges (hence the worst performer); in sharply rising markets they cost less premium than puts, and do not cap the upside like collars, or put spread collars (hence the best performer).
- In contrast, it might appear (again) that put spread collars are consistently the most effective strategy. But, in fact, put spreads also flip regularly from being the best strategy to the worst strategy: in severe down markets they provide less protection for the same premium as other hedges (hence the worst performer); in sharply rising markets they cost less premium than puts, and do not cap the upside like collars, or put spread collars (hence the best performer).

The distribution of returns is therefore important to get a realistic sense of each risk and reward.

- Put spread collars, while outperforming "on average", and while seeming to be the best-performing hedge for much of the past 20 years, expose investors to extreme "tail risks": in ~5% of the historic start dates a put spread collar loses 10 to 30% versus the underlying, as a consequence of the short call. Put spreads are also notable by their absence on "big payoffs": there are no observations above 20% returns. This is a function of the short put leg limiting the maximum payoff.
- For average payoffs, between 0 and 15%, the put spread collar is clearly dominant. However, such an "average" world may not be the path we are moving along.
- Only puts and collars provide large returns above 25%, by virtue of the unlimited long put payoff. Puts in particular suffer a high proportion of losses in the 2.5 to 7.5% range, largely as a function of the high premium during especially high-volatility environments. This reinforces the point that they need to be used sparingly.

So, what does this mean for hedging? Simply, that there is no right answer. The preferred hedge will depend on an investor's individual utility function – i.e. either her requirement for big positive hedge returns on the downside, or a desire not to suffer large underperformance if the market rallies or falls significantly.

Mark-to-market analysis during the life of the hedges is the next natural step for analysing hedging strategies, which throws up many important differences. This is a complex task, as there are many factors at work, such as spot moves (delta), volatility changes (vega and skew sensitivity) and time decay (theta), not to mention strike risk for different strategies (e.g. collars versus puts). Not only this, but sensitivities change through time.

- After two weeks, small changes in spot seem to show very little difference in behaviour between the strategies.
- For relatively large down moves (more than –10%), puts and collars perform strongly, as the embedded puts retain time value. Put spreads (and to some extent put spread collars) underperform, due to the lower delta of the positions. For large up moves (over 10%), put spread collars and collars underperform, as the embedded short calls cap the upside. It is worth noting, though, that, for both these extreme moves up and down, the probability of such events is historically low.
- After five months – i.e. close to maturity – things look very different. For small spot moves, put and put spreads clearly underperform (particularly puts, as there has been a lot of time decay). Put spread collars outperform between –10% and +10%, and this is because of the very low initial premium spend, and because indices do not cause losses on the short option legs. The latter point is true because of the time decay on these components of the hedge.
- Only for large spot declines (greater than –20%) do we see puts and collars start to perform well versus the other strategies, as the embedded puts start to payoff strongly. Only for large up moves (over +15%) do puts and put spreads outperform collars (and, for even larger moves higher, they also outperform put spread collars). In the case of large index rallies, put spreads are the notable outperformers (rather than puts), due to a lower initial premium spend versus puts. Note that the probability of these extreme moves higher or lower becomes more likely through time.
- It is even more interesting to look at each strategy in turn, and highlight the volatility risks that were observed during the life of each strategy. This is because volatility changes can be quite significant for any spot move: if this volatility decline is significant, in a "worst case" a put could have a far lower value than in the "average case", while collars and put spread collars should benefit, as the short-call leg decreases in value, providing a mark-to-market benefit.

- Put spreads have a much tighter return band (the max and min bars are closer together), as they are less sensitive to volatility than puts and they do not cap upside returns (collars and put spread collars suffer on a mark-to-market basis on the upside before expiry). Overall, put spreads are stable performers: they do not have the extreme protection of puts or collars, but, on the positive side, they do not have the same "worst case" as puts (less exposure to volatility slumps, and a lower premium loss than puts if spot prices rally sharply).
- In all markets, puts had most variability of returns – i.e. a wide band between the best and worst losses. Puts are volatility-sensitive: if volatility spikes on the downside, then the payoff is made more attractive on top of the full downside protection. If volatility slumps in down markets, or if the option suffers significant time decay, then the worst-case losses for puts in most scenarios is quite severe. Puts are clearly high return/high risk hedges – e.g. they have an attractive "average payoff" in bear markets, but even here their worst-case losses are quite severe.
- Put spread collars perform strongly, on average, in most market conditions, largely due to the lower premium spend versus other hedges. However, in most markets their return band is surprisingly wide, due to mark-to-market behaviour: the maximum return can be very attractive in many markets; but they can also underperform quite strongly. For example, in bull markets the upside call causes mark-to-market issues (causing some low worst-case losses versus put spreads or even puts). In bear markets the lack of full downside protection becomes less of a drag towards expiry (the best case is above a collar, due to the lower premium paid). However, the "worst case" and average return are below a collar, which has an uncapped embedded put.
- Collars and put spread collars perform quite similarly, but with a few differences. In bull markets collars can underperform, as the higher premium spend versus put spread collars can add drag to a portfolio. In bear markets the collar looks more compelling versus put spread collars (despite paying a higher premium), given that the put spread collar has a weaker mark-to-market performance due to the embedded put spread.

These results highlight how much the behaviours change as the hedges evolve. Ultimately, there is no scientific law that determines the best hedge.

In order to reduce the drag on collars, an alternative downside protection instrument lies in calendar put spread collars: instead of selling a six-month call (as in a normal put spread collar), we systematically sell one-month calls, and roll into a new one-month call every month, during the life of the six-month put spread. Selling one-month calls has several potential benefits.

- It is possible to take advantage of the risk-adjusted benefits of overwriting with short-dated calls.
- A major disadvantage of selling six-month calls is the capped upside, at one specific strike for a six-month period, which could cause major underperformance in rising markets. By selling one-month calls it is possible to re-strike as markets move higher, reducing the upside drag on a portfolio.
- Selling six one-month (close to ATM) calls will potentially generate a greater option premium than selling one six-month call.
- Investors can choose not to overwrite each month, if there is a risk of a market spike. They can therefore apply their timing skill to the strategy.
- In many cases (ATM to 105% strike one-month calls), the returns are superior to standard put spread collars, and versus the index over the full period.

- The volatility of the returns for all strikes is below that of the index, and for most is below that of the put spread collar.
- The one major drawback is that the maximum underperformance is greater for all the strikes, compared to the put spread collar, because some one-month calls would have suffered heavy call-away losses if they had been struck just before a major market rebound. On the positive side, the maximum hedge was greater for all the calendar strategies, partly due to the greater option premium received by the repeated one-month call selling.

The fact that a few one-month calls caused significant losses during the life of the trades (but, overall, the strategy outperformed put spread collars) means that investors who apply timing skill on selling the one-month calls could access additional outperformance – i.e. they have the flexibility to choose *not* to overwrite when they see the risk of a major market rebound.

Whether long-dated or short-dated puts offer better protection is a tough question to answer; it depends on many factors, such as the strikes chosen, the path of the market (e.g. a 70% two-year put can become a 50% put if markets rally +20%, and so provide very little protection for a subsequent 30% market drop).

- The long-dated puts provide a superior return, and lower volatility compared to short-dated puts over the full 12-year period. The lower volatility of long-dated puts is expected, as rolling short-dated puts caused frequent rolling to lower put strikes, which in bear markets will potentially remove protection just before another market slump.
- This latter point also explains why the long-dated puts provided a greater maximum hedge – as they continue to provide protection with a high strike put in long-drawn-out bear markets. The benefits of a long-dated floor are clear in the 2002 and 2008 sell-offs, when the downside was strongly restricted. There is clearly a timing risk here for longer-dated puts: a hedge started at a low point in the market will have a far lower OTM strike than if the market rallies and then collapses from a higher starting point.
- There are often drawbacks to long-dated puts; for example, the maximum underperformance is far greater, mainly due to the underperformance in large rallies, which occurs when the puts are very ITM (hence the hedged position has a net very low delta). This was very clear around the time of the 2003 and 2009 rebounds. The maximum hedge in any 12-month period can be less for a long-dated put if, for example, a short-dated put is struck at the top of a large market rebound and then provides a strong payoff in any subsequent sell-off.

When comparing puts versus put spreads, a common criticism of put spreads is the lower amount of protection they give to puts prior to expiry, even for small sell-offs (the latter is because of the relatively low delta compared to puts, due to mark-to-market effects prior to expiry). The issue is that the strategies are often compared on a fixed notional basis. We think this is an unfair comparison for put spreads: they have a lower delta (hence the lower protection) but also a lower initial premium. A fairer method is to adjust the notional such that the initial premium spent is the same for both strategies (i.e. a "leveraged put spread"). The performance of the leveraged put spread is then indeed compelling versus puts: the return is better than with a put hedge, and the volatility reduction is much closer to a put than a non-leveraged put spread. One major risk is that the maximum hedge is clearly lower for a leveraged put spread when there are extreme sell-offs.

The different KPIs of the hedge strategies (e.g. absolute returns, downside volatility, maximum drawdown of hedge + index, maximum underperformance of hedge), and a changing (riskier) world, reinforce the need to separate hedging strategies into (a) "core" hedges, for normal market conditions – so-called "downside protection overlays", which provide payoffs in less extreme markets – and (b) tail risk hedges, which provide significant payoffs only in extreme markets (either up or down markets).

5.2.3.1.4 Hedge Effectiveness and Hedge Cost of Target Volatility Options (TVOs)

5.2.3.1.4.1 Significantly Lower Volatility Exposure As introduced in Chapter 4, a target volatility fund is very efficient in controlling the volatility risk exposure.

- The one-year realized volatility of the TVF is, more than 90% of the time, below its target.
- The one-year realized volatility 99.5th percentile is very close to the target.
- The vol of vol risk is significantly reduced (the standard deviation is significantly lower).

There are several investment applications for which this volatility-stabilizing feature is valuable, and it has emerged as an increasing popular hedge asset class. The primary example is the variable annuity industry, in which TVFs can significantly reduce the hedging costs to provide the capital guarantee versus the use of put options (roughly by a percentage equal to the ratio of the realized managed volatility to the put implied volatility), implying lower capital requirements. The variable annuity AuM invested in such funds rose from about $55 billion in 2011 to more than $400 billion by the end of the 2010s, providing much cheaper funds (0.25% as opposed to 1.5 to 3% for mutual funds' annual fees).

As a result, intuitively it should be more efficient to hedge liabilities linked to target volatility funds using puts on target volatility indices so as to avoid overpaying for implied volatility and skew.

They are like vanilla puts, but with a floating premium, with the up-front premium based on the buyer's chosen "target volatility", with an additional premium at maturity if the realized volatility is below the target volatility.

5.2.3.1.4.2 Significantly Lower Risks Although the delta/rho Greeks are hardly impacted, options on such managed-volatility funds tend to divide the volatility risk exposure by a very large factor.

- There is significantly lower convexity and volatility risk (thus net income), and target volatility strategies are likely to exhibit long-term volatilities close to their targets, neutralizing the vega exposure of VA guarantees; indeed, exponential smoothing almost completely removes the long-term vega P&L vega, though some short-term vega remains. The TVF mechanism removes the volatility term premium (the spread long-term implied volatility minus the short-term implied volatility), but it does not remove the short-term implied-realized volatility spread.

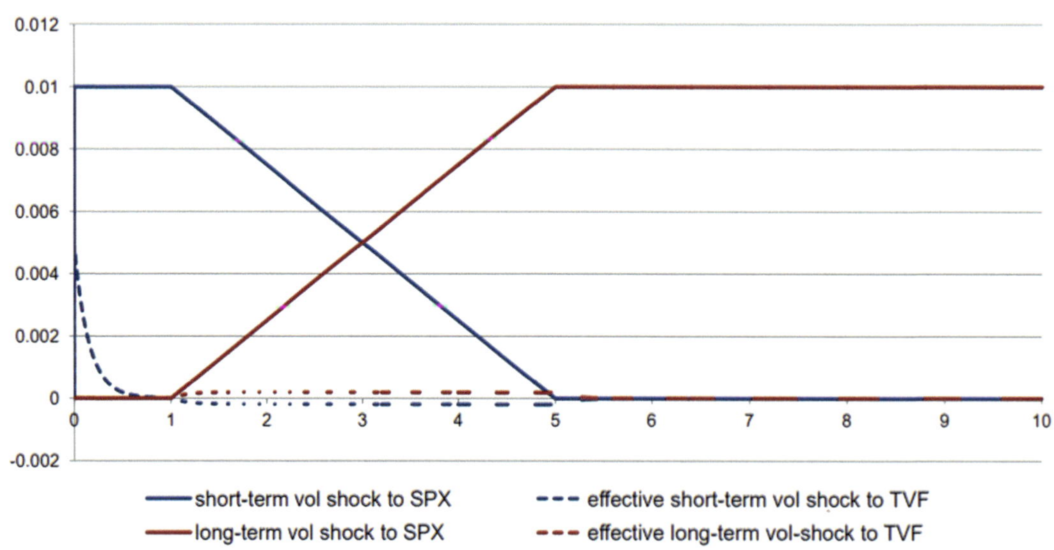

FIGURE 5.91 Vega profiles

- There is significantly lower skew risk, as target volatility strategies flatten the volatility skew, providing a return distribution that is closer to Gaussian.
- There is significantly lower basis risk through the use of TRSs, with significantly reduced P&L volatility (track error volatility divided by about seven).
- There are significantly lower transaction costs, since the realized volatility remains below the hedging volatility.
- Overall, TVF mechanism usage provides a discount of around 30% in capital requirements and a decrease of some 66% in transaction costs.

Indeed, the impact on reserve Greeks between the use of vanilla options versus options on target volatility funds using a VIX or an EWMA volatility target is massive (see Figure 5.92).

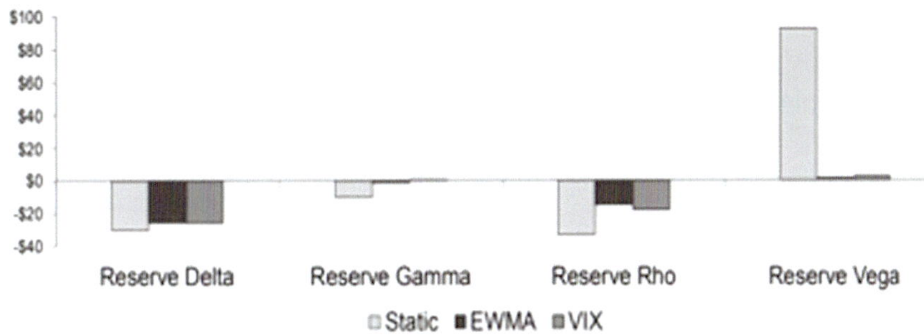

FIGURE 5.92 Reserve Greek comparisons

5.2.3.1.4.3 Significantly Improved P&L We compare the P&L from buying puts on target volatility indices with the hedging P&L from buying comparable vanilla puts, for each trade date. Comparing target volatility puts with vanilla equity puts requires careful consideration, as the vanilla equity put embedded in a TV put has changing notional and strike over the life of the trade.

We adjust the vanilla strike and notional to have equivalent intrinsic value on the trade date for all equity scenarios ("equal crash protection"), through the following "static" and "dynamic" adjustments.

- Dynamic adjustment: compare the TV put to the scaled vanilla put, adjusted using the most recent TV participation level.
- Static adjustment: compare the TV put to the scaled vanilla put using the fixed long-term average TV participation figure.

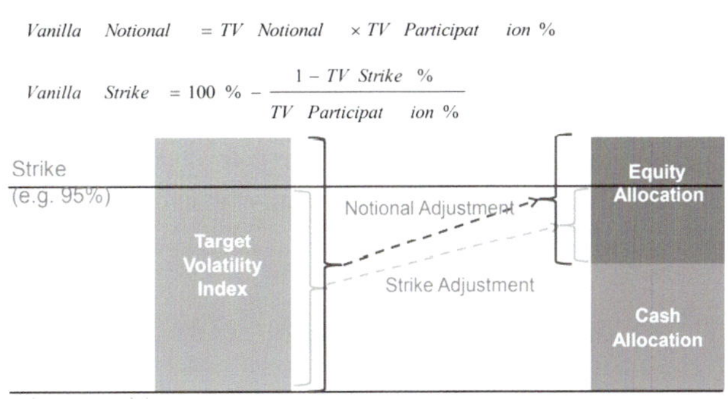

FIGURE 5.93 Adjustment of TVO

FIGURE 5.94 Price and payoff by trade date year, 2000–2016

We use here target volatility puts versus vanilla SPXT10UT six-month 95% puts with dynamic adjustment, which we compare to static vanilla put strikes and notionals using the same long-term participation rate of 67%.

As a result, such target volatility options provide significant savings, split as shown in Figure 5.95.

Source of Savings	Annual Basis Point Savings
Reduction in expected hedge costs	40
Cost of capital savings	36
Savings on quarterly earnings volatility mitigation	20
Total	96

FIGURE 5.95 Savings (bps)

5.2.3.1.5 Hedge Effectiveness and Hedge Cost of Exotic Options For example, knock-out puts provide downside protection unless a trigger market level has been reached (e.g. a one-year 100% SPX put that knocks out, or ceases to exist, if the S&P 500 closes at or above 103% of its spot level on the trade date). Another product is the "option on an option", in which the issuer purchases, for example, a call on a put.

Pricing on these options can be expensive and often does not meet the insurer's hedging budget, but the structures are worth tracking to see if relative value opportunities arise. For example, a life insurance guarantee issuer could pay a dealer 2% up-front to purchase a four-year call struck at 12% on a ten-year 90% strike SPX put. Compare this to a straight ten-year 90% strike put, offered at 8.75%. Obviously, the insurer pays little premium today, but, in this example, must be willing to pay a total of 14% (2% plus 12%) for protection if a severe downside scenario arises.

Hybrid products represent another variation, and may encompass puts with strikes that adjust according to reference interest rate levels or puts that pay out on a deferred basis so that reinvestment rate risk may be effectively locked in for the VA hedger. These products naturally pose correlation risk to the selling dealer and are quite sensitive to pricing assumptions.

5.2.3.1.5.1 Rent-a-Put Strategy The idea of this strategy is to buy today a ten-year put on S&P 500 total return and sell back in one year.

It is a cost-efficient trade.

- You pay today a premium based on high ten-year implied volatility for one year only.
- You receive a high premium in one year based on (the pre-agreed) one-year–nine-year forward implied volatility, which experiences similar behaviour, capitalizing on the progressively de-steepening implied volatility term structure.
- You therefore lock in the hedging costs for the coming year at a cheap level (5%) compared to a one-year put (15%) or a ten-year put (32%).

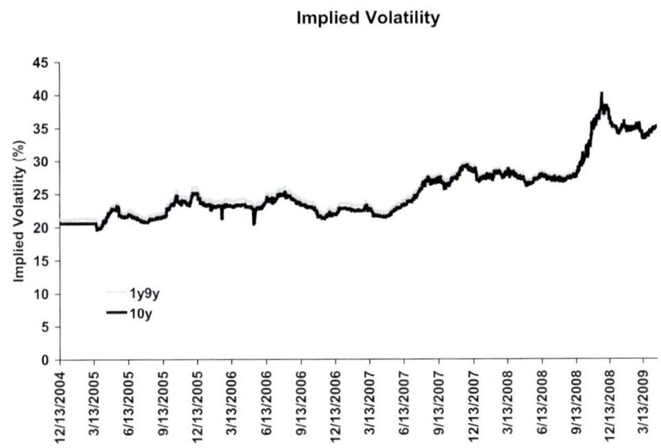

FIGURE 5.96 Ten-year and one-year–nine-year implied volatility similarity of behaviour, 13 December 2004–13 March 2009

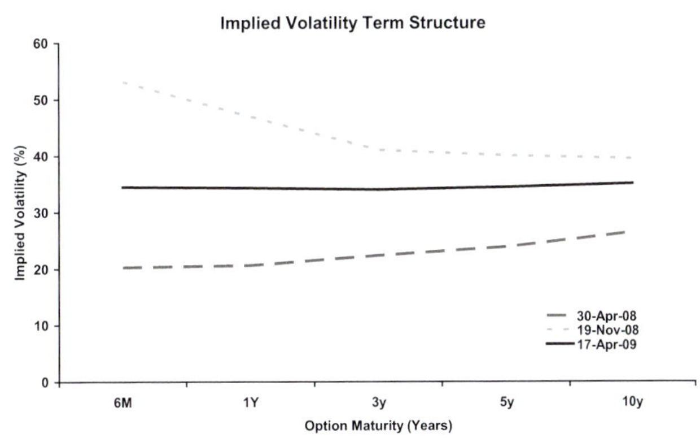

FIGURE 5.97 Flattening of implied volatility term structure

It is, to a large extent, a risk-mitigating trade.

- It has a long maturity (ten-year) strategy, consistent with VAs, in contrast with a one-year put.
- There is exposure to (long-dated) volatility risk for one year only.
- There is reduced exposure to volatility risk due to the selling back of the put.
- There is no exposure to the expected rise in realized volatility.

However, such a strategy is liable to potential tail risks: a potential increase in long-dated implied volatility by the next year, in contrast with the current historically low volatility term structure and de-steepening skew, would increase risks and reduce cost efficiency (e.g. consumption credit risks or the feedback effects of rising unemployment).

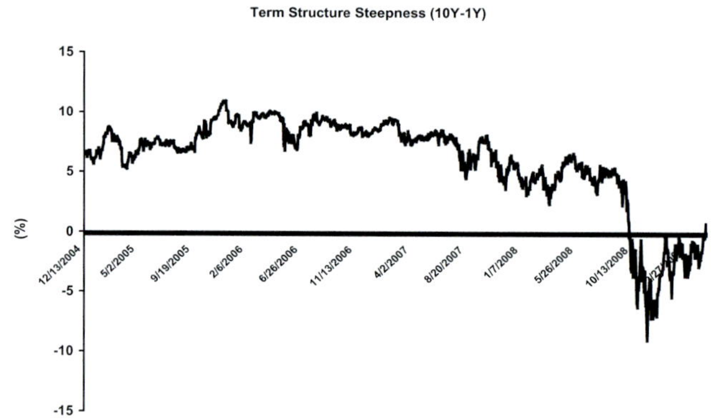

FIGURE 5.98 Volatility term structure at historical lows

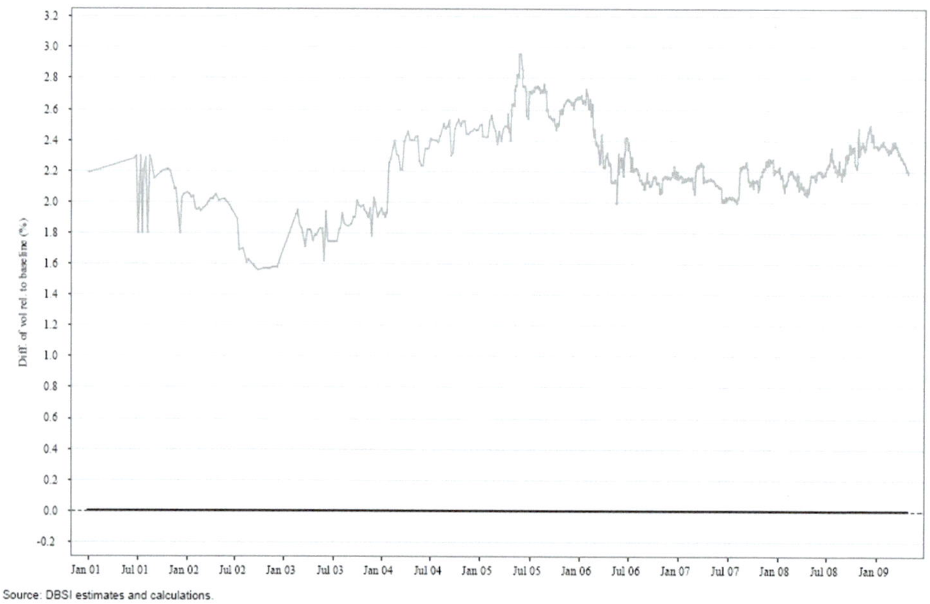

FIGURE 5.99 De-steepening skew, January 2001–December 2009

Source: **Deutsche Bank Securities Inc. estimates and calculations.**

Overall, the SWOT analysis shown in Figure 5.100 can be drawn.

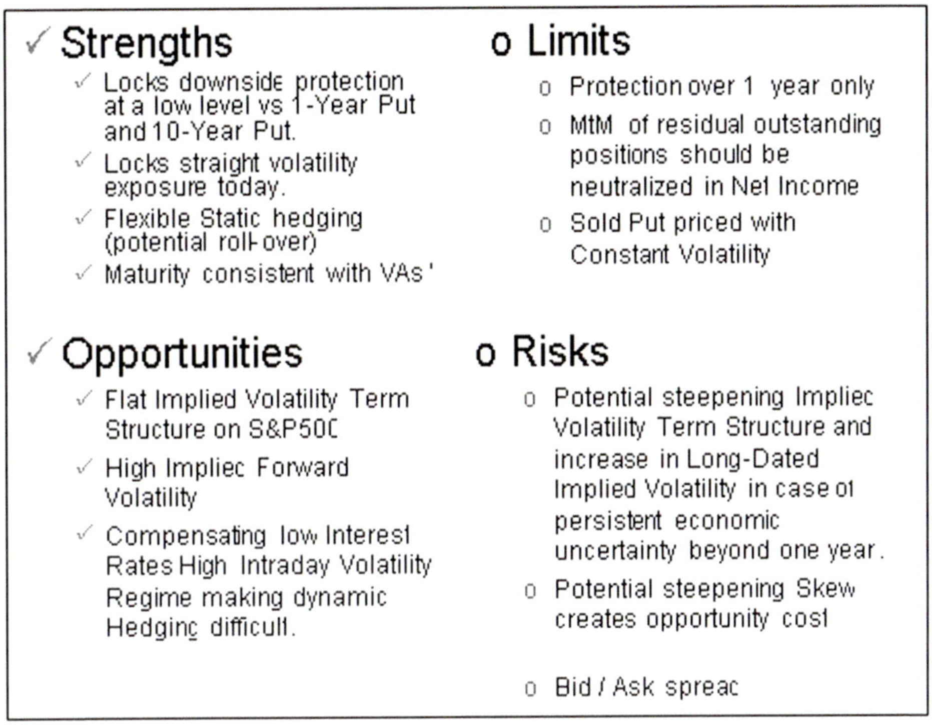

FIGURE 5.100 SWOT analysis

5.2.3.1.5.2 Hedge Effectiveness and Hedge Cost of VIX Products The liquidity of VIX products continues to increase, as most vega is being traded in VIX-related instruments, although SPX options often dominate in terms of open interests outstanding.

VIX future returns are convex relative to SPX returns – but not for free.

- VIX futures can actually lose money even if VIX does not change.
- VIX futures are typically higher than VIX during rallying markets and lose value as they approach maturity.
- This "roll cost" is equivalent to the up-front premium for vanilla puts.
- Short-term VIX futures have more convexity and so a higher cost (10 to 15% per month) relative to mid-term VIX futures (2 to 5% per month) during rallying markets.

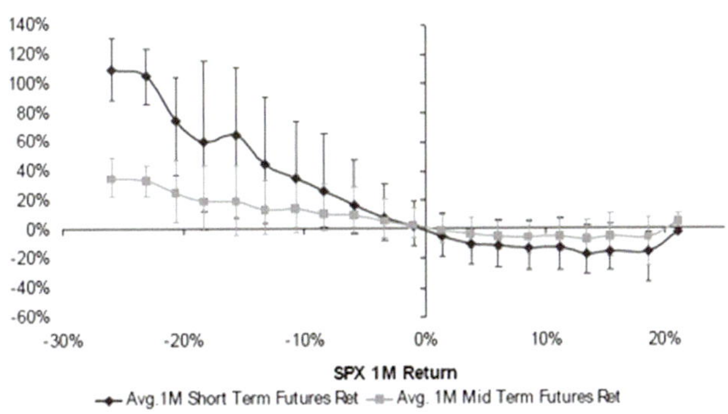

FIGURE 5.101 VIX futures payoff looks like a put

As a result, in order to get both low cost *and* high reactivity during crises, one strategy is to sell small amounts of short-term VIX futures to finance the cost of a long position in mid-term VIX futures, while reversing this short position when the probability of a risk flare-up is higher; in other words, going long on short-term VIX futures, based on the ratio of VIX and VXV (the symbol for the CBOE S&P 500 three-month volatility index, and so equivalent to the three-month VIX), which is a good predictor signal of VIX future returns. Regarding sizing, it is based on the beta of short-term versus mid-term futures, which is around two.

The aim of such a strategy is to minimize the carry cost but provide convexity during risk flare-ups.

5.2.3.2 From Systematic to Tactical Hedging

One alternative to the systematic hedging strategies developed above lies in tactical hedging programmes, which may face less time decay with hedges not running continuously, although there is still a negative impact on expected returns. The challenge is to add value in the timing of hedging decisions and to minimize drag on expected returns as much as possible.

To improve the chances of success in a tactical hedging programme, 1) diversify the timing of hedge trades and diversify execution points and equity exposures hedged; 2) align executions with forward-looking market views; and 3) alter the specifics of hedges according to the volatility environment.

Active investing within a solid investment decision framework with in-depth knowledge of options and volatility markets can add value over time. Timing is as important as direction. Success requires a sound, disciplined investment process to drive the size and timing of hedging transactions; otherwise the influence of human behavioural biases (in this case fear and greed) in the face of turmoil can be very counterproductive.

Two examples are presented below applied to equity, then fixed-income, volatility risk.

5.2.3.2.1 Equity Volatility Triggers for Sustainable Convexity Hedging under Budget Constraints

Alternatively, some investors may follow a counter-cyclical approach, setting a fixed budget and spreading the amount of hedges purchased based on the cost (proportional to the level of implied volatility) over time and across strikes, called a "laddered hedge purchase programme".

Spreading hedge purchases over time obtains more exposure when the cost of hedging is low. In fact, the cost of hedge instruments can sometimes eat more than 10% of the operating earnings for variable annuities, due to the non-insignificant transaction costs implied by convexity hedging (mostly from equities options, whose transactions costs are usually around four times swaption transaction costs).

As a result, there is a need for a gradual increase of hedging coverage in line with risks, to protect earnings volatility while trying to cheapen the premium, for which two different systematic rules-based triggers are possible:

- triggers that impose gradually increasing coverage with time (all else equal); and
- stationary triggers that remain fixed over time.

In both cases trigger schedules break down targeted whole trades into a series of tranches: tables indicate the amount of target coverage (as a percentage of the total in force) versus the market volatility levels that could be executed based on these trigger volatility levels, across maturities.

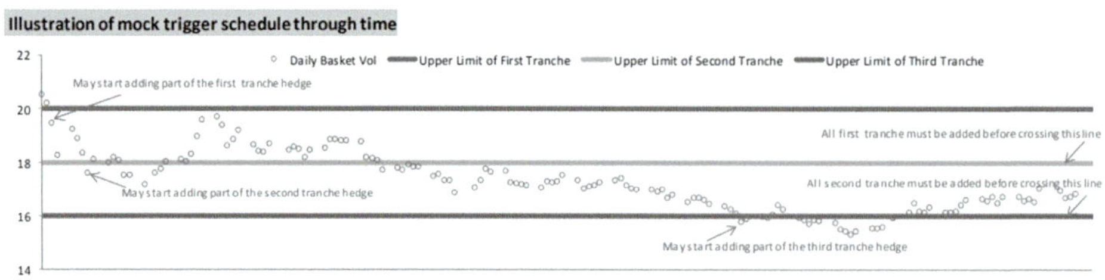

FIGURE 5.102 Mock trigger schedule through time

	Target Coverage*	Target Hedge Vol
Hedge Tranche 1	8%	18-19
Hedge Tranche 2	12%	16-18
Hedge Tranche 3	20%	<16
Total Hedge	40%	

FIGURE 5.103 Target coverage and hedge volatility

The choice of volatility triggers must be in line not only with the four risk appetite dimensions but also with some budget constraints. In this respect, a longer-term hedge is more costly and less protective of spikes, while cost rises with the square of the strike level (see Figure 5.104).

Hedge	Tenor	Coverage	Gamma	Strike	FY Cost Vs. 16 Vol	P&L from hedges (realized vs. strike) Scenario A: vol realizes 30 for 1 mo., 16 thereafter (19 overall) month 1	months 2-6	overall	Scenario B: vol realizes 30 for 6 mos.
"smaller for longer"	6 mos	20%	0.5	19	-25	11	-11	0	65
"shorter for larger"	2 mos	40%	1.0	16	-6	26	-1	25	51
difference				*3*	*-19*	*-15*	*-10*	*-25*	*14*

FIGURE 5.104 Hedging P&L

Moreover, tranche sizing, expressed as a percentage of total gamma, increases as liability gamma grows (see Figure 5.105).

ATM Tranche	Liability Gamma: 1.5m		Liability Gamma: 2.5m		Liability Gamma: 3.0m	
	Gamma	Pctg. of Total Liability	Gamma	Pctg. of Total Liability	Gamma	Pctg. of Total Liability
Hedge Tranche 1	0.1	8%	0.2	8%	0.3	9%
Hedge Tranche 2	0.2	12%	0.3	12%	0.4	14%
Hedge Tranche 3	0.3	20%	0.5	20%	0.7	23%

FIGURE 5.105 Liability gamma

It is recommended to increase the concentration of hedges in shorter maturities given that the majority of the risk is in shorter tenors. In addition, this could result in not paying more for volatility protection given the current upward-sloping volatility term structure.

The downside of such a short-dated option programme is the small amount of time, which does not provide as much hedge effectiveness as in long-dated options if a significant market event takes place. As a result, a 12-month rolling strategy gives better results in terms of mean, variance and tail risks than a three-month rolling strategy.

- A rolling three-month volatility is often expensive as it occurs around volatile financial year and third quarter earnings seasons with three- to six-month persistence.
- Buying a 12-month varswap is usually cheaper than a rolling three-month over one year, specifically in highly volatile markets, while providing lower tail risks and P&L variance as the skew is far less steep than for a three-month volatility. In fact, the volatility cost is proportional to the square root of the maturity, despite the usual upward-sloping term structure.
- Long-dated volatility is less risky: a volatility shock tends to last more than three months but not up to 12 months. Three-month volatility experiences larger extreme risks (skews) than 12-month volatility.

FIGURE 5.106 12-month volatility lower than three-month during extreme market conditions, July 2007–January 2015

The risk in setting triggers also lies in the fact that, potentially, they may be too high and so may rarely be experienced, implying under-hedged situations and thus a lack of hedge effectiveness.

Risks are also dependent on the moneyness, as the liability holds asymmetric dollar gamma profiles depending on the moneyness. The inflexion point is characterized by the change in sign in the derivatives of dollar gamma with respect to the stock level, which roughly coincides with the change in moneyness.

- There is a decreasing shape as the liability gets close to being ITM, which is in favour of lower hedge coverage.
- There is a flat shape as the liability gets close to being OTM, which is all the greater when interest rates rise at the same time, and this is in favour of higher coverage.

In addition, a rising interest rate environment would flatten the dollar gamma, in favour of higher coverage ratios.

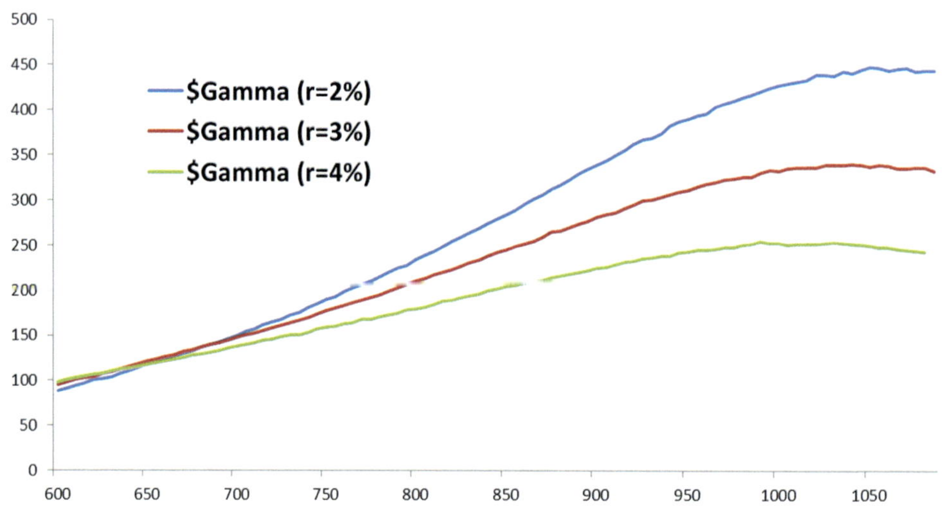

FIGURE 5.107 Gamma versus stock level, depending on interest rate level

5.2.3.2.2 Traffic Light System Aligned with Risk Appetite for Rho Convexity Rebalancing The purpose here is to outline an approach to swaption rebalancing that meets two specific requirements.

(a) The approach needs to be adequately defined so that it allows for regular action to be taken to keep rho convexity exposures at a certain defined and agreed level. Such a defined process will therefore allow some prudent credit for rho convexity protection to be taken when reserving for this business.

(b) The approach also needs to give adequate time and flexibility to define the most appropriate portfolio of swaptions to be held at a point in time. Too rigorous an approach could result in the purchase of a portfolio that is not appropriate for the business either from a cost viewpoint or a protection viewpoint.

The general approach is to hedge certain risks arising from guarantees on a systematic basis. These risks are first-order risks for which the used hedge instruments do not contain optionality and are not depended on market volatility levels (futures, swaps).

If realized market volatility is higher than the volatility assumptions used in pricing this business, this can lead to losses coming from second-order risks.

The insurer has decided here to hedge some of these risks on an opportunistic basis – i.e. when market conditions seem to be favourable and/or when the resulting risks seem to be not acceptable given the risk appetite framework.

Convexity is the second-order risk with respect to interest rates. In a dynamic hedging context, we usually have

$$\text{HedgeP\&L} = \text{First-order P\&L} + \text{Theta P\&L} + \text{Second-order P\&L}$$

And, if market volatility is not exceedingly high, theta and second-order P&L effects should compensate each other. First-order P&L is kept low by dynamically reducing first-order sensitivities.

By hedging convexity, the second-order P&L effect can be further reduced. This comes at the cost of locking in market volatility levels.

Convexity hedging is performed here on opportunistic basis: if market conditions are favourable, the insurer might decide to purchase swaptions.

Convexity exposure should be limited by limits to be back-tested and be checked with the risk appetite framework, such as these.

- The absolute value of the global net convexity exposure is less than [X%] of the total liability rho convexity.
- The absolute value of the net global convexity exposure is, at most, [€Y].
- It is based on rho and rho convexity sensitivities and capital requirement shocks.

In this respect, on a daily basis a one-day VaR metric is computed, based on the daily rho and rho convexity.

Given the residual number of days for the year, n, corresponding n-day VaR metrics are computed.

- If the VaR metrics < the risk appetite proxy budget, the risk appetite proxy budget is just reduced by the VaR.
- If the VaR metrics > the risk appetite proxy budget, the rho exposure is brought down to a “green light” boundary.

A process to monitor and review the convexity risks is set up in order to remain within these limits.

- The net ("global") convexity exposure is monitored on a daily basis.
- Rebalancing action is triggered if convexity exposure is outside the limits for z consecutive days.
- Once this trigger point has been reached, an appropriate portfolio of assets is analysed for, at most, a period of w months from the trigger date.
- The calibrated portfolio should result in the following: net ("global") convexity exposure is, at most, [X%] of the total liability rho convexity.
- Once a portfolio is defined, a meeting is required within a week to decide whether the portfolio should be implemented:
  - with respect to cost/risk aspects;
  - subject to the capital being available to support the capital impact in terms of volatility mismatches; and
  - if net convexity reverts back to within the limits, the insurer might decide not to trade any swaptions.
- Once the portfolio is agreed, the assets should be purchased within a one-month window.

5.2.3.3 Tail Risk Hedging

The S&P 500 posted 34 gap events (defined as a greater than 5% difference between the intraday low with the prior day's market close) over the 50 years ended July 2017. This represents just 0.27% of trading days, but some were unforgettable, such as the near-25% single-day fall on 20 October 1987. While most exchanges now implement circuit breakers to prevent falls of that size, the chance of sudden intraday falls of more than 5% still exceeds some investors' risk tolerance.

Tail risk hedging (e.g. variance swap, or call options on VSTOXX or VIX) refers to a strategy of buying derivative contracts that are expected to perform well in times of significant market stress. Allocations to tail-risk-hedging strategies do not necessarily have a directly proportional relationship to an investor's equity allocation, since the depth of a market correction is not sufficient information to predict the size of the offsetting gain from a tail risk hedge.

The tail risk protection is mostly provided by exposure to jumps in implied volatility, which is related to the speed and severity of a correction but not explicitly to the depth of a drawdown: position sizing requires an estimation of the market volatility increase as well as the extent of the market decline. As a result, sizing tail risk positions is a bit of an inexact science, based on quantifying sensitivity to increases in implied volatility (vega).

A tail risk mandate serves as a diversifier within the total multi-asset portfolio, rather than an overlay to an equity allocation. Such strategies can be split into two broad camps: rules-based strategies (medium fee levels) and active discretionary strategies (often hedge fund mandates, with much higher fees).

There is some debate about the definition of a "tail event". Most academic observers describe it as a three-sigma event or the worst 1% of potential investment outcomes. Unfortunately, that leaves out the aspect of fear that accompanies a tail risk event. A more pragmatic description is "something sudden and troubling that does not happen very often".

The size of a tail-risk-hedging mandate should be based on one of the following considerations: impact (the desired loss offset if a major event occurs – a coverage amount); and cost (an acceptable rate of loss in the scenario that no event occurs, which is similar to the concept of an insurance premium).

Regarding cost, tail risk hedges are generally less affordable than downside-protected equity due to their higher complexity. Over time, protecting against a tail event is much more expensive with downside protection (e.g. puts or put spreads) than it is with a tail risk strategy (e.g. well-structured volatility trade), but downside protection can be useful for tactical purposes, when the investor places value on a lower return with less volatility and downside risk.

5.2.3.3.1 Option Tail Risk Strategies Lack Efficiencies in the Long Term due to Costly Carry Theoretically, if the investment horizon were infinite, hedging might not be necessary, as an investor could incur positive returns from many sources of risk over long periods of time. However, reporting to shareholders usually takes place at specific times, during which there can be considerable volatility. Just-in-time hedging is impossible, and, by the time an investor decides to hedge, the market correction may have already begun, and hedging may have already become expensive.

Most tail risk option strategies have excessively high costs due to high fees, underperformance during normal markets and limits to upside potential. In fact, they fail to meet the following shape of an ideal tail risk hedge, as their costs usually exceed their benefits, as shown in Figure 5.108.

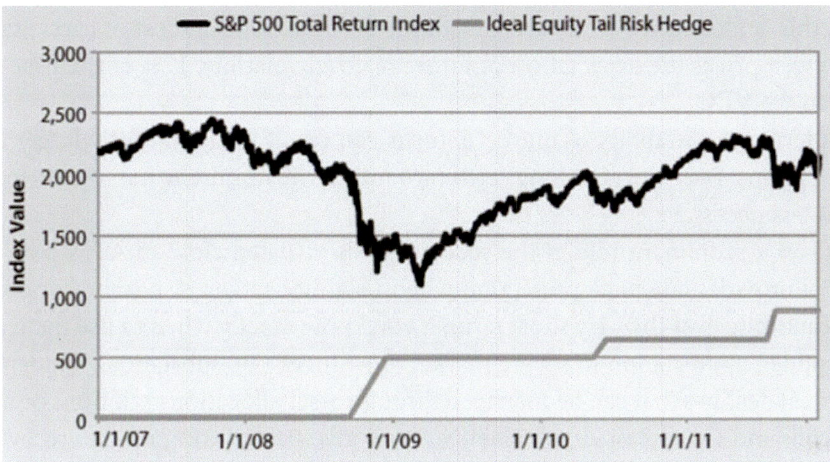

FIGURE 5.108 The shape of an ideal tail risk hedge, January 2007–January 2012

For example, investors using put options to protect against gap risk indeed face significant challenges. It is expensive to maintain ongoing protection, it is difficult to time and often it does not pay off. Meanwhile, put spreads and deep out-of-the-money option purchases still leave many portfolios significantly exposed to gap events. For example, consider a gap event of 10%. A hypothetical investor has bought a three-month 97%–90% put spread at the top of the market with a cost of 1.5% (6% to run annually). This strategy would still leave the portfolio participating in 45% of the market fall at expiry. However, if the gap event were 20%, this strategy would leave the portfolio exposed to 72.5% of the fall (both assuming the strike was from the top of the market) at expiry.

The level of instantaneous gap protection is highly variable and heavily dependent on where the option was struck and the remaining tenor, as illustrated in Figure 5.109. For example, holding a three-month 90% put option will generate 1.8% return on the notional – or provide just 18% protection against the gap event (assuming the option notional amount was equal to the underlying holding). If markets trend lower as the option matures, then the hard floor will pay off. However, markets have historically recovered quickly after large drawdowns.

Proportion of a -10% gap event, protected by a vanilla put option (hard floor)																
Option Expiry	Option Strike															
	85%	86%	87%	88%	89%	90%	91%	92%	93%	94%	95%	96%	97%	98%	99%	100%
6 Months	7%	10%	13%	16%	20%	24%	29%	34%	39%	44%	50%	55%	60%	65%	70%	74%
3 Months	3%	4%	7%	10%	14%	18%	23%	29%	36%	43%	50%	57%	64%	70%	76%	81%
2 Months	1%	2%	4%	7%	10%	15%	21%	27%	34%	42%	50%	58%	66%	72%	79%	84%
1 Month	0%	1%	2%	3%	6%	11%	16%	24%	32%	41%	50%	59%	68%	76%	83%	89%
1 Week	0%	0%	0%	0%	1%	5%	12%	20%	30%	40%	50%	60%	70%	80%	88%	94%
1 Day	0%	0%	0%	0%	0%	2%	10%	20%	30%	40%	50%	60%	70%	80%	90%	98%

Based on 2017/07/31 market levels on the S&P 500 index

FIGURE 5.109 Proportion of a –10% gap event, protected by a vanilla put option (hard floor)

There are two sources of negative expected return. The first is the option time decay, as an option's time value is eroded each day an option remains out of the money, illustrative of the first source of the "significant cost of carry". The second component of negative expected return is the negative delta of put options, as they lose value as equities rise.

Implied volatility is usually higher than realized volatility (the volatility risk premium) priced within the put option, and, again, this is illustrative of the second source of the significant cost of carry; since the inception of the VIX index, which prices the expectation of future realized volatility, 12% of the time realized volatility has been higher than the VIX.

If over the short term the instability of market returns can result in significantly larger losses that can be mitigated with put options, over the very long term (around ten years) the worst loss often approaches zero, making put options less necessary.

Market timing plays a prominent role in the success, as short-dated close to ATM put options offer high levels of gamma that provides adequate protection when volatility spikes. A number of portfolio protection techniques work beautifully over the very short term, if you know exactly when to use them and when to stop, translating into a sudden appreciation in severe market downturns (i.e. substantial, sudden and unexpected, and not minor, gradual declines that can be managed through asset allocations rotations or momentum-based strategies: see Miccolis and Goodman 2012), albeit with no give-back during market recoveries, notably due to higher implied volatility than realized volatility (significant cost of carry).

Maintaining a long-term position in put options for hedging will pay at the outset of a crisis, but focusing on shorter-dated and more liquid volatility durations may be preferable in terms of risk and reward, whereby an investor would be required to roll short-term options to remain protected, leading to the risk of increased levels of volatility when puts are rolled. This can be partially mitigated by a renewing coverage of implied volatility on a one-year periodic basis, within which the volatility is then locked in at inception.

5.2.3.3.2 Option Tail Risk Strategies There has been a proliferation of more sophisticated tail risk options since the market collapse of 2008, some of which use the VIX as the underlying, as it exhibits a positive skewed distribution.

Start	End	SPX	3m ATM Put	3m 95% Put	1y ATM Put	1y 90% Put	Short Term VIX Future	Medium Term VIX Future	1m ATM Put	1m 95% Put
29-Apr-10	1-Jul-10	-15%	8%	7%	9%	8%	8%	14%	9%	5%
22-Jul-11	3-Oct-11	-18%	14%	14%	17%	17%	25%	19%	12%	11%
17-Aug-15	29-Sep-15	-10%	8%	6%	10%	6%	10%	10%	7%	3%
29-Dec-15	11-Feb-16	-12%	9%	8%	13%	9%	8%	9%	8%	4%

FIGURE 5.110 Hedging strategy performance during several equity market downturns

	Annualized Carry	3m ATM Put	3m 95% Put	1y ATM Put	1y 90% Put	Short Term VIX Future	Medium Term VIX Future	1m ATM Put	1m 95% Put
Since May 2014	7.0%	100%	160%	150%	230%	14%	33%	100%	220%
Since Dec 2016	8.2%	100%	250%	175%	350%	8%	15%	85%	450%

FIGURE 5.111 Hedging strategy costs benchmarked versus three-month put roll-over strategy, 2014–2018

5.2.3.3.2.1 Using the VIX

5.2.3.3.2.1.1 VIX Futures

- Regarding the carry between implied and realized volatility, it is possible to systematically trade proportionate rolling of monthly VIX futures strategies, which are closely related to monthly forward-starting variance (e.g. 30-day variance starting at the futures' expiry), thus expressing a constant one-month VIX play.
- Moreover, a way to partially mitigate the theta time decay is to add an opportunistic overlay by selling VIX futures when equity markets are not in distress (i.e. whenever the term structure is positive), which generates alpha and translates into a spot-starting variance butterfly position.

5.2.3.3.2.1.2 Using VIX Options

- Today the S&P 500 VIX Short-Term Futures Index (SPVXSTR) and S&P 500 Mid-Term Futures Index (SPVXMTR) offer constant-maturity tradable equity volatility exposure data going back over 15 years now. The market is deep enough for the largest investors (e.g. tens of millions of dollars of vega can result in hundreds of millions of dollars in P&L) and efficient enough for the smallest investor (e.g. 8 bps of volatility in a wide bid–ask spread in cash-based products).
- There is a ratio to the call spread on the VIX, with a long position in further out-of-the money contracts and a smaller (hence the "ratio") short position in closer to-the-money contracts. The option structure is maintained continuously, with three-month options that are rolled forward at one month prior to expiration.
- The time decay is more manageable, as it avoids the steepest rate of time decay in the last month of the option's life, and this helps focus the exposure on the primary objective of sensitivity to jumps in implied volatility, as it reacts net-positively to large volatility shocks.
- Even so, such a VIX-based hedge results in performance drag from time decay, as there is generally a negative cost of carry to the hedge; this can be thought of as paying an ongoing premium for the protection provided. That cost is evident over periods where there is no tail event and these premiums did not "pay off". This is illustrated by the fairly steadily declining performance from mid-2012 to 2015, when the corrections were too small to trigger hedging gains in an "always-on" strategy. While this strategy provided a large gain in the significant correction of 2008/9, in the absence of that jump the always-on tail hedge did not add consistent value.
- The tail risk hedge does not have anywhere near a normal distribution of returns – it has a longer distorted right tail from those cases when the hedge was beneficial – so the term "standard deviation" can be a little misleading. That said, it is still a good diversifier of risk, as illustrated by the significant negative correlation to the market; it was designed with that criterion in mind. In fact, during the period from 2007 to 2015 the correlation of monthly returns was −34%.

	Annualised return	Standard deviation	Sharpe Ratio
SPTR index	6.49%	16.40%	0.40
VIX call spread - always on	5.85%	28.10%	0.21
70/30 blend	7.64%	11.70%	0.65

FIGURE 5.112 Comparative performance

One of the most interesting aspects of this particular structure is the spike in performance with a jump in implied volatility, with a tendency to give gains straight back as the volatility term structure normalizes. This relatively quick mean reversion in performance happens for all but the biggest of corrections (e.g. several months for the 2008/9 crash).

A fixed 15% asset allocation of a portfolio into volatility indices may enhance portfolio performance return distribution, as it can make return distribution more positively skewed, while smart allocation strategies that differentiate between low- and high-volatility environments can bring further enhancement to performance during high-volatility spike periods.

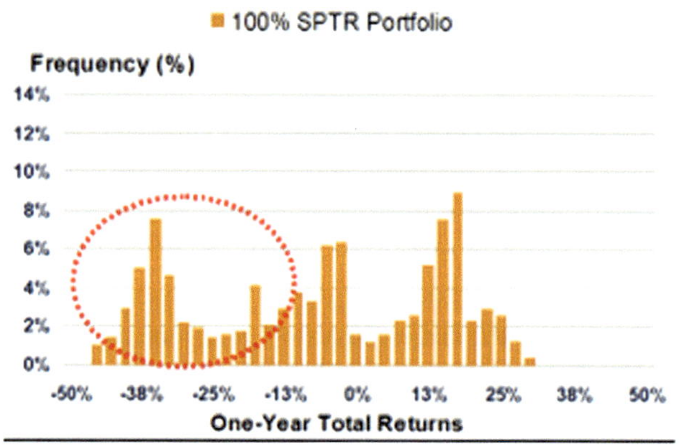

FIGURE 5.113 Returns distribution before volatility allocation

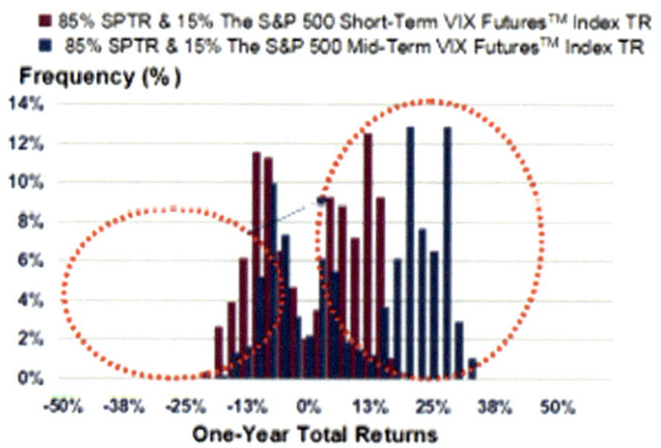

FIGURE 5.114 Returns distribution after volatility allocation

However, it underperforms long equity-only portfolio in low-volatility environments, due to the negative carry cost associated with the upward-sloping VIX futures term structure. More optimized equity and long volatility allocation is required to achieve consistently better performance in various volatility regimes.

Overall, managing a VIX tail risk mandate requires daily, if not continuous, oversight. The VIX cannot be directly traded, and the instruments that are based on VIX are forward exposures. Given that the structure of the volatility curve does not always react the same way as the spot VIX index (i.e. the relationship between the actual volatility of the index and the implied volatility of an option is not constant through time), the management of such strategies cannot be run on a "set and forget" basis.

5.2.3.3.2.2 *Using Exotic Tail Risk Put Options*

5.2.3.3.2.2.1 Using Best-of Puts Given a high – and even close to 100% – correlation between puts during market crashes, using a best-of put instead of a vanilla put cheapens the cost of the downside protection, combined with the following.

- A condition on certain events (e.g. a 30% market correction), so as to activate the best-of feature.
- A variable notional on the best-of, increasing as the market goes up before a crash, will further lower the cost, by more than 50%.
- The writing of options is contingent on the increase in the best-performing asset (e.g. +20%) or in the realized volatility of the basket; however, while worst-of options are popular, there is relatively little demand for best-of options.
- There are also resettable puts, within which the strike follows the market (reset every two month to 90% of the maximum level of the underlying on the reset dates) and keeps the same protection level, monetizing the view that the market is volatile and expected to make new highs and avoiding the cost of dynamic hedging.

5.2.3.3.2.2.2 Using Hybrid Equity: Interest Rate Options Clients who manage equity assets to fund liabilities are most exposed when both equities and long rates are down: their assets decline, while the PV of their liabilities rises. Empirically, equity–rate correlation has been strongly positive during recent equity sell-offs; thus equity–rate puts significantly outperform purchasing protection on equity and rates separately, notably through lower cost, as these examples illustrate.

- Vanilla equity–rate puts: a put option on a basket of equity and a ten-year swap rate costs about 54% of the average cost of individual puts.
- Knock-out equity–rate puts: a one-year SPX 95% put that exists only if the ten-year swap rate is below 4.75% at maturity with an indicative offer of 2.80%. Rate observation occurs only at maturity. This compares to a vanilla one-year 95% put cost of 6.50%.
- Knock-in equity–rate puts: a one-year SPX 95% put with knock-in if the ten-year swap rate touches 4.75%, with an indicative offer of 4.70%. Rate observation occurs throughout the transaction's life. This compares to a vanilla one-year 95% put cost of 6.50%.

5.2.3.3.2.2.3 Using Timer Puts These are puts that mature once the variance budget has been consumed. The variance budget corresponds to the reasonable volatility expected (V) on an investment horizon (T).

At inception, investors define the variance budget they estimate to be the most appropriate. The price of the timer put is a function of the variance budget chosen. Once the variance budget has been consumed, the put terminates

Timer put formula:

$$\left(K - \frac{S_T}{S_0}\right)^+$$

With

S_0 = Underlying price at inception

S_T = Underlying price when all the variance budget is consumed

K = strike

5.2.3.3.2.2.4 Reducing Risk Premiums Whenever Realized Volatility Is below High Levels of Implieds These options are available.

- Volatility strike puts: the buyer of the put compensates the put seller for the gamma P&L resulting from the realized volatility being greater than the target volatility.
- Strike-adjusted puts: the buyer of the put has a sliding strike, which depends on the ratio of the spread between the target and the realized variance versus the target volatility.
- Realized volatility puts: the buyer of the put will pay for the option at maturity, using a Black–Scholes pricing formula with the realized volatility.
- Mileage options: the buyer of the put is paid downside protection as soon as the realized variance exceeds a budget based on the targeted realized volatility.

In conclusion, tail-risk-hedging positions are structured to have large payoffs in certain conditions. However, after an event these positions can lose time value at an accelerated pace when markets start to calm down, so knowing when to close or roll a position (via active investment decisions or well-designed rules) is as critical to success as choosing the right structure in the first place. Performance can be improved in this regard by applying switching rules to monetize or take off the hedge until conditions return to a more normal state.

Given the complexity of these tail risk mandates, the potential for management fees to be higher (than simpler index-option-based hedges) and the propensity for long periods with zero to negative absolute returns, fiduciaries may find it challenging to keep such a strategy alive long enough to do its job. When it is implemented, we would recommend regularly educating boards to ensure that the role of these mandates is understood as members change. It is also important to diversify across more than one strategy, as there are challenging return paths for any given strategy.

5.2.3.3.3 Emerging Gap Risk Options Offer an Agnostic Cheap Alternative to Capture the Velocity of Changes

5.2.3.3.3.1 Definition There is a risk that the price of an underlying portfolio investment will change from one level to another with gap risk and no trading in between.

Such movements occur when the underlying portfolio drops substantially from the previous day's closing price. A gap event will occur if the index level drops in value by more than the gap strike (e.g. a performance of –25%) over the gap observation period, based on realized redemption proceeds. Gap risk is then defined as the drop in the underlying portfolio over the observation period. The minimum period of gap risk associated with an underlying portfolio depends on its liquidity profile: the frequency of valuation dates; the notice period required to submit a valid redemption request in respect of a dealing date; and the settlement period for redemption proceeds.

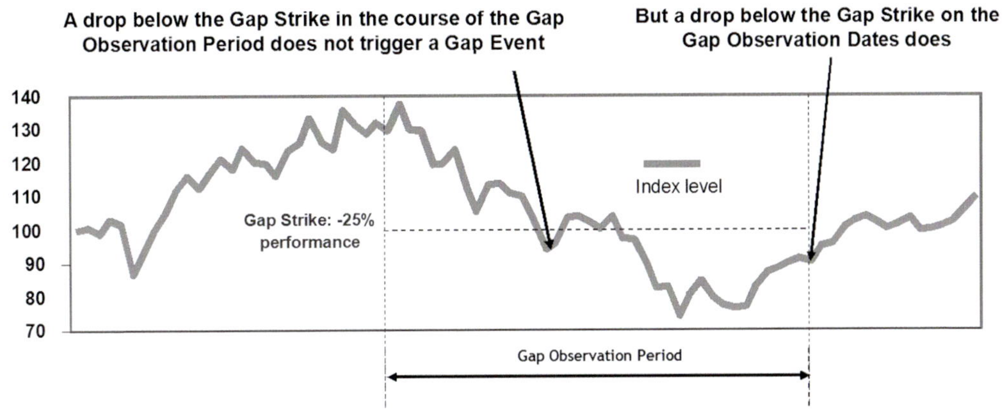

FIGURE 5.115 Gap risk illustration

Realized volatility on the S&P 500 index reached 89% at the height of the global financial crisis in 2008. An investment in the index starting on 10 October 1987 would have experienced a 55% drawdown lasting 510 days from the peak to the trough, and another 1,256 days to break even. Statistical models assuming a Gaussian distribution would predict such events to occur once every 13,457 years.

Date	Closing	Log Return	Average frequency under Normal law
19-Oct-87	1738.74	-0.2563	once in 10^{53} years
26-Oct-87	1793.93	-0.0838	once in 72503 years
15-Oct-08	8577.91	-0.0820	once in 41318 years
01-Dec-08	8149.09	-0.0801	once in 21725 years
09-Oct-08	8579.19	-0.0762	once in 6068 years
27-Oct-97	7161.15	-0.0745	once in 3402 years
17-Sep-01	8920.70	-0.0740	once in 2914 years
29-Sep-08	10365.45	-0.0723	once in 1798 years
13-Oct-89	2569.26	-0.0716	once in 1405 years
08-Jan-88	1911.31	-0.0710	once in 1173 years

FIGURE 5.116 Worst downside events of S&P 500 index, October 1987–October 2008

For instance, for a stock with a 25% volatility, the probability of having a 10% gap on any one day during one year is 3×10^{-8}, and the probability of a 20% gap is entirely negligible. We therefore suggest pricing and hedging gap options using models based on processes with discontinuous trajectories.

The gap options are indeed designed to capture stock jumps, and, clearly, cannot be priced within a diffusion model with continuous paths, since any such model will largely underestimate the gap risk.

The payoff of a gap option is linked to the occurrence of a gap event – that is, a one-day downside move of sufficient size in the underlying. There is ample evidence for crash fears and jump risk premiums in quoted European option prices, and many authors have argued that jump models allow a precise calibration to short-term European calls and puts and provide an adequate vision of short-term crash risk. Gap options capture exactly the same kind of risk. Further, as an approximate hedge of a gap option can be constructed using out-of-the-money puts (see below), it is therefore natural to price and risk manage gap options within a model with jumps, calibrated to market-quoted near-expiry Europeans.

The following gap option was commercialized by a big international bank in 2007 under the name of "gap risk swap": the protection seller pays the notional amount N to the protection buyer at inception and receives LIBOR + spread monthly until maturity or the first occurrence of the gap event, whichever comes first, plus the notional at maturity if no gap event occurs. The gap event is defined as a downside move of over 10% in the underlying within one day (close to close). If a gap event occurs between dates t – 1 and t, the protection seller immediately receives the reduced notional $N(1-10 \times (0.9-R))^{+}$, where $R = S_t / S_{t-1}$ is the index performance at gap, after which the product terminates.

As a result, the gap option payoff can be defined as follows. Suppose that the time to maturity T of a gap option is subdivided into N periods of length Δ (e.g. days): T = NΔ. The return of the k-th period will be denoted by $R\Delta\, k = S_{k\Delta} / S_{(k-1)\Delta}$. Let α denote the return level that triggers the gap event and k^* be the time of first gap expressed in the units of Δ: $k^* := \inf\{k: R^{\Delta}{}_{k} \le \alpha\}$. The gap option is an option that pays to its holder the amount $f(R^{\Delta}{}_{k^*})$ at time Δk^*, if $k^* \le N$, and nothing otherwise.

5.2.3.3.3.2 Pricing of Gap Options We first treat the case in which the log-returns are independent and stationary. We follow Tankov (2010).

Let the log-returns be i.i.d. and denote the distribution by $p\Delta(dx)$. Then the price of a gap option is given by

$$G_\Delta = e^{-r\Delta}\int_{-\infty}^{\beta} f\left(e^x\right)p_\Delta(dx)\frac{1-e^{-rT}\left(\int_\beta^\infty p_\Delta(dx)\right)^N}{1-e^{-r\Delta}\int_\beta^\infty p_\Delta(dx)},$$

Proof: see Tankov (2010).

$$\begin{aligned}
G_\Delta &= E\left[e^{-\Delta k^* r} f\left(R_{k^*}^{\Delta}\right)1_{k^*\le N}\right],\\
&= \sum_{n=1}^{N}\mathbb{P}\left[k^*=n\right]E\left[f\left(R_n^{\Delta}\right)\# k^*=n\right]e^{-\Delta nr},\\
&= \sum_{n=1}^{N}\mathbb{P}\left[R_n^{\Delta}\le\alpha\right]E\left[f\left(R_n^{\Delta}\right)\# R_n^{\Delta}\le\alpha\right]e^{-\Delta nr}\prod_{l=1}^{n-1}\mathbb{P}\left[R_l^{\Delta}>\alpha\right],\\
&= e^{-r\Delta}\int_{-\infty}^{\beta} f\left(e^x\right)p_\Delta(dx)\frac{1-e^{-rT}\left(\int_\beta^\infty p_\Delta(dx)\right)^N}{1-e^{-r\Delta}\int_\beta^\infty p_\Delta(dx)}
\end{aligned}$$

This formula allows us to compute gap option prices by Fourier inversion. For this, we need to be able to evaluate the cumulative distribution function,

$$F_\Delta(x) := \int_{-\infty}^{x} p_\Delta(d\xi),$$

and the integral,

$$\int_{-\infty}^{\beta} f\left(e^x\right)p_\Delta(dx).$$

- Let φ_Δ be the characteristic function of p_Δ, F' be the CDF and φ' the characteristic function of a Gaussian random variable with zero mean and standard deviation $\sigma' > 0$. Then

$$F_\Delta(x) = F'(x) + \frac{1}{2\pi}\int_{\mathbb{R}} e^{-iux}\frac{\phi'(u)-\phi_\Delta(u)}{iu}du.$$

- The integral (2) is nothing but the price of a European option with payoff function f and maturity Δ. So the price of a put gap option with log-forward moneyness k = log(K/S) – rΔ is given by

$$P_{\Delta}(k) = P_{\Delta}^{BS}(k) + \frac{S_0}{2\pi}\int_{\mathrm{R}} e^{-ivk}\tilde{\zeta}\Delta(v)\,dv,$$

$$\tilde{\zeta}_{\Delta}(v) = \frac{\phi_{\Delta}(v-i) - \phi_{\Delta}^{\sigma}(v-i)}{iv(1+iv)},$$

$$\phi_{\Delta}^{\sigma}(v) = \exp\left(-\frac{\sigma^2 T}{2}\left(v^2 + iv\right)\right).$$

$P_{\Delta}^{BS}(k)$ is the price of a European put option with log-moneyness k.

In practice, the corresponding integrals will be truncated to a finite interval [–L, L]. Since Δ is small, the characteristic function decays slowly at infinity, which means that L must be sufficiently big (typically L ~ 100), and the computation of the integrals will be costly. On the other hand, precisely the fact that Δ is small allows us, in exponential Lévy models, to construct an accurate approximation of the gap option price as follows.

- $S_t = S_0 e^{X_t}$, where X is a Lévy process, which means that p_{Δ} as defined above is the distribution of X.

- Since rΔ ~ 10 – 4 and the probability of having a gap on a given day $\int_{-\infty}^{\beta} p_{\Delta}(dx)$ is also extremely small, with very high precision

$$G_{\Delta} \approx \int_{-\infty}^{\beta} f\left(e^x\right) p_{\Delta}(dx) \frac{1 - e^{-rT - N\int_{-\infty}^{\beta} p_{\Delta}(dx)}}{r\Delta + \int_{-\infty}^{\beta} p_{\Delta}(dx)}.$$

Our second approximation is less trivial. For all Lévy processes, and under very mild hypotheses on the function f, we have

$$\int_{-\infty}^{\beta} g(x) p_{\Delta}(dx) \sim \Delta \int_{-\infty}^{\beta} g(x)\nu(dx),$$

as Δ → 0, where ν is the Lévy measure of X. Consequently, when Δ is non-zero but small, we can replace the integrals with respect to the density with the integrals with respect to the Lévy measure, obtaining an approximate but explicit expression for the gap option price as G_0:

$$G_{\Delta} \approx G_0 = \lim_{\Delta \to 0} G_{\Delta} = \int_{-\infty}^{\beta} f\left(e^x\right)\nu(dx) \frac{1 - e^{-rT - T\int_{-\infty}^{\beta}\nu(dx)}}{r + \int_{-\infty}^{\beta}\nu(dx)}.$$

Moreover, if we consider the following modified gap option payoff definition as

$$f\left(\frac{S_{\tau}}{S_{\tau-}}\right) = f\left(e^{\Delta X_{\tau}}\right) \text{if } \tau \le T \text{ and zero otherwise},$$

where τ = inf{t: ΔXt ≤ β} be the time of the first jump of X smaller than β, then the price is given by and easily seen to be equal to the above G_0:

$$G = E^Q\left[e^{-r\tau} f\left(e^{\Delta X_\tau}\right) 1_{\tau \le T}\right].$$

Remark: the gap option then arises as a pure jump risk product, which is only sensitive to negative jumps larger than β in absolute value, but not to small fluctuations of the underlying. In particular, it has zero delta.

Consider Kou's model, where the Lévy density holds the following form:

$$\nu(x) = \frac{\lambda(1-p)}{\eta_+} e^{-x/\eta_+} 1_{x>0} + \frac{\lambda p}{\eta_-} e^{-|x|/\eta_-} 1_{x<0}.$$

Here λ is the total intensity of positive and negative jumps, p is the probability that a given jump is negative and η− and η+ are characteristic lengths of respectively negative and positive jumps. In this case, for most common choices of f, the integrals can be computed explicitly:

$$\int_{-\infty}^{\beta} \nu(dx) = \lambda p e^{\beta/\eta_-}.$$

Then, for an OTM gap put option, the price is

$$\int_{-\infty}^{\beta} f\left(e^x\right) \nu(dx) = \frac{\lambda p \eta_-}{1+\eta_-} K^{1+1/\eta_-}.$$

Remark: the model parameter estimation is a tricky issue here: it is next to impossible to estimate the probability of a 10% gap from historical data, since the historical data simply do not contain negative daily returns of this size. Still, even though a 10% negative return has never occurred yet, the market participants believe that it has a positive probability of occurrence in the future. This suggests extracting the information about the probability of sharp downside moves from short-maturity OTM put prices by calibrating an exponential Lévy model to market option quotes and use it to price gap options.

As illustrated above, the gap option is close to a zero-delta product, so the associated risk cannot be delta-hedged and, more generally, it is hopeless to try to hedge it with the underlying. Moreover, the gap options are designed to offset jump risk, and the markets with jumps are typically incomplete. Therefore, one can try only to approximately hedge a gap option, for example, in the sense of L^2 approximation, and even then one would need to find a suitable hedging product, which is sensitive to extreme downside moves of the underlying and has little sensitivity to the small everyday movements.

5.2.3.3.3.3 Replicating Gap Options with Short-Dated European Options A natural example of such a product is a short-dated (one- to two-week maturities) deep out-of-the-money put option, for which the market share has grown exponentially in the United States since 2011 (from 5% to 30 to 50% of the S&P 500 options since 2016). These short-dated deep OTM put options offer protection against jumps – that is, against the same kind of risk as the gap option.

Our aim is now to compute the optimal quadratic hedge ratio for hedging a gap option with an OTM put – that is, the hedge ratio minimizing the expected squared hedging error. We presume that this expected squared error is computed under the martingale probability. We start by expressing the martingale dynamics of a gap option price.

We denote by Gt the price of a gap option evaluated at time t, defined as the second proxy developed above. The terminal value of a gap option can be then expressed as an integral with respect to J defined as the jump measure of X: J([s, t] × A): = #{r ∈ [s, t]: ΔXr ∈ A}. with compensated version J˜: J˜([s, t] × A): = J([s, t] × A) – (t – s)ν(A).:

$$G_T = \int_0^T \int_{\mathbb{R}} f\left(e^x\right) 1_{t \le \tau} J(dt \times dx).$$

Taking the conditional expectation under the risk-neutral probability, we can compute the price of a gap option at any time t:

$$G_t = E^Q\left[G_T \mid \mathcal{F}_t\right] = 1_{\tau \le t} f\left(e^{\Delta X_\tau}\right) + 1_{\tau > t} \int_{-\infty}^{\beta} f\left(e^x\right) \nu(dx) \frac{1 - e^{-\lambda^*(T-t)}}{\lambda^*}.$$

The interpretation of this formula is very simple: if, at time t, the gap event has already occurred, then the price of a gap option is constant and equal to its payoff; otherwise, it is given by the G_0 above applied to the remainder of the interval.

Alternatively, this can be rewritten as a stochastic integral with respect to J˜:

$$G_t = G_0 + \int_0^t \int_{-\infty}^{\beta} 1_{s \le \tau} e^{-\lambda^*(T-s)} f\left(e^x\right) \tilde{J}(ds \times dx).$$

Let P(t, S) denote the price of a European put option evaluated at time t

$$P(t,S) = E^Q\left[\left(K - S_T\right)^+ \mid S_t = S\right],$$

$$P_t \equiv P\left(t, S_t\right) = P\left(0, S_0\right) + \int_0^t \sigma S_u \frac{\partial P\left(u, S_u\right)}{\partial S} dW_u + \int_0^t \int_{\mathbb{R}} \left\{P\left(u, S_{u-} e^z\right) - P\left(u, S_{u-}\right)\right\} \tilde{J}(du \times dz).$$

A self-financing portfolio containing φt units of the put and the risk-free asset has value

$$V_t = c + \int_0^t \phi_s dP_s,$$

where c is the initial cost of the portfolio. The hedging strategy minimizing the risk-neutral L^2 hedging error,

$$E^Q\left[\left(c + \int_0^T \phi_t dP_t - G_T\right)^2\right],$$

is then

$$\hat{c} = E^Q[G_T] = G_0,$$

$$\hat{\phi}_t = 1_{t \le \tau} \frac{\int_{-\infty}^{\beta} V(dz) f(e^z) e^{-\lambda^*(T-t)} \{P(t, S_t e^z) - P(t, S_t)\}}{\sigma^2 S_t^2 \left(\frac{\partial P}{\partial S}\right)^2 + \int_{\mathbb{R}} \nu(dz) \left\{P(t, S_t e^z) - P(t, S_t)\right\}^2}$$

Note that $\hat{\varphi}_t$ is nothing but the local regression coefficient of G_t on P_t.

The cost of the hedging strategy, $\hat{c}$, coincides with the price of the gap option.

The strategy $\hat{\varphi}_t$ is optimal but is does not allow perfect hedging (there is always a residual risk), and it is not feasible, because it requires continuous rebalancing of an option portfolio.

In practice, due to the relatively low liquidity of the option market, the portfolio will be rebalanced rather seldom, say once a week or once every two weeks, as the hedging options arriving to maturity are replaced with longer-dated ones.

To test the efficiency of out-of-the-money puts for hedging gap options, we simulate the L^2 hedging error (the variance of the terminal P&L) over one rebalancing period (one week or two weeks) using two feasible hedging strategies: the trader buys $\hat{\varphi}_0$ options in the beginning of the period and keeps the number of the options constant until the end of the period; the trader buys $\hat{\varphi}_0$ options in the beginning of the period and keeps the number of the options constant until the end of the period unless a gap event occurs, in which case the options are sold immediately.

This strategy achieves a fourfold reduction compared to no hedging at all if one-week options are used and a twofold reduction if two-week options are used.

In addition, such hedging modifies considerably the shape of the distribution of the terminal P&L, reducing, in particular, the probability of extreme negative payoffs (Tankov 2010).

- Without hedging, the distribution of the terminal payoff of the gap option has an important peak at −1, corresponding to the maximum possible payoff (the graphs in Figure 5.117 are drawn from the point of view of the gap option seller). In the presence of hedging, this peak is absent and the distribution is concentrated around zero.

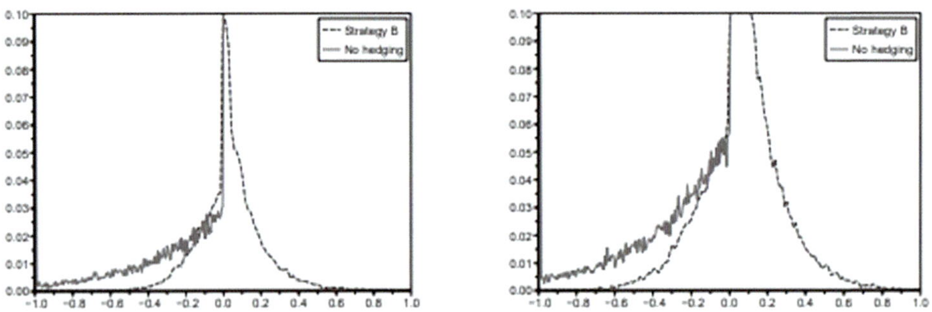

FIGURE 5.117 Distribution of the terminal payoff

- If one-week options are used for hedging (the left hand panel of Figure 5.117), the value at risk of the portfolio for the horizon of one week and at the level of 0.1% is equal to 0.85 without hedging, and only to 0.23 in the presence of hedging; this means that the hedging will allow a reduction in the regulatory capital by a factor of four.

- If two-week options are used (the right hand panel), the two week VaR at the level of 0.1% is 0.38 with hedging, and without hedging it is equal to 0.99 (the probability of having a gap event within two weeks is slightly greater than 0.1%).

We conclude that hedging gap options with OTM puts is feasible, but it is best to use the shortest available maturity.

5.2.3.3.3.4 Gap Options Empirically Outperform Vanilla Options The tail risk is defined here as the conditional mean monthly portfolio return during any month the equity loses in excess of 5%.

We compare here a roll-over put strategy with a roll-over gap put strategy with a T-bill strategy. The effectiveness of such a tail risk strategy is assessed here through two main criteria.

- The first is the tail risk protection efficiency measured by the performance drag, which is the reduction in annual return when adopting the tail risk strategy (here reducing the tail risk defined above by 20%) versus the long equity strategy used as benchmark. Such tail risk protection efficiency for a gap option is, naturally, significantly impacted by the choice of the activation threshold (here 95%).
- The second criterion is a certainty measure that looks at the consistency with which it outperforms during a crash, which is here the conditional information ratio, which measures the average excess return per unit of volatility during events when the equity falls by more than 5%.

In fact, thanks to their significantly lower premium, and despite their activation only below a specific threshold, gap options have better performance drag than vanilla put options (here both with strikes of 90%), with similar information ratios.

Such outperformance has been tested across one- and two- to three-year periods since 2007, so that they cover both quiet spells and some mini-crashes, over which gap put options offer similar average performance drag while better, with the following consistent pattern:

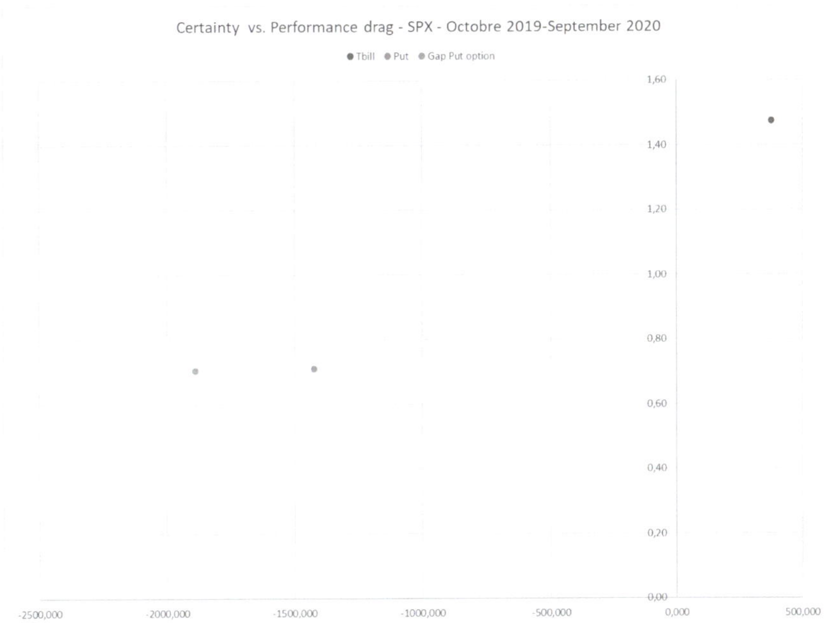

FIGURE 5.118 Certainty measure versus tail risk measure, October 2019–September 2020

5.2.3.3.3.4 Conclusion on Gap Risk Options Gap options turn out to be resilient and effective for hedging.

- They have high hedging power, based on tail risk metrics (conditional mean monthly return during months the S&P 500 loses in excess of 5%).
- They have performance similar to or higher than OTM puts during massive drawdowns (albeit weaker performance during a series of small to medium-sized market swings).
- They clearly outperform whatever the negative returns threshold.
- They have a high certainty of tail risk protection based on the conditional "information ratio" during tail risk events.

Gap options prove to be cheap and simple to use.

- They have a very low premium, with a 10 to 15% cost versus OTM puts.
- They have low performance drag when added to an equity portfolio.
- They enable an agnostic view on equity markets.
- They are customizable through the choice of the activating gap barrier.
- They are easily replicable using short-dated (weekly liquid) OTM puts.

5.3 MANAGING HEDGING LIQUIDITY CONSTRAINTS

The recent crisis has accelerated and deepened the interest of regulators in market risk positions. Furthermore, the recognition of the role of liquidity risk in the crisis has further broadened the range of risks that regulators and risk managers take into consideration.

What happens when and if guarantees move fully out of the money? Future market movements would then lead to net cash transfers from insurers to their counterparties, backed only by fair-value assets in the hands of the insurer, thus creating a net cash strain. This situation introduces a borrowing requirement for insurers, or, in a general sense, introduces an additional consideration for funding.

Therefore, by virtue of risk management alone, illiquid policyholder liabilities become liquid obligations in the hands of the insurer. It is thus critical that insurers appreciate why, how and to what degree market microstructure in liquidity affects option positions to enable them to assess their risk appetite and pricing expectations of the same.

When an insurance company sells an option on an illiquid index and seeks to hedge with a liquid index, what then is the correct price for the option? Ignoring all other elements, the sold position should reflect the illiquidity of the underlying fund, on the basis that the insurance company will incur either expensive replication costs or increased hedging errors in providing the protection. In this regard, the insurer is transforming the illiquid obligation, and requires recompense and reward for this.

Thus, the volatility parameter should have regard to the sold obligation in addition to the replication strategy, as illustrated below on two key life insurance products: unit-linked hybrid investments and long-term equity guaranteed investments.

5.3.1 Managing Market "Gap Risks" for Unit-Linked Hybrids (iCPPI/TIPPs)

Individualized constant proportion portfolio insurance products are attractive alternatives to traditional unit-linked products offering a guaranteed minimum return, such as variable annuities. They offer high potential returns while limiting the downside risk by implementing a dynamic allocation strategy between risky and risk-free assets tailored to the risk appetite of the beneficiary. But the performance evaluation of iCPPI products should not rely on the unrealistic assumptions of continuous market price variation and continuous rebalancing of asset allocations. We adopt a more general and realistic pricing jump model and examine several dynamic strategies, as well as put options, to mitigate the risk that the value of the product falls below the guaranteed minimum (so-called "gap risk").

With rising life expectancies, current provisions for retirement may not be sufficient for people to secure an acceptable standard of living after retirement. To achieve sufficiently high investment returns together with low risks over the long term, customers' funds should remain invested in risky assets as well as in safer bonds over an extended period well into retirement. The design of long-term investment products should also reflect the requirements and risk appetite of individual investors.

From the point of view of the provider as well, CPPI products provide an attractive alternative to many traditional long-term investment products, offering a guaranteed minimum return for several key reasons:

- lower exposure to volatility and extreme market price movements, along with slightly lower returns;
- lower costs; and
- lower regulatory capital requirements.

Besides their price transparency, open time horizon and no early redemption penalty, CPPI products offer a wide range of alternative investments for the risky asset, and flexibility to add other guarantees such as ratchets.

The CPPI investment strategy provides a minimum guaranteed return, the "floor" (usually defined as the discounted value of the final capital guarantee), and aims to maintain a risk asset exposure equal to a constant multiple of the "cushion" (defined as the excess value of the fund above the floor) at all times. The capital guarantee at maturity and the multiplier are customized to the customer's risk appetite, usually between three and six (may be constant or not depending on the contract).

However, implementation comes with many of the concrete challenges raised in section 1. The rebalancing of the asset allocation can be made only at discrete times. There are transaction costs, and risky asset prices may jump. There is likely to be a difference between the realized compared to the value of a CPPI strategy computed under traditional unrealistic theoretical conditions of continuous price movements, unfettered zero-cost trading and continuous rebalancing. In particular, there is a non-zero probability for the value of the fund to fall below the guaranteed floor, called the gap risk, as illustrated by the impact of introducing discontinuous jump processes in the modeling within the risky asset dynamics.

Section 5.3.1.2 specifically deals with concrete strategies that, at least partially, mitigate such gap risk, through a dynamically risk-adjusted multiplier and the use of put options.

5.3.1.1 CPPI Management: from Theory to Practice

5.3.1.1.1 CPPI Mechanism Basics Consider at time t a risky asset (e.g. a share) with price S and a risk-free asset (e.g. a Treasury bond) with price B returning a constant rate r. The CPPI fund is invested into these two assets so that part of its value, called the "floor", Ft, is guaranteed, while the excess value above the floor (the "cushion"), Ct = Vt – Ft, remains exposed to the risky asset price fluctuations. At any one time the exposure to the risky asset is kept at a constant multiple m of the cushion – i.e. m × Ct (where m is usually held in practice between three and six, implying that the asset manager borrows dynamically to buy the risky asset, or may buy in practice the non-risky part only close to the expiry of the contract).

The risky asset S is defined by the usual log-normal continuous-time diffusion equation with drift μ and volatility σ:

$$\frac{dS_t}{S_t} = \mu dt + \sigma dW_t \quad \text{and} \quad \frac{dB_t}{B_t} = r dt,$$

$$dV_t = m(V_t - F_t)\frac{dS_t}{S_t} + \big(V_t - m(V_t - F_t)\big) r dt,$$

$$V_t = F_t + (V_0 - F_0)\exp\left[\left[m(\mu - r) + r - \frac{m^2\sigma^2}{2}\right]t + m\sigma W_t\right].$$

This makes the portfolio value V independent on the path followed by the underlying S, while the probability of touching the floor is zero (see Black & Jones 1987).

Then the cushion Ct is also log-normally distributed:

$$\frac{dC^t}{C_t} = (m\mu + (1-m)r)dt + m\sigma dW_t \cdots -> C_t = C_0 \exp\left(\left(m(\mu - r) + r - \frac{m^2\sigma^2}{2}\right)t + m\sigma W_t\right).$$

However, such statistical assumptions are unrealistic and not consistent with market practice. As remediation, two alternatives are studied: modelling in a discrete-time framework and using discontinuous "jump" processes (e.g. the Kou model).

5.3.1.1.2 Discrete-Time CPPI A sequence of equidistant points in the interval [0, T] is defined, between which the portfolio asset allocation is updated. The first time the portfolio value touches the floor is defined by the following formula;

$$t_s = \min\{t_k \epsilon \Theta \mid V_{tk} - F_{tk} \le 0\}.$$

The probability of touching the floor now becomes greater than zero, assuming that the portfolio has not breached the floor up to time t_k. The probability of breaching the floor at time t_{k+1} is that of a downside jump in the risky asset of more than about 1/m, as evidenced below:

$$V^{\tau}_{t_{k+1}} - F_{t_{k+1}} = \begin{cases} \left(V_{t_k} - F_{t_k}\right)\left(m\frac{S_{t_i}}{S_{t_{i-1}}} - (m-1)e^{r\frac{T}{n}}\right) & \text{if } V_{t_k} - F_{t_k} > 0 \\ \left(V_{t_k} - F_{t_k}\right)e^{r\frac{T}{n}} & \text{if } V_{t_k} - F_{t_k} \le 0. \end{cases}$$

Assuming that the breach of the floor did not occur until tk,

$$V^{\tau}_{t_{k+1}} > F_{t_{k+1}} \Leftrightarrow \left(m\frac{S_{t_i}}{S_{t_{i-1}}} - (m-1)e^{r\frac{T}{n}}\right) > 0,$$

$$\Leftrightarrow \frac{S_{t_{k+1}}}{S_{t_k}} > \frac{m-1}{m}e^{r\frac{T}{n}}.$$

As the interest rate return is close to zero over one day, we get the result

$$\frac{S_{t_{k+1}}}{S_{t_k}} - 1 > -\frac{1}{m}.$$

Back-testing on three rebalancing frequencies (daily, weekly and monthly) from the first quarter of 2006 to the fourth quarter of 2010 (S&P 500 index) illustrates that the CPPI strategy under daily rebalancing performs better than the weekly and monthly ones within bearish markets. We tested 10,000 simulations paths using the Black–Scholes model with a three-month realized volatility, a constant asset return m = 8%, a risk-free rate r = 4%, duration of five years and 10 bps transaction costs. This result is due to the way daily rebalancing is highly responsive to decrease the risk exposure, which prevents the bond floor from being breached and thus ensures the capital guarantee at maturity (as illustrated by fatter left tails in Figure 5.120). On the other hand, the 5% and 0.5% quantiles show that the CPPI with m = 6 has a larger right tail. It performs better than the two others in bullish market even though the mean return is similar to CPPI with m = 3.

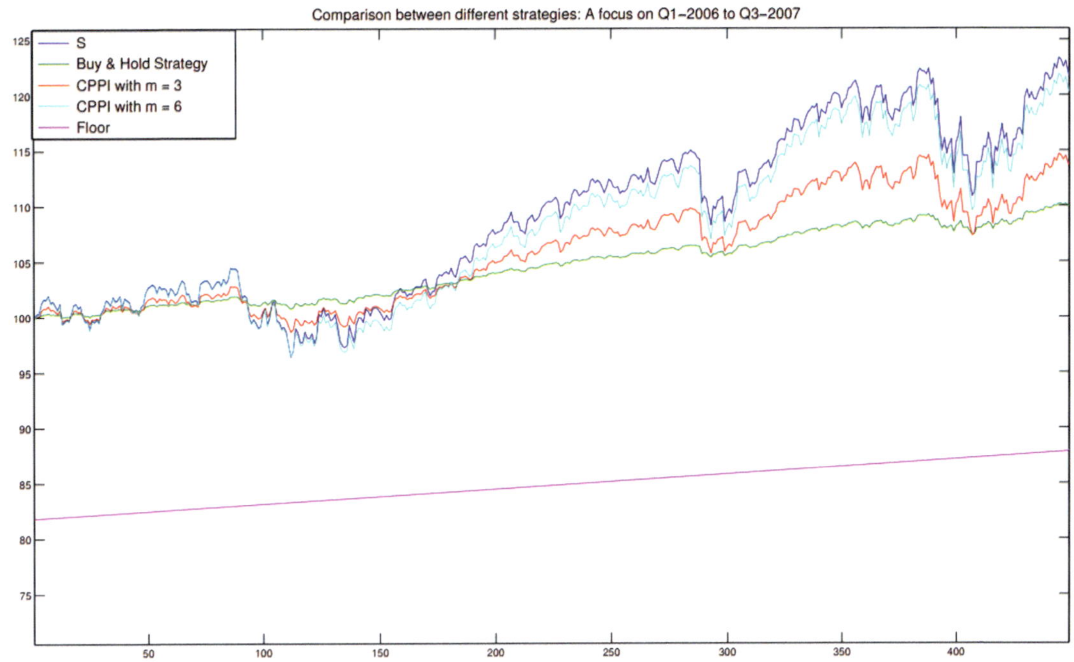

FIGURE 5.119 Performance depending on multiplier versus buy and hold strategies, first quarter 2006–third quarter 2007

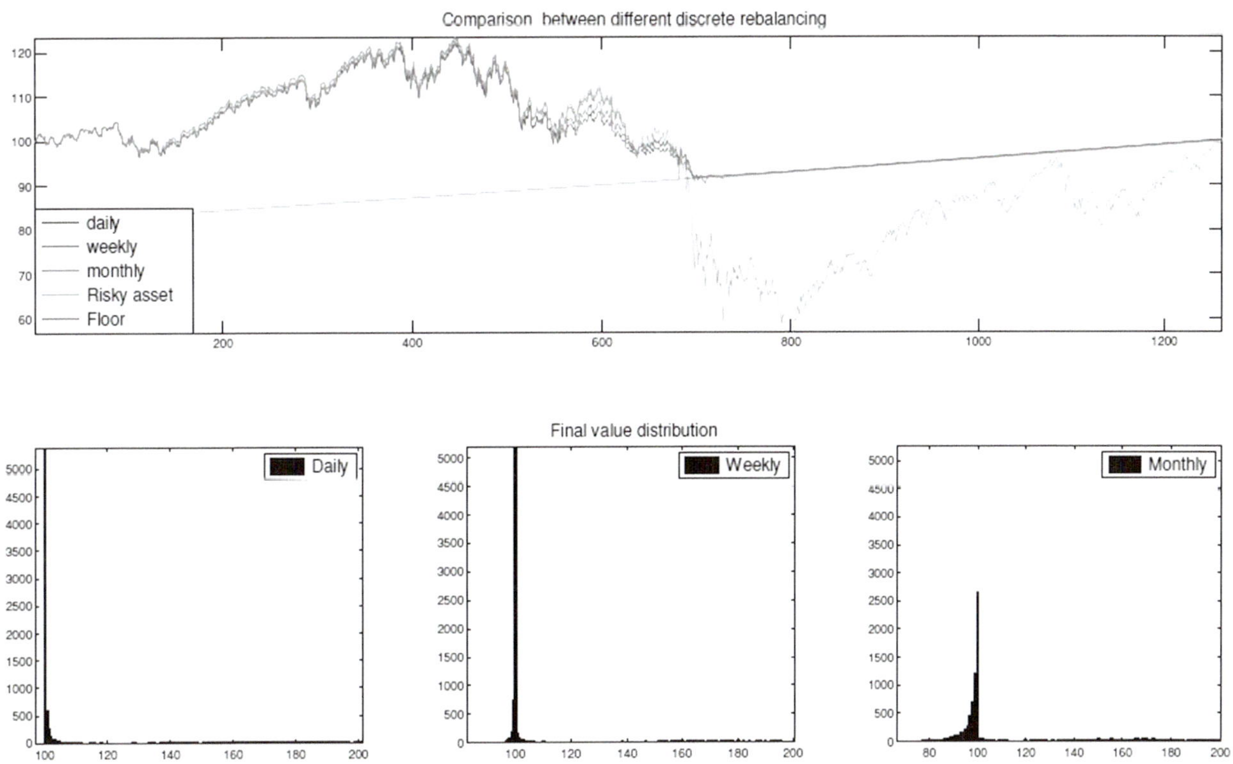

FIGURE 5.120 Performance between different discrete rebalancing frequencies

	CPPI with m=3			CPPI with m=6		
	Daily	Weekly	Monthly	Daily	Weekly	Monthly
Mean	123.31	123.39	119.75	124.10	124.87	125.01
Std-dev	31.58	32.66	36.86	42.62	43.88	48.10
95% quantile	100.48	99.98	97.01	99.99	99.13	89.69
99.5% quantile	100.02	99.98	91.47	99.98	95.20	74.28
5% quantile	194.37	195.23	197.94	216.51	218.50	225.46
0.5% quantile	226.47	284.07	282.58	291.49	293.75	311.46
Rebalancing cost	0.91	0.44	0.26	0.78	0.46	0.31
PBF	0.0018	0.0947	0.5289	0.2016	0.5730	0.6555

FIGURE 5.121 Statistical metrics depending on multiplier and rebalancing frequency

Note: PBF = probability of breaching the floor.

However, using a constant volatility and log-normal distribution modelling is not consistent with empirically observed jumps during extreme market moves. They are likely to breach the bond floor. Jumps are thus added in the next section.

5.3.1.1.3 Jump Modelling For computational tractability, we choose the double exponential Kou model (see Kou 2002). The Kou model introduces jumps into the stochastic process for stock returns as a set of random Poisson processes. The Kou model is defined as follows:

$$\frac{dS_t}{S_{t^-}} = \mu dt + \sigma dW + d\left(\sum_{i=1}^{N_t} e^{Y_i} - 1\right),$$

where W is a standard Brownian motion, N is the added (Poisson) jump process, where the jump sizes {Y1, Y2,...} are i.i.d. random variables with a common asymmetric double exponential density, and

$$f(y) = (1-p)\gamma^+ e^{-\gamma^+ y} 1_{y\geq 0} + p\gamma^- e^{-\gamma^- y} 1_{y<0}.$$

$\gamma+/\gamma-$ are the intensity of positive/negative jumps while (1 – p) and p are the likelihood of positive and negative jumps respectively. The calibration has been carried out by minimizing the quadratic error on options prices with one-month maturity, and strikes 80 to 110% (see Figure 5.122).

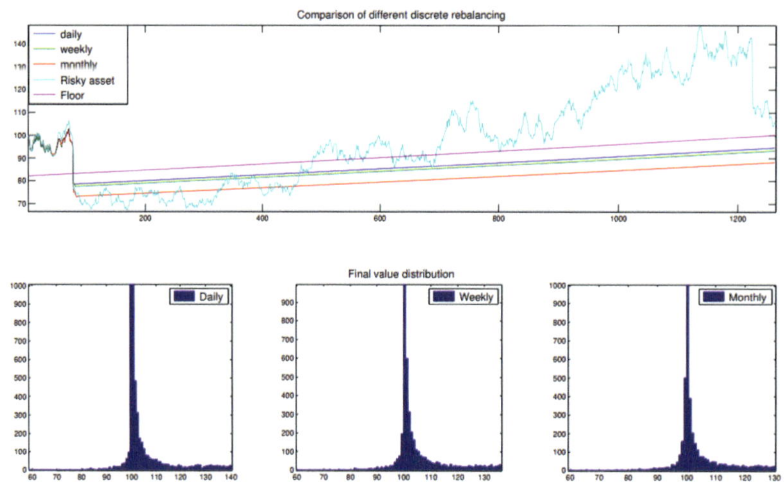

FIGURE 5.122 Statistical metrics of the Kou model

	Daily	Weekly	Monthly
Mean	146.28	147.10	147.57
Stdev	52.84	52.93	53.11
95% quantile	92.19	92.21	92.03
99.5% quantile	59.38	59.08	59.23
5% quantile	238.12	238.67	239.41
0.5% quantile	349.41	350.92	350.37
Rebalancing cost	0.92	0.45	0.26

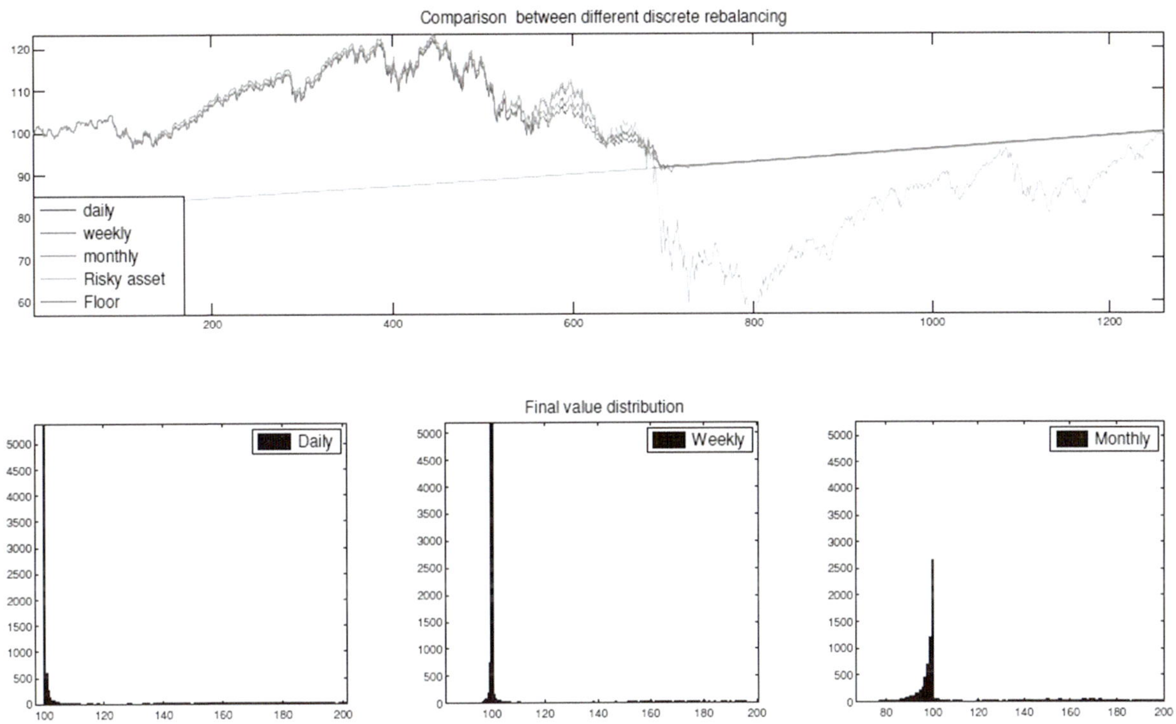

FIGURE 5.123 Performance between different discrete rebalancing frequencies

The results in Figure 5.124 illustrate that, whereas the probability of breaching the floor (so-called "gap risk") significantly decreases down to being negligible under the traditional unrealistic assumption of continuous price movements (Black–Scholes model assumptions) as the rebalancing frequency increases to daily, this is no longer the case under more realistic discontinuous modeling assumptions (the Kou model), even with continuous rebalancing frequency (0.4%, so no more negligible in terms of occurrence for such extreme-magnitude events).

Model	Frequency	PBF
B&S	Monthly	9.07×10^{-6}
	Weekly	1.2×10^{-10}
	Daily	~
Kou	Continuous	0.00410

FIGURE 5.124 Probability of breaching the floor depending on asset dynamics modelling and rebalancing frequency

The next section deals specifically with concrete strategies that at least partially mitigate such gap risk, through a dynamically risk-adjusted multiplier, and the use of put options.

5.3.1.2 Mitigating the Downside Risk by Preventing Breaching of the Floor (Gap Risk)

5.3.1.2.1 Adjusting the Multiplier to Market Conditions The manager usually sets the multiplier at the beginning of the period. Even so, the probability of breaching the floor may surge in a market crash, or the manager might miss the subsequent market recovery. Thus the multiplier needs to be adjusted so as to be consistent with the current market conditions.

A first approach to define a dynamic multiplier is the choice of an "optimal" m* (e.g. using optimal certainty-equivalent returns with hyperbolic absolute risk aversion utilities and log-normal distribution). m* is defined by the following formula:

$$m^* = \eta \frac{(\mu - r)}{\sigma^2}$$

where η is the investor's sensitivity of risk tolerance to wealth

An alternative is a value-at-risk-based multiplier for which investors choose the confidence level according to their tolerance for tail risks; see Jiang, Ma & An (2009). m_t is defined as follows:

$$m_t = \frac{1}{1 - \exp\left(\left(\mu - r - \frac{1}{2}\sigma^2\right)(T-t) - z_p \sigma \sqrt{T-t}\right)}.$$

These two approaches offer an interesting alternative to the constant multiplier, which lacks flexibility depending on market conditions. Based on the back-testing of data from 2006 to 2011, the VaR-based multiplier performs better than the "optimal" one in bullish and recovery markets: the multiplier decreases more in periods of market turmoil, reducing the risk exposure, and increases back during periods of market recovery. In contrast, during bear markets, using the "optimal" multiplier (through m < 1) helps keep a somewhat higher cushion, but it misses the recovery, as it makes no provision for high leverage.

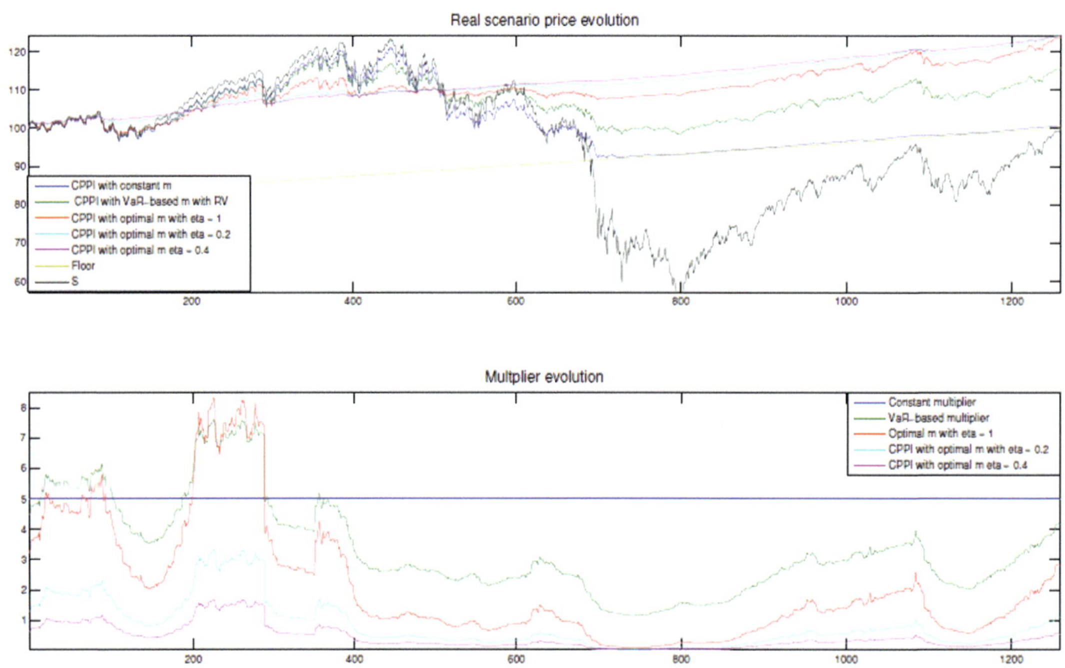

FIGURE 5.125 Comparison between different multipliers (VaR-based with p = 99.5% versus the optimal one with γ = 0.2, 0.4) and based on realized volatility

Note: RV = realized volatility.

In order to allow for a higher level of participation in the market recovery, the multiplier is adjusted with a modified volatility estimator. This is done either through a short-term weighted moving average realized volatility (EWMA with $\lambda = 0.94$) or through an estimator based on the implied volatility of the strike consistent with the latest market returns. For example, if the underlying jumped 5% downwards, the implied volatility with strike 95% will be chosen. This adjustment enables the model to capture more of the upside return when markets rebound. For example, reinvesting in the risky asset as soon as the third quarter of 2009 in the back test results in higher returns, as illustrated with the rising ongoing performance after the rebound in the stock in Figure 5.126, consistent with a dynamic multiplier quickly adjusting to the stock performance.

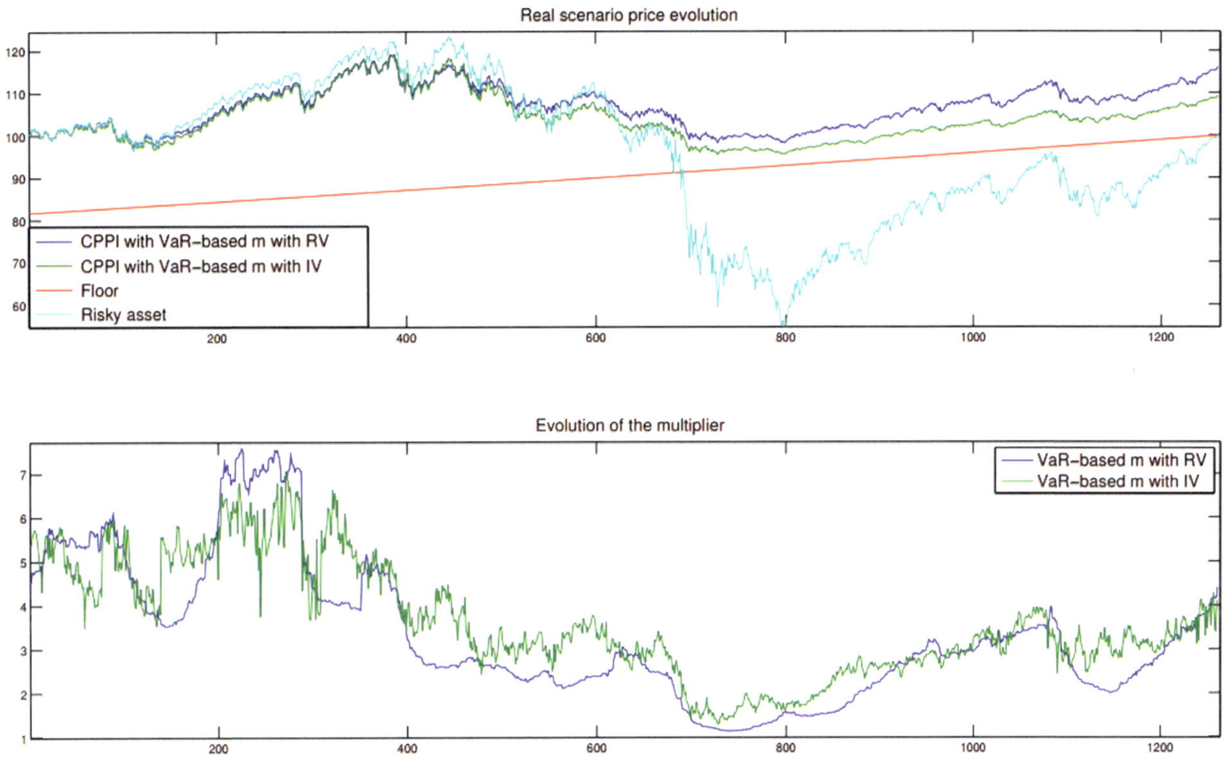

FIGURE 5.126 Comparison between dynamic multipliers based on realized volatility and implied volatility

Note: IV = implied volatility.

Finally, the fixed-frequency rebalancing may be switched to trigger rebalancing when the multiplier is out of a specific range chosen by the portfolio manager, as illustrated by the higher performance in Figure 5.127, in which the trigger-based management (green line) consistently outperforms the daily rebalancing management (red line). In this case, on average the rebalancing frequency becomes every other day, which is consistent with the usual practice in CPPI asset management, while the cost of rebalancing is cut by a half in comparison to a daily rebalancing (i.e. as low as a weekly or monthly rebalancing).

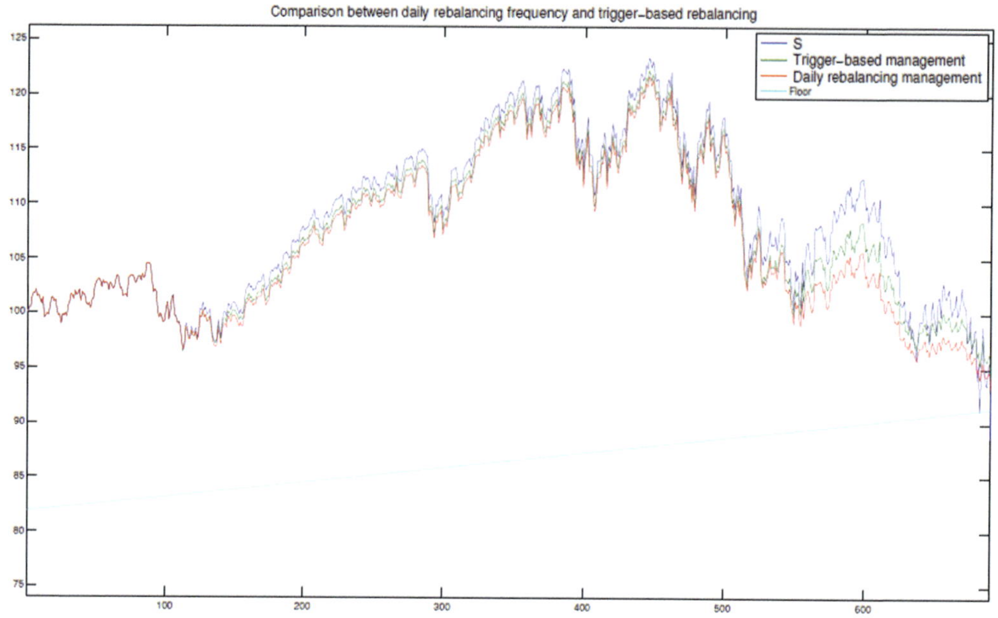

FIGURE 5.127 Comparison between trigger rebalancing and fixed-frequency (daily) rebalancing

Adjusting the multiplier dynamically allows it to be more reactive to market conditions and explicitly dependent on the investor's risk aversion. However, it is still exposed to the downside risk in case of sudden jumps ("black swan" events, such as a market crash of 20% in a single day), when options may be useful to hedge such gap risks.

5.3.1.2.2 Hedging the Gap Risk A simple hedging strategy for the CPPI can be constructed using short maturity put options. Touching the bond floor is mathematically equivalent to the cushion becoming negative. Assuming that the event has not occurred up to time t_k, the "gap risk" is defined by

$$C_{t_{k+1}} < 0 \Leftrightarrow m\frac{S_{t_{k+1}}}{S_{t_k}} - (m-1)e^{r\frac{T}{n}} < 0.$$

Hedging this risk can be done by buying put options at each rebalancing period with strike price

$$\left(1 - \frac{1}{m}e^{r\frac{T}{n}} S_{t_k}\right)$$

and with maturity equal to the CPPI rebalancing frequency. To hedge the whole portfolio, the manager needs a number of puts equal to

$$m\frac{C_{t_k}}{S_{t_k}},$$

which is the risky asset exposure. The discounted payoff in this case is then

$$e^{-r\frac{T}{N}}C_{t_k}\left((m-1)e^{r\frac{T}{N}}-m\frac{S_{t_{k+1}}}{S_{t_k}}\right)^+$$

while the hedging cost is

$$\text{cost}_{t_k}=m\frac{C_{t_k}}{S_{t_k}}\mathbb{E}^{\mathbb{Q}}\left[\left(\left(1-\frac{1}{m}\right)e^{r\frac{T}{N}S_{t_k}}-S_{t_{k+1}}\right)^+\right].$$

Figure 5.128 illustrates the following impacts of hedging with puts: as there is no final value below 100 in the "Put hedge" P&L distribution, the guarantee is, as expected, ensured, and the manager no longer holds the risk of breaching the floor. In terms of P&L distributions, the CPPI distribution with put option hedging is indeed a truncation of the classical CPPI, where losses are cut (the left tail is limited by the guarantee). However, once the put is exercised and the floor recovered, the manager needs to monetize that option in order to keep the guarantee until maturity.

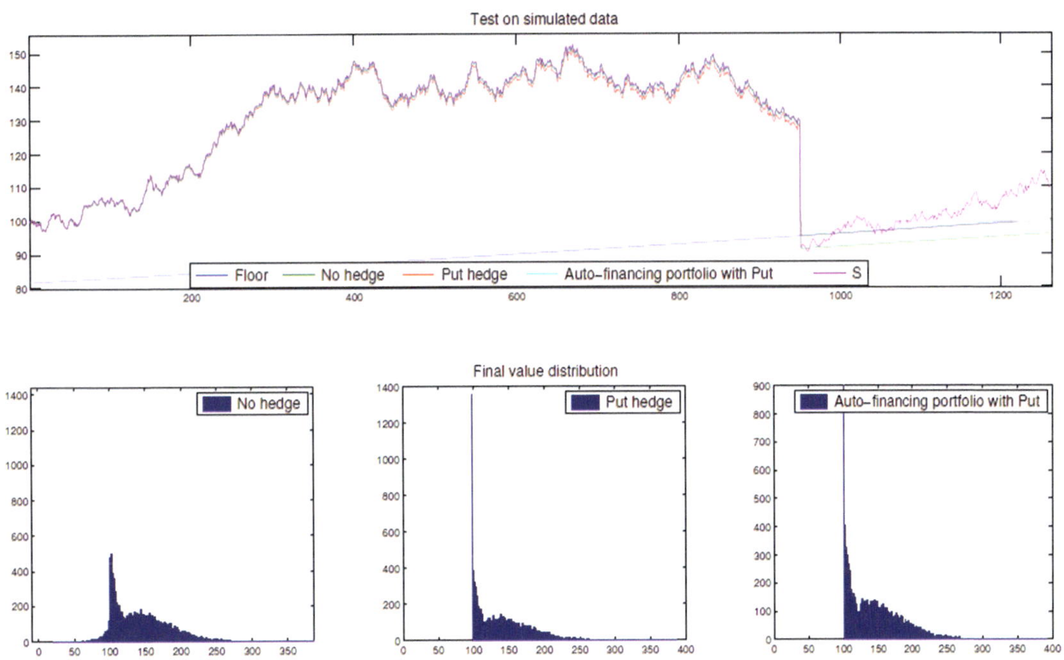

FIGURE 5.128 Comparison between no hedging and put hedging

5.3.1.2.3 Conclusions The main conclusions regarding modeling and hedging issues from a risk/return/cost perspective are as follows.

- Jump modelling is an essential element of CPPI modelling. It allows the model to measure the non-zero probability of breaching the floor.

- Correctly choosing and adjusting the multiplier dynamically significantly reduces the downside risk according to a value-at-risk indicator.
- Hedging the gap risk is possible through vanilla puts.

5.3.2 Minimizing the Market Impact of Hedging Life Insurance Liabilities within Risk Appetite

5.3.2.1 Market Impact as a Growing Issue

As a long-term equity guaranteed investment can be decomposed into the equity asset and a put option, hedging the risks embedded within the downside protection can be done through buying put options available in limited quantity in the financial market. However, buying a large number of equity put options may impose pressures on the unit price, as equity put options are highly sensitive to the supply and demand balance, which implies a higher cost of hedging than initially expected, illustrative of market liquidity risk.

There has indeed been significant evidence of growing costs stemming from an imbalance of supply and demand for options, driven by the derivatives hedging activity of large players, notably near the market close, in particular during equity sell-offs, which have been occurring with growing frequency over the past decade (May 2010, August 2011, August 2015, January 2016, June 2016, February 2018, October–December 2018, March–April 2020).

This might help exacerbate the acuity of such volatility spikes from the illiquidity premium in option markets since the 1990s (the late 1990s bull market, with overwriting-supplied volatility from pressure for income generation, or the period from 1997 to 2000, when there was strong demand for bond and call participation rates).

An empirical study conducted by Christoffersen et al. (2018) has presented strong evidence for this, stemming from the structural imbalance between supply and demand in derivatives, as illustrated by the following.

- There have been some $5 trillion outstanding in equity derivatives in the United States: $3 trillion are listed, representing a 20-fold increase since 2000, above the cash equity volume from 2005 to 2007; and there were also $2 trillion in the OTC market in 2005, before a 40% drop since 2008.
- There is a structural supply–demand imbalance in short-dated volatility: there are 70% more put options outstanding than calls outstanding, with directional users buying puts for portfolio protection; this has been matched by growth of the variance swap market requiring short positions in deep out-of-the money puts.

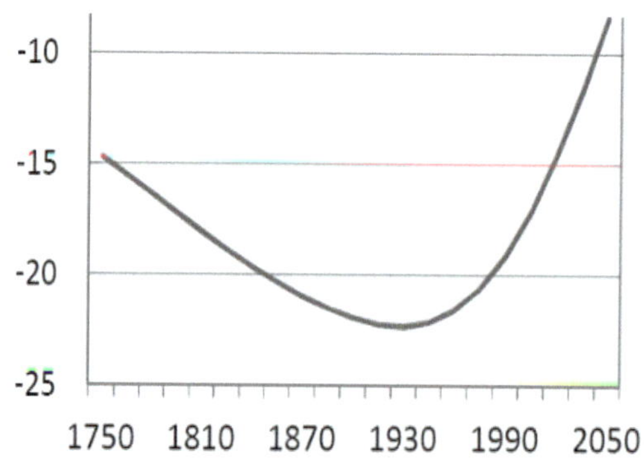

FIGURE 5.129 Put–call gamma imbalance, 1750–2020

- There is also a structural supply–demand imbalance in long-dated volatility, again implying growing cost in terms of hedging.
  - First, from US variable annuity and Asian structured product hedging, translating into a steepening of the term structure.

- Second, from hedging US mortgage convexity (homeowners own the option to prepay, which is valuable when interest rates fall, leaving the market net short volatility): there were $900 billion in option hedges in 2005 (as against $200 billion in 1993).
- Third, from the life insurance hedging of long-term guarantees (e.g. embedded within variable and fixed indexed annuities), characterized by three main features: a medium- to long-term duration, large volumes and significant market risk convexity exposure. Some 97% of the derivatives used by the US insurance industry are used by life insurance, 95% of which are used for hedging (49% swaps, 45% options, 7% futures and forwards), the total rising from $786 billion as of FY 2010 to $2,600 billion as of FY 2018.

Specifically regarding the hedging of large life insurance liabilities such as variable annuities, the impact on derivatives markets was already significant even before the financial crisis, in terms of both gamma and vega.

- Gamma hedging needs can be met with listed options (based on 2007 cash index options statistics from the CBOE).
  - Annual sales have about 25,000 of gamma; gamma is about 0.425 per $1 million of account value.
  - One year of sales volume would be equivalent to ½% of the annual dollar volume of cash index options, while multiple years of sales could result in 5+% of the dollar volume.
  - One year of sales volume would be equal to 1% of the open interest of cash index options, while multiple years of sales could result in 10+% of the open interest of cash index options.
- The vega hedging needs exceed supply, though they are not all hedged in practice by most annuity carriers (as the implied volatility is modelled as constant, thus implying no vega P&L, and consistent with the fact that there is no market at all for such long 50-year liabilities while prices exceed what insurers are willing to pay). The outstanding vega hedges represent less than 20% of total needed hedges. Annual sales stand at about $250 million of vega; vega is about $4.2 per $1,000 of account value.

Given such a supply–demand imbalance in derivatives, the cost of placing one large order to close a position will be far greater than the sum of infinitely small orders placed at different times.

The usual transaction costs are as follows:

- execution commission: roughly $1.00/contract;
- exchange commission: the CBOE charges up to $0.80/contract (the average is $0.35 to $0.55/contract);
- clearing commission: roughly $1.00/contract;
- clearing fees: the Options Clearing Corporation charges up to $0.055/contract;
- options regulatory fee: $0.0388/contract; and
- bid–ask spread: at least a few basis points.

There are some ways to estimate the total market gamma needs, by doing the following.

- Using the data related to options volumes and options gamma.
- Making some assumptions, such as: 1) dealers hold 50% of the gross total gamma; 2) out of which they are short 75% and long 25%; 3) hence dealers are net short gamma = (75 – 25)% × 50% = 25% of total gross gamma.

								Call				Put			
VOLUME_TOTAL_CALL	377 073	Call	Put	Strike	Exipry	Type	Size	OI	Gamma for 1% move	OI-Adj $Gamma	Volume-Adj Gamma	OI	Gamma for 1% move	OI-Adj $Gamma	Volume-Adj Gamma
VOLUME_TOTAL_PUT	811 006	SPX US 03/16/13 C100 Index	SPX US 03/16/13 P100 Index	100	16/03/2013	E	100	0	2,03779E-05	$0	$0	434	6,94432E-06	$5 722	$585
		SPX US 01/18/14 C100 Index	SPX US 01/18/14 P100 Index	100	18/01/2014	E	100	0	0,000100742	$0	$0	712	8,59204E-06	$11 614	$1 187
		SPX US 06/21/14 C100 Index	SPX US 06/21/14 P100 Index	100	21/06/2014	E	100	0	1,70557E-06	$0	$0	158	1,05342E-05	$3 160	$323
		SPX US 12/22/12 C100 Index	SPX US 12/22/12 P100 Index	100	22/12/2012	E	100	108	1,11891E-05	$2 294	$182	61340	6,93118E-06	$807 139	$82 470
		SPX US 02/16/13 C100 Index	SPX US 02/16/13 P100 Index	100	16/02/2013	E	100	0	0	$0	$0	0	6,70275E-06	$0	$0
Call Gamma	$17 957 138 045	SPX US 06/22/13 C100 Index	SPX US 06/22/13 P100 Index	100	22/06/2013	E	100	0	8,33421E-05	$0	$0	7072	1,131E-05	$151 845	$15 515
Put Gamma	$42 404 730 243	SPX US 09/21/13 C100 Index	SPX US 09/21/13 P100 Index	100	21/09/2013	E	100	0	0,000101184	$0	$0	113	1,04573E-05	$2 243	$229
Total Gamma	$60 361 868 289	SPX US 12/21/13 C100 Index	SPX US 12/21/13 P100 Index	100	21/12/2013	E	100	1735	1,52411E-05	$50 201	$3 976	28145	8,58781E-06	$458 861	$46 884
		SPX US 12/20/14 C100 Index	SPX US 12/20/14 P100 Index	100	20/12/2014	E	100	8	6,62331E-06	$101	$8	6051	2,18128E-05	$250 574	$25 603
Net Short Gamma	$18 108 560 487	SPX US 01/19/13 C100 Index	SPX US 01/19/13 P100 Index	100	19/01/2013	E	100	0	0	$0	$0	0	6,69685E-06	$0	$0
# Shares to sell on a 1% down move	13 142 716	SPX US 11/17/12 C100 Index	SPX US 11/17/12 P100 Index	100	17/11/2012	E	100	0	1,75157E-05	$0	$0	0	6,92597E-06	$0	$0

FIGURE 5.130 Portfolio gamma

Note: OI = open interest.

In addition, recent regulatory framework changes (e.g. Solvency II in Europe, NAIC reform and US GAAP LDTI in the United States) force insurers to hold sufficient capital requirements offsetting the economic (mark-to-market) risks embedded within those downside protection guarantees, in order to remain solvent during periods of market stress. Those capital requirements can be reduced if the economic risks are mitigated through buying put options, whose prices will increase as a result of the supply–demand imbalance resulting into some market liquidity risk.

In fact, the market liquidity risk refers to the degree to which large transactions can be carried out in a timely manner with a minimal impact on prices. Fundamentally, market impact is an illustration of market inefficiency: theories of efficient markets typically expect that investors are buying and selling assets based on assessments of their value; in contrast, large derivatives players buy and sell based on observations of market price movements, and so have a utility function quite different from that of a value investor.

As a result, the cost of placing one large order to close a position becomes significantly greater than the sum of infinitely small orders placed at differing times. For this reason, explicit modelling of the market impact function driven by the transaction size is required (something that is not considered by traditional pricing models). Such market impact function depends on the "temporary impact strength", which is proportional to the main empirically observed drivers: the speed of option trading (i.e. the number of options per unit of time), the equity stock level and the option sensitivity to the equity stock.

In this context, the best-order execution (i.e. to minimize the market impact) cannot be defined as a single trade, but turns out to be a sequence of small trades over the course of several days, thus optimizing the target. Minimizing the market impact will be beneficial not only to the insurer, through a lower hedging cost, but also to the customer, through a lower price.

Such an order execution strategy will also depend on the insurer's risk appetite (e.g. minimization of the mean cost or of the mean variance, which also penalizes the dispersion of the profit and loss). The standard procedure in stochastic control problems, namely the Hamilton–Jacobi–Bellman (HJB) framework, is applied, coupled with numerical schemes.

In case the equity put option becomes too expensive (due to higher market uncertainties or too large a market impact), the insurer will replicate them synthetically and dynamically through equities (or equity futures) that benefit from lower transaction costs. Such a synthetic replication needs to be dynamic in order for linear equities to piecewise match the convex shapes of the equity puts as the market moves, which provides higher replication costs and optimal transaction size resulting from the "feedback" market impact of the large transaction size of the equities on the equity asset price.

In the next section the equity put price consistent with market impact is modelled and best-order execution strategies consistent with the insurer's risk appetite are illustrated. In Section 5.3.2.3 the proxy synthetic replication through using a large quantity of equity assets is modelled within a specific no-arbitrage framework, providing a higher cost of replication and larger execution order size as compared with no market impact.

5.3.2.2 Dynamic Risk Management Reduces the Cost to the Insurer and the Price to the Insured by Minimizing the Market Impact with respect to the Insurer's Risk Aversion

5.3.2.2.1 An Equity Puts Pricing Model including Market Impact The model below is inspired from Leland's option replication with transaction costs (see Leland 1985), where the market impact is incorporated into the volatility of the asset σ as a market impact function f (dependent on time, volatility, inventory and trading rate):

$$\tilde{\sigma}^2 = \sigma^2 + f\left(t, \dot{x}_t, x_t, \sigma\right).$$

We follow the approach by Almgren (see Lorenz & Almgren 2011 and Gatheral & Schied 2011), where the market impact function is a combination of two components: a permanent component, which reflects the information transmitted to the market by the buy–sell imbalance; and a temporary component, which reflects the price concession needed to attract counterparties within a specified short time interval. We adapt such approach to derivatives through the "enlarged volatility" expression, as follows:

$$\tilde{\sigma}_t^2 = \sigma^2 + \left(\tilde{\eta}\dot{x}_t + \tilde{\gamma}\left(x_t - x_0\right)\right)\sqrt{\check{T} - t}\,\sigma, \text{where } \tilde{\eta} = \eta\sqrt{\frac{8}{h\pi}} \textit{ and}$$

where $\tilde{\eta} = \eta\sqrt{\frac{8}{h\pi}}$, and where η and γ are constants. The number of shares is x(t), while $\dot{x}_t$ (its derivative with regard to time) is the speed of trading of the security. The term $\eta\dot{x}_t$ corresponds to the temporary or instantaneous impact of trading $\dot{x}_t dt$ shares at time t, which affects only this current order. The term $\gamma\left(x_t - x_0\right)$ is the permanent price impact that was accumulated by all transactions until time t.

Let $(\Omega, \mathcal{F}, \mathbb{P})$ be the usual probability space on the filtration $\left(\mathcal{F}_t\right)_{t\in\mathbb{R}_+}$. In the absence of market impact and under a zero risk-free rate, the no-arbitrage price of a put option is defined by $P_t = \mathbb{E}_{\mathbb{Q}}\left[\left(K - S_{\check{T}}\right)^+ \mid \mathcal{F}_t\right]$ under the risk-neutral probability measure Q, under which the discounted asset price is a martingale.

The equity put option price including market impact is then expressed through a Black–Scholes-like partial differential equation using the "enlarged volatility" illustrative of the market impact, such that buying the option will typically lead to increasing its price. The higher the trading speed and quantity, the higher the volatility and, thus, the higher the option price:

$$\begin{cases} \partial_u \tilde{P}(u,S) + \frac{1}{2}\tilde{\sigma}_t^2 S^2 \partial_{SS}\tilde{P}(u,S) = 0, & (u,S) \in [t, \check{T}[\times]0,\infty] \\ \tilde{P}(\check{T}, s) = (K - s)^+. \end{cases}$$

Using a simple Taylor approximation to the first order, the expression can be expressed as a sum of the Black–Scholes option price without market impact and an additional term corresponding to the option market impact:

$$\tilde{P}\left(t,S_t\right) \approx P\left(t,S_t\right) + \left(\tilde{\sigma}_t^2 - \sigma^2\right)\partial_v P\left(t,S_t\right) \approx P\left(t,S_t\right) + \frac{1}{2}\left\{\tilde{\eta}\cdot\dot{x}_t + \tilde{\gamma}\cdot\left(x_t - x_0\right)\right\}\cdot\sqrt{\check{T} - t}v\left(t,S_t\right),$$

where $\sum(\Delta_A \times R_A) - \sum(\Delta_L \times R_L)$ is the Black–Scholes vega of the option:

$$v(t,S_t) = \sqrt{\breve{T}-t}S_t N'(d_1) = \sqrt{\breve{T}-t}KN'(d_2),$$

$$N'(x) = \frac{1}{\sqrt{2\pi}}e^{-x^2/2},$$

$$d_1 = \frac{\log\frac{S_t}{K}+\frac{1}{2}\sigma^2(\breve{T}-t)}{\sigma\sqrt{\breve{T}-t}},$$

$$d_2 = \frac{\log\frac{S_t}{K}-\frac{1}{2}\sigma^2(\breve{T}-t)}{\sqrt{\breve{T}t}} = d_1 - \sigma\sqrt{\breve{T}-t}.$$

For simplicity of notation, the permanent impact is excluded in the following sections (i.e. $\tilde{\gamma}=0$). Using the vega–gamma relationship (gamma is the second-order sensitivity of the option price with respect to the underlying $\frac{d^2P}{d^2S}$, which is equal to the optimal change in the quantity of assets required to hedge the put),

$$v = \sigma\tau S^2\Gamma,$$

$$\text{where}\quad \Gamma(t,S_t) = \frac{N'(d_1)}{S_t\sigma\sqrt{\breve{T}-t}} = \frac{KN'(d_2)}{S_t^2\sigma\sqrt{\breve{T}-t}},$$

and the price including market impact is then

$$\tilde{P}_t = P_t + \frac{1}{2}\tilde{\eta}\dot{x}_t\sigma S_t^2(\breve{T}-t)^{3/2}\Gamma(t,S_t).$$

5.3.2.2.2 The Optimal Order Execution Minimizes the Market Impact with respect to the Insurer's Risk Appetite The optimal dynamic order execution strategy unfolds over the course of several days [0,T], consistent with changing market conditions and according market impact. Let us consider a buying order execution strategy x(t) in which an amount X of options with fixed strike K and maturity T are bought by a fixed time horizon [0,T] with the conditions x(0) = X and x(T) = 0.

At each time t, $\dot{x}_t dt$ options are bought at price $\tilde{P}_t$, which is the option impact price defined by the price equation above. Thus, the cost arising from the strategy x is

$$C(x) := \int_0^T \tilde{P}_t\dot{x}_t dt,$$

$$C(x) = \int_0^T P_t\dot{x}_t dt + \frac{1}{2}\tilde{\eta}\int_0^T \dot{x}_t^2\sigma S_t^2(\breve{T}-t)^{3/2}\Gamma(t,S_t)dt,$$

$$\mathcal{C}(x) = -XP_0 - \int_0^T \sigma x_t S_t \Delta(t, S_t) dW_t + \frac{1}{2}\tilde{\eta}\sigma \int_0^T \dot{x}_t^2 S_t^2 (\check{T} - t)^{3/2} \Gamma(t, S_t) dt,$$

where $\Delta = \frac{dP}{dS}$ the first-order sensitivity of the option price with respect to the underlying, so-called Black–Scholes delta of the option, which is equal to the quantity of assets required to hedge the put.

The insurer's objective is then to minimize a certain cost objective function, which may take into account her risk aversion. Here we consider two risk appetite cases:

- the mean cost $\mathbb{E}[\mathcal{C}(x)]$, which corresponds to the risk-neutral case (the insurer has no risk aversion); and
- the mean-variance cost $|\mathbb{E}[\mathcal{C}(x)] + \lambda \mathbb{V}ar[\mathcal{C}(x)]$ – i.e. a risk–reward criterion (which includes the mean case if $\lambda = 0$), where λ is the variance penalty

5.3.2.2.2.1 Optimal Order Execution Minimizes the Market Impact with No Insurer Risk Aversion The mean cost is usually used for an agent who does not monitor the risk of her strategy:

$$\mathbb{E}[\mathcal{C}(x)] = -XP_0 + \frac{1}{2}\tilde{\eta}\mathbb{E}\left[\int_0^T \dot{x}_t^2 S_t^2 (\check{T} - t)^{3/2} \Gamma(t, S_t) dt\right]$$

Theorem (Kalife & Mouti 2017): the optimal transaction size and pace strategies resulting in minimizing the mean cost under the Black–Scholes framework are illustrated in Figure 5.131 and characterized by

$$\dot{x}^*(t) = \frac{K_1}{(\check{T} - t)^{3/2}},$$
$$x^*(t) = \frac{K_1}{(\check{T} - t)^{1/2}} + K_2$$

$$\text{where} \quad K_1 = \frac{X}{2\left(\check{T}^{-\frac{1}{2}} - (\check{T} - T)^{-\frac{1}{2}}\right)}$$

$$\text{and} \quad K_2 = -2K_1(\check{T} - T)^{-1/2}.$$

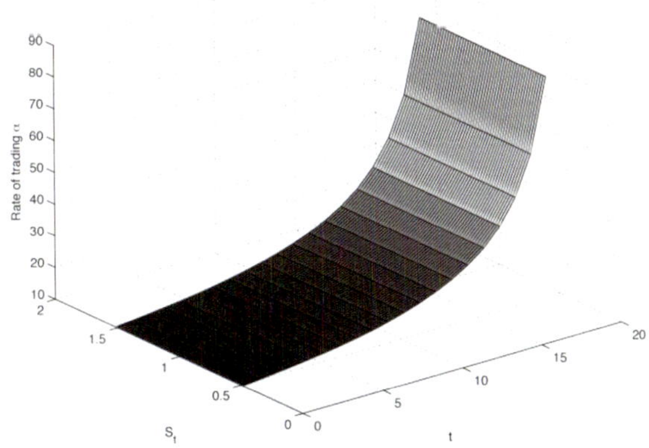

FIGURE 5.131 Optimal rate of trading depending on stock level and time passing

Under the mean cost case, the optimal order execution strategy provides a rather stable pace of trading, depending only mildly on the stock price path. Besides, the insurer must acquire at a gradual faster rate as time passes, as expected given the fixed quantity to buy within a fixed time period.

5.3.2.2.2.2 Optimal Order Execution Minimizing the Market Impact with Risk Aversion We now develop the optimal order execution framework under the mean-variance case (see Tse et al. 2013 and Lorenz & Almgren 2011), where the optimal strategy turns out to be more sensitive to the underlying price evolution. In fact, the mean-variance objective function can be approximated as

$$E[C(x)]+\lambda \operatorname{Var}[C(x)] \approx E\left[\int_0^T \frac{1}{2}\tilde{\eta}\sigma\dot{x}_t^2 S_t^2(\check{T}-t)^{\frac{3}{2}}\Gamma(t,S_t)dt+\tilde{\lambda}\int_0^T x_t^2\sigma^2 S_t^2\Delta^2(t,S_t)dt\right].$$

We then set up the dynamic programming problem, in which we parameterize as before the strategies x by their trading speed or trading rate α, defined as

$$-\dot{x}_t : x_t^{\alpha} := X-\int_0^t \alpha_s ds,\quad 0\le t\le T.$$

We restrict our framework to a Markovian trading rate (i.e. the agent's optimal trading speed at time t is completely determined by the current state). Using the standard procedure of deriving the Hamilton–Jacobi–Bellman equation in stochastic control problems, the solution to the optimization problem solves the following PDE,

$$\partial_t U+\frac{1}{2}\sigma^2 S^2\partial_{SS}U+\tilde{\lambda}x^2S^2\Delta^2(t,S)+\inf_{\alpha\in\mathbb{R}}\left\{\alpha^2S^2(\check{T}-t)^{3/2}\Gamma(t,S)-\alpha\partial_x U\right\}=0$$

combined with the so-called "finite-fuel constraint" (i.e. $\int_0^T \alpha_t dt = X$).

Although this minimization problem does not admit a closed-form solution, this quasi-linear PDE can be solved numerically using finite differences methods. Figure 5.132 illustrates the optimal-order execution strategy through the rate of trading as a function of the underlying price S and time t.

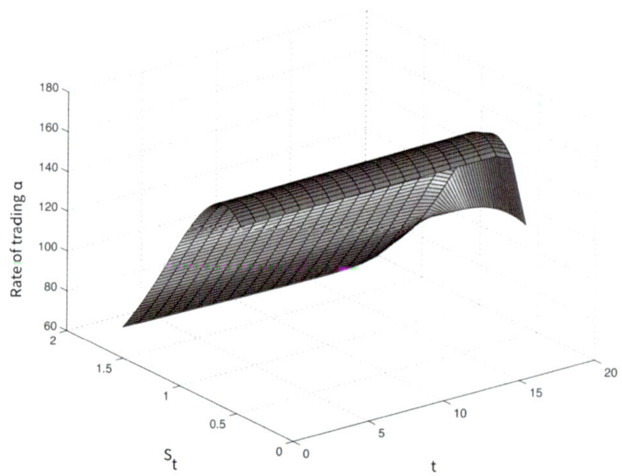

FIGURE 5.132 Optimal rate of trading depending on stock level and time passing

Note: Mean objective (λ = 0, top left) or mean variance (λ = 1, top right; 10, bottom left; 100, bottom right).

In contrast to the mean cost case, as the risk aversion (dispersion) λ increases, the optimal-order execution increasingly depends on the stock path, with a faster pace as the stock level decreases, consistent with a higher put option cost, which implies the need to trade as soon as possible. Moreover, as time passes the trading pace decreases, as the mean variance prevents the insurance company from waiting too long to buy a large quantity.

5.3.2.2.3 The Potential Lack of Puts Calls for a Dynamic Synthetic Replication that Minimizes the "Feedback" Effects of the Stock Sales on the Cost of Replication In the event that the equity put options become too expensive due to higher market uncertainties or too large a market impact, or just unavailable due to the long-term maturities of retirement products, the insurer will replicate them synthetically and dynamically through equities (or equity futures) benefiting from lower transaction costs.

The quality of such a piecewise synthetic approximation of the equity puts lies in the ability to dynamically fit the convex shape of the equity put by a series of subsequent dynamic trades in linear equity assets as the equity market changes, which involves a regularly updated quantity of equity assets, equal to the sensitivity of the equity put to the equity underlying (so-called "delta").

Unfortunately, due to the market impact associated with the large equities transaction size, such a dynamic sequence of equity trades involves some market inefficiencies (so-called "feedback effects"), as the insurer sells equity as the market drops or buys equity as the market rallies, which exacerbates the equity market shocks and increases the replication cost, as illustrated over the past decade.

5.3.2.2.3.1 Illustrations of Feedback Effects

5.3.2.2.3.1.1 Illustrations on Options, from Short-Term to Long-Term **Short-Dated Options** Investors typically (i.e. 65 to 80% of the time) buy index puts for downside protection, which are thus market-making short put options, which they hedge by selling futures to be market neutral. If the market suddenly drops, they would need to sell further to adjust, which amplifies the down-market move. Options dealers (delta hedgers) are net short options against outright investors (non-delta hedgers), and are large enough to move the equity market, which might further exacerbate market moves, and so the volatility. Their total risk exposure amounts to 40% of the total gamma – i.e. $7 billion sold near the market close for a 1% move per day.

Long-Dated Options and Structured Products

The capital guarantee of many structured products leaves the seller of the product effectively short an OTM put. A short OTM put is short skew and short vega convexity (or volga). This is a simplification, as structured products tend to buy visually cheap options (i.e. OTM options) and sell visually expensive options (i.e. ATM options), leaving the seller with a long ATM and short OTM volatility position. As OTM options have more volga (or vega convexity) than ATM options, the seller is short volga. The embedded option in structured products is floored, which causes the seller to be short skew (as the position is similar to being short an OTM put).

As market makers are short long-dated skew and short long-dated vega convexity, they buy long-dated volatility when it rises and vice versa, exaggerating any move and starting a vicious circle. As similar products are sold by all banks, all participants have the same position, leading to an imbalance due to the lack of natural counterparties, up to ten-year maturities, concentrated around the five-year maturity.

To demonstrate how this occurs, we examine a short skew position through a vega flat risk reversal (short 90% put, long 110% call): If there is a 10% decline in equity markets, the 90% put becomes ATM and increases in vega. As the risk reversal is short the 90% put, the position becomes short vega (or short implied volatility). In addition, the 110% call option becomes more OTM and further decreases the vega of the position (increasing the value of the short-dated implied volatility position).

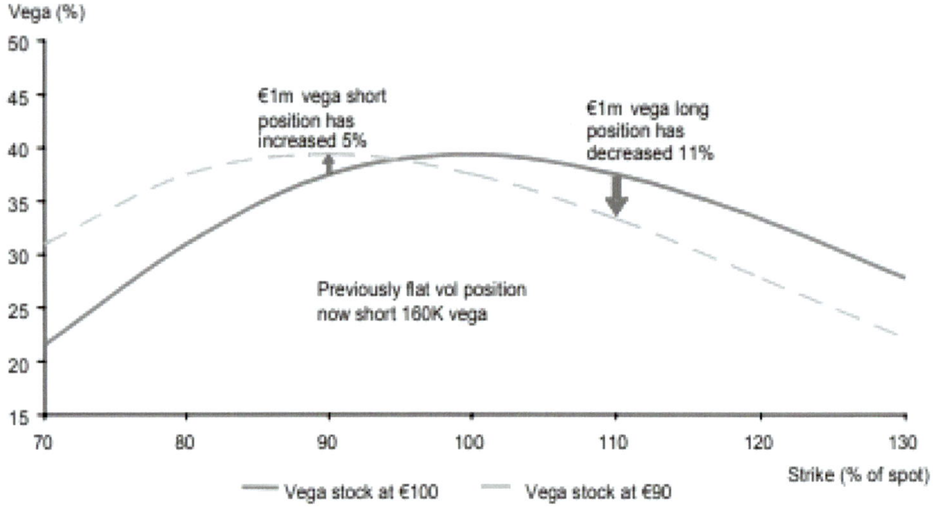

FIGURE 5.133 Vega versus strike

It is the fact that options become more or less ATM that causes the change in vega; it is not the fact that downside put options have a higher implied than upside call options. If skew were flat (or even if puts traded at a lower implied than calls), the above argument would still hold. Vanna (= dVega / dSpot) can be thought of as the size of the skew position (in a similar way that vega is the size of a volatility position), while skew (e.g. 90%–100% skew) measures the value of skew (in a similar way that implied volatility measures the value of a volatility position).

As the size of trading desks' short vega position increases during equity market declines, this position is likely to be covered. As all trading desks have similar positions, this buying pressure causes an increase in implied volatility. This flow is in addition to any buying pressure due to an increase in realized volatility, and hence it can cause an overshoot in implied volatility.

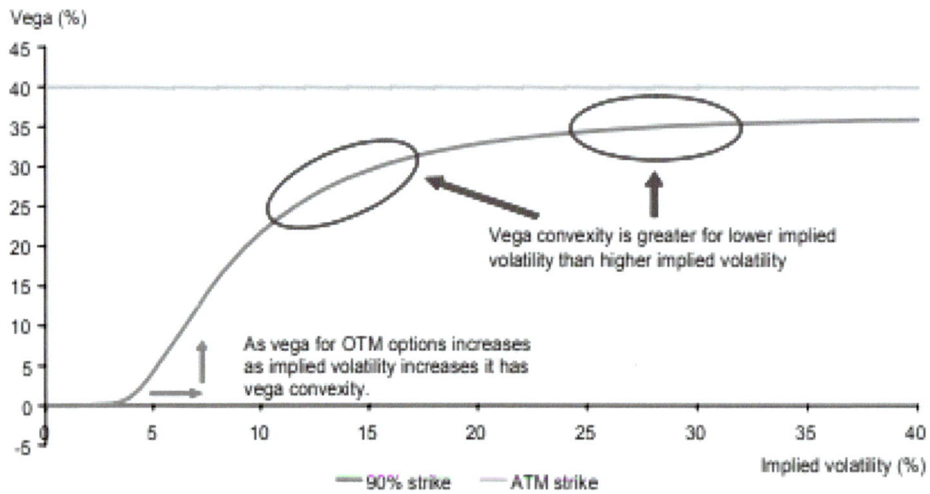

FIGURE 5.134 Vega versus 90% and ATM strike implied volatility

Options have their peak vega when they are (approximately) ATM. As implied volatility increases, the vega of OTM options increases and converges with the vega of the peak ATM option. Therefore, as implied volatility increases, the vega of OTM options increases, and so positive vega convexity = dVega / dVol (also called volga or vomma).

As the vega of options rises as volatility increases, this increases the size of the short volatility position that needs to be hedged. As trading desks' volatility short position has now increased, they have to buy volatility to cover this increased short position, which leads to further gains in implied volatility. This starts a vicious circle of increasing volatility, which we call the "structured products vicious circle".

The slope of vega against volatility is steepest (i.e. vega convexity is highest) for low to medium implied volatilities. This effect of vega convexity is therefore more important in volatility regimes of around 20% or less; hence, the effect of structured products can have a similar effect when markets rise and volatilities decline. In this case, trading desks become long vega, due to skew, and have to sell volatility. Vega convexity causes this long position to increase as volatility declines, causing further volatility sellings. This is typically seen when a market recovers after a volatile decline (e.g. in 2003 and 2009, following the end of the tech bubble and the credit crunch respectively).

While this position has the greatest impact at the far end of volatility surfaces, a rise in far-dated term volatility and skew tends to be mirrored to a lesser extent for nearer-dated expiries. If there is a disconnect between near- and far-dated implied volatilities, this can cause a significant change in term structure.

- The 2002 sell-off induced large-scale hedging amid extraordinary volatility and a spike in option prices. The sudden collapse of implied volatility in early 2003 from an overwhelming issuance of structured products (Asia) reacting to a huge rise in stock markets, for yield enhancement or upside participation, led to implied volatility being depressed to historically low levels.
- In April 2005 a large fall in the equity markets caused significant hedging issues for banks hedging the short skew position dynamically, lifting longer-dated implieds and hence the term structure.
- The huge downward moves in October 2008 triggered the short 50 to 70% knock-in barrier put option embedded within massively autocallable notes on the Nikkei ($6 billion, with KI barriers between the 7,000 and 9,000 levels) and the Korea Composite Stock Price Index (KOSPI) ($3 billion, barrier at the 160 level), whose hedging required the unwinding of hedging by selling the index as it crossed the barrier -> exacerbating the downward move of the market by 15%, from sticky to choppy, as both gamma and vega flipped sign, with jumps in volatility and bid–offer spreads.

  With maturity gamma concentrating around the strike -> barriers can cause gamma to reverse in large size across very small movements as spot rallies towards a knock-out barrier-> large hedgers will sell large amounts of spot as it approaches the barrier, preventing the spot from hitting the barrier. Even so, the barrier may break at some point -> the option disappears, and the trader is left only with naked hedges, which he must cover by buying back spot and gamma, driving the spot even higher -> choppy trading and market gaps.

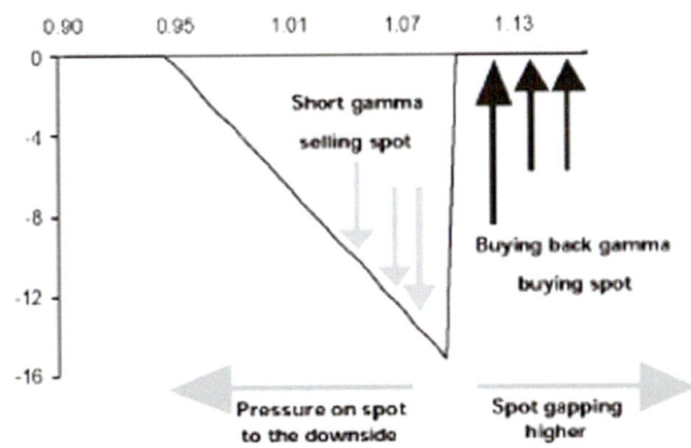

FIGURE 5.135 Feedback effects on barrier options

To hedge the long vega exposure from the autocallable issuance, the issuers short volatility using variance swaps and put options. When the structure is knocked in on the downside, the embedded knock-in put becomes a vanilla option that is very much in-the-money with very little vega exposure, forcing the issuers to buy back their short volatility positions to rehedge.

- The hedging of variable annuity products lifts the index term structure and skew: the hedging of variable annuity involves the purchase of downside protection for long maturities. Often the products are 20+ years long, but, as the maximum maturity with sufficient liquidity available on indices can only be three to five years, the position has to be dynamically hedged with the shorter-dated option. This constant bid for long-dated protection lifts the index term structure and skew, particularly for the S&P 500, but it also affects other major indices (potentially due to relative value trading). The demand for protection (from viable annuity providers or other investors), particularly on the downside and for longer maturities, could be considered to be the reason why volatility (of all strikes and maturities), skew (for all maturities) and term structure are usually overpriced.

5.3.2.2.3.1.2 Illustrations of Feedback Effects on Variance Swaps Hedging variance swaps involves the trading of a strip of options of all strikes and delta hedging at the close.

The short end of volatility surfaces is now pinned to realized: if there is a divergence between short-dated variance swaps and realized volatility, hedge funds will put on variance swap trades to profit from this divergence. This puts pressure on the short-dated end of volatility surfaces to trade close to the current levels of realized volatility. Due to the greater risk of unexpected events, it is riskier to attempt a similar trade at the longer-dated end of volatility surfaces.

Skew levels are affected by the direction of volatility trading: as variance swaps became a popular way to express a view of the direction of implied volatility, they impact the levels of skew. This occurs as variance swaps are long skew (explained below), and, if volatility is being sold through variance swaps, this weighs on skew. This occurred between 2003 and 2005, which pushed skew to a multiple-year low. As volatility bottomed, the pressure from variance swap selling abated and skew recovered.

Delta hedging can suppress or exaggerate market moves: as the payout of variance swaps is based on the close-to-close return, they all have an intraday delta (which is equal to zero if the spot is equal to the previous day's close). As this intraday delta resets to zero at the end of the day, the hedging of these products requires a delta hedge at the cash close. A rule of thumb is that the direction of hedging flow is in the direction that makes the trade the least profit (ensuring that, if a trade is crowded, it makes less money). This flow can be hundreds of millions of dollars or euros per day, especially when structured products based on selling short-dated variance are popular (as they were in 2006 and 2007, but less so since the high volatility of the credit crunch).

Although fears about growth or sovereign debt sustainability were drivers of significant declines in equities, as experienced in May 2006, May 2010 and August 2011, they did not fully explain 1) the extreme magnitude of the shocks and 2) their repeated occurrence at the close in European and US markets. The derivative hedging activity of long-dated derivatives by large players might help exacerbate the acuity of such volatility spikes:

In fact, most investors buy puts for portfolio protection while selling variance swaps, which implies a significant 30 to 60% short option convexity imbalance for market makers, then hedged with large transactions in the spot; they sell more index as the market drops or buy more as the market rallies, exacerbating the initial move and realized volatility.

Variance selling exaggerates equity market moves. Should clients be predominantly selling variance swaps, the hedging of these products exaggerates market moves. The argument is simply the inverse of the argument above. The trader who is long a variance swap (as the client is short) has hedged by selling a portfolio of options. If markets rise, the delta of the position is negative and, as the variance swap delta is reset to zero at the end of the day, the trader has to buy equities at the same time (causing the close to be lifted for underlyings that have increased in value over the day). If markets fall, then the trader has to sell equities at the end of the day (as the delta of a short portfolio of options is positive). Movements are therefore exaggerated, and realized volatility increases if clients have sold variance swaps.

Prolonged periods of calm markets may encourage hedge funds to bet that equity market volatility will remain low, selling volatility to banks through "variance swaps". The banks then hedge against the variance swaps in two ways: 1) selling vanilla options to hedge their volatility exposure; and 2) delta-hedging these vanilla options using cash equities. As a result, they trade in the direction of the day's trend, and thus exacerbate the decline in market and rise in volatility, which is illustrative of pro-cyclical feedback hedging effects. As variance swaps are priced off the market's closing price, these feedback hedging effects are concentrated at the end of the day, implying drying in market liquidity.

Basis risk between cash and futures can cause traders problems: the payout of variance swaps is based on the cash close, but traders normally delta hedge using futures. The difference between the cash and futures price is called the basis, and the risk due to a change in basis is called basis risk. Traders just have to take this basis risk between the cash close and futures close, which can be significant, as liquidity in the futures market tends to be reduced after the cash market closes.

5.3.2.2.3.1.3 Illustrations of Feedback Effects on Variable Annuities through Managed Volatility Funds Volatility target (or volatility control) strategies provide the most immediate selling as a reaction to the increase in volatility. These strategies adjust equity leverage based on short-term realized volatility. Typical signals are one-, two- or three-month realized volatility. Volatility target products are provided by many dealers, index providers and asset managers. Volatility-targeting strategies also became very popular in the insurance industry.

After the 2008 financial crisis many variable annuity providers moved from hedging their equity exposure with options to investing directly in volatility target indices (e.g. a 10% volatility target for the S&P 500). It is estimated that VA issuers have some $360 billion in strategies that are managing volatility; some of these use options to manage tail risk, some buy low-volatility stocks and some invest in volatility target strategies. We estimate that strategies that are targeting a particular level of volatility or managing to an equity floor could have $100 to 200 billion of assets.

Assuming that, on average, these strategies follow a two-month realized volatility signal, we can estimate their selling pressure. Two-month realized volatility increased in early February 2018 from ~10% to ~20% (i.e. doubled), so these strategies need to reduce equity exposure by up to ~50% to keep volatility constant. This could lead to $50 to 100 billion of selling. There is often a delay of one to three days between when a signal is triggered and trade implementation, and positions are often reduced over several days. We think this could have contributed to the "unexpected" sell-off that happened in the last hour of the trading session on Tuesday 6 February 2018. Although these flows may have continued to have a negative impact over the next few days, they would be the first to reverse (start buying the market) when volatility declined.

It can be seen that the hedging of variable annuities can cause volatility overshoot. Sales of variable annuities cause insurance companies to be short skew and short vega convexity; in other words, as the market decline the short skew position causes them to be short volatility, which requires the buying of volatility, lifting implied volatility, which as a feedback increases the short volatility position due to the short vega convexity position. This causes a vicious circle and an implied volatility overshoot.

5.3.2.2.3.2 Allowing for Feedback Effects through a Pricing Model of Synthetic Replication with Market Impact The objective of the optimal-order replication strategy is to minimize such a market impact. It is determined by a no-arbitrage framework that incorporates the impact of the large trader's large transactions on equities. We consider here the interaction of one "large trader" whose action affects prices versus many price takers or "small traders"; the usual no-arbitrage condition does not apply. We use a continuous-time version of

Jarrow's "no market manipulation strategies" (see Jarrow 1994 and Bierbaum 1997), which requires additional assumptions:

- the asset price is independent of the large trader's past holdings;
- real wealth (as if the holdings were liquidated);
- synchronous markets condition;
- prices adjust instantaneously across the underlying and derivatives; and
- an absence of corners.

The dynamics of the asset price return with impact can be modelled as a modified version of the usual dynamics (under the assumption of zero interest rates):

$$\frac{dS_t}{S_t} = \sigma_t dW + \rho_t d\alpha_t,$$

where

- $\sigma_t dW_t$ is the dynamics of the asset return without market impact; and
- $\rho_t d\alpha_t$ is the specific market impact contribution, where α is the transaction size in assets, while ρ represents the intensity of the market impact (e.g. the ratio of change in the price of the underlying to the quantity traded); $\frac{1}{\rho_t s_t}$ therefore represents the market depth at time t, (the quantity of assets required to move prices by one unit).

Under a zero risk-free interest rate (for simplicity of notation), the no-arbitrage cost of the synthetic replication of a put is the expected discounted future payoff under the associated martingale measure

$$\mathbb{Q}', P_t = \mathbb{E}_{\mathbb{Q}}\left[\left\{(K - S_T)^+ \# \mathcal{F}_t\right\}\right],$$

which provides the following pricing equation, as we now apply the Black–Scholes continuous-time delta-hedging replication framework to the modified asset return dynamics under $\mathbb{Q}'$ (see Platen & Schweizer 1998 and Frey 1998):

$$\begin{cases} \dfrac{dP}{dt} + \dfrac{1}{2}\left(\dfrac{\sigma S}{1 - \varrho S \dfrac{d^2 P}{d^2 S}}\right)^2 \dfrac{d^2 P}{d^2 S} = 0 \\ P_T = n(K - S_T)^+ \end{cases},$$

where n is the number of equity put options to hedge. Such a cost including the market impact is equivalent to the cost without the market impact but with so-called "feedback" volatility, as below, increasing with both the market impact intensity ρ and the increase in the sales of assets required hedging the put:

$$\tilde{\sigma} = \frac{\sigma}{\left(1 - \varrho S \frac{d^2 P}{d^2 S}\right)}.$$

This modified Black–Scholes equation is a fully nonlinear parabolic PDE, requiring specific numerical implementation to ensure accuracy, flexibility and stability (see Fahim, Touzi & Warin 2011 and Kalife, Tan & Wang 2012), which can be summarized below.

- The method of Lie uses a symmetry reduction technique to translate a PDE into some ordinary differential equations (ODEs). However, the solution may not fit our terminal and boundary conditions, despite giving a reference.
- Finite difference: Kilianová and Ševčovič (2013) propose a transformation method, originally for pricing American-style options in a nonlinear model:

$$S = S_0 e^{-x}, \quad \tau = T - t \quad \text{and} \quad \Pi(\tau, x) = u(t, S) - S\frac{\partial u}{\partial S}(t, S),$$

$$\begin{aligned}
&\frac{\partial \Pi}{\partial \tau} + \left(r - q - \frac{\sigma^2}{2}\right)\frac{\partial}{\partial x} - \frac{1}{2}\frac{\partial}{\partial x}\left(\sigma^2 \frac{\partial \text{I}}{\partial x}\right) + r\Pi \\
&= \left(S\frac{\partial}{\partial S} - 1\right)\left(\frac{\partial u}{\partial t} + (r-q)S\frac{\partial u}{\partial S} + \frac{1}{2}\sigma^2\left(t, S, u_S, u_{SS}\right)S^2\frac{\partial^2 u}{\partial S^2} - ru\right) \\
&= 0.
\end{aligned}$$

✓ For the convective part, we use the exact scheme

$$\Pi\left(t_{n+\frac{1}{2}}, x_i\right) = \Pi\left(t_n, x_i - (r-q)\Delta t\right).$$

✓ For the diffusion part, we use the implicit scheme

$$\begin{aligned}
\frac{\Pi_i^{n+1} - \Pi_i^{n+\frac{1}{2}}}{\Delta t} \quad & - \frac{\tilde{\sigma}_i^2}{2}\frac{\Pi_{i+1}^{n+1} - \Pi_{i-1}^{n+1}}{2\Delta x} + r\prod_i^{n+1} \\
& - \frac{1}{2}\left(\tilde{\sigma}_i^2 \frac{\Pi_{i+1}^{n+1} - \Pi_i^{n+1}}{\Delta x^2} - \tilde{\sigma}_{i-1}^2 \frac{\Pi_i^{n+1} - \Pi_{i-1}^{n+1}}{\Delta x^2}\right) = 0.
\end{aligned}$$

- BSDE as an extension of the Feynman–Kac formula:

$$\begin{aligned}
dX_t &= b\left(t, X_t\right)dt + \sigma\left(t, X_t\right)dW_t \quad \text{with } X_0 = x, \\
-dY_t &= f\left(t, X_t, Y_t, Z_t\right)dt - Z_t dW_t \quad \text{with } Y_T = \Phi\left(X_T\right).
\end{aligned}$$

✓ Semilinear parabolic equation:

$$\partial_t u(t,s)+b(t,x)D_x u+\frac{1}{2}\sigma\sigma^T\cdot D^2_{xx}u(t,s)+f\left(t,s,u,\sigma(t,s)*\partial_s u\right)=0,$$

$$\left(Y_s,Z_s\right)=\left(u\left(s,X_s\right),\partial_x u\left(s,X_s\right)\sigma\left(s,X_s\right)\right),$$

$$\begin{aligned}Y^N_{t_k}&=E_{t_k}\left(Y^N_{t_{k+1}}\right)+\Delta t\cdot E_{t_k}f\left(t_k,X^N_{t_k},Y^N_{t_{k+1}},Z^N_{t_k}\right),\\ \Delta t\cdot Z^N_{t_k}&=E_{t_k}\left(Y^N_{t_{k+1}}\Delta W_k\right).\end{aligned}$$

✓ Fully nonlinear parabolic equation (Fahim, Touzi & Warin 2011):

$$-\mathcal{L}v-F'\left(\cdot,v,Dv,D^2v\right)=0,\quad \mathcal{L}v:=\partial_t v+\mu\cdot Dv+\frac{1}{2}\sigma\sigma^T\cdot D^2v,$$

$$v^h(t,x)=E\left(v^h\left(t+h,\check{X}^{t,x}_h\right)\right)+F\left(t,x,\mathcal{D}_h v\right).\quad \mathcal{D}_h v=E\left[v\left(\check{X}^{t,x}_h\right)H^h_i\right],i=0,1,2,$$

$$H_0=1,H_1=\left(\sigma^T\right)^{-1}\frac{W_h}{h},H_2=\left(\sigma^T\right)^{-1}\frac{W_hW_h^T-h1_d}{h^2}\sigma^{-1}.$$

✓ For conditional expectations, Gobet, Lemor and Warin (2005) propose a regression technique:

$$E\left(G\,|\,\mathcal{F}_t\right)=\arg\min_{k\in\mathcal{F}_t}E\left((G-k)^2\right).$$

✓ In our case, we solve as above with

$$F(\cdots,\Gamma)=\frac{1}{2}\sigma^2S^21'\left(\frac{1}{(1-\rho S\Gamma)^2}-1\right).$$

5.3.2.2.3.3 An Optimal Synthetic Replication with Market Impact Minimizes the "Feedback Effects" Stemming from the Stock Sales on the Replication Cost In fact, as the insurer synthetically replicates the long equity put position by selling a quantity of stocks equal to the sensitivity of the put price to the stock, whenever the market falls he has to sell further, which as a "feedback effect" makes the stock price fall further still, compounded by a larger volatility, and thus a larger replication cost.

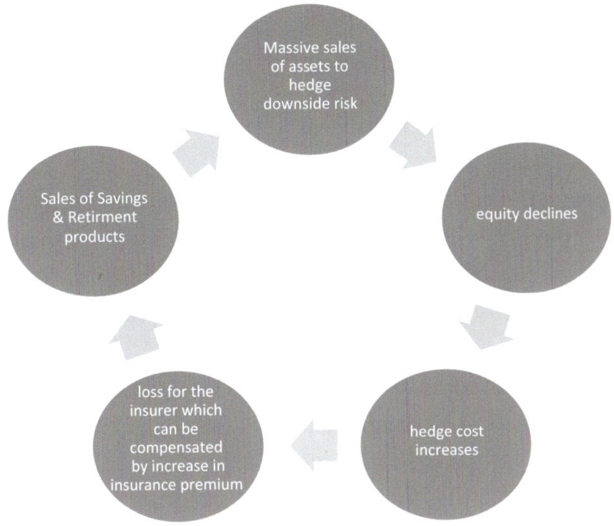

FIGURE 5.136 The feedback effect process

As illustrated in Figure 5.137, the synthetic dynamic strategy presented above enables us to minimize such an impact into an optimal replication cost that depends on the market impact intensity ("rho"). The associated optimal dynamic strategy in stocks is provided by the sensitivity of such optimal replication cost to the equity level, as illustrated in Figure 138.

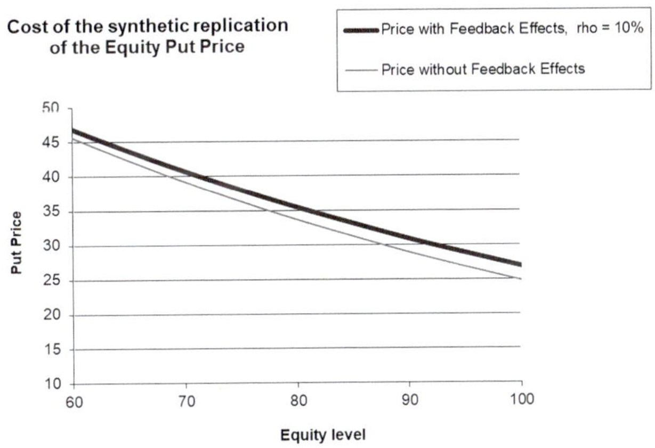

FIGURE 5.137 Optimal replication cost

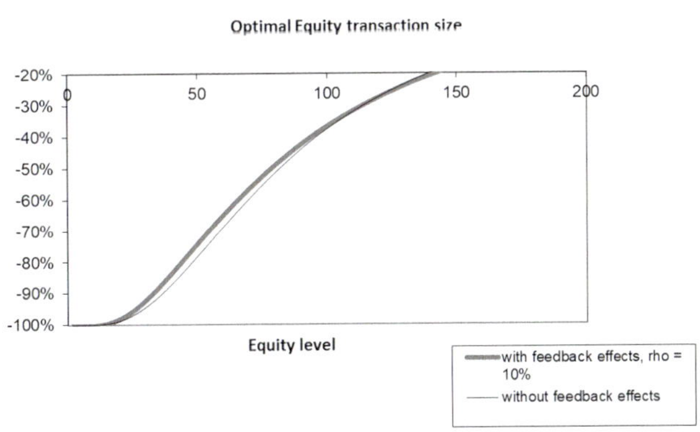

FIGURE 5.138 Optimal equity transaction size

5.3.2.2.3.4 Conclusion Within the context of a rapidly ageing population with increasing life expectancy, reductions in social security to meet retirement income needs and the zero interest rate environment compounded by recurrent market crashes, sustainable long-term savings and retirement need equity market upside combined with some capital guarantees after a waiting period, in order to embed downside protection in the manner of an equity put.

Given that the downside protection is subject to financial market downturns, insurance companies need to hedge their financial risks though buying large quantities of equity puts, thus incurring a market impact, which translates into higher hedging costs than expected. Dynamic risk management provides a best-order execution strategy that minimizes the market impact, and thus the hedging cost with respect to the insurer's risk appetite, which also benefits the customer by lowering the price.

In the event that the equity put option becomes too expensive or illiquid, the insurer replicates them synthetically and dynamically through selling large equity transactions, reinforcing equity market declines, which generates larger replication costs. Dynamic risk management provides the best-order execution strategy that minimizes such impacts on replication costs, through optimal dynamic equity-selling transactions depending on the equity levels.

Bibliography

Abergel, F., & Loeper, G. (2013). Pricing and hedging contingent claims with liquidity costs and market impact. DOI: 10.2139/ssrn.2239498.

Agarwal, N., & Ouaknine, Y. (2019). ESG rating and momentum. https://corpgov.law.harvard.edu/2019/03/21/esg-rating-and-momentum.

Albizzati, M., & Geman, H. (1994). Interest rate risk management and valuation of the surrender option in life insurance policies. *Journal of Risk and Insurance*, 61 (4), 616–37.

Alfonsi, A. (2005). On the discretization schemes for the CIR (and Bessel squared) processes. *Monte Carlo Methods and Applications*, 11 (4), 355–84.

Almgren, R. (2003). Optimal execution with nonlinear impact functions and trading-enhanced risk. *Applied Mathematical Finance*, 10 (1), 1–18.

Almgren, R. (2012). Optimal trading with stochastic liquidity and volatility. *SIAM Journal on Financial Mathematics*, 3 (1), 163–81.

Almgren, R., & Chriss, N. (2000). Optimal execution of portfolio transactions. *Journal of Risk*, 3 (2), 5–39.

Almgren, R., Thum, C., Hauptmann, E., & Li, H. (2005). Direct estimation of equity market impact. DOI: 10.1.1.146.1241.

Alòs, E., Chatterjee, R., Tudor, S., & Wang, T.-H. (2019). Target volatility option pricing in the lognormal fractional SABR model. *Quantitative Finance*, 19 (8), 1339–56.

Andersen, L. (2008). Simple and efficient simulation of the Heston stochastic volatility model. *Journal of Computational Finance*, 11 (3), 1–42.

Applebaum, D. (2009). *Lévy Processes and Stochastic Calculus*. New York: Cambridge University Press.

Ashwin Kumar, N., et al. (2016). ESG factors and risk-adjusted performance: a new quantitative model. *Journal of Sustainable Finance & Investment*, 6 (4), 292–300.

Bacinello, R., Millossovich, P., Olivieri, A., & Pitacco, E. (2011). Variable annuities: a unifying valuation approach. *Insurance: Mathematics and Economics*, 49 (3), 285–97.

Barraquand, J., & Martineau, D. (1995). Numerical valuation of high dimensional multivariate American securities. *Management Science*, 41 (12), 1882–91.

Barles, G., Daher, C., & Romano, M. (1995). Convergence of numerical schemes for parabolic equations arising in finance theory. *Mathematical Models and Methods in Applied Sciences*, 5 (1), 125–43.

Barone-Adesi, G., & Whaley, R. (1987). Efficient analytic approximation of American option value. *Journal of Finance*, 42 (2), 301–20.

Bauer, D., Bergmann, D., & Kiesel, R. (2010). On the risk-neutral valuation of life insurance contracts with numerical methods in view. *ASTIN Bulletin*, 40 (1), 65–95.

Bauer, D., Kling, A., & Russ, J. (2008). A universal pricing framework for guaranteed minimum benefits in variable annuities. *ASTIN Bulletin*, 38 (2), 621–51.

Bebchuck, L., Cohen, A., & Ferrell, A. (2009). What matters in corporate governance? *Review of Financial Studies*, 22 (2), 783–827.

Bennani, L., et al. (2018). How ESG investing has impacted the asset pricing in the equity market. DOI: 10.2139/ssrn.3316862.

Berger, K., & Lefebvre, D. (2013). Dynamiser l'épargne financière des ménages pour financer l'investissement et la compétitivité, report to the prime minister. Paris: Assemblée nationale.

Bierbaum, J. (1997). Uber die Rückwirkung von Handelsstrategienausgewählter Investoren auf Wertpapierpreisprozesse, PhD dissertation. Berlin University.

Bjerksund, P., & Stensland, G. (1993). Closed-form approximation of American options. *Scandinavian Journal of Management*, 9 (Supp. 1), 87–99.

Björk, T. (2009). *Arbitrage Theory in Continuous Time*. Oxford: Oxford University Press.

Black, F., & Jones, R. (1987). Simplifying portfolio insurance. *Journal of Portfolio Management*, 14 (1), 48–51.

Blanchard, A. (2012). The two-factor Hull-White model: pricing and calibration of interest rates derivatives, PhD thesis. Royal Institute of Technology, Stockholm.

Bohnert, A., & Gatzert, N. (2014). Fair valuation and risk assessment of dynamic hybrid products in life insurance: a portfolio consideration. *Geneva Papers on Risk and Insurance: Issues and Practice*, 39 (1), 148–72.

Bossu, S. (2006). Introduction to variance swap. *Wilmott Magazine*, March, 50–5.

Bouchard, B., & Warin, X. (2010). Monte-Carlo valorisation of American options: facts and new algorithms to improve existing methods, Research Report 10-07. Paris: Laboratoire de finance des marchés de l'énergie.

Bouchaud, J.-P. (2009). Price impact. DOI: 10.48550/arXiv.0903.2428.

Bouchaud, J.-P., Farmer, J. D., & Lillo, F. (2009). How markets slowly digest changes in supply and demand. In Hens, T., & Schenk-Hoppé, K. (eds.), *Handbook of Financial Markets: Dynamics and Evolution* (pp. 57–160). Amsterdam: North-Holland.

Bourgeois, B. (2014). Solvabilité II (course). Paris: École nationale de la statistique et de l'administration économique.

Bourhis, C. (2016). Taux de rendement interne Solvabilité 2: application à un fonds en euros. www.institutdesactuaires.com/global/gene/link.php?news_link=mem%2F3c78a3d07ee37b63e4ae8f3f9d11a159.pdf&fg=1. Supervised by Aymeric Kalife

Boyle, P., & Hardy, M. (2003). Guaranteed annuity options. *ASTIN Bulletin*, 33 (2), 125–52.

Boyle, P., & Tian, W. (2008). The design of equity-indexed annuities. *Insurance: Mathematics and Economics*, 43 (3), 303–15.

Brennan, M., & Schwartz, E. (1977). The valuation of American put options. *Journal of Finance*, 32 (2), 449–62.

Broadie, M., & Jain, A. (2008). The effect of jumps and discrete sampling on volatility and variance swaps. *International Journal of Theoretical and Applied Finance*, 11 (8), 761–97.

Broadie, M., & Kaya, O. (2006). Exact simulation of stochastic volatility and other affine jump diffusion processes. *Operations Research*, 54 (2), 217–31.

Carr, P., Geman, H., Madan, D., & Yor, M. (2002). The fine structure of asset returns: an empirical investigation. *Journal of Business*, 75 (2), 305–32.

Carr, P., Geman, H., Madan, D., & Yor, M. (2003). Stochastic volatility for Lévy processes. *Mathematical Finance*, 13 (3), 345–82.

Carr, P., Geman, H., Madan, D., & Yor, M. (2005). Pricing options on realized variance. *Finance and Stochastics*, 9 (4), 453–75.

Carr, P., & Lee, R. (2007). Realized volatility and variance: options via swaps. *Risk*, May, 76–83.

Carr, P., & Lee, R. (2008). Robust replication of volatility derivatives. DOI: 10.48550/arXiv.2107.00554.

Carr, P., & Lee, R. (2009). Volatility derivatives. *Annual Review of Finance and Economics*, 1, 1–21.

Carr, P., & Madan, D. (1999). Option valuation using the fast Fourier transform. *Journal of Computational Finance*, 2 (4), 61–73.

Carr, P., & Wu, L. (2004). Time-changed Lévy processes and option pricing. *Journal of Financial Economics*, 71 (1), 113–41.

Chan, L., & Lakonishok, J. (1995). The behavior of stock prices around institutional trades, *Journal of Finance*, 50 (4), 1147–74.

Chava, S. (2014). Environmental externalities and cost of capital. *Management Science*, 60 (9), 2223–47.

Chen, Z., Chen, B., Hu, Y., & Zhang, H. (2019). Three-fund constant proportion portfolio insurance strategy. Paper presented at European Financial Management Association annual meeting, Ponta Delgada, Portugal, 29 June.

Chen, Z., & Forsyth, P. (2008). A numerical scheme for the impulse control formulation for pricing variable annuities with a guaranteed minimum withdrawal benefit (GMWB). *Numerische Mathematik*, 109, 535–69.

Chen, Z., Vetzal, K., & Forsyth, P. (2008). The effect of modelling parameters on the value of GMWB guarantees. *Insurance: Mathematics and Economics*, 42 (1), 165–73.

Chong, J., Her, M., & Phillips, M. (2006). To sin or not to sin? Now that is the question. *Journal of Asset Management*, 6 (6), 406–17.

Chow, V. (2017). An examination of alternative ricing option hedging strategies in the presence of transaction costs, master's thesis. University of Waterloo.

Christoffersen, P., Goyenko, R., Jacobs, K., & Karoui, M. (2018). Illiquidity premia in the equity options market. *Review of Financial Studies*, 31 (3), 811–51.

Clark, G., Feiner, A., & Viehs, M. (2015). From the stockholder to the stakeholder: how sustainability can drive financial outperformance. DOI: 10.2139/ssrn.2508281.

Constantinides, G. (1986). Capital market equilibrium with transaction costs. *Journal of Political Economy*, 94 (4), 842–62.

Cont, R., & Tankov, P. (2003). *Financial Modelling with Jump Processes*. London: Chapman & Hall/CRC.

Dai, M., Kwok, Y., & Zong, J. (2008). Guaranteed minimum withdrawal benefit in variable annuities. *Mathematical Finance*, 18 (4), 595–611.

Damiano, B., & Mercurio, F. (2006). *Interest Rate Models – Theory and Practice: With Smile, Inflation and Credit*. Berlin: Springer.

Davis, M., & Norman, A. (1990). Portfolio selection with transaction costs. *Mathematics of Operations Research*, 15 (4), 676–713.

De Giovanni, D. (2010). Lapse rate modeling: a rational expectations approach. *Scandinavian Actuarial Journal*, 2010, 56–67.

De Spiegeleer, J., et al. (2021). ESG: a new dimension in portfolio allocation. DOI: 10.2139/ssrn.3712857.

Deelstra, G., & Rayée, G. (2013). Pricing variable annuity guarantees in a local volatility framework. *Insurance: Mathematics and Economics*, 53 (3), 650–63.

Dellinger, J. (2006). *The Handbook of Variable Income Annuities*. Hoboken, NJ: Wiley.

Demeterfi, K., Derman, E., Kamal, M., & Zou, J. (1999). A guide to volatility and variance swaps. *Journal of Derivatives*, 6 (4), 9–32.

Derman, E. (1999). When you cannot hedge continuously, the corrections of Black–Scholes. https://emanuelderman.com/when-you-cannot-hedge-continuously-the-corrections-of-black-scholes.

Derman, E., & Kani, I. (1994). Riding on a smile. *Risk*, February, 32–9.

Derwall, J., Koedijk, K., & Ter Horst, J. (2011). A tale of values-driven and profit-seeking social investors. *Journal of Banking & Finance*, 35 (8), 2137–47.

Dixit, A. (1991). *The Making of Economic Policy: A Transaction-Cost Politics Perspective*. Cambridge, MA: MIT Press.

Dupire, B. (1994). Pricing with a smile. *Risk*, January, 18–20.

Dupire, B. (2006). Fair skew: break-even volatility surface, technical report. New York: Bloomberg.

Eraker, B., Johannes, M., & Polson, N. (2003). The impact of jumps in volatility and return. *Journal of Finance*, 58 (3), 1269–300.

Fahim, A., Touzi, N., & Warin, X. (2011). Probabilistic numerical method for fully nonlinear parabolic PDEs. *Annals of Applied Probability*, 21 (4), 1322–64.

Feng, R., Kuznetsov, A., & Yang, F. (2019). Exponential functionals of Lévy processes and variable annuity guaranteed benefits. *Stochastic Processes and Their Applications*, 129 (2), 604–25.

Feng, R., & Vecer, J. (2017). Risk based capital for guaranteed minimum withdrawal benefit. *Quantitative Finance*, 17 (3), 471–8.

Föllmer, H., & Schweizer, M. (1991). Hedging of contingent claims under incomplete information. In Davis, M., & Elliott, R. (eds.), *Applied Stochastic Analysis* (pp. 389–414). London: Gordon & Breach.

Föllmer, H., & Sondermann, D. (1986). Hedging of non-redundant contingent claims. In Hildenbrand, W., & Mas-Colell, A. (eds.), *Contributions to Mathematical Economics: In Honour of Gérard Debreu* (pp. 205–23). Amsterdam: North-Holland.

Forsyth, P. (2011). A Hamilton–Jacobi–Bellman approach to optimal trade execution. *Applied Numerical Mathematics*, 61 (2), 241–65.

Forsyth, P., Kennedy, J., Tse, S., & Windcliff, H. (2012). Optimal trade execution: a mean quadratic variation approach. *Journal of Economic Dynamics and Control*, 36 (12), 1971–91.

Frey, R. (1998). Perfect option hedging for a large trader. *Finance and Stochastics*, 2 (2), 115–41.

Frey, R., & Stremme, A. (1997). Market volatility and feedback effects from dynamic hedging. *Mathematical Finance*, 7 (4), 351–74.

Friede, G., Busch, T., & Bassen, A. (2015). ESG and financial performance: aggregated evidence from more than 2000 empirical studies. *Journal of Sustainable Finance & Investment*, 5 (4), 210–33.

Fung, M., Ignatieva, K., & Sherris, M. (2014). Systematic mortality risk: an analysis of guaranteed life-time withdrawal benefits in variable annuities. *Insurance: Mathematics and Economics*, 58, 103–15.

Garcia, J., Goosens, S., & Schoutens, W. (2008). Let's jump together: pricing credit derivatives. *Risk*, September, 130–3.

Gao, J., & Ulm, E. (2012). Optimal consumption and allocation in variable annuities with guaranteed minimum death benefits. *Insurance: Mathematics and Economics*, 51 (3), 586–98.

Gatheral, J., & Schied, A. (2011). Optimal trade execution under geometric Brownian motion in the Almgren and Chriss framework. *International Journal of Theoretical and Applied Finance*, 14 (3), 353–68.

Geman, H. (1989). The importance of the forward neural probability measure in a stochastic approach to interest rates, working paper. Paris: École supérieure des sciences économiques et commerciales.

Geman, H., Madan, D., & Yor, M. (2001). Time changes for Lévy processes. *Mathematical Finance*, 11 (1), 79–96.

Gerber, H., & Shiu, E. (1994). Option pricing by Esscher transforms. *Transactions of the Society of Actuaries*, 46, 99–191.

Gerber, H., & Shiu, E. (1996). Actuarial bridges to dynamic hedging and option pricing. *Insurance: Mathematics and Economics*, 18 (3), 183–218.

Gerber, H., & Shiu, E. (2003). Pricing lookback options and dynamic guarantees. *North American Actuarial Journal*, 7 (1), 48–66.

Giles, M. 2008. Multilevel Monte Carlo path simulation. *Operations Research*, 56 (3), 607–17.

Giles, M. (2006). Improved multilevel Monte Carlo convergence using the Milstein scheme, Technical Report NA06/02. Oxford: Oxford University Computing Laboratory.

Glasserman, P. (2004). *Monte Carlo Methods in Financial Engineering*. Berlin: Springer.

Gobet, E., Lemor, J.-P., & Warin, X. (2005). A regression-based Monte Carlo method to solve backward stochastic differential equations. *Annals of Applied Probability*, 15 (3), 2172–202.

Gompers, P., Ishii, J., & Metrick, A. (2003). Corporate governance and equity prices. *Quarterly Journal of Economics*, 118 (1), 107–56.

Graf, S., Kling, A., & Ruß, J. (2011). Risk analysis and valuation of life insurance contracts: combining actuarial and financial approaches. *Insurance: Mathematics and Economics*, 49 (1), 115–25.

Grasselli, M., & Romo, J. (2015). Stochastic skew and target volatility options. *Journal of Futures Markets*, 36 (2), 174–93.

Grosen, A., & Jørgensen, P. (2000). Fair valuation of life insurance liabilities: the impact of interest rate guarantees, surrender options, and bonus policies. *Insurance: Mathematics and Economics*, 26 (1), 37–57.

Hagan, P., Kumar, D., Lesniewski, A., & Woodward, D. (2002). Managing smile risk. *Wilmott Magazine*, September, 84–108.

Hardy, M. (2003). *Investment Guarantees: Modeling and Risk Management for Equity-Linked Life Insurance*. Hoboken, NJ: Wiley.

Heinkel, R., Kraus, A., & Zechner, J. (2001). The effect of green investment on corporate behavior. *Journal of Financial and Quantitative Analysis*, 36 (4), 431–49.

Heston, S. (1993). A closed-form solution for options with stochastic volatility with applications to bond and currency options. *Review of Financial Studies*, 6 (2), 327–43.

Hodges, S., & Neuberger, A. (1989). Optimal replication of contingent claims under transaction costs. *Review of Futures Markets*, 8 (2), 222–42.

Hoepner, A. (2010). Corporate social responsibility and investment portfolio diversification. *Risk Management eJournal*. DOI: 10.2139/ssrn.3683170.

Hoepner, A., et al. (2018). ESG shareholder engagement and downside risk. DOI: 10.2139/ssrn.2874252.

Holz, D., Kling, A., & Ruß, J. (2012). GMWB for life: an analysis of lifelong withdrawal guarantees. *Zeitschrift für die gesamte Versicherungswissenschaft*, 101 (3), 305–25.

Huang, H., Milevsky, M., & Salisbury, T. (2014). Optimal initiation of a GLWB in a variable annuity: no arbitrage approach. *Insurance: Mathematics and Economics*, 56, 102–11.

Huang, Y., Forsyth, P., & Labahn, G. (2012). Iterative methods for the solution of a singular control formulation of a GMWB pricing problem. *Numerische Mathematik*, 122, 133–67.

Hull, J., & White, A. (1987). The pricing of options on assets with stochastic volatilities. *Journal of Finance*, 42 (2), 281–300.

Hvidkjaer, S. (2017). ESG investing: a literature review. Copenhagen: Dansif.

Jarrow, R. (1994). Derivative securities markets, market manipulation, and option pricing theory. *Journal of Financial and Quantitative Analysis*, 29 (2), 241–61.

Jeanblanc, M., Yor, M., & Chesney, M. (2009). *Mathematical Methods for Financial Markets*. London: Springer.

Jiang, C., Ma, Y., & An, Y. (2009). The effectiveness of the VaR-based portfolio insurance strategy: an empirical analysis. *International Review of Financial Analysis*, 18 (4), 185–97.

Jiang, G., & Tian, Y. (2005). The model-free implied volatility and its information content. *Review of Financial Studies*, 18 (4), 1305–42.

Jiang, G., & Tian, Y. (2007). Extracting model-free volatility from option prices: an examination of the VIX index. *Journal of Derivatives*, 14 (3), 18–36.

Kahl, C., & Jäckel, P. (2005). Not-so-complex logarithms in the Heston model. *Wilmott Magazine*, September, 94–103.

Kahl, C., & Jäckel, P. (2006). Fast strong approximation Monte Carlo schemes for stochastic volatility models. *Quantitative Finance*, 6 (6), 513–36.

Kaiser, L., & Welters, J. (2019). Risk-mitigating effect of ESG on momentum portfolios. *Journal of Risk Finance*, 20 (5), 542–55.

Kalife, A. (2004). Portfolio insurance strategies by a large player. DOI: 10.2139/ssrn.683705.

Kalife, A. (2017). Managing investment and liquidity risks for derivatives within a market impact perspective. *Insurance Markets and Companies*, 8 (1), 59–73.

Kalife, A., López Ruiz, G., Mouti, S., & Tan, X. (2018). Optimal behavior strategy in the GMIB product. *Insurance Markets and Companies*, 9 (1), 41–69.

Kalife, A., & Mouti, S. (2017). On optimal options book execution strategies with market impact. *Market Microstructure and Liquidity*, 2 (3). DOI: 10.1142/S2382626617500022.

Kalife, A., Mouti, S., & Goudenège, L. (2014). Managing gap risks in iCPPI for life insurance companies: a risk return cost analysis. *Insurance Markets and Companies*, 5 (2), 69–77.

Kalife, A., Mouti, S., & Tan, X. (2015). Minimizing market impact of hedging insurance liabilities within risk appetite constraints. *Insurance Markets and Companies*, 6 (2), 58–64.

Kalife, A., Tan, X., & Wang, L. (2012). Dynamic hedging by a large player: from theory to practical implementation. *Neural, Parallel, and Scientific Computations*, 20 (2), 191–214.

Karatzas, I., & Shreve, S. (1991). *Brownian Motion and Stochastic Calculus*. New York: Springer.

Kilianová, S., & Ševčovič, D. (2013). A transformation mdethod for solving the Hamilton–Jacobi–Bellman equation for a constrained dynamic stochastic optimal allocation problem. *ANZIAM Journal*, 55 (1), 14–38.

Kim, C. (2005). Modeling surrender and lapse rates with economic variables. *North American Actuarial Journal*, 9 (1), 56–70.

Kling, A., Ruez, F., and Russ, J. (2013). The impact of policyholder behavior on pricing, hedging, and hedge efficiency of withdrawal benefit guarantees in variable annuities. DOI: 10.1007/S13385-014-0093-0.

Kloeden, P. (2002). The systematic derivation of higher order numerical schemes for stochastic differential equations. *Milan Journal of Mathematics*, 70, 187–207.

Kochanski, M., & Karnarski, B. (2011). Solvency capital requirement for hybrid products. *European Actuarial Journal*, 1, 173–98.

Kou, S. (2002). A jump-diffusion model for option pricing. *Management Science*, 48 (8), 1086–101.

Kroll, Y., Levy, H., & Markowitz, H. (1984). Mean-variance versus direct utility maximization. *Journal of Finance*, 39 (1), 47–61.

Kuo, W., Tsai, C., & Chen, W.-K. (2003). An empirical study on the lapse rate: the cointegration approach. *Journal of Risk and Insurance*, 70 (3), 489–508.

Le Floc'h, F. (2018). Variance swap replication: discrete or continuous. *Journal of Risk and Financial Management*, 6, 11–15.

Leland, H. (1985). Option pricing and replication with transactions costs. *Journal of Finance*, 40 (5), 1283–301.

Levy, H., & Markowitz, H. (1979). Approximating expected utility by a function of mean and variance. *American Economic Review*, 69 (3), 308–17.

Longstaff, F., & Schwartz, E. (2001). Valuing American options by simulation: a simple least-squares approach. *Review of Financial Studies*, 14 (1), 113–47.

Lord, R., Koekkoek, R., & van Dijk, D. (2010). A comparison of biased simulation schemes for stochastic volatility models. *Quantitative Finance*, 10 (2), 177–94.

Lorenz, J., & Almgren, R. (2011). Mean-variance optimal adaptive execution. *Applied Mathematical Finance*, 18 (5), 395–422.

Lin, X., & Tan, K. (2003). Valuation of equity-indexed annuities under stochastic interest rates. *North American Actuarial Journal*, 7 (4), 72–91.

McKean, H. (1969). *Stochastic Integrals*. New York: Academic Press.

Madan, D., Carr, P., & Chang, E. (1998). The variance gamma process and option pricing. *European Finance Review*, 2 (1), 79–105.

Madan, D., & Milne, F. (1991). Option pricing with VG martingale components. *Mathematical Finance*, 1 (1), 39–55.

Madan, D., & Seneta, E. (1990). The variance-gamma (VG) model for share market returns. *Journal of Business*, 63 (4), 511–24.

Melnikov, A. & Nosrati, A. (2017). *Equity-Linked Life Insurance: Partial Hedging Methods*. London: Chapman & Hall/CRC.

Merton, R. (1971). Optimum consumption and portfolio rules in a continuous-time model. *Journal of Economic Theory*, 3 (4), 373–413.

Merton, R. (1976). Option pricing when underlying stock returns are discontinuous. *Journal of Financial Economics*, 3 (1/2), 125–44.

Miccolis, J., & Goodman, T. (2012). Dynamic asset allocation: using momentum to enhance portfolio risk management. *Journal of Financial Planning*, 25, 38–48.

Milevsky, M., & Salisbury, T. (2006). Financial valuation of guaranteed minimum withdrawal benefits. *Insurance: Mathematics and Economics*, 38 (1), 21–38.

Milstein, G. (1995). *Numerical Integration of Stochastic Differential Equations*. Dordrecht: Kluwer Academic [translated and revised from the Russian original].

Moenig, T., & Bauer, D. (2013). Revisiting the risk-neutral approach to optimal policyholder behavior: a study of withdrawal guarantees in variable annuities. DOI: 10.2139/ssrn.3797995.

Moody's Investors Service (2013). Unpredictable policyholder behavior challenges US life insurers' variable annuity business, press release, 24 June.

Moreira, A., & Muir, T. (2017). Volatility-managed portfolios. *Journal of Finance*, 72 (4), 1611–44.

Mougeot, N. (2005). Volatility investing handbook, topic paper. Paris: BNP Paribas.

Nagy, Z., Kassam, A., & Lee, L.-E. (2015). Can ESG add alpha? New York: MSCI.

Neuberger, A. (1990). Volatility trading, Working Paper 91. London: Institute of Finance and Accounting, London Business School.

Neuberger, A. (1994). The log contract. *Journal of Portfolio Management*, 20 (2), 74–80.

Papageorgiou, A. (2002). The Brownian bridge does not offer a consistent advantage in quasi-Monte Carlo integration. *Journal of Complexity*, 18 (1), 171–86.

Pavlova, I., & de Boyrie, M. (2022). ESG ETFs and the Covid-19 stock market crash of 2020: did clean funds fare better? *Finance Research Letters*, 44. DOI: 10.1016/j.frl.2021.102051.

Pedersen, L., Fitzgibbons, S., & Pomorski, L. (2021). Responsible investing: the ESG-efficient frontier. *Journal of Financial Economics*, 142 (2), 572–97.

Peng, J., Leung, K., & Kwok, Y. (2012). Pricing guaranteed minimum withdrawal benefits under stochastic interest rates. *Quantitative Finance*, 12 (6), 933–41.

Pinault, A. (2013). Equity Indexed-Annuities: Pricing and Optimization, Master Thesis, Paris Dauphine University, supervised by Aymeric Kalife.

Platen, E., & Schweizer, M. (1998). On feedback effects from hedging derivatives. *Mathematical Finance*, 8 (1), 67–84.

Renault, É., & Touzi, N. (1996). Option hedging and implied volatilities in a stochastic volatility model. *Mathematical Finance*, 6 (3), 279–302.

Revelli, C., & Viviani, J.-L. (2015). Financial performance of socially responsible investing (SRI): what have we learned? A meta-analysis. *Business Ethics, the Environment & Responsibility*, 24 (2), 158–85.

Roncalli, T., et al. (2021). The market measure of carbon risk and its impact on the minimum variance portfolio. DOI: 10.2139/ssrn.3772707.

Rudd, A. (1998). Social responsibility and portfolio performance. *California Management Review*, 23 (4), 55–61.

Sato, K.-I. (1999). *Lévy Processes and Infinitely Divisible Distributions*. Cambridge: Cambridge University Press.

Schmitz-Abe, K. (2007). Pricing exotic options using improved strong convergence, PhD thesis. St Catherine's College, University of Oxford.

Schmitz-Abe, K., & Giles, M. (2008). Pricing exotic options using strong convergence properties. In Bonilla, L., Moscoso, M., Platero, G., & Vega, J. (eds.), *Progress in Industrial Mathematics at ECMI 2006* (pp. 614–29). Berlin: Springer.

Schmitz-Abe, K., & Shaw, W. (2005). Measure order of convergence without an exact solution, Euler vs Milstein scheme. *International Journal of Pure and Applied Mathematics*, 24 (3), 361–77.

Schoutens, W. (2003). *Lévy Processes in Finance: Pricing Financial Derivatives*. Chichester: Wiley.

Seneta, E. (2007). The early years of variance-gamma process. In Fu, M., Jarrow, R., Yen, J.-Y., & Elliott, R. (eds.), *Advances in Mathematical Finance* (pp. 3–20). Boston: Springer.

Sepp, A. (2015). When you hedge discretely: optimization of Sharpe ratio for delta-hedging strategy under discrete hedging and transaction costs. *Journal of Investment Strategies*, 3 (1), 19–59.

Shevchenko, P., & Luo, X. (2016). A unified pricing of variable annuity guarantees under the optimal stochastic control framework. *Risks*, 4 (3), 1–31.

Singh, A. (2020). Covid-19 and safer investment bets. *Finance Research Letters*, 36. DOI: 10.1016/j.frl.2020.101729.

Statman, M. (2004). The diversification puzzle. *Financial Analysts Journal*, 60 (4), 44–53.

Stein, E., & Stein, J. (1991). Stock price distributions with stochastic volatility: an analytic approach. *Review of Financial Studies*, 4 (4), 727–52.

Suzuki, K., & Vyas, K. (2011). Break-even volatility: a hedging cost approach for option volatility, technical report. Tokyo: Nomura Securities.

Tankov, P. (2010). Pricing and hedging gap risk. *Journal of Computational Finance*, 13 (3), 33–59.

Tiong, S. (2000). Valuing equity-indexed annuities. *North American Actuarial Journal*, 4 (4), 149–70.

Tse, S., Forsyth, P., Kennedy, J., & Windcliff, H. (2013). Comparison between the mean-variance optimal and the mean-quadratic-variation optimal trading strategies. *Applied Mathematical Finance*, 20 (5), 415–49.

Ulm, E. (2006). The effect of the real option to transfer on the value of guaranteed minimum death benefits. *Journal of Risk and Insurance*, 73 (1), 43–69.

Van der Corput, J. G. 1935. Verteilungsfunktionen I. *Proceedings of the Koninklijke Nederlandse Akademie van Wetenschappen*, 38, 813–21.

Verheyden, T., Eccles, R., & Feiner, A. (2016). ESG for all? The impact of ESG screening on return, risk and diversification. *Journal of Applied Corporate Finance*, 28 (2), 47–55.

Wang, L., et al. (2015). Understanding guaranteed minimum withdrawal benefit: a study on financial risks and rational lapse strategy. *Insurance Markets and Companies*, 6 (1), 6–21.

Wang, X., & Wang, Y. (2014). Variance-optimal hedging for target volatility options. *Journal of Industrial and Management Optimization*, 10 (1), 207–18.

Whalley, E., & Wilmott, P. (1999). The best hedging strategy, Mathematical Finance Working Paper MF 1999-17. Oxford: University of Oxford.

Windcliff, H., Forsyth, P., & Vetzal, K. (2006). Pricing methods and hedging strategies for volatility derivatives. *Journal of Banking & Finance*, 30 (2), 409–31.

Wu, L. (2008). Modeling financial security returns using Lévy processes. In Birge, J., & Linetsky, V. (eds.), *Handbooks in Operations Research and Management Science*, vol. 15, *Financial Engineering* (pp. 117–62). New York: Elsevier.

Yong, J., & Zhou, X. (1999). *Stochastic Controls: Hamiltonian Systems and HJB Equations*. New York: Springer.

Zakamouline, V. (2006). Optimal hedging of option portfolios with transaction costs. *Journal of Economic Dynamics and Control*, 30 (1), 69–98.

Zhang, J., & Zhu, Y. (2006). VIX futures. *Journal of Futures Markets*, 26 (6), 521–31.

Zhu, S.-P., & Lian, G.-H. (2015). Pricing forward-start variance swaps with stochastic volatility. *Applied Mathematics and Computation*, 250, 920–33.

Zhu, S.-P., & Lian, G.-H. (2018). On the convexity correction approximation in pricing volatility swaps and VIX futures. *New Mathematics and Natural Computation*, 14 (3), 383–401.

Zhu, Y., & Zhang, J. (2007). Variance term structure and VIX futures pricing. *International Journal of Theoretical and Applied Finance*, 10 (1), 111–27.

Index

Note: Page locators in **bold** and *italics* represent tables and figures, respectively.

For Product Safety Concerns and Information please contact our EU representative GPSR@taylorandfrancis.com Taylor & Francis Verlag GmbH, Kaufingerstraße 24, 80331 München, Germany

Batch number: 10748876

Printed by Printforce, the Netherlands